Innovations in Materials Manufacturing, Fabrication, and Environmental Safety

Innovations in Materials Manufacturing, Fabrication, and Environmental Safety

Edited by
Mel Schwartz

CRC Press
Taylor & Francis Group
Boca Raton London New York

CRC Press is an imprint of the
Taylor & Francis Group, an **informa** business

CRC Press
Taylor & Francis Group
6000 Broken Sound Parkway NW, Suite 300
Boca Raton, FL 33487-2742

First issued in paperback 2019

© 2011 by Taylor and Francis Group, LLC
CRC Press is an imprint of Taylor & Francis Group, an Informa business

No claim to original U.S. Government works

ISBN-13: 978-1-4200-8215-9 (hbk)
ISBN-13: 978-0-367-38329-9 (pbk)

Visit the Taylor & Francis Web site at
http://www.taylorandfrancis.com

and the CRC Press Web site at
http://www.crcpress.com

To my family whose free-flowing love and encouragement have

endured for 40 years of my book publishing career.

Contents

Preface

Throughout industries, pressure for change now comes from all directions—technical, financial, environmental, political, and social. The roadmap for new and developing processes, technologies, and materials provides a vision that shows company-by-company efforts aren't enough—implementing such a vision requires extensive value/chain collaboration and public/private partnerships. To ensure success in a highly competitive environment, the government must be part of the business strategy, especially long-term funding of research and development. Technical societies could also be a prime mover and act as a catalyst/facilitator at times.

In the research and development of new materials or variations of new processes, the word "innovation" for the scientist means something altogether different from the interpretation by the general public. According to Margaret W. Hunt, editor of *Advanced Materials & Processes* magazine, "Innovation is really a state of mind, a fundamental attitude of willingness to try new and sometimes radically different approaches to problems." This is the state of mind of the engineers who help drive the innovations that prime the U.S. economy. No one can predict the wonders that will become commonplace in the future, but it is safe to predict that they will be based on unseen developments.

For the general public, innovation brings to mind such products as cell phones and laptop computers, not advanced materials and process technologies. However, the inventors of these devices know that without advanced materials, joining methods, processing, and testing technologies, clunky telephones and typewriters would still be the norm, not to mention heavy and inefficient automobiles. These innovations that save money and reduce costs are often hidden from the ultimate consumer, who might not realize that the improved mileage in a new car is partially due to a metal diecast or molded plastic part that weighs less, that better safety is provided by stronger steel, and that new coating methods make the car look better and last longer. These things are invisible and the engineers who design and develop them often do not receive the credit they deserve.

A roadmap that was created in 1998 to identify the robotics and intelligent machine (RIM) goals by the year 2020 is now in place. This national initiative called for focusing and strengthening research in intelligent systems to strengthen the entire industry—currently U.S. companies lead the world in sensory devices and algorithms. Intelligent machines are advanced sensory devices to collect information about their environment and use sophisticated algorithms to respond to the information. This RIM science will benefit industry and society.

Intelligent machines can provide value to companies in all segments of manufacturing. An important question to ask is: What can RIM help you do better? For example, the high-value consumer electronics manufacturers want intelligent production systems that can rapidly and easily accommodate new product lines; they want intelligent systems to ensure the manufacturability of a product line and autonomously reprogram themselves when a new design is introduced. The welding industry wants an intelligent machine that will result in a repeatable, high-quality, structurally reliable weldment.

Many of the manufacturing processes described in this book can, and likely will, benefit from the incorporation of an intelligent machine into the processing and fabrication cycle. For example, a project at the Idaho National Engineering & Environmental Laboratory (INEEL) currently involves intelligent welding machines that incorporate both knowledge of welding physics and empirical learning capabilities. Currently, welding is considered as an industrial art based on a welder's manual skills or very simple duplication of these skills by an automatic (but dumb) machine. A typical approach to a machine control problem is to have a central body of intelligence (and control) in the machine. However, researchers at INEEL have developed a conceptual design of a machine using distributed learning and intelligence. The design is loosely based on biological models of social insects. For example, in an ant colony each ant functions according to local rules of behavior. Thus methods of learning and behavior modification have been developed that ensure global stability and optimization of the total machine. Researchers believe a qualitative understanding of the relationships between local costs and global subcosts can be used to develop future models for a welding process, as well as in more traditional control systems for welding processes.

In this 21st century, there are numerous challenges that are being overcome and in some instances have been overcome. Increased emphasis has been put on new materials, new fuels and propulsion technologies, advanced manufacturing, and predictive engineering that will put tremendous demands on our educational infrastructure and necessitate changes in curricula. Various industrial learning centers have been actively engaged during the past 10 years with design-oriented and engineering schools, industry scientists and engineers who work with materials, and advanced processes in activities including seminars, workshops, forums, and national conferences.

The initiatives mentioned here require steadfast commitment to long-term research and development, advocacy, and communication. Individual and corporate participation is, of course, critical.

This book will discuss environmental issues and safety in regard to processes and processing. In fact, all products manufactured need energy for material and production. Professor Timothy G. Gutowski, Faculty of Mechanical Energy at MIT, has received a grant from the National Science Foundation (NSF) to conduct research into the energy use of manufacturing processes of materials. "Manufacturing processes can be thought of as products with a huge energy appetite," Gutowski says. "These processes contribute to global warming, but are not visible to the public, unlike gas-guzzling SUVs or images of melting polar ice caps." Many people are not aware of the energy requirement for a lot of manufacturing processes, claims Gutowski. For example, the whole of the Western composites industry should be concerned. Upcoming economies such as China, India, Korea, and Malaysia are rapidly developing composite technologies and production processing.

Composites and energy could well be the criteria for future success in public transport. The subject of composites and energy is strongly related to population growth, mobility, and transport systems. Countries that have new economies are less hindered by rules, regulations, laws, and standards, and they are able to start from a different playing field. These countries have the best of both worlds—their own low cost of manufacturing and a highly educated population, and the loan of the often rusted technology know-how from Western countries. Southeast Asia is going to be in strong competition with the Western world. This should be a major concern to industry in the future. However, being different may be the only way the Western world of industries can sustain, by being one step ahead of the rest of the world and focus on what is the most important area for future growth and development; that is, resin transfer molding (RTM) composites and energy.

Speaking about energy, fresh research out of Yale University concludes that the energy required to produce nickel-containing, austenitic stainless steel from scrap is less than a third of the energy used to produce stainless steel from virgin sources. As an additional environmental bonus, recycling produces just 30% of the CO_2 emissions. Already one of the most recycled materials in the world, stainless steel could, theoretically, be made entirely from scrap if there weren't serious limitations on the availability of this material. Ironically, one of the main benefits of the material—its durability—limits its recycling potential: stainless steel structures and products tend to last a long, long time.

Current recycling operations reduce primary energy use by about 33% and CO_2 by 32% compared with production from virgin sources alone. But if stainless steel were to be produced solely from scrap (a merely hypothetical scenario), about 67% of the energy could be saved and CO_2 emissions cut by 70%.

"It confirms common sense," says Barbara Reck, a research associate at the School of Forestry and Environmental Studies at Yale. "The biggest energy use is in the mining and smelting phase, and you don't have to go through this phase using scrap. But now calculations have shown this systematically and the hypothesis has been confirmed." These findings have implications for the thousands of end-users of nickel-containing stainless steels.

In today's environmentally conscientious marketplace, customers want assurance that the products they buy will not contribute to climate change. They prefer to contribute to a sustainable world. Today products are advertised as being made of materials that have been validated, by one association or another, as safe for the environment. For example, in Northbrook, Illinois, a U.S.-based Crate and Barrel reports that its sofas have wood frames "certified by the Sustainable Forestry Initiative" and that they are "guaranteed for life." What's more, its cushions are "created with revolutionary, biobased materials that are environmentally renewable." The advertisement concludes: "Sustainability is a beautiful thing."

The same can be said of products made of nickel-containing stainless steel. It is one of the world's most recycled materials. Austenitic stainless steel products are all around us. Our kitchen appliances and sinks are made of it, we cook our meals in it, we eat our meals with it, the material has been available for less than 100 years, and is increasingly recycled. More than 80% of all products made of austenitic stainless steel are recycled at the end of their useful life. That has significance for the environment and sustainability. It means less energy is needed and less CO_2 emitted in the manufacture of austenitic stainless steel than in the past, when virgin material was all that was available. As more scrap becomes available, the need for virgin material declines and the carbon footprint left by a ton of austenitic stainless steel becomes smaller. The production of austenitic stainless steel is more sustainable than ever.

Fabricators and original equipment manufacturers (OEMs) must address health, safety, and environmental concerns. Infusion processing technologies especially must maintain a safe workplace including periodic training, adherence to detailed handling procedures, maintenance of current toxicity information, use of protective equipment (gloves, aprons, dust-control systems, and respirators), and development of company monitoring policies. Both suppliers and OEMs are working to reduce emissions of highly volatile organic compounds (VOCs) by reformulating resins and prepregs (pre-impregnated) and switching to water-dispensable cleaning agents.

So where does all this lead? Advances in materials technology and the convergence of materials, information, and miniaturization will drive less resource intensity, more complexity, and smart products. Older, established technologies will continue sidewise development into new markets and applications. Innovations such as nanotechnologies, portable energy sources, multifuel products, miniaturization, and customizable intelligent materials will grow.

But how do we cope with these challenges and the opportunities that result?

We put the spotlight on innovation of new and exciting processing techniques, methods, and fabrication technologies developing the next business models, not only on new products, processes, and services within your own company. Identify novel segments and geographies, quickly spot new competitors and partners, and establish global positions quickly through acquisitions, alliances, and licensing. In addition, we must manage strategic risk—not just operational risk or political risk. Speed, flexibility, market knowledge, effective alliances and acquisitions, and innovation will be critical success factors required to win in an environment driven by these mega trends.

In conclusion, we have before us opportunities and huge challenges. They are coming at us simultaneously … we have no option to take them in sequence. We must play two hands.

Mel Schwartz

Editor

Mel Schwartz has degrees in metallurgy and engineering management and has studied law, metallurgical engineering, and education. His professional experience extends over 51 years serving as a metallurgist in the U.S. Bureau of Mines; metallurgist and producibility engineer, U.S. Chemical Corps; technical manufacturing manager, chief R&D Lab, research manufacturing engineering, and senior staff engineer, Martin-Marietta Corporation for 16 years; program director, manager and director of manufacturing for R&D, and chief metals researcher, Rohr Corp for 8 years; staff engineer and specification specialist, chief metals and metals processes, and manager of manufacturing technology, Sikorsky Aircraft for 21 years. While retired Mel has been a consultant for many industrial and commercial companies including Intel and Foster Wheeler, and was former editor for *SAMPE Journal of Advanced Materials*.

Mel's professional awards and honors include Inventor Achievement Awards and Inventor of the Year at Martin-Marietta; C. Adams Award & Lecture and R.D. Thomas Memorial Award from AWS; first recipient of the G. Lubin Award and an elected Fellow from SAMPE; an elected Fellow and Engineer of the Year in CT from ASM; and Jud Hall Award from SME.

Mel's other professional activities involve his appointment to ASM Technical Committees (Joining, Composites and Technical Books; Ceramics); manuscript board of review, *Journal of Metals Engineering* as peer reviewer; the Institute of Metals as peer reviewer as well as *Welding Journal*; U.S. Leader of IIW (International Institute of Welding) Commission I (Brazing & Related Processes) for 20 years and leader of IIW Commission IV (Electron Beam/ Laser and Other Specialized Processes) for 18 years.

Mel's considerable patent activity has resulted in the issuance of five patents, especially aluminum dip brazing paste commercially sold as Alumibraze.

Mel has authored 17 books and over 100 technical papers and articles. Internationally known as a lecturer in Europe, the Far East and Canada, Mel has taught in U.S. colleges (San Diego State, Yale University), ASM Institutes, McGraw-Hill Seminars, and in-house company courses.

List of Contributors

Jean-Michel Bergheau
Université de Lyon
École Nationale d'Ingénieurs de
  Saint-Étienne
Saint-Étienne, France

Thomas S. Bloodworth III
Department of Mechanical
  Engineering
Vanderbilt Welding Automation
  Laboratory
Nashville, Tennessee

Douglas E. Burkes
Nuclear Fuels and Materials
  Division
Idaho National Laboratory
Idaho Falls, Idaho

Kathy C. Chuang
Structures and Materials Division,
  Polymer Branch
NASA Glenn Research Center
Cleveland, Ohio

George E. Cook
Department of Electrical
  Engineering and Computer
  Science
Vanderbilt Welding Automation
  Laboratory
Nashville, Tennessee

Sylvain Drapier
Ecole des Mines de Saint-Etienne
Structures and Materials Science
  Division & LTDS UMR CNRS 5513
Saint-Étienne, France

Adrian P. Gerlich
Department of Chemical and
  Materials Engineering
University of Alberta
Edmonton, Alberta, Canada

James L. Glancey
Department of Mechanical
  Engineering
University of Delaware
Newark, Delaware

John F. Hinrichs
Friction Stir Link, Inc.
Brookfield, Wisconsin

Leanna Micona
The Boeing Company
Seattle, Washington

Thomas H. North
Department of
  Chemical and Materials
  Engineering
University of Alberta
Edmonton, Alberta, Canada

Arthur C. Nunes, Jr.
NASA Marshall Space Flight
  Center
Huntsville, Alabama

Dr.- Ing. Carolin Radscheit
Process Prototyping
INPRO
Berlin, Germany

Burke Reichlinger
The Boeing Company
Seattle, Washington

Francisco F. Roberto
Biological Systems Department
Idaho National Laboratory
Idaho Falls, Idaho

Mel Schwartz
Consultant
Clearwater, Florida

Heather G. Silverman
Biological Systems Department
Idaho National Laboratory
Idaho Falls, Idaho

Christopher B. Smith
Friction Stir Link, Inc.
Brookfield, Wisconsin

Alvin M. Strauss
Department of Mechanical
  Engineering
Vanderbilt Welding Automation
  Laboratory
Nashville, Tennessee

Myung-Keun Yoon
Mechanical Engineering
  Department
South Dakota School of Mines and
  Technology
Rapid City, South Dakota

1

Plasma Brazing

Dr.- Ing. Carolin Radscheit

CONTENTS

Why Plasma Brazing?

In the last 20 years, the weight of vehicles has been steadily increasing. This tendency can be applied to all makes and models of vehicles. The Volkswagen Golf I automobile was made from 1974 until 1983 and weighed only 870 kg, whereas the Golf 5 that started production in 2004 weighed about 1200 kg. This increase in weight is largely due to two factors:

- The increase in components concerning the safety of the occupants and the environment, for example, curtain and side airbags, antilock braking systems, belt tensioners, catalytic converters, and so on.

- The increase in components that increased driving comfort: electric windows, electrically operated sunroofs, power steering, air conditioning, and so on. For example, a maximum of five electric motors and control units were installed in the Golf 3 (from November 1991 until May 1997), whereas the Golf 5 possesses 25–30 such units with optimal equipment.

Parallel to this development, the construction of vehicles was changed so that safety could be provided through stable passenger compartments and consequently, the thickness of the body sheet metal was reduced. Most body panels today are thinner than 1.0 mm. Galvanization provides corrosion protection, which helps maintain the vehicle, its value, and enables the vehicle manufacturer, Volkswagen, to offer guarantees against rust perforation.

Another development was the influence of joining techniques in body construction. Previous welding procedures have caused spattering on galvanized metal, and if the sheet metal is thin, there is a danger that it will be burn-through. As a result, new or modified joining methods were needed and required development.

Several brazing methods have been examined and developments have produced several new techniques. Three of these methods appear to have promise; however, the third method is the most applicable and is covered in this chapter.

The first process that was examined was laser beam brazing, which is suited for producing visually demanding and completely splatter-free seams. Therefore, this process has been used for outer surface joints, the boot lid (trunk) or hatch, and roof/side panel joints in autos. However, this process depends on maintaining small gaps (g <1.0 mm), and can only be used in automated applications and also requires safety measures specific to laser beams.

The second method that was examined was MIG brazing (gas metal brazing), a metal inert gas process in which a copper-containing filler metal was used as a brazing material and is well suited for joining thin galvanized sheet metal. This process or ability to fill gaps well is especially advantageous when used with robots. However, this process is not splatter free and consequently has limited use for outer body surfaces. The splatter must be removed by grinding.

The third method and the subject of this chapter is plasma brazing, which offers an excellent compromise between the previous two modern brazing methods. Plasma brazing is very well suited for joining thin galvanized sheet metal, produces a narrow, visually appealing, and completely splatter-free seams and therefore has been used primarily on the outer body surfaces of autos. Plasma brazing can be performed manually as well as robotically and requires the same safety precautions as inert gas shielded welding.

How Plasma Brazing Works

A schematic diagram of the plasma brazing torch is shown in Figure 1.1. A plasma brazing machine consists of a plasma torch, a power source, a filler material wire feed, gas feed and, if required, a seam guide sensor, Figure 1.2. Plasma brazing is the equivalent of plasma welding with nonconducting filler wire. Just as with other welding and brazing processes with a side-feed

1. Work piece
2. Brazed seam
3. Brazed metal
4. Wire feed
5. Shielding gas nozzle
6. Plasma gas nozzle
7. Tungsten electrode
8. Shielding gas
9. Plasma gas
10. Wire feed nozzle
11. Ignition power source

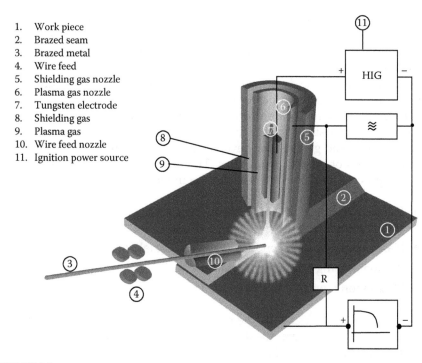

FIGURE 1.1
Schematic of a plasma brazing torch. (Courtesy of Fa. MIG-WELD, Landau Isar. Germany.)

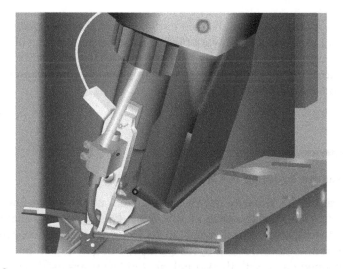

FIGURE 1.2
Plasma brazing head, including braze wire supply, brazing torch, and seam tracking sensor.
(Courtesy of IGM-Robotersysteme, Wiener Neudorf, Austria.)

of wire, the accessibility of the torch is limited and in robotic applications, there is one degree of freedom less than with MIG brazing. Just as in laser or MIG brazing, a copper-based filler material has been used successfully in plasma brazing.

As can be seen in the diagram in Figure 1.1, plasma brazing requires two power sources: an ignition power source and a plasma power source with dropping power source characteristics. Actually, only manual procedures have this characteristic. Through the use of a seam guide sensor or reproducible work piece and gap size, however, this process can be used without any problems in robot-guided procedures. There are a number of variations of this new process, in part still in the developmental stages:

- Plasma brazing with only one inert gas and without an ignition power source
- Plasma brazing with conductive filler material wire
- Plasma brazing with brazing powder
- Plasma torch with central wire feed.

The plasma arc is restricted by the plasma gas and has an opening angle of only about 7° [1]. Therefore, the delivery of energy is very concentrated and the thermal efficiency is high.

Degree of efficiency	Plasma brazing	73 80%
For comparison [2]:	TIG	ca. 30%

The results reflect both a high fusion efficiency of the filler material and a low thermal influence on the work piece. The separate selection of wire feed and arc current strength in plasma brazing creates the possibility, for example, of performing repair brazing by remelting the existing material without needing to feed more wire [3].

What Plasma Brazing is Capable of Doing

Plasma brazing is primarily used to join coated, unalloyed or low-alloyed, deep-drawn sheet steel less than 1.5 mm thick. Plasma brazing of aluminum is under development. Sheet metal up to 15 μm galvanization can generally be joined without trouble. If, for example, hot galvanized or individual hot zinc dipped parts with thicker galvanized layers are used, a subsequent examination should be performed. Even sheet metal with organic layers (on zinc) can be joined with plasma brazing.

Combinations of material	ASTM A620-1008, 1.0 mm	ASTM A620-1008, 1.5 mm	H 180 B, 0.8 mm	H 220 P, 1.0 mm	H 220 P, 2.0 mm	H 260 LA, 1.0 mm	H 260 LA, 1.2 mm	H 260 LA, 2.0 mm	H 340 LA, 1.5 mm
ASTM A620-1008, 1.0 mm	X	X	X	X	X	X	X	X	X
ASTM A620-1008, 1.5 mm	X		X	X					
H 180 B, 0.8mm	X	X	X	X	X	X	X	X	X
H 220 P, 1.0 mm	X	X	X	X	X	X	X	X	X
H 220 P, 2.0 mm	X		X	X		X			
H 260 LA, 1.0 mm	X		X	X	X	X	X	X	X
H 260 LA, 1.2 mm	X		X	X		X			
H 260 LA, 2.0 mm	X		X	X		X			
H 340 LA, 1.5 mm	X		X	X		X			

FIGURE 1.3
Brazed filler metal. Combinations of material and their suitability for plasma brazing (X), tensile shear test, R-CuSi-A (AWS/ASME).

The strength of the brazed joint depends on the filler metal used, the strength of the basic material used and the thickness of the material, and should be determined in individual cases, Figure 1.3. Currently, standardized copper-based brazing filler metal as well as nonstandardized copper-based brazing filler metal and welding wire have been used in plasma brazing. In car manufacturing, standard brazing filler metal R-CuSi-A (AWS/ASME, material number 2.1461) has proven to be an excellent and acceptable material of choice. This filler metal has been used in the form of wire, generally with diameters between 0.8 mm and 1.2 mm. Other available forms that have been used include powder, paste, cored wire, and solid, and cored wire with square or rectangular cross sections.

The surfaces of all work pieces and filler metal wire should be grease-free. Generally, welding argon Ar has been used as a shielding and plasma gas. The attainable brazing speed depends on the contours of the parts to be joined and, for a manual operation is about $v_B = 0.5 + 0.7$ m/min and for automatic operation, from about 7 m/min to about 1.0 m/min for $v_B = 0.3$ [2]. The holding fixture is crucial for the quality of brazing. The parts should be clamped close to the edge so that the parts cannot move during the brazing process. Gap widths of 0 is less than or equal to gap and is less than or equal to 1.5 t_{min} and has produced good results.

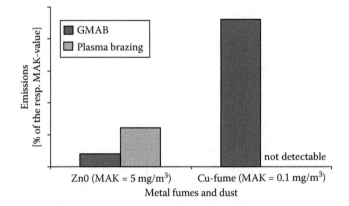

FIGURE 1.4
Comparison of emissions, GMAB versus plasma brazing, MAK = maximum work environment concentration.

FIGURE 1.5
Manual plasma brazing, carried out at the sidewall of a 31-Lupo vehicle.

A further characteristic of plasma brazing is that, in comparison to MIG brazing, there are no measurable Cu emissions, Figure 1.4. A few examples demonstrate the possible uses of plasma brazing. Manual plasma brazing is shown in Figure 1.5, which is appropriate for a small series of parts or prototype construction. Because the process is splatter free, it is also possible to make seams directly next to threads, Figure 1.6. In Figure 1.7, a robot-guided implementation series can be seen. The roof channel drain of the Volkswagen

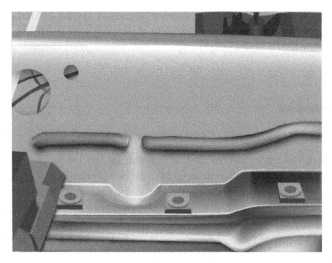

FIGURE 1.6
Plasma brazing at the pivot reinforcement. (Courtesy of VA-Tech, Linz, Austria.)

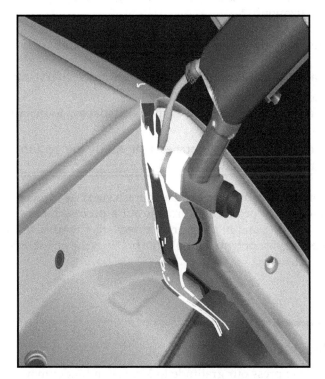

FIGURE 1.7
Robot guided plasma brazing at the car body of a GOLF 4 Volkswagen.

Golf 4 is welded with the aid of a seam sensor. Plasma brazing was chosen for this application because seams on the outside of the body would not require further work (grinding) [3,4].

Expected Developments, Future Potential, and Conclusions

Plasma brazing is appropriate on the exterior of the body of a vehicle. It is ideal for seams, which should not require further work but nevertheless must meet stringent visual standards. Currently, seams having these requirements are also made with laser brazing. Therefore, plasma brazing offers a possible alternative to the very elaborate and expensive laser process. Because plasma brazing is still a rather new process, some questions remain to be answered, for example:

- Do plasma brazers require special training?
- Will the process be classified as a welding or a brazing process in the standardization?

Results expected in the future from research and development include:

- Plasma brazing of aluminum, magnesium, and mixed combinations: Al–Mg or Al–Steel.
- Materials including filler metal specially adapted to this process as well as coating material.
- Robot-operated welding torches that fulfill the requirements of industrial series production.

In conclusion, plasma brazing can be characterized as a process, which is not a comprehensive production process but a process for special tasks at which other welding and brazing procedures fail. The uses, until now, also demonstrate the great potential for plasma brazing, the possibilities of which are far from exhausted.

Acknowledgments

I would like to express our gratitude to Mr. Stefan Lindenmayer and to Mr. Uwe Lahrmann for their help.

References

1. Eichhorn, F. *Schweiss Technische Fertigungsverfahren, Düsseldorf*, Germany: VDI-Verlag, 1983, ISBN 3-18-400603-4.
2. Kallabis, M. (2003) Einsatz der Plasmatron-Technologie im Karosserie- und Anlagenbau, http://www.auscartec.com/news/powerslave,id,33,nodeid,89,_language,de.html
3. N.N. Lichtbogenlen Grundlagen, Verfahren, Anforderungen an die Anlagentechnik Merkblatt DVS 0938-1, Düsseldorf, Germany: DVS-Verlag, 2001.
4. N.N. Lichtbogenlen Anwendungshinweise Merkblatt DVS 0938-2, Düsseldorf, Germany: DVS-Verlag, 2004.

2

Adhesive Bonding

Mel Schwartz

CONTENTS

Introduction and Brief History

Adhesive bonding as a joining technology was utilized in ancient times using natural materials. The first known application of adhesive in a structural application was the use of bitumen about 36,000 years ago [1] to bond wooden tool handles to flint stone. Various adhesive materials of animal or vegetable origin were used in ancient Egypt 3,300 years ago. The adhesive bonding of thin layers of a material to build up a laminate was used with various types of wood (plywood) for thousands of years. The earliest known occurrence of plywood was in ancient Egypt around 3500 B.C. when wooden articles were made from sawed veneers glued together cross-wise. This was originally done due to a shortage of fine wood. Thin sheets of high-quality wood were glued over a substrate of lower quality wood for cosmetic effect, with incidental structural benefits. This manner of inventing plywood has occured repeatedly throughout history.

In early aircraft, adhesive bonding was used to build up wood structures for frames, contoured ribs, and spars. Wooden parts were generally laminated with various fiber directions to create sufficient strength. By laminating wood, the best properties were obtained by using the highest quality materials available and improved the stability of structures. The irregularities of grain often resulted in uncontrollable warping and cracking if any attempt is made to use the wood in thicknesses much greater than the typical veneer thicknesses (1–2 mm). High-strength plywood was used for aircraft skin molding in the 1930s and 1940s and made from birch. By combining laminated wood fiber orientations, properties could be tailored as in advanced composites and hybrid laminates today.

Protein-based adhesives (hide glues) were used in the early days with the disadvantage that these structures were sensitive to moisture. This problem was solved with the introduction of the synthetic polymers (i.e., urea-formaldehyde and phenol-formaldehyde-based adhesive systems. With the development of synthetic polymeric materials, higher loaded joints in more demanding applications became possible.

Around 1938, Aero Research Ltd. (Duxford, England) started investigating metal bonding that led to the introduction of the Redux adhesive system. This system, based on phenol-formaldehyde resin toughened with polyvinylformaldehyde thermoplastic particles and is still available in the market today.

The adhesive, Redux 775, was used by DeHavilland in the first application of metal adhesive bonding in the Hornet fighter aircraft, a derivative of the famous Mosquito aircraft. The aircraft first flew in 1944 and had a mixed wood/aluminum structure. The wing had a lower skin, leading and trailing edge, and spar caps on aluminum alloy. To ease the manufacturing, a layer of veneer was bonded to the aluminum parts (high pressure and temperature) and as an assembly step, the wood-on-wood joint was bonded with urea formaldehyde (Aerolite) adhesive cured at room temperature.

The first large-scale application of structural adhesive bonding was in the fuselage (stiffened panels) of the first jet airliner, the De Havilland Comet. Along in this time period, adhesive bonding came into its own with the first of many in a family of all-aluminum sheet metal and the honeycomb-guided missile, Matador, followed by Mace, Bullpup, Lacrosse, Pershing, Sargent, Sprint, and others. Since these early applications, adhesive bonding technology has developed much further over the years. This has led to many examples ranging from adhesive-bonded doublers on fuselage skins to large fully integrated built-up wings and fuselage structures in which stringers, doublers, and skin are cured in one single cycle. More on these examples under the applications section.

Basics

In order to produce successful bonds it is critical, in addition to surface treatment, to do the following:

1. Thoroughly mix and precisely proportion resin and hardener before application.
2. Curing temperature and times must be correct.
3. Jigs and other fixtures should securely hold bonded surfaces during curing and eliminate any movement before the part attains handling strength.
4. Bonded surfaces need only light pressure during assembly as the adhesive cures, but pressure should be applied as evenly as possible over the entire bond area. Excessive pressure can force adhesive to run out, leaving a joint starved of adhesive. Under lab conditions, pressure applied is generally 30–50 N for a bonding area 312.5 mm^2 (0.625 in^2). This figure may change for high-viscosity adhesives.
5. Durable joints must be designed properly and engineers must evaluate all the forces that will act on the bond line such as shear, tension, or compression. Butt joints and bond lines that will see peel and/or cleavage forces should be avoided. In addition, a glue line of 0.05–0.2

mm (0.002–0.008 in.) is best for optimum adhesive strength, although many thixotropic/paste adhesives can fill some of the gap between two substrates.

6. Adhesive selection depends on operating conditions. An assembly subjected to vibration or impact, for example, should use a toughened adhesive. Adhesive choice should also account for service conditions. Loading, temperature, and other environmental factors can all have major effects on bond strength.

Comparison of Joints

In contrast to other joining methods, such as riveting and bolting, adhesive bonding has no adverse effect on the material characteristics of the surfaces to be joined, for example, drilling of holes damaging the joined parts and creating stress concentrations (Figure 2.1). Due to this, bonded joints will have good fatigue properties. There is also no adverse effect on mechanical properties of high temperatures or distortion by local heating, as is the case with most welding or soldering processes.

In the manufacturing environment, bonding technology makes sure that the characteristic material properties are utilized to the utmost. For example, sandwich structures, in which high-strength materials are used as face sheets with a relative weak material, such as foam that is used as a core, show an excellent bending stiffness over weight ratio. Furthermore, adhesive-bonded joints create the possibility for new structural design solutions. By laminating, materials can be created with improved, often tailored characteristics compared to the properties of individual layers.

Next to the advantages mentioned, some disadvantages have to be considered. Adhesive-bonded joints are difficult to disassemble. If, for any reason, disassembly of components is desirable during the lifetime of a product then mechanically fastened joints are the preferred option. As the adhesive materials are often weaker than the substrate materials, adhesive-bonded joints

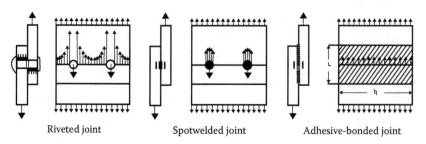

Riveted joint   Spotwelded joint   Adhesive-bonded joint

FIGURE 2.1
Comparison of stress distributions of various types of joints.

have to be designed mainly as shear loaded joints (like in riveted joints). This requires different design solutions than for bolted or welded joints.

Preparation of Surfaces for Adhesive Bonding

All substrates should be degreased and, typically, lightly abraded before bonding. But the way to get the maximum strength and long-term resistance to deterioration is through a chemical or electrolytic pretreatment, particularly on thermoset and thermoplastic surfaces. Bonds are stronger with properly prepared surfaces. Pretreatment removes low-surface energy contaminants such as waxes, oils, greases, plasticizers, and release agents. Proper preparation also reduces dust, dirt, and loose particles resulting from abrasion. Substrate pretreatment expands the surface area available for bonding by creating textured surfaces. In addition, surface preparation promotes the thorough adhesive wetting needed for strong bonds. It does so by giving the bonded material higher surface energy than that of the adhesive.

Metal substrates exhibit much higher surface energies than polymers. As a result, they are easier to wet-out and bond with adhesive-like epoxies that have surface energies around 40 ml/m^2. Conversely, polymers don't wet-out as well because their surface energies are slightly lower than those of epoxy adhesives. That's why their surface energy must be raised to get a strong bond. A variety of techniques can accomplish this. These methods also help produce joints that better withstand moisture and aggressive chemicals without disbonding. In most cases, abrading surfaces and wiping them with a solvent such as acetone, ethanol, or isopropanol will be enough. But certain plastics may need flame, corona, or other chemical pretreatments to change their surface textures.

Surface Preparation for Various Materials

A key element in achieving maximum joint strength is good preparation of the substrate surface of a material. Designers must include this preparation process when selecting the substrate and adhesive that provide desired joint properties. Poor surface preparation often results in reduced joint strength. This region on the bond depends on the substrate's chemical composition and processing methods that may produce undesirable surface conditions. For example, metal oxides often form on metallic substrates during processing or storage. Plastics can have impurities on their surface related to processing or as a result of the bonding process.

The mechanical and physical properties imposed by the application often dictate the substrate composition. But all metal, plastic, elastomer, and ceramic substrates require cleaning or surface modification. This preparation boosts the surface energy that promotes wetting of the adhesive and subsequent adhesion. A surface that produces the best bond is clean and free of contamination, uniform and continuous in finish, and stable with high-surface energy. See Tables 2.1 through 2.4.

TABLE 2.1

Adhesive Selection

	Acrylic	Anaerobic	Cyanoacrylate	Epoxy	Hot Melt	Polyurethane	Polysulfide	Silicone	Solvent Base	Water Base	UV
Viscosity	Medium	Low	Low	Medium to thick	Thick	Medium	Thick	Thick	Low to medium	Low to medium	Low to medium
Void filling	Good	Poor to fair	Poor to fair	Very good	Very good	Good	Very good	Very good	Fair	Poor	Low to medium
Heat resistance	Good	Good	Fair	Good	Poor to fair	Fair	Good	Very good	Good	Fair	Fair
Cold resistance	Good	Good	Fair	Fair	Fair	Good	Good	Very good	Good	Fair	Good
Flexibility	Good	Good	Poor to fair	Fair	Fair to good	Good	Good	Very good	Good	Poor	Good
Chemical resistance	Good	Good	Good	Good	Fair	Good	Very good	Very good	Good	Poor	Fair
Humidity resistance	Good	Good	Fair	Good	Good	Fair	Good	Very good	Good	Poor	Good
Work time	Medium to fast	Medium	Fast	Slow to medium	Fast	Slow to medium	Medium	Slow to medium	Slow to medium	Medium	Slow
Cure time	Medium to fast	Medium	Fast	Slow	Fast	Medium	Medium	Medium	Medium	Medium	Fast
Metal bond (steel, aluminum)	Good	Fair	Good	Good	Fair	Good	Good	Fair	Good	Poor	Good
Plastic bonding (ABS, styrene)	Very good	Poor	Very good	Fair	Fair	Very good	Fair	Fair	Fair	Poor	Good
Polvolefin	Fair	Not suggested	Good	Poor	Poor	Good	Not suggested	Fair	Fair	Poor	Fair
Wood	Not suggested	Not suggested	Not suggested	Good	Very good	Not suggested	Not suggested	Not suggested	Good	Very good	Fair
Paper cardboard	Not suggested	Not suggested	Not suggested	Not suggested	Very good	Not suggested	Not suggested	Not suggested	Good	Very good	Fair

Source: Ellsworth Adhesives, Germantown, Wisconsin

TABLE 2.2

Surface Pretreatments of Common Metals and Ceramics

Substrate	Solvent Cleaning[a]	Intermediate Cleaning	Chemical Treatment or Other
Aluminum	Chlorinated solvent, ketone, or mineral spirits[b]	Detergent scrub	Sulfuric (96%) acid/sodium dichromate (77.8/22.2 pbw) solution at 25°C for 20 min; rinse with tap water followed by distilled water, dry for 30 min at 70°C
Beryllium/ copper	Chlorinated solvent, ketone, or mineral spirits[b]	Wet abrasive blast	—
Copper	Chlorinated solvent, ketone, or mineral spirits[b]	Dry abrasion or wire brushing	Nitric (69%) acid/ferric chloride/ distilled water (12.4/6.2/81.4 pbw) solution at 21–32°C for 1–2 min; rinse in tap water followed by distilled water; dry at 65°C maximum
Steel (stainless)	Chlorinated or aromatic solvent[b]	Heavy-duty alkaline cleaner	Nitric (69%) acid/distilled water (20/80 pbw) solution at 21–32°C for 25–35 min; rinse with tap water followed by distilled water; dry at 65°C maximum
Steel (mild)	Same as stainless	Same as stainless	Ethyl alcohol (denatured)/ orthophosphoric (85%) acid (66.7/33.3 pbw) solution at 60°C for 10 min; rinse in tap water followed by distilled water; dry for 60 min at 120°C
Titanium, titanium alloys	Ketone or aromatic solvent	Mild alkaline cleaner or wet abrasive scour	Nitric (69%) acid/hydrofluoric (60%) acid/distilled-water (28.8/3.4/67.8 pbw) solution at 38–52°C for 10 to 15 min; rinse with tap water followed by distilled water; dry 15 min at 71–82°C; brush off carbon residue with nylon brush while rinsing
Ceramic	Ketone solvent	—	Sulfuric (96%) acid/sodium dichromate/distilled water (96.6/1.7/1.7 pbw) at 20°C for 15 min; rinse with tap water followed by distilled water; dry at 65°C maximum

Source: Ellsworth Adhesives, Germantown, Wisconsin

[a] Immerse, spray, or wipe.

[b] Or vapor degrease with chlorinated solvents.

TABLE 2.3

Surface Pretreatments of Common Plastics

Substrate	Solvent Cleaning[a]	Intermediate Cleaning	Chemical Treatment or Other
Acetal (Delrin)	Ketone solvent	Dry abrasion or wet or dry abrasive blast	Sulfuric (96%) acid/potassium dichromate/distilled water (88.5/4.4/7.1 pbw) solution at 25°C for 10 sec; rinse with tap water followed by distilled water and dry at room temperature
ABS	Ketone solvent	Dry abrasion or wet or dry abrasive blast	Sulfuric (96%) acid/potassium chromate/distilled water (65/7.5/27.5 pbw) solution at 60°C for 20 min; rinse with tap water followed by distilled water; dry with warm air
Polycarbonate (PC)	Alcohol	Dry abrasion or wet or dry abrasive blast	—
Polyethylene (PE) and polypropylene (PP)	Ketone solvent	—	Sulfuric (96%) acid/sodium dichromate/distilled water (88.5/4.4/7.1 pbw) at 70°C for 60 sec; expose surface to gas burner flame until the substrate is glossy. Can also be treated with corona discharge or flame.

Source: Ellsworth Adhesives, Germantown, Wisconsin
[a]Immerse, spray, or wipe.

Surface energy defines the ability of adhesives and pressure-sensitive adhesive tapes to wet-out on to the substrate surfaces and promote adhesion. Materials with a low surface energy—including olefin-based thermoplastics and polypropylene—may need priming or flame or corona treating before they can take an adhesive. These pretreatments convert low surface energy substrates to a higher surface energy better suited for strong adhesive bonds. Likewise, anodizing aluminum surfaces is one method that will boost reliability of adhesive-bonded joints incorporating those materials. Anodizing increases surface roughness, resulting in a significant improvement in mechanical adhesion behavior of the substrate.

Solvent Wipe

This is the simplest form of surface preparation and is ideal for removing waxes, oils, and other low-molecular weight contaminants from substrates. The technique relies on the contaminants being soluble in the solvent and the solvent itself being free of dissolved contaminants. But some solvents aren't

TABLE 2.4

Prepped to Bond

Substrate	Recommended Pretreatments
Aluminum	Solvent degrease and 20–40 min etch in 140–150°F sulfuric acid and sodium dichromate or ceramic grit blast
Copper, brass, and their alloys	Ferric chloride/nitric acid etch or ammonium sulfate/water etch
Steels (nonstainless)	Solvent degrease and etch in diluted sulfuric or hydrochloric acid
Stainless	Solvent degrease, etch in concentrated sulfuric acid 20–30 min
Nickel and its alloys	Abrasive or solvent clean, etch in sulfuric acid or sulfuric/nitric acid pickle
Titanium	Abrasive or solvent clean, acid etch at room temperature
Zinc	Abrade then degrease or vapor clean, or acid etching
Silica ceramics and glasses	Abrade, apply silicone primers, solvent clean
Silicon carbide	Abrade, solvent clean
Carbons and carbon fibers, zirconia ceramics	Abrade, solvent clean, apply primers
Epoxies, phenolics, and polyesters	Abrade, solvent clean, apply primers
Polyimides and polyether ketones	Abrade, solvent clean, and degrease
Fluoropolymers (Teflon, PTFE, etc.), silicones	Apply special primers

Note: Surface preparation is the key to an adhesive bond that will last, even at high temperatures. These are the recommended treatments for some common substrates.

compatible with particular polymeric substrates. Some solvents dissolve thermoplastics or will create stress cracking or crazing on surfaces. Common solvents are acetone, MEK (methyl ethyl ketone), MIBK (methyl isobutyl ketone), xylene, TCE (trichloroethylene), ethanol, and IPA (isopropyl alcohol).

Solvents should be applied with clean, lint-free cloths, or paper towels. Take care to prevent cross-contamination from sample to sample by not reusing cloths or dipping a contaminated cloth dipped into the solvent. It's also important to monitor for hazardous and toxis-vapor buildup when using solvents. Solvent wiping is not suitable for large-scale bonding projects. Instead, vapor degreasing or ultrasonic vapor degreasing are more appropriate.

Abrasion

This can be accomplished several ways. Manual abrasion includes wire brushing, paper sanding, and filing. Automated techniques include high speed sanding, grinding, and shot/grit blasting. Mechanical/automated abrasion processes are relatively quick and economical. They are also reproducible and depend less on skilled labor than manual methods.

Flame Treatment

This technique partially oxidizes surfaces, producing polar groups that raise the polymer's surface energy. This technique uses a gas or gas/oxygen flame. It works well for uneven profiles and thick substrates. It is easy to adjust and control the gas/oxygen ratio, flow rate, exposure time, and proximity of flame to the substrate. It is an effective method for use on PE (polyethylene) and PP (polypropylene). Thinner substrates are better suited to corona pretreatment.

Plasma Treatment

The process uses a plasma created by charging a gas with a high amount of energy. The free ions and electrons in the plasma clean the surfaces of any material it touches. On organic surfaces, plasmas, like flames, create polar groups/active radicals that boost surface energy and aid adhesion. Low pressure plasma is a technique that involves exciting a gas with high frequency and high voltage between two electrodes in a low pressure chamber. The process uses various plasmas of argon, ammonia, nitrogen, or oxygen, which make it suitable for a wide range of substrates.

With low pressure plasmas, air can serve as the plasma source. In this case, oxygen in air produces the greatest results because it reacts with carbohydrate-based contaminants and breaks up large-chain molecules. Handheld and automated devices are available for the method. Some devices can be fit to a robotic head so adhesives can go on immediately after pretreatment. The process creates surfaces with energies greater than 70 mJ/m^2 in many cases.

Low pressure plasma has an active bandwidth of 1.27 cm (0.5 in.). It can be used at up to 3000 fpm and its a candidate for large industrial applications. Among the plastics suitable for treatment with low pressure plasma are PP, PE, polyamide, polyethylene terephthalate (PET), acrylonitrile butadiene styrene (ABS), polycarbonate (PC), rubber, and composites.

Corona Discharge

This effect works like low pressure plasma in principle, but events take place in the air at atmospheric pressure. A corona is generated by applying high voltage (up to 30 kV) at frequencies ranging from 9 to 50 kHz to an electrode separated from a grounded table by an air gap. At 3–5 kV/mm, a current passes through the air gap to generate free electrons that move toward the positive electrode with great energy. The free electrons displace electrons from molecules in the air gap and, in turn, create more free electrons. The corresponding ions contribute to a current that flows across the gap. As ionization currents rise, the corona discharge rate also increases (i.e., the particles move faster). The resulting plasma then activates the surface of the plastic part onto which the discharge is directed. Corona discharge is especially well suited for thin films and composite laminates.

Chemical Treatment

It works well for bonding polymeric substrates. These treatments are usually applied by polymer manufacturers or companies specializing in chemical treating. Treatments used for polymers include etchants for polytetrafluoroethylene (PTFE), caustic soda etching for polyesters, proprietary primers for PP, sulfuric acid for polysulfide (PS), and the Sicor process for thermoplastics. Sicor was developed by the Commonwealth Scientific and Industrial Research Organisation (CSIRO) in Australia, primarily for the automotive industry.

Laser

Recently demonstrated was the use of a low power handheld laser system that could be used to pretreat aircraft surfaces prior to adhesive bonding surface preparation via a sol-gel process that uses a specific chemistry. Engineers conducted tests with a prototype laser system to validate this promising method for removing organic coatings, contaminants, and an oxide layer from aluminum surfaces in an environmentally friendly manner. Scientists also studied the surface morphology and chemistry produced by the laser pretreatment to determine how bond strength and moisture durability correlate with surface charateristics, and to compare laser pretreatment with other pretreatment methods.

Chemical paint stripping methods and sandblasting techniques create large amounts of waste and residue that require careful handling and disposal. Due to increasing restrictions and safety concerns raised by government agencies and commercial industry regarding these processes, material scientists have identified laser technology as a potential, environmentally friendly alternative. Engineers demonstrated that the use of a handheld laser can also eliminate laborious and time-consuming surface pretreatment processes required for conventional aircraft metal bonding techniques. A prototype handheld laser end effector using a commercially available, 10 W, neodymium (Nd): yttrium aluminum garnet (YAG) laser system was previously researched by laboratory engineers who were investigating laser cleaning of aircraft oxygen tubes and recognized the laser technology's potential for environmentally friendly surface cleaning and paint stripping.

During an initial laser surface pretreatment for metal adhesive bonding project, engineers optimized and evaluated the system's ability to remove organic coatings and to clean, deoxidize, and texturize the surface for bonding. They demonstrated the technology as an environmentally favorable alternative to chemically stripping, hand sanding, grit blasting, and chemically cleaning metallic surfaces. They also proved through lap shear, 90°, peel, and wedge crack extension testing that acceptable bond strength and moisture durability are attainable on a surface contaminated by baked-on hydraulic fluid prior to laser pretreatment and sol-gel surface preparation.

The Air Force Research Laboratory initiated a follow-up project to determine if commercial, handheld laser systems can properly strip, clean, deoxidize, and provide the surface morphology required for acceptable adhesively bonded aircraft repairs. Once again, engineers utilized the water-based, nonchromated sol-gel technology as the bonding treatment. In the final phase of the project, engineers plan to evaluate the Nd:YAG laser's capability to prepare titanium (using sol-gel) and composite surfaces for adhesive bonding.

Joint Design and Proper Adhesive Selection

The two most important considerations needed to achieve the following properties are careful joint design and proper adhesive selection.

- Uniform stress distribution and larger stress-bearing area.
- Outstanding fatigue as well as mechanical and thermal shock resistance.
- Continuous contact between substrates, producing good seals, and load-bearing properties.
- Can bond dissimilar substrate materials including metals, plastics, elastomers, ceramics, glass, and wood.
- Can bond different substrates at low or elevated temperatures.
- Smooth, contour-free surfaces, free from external projection and gaps.
- Less critical tolerances when components are joined because of gap-filling capability.
- Minimized galvanic (electromechanical) corrosion between dissimilar metal substrates.
- Thermal and/or electrical insulation or conductivity.
- Wide service temperature range capability.
- Long-term durability.
- Resistance to thermal or mechanical shock and vibrations.

Joint design requires the selection of an appropriate style, proper surface preparation, and use of careful application and assembly procedures. Special attention must also be given to the following recommended adhesive curing procedures. Joint design should minimize stress concentrations so the load is distributed over the entire area.

Joint design should be selected to enhance bond strength for the specific use under consideration. Popular joints used in structural bonding applications include butt, scarf, lap, and offset lap. Butt joints are used when stress concentrations are concentrated along the bond line and when forces perpendicular to the bond are minimal. Scarf joints allow for an ample adhesive bond area, but parts joined in this way must maintain closer fits. Lap and lap offset joints are recommended for thin sections and rigid parts. In lap joints, the bonded parts are slightly offset, thus peel and cleavage forces develop when the joints are under load. These forces can be minimized by using the offset lap joint design.

Surface preparation is critical for achieving high-strength structural bonds. Surfaces contaminated with oils, greases, dirt, moisture, and other contaminants must be carefully cleaned prior to adhesive application. Certain forms of oxidation, notably loose rust, must be completely removed. Other metallic substrates are preferred. The adhesive supplier's recommendation should be carefully followed to assure good results. Curing conditions must be duplicated to achieve desired performance.

Adhesive Types

The choice of adhesive makes a difference thus it is critical to produce durable joints.

Nanoglue

An adhesive based on self-assembling nanoscale chains for bonding materials that do not normally stick together is under development at Rensselaer Polytechnic Institute, Troy, New York. The glue material is already commercially available, but the research team's method of treating the glue to dramatically enhance its "stickiness" and heat resistance is completely new. The nanoglue nanolayer is sandwiched between a thin film of copper and silica, which strengthens the bond and boosts adhesive properties. Less than a nanometer thick, the nanoglue is inexpensive to make and can reportedly withstand temperatures far higher than what was previously envisioned. See Figure 2.2. When exposed to heat, the middle layer of the "nanosandwich" does not break down or fall off, as it has nowhere to go. Constrained between the copper and silica, the nanolayer's molecules hook onto an adjoining surface with unexpectedly strong chemical bonds. In addition, the nanolayer's bonds grow stronger and more adhesive when exposed to temperatures above 400°C (750°F) [2].

In another study iron oxide nanoparticles were inbedded in nanoparticles of silicon dioxide in a variety of adhesives that enables both bonding and disbonding by application of high-frequency alternating magnetic fields. The adhesives contain the MagSilica filler, which is nanostructured and contains a powder of superparamagnetic particles. When the adhesives are exposed to a high-frequency alternating magnetic field, the particles oscillate and heat the adhesive. This causes one- and two-component adhesives to harden

FIGURE 2.2
Self-assembling nanoglue can bond nearly anything together.

within seconds. The bonds can be dissolved by exposing the adhesive bond to another high-frequency magnetic field. This field has the same frequency as during hardening, but a higher intensity. For the method to work, at least one of the components to be bonded must be electrically nonconducting.

Flexible

Flexible adhesives provide joints with a more uniform distribution and less of a difference between average and maximum stress. These adhesives distribute peel and shear stresses over a larger area thereby improving joint efficiency. However, since adhesives with high flexibility and elongation typically have lower cohesive strength than more rigid adhesives, the advantage of flexibility and high elongation is usually compromised. In order to transfer the same load a much larger overlap is needed.

Three structural adhesives have been designed to produce strong, flexible bonds on composites and offer medium to long working times for large assemblies. The products are said to be ideal for use on boats, trucks, buses, architectural ornamentation, wind turbine blades, and composite bridges and decks. MA530, MA560-1, and MA-590 are two-part methacrylates that can be dispensed manually or with standard meter-mix equipment. The 100% reactive, nonsagging gels have been recommended for joining composite assemblies, ABS, acrylics, gel-coats, polyesters including DCPD modified (dicyclopentadiene), polyvinyl chloride (PVC), styrenics, and vinyl esters. They require little or no surface preparation and cure at room temperature to produce bonds that withstand service temperatures of –40–82°C (–40°F–180°F).

MA530 has a working time of 30–40 minutes and fills gaps up to 17.78 mm (0.70 in.). A 0.76 mm (0.03 in.) bond line achieves approximately 75% of its ultimate shear strength in 90–160 minutes at 23.3°C (74°F), according to preliminary tests. MA560-1 has a longer working time (55–70 minutes) and fills gaps up to 2.54 cm (1 in.). It is recommended for large structures as well as small part assembly where fewer workers are required. Fixture time for a 0.76 mm (0.03 in.) bond line is 220–240 minutes at 23.3°C (74°F). MA590 has a long working time of 90–105 minutes and was formulated for building large fiber reinforced plastics (FRP) boats. A 0.76 mm (0.03 in.) bond line achieves approximately 75% of its ultimate shear strength in 180–240 minutes at 23.3°C (74°F). This adhesive fills gaps up to 2.54 cm (1 in.).

For flexible joints, the latest polyurethanes and modified silanes withstand vibration and provide long-term durability. They are used for window glazing and roof-panel bonding on heavy-duty construction equipment, trailers, utility trucks, agricultural equipment, and school buses.

Pressure-Sensitive

A manufacturer of pressure-sensitive adhesives, says patient comfort can be a difficult goal when it comes to adhesives used on bandages for covering wounds or holding catheters in place. Sometimes medical devices must stick to skin for long periods of time while withstanding showering and exercising. New silk adhesives have been developed with good skin adhesion that can be removed cleanly and gently without hurting the skin. In addition, the adhesives are okay for sustained skin contact and remain suitable after sterilization. These new silk adhesive systems allow engineers to select adhesives with strength, quick tack, peel strength from skin, shear strength, and removability.

Waterproof Bandages

A newly developed waterproof adhesive bandage inspired by Gecko lizards may soon join sutures and staples as a basic operating room tool for patching up surgical wounds or internal injuries reported by MIT researchers. The MIT reseachers built the adhesive with a biorubber and, using micropatterning technology, shaped the biorubber into different hill and valley profiles at nanoscale dimensions.

The surface of the bandage has the same kind of nanoscale hills and valleys that allow the lizards to cling to walls and ceilings. Applied over this landscape is a thin coating of sugar-based glue that helps the biodegradable bandage stick in wet environments, such as to heart, bladder, or lung tissue. Because it can be folded or unfolded, the bandage is potentially suitable for minimally invasive surgical procedures that are performed through a very small incision. The adhesive could also be infused with drugs designed to release as the biorubber degrades. The elasticity and degradation rate of the biorubber are tunable as is the pillared landscape—allowing for customizable elasticity, resilience, and grip for different medical applications [3].

Electrically Conductive Adhesives

Electrically conductive adhesives (ECAs) are composites of polymeric matrices and electrically conductive fillers. The polymeric resin, such as an epoxy, silicone, or polyimide, provides physical and mechanical properties such as adhesion, mechanical strength, impact strength, and the metal filler (such as silver, gold, nickel, or copper) conducts electricity. Recently, ECA materials have been identified as one of the major alternatives for lead-containing solders for microelectronics packaging applications. ECAs offer numerous advantages over conventional solder technology, such as environmental friendliness, and mild processing conditions (enabling the miniaturization of electronic devices) [4–7]. Therefore, conductive adhesives have been used in flat panel displays such as liquid crystal display (LCD), and smart card applications as an interconnect material and in flip-chip assembly, chip scale package (CSP) and ball grid array (BGA) applications in the replacement of solder. However, no currently commercialized ECAs can replace tin–lead metal solders in all applications due to some challenging issues such as lower electrical conductivity, conductivity fatigue (decreased conductivity at elevated temperature and humidity aging or normal use condition) in reliability testing, limited current-carrying capability, and poor impact strength. Table 2.5 gives a general comparison between tin–lead solder and generic commercialized ECAs [8,9].

Depending on the conductive filler loading level, ECAs are divided into isotropically conductive adhesives (ICAs), anisotropically conductive adhesives (ACAs), and nonconductive adhesives (NCAs). For ICAs, the electrical conductivity in all x-, y-, and z-directions is provided due to high filler content exceeding the percolation threshold. For ACAs or NCAs, the electrical

TABLE 2.5

Conductive Adhesives Compared With Solder

Characteristic	Sn/Pb Solder	ECA
Volume resistivity	0.000015 Ohm-cm	0.00035 Ohm-cm
Typical Junction R	10–15 mW	<25 mW
Thermal conductivity	30 W/m°K	3.5 W/m°K
Shear strength	2200 psi	2000 psi
Finest pitch	300 μm	<150–200 μm
Minimum processing temperature	215°C	150–170°C
Environmental impact	negative	very minor
Thermal fatigue	yes	minimal

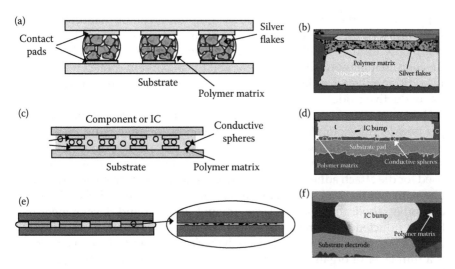

FIGURE 2.3
Schematic illustrations and cross-sectional views of (a,b) ICA, (c,d) ACA, and (e,f) NCA flip chip bomdings.

conductivity is provided only in the z-direction between the electrodes of the assembly. Figure 2.3 shows the schematics of the interconnect structures and typical cross-sectional images of flip-chip joints by ICA, ACA, and NCA materials illustrating the bonding mechanism for all three adhesives.

Isotropically Conductive Adhesives (ICAs)

Isotropic conductive adhesives, also called polymer solder, are composites of polymer resin and conductive fillers. The adhesive matrix is used to form a mechanical bond for the interconnects. Both thermosetting and thermoplastic materials are used as the polymer matrix. Epoxy, cyanate ester, silicone, polyurethane, and so on are widely used thermosets; and phenolic epoxy; maleimide, acrylic, preimidized polyimide, and so on are the commonly used thermoplastics. An attractive advantage of thermoplastic ICAs is that they are reworkable (i.e., can easily be repaired). A major drawback of thermoplastic ICAs, however, is the degradation of adhesion at high temperatures. Another drawback of polyimide-based ICAs is that they generally contain solvents. During heating, voids are formed when the solvent evaporates. Most commercial ICAs are based on thermosetting resins. Thermoset epoxies are by far the most common binders due to the superior balanced properties, such as excellent adhesive strength, good chemical and corrosion resistances, and low cost, while thermoplastics are usually added to allow softening and rework under moderate heat. The conductive fillers provide the composite with electrical conductivity through contact between the conductive particles. The possible conductive fillers include silver, gold, nickel, copper, and carbon in various forms (graphites, carbon nanotubes, etc.), sizes, and shapes. Among different metal particles, silver flakes are the most commonly used conductive fillers for current commercial ICAs because of the high conductivity, simple process, and the maximum contact with flakes. In addition, silver is unique among all the cost-effective metals by nature of its conductive oxide.

Oxides of most common metals are good electrical insulators, and copper powder, for example, becomes a poor conductor after aging. Nickel and copper-based conductive adhesives generally do not have good resistance stability, because both nickel and copper are easily oxidized. ICAs have been used for die attach adhesives. Recently, ICAs have also been considered as an alternative to tin/lead solders in surface mount technology (SMT), flip-chip, and other applications and a large amount of effort and development has been conducted to improve the properties of ICAs in the past years [10–14]. An ICA is generally composed of a polymeric binder and silver flakes. There is a thin layer of organic lubricant on the silver flake surface. This lubricant layer plays an important role for the performance of ICAs, including the dispersion of the silver flakes in the adhesives and the rheology of the adhesive formulations [15,16].

Anisotropic Conductive Adhesives (ACAs/ACFs)

Anisotropic conductive adhesives (ACAs) or anisotropic conductive films (ACFs) provide unidirectional electrical conductivity in the vertical or Z-axis. This directional conductivity is achieved by using a relatively low volume loading of conductive fillers (5–20 vol.%). The Z-axis adhesive, in film or paste form, is interposed between the surfaces to be connected. Heat and pressure are simultaneously applied to this stack-up until the particles bridge the two connector surfaces.

The ACF bonding method is a thermocompression bonding process shown in Figure 2.4. Interconnection technologies using ACFs are major

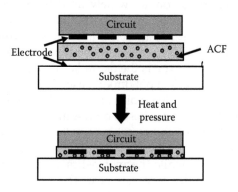

FIGURE 2.4
Thermocompression bonding using ACF.

packaging methods for flat panel display modules and have high resolution, light weight, thin profile, and low consumption power. In addition to the LCD industry, ACA/ACF is now finding applications in flex circuits and SMT for chip-scale package (CSP), application specific integrated circuit (ASIC), and flip-chip attachment for cell phones, radios, personal digital assistants (PDAs), sensor chip in digital cameras, and memory chip in laptop computers. The ACA/ACF joints generally have lower electrical conductivity, poor current carrying capability, and electrical failure during thermal cycling.

Nonconductive Adhesive/Films (NCAs/NCFs)

Electrically conductive adhesive joints can be formed without any conductive fillers. The electrical connection of NCAs is achieved by sealing the two contact partners under pressure and heat. Thus, the small gap contact is created, approaching the two surfaces to the distance of the surface asperities, as illustrated in Figure 2.3e. The formation of contact spots depends on the surface roughness of the contact partners. Approaching the two surfaces enables a small number of contact spots to form that allows the electric current to flow. When the parts are pressed together during the sealing process, the number and area of the single contact spots are increased according to the macroscopic elasticity or flexibility of the parts and the microhardness and plasticity of the surfaces, respectively.

Conductive joints with NCA/NCF provide a number of advantages compared to other adhesive bonding techniques. NCA/NCF joints avoid short-circuiting and are not limited, in terms of particle size or percolation phenomena, to a reduction of connector pitches. The size reduction of electronic devices can be realized by the shrinkage of the package and the chip. Further advantages include cost-effectiveness, ease of processing regarding the possibility of nonstructures adhesive application, good compatibility with a wide range of contact materials, and low temperature cure. In fact, the pitch size of the NCF joints can be limited only by the pitch pattern of the bond pad, rather than the adhesive materials.

Summary

Polymer metal-based, electrically conductive adhesive has been evolved to meet the high electrical/mechanical/thermal performance, fine pitch capability, low temperature process, and strong adhesion/reliability requirements for electronic packaging module and assemblies. As one of the most promising, lead-free alternatives in electronic and optoelectronic device packaging interconnect materials, electrically conductive adhesives have shown remarkable advantages and attracted many research interests. Significant improvements in the electrical, mechanical, thermal properties of different types of conductive adhesives have been achieved. Electrically conductive adhesives with high performance and acceptable reliability have been developed for applications in die attach, flip-chip, and surface-mount interconnect applications. These lead-free materials and processing methods have great potential to replace the lead-containing solders in the electronics industry.

Conductive

A new conductive adhesive technology has recently been introduced to the marketplace, delivering dramatic improvements in thermal performance, drop test reliability, and providing significant personnel and equipment savings. With this new technology, the traditional mixing and measuring of previous systems (not to mention the issues associated with managing pot life) are completely eliminated. The advanced adhesive technology, called Bead-On-Bead, is a remarkable new product that has the potential to completely change the way the market thinks about adhesive systems.

Loctite Bead-On-Bead technology is a two-part system based on proprietary technology that enables simple, yet incredibly strong, bonding of heat-dissipating devices to heat sinks and spreaders. Ease of use is one of the premiere benefits of this extraordinary system. Bead-On-Bead requires no meter mixing, no volumetric measuring, no management of pot life and, because of its color characteristics, visual inspection is made easy. Each part (parts A and B) are applied to the electronic device and heat dissipating device separately, so no mixing is required. When parts A and B combine during assembly, cure is achieved and fixture strength is realized within seconds. All of these process enhancements come with the added benefit of improved thermal performance. The product has a thermal conductivity of 1.75W/ m°K, which is a 40% improvement over the older generation products.

The Bead-On-Bead adhesive technology has been applied to an light emitting diode (LED) and a heat sink in a flashlight and by adjusting the pressure for the seating of the LEDs to the heat sink combined with the robust Bead-On-Bead adhesive system, bond lines yielded joints close to 0.13 mm (0.005 in.) thick. In addition, because these flashlights are often dropped, a stringent drop test requirement was imposed of 1500 cm (50 ft), which Bead-On-Bead satisfied, delivering very high reliability.

This new technology also provided a solution for inspection.The technology provided a solution where Bead-On-Bead's unique color coded system delivered increased manufacturing control. The adhesives are colored part yellow and part blue and when the parts are assembled, the bond line becomes green, allowing for very simple visual inspection.

Another new high temperature resistance conductive glue offers an adhesive that is suitable for making an electrically conductive connection to a material or system that is operated at high temperatures or can even be a heating element. The technology was developed by Philips and SIMTech as part of the flat heating technology for clothes irons, where a need for a reliable, strong, and cost-efficient glue was required as techniques such as soft solder-ing had too low a melting point and hard solder too high a melting point for use with aluminum plates. The glue developed was a super glue based on sol-gel precursors that had all the properties that were required. The product is currently commercially available in clothes irons from Philips, and is also proven in flat panel heaters where glues and adhesives were not suitable due to high temperatures and high power densities. This new adhesive utilizes complex sol-gel precursors resulting in a hybrid sol-gel material.

Structural

Recent advances in structural adhesives deliver the same strength as welds, or go through paint baking unscathed. Other new adhesives require no surface cleaning, a process that was once a given before any bonding could commence. Today's structural adhesives are fomulated to withstand severe shock, peel, and impact. They bear heavy loads, withstand chemicals, endure extreme temperatures, absorb energy, and can take large deformations with-out rupturing.

Specialty-vehicle manufacturers, for example, now use structural adhe-sives to replace or augment rivets, bolts, welding, and other traditional fas-tening methods. Adhesives rather than fasteners assemble frames, panels, booms, and cabs. Better product aesthetics result from adhesive assembly and, thus, can greatly boost vehicle value. Structural adhesives also simplify production by eliminating the need to precisely bore holes for installation of rivets or other mechanical fasteners. Likewise, in most environments such as bathtubs and spas, structural adhesives attach and bond galvanized steel frames to fiberglass and ABS. Commercial furniture manufacturers bond plain, painted, or powder-coated metals and plastics in chairs, desks, and cabinets.

Structural-adhesive technologies include epoxies, acrylics, methylmetha-crylates (MMAs), modified silanes, and polyurethanes. Adhesives can be tailored to deliver processing and performance benefits. For example, adhe-sives come as one-part no-mix or two-part mix systems. They can cure under ambient room temperature or at elevated temperatures in cure ovens. For high-speed assembly, there are fast-curing adhesives as well as those with long work lives for parts that need alignment once they are in place. High, medium, and low viscosity formulations can fill large gaps or provide thin, virtually invisible bond lines.

It is usual practice to add dispensing and curing equipment to a specific assembly process. This lets manufacturers optimize process efficiency, keep down waste, and make dispensing and curing repeatable and consistent. Advances in structural-adhesive technology have dramatically expanded the scope of potential bonding applications. Over time, traditional structural adhesives lose strength on substrates such as galvanized steel. Here, the zinc coatings used on the galvanized steel eventually delaminates the bond.

New structural adhesives are providing long-term durability on such hard-to-bond substrates. Prefabricated building components including galvanized trusses, joists, shear walls, headers, studs, and cantilevered beams are also adhesively bonded. Traditional structural adhesives can also have trouble maintaining long-term strength when exposed to elevated temperatures. This data is significant for a manufacturer of electric motors trying to decide whether to attach magnets using mechanical clips or adhesives. Electric motors continuously operate at elevated temperatures. Adhesive joints in the magnet assembly will gain strength over time and will withstand vibration better than mechanical fasteners.

Adhesives are also making inroads in loudspeakers that are getting smaller with each new generation. Speaker manufacturers are demanding impact and temperature-resistant structural adhesives that cure rapidly. As speakers get smaller, adhesives must also withstand high temperatures that result from tighter operating spaces. The demand for more portable speakers also means adhesives must have higher resistance to impacts. And to improve the bottom line, speaker manufacturers demand fast-fixturing adhesives to speed production, reduce work-in-process, and eliminate mechanical-fastener inventories.

Environmental considerations can also greatly affect the long-term performance of an assembly. A two-part MMA, for example, was exposed to salt, fog, and humidity for 1,000 hours. Here the adhesive maintains more than 95% of its strength when compared to the control sample, see Table 2.6. Recently introduced structural adhesives can withstand welding, phosphate pickling, and powder-coating processes while maintaining strength. And some can replace costly metallic inert gas (MIG) or tungsten inert gas (TIG) welds by bonding two substrates and supporting the bond with low-cost spot welds that hold the assembly together while the adhesive cures. These fast-fixturing, weld-tolerant adhesives let asemblies go in pickling and paint-bake processes within 20 minutes.

The MMA adhesives can bite through surface oils and contaminants, eliminating the need for surface preparation and activators. For traditional adhesives, surface preparation has always been critical to ensure a long-lasting, reliable bond. See the section earlier on surface preparation. Surface treatment requirements depend on the level of contamination, the substrates, the initial and long-term bond performance, and the financial practicality of the treatment process. For difficult-to-bond substrates such as polyolefin plastics, new adhesives bond exceptionally well within the need for primers or surface treatments.

Recently developed structural adhesives seal out moisture and gases in applications where substrates have different coefficients of thermal expansion (CTE). For example, paintable, clear, UV-resistant modified silanes are being used for bonding appliance assemblies, truck bodies, and in HVAC applications. Structural adhesive bonding offers significant technical as well as economic advantages over mechanical fasteners as well as joining

TABLE 2.6

Structural Adhesives

	Epoxy	Two-Part MMA	Two-Part Acrylic	Two-Part Urethane	Modified Silane
Benefits	High-temperature resistance, chemical resistance, low shrinkage, high shear and tensile strength, bond glass	Impact resistance; good elongation/flexibility; fills gaps; high peel, tensile, and shear strength; bonds metals, composites, and plastics	Impact resistance, high-temperature resistance, high shear and tensile strength, no mixing required, rapid cure, bonds metal and glass	High elongation/flexibility, toughness	Flexibility, paintable, sealing properties; no mixing required
Limitations	Low elongation/flexibility, low adhesion to plastics and rubbers, mixing required for two-part systems	Low adhesion to rubber, mixing required, strong odor	Need an activator, low adhesion to plastics and rubber, limited gap filling	Sensitive to moisture, mixing required, contains isocyanates	Limited temperature resistance, low resistance to nonpolar solvents, low adhesion to rubber, extended cure time

techniques such as soldering, brazing, or welding for many demanding assembly constructions. Epoxy polymer-based resin systems have become the engine for the growth of structural bonding because of their unmatched processing versatility, high-strength, low-weight, and wide-service temperature capabilities, even when exposed to hostile environmental conditions. Their strength properties can be further improved by compounding with glass, carbon, or polyimide reinforcing fibers. Both thermally conductive and also electrically conductive adhesive formulations are available and are widely used for many electrical, mechanical, and medical devices.

Fiber-reinforced, epoxy-resin composites can compete economically with steadily increasing success against conventional metal constructions both on the basis of a more favorable strength/weight ratio and enhanced corrosion-free service life. Their industrial success is demonstrated by the fact that some 60% of today's aerospace structures are manufactured with epoxy-resin adhesive systems. Other polymer adhesive candidates, such as polyimides, are mainly considered for specialty demanding applications. Thermal stability and chemical resistance as well as specific curing conditions are mainly determined by the choice of the base resin and the hardener components. The workhorse of the structural adhesives field are the so-called bisphenol A-type epoxy resins and amine type hardeners. Epoxy-base resin and hardeners can be either liquid or solid with liquids generally preferred. They are also available in the form of both supported and unsupported semi-cured films.

Depending on the choice of base resin and hardener, an epoxy-adhesive composition can cure as fast as 45 seconds or as slowly as 48 hours or more at ambient or quickly elevated temperatures. The bond strength, mechanical strength properties, toughness, flexibility, electrical and thermal conductivity, thermal and mechanical shock resistance, thermal expansion coefficients, and so on can be varied greatly depending on the selection and amount of the various other ingredients added to the base resin-hardener system. The number of such additives that can be added to optimize specific desirable processing and/or performance properties is steadily increasing. The performance of epoxy adhesives can thus be designed to maximize such characteristics as physical strength, heat stability, chemical resistance, thermal shock, and electrical insulation for different end product applications. They are used for bonding metal and nonmetallic substrates, including most plastics, ceramics, glass, and wood, as well as many elastomers.

Structural epoxy-adhesive systems feature the highest tensile strength of all commercially available bonding agents. Resistance to moisture, fuels, oils, acids, bases, and many other aggressive chemicals is of a very high order over a wide service temperature range. They can safely be operated at service temperatures from as high as 300°C (572°F) to cryogenic conditions. Most advantageous for structural design reliability are their outstanding gap-filling properties. Only positive contact pressure is generally required to effect cures at the temperature/time schedules recommended by the adhesive manufacturer. When bonding dissimilar metals, the epoxy-adhesive bond also functions as protection against corrosion.

The most widely used epoxy adhesives are one-and two-component liquids or pastes. The two-component systems may be cured at ambient or, more quickly, at elevated temperatures. One-component formulations require a cure at elevated temperatures. Recently, one-component epoxy-types of adhesives have been introduced that can be cured in seconds by exposure to UV light. Dual cure adhesive systems capable of use by either exposure to UV light or heat have been successfully developed for certain sensitive plastic substances as well as assemblies where an ingress of UV light in selected areas is limited. Substantial technical progress has been made to enhance toughness, shear strength, thermal stability, and chemical resistance even upon prolonged exposure to hostile environments. The demands of the aerospace and defense industries for lighter-weight, high-strength, and heat-resistant structures have stimulated these developments. Manual, semiautomatic, and automatic equipment for dispensing such one-and two-component epoxy-resin adhesive composites is readily available commercially from many manufacturers.

Major industrial and product applications for structural epoxies include:

- Aerspace and defense
- Oil-servicing industry
- Sporting goods
- Civil engineering structures
- Electronic assemblies
- Medical devices
- Industrial processing
- Automotive components

The number of applications are steadily growing, fueled by the increasing need for lighter weight corrosion-resistant and more cost-effective assemblies.

Animal

The race for the best "Gecko foot" dry adhesive has a new competitor with a stronger and more practical material reported in the journal *Science* [17]. Scientists have long been interested in the ability of Gecko lizards to scurry up walls and cling to ceilings by their toes. The creatures owe this amazing ability to microscopic branched elastic hairs in their toes that take advantage of atomic-scale attractive forces to grip surfaces and support surprisingly heavy loads (Figure 2.5) [17]. Several research groups have attempted to mimic those hairs with structures made of polymers or carbon nanotubes.

FIGURE 2.5
Closeup of a gecko foot, showing the pads that bear microscopic branched elastic hairs that use atomic-scale forces to grip surfaces.

Scientists and researchers have developed and improved carbon nanotube-based material that for the first time creates directionally varied (anisotropic) adhesive force. With a gripping ability nearly three times the previous record—and 10 times better than a real gecko at resisting perpendicular shear forces—the new carbon nanotube array could give artificial gecko feet the ability to tightly grip vertical surfaces while being easily lifted off when desired. Beyond the ability to walk on walls, the material could have many technological applications, including connecting electronic devices and substituting for conventional adhesives in the dry vacuum of space. The resistance to shear force keeps the nanotube adhesive attached strongly to the vertical surface, but you can still remove it from the surface by pulling away from the surface in a normal direction. This directional difference in the adhesion force is a significant improvement that could help make this material useful as a transient adhesive.

The key to the new material is the use of rationally designed, multiwalled carbon nanotubes formed into arrays with "curly entangled tops." The tops (can be compared to spaghetti or a jungle of vines) mimic the hierarchical structure of real gecko feet, which include branching hairs of different diameters. When pressed onto a vertical surface, the tangled portion of the nanotubes becomes aligned in contact with the surface. That dramatically increases the amount of contact between the nanotubes and the surface, maximizing the van der Waals forces that occur at the atomic scale. When lifted off the surface in a direction parallel to the main body of the nanotubes, only the tips remain in contact, minimizing the attraction forces. In tests done on a variety of surfaces—including glass, a polymer sheet, Teflon, and even rough sandpaper—the researchers measured adhesive forces of up to 100 Newtons per square centimeter in the shear direction. In the normal direction, the adhesive forces were 10 Newtons per square centimeter—about the same as a real Gecko lizard.

The resistance to shear increased with the length of the nanotubes, while the resistance to normal force was independent of tube length. Though the material might seem most appropriate for use by the comic book hero, Spider-Man, the real applications may be less glamorous. Because carbon nanotubes conduct heat and electrical current, the dry adhesive arrays could be used to connect electronic devices.

Thermal management is a real problem today in electronics, and if you could use a nanotube dry adhesive, you could simply apply the devices and allow van der Waals forces to hold them together and that would eliminate the heat required for soldering. Another application might be for adhesives that work long-term in space. In space, there is a vacuum and traditional kinds of adhesives dry out. But nanotube dry adhesives would not be bothered by the space environment.

For the future, researchers hope to learn more about the surface interactions so they can further increase the adhesive force. They also want to study the long-term durability of the adhesive, which in a small number of tests became stronger with each attachment. And they may also determine how much adhesive might be necessary to support a human wearing tights and red mask. Because the surfaces may not be uniform, the adhesive force produced by a larger patch may not increase linearly with the size. There is much to learn about the contact between nanotubes and different surfaces.

Low Energy Substrate Technology (LESA)

Advances in low energy substrate adhesive (LESA) technology are helping spur interest in hybrid systems. The materials were developed to bond hybrids and all-plastic structures together without mechanical fasteners. The two-component structural adhesives graft to low energy surfaces and create a sturdy durable bond. And they are suited for materials that can be tough to bond with other types of adhesives. For instance, LESA bonds thermoplastic and thermoset resins, including polypropylene, polyethylene, polystyrene, PTFE, PET, ABS, and others without any surface treatment. It also bonds many thermoplastic-olefin elastomers (TPOs), as well as coated metallic substrates and glass.

Overall adhesive properties and long-term durability make LESA a good candidate for structural bonding of metal/plastic hybrids. For instance, one commercial adhesive performs well at automobile temperatures −40–120°C (−40–248°F). It has a tensile strength of 22.1 MPa (3250 psi) and, when subjected to extreme loads, the plastic substrate typically fails before the adhesive interface. Environmental weathering does not adversely affect performance. As with other adhesives, LESA can replace or supplement welds and mechanical

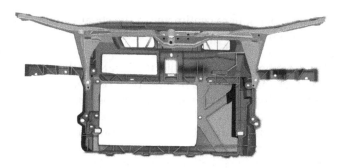

FIGURE 2.6
The front-end carrier on Volkswagen's new Polo features steel reinforcement bonded to a plastic framework with Betamate low-energy-substrate adhesive from Dow Automotive. The resulting structure reduces weight by 25% compared with previous versions, lowers costs by 10%, and meets structural requirements.

fasteners. It applies and cures at room temperature and requires no extra pressure during curing—the only need is to maintain contact between parts. The benefits of adhesive-bonded hybrid structures has recently been applied in the automobile front-end carrier by Volkswagen, (Figure 2.6). A front-end carrier (FEC) supports most of the cooling system, headlights, hood/trunk latch, and various other components. It connects the upper and lower longitudinal body rails and plays a role in vehicle structural stiffness. Traditional steel FECs are bolted or welded to the body. Adding subcomponents, however, requires welding attachment brackets that increases weight and cost. A recent trend in the automobile industry has been the move to modular systems, which has opened the door to plastics. Molded-plastic FECs consolidate parts as well as reduce weight and cost. However, in most cases, it still takes a hybrid structure of metal and plastic to resist loads on the carrier, particularly in a crash. In traditional hybrid front-end carriers, metal reinforcements attach to the plastic using rivets or heat stakes, or through overmolding. This new solution uses a Dow Automotive adhesive, Betamate, and LESA technology to join the plastic molding and metal reinforcement. The adhesive produces a continuous load-bearing joint between metal and plastic parts, improving stiffness and reducing the high-stress point loads associated with typical mechanical fasteners. It also enables bonding without pretreatment. The assembly techmique lets designers use molded box sections, which maintain structural integrity while permitting more cost-effective thermoplastics, such as long-glass, fiber-reinforced polypropylene and thiner wall sections. The FEC bonded with Betamate LESA is now in production in the new Volkswagen Polo in Europe. The Polo's new front end reduces weight by 25%, compared with previous solutions, and meets rigidity demands at reduced cost. The FEC also meets or exceeds requirements for vibration, hood slam, and latch pull under various temperature and environmental conditions. It offers opportunities to integrate components such as air-intake ducting into the carrier, and facilitates a design that improves pedestrian safety in terms of leg impact.

Hot Melt

A hot melt, solvent-free alternative to the aerosol adhesives commonly used to position and secure reinforcement materials during the vacuum infusion process has been developed. The VAC-TAC adhesive system is designed to hold in place materials such as reinforcement fabrics, core materials, breathers, peel plies, and release films. Some boat builders who use the vacuum infusion process to manufacture one-piece hulls and lower decks have switched from using a traditional aerosol contact adhesive to this new adhesive system. Manufacturers found that they could build the composite structures much faster, saving time and reducing production cost as well as eliminating health and safety issues because the old system depended on the use of a solvent-based contact adhesive.

The VAC-TAC system incorporates a hot melt glue gun, 6 m (.2 ft) of hose, adhesive and toolbox carry case. The adhesive is applied in a hot melt cartridge and the glue gun is connected to an air and electrical supply. The adhesive is melted inside the gun chamber and sprayed directly onto one of the substrates. The materials can be bonded instantly although the adhesive remains sticky for up to 5 minutes, allowing for positioning of the substrates.

Protein

The use of protein-based adhesive materials is being developed for surgical adhesives and implant coatings based on mussels and other marine organism's secretions for adherence to the substrates upon which they reside. Secreted as fluids, these protein adhesives undergo an in-situ cross-linking or hardening reaction leading to the formation of a solid adhesive plaque, which mediates the attachment of the organism to a variety of substrates. By combining key components with a variety of polymers, novel synthetic constructs have been created for both adhesives and antifouling coatings [18].

Epoxy

When you look out your airplane window in a few years, you may not see the familiar rows of rivets. Believe it or not, adhesive bonding has advanced to the point where it can compete with high-performance traditional fastening technologies. Advanced adhesive-bonding technologies are increasingly pulling their weight in high-performance structural assemblies. Besides replacing traditional mechanical fasteners like rivets, nuts and bolts, and screws, they are competing with other bonding techniques like brazing and welding.

Structural adhesives have tensile strengths in the 54.4–74.8 MPa (8000–11,000 psi) range and compressive strengths between 136 and 272 MPa (20,000 to 40,000 psi). Shear stresses act in plane with the two substrates to move them in opposing directions. Most structural adhesives can take 17–23.8 MPa (2500–3500 psi) of shear stress at room temperature. This property is the one most commonly affected by water or chemical attack, so screening tests often take place both with and without water or fluid immersion. Structural epoxies, like Master Bond's EP17HT, can resist moisture, acids, bases, and temperatures up to 316°C (600°F). The epoxy has been used to bond stainless steel bolts to a stainless steel ring that will see 232°C (450°F) service in an acidic environment.

Designers must weigh the stresses the joint will see against assembly considerations to choose the joint best suited to a particular application. The butt joint is the simplest, with two parts bonded end-to-end. Scarf joints are similar, but have the ends of the joining parts beveled at matching angles for more surface area and shear resistance. Lap joints allow even more bonding area, but result in offfset surfaces that can be susceptable to peel. This can be resolved with an offset or joggle lap where one part is formed to compensate for the bond line thickness. Machined laps and double laps are other options, but they entail additional manufacturing steps. A strap joint combines the end-to-end butt joint with the greater surface area of a double lap joint.

Anisotropic Conductive Paste (ACP)

For nearly two decades [19], flip-chip technology has been widely used and accepted for several mainstream and high-end applications including flat panel displays and semiconductor modules in the form of chip (die) on glass (COG) and chip (die) on flex (COF). To facilitate the interconnect within these applications, ACFs have generally been the preferred material for several reasons: ACFs offer excellent low contact resistance and compatibility with noble metallization, they have outstanding adhesion to glass and offer highly reliable interconnection on fine resolution lines. But, the higher materials costs of ACFs combined with their multiple step processing requirements and higher cure temperatures make them a less than ideal solution for some of today's lower cost consumer applications.

As the benefits of flip-chip technology have been realized for high-end applications, this process is also finding favor among manufacturers in the consumer products realm. In particular, these devices have proliferated in applications such as radio frequency identification (RFID) and mobile phones. But lower cost consumer electronic products pose different challenges than their high-end counterparts and, therefore, require alternative and more cost-effective materials to facilitate high yield, low-cost production. Particularly in the case of RFID devices, which are found in everything from pet ID tags to department store inventory oversight, controlling the production cost and enabling extremely high-throughput rates is essential to widespread use of the technology. For RFID assembly and many other low-cost applications, the interconnect adhesive must be less costly than alternative materials, deliver faster throughput, be compatible with lower processing temperatures and provide very good compatibility with nonnoble metals such as etched copper and etched aluminum. Because ACFs don't deliver on these requirements, anisotropic conductive pastes have emerged as the most cost-effective interconnect material for RFID assembly. But, as RFID inlay manufacturers continue to reduce the cost of the tags by using lower cost substrates such as PET or also known as polyester and antennae metallizations like etched aluminum, die cut aluminum, and etched copper, traditional anisotropic conductive paste (ACP) materials aren't offering the robust performance these devices demand. The most common issue with conventional ACPs when used with nonnoble RFID antenna metallizations is that they are prone to galvanic corrosion when subjected to high humidity and high temperature environments.

Because of this, material specialists have developed a newer generation ACP that delivered the known benefits of existing ACP formulations—snap cure capability at low temperature, strong adhesion to etched aluminum and etched copper, long work life at ambient temperature, and a low interconnect contact electric resistance—and extend them to include corrosion resistance. Building on an existing and well-proven Henkel ACP material [19], chemical scientist's applied design of experiment (DOE) methodology to modify the formula, making it corrosion-resistant in humid, higher temperature environments and its rheology more suitable to jet dispensing. Through filler modification and mixture optimization, a robust ACP formulation that addressed all of the needs of modern, low-cost RFID assembly was created and engineered.

The new material, called Acheson CE-3126 [19], not only meets all of the processability requirements of good adhesion strength, snap cure capability, and a long work life, but also has a much improved reliability performance on nonnoble metal substrates as compared to older generation versions. Three materials were evaluated: etched aluminum, etched copper, and vapor deposited copper. In all three cases, the above adhesive's contact joint resistance remained less than 1 ohm after aging in both 85°C (185°F) at 85% relative humidity (RH) up to 168 hours and after 200 thermal shock cycles. The material also offers exceptional process versatility, as it can be screen printed, dispensed, or jetted to address a variety of manufacturing preferences.

RFIDs and other low-cost electronic devices are part of our everyday life—sometimes even unbeknownst to us. The technology that affords our common conveniences must also address consumer-driven cost pressures and RFID specialists, who have turned to lower cost substrate and antenna metallizations now require complementary materials to enable robust assembly, performance, and reliability.

Methacrylate Adhesives

Building a sheet-metal structure? It's likely you might first think of welding when picking an assembly technique. But for some applications, durable adhesives deliver the desired aesthetic.

Tooling Research Inc. (TRI) [20], Walpole, Massachusetts, primarily designs and builds custom electrical, mechanical, and pneumatic equipment and controls as well as machining metal and plastic components. Recently, TRI was tapped to provide outdoor signs for a college campus, the plans called for thin sheets of aluminum to be mounted on a structural framework. The finished sign panels had to be lightweight and removable as well as weather resistant.

Typically, such sheets are welded to a structural backing. But welds distort the sheet metal, leaving divots on the surface. Sign makers must repair the dents with body filler. Filling and smoothing can take several applications and add time and expense to the project. Mechanical fasteners also deform sheet metal. Sheets must be drilled or punched before joining, and installers still have filling and smoothing tasks to perform afterward. Fasteners are also prone to leak or loosen from weather and vandalism.

For the college campus signs, TRI used an adhesive from Devcon, Danvers, Massachusetts. The two-part, 10:1 methacrylate mixes as workers dispense it from manual or pneumatic mixing nozzles. There's no need for a primer, and bonds attain functional cure in an hour at room temperature. Once the adhesive is cured, which takes 24 hours, it can hold 16.6 MPa (2450 psi) in tensile shear using a 0.05 cm (0.02 in.) bond line on grit-blasted substrates. The adhesive spreads loads over the structural framework instead of concentrating them at fastening points. This adds strength. The adhesive also acts to separate substrates electrically, diminishing the likehood of corrosive electrolysis. The adhesives use 0.076 cm (0.03 in.) diameter glass beads to control bond thickness. "The adhesive works particularly well on panels and thin sections where clean lines and surfaces are the ultimate goal," says TRI President M. Florst.

Keeping that clean look can involve tight tolerances. For this level of precision, TRI uses an adhesive that permits bonds as thin as 0.03 cm (0.01 in.). This Metal Welder I adhesive has a working time of 5–6 minutes, while Metal Welder II sets up in 14–16 minutes. Plastic Welder II has an 18-minute working time. This adhesive, Plastic Welder II, which is a 1:1 two-part mix that cures in 2 hours at room temperature and can join nylon alloys and polyesters, in addition to standard epoxy composites, metals, and wood. It resists UV aging and weathering.

An adhesive's working time is also important when building complex signs. There could be three or four people working at the same time—one gluing, one holding, and one clamping—to complete an assembly before the adhesive sets up.

Some Significant Applications and Future Potential

1. Under the U.S. Navy's Surface Strike Affordability Initiative, the Navy Joining Center is leading a project team that is developing an adhesive bonding technology for composite-to-steel structures [21]. While this project is directly supporting the design of the DD(X) multimission destroyer, the technology will have applications for other surface combatant ships including aircraft carriers. The project is being performed by a team with participants from shipyards, aerospace companies, research institutes, and academic universities. The DD(X)s composite deckhouse and helicopter hanger are key components that will help the destroyer to achieve the operational requirements for improved performance, increased survivability, and low ownership cost. These structures take advantage of lightweight, nonmetallic, composite materials to reduce structural weight, ship signature, and in-service maintenance. These new joining techniques are needed to take full advantage of the benefits of composite materials. Currently, mechanical fasteners are used to join composites to steel on navy ships. These fasteners are expensive and their installation procedures are labor intensive. Mechanical fasteners also can become sites for corrosion that often require maintenance during the life of the ship. This project has developed cost-effective adhesive joining technology in response to the DD(X) program's need for improved connections. The joints have been designed to carry both structural and combat loads between major composite and steel ship structures.

   Adhesive bonding the composite and steel structures not only eliminates the labor costs associated with mechanical fasteners, but the adhesive-bonded joints have been demonstrated to be 40% lighter and 50% less expensive to produce than the existing bolted joint configuration. The new joints also improve the ship's signature and require less in-service maintenance. Successful designs, analyses, manufacturing development, and testing have resulted in adhesive-bonded joints being selected as the baseline method of attaching the composite deckhouse to the steel hull of the DD(X). This newly developed adhesive bonding technology has already been used to manufacture demonstration articles and test components. The destructive tests of these components have verified the ability of bonded joints to meet or exceed DD(X) design requirements. The project also included development of methods for nondestructuve inspection and repair of the joints. These techniques can be applied to both newly manufactured joints as well as joints that have been in service. The vital technology emerging from this effort will serve as a building block for all future composite ship designs and will help meet the needs of the twenty-first-century U.S. Navy.

2. A process for hydroxide-assisted bonding [22] has been developed as a means of joining optical components made of ultra low expansion (ULE) glass, while maintaining sufficiently precise alignment between. The process is intended mainly for use in applications in which (1) bonding of glass optical components by use of epoxy does not enable attainment of the required accuracy and dimensional stability and (2) conventional optical contacting (which affords the required accuracy and stability) does not afford adequate bond strength. The basic concept of hydroxide-assisted bonding is not new. The development of the present process was prompted by two considerations: (1) The expertise in hydroxide-assisted bonding has resided in very few places and the experts have not been willing to reveal the details of their processes and (2) data on the reliability and strength atainable by hydroxide-assisted bonding have been scarce. The first and most critical phase of the present hydroxide-assisted bonding proess is the preparation of the surfaces to be bonded. This phase includes the following steps:

- Ultrasonic cleaning in successive baths of acetone, methanol, and propanol, using an ultrasound cleaner that operates at several megahertz.

- Treatment in a solution of potassium hydroxide and ammonium hydroxide in an ultrasonic cleaner, at megahertz frequencies.

Thorough rinsing with deionized water is carried out after each of the above-mentioned steps. The last rinse is followed by ultrasonic cleaning in deionized water, then the cleaned surfaces are blow dried with ionized air. Next, a droplet of a dilute solution of potassium hydroxide is placed on one of the surfaces, then the surfaces are placed in contact and gently squeezed together (see Figure 2.7). The resulting assembly is allowed to sit at room temperature for 24 hours and is then baked at a temperature of 200°C (392°F) for 24 hours. In mechanical tests, sample bonds made by this process were found to have tensile strengths of at least ~9 MPa (1.3 kpsi), where the epoxy bond used to attach the sample to the tensile stress test apparatus broke.3. Scientists have long dreamed about putting telescopes on the far side of the moon where they would be shielded from the interference that affects telescopes on the earth's surface and also in the earth's orbit. One big problem, though, is that the bulk and mass of telescopes make them very expensive to transport to a distant location such as the moon. But researchers at Goddard Space Flight Center in Greenbelt, Maryland, are working on a technique that would make it possible to build a telescope from moon dust and a few common materials that could be transported from Earth. "You can go to the Moon with a few buckets, and build something

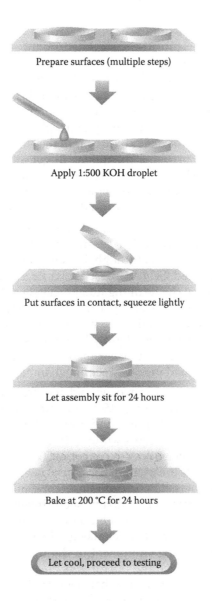

FIGURE 2.7
Two proposed surfaces are placed in contact with a small amount of a hydroxide solution at the interface. The assembly is allowed to sit and is then baked. The resulting bond is at least as strong as an epoxy bond.

far larger than anything a rocket can carry," Goddard physicist P. Chen has said.

This work could have an impact on a long-standing debate among astronomers: whether it will be better in the future to build additional orbiting space telescopes such as the Hubble or set them up on the moon. Both types of telescopes would be beyond the interference and distortions created by the earth's atmosphere, but the moon has the added advantage of being a stable platform with a far side that also would shield a telescope from radio interference from the earth's cluttered radio background. Creating gigantic lunar telescopes would normally carry an astronomical price tag because the earth's gravity places both size and cost limits on what rockets can carry into the earth's orbit, much less a trip to the surface of the moon. To get around that problem NASA researchers used a mix of epoxy, simulated lunar dust, and carbon nanotubes to demonstrate how materials already found on the moon could be used to manufacture a telescope.

Chen has already worked with epoxy and carbon nanotubes to create "smart" materials that can flex or change shape when an electric current passes through, but ended up adding simulated lunar dust, called JSC-1A Coarse Lunar Regolith Simulant, to the mix. Chen used the resulting lunar concrete as the foundation of a 30 cm (11.8 in.) disk and poured more plastic epoxy on top of it. Then he spun the mirror at a constant speed that formed the epoxy into a parabolic, slightly bowl-like shape as it hardened. The mirror's finishing touch came with a thin layer of reflective aluminum applied inside a vacuum chamber.

Making a Hubble-sized mirror would require transporting 60 kg (133.3 lb) of epoxy, 1.3 kg (2.9 lb) of carbon nanotubes and less than 1 gram (0.45 kg) of aluminum to the moon, according to Chen's calculations. Meanwhile, 600 kg (1333 lb) of lunar dust could provide the bulk of the material. The moon's lack of atmosphere also suits the vacuum conditions needed to make the mirror. Astronomers may imagine telescope mirrors half the size of a football field, but realizing such dreams depends heavily on whenever NASA returns human explorers to the moon and sets up a base there. Other challenges include getting the necessary manufacturing equipment to the moon, such as the spinning table on which the mirror gets created. Future astronauts also would have to ensure that none of the free-floating lunar dust contaminates the mirror. "It's a great idea in principle, but nothing is simple on the Moon," J. Spann, a physicist heading the Space and Exploration Research Office at the Marshall Space Flight Center (MSFC), was quoted as saying in an article about the lunar mirror. Spann said the moon's dusty environment would pose challenges, noting that the mirror coating on the earth is done in clean environments. But he also has stated that making simple composite structures has a lot of promise and could be useful for things such as building habitats for lunar bases.

Future efforts by Chen and his colleagues will try to scale up their demonstration by creating 0.5 m (19.68 in.) and 1 m (39.37 in.) mirrors using the simulated lunar dust. They also plan to figure out ways to hone the quality of the finished mirror's surface and are already speculating about ways future explorers and robots could build even larger telescope mirrors on the moon—perhaps within an impact crater.

4. The U.S. Navy Joining Center (NJC) [23] has completed a project to characterize manufacturing variabilities that effect adhesive bonding for composite joints used in the fabrication of primary aircraft structures. The project developed adhesive joint mechanical property data that quantifies the effects of manufacturing variability in adhesive-bonded systems to meet U.S. Navy aircraft service requirements for bonding composite structures. Adhesive bonding of primary aircraft structures has the potential to save significant weight and cost over conventional bonded, riveted, or bolted joint designs with bolts to insure that premature failure will not occur should the bond fail.

As a result of the above-mentioned program, new bonded joint designs have been developed, which minimize peel stresses and provide a more robust joint. The joint, commonly referred to as Pi (P) or a tongue-and-groove joint included a male and female section bonded with an adhesive. This type of joint is more tolerant to damage and flaws than other composite designs. The joints are lighter and less expensive to fabricate than conventional mechanical fastened joints. The target applications of this joint include skin-to-ribbed box composite subassemblies requiring both structural and closeout connections. Even though the Pi-joint configuration has the potential to improve the performance of adhesively bonded joints, fabrication issues were also addressed in the project to insure that the bonded joints can be reproducibly manufactured for primary structure. Large Pi-joint fabrication test elements were manufactured using a range of process variables. Test specimens were produced from the large Pi-joint test elements in sufficient number to provide data on shear loading, combined angle loading, and fatigue performance.

A number of manufacturing variables were identified by airframe manufacturers and tests were performed quantifying their effects on adhesive-bonded joints. Manufacturing variables investigated for adhesive bonding included surface preparation, bond line thickness, offset web, porosity, and impact damage. Results have been released to goverment agencies as well as industrial aircraft firms [23]. Some of the early data developed in the project was used by designers for bonding of composite structures in the Joint Strike Fighter (JSF) F-35 program. The application of bonding technology on the inlet duct assembly for the JSF is one example. The results of this project support the U.S. Navy and Air Force requirements for future adhesive-bonded aircraft structures for enhanced performance and operation costs. Cost savings of 60–70% have been realized in assembly operations in bonded structures. As the reliability of adhesive bonding is confirmed, both the Navy and the airframe manufacturers will encourage the use of this joining technology for primary structures.

5. Recently, a new fiber metal laminate (FML) called centrally reinforced aluminum (CentrAl) was introduced for application in aircraft wings [24–27]. The CentrAl concept comprises a central layer of fiber metal laminate (Glare), sandwiched between one or more thick layers of new generation damage tolerant aluminum alloys. The outer layers are also bonded by fiber reinforced adhesive layers, called bondpreg. This creates a robust structural material that is not only exceptionally strong but also insensitive to fatigue. Because the hybrid material is practically immune to fatigue, wing panels can be designed that neeed no frequent inspection and repair of cracks during the life of the aircraft (i.e., have a "care-free" economic life). The new CentrAl structures are stronger than carbon fiber reinforced plastic (CFRP) structures [24]. CentrAl allows higher stress levels and by using it in lower wing strcutures, the weight can be reduced by 20% compared to CFRP structures. The application of CentrAl will result in considerably lower manufacturing and maintenance costs.

6. A fast dry adhesive has been used on all FRP components in a new Stark 4-wheel drive vehicle produced by Brazilian vehicle manufacturer, Technologia Automotiva Catarinense (TAC). According to Leandro Correia, TAC's engineer, the need to reduce noise and vibrations was an essential factor in the decision to join FRP components with a fast drying adhesive. Fusor is a two-component polyurethane-based solution from the adhesive manufacturer, Lord Corp., and Lord Corp. was selected to replace the Stark's mechanical fasteners and act as a substitute for the lamination of the FRP. With an average time for component handling of 30–35 minutes before dryness, the adhesive proved flexible and, according to Correia, improved the vehicles visual style by not using screws and clinches. Future possibilites and applications include replacing the glue currently used on the PP dashboard of the Stark. The Stark is air-conditioned, with a hydraulic wheel, leather seats, electric windows, and a 1.8 VW engine. It is an off-road vehicle designed for muddy and curvy trails.

7. Automaker BMW stepped outside the box when it recently used adhesive-bonded carbon fiber/epoxy roofs to 1500 limited edition BMW M3 CSL sedans. The automaker wanted to enhance the car's performance by shedding as much weight as possible—other composite components include the doors, front skirt, trunk lid, and rear bumper support. The 6 kg (13 lb) carbon roof not only reduces vehicle weight but also lowers its center of gravity, improving stability and handling. Yet, a bonded solution meant that the composite/adhesive combination had to possess stiffness, strength, torsional rigidity, crash impact resistance, and vibration performance equal to a welded steel roof. Manufacturing goals would include ease of application, adequate open time, and a cure time of less than two hours.

   BMW conducted extensive adhesive validation studies, including dynamic and static torsion tests, lap shear tests, dynamic mechanical analysis, peel tests, aging/climatic stress tests, and an actual crash test, using different steel paints, surface prep methods, and curing times. While phenolics, epoxies, and acrylics were considered, a two-part polyurethane, PLIOGRIP, manufactured by Ashland Specialty Polymers and Adhesives, Dublin, Ohio, was selected that meets the structural performance requirements for bonds betwen dissimilar materials at service temperatures of 80°C (176°F), which is the highest temperature the roof will see in hot, sunny weather, yet has sufficient elasticity (~50% elongation) to handle vibration loads.

   At installation, the M3 is taken off the production line, workers manually clean the roof with alcohol and water, roughen the mating surfaces, and then hand place a continuous bead of adhesive onto the car's steel roof opening. An assembly jig assists technicians in lowering the roof into place. Pressure is applied with clamps to ensure a tight bond, and "squeeze-out" of excess adhesive is visually confirmed. After a 100 minute room temperature cure, excess adhesive is trimmed and the M3 goes back on line. The adhesives even cure speed, relatively low exotherm, and balance of strength and elasticity were key in its choice: While epoxies may have higher modulus of shear, the polyurethans have adequate shear to maintain vehicle rigidity—the advantage of higher elasticity is by far the predominant factor.

8. Adhesive tapes, the adhesive resins of which can be cured (and thereby rigidized) by exposure to ultraviolet and/or visible light, are being developed at the Marshall Space Flight Center as repair patch materials. The tapes, including their resin components, consist entirely of solid, low out gassing, nonhazardous or minimally hazardous materials. They can be used in air or in a vacuum and can be cured rapidly, even at temperatures as low as –20°C (–4°F).

Although these tapes were originally intended for use in repairing structures in outer space, they can also be used on Earth for quickly repairing a wide variety of structures. They can be expected to be especially useful in situations in which it is necessary to rigidize tapes after wrapping them around or pressing them into the parts to be repaired.

9. Nanogrip, a Carnegie-Mellon University spin-off [28], has recently been funded to commercialize a fibrous adhesive technology based on the foot hairs of Geckos. The materials mimic the nano- and micro-fibers that provide Geckos and a number of other animals with their ability to grip strongly and repeatedly to smooth and rough surfaces, even in dry and wet outdoor conditions. The goal of this project will assist companies in the design, manufacturing, material selection, and testing of the adhesive for new commercial sportswear applications, in collaboration with leading companies in the sporting goods and materials industries.

References

1. Boëda, E., J. Connan, D. Dessort, S. Muhesen, N. Mercier, H. Valladas, and N. Tisnerat. Bitumen as Hafting Material on Middle Palaeolithic Artefacts, *Nature* 380 (1996): 336–8.
2. Ramanath, G., Rensselaer Polytechnic Institute, Troy, NY, ramanath@rpi.edu; www.rpi.edu; and *AM&P eNews* (August 2007): 24.
3. http://link.alpi.net/l.php?20080221A2
4. Li, Y., K. Moon, and C. P. Wong. *Science* 308 (2005): 1419–20.
5. Hwang, J. S. ed. Chapter 1 in *Environment-Friendly Electronics: Lead-Free Technology*, 4–10. Port Erin, UK: Electrochemical Publications, 2001.
6. Li, Y., and C. P. Wong. *Materials Science and Engineering* R51 (2006): 1–35.
7. Lau, J., C. P. Wong, N. C. Lee, S. and W. R. Lee. In *Electronics Manufacturing: With Lead-Free, Halogen-Free, and Conductive-Adhesive Materials*. New York: McGraw-Hill, 2002.
8. Gilleo, K. Chapter 24 in *Environment-Friendly Electronics: Lead-Free Technology*. Edited by J. S. Hwang. Port Erin, UK: Electrochemical Publications, 2001.
9. Li, Y., M. J.Yim, K.-S. Moon, and C. P. Wong. Chapter 11.3 in *Electrically Conductive Adhesives, Smart Materials*, 11.12–11.25. Edited by M. Schwartz. Boca Raton, FL: CRC Press, Taylor & Francis Group, 2009.
10. Dietz, R. L. et al., *Soldering & Surface Mount Technology* 9 (1997): 55.
11. Miragliotta, J., R. C. Benson, T. E. Phillips, and J. A. Emerson. In *Proceedings of the Materials Research Society Symposium* 515 (1998): 245.
12. Cavasin, D., K. Brice-Heans, and A. Arab. In *Proceedings of the 53rd Electronic Components and Technology Conference* (2003): 1404–7.
13. Kisiel, R. *Journal of Electronic Packaging* 124 (2002): 367.

14. de Vries, H., J. van Delft, and K. Slob. IEEE Transactions on Components and Packaging Technologies 28 (2005): 499.
15. Jost, E. M., and K. McNeilly. Proceedings of International Society for Hybrids and Microelectronics Society (1987): 548–53.
16. Lu, D., Q. K. Tong, and C. P. Wong. IEEE Transactions on Components, Packaging and Manufacturing Technology, Part A, 22-23 (1999): 365–71.
17. Mimicking Gecko Feet: Dry Adhesive Based on Carbon Nanotubes Gets Stronger, with Directional Gripping Ability, *Science* October 10, 2008; University of Dayton; Georgia Institute of Technology; AFRL; University of Akron; http://gtresearchnews.gatech.edu/newsrelease/gecko-fee: 10/23/08.
18. Adhesives from Mussels Combined with Polymers for Implant Coatings, Nerites Corp., 525 Science Drive, University Research Park, Madison, WI, 5371: www.nerites.com
19. Xia, B., J. Shan, and W. O'Hara. *New Corrosion-Resistant Conductive Adhesive for Consumer Applications Delivers Cost-Efficient Alternatives to Current Technologies*, Loctite Industrial Adhesives, Henkel Corp., Rocky Hill, CT.
20. Sign Builder Sticks with Adhesives, *Machine Design* (November 20, 2008): 26–27.
21. Bonded Joint Design Selected for the Baseline for DD(X) Multi-Mission Destroyer, *Welding Journal* (December 2005): 75; and Navy Joining Center (NJC), Columbus, OH (L. Brown, NJC or G. Ritter, EWI).
22. Abramovici, A., and V. White. Hydroxide-Assisted Bonding of Ultra-Low-Expansion Glass, Cal Tech, NASA Jet Propulsion Laboratory, Pasadena, CA and NASA Tech Briefs, April 2008, pp 55–56.
23. NJC Completes Adhesive Bonding Project for Primary Aircraft Strcutures, NJC and EWI, Columbus, OH (L. Brown, NJC or G. Ritter, EWI).
24. Kwakernaak, A., and J. C. J. Hofstede. Adhesive Bonding: Providing Improved Fatigue Resistance and Damage Tolerance at Lower Costs, *SAMPE Journal* 44, no. 5 (Sept/Oct 2006): 6–15.
25. Bucci, R. J. Advanced Metallic and Hybrid Structural Concepts, *Proceedings USAF Structural Integrity Program Conf. (ASIP 2006)*, San Antonio, TX 2006.
26. Roebroeks, G. H. J. J., P. A. Hooijmeijer, E. J. Kroon, and M. Heinimann. The Development of CentrAl, *Proceedings First International Conference on Damage Tolerance of Aircraft Structures*. The Netherlands: TU Delft, 2007.
27. Fredell, R. S., J. W. Funnink, R. J. Bucci, and J. Hinrichsen. "Carefree" Hybrid Wing Structures for Aging USAF Transports, *Proceedings First International Conference on Damage Tolerance of Aircraft Structures*. The Netherlands: TU Delft, 2007.
28. http://www.compositesworld.com/news/arkema-nanogriptech-receive-nanomaterial-funding.aspx: p 1 of 1, 8/11/2009.

Bibliography

Advance High-Performance Epoxy Adhesives Revolutionize Structural Bonding, Master Bond, Inc., 154 Hobart St., Hackensack, NJ 07601.

Aleksic, S., H. Schweigart, and H. Wack. How Does Cleaning Affect Conformal Coatings, *Zestron America and SMT*, July 13, 2008.

Bieniak, D., Secrets of Bond Strength, *Machine Design* (July 19, 2008): 60–65.

Epoxy Adhesives Hold Their Own, *Machine Design* (September 25, 2008): 56–58.

Fisher, L. W., Better Bonds, *Machine Design* (January 12, 2006): 82–86.

Hsu, J., NASA Scentists Study Using Moon Dust to Build Telescopes, *Space News* (July 21, 2008): 17.

Kadioglu, F., M. Es-souni, and S. Hinisliouglu, The Effect of Temperature Increase on the Stress Concentrations of Adhesive Joints, *SAMPE Journal* 37, no. 3 (July 2005): 21–24.

Naguy, T. A., J. R. Kolek, and T. R. Anderl. *Laser Surface Pretreatment Method for Metal Adhesive Bonding, AFRL Technology Horizons*, 41–42. Dayton, OH: AFRL Materials and Manufacturing, Directorate and Anteon Corp, February 2005.

Recktenwald, D. Advanced Adhesives Foster Hybrid Structures, *Machine Design* (November 3, 2005): 124–26.

Rosselli, F. Making the Move From Conventional Joining to Structural Adhesives, *Reinforced Plastics* (February 2006): 42–46.

Toleno, B. Step 5: Adhesives/Epoxies & Dispensing, *Henkel and SMT*, September 18, 2008.

Wong, E. W., M. J. Bronikowski, and R. S. Kowalczyk. Using ALD to Bond CNTs to Substrates and Matrices, Cal Tech, NASA Jet Propulsion Laboratory, Pasadena, CA and NASA Tech Briefs, July 2008, pp 50–51.

3

Adhesive Proteins from Mussels

Francisco F. Roberto and Heather G. Silverman

CONTENTS

Preface

Mussels, or mollusks, adhere to inanimate and living surfaces in both freshwater and marine environments. The byssus—an exogenous proteinaceous secretion that enables strong, durable adhesion to many different surfaces in both dry and wet environments—has been the focus of substantial biomaterials development research within the last decade. Understanding the complex mechanisms involved in adhesion by these invertebrates is vital to advancing the formulation and production of mussel-derived bioadhesives. The biomaterial development field is actively pursuing the use of various synthetic and naturally derived mussel adhesive proteins for adhesive applications. Potential trends in the adhesives industry, as well as challenges to commercialization of a new field of ecologically friendly adhesives will be discussed.

Background

Marine and freshwater bivalves often use a holdfast termed the byssus to either attach temporarily or permanently in an aqueous environment to a variety of surfaces (Bromley and Heinberg 2005). Individual proteins isolated from mussel byssi have been shown to bond with numerous substrates including glass, Teflon® (E.I. Du Pont de Nemours and Company, Inc., Wilmington, DE, USA) wood, concrete, plastics, metals, biological cell lines, bone, teeth, and others. Mussel byssi contain a mixture of at least 11 proteins that are stockpiled in the foot organ of the animal before being secreted and self-assembled in a process unique in nature. These adhesive-related proteins have been classified as (a) foot proteins (fp): designated fp-1, fp-2, fp-3, fp-4, fp-5, and fp-6; (b) collagens: designated distal collagen (precol-D, furthest from the animal), proximal collagen (precol-P, closest to the animal), and pepsin-resistant nongradient collagen (precol-NG); (c) a proximal thread matrix protein (PTMP-1); and (d) a polyphenol oxidase enzyme. See Table 3.1 for the relationships between these proteins and their materials properties. The adhesive strength of the mussel byssus is determined by the properties of the individual proteins secreted by the animal, the interaction of these proteins with each other, the bonding

TABLE 3.1

Byssal Protein Adhesive and Mechanical Properties

Biological Classification			Adhesive Materials Classification	
Byssus	Region	Mussel Proteins Identified	Function	Curing Mechanism
Plaque	Primer layer	fp-3, fp-5, fp-6	Primer Plaque/primer	Couple with inorganic (i.e., metals) and organics
	Plaque foam	fp-2	Stabilize	Inter- and intramolecular cross-linking
	Thread collagen anchor	fp-4		
Thread	Thread sheath (cuticle)	fp-1	Varnish/coating	Inter- and intramolecular cross-linking
		polyphenol oxidase		Oxidation
	Thread core	distal precol-D	Rigid	Inter- and intramolecular cross-linking
		proximal precol-P, PTMP-1	Elastic Rigid	
		nongradient precol-NG	Rigid and elastic	

of these proteins to substrates, the environmental conditions under which adhesion occurs (e.g., salinity, temperature, pH, season), the biological state of the animal (e.g., age and metabolic state), and the mussel species (Van Winkle 1970; Crisp et al. 1985; Carrington 2002; Holten-Anderson and Waite 2008).

The Mussel Byssus

Freshwater mussels with byssi are found in the subfamily *Dreissena*. They inhabit lakes, ponds, rivers, creeks, canals, and similar habitats. Marine mussels live in intertidal and subtidal areas along coastlines throughout the world, and have also been found in hydrothermal vents and deep seeps on the ocean floor (Nelson and Fisher 1995; MacDonald and Fisher 1996). Mussel byssi from the marine subfamilies *Bathymodiolinae* and *Mytilinae* have been studied extensively and comprise the majority of the fps and collagens studied to date in mussel adhesion. Mussel byssi originate from a root that is attached to the byssal retractor muscle and a stem that extends from the root. Byssal threads terminate into byssal plaques. Figure 3.1 illustrates the anatomy of the common marine mussel, *Mytilus edulis*, also referred to as the "blue mussel" because of its color, or the "bearded mussel" because of the threads produced by the animal.

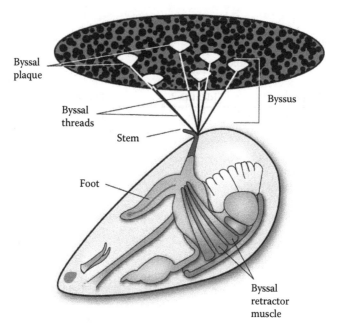

FIGURE 3.1

Anatomy of the common marine mussel, *Mytilus edulis*, also referred to as the "blue mussel" or "bearded mussel." (From Silverman, H. G. and Roberto, F. F., *Marine Biotechnol.*, 9, 661–8, Figure 3, 2007. With permission from Springer Science+Business Media.)

Adhesive Properties of Byssus Proteins

The first description of the mollusk byssus occurred in 1711 (Brown 1952). Brown's work nearly 60 years ago noted that byssal threads of *M. edulis* exhibited different morphologies throughout their length, as they emerged from the mussel (proximal thread) and attached to surfaces (distal thread and plaque). Today we know that the distribution of proteins within the thread and plaque are important to the overall mechanical behavior of the threads and their utility to the organism.

As the biochemistry of the byssal thread has been examined, polyphenolic proteins containing 3,4-dihydroxyphenylalanine (DOPA), collagen, and a catechol oxidase that oxidizes tyrosine residues to DOPA (Figure 3.2) have been revealed. DOPA has proven to be a critical chemical contributor to the adhesive properties of proteins making up the byssal thread, and while the exact mechanisms of surface binding and protein cross-linking are still unclear, the ability of this reactive dicatechol to engage in coordination with metals and metal oxides, form covalent linkages, as well as hydrogen bonds, suggest likely molecular interactions that can be explored and exploited. Repetitive amino acid motifs (fp-1, fp-2), structural amino acid motifs (the collagens), and posttranslational modifications (hydroxylation in fp-1, fp-2, fp-3, fp-5, phosphorylation of fp-5) are additional

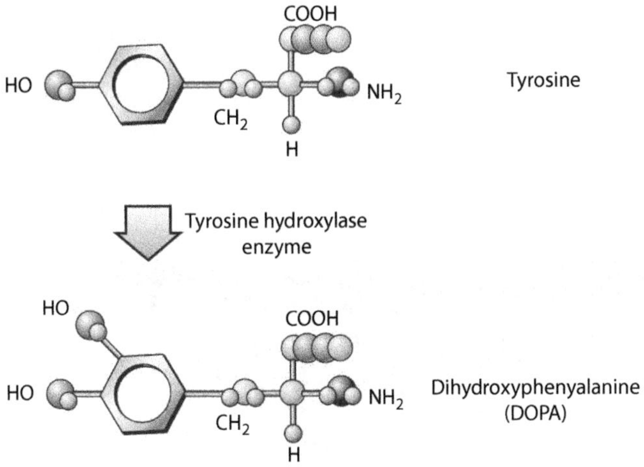

FIGURE 3.2
Hydroxylation of tyrosine residues in polyphenolic mussel adhesive proteins. (From Silverman, H. G. and Roberto, F. F., *Marine Biotechnol.*, 9, 661–8, Figure 2, 2007. With permission from Springer Science+Business Media.)

biochemical characteristics of the proteins in the byssus (Waite and Tanzer 1981; Rzepecki, Hansen, and Waite 1992; Burzio et al. 1996; Zhao and Waite. 2006a, 2006b, 2006c). The organization of these proteins in the byssal thread, and some of the chemical interactions of DOPA-containing proteins with surfaces and other proteins are shown in Figure 3.3. The importance of metal ions to the polymerization of DOPA-containing byssal proteins was recently demonstrated (Sever et al. 2004), and while ferric iron is clearly important, other metals may also play a role. We discuss additional details of the byssal proteins below.

Byssal Thread Proteins

Byssal threads vary in dimensions (~0.1 mm diameter, 2–4 cm length) and consist of a flexible core encased by a hardened sheath or cuticle. The core contains three unique collagens, precol-P, precol-D, and precol-NG, extending the length of the threads in a gradient fashion (Coyne and Waite 2000; Lucas, Vaccaro, and Waite 2002; Hassenkam et al. 2004). The fp-1 encapsulates the collagens and forms a protective coating after interacting with metals (Sever et al. 2004) and a polyphenol oxidase (Waite and Tanzer 1981).

The Precollagens

Precol-P is ~95 kDa and contains seven separate protein domains: the customary amino and carboxyl termini found in all collagens, and a collagenous domain flanked by two elastin-like domains, which in turn are flanked

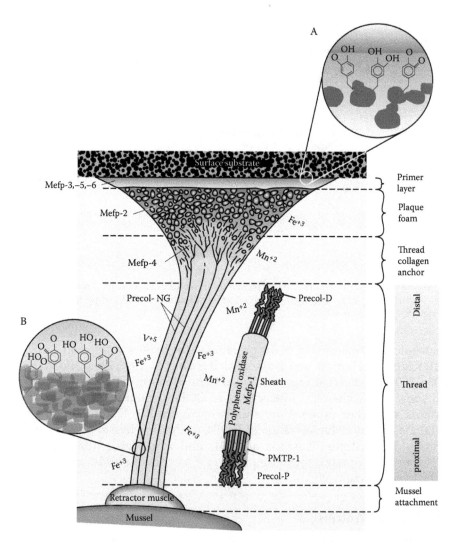

FIGURE 3.3

Organization and location of byssal proteins in the mollusk byssus. Note the gradients of proteins conferring different material properties on the thread between the proximal (adjacent to the mollusk) and distal (surface attachment) ends of the thread. These gradients are described in more detail in the text. Inset A depicts the possible role of DOPA-containing proteins fp-3 and fp-5 at the interface between the plaque and the surface, with catecholic and quinone moieties available for interaction with the surface. Inset B interprets/depicts recent information from the Waite laboratory (Adapted from Holten-Andersen, N. and Waite, J. H., *J. Dent. Res.*, 87, 701–9, 2008.) indicating that fp-1 organizes in protein granules to form a protective coating that resists failure in part due to this organization. Fp-1 also contains large amounts of DOPA that is involved in protein cross-linking. (Adapted from Silverman, H. G. and Roberto, F. F., *Marine Biotechnol.*, 9, 661–8, Figure 4, 2007. With permission from Springer Science+Business Media.)

by two histidine-rich domains. Amino acid sequence variants have been observed. The elastic proximal region of the thread is arranged in a coiled configuration, allowing for tough and extensible properties (Qin and Waite 1995; Coyne, Qin, and Waite 1997). Precol-P interacts with metals, suggesting metal-mediated cross-linking may occur (Coyne and Waite 2000). Precol-D is ~97 kDa and also contains seven separate protein domains. Silk fibroin-like domains rather than the elastin-like domains of precol-P flank the central collagenous region, imparting a strong, stiff extensible property to the distal end of the thread (Waite, Qin, and Coyne 1998). The collagenous distal region is configured in straight bundles rather than the coiled nature of precol-P. Precol-NG is ~76 kDa and contains a combination of protein domains found in precol-P and precol-D. It functions as a mediator between these two collagens along the full length of the byssal thread (i.e., nongradient fashion). The protein domains include the amino and carboxyl termini with a larger distribution of tyrosine than the other precols. A central collagenous region is flanked by plant cell wall-like domains, which in turn are flanked by the fibroin-like domains found in precol-D (Qin and Waite 1995).

Foot Protein-1

Fp-1 was the first mussel adhesive protein (MAP) to be identified by J. H. Waite and colleagues (1981, 1983), and has been described in the literature as MAP or *Mytilus edulis* foot protein-1 (Mefp-1). It is ~115 kDa and contains 10–15 mol% DOPA and substantial posttranslational modification of its amino acids. Decapeptide and hexapeptide repeats present in fp-1 contain tyrosine residues that are readily hydroxylated to DOPA and proline residues that are hydroxylated to trans-2,3-*cis*-3,4-dihydroxyproline or trans-4-hydroxy-L-proline. The quinone-tanning reaction (tyrosine to DOPA) is catalyzed by a polyphenol oxidase to form the hardened, protective sheath that surrounds the collagen core. It has recently been shown that these proteins polymerize in granules lend mechanical strength and durability, including an ability to self-heal upon strain, to the sheath (also referred to as the cuticle). Analogous fp-1 sheath proteins appear in many other mussel species, with differences in decapeptide repeat frequencies, amino acid composition and nonrepetitive coding regions [*M. galloprovincialis* (Mgfp-1, Inoue and Odo 1994); *M. coruscus* (Mcfp-1, Inoue et al. 1996b); *M. trossulus* (Inoue et al. 1995b); *M. californianus* Conrad (Waite 1986); *M. chilensis* (Pardo et al. 1990); *M.* sp. JHX-2002 (Wang et al. 2004); *Dreissena polymorpha*, zebra mussel (Dpfp-1, Rzepecki and Waite 1993; Anderson and Waite 2000); *D. bugensis*, quagga mussel (Dbfp-1, Anderson and Waite 2002); *Perna viridis*, green mussel (Pvfp-1, Ohkawa et al. 2004); *P. canaliculus*, green shell mussel (Pcfp-1, Zhao and Waite 2005); *Guekensia demissa*, ribbed mussel (Waite, Hansen, and Little 1989); *Limnoperna fortunei*, Asian freshwater mussel (Lffp-1, Ohkawa et al. 1999); *Aulacomya ater* (Saez et al. 1991); *Choromytilus chorus* (Pardo et al. 1990)].

Polyphenol Oxidase

A specific polyphenol oxidase (an oxidoreductase) responsible for cross-linking of fp-1 has not been identified. However, catechol oxidase activity has been measured in the foot organ and byssus (Zuccarello 1981; Waite 1985) and nonhomogeneous preparations of enzymes classified as catechol oxidases have been isolated from the mussel foot and regions of the byssus (Burzio 1996; Hellio, Bourgougnon, and Le Gal 2000). Enzyme subunits have ranged from 34 to 174 kDa. Hydroxylation of native (chemically extracted) or synthetically derived fp-1 can be accomplished in vitro with chemical oxidants (e.g., periodate), or enzymatically (mushroom tyrosinase, peroxidase; Marumo and Waite 1986), alkaline environments (pH > 8.5) or dissolved oxygen.

Proximal Thread Matrix Protein

PTMP-1 is a ~50 kDa, water-soluble, noncollagenous protein present in the proximal region of the byssal thread, functioning as a collagen-binding protein and containing amino acid sequences that exhibit a stiffening effect (Sun, Lucas, and Waite 2002). This protein may aid in stabilizing the more flexible precol-P portion of the byssal thread where it attaches to the animal.

Byssal Plaque Proteins

Byssal plaques vary in size, but are usually ~0.15 mm in diameter at the distal end of the byssal thread and ~2–3 mm diameter where they adhere to a substrate (Crisp et al. 1985). Six proteins have been identified in byssal plaques: fp-3, -5, and -6 constitute a region that is believed to function as a primer layer; fp-2 comprises the plaque foam; and fp-4 occupies the thread–plaque junction region between the byssal thread and the byssal plaque. Fp-1 functions as a hardened protective sheath encompassing the plaque—similar to its role in byssal thread formation.

Primer Layer

Fp-3 is ~5–7 kDa and contains 20–25 mol% DOPA as well as 4-hydroxyarginine and tryptophan (Papov et al. 1995; Warner and Waite 1999; Zhou et al. 2006a). Gene families of fp-3 variants have been identified in *M. edulis, M. californianus,* and *M. galloprovincialis* with differing conclusions as to why and when the animals secrete specific variants (Inoue et al. 1996a; Floriolli, von Langen, and Waite 2000; Zhou et al. 2006a). Fp-5 is ~9.5 kDa and contains 27 mol% DOPA as well as phosphoserine. Fp-5 analogs have been identified in *M. edulis* (Waite and Qin 2001), *M. californianus* (with two variants; Zhao and Waite 2006c), and *M. galloprovincialis* (Hwang et al. 2004).

Fp-6 is ~11.6 kDa and contains a small amount of DOPA although tyrosine is present in large quantities. Two fp-6 variants have been isolated from *M. californianus* (Zhao and Waite 2006c). Fp-6 may be involved in adhesion in a number of ways: bonding via DOPA residues with fp-3 and fp-5 in the plaque and/or fp-1 in the byssal thread sheath and direct bonding to substrate surfaces.

Plaque Foam

Fp-2 is ~42–47 kDa and contains only 2–3 mol% DOPA as well as 6–7 mol% cysteine. The presence of cysteine and tandem, repetitive amino acid repeats suggest a stabilization role for fp-2 in the byssal plaque. Fp-2 is resistant to proteases, has substantial secondary structure, and constitutes approximately 25–40% of the total plaque proteins. Fp-2 analogs have been identified in *M. edulis* (Rzepecki, Hansen, and Waite 1992; Inoue et al. 1995a), *M. coruscus* (Inoue et al. 2000), and *D. polymorpha* (Rzepecki and Waite 1993). A fp-2 multigene family (rather than the concept of variants) is supported by the identification of varying repetitive motifs and distinct complementary DNA (cDNA) sequences in *M. galloprovincialis* and *M. edulis* (Rzepecki, Hansen, and Waite 1992; Silverman and Roberto 2006a, 2006b; Silverman and Roberto 2007).

Thread–Plaque Junction

Fp-4 is ~79 kDa and contains 4 mol% DOPA as well as elevated levels of glycine, histidine, and arginine and a tyrosine-rich octapeptide repeat. Fp-4 is hypothesized to function as a coupling agent between the precollagens of the thread and fp-2 located in the plaque. Fp-4 has been isolated from two mussel species, *M. edulis* and *M. californianus* (Weaver 1998; Warner and Waite 1999; Zhao and Waite 2006b). Table 3.2 summarizes common amino acid repeats, DOPA concentration, protein variants, and gene families identified in the various adhesive proteins involved in mussel adhesion.

Materials Testing and Characterization of Mussel Adhesive Proteins and Hybrids

Mussel underwater attachment is a complex combination of biochemical and physical processes that is inspiring biologically related materials development (Tonegawa et al. 2004; Broomell et al. 2007; Kamino 2008). Adhesive testing techniques have evolved to enable the mechanical evaluation of mussel adhesive proteins and to offer insight into the surface and cohesive interactions relative to structure and function (Berglin et al. 2005). Techniques to study live animals are also being used to understand the relationship between life cycle and byssus formation (Gao et al. 2007; Petrone, Ragg, and McQuillan 2008).

TABLE 3.2

Amino Acid Repeats, DOPA Concentration, Protein Variants, and Gene Families in Adhesive Proteins from Mussels

Protein	Mass (kDa)	Amino Acid Repeat Type			DOPA (%mol)	Variants Within a Species	Gene Families
		Deca	Hexa	Other			
Fp-1	~115	√	√	—	~13	√	√
Fp-2	~42–47	—	—	Extensive	~2–3	√	√
Fp-3	~5–7	—	—	—	~20–25	√	—
Fp-4	~79	—	—	Tyrosine rich octapeptide	~4	√	√
Fp-5	~9.5	—	—	—	~27	√	—
Fp-6	~11.6	—	—	—	~4	—	—
Precol-P	~95	—	—	[Gly–X–Y]$_n$	—	√	—
Precol-D	~97	—	—	[Gly–X–Y]$_n$	—	√	—
Precol-NG	~76	—	—	[Gly–X–Y]$_n$	—	—	—
PMTP-1	~50	—	—	—	—	√	—
Polyphenol oxidase	~34–174	—	—	—	—	√	—

Gly	glycine
X	proline
Y	hydroxyproline
—	not determined to date

The material properties of whole mussel byssal threads have been determined for a range of mussel species (Brazee and Carrington, 2006). Strength ranged from 13 to 48 MPa, extensibility between 0.50 and 0.76, and modulus from 35 to 137 MPa. In another recent study (Harrington and Waite 2008), the Young's modulus of hand-drawn fibers formed from purified byssal precols (70.8 MPa) was compared with the distal (868.6 MPa) and proximal (15.6 MPa) measurements of intact threads of *M. californianus*. Perhaps the most interesting recent comparison (Holten-Anderson and Waite 2008) for materials scientists is the determination of the ultimate strain tolerance of *M. galloprovincialis* thread cuticle (70%), collagen core (75%), with synthetic epoxies (<5%). These studies highlight the remarkable strength of intact byssal threads and point to the potential for use of purified byssal thread components to elucidate the component contributions of the various proteins to the final biological structure and as future new materials.

Conventional approaches and more recent nanocharacterization strategies for the evaluation of mussel adhesion are presented in Table 3.3. As the ability to perform sensitive measurements with small amounts of protein improves, along with the increased resolution of the techniques used, we can anticipate that a greater understanding of the interactions of adhesive proteins necessary to achieve robust underwater adhesion will increase. Materials testing of predominantly fp-1 or fp-1 repetitive motifs with surfaces such as glass,

TABLE 3.3

Techniques for Materials Characterization and Mechanisms for Adhesion from Mussel Adhesive Proteins

Surface (S) Cohesive (C) Biological (B) Apparatus (A) Fabrication (F)	Technique	Description/Use
B	**Amino acid sequencing** **[N-terminal Edman sequencing]**	Determination of individual amino acids within a protein sequence
B	**Biochemical characterization** colorimetric, fluorescent, radiochemical, immunocyto- chemical assays	Specific assays and methods used to identify characteristics and properties of proteins or specific amino acids
B	**Cell** metabolic viability, morphology, adhesion assay	Effects of adhesive proteins and/or protein analogs or derivatives on various cell lines
B, C	**Gel electrophoresis** SDS-PAGE: sodium dodecyl sulfate polyacrylamide gel electrophoresis IEF: isoelectric focusing	Technique(s) by which proteins are separated based on size, shape, charge, etc.; one- and two-dimensional
S	**AFM: Atomic force microscopy** Tapping mode: hydrated Contact mode: dehydrated modified cantilever with microglass bead	Surface topography with atomic resolution, interatomic force; "profilometry"; direct force measurements
S	**ATR: Attenuated total reflection** **transform infrared spectroscopy** **ATR/FT-IR: Attenuated total** **reflection Fourier transform** **infrared spectroscopy** with flow cell	For depth profiles; information on molecular composition, bonding, conformation, and orientation with respect to the interface
C	**CD (circular dichroism)** **spectroscopy**	Electronic transitions and vibrational modes of molecules
S	**Contact angle analysis** **(goniometer)**	A goniometer is used to measure contact angles of droplets on a substrate surface to indicate surface energies/wetting capabilities
C	**Dynamic light scattering or** **photon correlation spectroscopy** **(PCS) or quasi elastic light** **scattering (QELS)**	Study of fluctuations in frequency and intensity of scattered light in a sample; used for particle and macromolecule sizing
S	**EDS: Energy dispersive X ray** **spectroscopy**	Detection of specific element distributions
S	**Ellipsometry**	Optical technique used to study thin films (layer thickness, morphology, chemical composition); optical mass

(Continued)

TABLE 3.3 (CONTINUED)

Techniques for Materials Characterization and Mechanisms for Adhesion from Mussel Adhesive Proteins

Surface (S) Cohesive (C) Biological (B) Apparatus (A) Fabrication (F)	Technique	Description/Use
C	**Flourescence microscopy**	Autoflourescence monitored for tyrosine, DOPA, and DOPA-quinone
C	**IRS: Infrared spectroscopy** FT-IRAS: Fourier transform infrared reflection analysis MAIR-IR: Multiple-attenuated internal reflection infrared	Electronic transitions and vibrational modes in crystals and molecules
B	**Gel electrophoresis** SDS-PAGE: Sodium dodecyl sulfate polyacrylamide gel electrophoresis IEF: Isoelectric focusing	Technique(s) by which proteins are separated based on size, shape, charge, etc.; one- and two-dimensional
B	**HPLC: High performance liquid chromatography**	Method to characterize proteins: separation, quantification, purification, and identification
F	**Matrix assisted pulsed laser evaporation (MAPLE)**	Deposition of thin films; a physical vapor deposition process that provides control over thickness, roughness, homogeneity, and other film parameters
C	**MS: Mass spectrometry** ESIT-IT: Electrospray ionization–ion trap (and nano) MALDI-TOF: Matrix-assisted laser desorption ionization–time of flight MS/MS: Tandem mass spectrometry	Determining the mass of molecules by producing and analyzing charged species (ions); many techniques and instruments available
C	**NMR: Nuclear magnetic resonance or magnetic resonance imaging (MRI)** Rotational echo double-resonance NMR Dipolar dephasing method	Local magnetic environment; conformational analyses (e.g., random coil, beta sheet)
C	**Penetration test**	Measure direct force as rod is penetrated into a sample
C	**Photon correlation spectroscopy**	Monitoring of oxidation and autooxidation of reactive groups
C	**QCM-D: Quartz crystal microbalance with dissipation**	Monitoring changes in mass measured as a function of resonance frequency

TABLE 3.3 (CONTINUED)

Techniques for Materials Characterization and Mechanisms for Adhesion from Mussel Adhesive Proteins

Surface (S) Cohesive (C) Biological (B) Apparatus (A) Fabrication (F)	Technique	Description/Use
S	**Small angle scattering** SANS: Neutron SAXS: X ray SALS or LS: Light	Radiation is elastically scattered by a sample; information about size, shape and orientation of some component of the sample
B, S	**SEM: Scanning electron microscopy** **TEM: Transmission electron microscopy**	Surface image, detect fine structure
S	**SERS: Surface enhanced Raman spectroscopy**	Vibrational spectroscopic technique using electromagnetic and chemical enhancement mechanisms
S and C	**SIMS: Secondary ion mass spectroscopy**	Elemental analysis (e.g., metals)
S	**SPR: Surface plasmon resonance**	Uses reflected light from a conducting film to study interactions of sample and surface
S	**Sum frequency generation (SFG) vibrational spectroscopy**	Track individual/single amino acid conformations on surfaces
A	**Wall jet**	Apparatus used to study adhesion strength
S	**XPS: X ray photoelectron microscopy** Angle-dependent Hydrated and nonhydrated	Secondary mineral/elemental composition; can identify chemical compounds and give information about the short-range chemical bond structure
A	**Instron**	Tensile strength
S	**Surface forces apparatus (SFA)**	Measures the normal (attractive adhesion or repulsion) forces as a function of surface separation between two initally curved elastic surfaces

metals, plastics, concrete, and wood have been presented in a recent review (Silverman and Roberto 2007). New mechanistic studies have investigated fp-1 from mussels other than *Mytilus*: *Limnoperna fortunei* (Matsui et al. 2001; Nagaya et al. 2001; Ohkawa et al. 2001) and *Geukensia demissa* (Yamamoto, Sakai, and K. Ohkawa 2000), as well as fp-3 (Lin et al. 2007; Even, Wang, and Chen 2008), and fp-5 (Kim et al. 2008).

Challenges in Applications and Commercialization

Biochemical, structural, and mechanical evaluation of mussel adhesive-derived materials requires ample quantities of sample. Fabrication, formulation, and ultimately manufacturing of mussel-inspired adhesive products will require processes that produce high yields of pure products, and are rapid, environmentally friendly, biocompatible (in some industries), and both functionally and economically competitive with current adhesives available today.

Chemical Extraction

The original method to obtain fp-1 and fp-2 involved excision of the mussel foot organ followed by protein purification via acid extraction (Waite 1983). Chemical extraction is costly; from a monetary perspective due to wet chemistry reagents, equipment and time, and from an environmental perspective due to both the number of animals (estimated to be about 10,000) needed to produce 1 gram of mixed adhesive fps (Strausberg and Link 1990) and the wastes associated with chemical use. In addition, chemical extraction does not easily yield individual or pure proteins—an example being Cell-Tak™, a commercial cell tissue adhesion product composed of fp-1 and fp-2 from *M. edulis*. Chemical extraction of the fps, collagens, polyphenol oxidase, and other proteins from the foot organ or regions of the byssus is still used today for materials characterization and biochemical studies requiring only small quantities of protein.

Recombinant Protein Production

Heterologous recombinant protein (rPRO) production is a molecular biology technique that involves the expression of deoxyribonucleic acids (DNA) that code for foreign amino acids as well as a host's suite of native proteins. Prokaryotic or eukaryotic micro- and macroorganisms can be used as host systems for rPRO production (e.g., insect or mammalian cells, plant plastids or chloroplasts, bacteria, yeast, etc.). Mussel adhesive rPROs have used DNA sequences chosen to express partial or full-length proteins, specific amino acid motifs or patterns that are known to incur biochemical and/or function properties of interest, or a combination of the two. "Native" DNA derived from the organism(s) of interest or "synthetic" DNA obtained by chemical means may be employed. As discussed in a recent review encompassing recombinant methodology for the expression of marine mussel adhesive proteins (Silverman and Roberto 2007), codon usage compatible hosts, growth and induction methods, regulators (for promotion, transcription, and translation), strategies to aid in purification (e.g., protein tags), and appropriate posttranslational modification of amino acids are additional variables important in approaching expression systems for biological mimics. Protein aggregation and the formation of inclusion bodies is also a common occurrence in rPRO production. This imbalance between protein production and release within a cell presents additional difficulties that must often be overcome during the purification process (Villaverde and Carrio 2003; Cha, Hwang, and Lim 2008). Recent molecular-based approaches to improving mussel adhesive-related rPRO production include the use of overlapping polymerase chain reaction (PCR) to design gene constructs (Platko et al. 2008), a truncated OmpA signal peptide fusion tag for directing secretion (Lee et al. 2008b), fusion with amino acid residues identified with the cell attachment site (RGD) of fibronectin to improve surface cell attachment and spreadability (Hwang, Sim, and Cha 2007b), and coexpression of bacterial hemoglobin using the oxygen-dependent bacterial promoter *nar* to improve soluble product yield (Kim et al. 2008). A relatively recent technique, the modules of degenerate codons (MDC) approach has been proposed as an alternative for synthesizing repetitive protein-based products. This method incorporates degeneracy in the nucleotides coding for amino acids, allowing for higher protein yields, less error-prone PCR with repetitive sequences, directed individual base or amino acid mutation (should this be desired), and the option of "scrambling" the expression of repeats when desired (modular codon scrambling strategy; MCS; Mi 2006). Achmuller et al. (2007) describe a PCR-based method for the synthesis of multiple domains that conceivably alleviates the potential exhaustion of oligonucleotide (primer) options that could occur with the MDC technique. Combining natural amino acid motifs with specific, desired functional properties is the current focus of an approach coined repeat sequence protein polymer (RSPP) technology (Kumar et al. 2006). Recombinant copolymers with antimicrobial, textile targeting, and UV protection have been successfully produced in *E. coli* by assembling protein "blocks" with functional amino acid motifs and RSSP.

Individual Recombinant Mussel Foot Proteins

Recombinant mussel fps have been produced in yeast and bacteria with varying degrees of success with respect to production yield, solubility, and adhesive properties of the protein. Fp-1 decapeptide repeats from *M. edulis* have been expressed in *S. cerevisiae* (Filpula et al. 1990) and *E. coli* (Salerno and Goldberg 1993; Kitamura et al. 1999; Lee, Rho, and Messersmith 2008a). Recently Lee, Rho, and Messersmith (2008a) reported the addition of a histidine tag and a truncated signal peptide (OmpA) resulting in increased expression levels of soluble rMefp-1. Full length fp-1 proteins have also been expressed (*M. edulis,* Silverman and Roberto 2006a; *D. polymorpha,* Anderson and Waite 2000).

A fusion protein consisting of the human influenza virus hemagglutinin (HA epitope tag) and fp-3 from *M. californianus* was expressed in the yeast *Kluyveromyces lactis* (Platko et al. 2008). Preferential codon usage for *K. lactis* was achieved by an overlapping PCR technique (mentioned previously). The fp-3 was shown to form higher molecular weight aggregates than expected from a single rPRO, supporting the ability of fp-3 to self-assemble and organize in a manner beneficial to surface adhesion.

Recombinant Hybrids With Fp Components

Within the past 5 years, the Cha group successfully produced *M. galloprovincialis* fp-3 and fp-5 in *E. coli*. Fusion proteins with a histidine tag demonstrated relatively high yields and soluble protein with adhesive ability (Hwang et al. 2004; Hwang, Gim, and Cha 2005). Subsequently, two hybrid "fusion" proteins containing both fp-1 and fp-5 were designed and produced. The first, fp-151, consisted of six fp-1 repeats at both the N- and C-termini of fp-5 (Hwang et al. 2007a). The second hybrid, fp-151-RGD, included a fusion with the fibronectin binding domain designated RGD (Hwang, Sim, and Cha 2007b). The RGD tag improved solubility, cell adhesion and cell spreadability of the hybrid rPRO. More recently, the production of fp-151 by coexpression of bacterial hemoglobin (*Vitreoscilla* hemoglobin, VHb) was directed by the oxygen-dependent *nar* promoter rather than the original *vhb* promoter that is customarily used for VHb expression (Kim et al. 2008). This expression strategy enhanced the production of rPRO yet again, providing another strategy to improve the overall efficiency of the *E. coli* system for use with the production of recombinant mussel fps. Yang and colleagues (2007) recently engineered a silk-like protein containing the fp-1 decapeptide repeat from *M. edulis*. The resulting rPRO, expressed in *E. coli*, demonstrated promising cell adhesion activities as well.

Transgenic Plant Systems

Transgenic plant systems are potential hosts for the economical production of recombinant mussel adhesive-related proteins (Doran 2000; Giddings 2001). Plant leaves, seed endosperm, plant tissue culture, whole plants, and plant organ culture (e.g., roots) are among the strategies currently used for the production of various vaccines, antibodies, and enzymes useful to industrial processes. Plant cell walls contain the large, repetitive proteins expansion and extension that are analogous in many ways to fp-1, fp-2, and collagens found in the mussel byssus. Patel et al. (2007) reported the use of an elastin-like fusion tag to improve rPRO yield in transgenic tobacco plants. McQueen-Mason (2009) has investigated spider silk and fp-2 production in plants. Procollagens have also been produced successfully in tobacco (Ruggiero, Chanut, and Fichard 2000a; Ruggiero et al. 2000b).

In Vitro Protein Production

Takeuchi et al. (1999) demonstrated the successful in vitro expression of fp-1, fp-2, and fp-3 from *M. galloprovincialis* cultured mussel foot cells, presenting the possibility of producing fps from mussel foot cell lines rather than via recombinant DNA approaches. Commercial in vitro protein production products are available and may be another alternative.

Synthetic Constructs and Novel Biomaterials

Biomaterials have been constructed synthetically with mussel adhesive mechanism concepts in mind. DOPA-based nanocomposite layered thin films consisting of fused nacre (clay) with DOPA lysine polyethylene glycol (PEG) and Fe^{3+} cross-linking are reported by Podsiadlo et al. (2007). Hydrogels comprised of DOPA and PEG (Lee et al. 2004) or gelatin and chitosan polymerized with tyrosinase (Chen et al. 2002) are suggested as new adhesives to be used for medical applications. And peptide-conjugated polypeptide fibers incorporating poly-L-lysine grafted with a tyrosine-catalyst and gellan have been designed (Tonegawa et al. 2004). These various biomimics have been shown to exhibit mechanical properties such as breaking stress, breaking strain, strain energy, and total strength adequate for use in numerous industries.

The use of biological targets such as keratin, elastin, collagen, silks, fibrin, and mussel adhesive proteins, for example, continue to offer new approaches to biomaterials development (Guthold et al. 2007).

Current and Potential Applications of Mussel Adhesive Proteins and Mussel-Inspired Adhesives

Benedict and Picciano (1991) received the first U.S. patent describing the preparation and use of mussel adhesive proteins as adhesives. In this key patent, virtually all uses for mussel adhesives contemplated today were anticipated. These included: bonding of inert materials, such as wood, cellulose, and polyurethane; use as orthopedic surgical adhesives for bone–bone, bone–tendon, bone–muscle, and other combinations; use as an ophthalmic adhesive for retinal reattachment, corneal reconstruction, lacerations, lens attachment; use as a dental adhesive for tooth reattachment, coating, filling; use as a veterinary adhesive for split hooves, bone; use in antifouling and anticorrosion coatings; use as a plant and tree graft adhesive; use as a filter coating to prepare media for column chromatography; use to prepare drug delivery systems; use as a cell adhesive for tissue grafts; use as a paint or adhesive primer. The key features of the adhesive as patented included a mussel-derived protein containing 10–400 decapeptide repeats, with additional components to include: bifunctional spacers; cross-linking agents including glutaraldehyde, formaldehyde, oxygen, peroxide, mushroom tyrosinase, catechol oxidase; additives including SDS, protein, mucoprotein, conductive metals; fillers such as collagen albumin, hydroxyapatite, chitin, and chitosan. Many of the biomedical applications have been explored with crude mussel adhesive preparations, or the commercially available, purified preparation from *Mytilus edulis* sold as Cell-Tak™ (BD Biosciences, San Jose, California). The broader, nonmedical applications have seen consideration of biomimetic formulations comprised of phenolic and amine compounds, often using a tyrosinase or chemical oxidant to promote cross-linking. The most exciting developments today, which will be considered in more detail below, have largely focused on the DOPA moiety and means of creating bifunctional reagents that exploit the ability of DOPA to bind to surfaces, and create reactive groups to promote addition of other functional groups and organic polymers, with or without additional cross-linking. While the ability of mussel adhesive proteins to bond to surfaces in wet environments is still of interest, new applications that allow DOPA-containing thin film coating of metals, and other inorganic and organic surfaces, appear to be the most promising uses of mussel-inspired adhesives in the future.

Dental Adhesives and Composites

As pointed out earlier in this review, the similarities between the marine environment in which mussels attach to surfaces and the oral environment in which dentists seek to bond a variety of substrates to teeth have not escaped U.S. research funding agencies. This research support continues to this day, although the promise of such applications has yet to be achieved (Dove and Sheridan 1986; Benedict and Picciano 1991; Holten-Anderson and Waite, 2008).

Protective, esthetic, and restorative dental procedures require waterproof adhesives. These include the application of protective coatings, veneers, caps, orthodontic apparatus, and crowns (consisting of a range of polymeric, ceramic, and metallic materials) to the teeth of adults and children; the restorative filling of dental caries after removal of deteriorated tissue using ceramic or metallic fillings; the obturation or filling of the root canal prior to closure with a crown during root canal procedures; and the sealing and/or implanting of soft tissues, appliances, and bone associated with periodontal procedures. Metal amalgams, particularly those using mercury, are a concern for both the environmental and health impacts associated with their use. Fibrin adhesives mentioned earlier have been tested for use in periodontal treatments (Becker 2005). The natural latex isoprene polymer produced from the sap of *Palaquium gutta* (colloquially, the Gutta-percha tree) known as gutta-percha continues to be widely used for obturating root canals, and in fact, has been shown to be superior to more modern polymeric materials (e.g., Resilon®/Resilon Research, LLC, Madison, WI USA; Gesi et al. 2005). A long-standing dental adhesive often used as a control in comparisons with other adhesive candidates is a phenolate cement, comprised of a mixture of zinc oxide and eugenol (eugenol is the primary aromatic component of clove oil). This adhesive has been preferred by many dentists for its sedative properties on the tooth pulp, its good sealing properties, and is often used in temporary restorations. Other adhesives in broad use today include glass ionomer cements (consisting of a glass powder mixed with polyacrylic acid in distilled water.) These cements may be modified with reactive methacrylate groups mixed with a photosensitive initiator to produce light-curable adhesives that set rapidly. Urethane acrylate cements have also been used, but may experience shrinking during polymerization unless space-filling materials such as glass are added. A very popular filling material is mineral trioxide aggregate (MTA), composed of 75% Portland cement, 20% bismuth oxide, and 5% gypsum (Al-Hezaimi et al. 2005), sold in gray and white formulations under the trade name ProRoot™ MTA (Dentsply/Tulsa Dental, Tulsa, OK, USA).

In spite of the lack of progress in testing mussel adhesive proteins for dental applications, it seems likely that formulations will be developed in the future that employ individual or combinations of adhesive proteins mixed with current materials to meet specific dental applications in the oral environment.

Biomedical Adhesives

Adhesives play an increasing role in medicine today (Reece, Maxey, and Kron 2001). Desirable features in biomedical adhesives include a lack of toxic effects, biocompatibility (no immunogenic or foreign body reaction), bioabsorbability, bonding under wet and dry conditions, and the ability to promote or fail to interfere with natural healing (Strausberg and Link 1990). Fibrin adhesives (Tisseel™, Baxter International, Inc., Deerfield, IL USA) are used in cardiovascular surgery to control bleeding, seal air leaks in lungs, seal dural closures, CSF leaks, control bleeding after debridement, knee replacements, sealing lymphatic leaks after radical neck surgery, and repair spleen and liver lacerations. Albumin glues include bovine albumin-glutaraldehyde (BioGlue®, CryoLife, Inc., Kennesaw, GA, USA) for aortic dissection, which is known to have inflammatory effects. At this time no formaldehyde-based albumin glues have been approved in the United States. Cyanoacrylate medical adhesives include DermaBond® (Ethicon, Inc., San Angelo, TX, USA) and Trufill n-BCA® (Cordis Corp., Warren, NJ, USA). These are stronger than fibrin adhesives, but are not bioabsorbable, can induce inflammation, tissue necrosis, and infection. However, they require no dressing, and can be used to avoid sutures in plastic surgery, close small cuts, and surfaces above support sutures. Their use is restricted to external or temporary applications. PEG hydrogels (FocalSeal-L®, Focal, Inc., Lexington, MA, USA) may reduce leaks after thoracic surgery, but the required polymerization process is too long for immediate or critical use. Collagen adhesives include FloSeal and Proceed from bovine collagen and thrombin. CoStasis® (Angiotech Pharmaceuticals, Inc., Vancouver, British Columbia) is a collagen adhesive that includes autologous plasma for surgical bleeding control. Fibrin adhesives utilizing human fibrinogen and thrombin, like all human blood-derived products, are considered to have a risk of potential infectious viral contamination. This risk can be mitigated through the use of autologous production, although these preparations have been shown experimentally to have reduced bonding effectiveness (Siedentop et al. 2001) and can only be produced in limited quantities. It should be noted that fibrin adhesives are the only class of current medical adhesives that promote healing.

The first U.S. Government-sponsored research we are aware of to study mussel adhesive proteins was focused on the potential application of these materials to dentistry (Bowen 1971). Holten-Andersen and Waite (2008) in a review published in the *Journal of Dental Research* remarked on how it has taken nearly 40 years for this field to begin bearing fruit. Several commercial firms clearly saw the biomedical utility of water impervious natural adhesives early on (Benedict and Picciano 1991), and such applications, for oral, ophthalmic, or general surgery were incorporated into the range of uses envisioned. After a promising initial demonstration in ocular surgery (Robin et al. 1988) where the use of sutures could be reduced along with the ability to remove those used more quickly, more recent studies have failed to show appreciable efficacy of mussel adhesive preparations in bonding a variety of tissues. Chivers and Wolowacz (1997) compared the ability of cyanoacrylate, fibrin, and mussel adhesives to bond skin, bone, and cartilage in butt and lap joints. They observed "scarcely perceptible" bonding after 22 hours of curing with a mussel adhesive preparation. Ninan et al. evaluated bonding of porcine skin (2003) and small intestinal submucosal (SIS) tissues (Ninan et al. 2007) with mussel adhesive proteins in comparison to fibrin and cyanoacrylate adhesives.

Cyanoacrylate adhesives were superior in both cases, but significant, unacceptable tissue hardening was observed. Fibrin was also superior to the mussel adhesive proteins in both studies, although an apparent 200-fold increase in bond strength of the MAP-bonded tissue was seen when the SIS tissues were tested, as compared with the earlier study's measurements with porcine skin. This may be due to their use of recent information on the role of metal ions in promoting MAP cross-linking, by adding various metals during the bond curing step of the experiment. The authors also comment on previous studies that showed little or no bonding with mussel adhesive preparations, concluding that excessive moisture could have played a role. It should be noted that many studies have employed crude preparations of mussel adhesive proteins (often referred to as "MAP") from mussel feet, while some studies have used Cell-Tak™, a purified mixture of Mefp-1 and -2 from *M. edulis*. Undoubtedly the variation in the nature of the adhesive protein preparations plays a role in the variable results that have been observed. Cell-Tak™, has been compared to BioGlue® and fibrin adhesives in a rat mastectomy model (Chung et al. 2006) and shown to inhibit seroma formation by 66% without eliciting a foreign body reaction and causing only a mild inflammatory response, so was considered to be promising for surgical applications where adhesion between adjacent layers of tissue is desired.

Future surgical adhesives inspired by mussel adhesive proteins may come from studies that have produced simple peptide mimics that are adhesive and can be cross-linked (Tatehata et al. 2001), or hybrid recombinant products combining motifs from spider silk with the decapeptide of *M. edulis*, to produce a protein that may be strong, lightweight, and adhesive under wet conditions (Yang et al. 2007). The creation of a hybrid protein containing regions of *M. galloprovincialis* fp-1 and fp-5 has been described to have four times greater shear strength than fibrin adhesive, and to be more easy to purify, with greater yields, than the cloned native proteins (Cha, Hwang, and Lim 2008). This product has been commercialized by Kollodis Biosciences (2009).

Orthopedic Applications

Adhesives play an important role in modern orthopedic surgery, where dissimilar materials (metal, bone, organic polymers, tendons, muscle) must be joined together to form durable, robust interfaces and the bonds must remain stable under mechanical stress and the moist conditions of the patient's body. Reattachment of connective tissues to bone has been a particular challenge, and a range of approaches, from the use of autologous tendon grafts, to incorporation of high-strength, lightweight polymers such as nylon, Dacron, and even carbon fiber have been employed to reinforce such repairs without great success. Using the MAP model, a dicatechol, nordihydroguaiaretic acid (NDGA) has been used to polymerize collagen fibers to the bovine tendon, producing a material strength that matches or exceeds that of the bovine tendon (Koob and Hernandez 2002).

Antifouling Coatings of Medical Devices

One of the more interesting uses of adhesive proteins from a marine fouling organism is the application of coatings to medical devices to inhibit attachment and fouling by potentially pathogenic microorganisms. Catheters, for example, are routinely used in hospitalized patients, but infections arising from long-term placement or contamination can be life threatening, or at least an additional complication of a hospital stay. Coatings inspired by mussel adhesive proteins—poly(ethylene glycol) conjugated to DOPA—were found to inhibit attachment of common uropathogenic bacteria incubated with coated silicon disks after 24 hours in human urine (Ko et al. 2008). A 94–97% reduction in adherent bacteria was observed from the initial bacterial inoculum concentration of 10^5 cells/ml. Chemically similar coatings produced by surface-initiated polymerization (SIP) were shown to inhibit the binding of osteoblasts (Dalsin et al. 2003) and fibroblasts (Fan et al. 2005), which may be useful in orthopedic surgeries where invasion of the implanted device is undesirable. This behavior is interesting in light of other studies mentioned earlier where adhesion of cells to bone or dentin was considered desirable. It should also be noted that there is at least one study where Cell-Tak™ was observed to cause almost immediate apoptosis (programmed cell death) in osteoblasts, a potentially undesirable effect when cell regeneration surrounding an implant might be desired (Benthien, Russlies, and Behrens 2004). This result conflicts with earlier work indicating no cytotoxic effects on osteoblasts, which may have been a somewhat different preparation of mussel adhesive proteins (Fulkerson et al. 1990).

Wood, Pulp, and Paper Products

Adhesives are used extensively throughout this industry. Paper and fiber-board products are assembled almost entirely with adhesives. Specialty adhesives, such as reversible adhesives found in Post-It Notes® (3M, St. Paul, MN, USA) also play a significant role in this area. Structural wood products make use of formaldehyde adhesives that have many environmental concerns. It has been estimated that 7.2 billion pounds of formaldehyde adhesives were used in the United States and Canada in 1999. Wood composite products use 2–10% adhesive by weight (Peshkova and Li 2003). While polyphenolic mussel adhesive proteins have not yet been tested in producing adhesives for bonded wood products, they have inspired the consideration of other polyphenolic compounds. For example, a durable, effective hot press adhesive has been developed from wood-processing derived tannins (natural polyphenolic compounds). Laccase-polymerized chitosan-phenolic mixtures have been investigated, and the need for polyphenolic starting materials has been shown to be critical to creating an adhesive mixture (Peshkova and Li 2003). Li et al. (2004) demonstrated that a tannin fraction could be combined with polyamines to form hot set adhesives for bonding composite wood products. This adhesive formulation could accommodate 12–24% solids, and was most effective when freshly formulated and applied at >100°C for 2 minutes.

Electronics

Products produced in this industry sector rely increasingly on adhesives to miniaturize equipment and bond small parts and layers. As these products are made smaller and lighter, mechanical connections, including conductive wiring, are being replaced with adhesives to reduce, weight, size, and cost. DVDs are produced by bonding the optically encoded film to a clear, protective plastic layer. While not designed generally to operate under wet conditions, moisture can have detrimental effects on bonding with conventional adhesives that could be reduced or eliminated with an adhesive that continued to function under such conditions.

Personal Products

Adhesives can be found in virtually every class of personal products, from clothing, shoes, and sports equipment, to cosmetics, infant care products such as diapers and baby wipes, and first aid, such as bandages and wound sealants (the cyanoacrylate-based DermaBond® is available in an over-the-counter preparation for sealing shallow skin lacerations, for example). Many of these applications require durable, water-resistant bonds using low-cost adhesives.

Home Construction

As already noted, there are a variety of structural wood products that rely on the use of adhesives. Beyond the structural components of homes, adhesives play a role throughout the construction process, from adhesives used in other structural components, such as drywall, to flooring (bonding of wood, ceramic, and vinyl finish materials to underlayment), to the application of carpeting, roofing, and wall coverings. Kitchen and bathroom sealants and adhesives have an obvious need for water resistance to protect structural components under finish materials, and to provide sealing of plumbing connections and fixtures. Adhesives are also used extensively in assembling decorative composites for furniture, doors, trim, and cabinets.

Automotive

It has been estimated that the automotive industry accounts for consumption of 9% of all adhesives used today, and that the modern automobile is constructed with 40 lb of adhesives. These adhesives are used not only in the attachment of vehicle trim parts, but in the assembly of structural body parts, where they impart not only lighter weight over mechanical connectors, but desirable flexibility during assembly, and under impact conditions. Structural adhesives must survive for the life of the vehicle, and resist temperature extremes, physical shear forces and vibration, as well as exposure to water, gasoline, and oil. Extremely durable, tenacious, and water-impervious adhesives are used in attaching windshields that are subject to large wind shear forces, hot and cold conditions, and must remain watertight in rapidly moving vehicles. Mussel-inspired adhesives could be used in the future to provide not only water-resistant sealants, but protective coatings and perhaps even in the manufacture of composite parts on the interior and exterior of the vehicle.

Aerospace

Adhesives have long played a role in this industry, as reducing weight of the airframe and fittings of aircraft are critical to their load-bearing capacity. The earliest aircraft utilized dopes, solvent-base adhesives in order to provide rigidity and durability to fabric coatings of airfoil surfaces. Glue was also used to assemble the wooden frames of early aircraft. In modern aircraft, adhesives are used throughout to eliminate mechanical fasteners, connect subframes and components, seal joints to prevent degradation from leakage and corrosion, and to create lightweight composites comprised of metals, plastics, and carbon fibers. Many of the same considerations that apply to automotive adhesives apply to aircraft, but with the obvious additional need for demonstrated reliability and performance under extremely low temperatures and high loadings.

Recent Developments in Mussel-Inspired Adhesives

The past 3 years have seen a dramatic increase in the number of papers reporting exciting progress in new adhesive materials inspired by mussel adhesive proteins. It is interesting that the key developments in our understanding of the natural system now date back over 5 years, first to a series of papers where adhesive organic polymers were created through the covalent linking of DOPA alone (Huang et al. 2002; Lee, Dalsin, and Messersmith 2002; Dalsin et al. 2003), and then to the discovery of a key role of metal ions in cross-linking of DOPA-containing proteins (Sever et al. 2004). The latter work was conducted with crude preparations of mussel adhesive proteins extracted from the mussel foot, so suggests, rather than demonstrates how metal coordination and oxidation may direct cross-linking and polymerization of the native adhesive proteins. The work of Messersmith's group, on the other hand, builds from the early realization that the presence of large amounts of DOPA in the mussel byssus, and in key protein components of the byssus, must be involved in their adhesive properties. Recent studies showing that the surface active proteins of the mussel byssus, Mefp-3 and Mefp-5 contain large amounts of DOPA further emphasized the importance of DOPA to surface binding. Perhaps the most significant result to date has been the demonstration that a simple DOPA protein mimic, dopamine, can be used to coat even irregular surfaces by dip coating (Lee et al. 2007a; Lee, Rho, and Messersmith 2008a). This process is depicted in Figure 3.4. A wide range of surfaces can be coated, including gold, copper, glass, silicon and titanium dioxide, polycarbonate, PEEK, and polystyrene without the use of vapor or electrodeposition techniques. The dip coating process produces uniform layers of 10–50 nm thickness, depending on the immersion time. Further derivatization with alkanethiols produced coatings that inhibited attachment of fouling microorganisms (Lee et al. 2007a). Nucleophilic attack of polydopamine could be tailored by adjusting the pH, permitting the attachment of the proteolytic enzyme trypsin to the dopamine-coated surfaces (Lee, Rho, and Messersmith 2008a). It should be noted that the exact nature of dopamine binding to the surface, and the formation of polydopamine is not yet clear. It appears that both amine and catechol functions remain available to react with metals and organic molecules. From this simple coating process, surface-initiate polymerization has also been shown (Hamming et al. 2008), as well as bonding of self-assembling polymers to metallic surfaces using the base dopamine molecule functionalized with methacrylate hydrogels (Guvendiren, Messersmith, and Shuli 2008). In elegant demonstrations of the technology, a synthetic lamellar material resembling seashell nacre was produced through layer-by-layer (LbL) assembly by coating clay layers with DOPA-modified polymers (Podsiadlo et al. 2007), and a reversible, water-resistant, "geckel" adhesive inspired by Geckos and mussels was produced by coating nanofabricated polymer pillars with poly(dopamine methacrylamide-comethoxyethyl acrylate; Lee, Lee, and Messersmith 2007b).

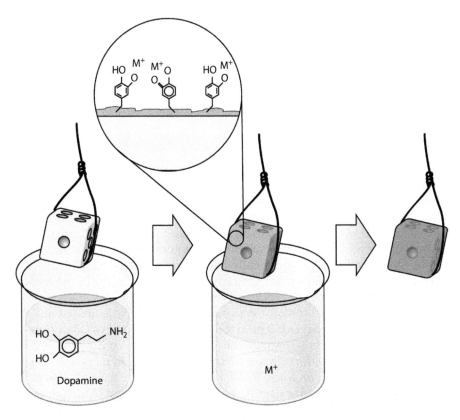

FIGURE 3.4
Schematic representation of surface modification using the mussel-inspired adhesive molecule dopamine. Simple dip-coating of a variety of surfaces and materials has been demonstrated. While it is not yet clear the nature of the binding and formation of polydopamine to the surface, both the terminal amine and catechol functions appear to remain active to promote additional binding of metals and proteins to the coated surface.

Use of semiconductor industry techniques such as laser deposition (Doraiswamy et al. 2007a, 2007b) and piezoelectric injket printer technology (Doraiswamy et al. 2009) have also been employed to apply and pattern natural adhesive proteins and DOPA-containing mimics on surfaces.

Recent work from J. H. Waite's group, who has pioneered much of our understanding of mussel adhesion and the role of polyphenolic proteins, emphasizes that the natural system still has much to teach us. It has now been shown that Mgfp-1 from *M. galloprovincialis*, resists mechanical damage by organization about the flexible collagen core of the byssal thread in a self-healing, granular arrangement. The hardness and stiffness of the Mefp-1 protective sheath is comparable to engineering epoxies, and in some ways, vastly superior. For example, tensile failure strain of 70% can be tolerated, due to the behavior of granules, which promote the formation of self-healing microtears, rather than by catastrophic failure that might result if a more rigid polymer were formed (in contrast, epoxies can tolerate strain less than 5%). Self-assembling collagen fibers with mechanical properties resembling native collagen fibrils have also been produced from purified mussel precol, demonstrating pH dependence that suggests a role for histidine in proper configuration of the protein monomers (Harrington and Waite 2008).

Environmental Considerations

There is increasing pressure to reduce the use and emissions of volatile organic compounds (VOCs) as well as the use of nonrenewable, petroleum-based adhesives. Mussel adhesive proteins and adhesives inspired by these natural adhesives are of obvious interest as they are water-based materials. Formaldehyde-based adhesives, such as those used in the production of composite wood products, and in insulation, are a good example of a target for substitution by more environmentally benign materials. Formaldehyde is a listed carcinogen responsible for a variety of health effects, including leukemia, nasopharyngeal cancer, sinonasal adenocarcinoma, and squamous cell carcinoma. Regulatory limits for exposure to formaldehyde are set at 0.75 ppm over an 8-hour period, or 2 ppm for short-term exposures (15 minutes). Some plywood products (UF grade) have been noted to continue to offgas formaldehyde, despite efforts to reduce the amount used in its production.

Interest in fabricating products that can be completely recycled is also generating more interest in adhesives that are water soluble and biodegradable. From computers and electronics that are manufactured without toxic metals, to automobiles that can be completely recycled after disassembling their components, to dismantlable homes, protein and/or biomimetic adhesives could be envisioned to enable such products, and substantially reduce the environmental impacts of manufacturing and disposal (the cradle to grave paradigm).

Conclusions

The mussel byssus and its component proteins have been the subject of great interest for many years. The potential for using mussel adhesive proteins in various applications has intrigued scientists for many years, but it is only now, after careful dissection of the byssus, and extrapolation of chemical features of the adhesive proteins that our understanding provides a means of exploiting mussel-inspired adhesion in useful ways. Commercialization of mussel-based adhesives continues to be plagued by the challenge of efficiently obtaining native or recombinant adhesive proteins, but mussel-inspired chemistries may overcome this challenge.

Acknowledgments

The authors would like to acknowledge the assistance of the Idaho National Laboratory (INL) Research Library staff for providing access to the extensive published literature referenced in this chapter, graphic artist Allen Haroldsen for the excellent technical illustrations, and Gordon Holt for technical writing and editing support.

References

Achmuller, C., E. Werther, P. Wechner, and B. Auer. 2007. Synthesis of genes with multiple identical domains. *BioTechniques* 42 (1):43–6.

Anderson, K. E., and J. H. Waite. 2000. Immunolocalization of Dpfp-1, a byssal protein of the Zebra Mussel, *Dreissena polymorpha*. *Journal of Experimental Biology* 203:3065.

Anderson, K. E., and J. H. Waite. 2002. Biochemical characterization of a byssal protein from *Dreissena bugensis* (Andrusov). *Biofouling* 18 (1):37.

Al-Hezaimi, K., J. Naghshbandi, S. Oglesby, J. H. S. Simon, and I. Rotstein. 2005. Human saliva penetration of root canals obturated with two types of mineral trioxide aggregate cements. *Journal of Endodontics* 31:453–56.

Becker, W. 2005. Fibrin sealants in implant and periodontal treatment: Case presentations. *Compendium* 26:539–44.

Benedict, C. V., and P. T. Picciano. 1991. Adhesives derived from bioadhesive polyphenolic proteins. U.S. Patent No. 5,105,677.

Benthien, J. P., M. Russlies, and P. Behrens. 2004. Investigating the effects of bone cement, cyanoacrylate glue and marine mussel adhesive protein from *Mytilus edulis* on human osteoblasts and fibroblasts in vitro. *Annals of Anatomy* 186:561–6.

Berglin, J. H., J. Hedlund, C. Fant, and H. Elwing. 2005. Use of surface-sensitive methods for the study of adsorption and cross-linking of marine bioadhesives. *The Journal of Adhesion* 81:805–22.

Bowen, J. H. 1971. Dental cement from marine sources. NIH Technical Report 70-2238.

Brazee, S. L., and E. Carrington. 2006. Interspecific comparison of the mechanical properties of mussel byssus. *Biological Bulletin* 211:263–74.

Bromley, R. G., and C. Heinberg. 2005. Attachment strategies of organisms on hard substrates: A palaeontological view. *Palaeogeography, Paleoclimatology, Palaeoecology* 232:429–53.

Broomell, C. C., R. K. Khan, D. N. Moses, A. Miserez, M. G. Pontin, G. D. Stucky, F. W. Zok, and J. H. Waite. 2007. Mineral minimization in nature's alternative teeth. *Journal of the Royal Society Interface [Abbreviated as J. R. Soc. Interface]* 4:19–31.

Brown, C. H. 1952. Some structural proteins of *Mytilus edulis*. *Quarterly Journal of Microscopical Science* 93 (4):487.

Burzio, L. A. 1996. Catechol oxidases associated with byssus formation in the Blue Mussel, *Mytilus edulis*. Master's Thesis. Newark, DE: University of Delaware.

Carrington, E. 2002. Seasonal variation in the attachment strength of Blue Mussels: Causes and consequences. *Limnology and Oceanography* 47 (6):1723.

Cha, H. J., D. S. Hwang, and S. Lim. 2008. Development of bioadhesives from marine mussels. *Biotechnology Journal* 3:631–8.

Chen, T., H. D. Embree, L. Wu, and G. F. Payne. 2002. In vitro protein-polysaccharide conjugation: Tyrosinase-catalyzed conjugation of gelatin and chitosan. *Biopolymers* 64:292–302.

Chivers, R. A., and R. G. Wolowacz. 1997. The strength of adhesive-bonded tissue joints. *International Journal of Adhesion and Adhesives* 17:127–32.

Chung, T. L., L. H. Holton, III, N. H. Goldberg, and R. P. Silverman. 2006. Seroma prevention using *Mytilus edulis* protein in a rat mastectomy model. *Breast Journal* 12:442–5.

Coyne, K. J., X. X. Qin, and J. H. Waite. 1997. Extensible collagen in mussel byssus: A natural block copolymer. *Science* 277 (5333):1785.

Coyne, K. J., and J. H. Waite. 2000. In search of molecular dovetails in mussel byssus: From the threads to the stem. *Journal of Experimental Biology* 203 (9):1425.

Crisp, D. J., G. Walker, G. A. Young, and A. B.Yule.1985. Adhesion and substrate choice in mussels and barnacles. *Journal of Colloid and Interface Science* 104 (1):40.

Dalsin, J. L., B.-H. Hu, B. P. Lee, and P. B. Messersmith. 2003. Mussel adhesive protein mimetic polymers for the preparation of nonfouling surfaces. *Journal of the American Chemical Society* 125:4253–8.

Doran, P. M. 2000. Foreign protein production in plant tissue cultures. *Current Opinion in Biotechnology* 11 (2):199–204.

Doraiswamy, A., C. Dinu, R. Cristescu, P. B. Messersmith, B. J. Chisholm, S. J. Stafslien, D. B. Chrisey, and R. J. Narayan. 2007a. Matrix-assisted pulsed-laser evaporation of DOPA-modified poly(ethylene glycol) thin films. *Journal of Adhesion Science and Technology* 21 (3–4): 287–99.

Doraiswamy, A., R. J. Narayan, R. Cristescu, I. N. Mihailescu, and D. B. Chrisey. 2007b. Laser processing of natural mussel adhesive protein thin films. *Materials Science and Engineering C* 27:409–13.

Doraiswamy, A., T. M. Dunaway, J. J. Wilker, and R. J. Narayan. 2009. Inkjet printing of bioadhesives. *Journal of Biomedical Materials Research B: Applied Biomaterials* 89 (1): 28–35.

Dove, J. and P. Sheridan. 1986. Adhesive proteins from mussels: possibilities for dentistry, medicine, and industry. *Journal of the American Dental Association* 112:879.

Even, M. A., J. Wang, and Z. Chen. 2008. Structural information of mussel adhesive protein mefp-3 acquired at various polymer/mefp-3 solution interfaces. *Langmuir* 24:5796–801.

Fan, X., L. Lin, J. L. Dalsin, and P. B. Messersmith. 2005. Biomimetic anchor for surface-initiated polymerization from metal substrates. *Journal of the American Chemical Society* 127:15843–7.

Filpula, D. R., S. M. Lee, R. P. Link, S. L. Strausberg, and R. L. Strausberg. 1990. Structural and functional repetition in a marine mussel adhesive protein. *Biotechnology Progress* 6 (3):171.

Floriolli, R. Y., J. von Langen, and J. H. Waite. 2000. Marine surfaces and the expression of specific byssal adhesive protein variants in *Mytilus*. *Marine Biotechnology* 2:352.

Fulkerson, J. P., L. A. Norton, G. Gronowicz, P. Picciano, J. M. Massicotte, and C. W. Nissen. 1990. Attachment of epiphyseal cartilage cells and 17/28 rat osteosarcoma osteoblasts using mussel adhesive protein. *Journal of Orthopaedic Research* 8:793–8.

Gao, Z., P. J. Bremer, M. F. Barker, E. W. Tan, and A. J. McQuillan. 2007. Adhesive secretions of live mussels observed in situ by attenuated total reflection-infrared spectroscopy. *Applied Spectroscopy* 61 (1):55–9.

Gesi, A., O. Raffaelli, C. Goracci, D. H. Pasley, F. R. Tay, and M. Ferrari. 2005. Interfacial strength of resilon and gutta-percha to intraradicular dentin. *Journal of Endodontics* 31:809–13.

Giddings, G. 2001. Transgenic plants as protein factories. *Current Opinion in Biotechnology* 12:450–4.

Guthold, M., W. Liu, E. A. Sparks, L. M. Jawerth, L. Peng, M. Falvo, R. Superfine, R. R. Hantgan, and S. T. Lord. 2007. A comparison of the mechanical and structural properties of fibrin fibers with other protein fibers. *Cell Biochemistry and Biophysics* [Abbreviated as *Cell Biochem Biophys* 49:165–81.

Guvendiren, M., P. B. Messersmith, and K. R. Shuli. 2008. Self-assembly and adhesion of DOPA-modified methacrylic triblock hydrogels. *Biomacromolecules* 9:122–8.

Hamming, L. M., X. W. Fan, P. B. Messersmith, and L. C. Brinson. 2008. Mimicking mussel adhesion to improve interfacial properties in composites. *Composites Science and Technology* 68:2042–8.

Harrington, M. J., and J. H. Waite. 2008. pH-dependent locking of giant mesogens in fibers drawn from mussel byssal collagens. *Biomacromolecules* 9:1480–6.

Hassenkam, T., T. Gutsmann, P. Hansma, J. Sagert, and J. H. Waite. 2004. Giant bent-core mesogens in the thread forming process of marine mussels. *Biomacromolecules* 5:1351.

Hellio, C., N. Bourgougnon, and Y. Le Gal. 2000. Phenoloxidase (E.C. 1.14.18.1) from the byssus gland of *Mytilus edulis*: Purification, partial characterization, and application for screening products with potential antifouling activities. *Biofouling* 16 (2–4): 235–44.

Holten-Andersen, N., and J. H. Waite. 2008. Mussel-designed protective coatings for compliant substrates. *Journal of Dental Research* 87:701–9.

Huang, K., B. P. Lee, D. R. Ingram, and P. B. Messersmith. 2002. Synthesis and characterization of self-assembling block copolymers containing bioadhesive end groups. *Biomacromolecules* 3:397–406.

Hwang, D.S., Y. Gim, and H. J. Cha. 2005. Expression of functional recombinant mussel adhesive protein type 3A in *Escherichia coli. Biotechnol Progress* 21:965.

Hwang, D. S., Y. Gim, H. J. Yoo, and H. J. Cha. 2007a. Practical recombinant hybrid mussel bioadhesive fp-151. *Biomaterials* 28:3560.

Hwang, D. S., S. B. Sim, and H. J. Cha. 2007b. Cell adhesion biomaterial based on mussel adhesive protein fused with RGD peptide. *Biomaterials* 28:4039–46.

Hwang, D. S., H. J. Yoo, J. J. Jun, W. K. Moon, and H. J. Cha. 2004. Expression of functional recombinant mussel adhesive protein Mgfp-5 in *Escherichia coli. Applied Environmental Microbiology* 70 (6):3352.

Inoue, K., K. Kamino, F. Sasaki, S. Odo, and S. Harayama. 2000. Conservative structure of the plaque matrix protein of mussels in the genus *Mytilus. Marine Biotechnology* 2:348.

Inoue, K., and S. Odo. 1994. The adhesive protein cDNA of *Mytilus galloprovincialis* encodes decapeptide repeats but no hexapeptide motif. *Biology Bulletin* 186 (3): 349.

Inoue, K., Y. Takeuchi, D. Miki, and S. Odo. 1995a. Mussel adhesive plaque protein gene is a novel member of epidermal growth factor-like gene family. *Journal of Biological Chemistry* 270 (12):6698.

Inoue, K., Y. Takeuchi, D. Miki, S. Odo, S. Harayama, and J. H. Waite. 1996a. Cloning, sequencing, and sites of expression of genes for the hydroxyarginine-containing adhesive-plaque protein of the mussel *Mytilus galloprovincialis*. *European Journal of Biochemistry* 239 (1):172.

Inoue, K., Y. Takeuchi, S. Takeyama, E. Yamaha, F. Yamazaki, S. Odo, S. Harayama. 1996b. Adhesive protein cDNA sequence of the mussel *Mytilus coruscus* and its evolutionary implications. *Journal of Molecular Evolution* 43 (4):348–56.

Inoue, K., J. H. Waite, M. Matsuoka, S. Odo, and S. Harayama. 1995b. Interspecific variations in adhesive protein sequences of *Mytilus edulis, M. galloprovincialis*, and *M. trossulus*. *Biology Bulletin* 189 (3):370.

Kamino, K. 2008. Underwater adhesive of marine organisms as the vital link between biological science and material science. *Marine Biotechnology* 10:111–21.

Kim, D., D. S. Hwang, D. G. Kang, J.Y.H. Kim, and H. J. Cha. 2008. Enhancement of mussel adhesive protein production in *Escherichia coli* by co-expression of bacterial hemoglobin. *Biotechnology Progress* 24:663–6.

Kitamura, M., K. Kawakami, N. Nakamura, K. Tsumoto, H. Uchiyama, Y. Ueda, I. Kumagai, and T. Nakaya. 1999. Expression of a model peptide of a marine mussel adhesive protein in *Escherichia coli* and characterization of its structural and functional properties. *Journal of Polymer Science* 37:729.

Ko, R., P. A. Cadieux, J. L. Dalsin, B. P. Lee, C. N. Elwood, and H. Razvi. 2008. Novel uropathogen-resistant coatings inspired by marine mussels. *Journal of Endourology* 22:1153–60.

Kollodis Biosciences. 2009. Recombinant MAP. http://www.kollodis.com/sub0201. html (Accessed September 21, 2010).

Koob, T. J., and D. J. Hernandez. 2002. Material properties of polymerized NDGA-collagen composite fibers: Development of biologically based tendon constructs. *Biomaterials* 23:203–12.

Kumar, M., K. J. Sanford, W. A. Cuevas, M. Du, K. D. Collier, and N. Chow. 2006. Designer protein-based performance materials. *Biomacromolecules* 7:2543–51.

Lee, B. P., J. L. Dalsin, and P. B. Messersmith. 2002. Synthesis and gelation of DOPA-modified poly(ethylene glycol) hydrogels. *Biomacromolecules* 3:1038–47.

Lee, B. P., K. Huang, F. N. Nunalee, K. R. Shull, and P. B. Messersmith. 2004. Synthesis of 3,4-dihydroxyphenylalanine (DOPA) containing monomoers and their co-polymerization with PEG-diacrylate to form hydrogels. *Journal of Biomaterials Science, Polymer Edition* 15 (4): 449–64.

Lee, H., B. P. Lee, and P. B. Messersmith. 2007b. A reversible wet/dry adhesive inspired by mussels and geckos. *Nature* 448:338–42.

Lee, H., J. Rho, and P. B. Messersmith. 2008a. Facile conjugation of biomolecules onto surfaces via mussel adhesive protein inspired coatings. *Advanced Materials* 20:1–4.

Lee, H., S. M. Dellatore, W. M. Miller, and P. B. Messersmith. 2007a. Mussel-inspired surface chemistry for multifunctional coatings. *Science* 318:426–30.

Lee, S. J., Y. H. Han, B. H. Nam, Y. O. Kim, and P. R. Reeves. 2008b. A novel expression system for recombinant marine mussel adhesive protein Mefp1 using a truncated OmpA signal peptide. *Molecules and Cells* 26:34–40.

Li, K., X. Geng, J. Simonsen, and J. Karchesy. 2004. Novel wood adhesives from condensed tannins and polyethylenimine. *International Journal of Adhesion and Adhesives* 24:327–33.

Lin, Q., D. Gourdon, C. Sun, N. Holten-Andersen, T. H. Anderson, J. Waite, and J. N. Israelachvili. 2007. Adhesion mechanisms of the mussel foot proteins. *Proceedings of the National Academy of Sciences of the United States of America* 104 (10):3782–6.

Lucas, J. M., E. Vaccaro, and J. H. Waite. 2002. A molecular, morphometric, and mechanical comparison of the structural elements of byssus from *Mytilus edulis* and *Mytilus galloprovincialis*. *Journal of Experimental Biology* 205 (12):1807.

MacDonald, I. R., and C. R. Fisher. 1996. Life without light. *National Geographic* 190 (4):313.

Marumo, K., and J. H. Waite. 1986. Optimization of hydroxylation of tyrosine and tyrosine containing peptides by mushroom tyrosinase. *Biochimica et Biophysica Acta* 872:98–103.

Matsui, Y., K. Nagaya, A. Yuasa, H. Naruto, H. Yamamoto, K. Ohkawa, and Y. Magara. 2001. Attachment strength of *Limnoperna fortunei* on substrates, and their surface properties. *Biofouling* 17 (1):29–39.

McQueen-Mason, S. 2009. Investigating the roles of biological materials and harnessing their potential for industrial applications. Mussel adhesive protein—A potential medical glue. http://www.cnap.org.uk (Accessed January 20, 2009).

Mi, L. 2006. Molecular cloning of protein-based polymers. *Biomacromolecules* 7:2099–107.

Nagaya, K., Y. Matsui, H. Ohira, A. Yuasa, H. Yamamoto, K. Ohkawa, and Y. Magara. 2001. Attachment strength of adhesive nuisance mussel, *Limnoperna fortunei*, against water flow. *Biofouling* 17:263–74.

Nelson, D. C., and C. R. Fisher. 1995. Chamoautotrophic and methanotrophic endosymbiotic bacteria at deep-sea vents and seeps. Chapter 3 in *Microbiology of deep-sea hydrothermal vent habitats*, 125–67. Boca Raton, FL: CRC Press.

Ninan, L., J. Monahan, R. L. Stroshine, J. J. Wilker, and R. Shi. 2003. Adhesive strength of marine mussel extracts on porcine skin. *Biomaterials* 24:4091–9.

Ninan, L., R. L. Stroshine, J. J. Wilker, and R. Shi. 2007. Adhesive strength and curing rate of marine mussel protein extracts on porcine small intestinal submucosa. *Acta Biomaterialia* 3:687–94.

Ohkawa, K., A. Nishida, K. Ichimiya, Y. Matsui, K. Nagaya, A. Yuasa, and H. Yamamoto. 1999. Purification and characterization of a DOPA-containing protein from the foot of the Asian Freshwater Mussel, *Limnoperna fortunei*. *Biofouling* 14 (3):181.

Ohkawa, K., A. Nishida, H. Yamamoto, and J. H. Waite. 2004. A glycosylated byssal precursor protein from the Green Mussel, *Perna viridis*, with modified DOPA side-chains. *Biofouling* 20 (2):101.

Ohkawa, K., K. Ichimiya, A. Nishida, and H. Yamamoto. 2001. Synthesis and surface chemical properties of adhesive protein of the Asian freshwater mussel, *Limnoperna fortunei*. *Macromolecular Bioscience* 1:376–86.

Papov, V. V., T. V. Diamond, K. Biemann, and J. H. Waite. 1995. Hydroxyarginine-containing polyphenolic proteins in the adhesive plaques of the marine mussel *Mytilus edulis*. *Journal of Biological Chemistry* 270 (34):20183.

Pardo, J., E. Gutierrez, C. Saez, M. Brito, and L. O. Burzio. 1990. Purification of adhesive proteins from mussels. *Protein Expression and Purification* 1:147.

Patel, J., H. Zhu, R. Menassa, L. Gyenis, A. Richman, and J. Brandle. 2007. Elastin-like polypeptide fusions enhance the accumulation of recombinant proteins in tobacco leaves. *Transgenic Research* 16:239.

Peshkova, S., and K. Li. 2003. Investigation of chitosan-phenolics systems as wood adhesives. *Journal of Biotechnology* 102:199–207.

Petrone, L., N. L. C. Ragg, and A. J. McQuillan. 2008. *In situ* infrared spectroscopic investigation of *Perna canaliculus* mussel larvae primary settlement. *Biofouling* 24 (6):405–13.

Platko, J. D., M. Deeg, V. Thompson, Z. Al-Hinai, H. Glick, K. Pontius, P. Colussi, C. Taron, and D. L. Kaplan. 2008. Heterologous expression of *Mytilus californianus* foot protein three (Mcfp-3) in *Kluyveromyces lactis*. *Protein Expression and Purification* 57:57–62.

Podsiadlo, P., Z. Liu, D. Paterson, P. B. Messersmith, and N. A. Kotov. 2007. Fusion of seashell nacre and marine bioadhesive analogs: High-strength nanocomposite by layer-by-layer assembly of clay and L-dihydroxyphenylalanine polymer. *Advanced Materials* 19:949–55.

Qin, X. X. and J. H. Waite. 1995. Exotic collagen gradients in the byssus of the mussel *Mytilus edulis*. *Journal of Experimental Biology* 198 (3):633.

Reece, T. B., T. S. Maxey, and I. L. Kron. 2001. A prospectus on tissue adhesives. *American Journal of Surgery* 182:40S–4S.

Robin, J. B., P. Picciano, R. S. Kusleika, J. Salazar, and C. Benedict. 1988. Preliminary evaluation of the use of mussel adhesive protein in experimental epikeratoplasty. *Archives of Opthalmology* 106:973–7.

Ruggiero, F., H. Chanut, and A. Fichard. 2000a. Production of recombinant collagen for biomedical devices. *BioPharm Journal* 13:32–7.

Ruggiero, F., J. Y. Exposito, P. Bournat, V. Gruber, S. Perret, J. Comte, B. Olagnier, R. Garrone, and M. Theisen. 2000b. Triple helix assembly and processing of human collagen produced in transgenic tobacco plants. *FEBS Letters* 469:132–6.

Rzepecki, L. M., K. M. Hansen, and J. H. Waite. 1992. Characterization of cystine-rich polyphenolic protein family from the Blue Mussel, *Mytilus edulis-L*. *Biology Bulletin* 183 (1):123.

Rzepecki, L. M. and J. H. Waite. 1993. The byssus of the Zebra Mussel, *Dreissena polymorpha II*: Structure and polymorphism of byssal polyphenolic protein families. *Molecular Marine Biology and Biotechnology* 2 (5):267.

Saez, C., J. Pardo, E. Gutierrez, M. Brito, and L. O. Burzio. 1991. Immunological studies of the polyphenolic proteins of mussels. *Comparative Biochemistry and Physiology B* 98 (4):569.

Salerno, A. J. and I. Goldberg. 1993. Cloning, expression, and characterization of a synthetic analog to the bioadhesive precursor protein of the sea mussel *Mytilus edulis*. *Applied Microbiology and Biotechnology* 39 (2):221.

Sever, M. J., J. T. Weisser, J. Monahan, S. Srinivasan, and J. J. Wilker. 2004. Metal-mediated crosslinking in the generation of a marine-mussel adhesive. *Angewandte Chemie* 43:448–50.

Siedentop, K. H. J. J. Park, A. N. Shah, T. K. Bhattacharyya, and K. M. O'Grady. 2001. Safety and efficacy of currently available fibrin tissue adhesives. *American Journal of Otolaryngology* 22:230–5.

Silverman, H. G., and F. F. Roberto. 2006a. Cloning and expression of recombinant adhesive protein Mefp-1 of the Blue Mussel, *Mytilus edulis*. US6987170B. Patent to Idaho National Laboratory.

Silverman, H. G., and F. F. Roberto. 2006b. Cloning and expression of recombinant adhesive protein Mefp-2 of the Blue Mussel, *Mytilus edulis*. US6995012B1. Patent to Idaho National Laboratory.

Silverman, H. G., and F. F. Roberto. 2007. Understanding marine mussel adhesion. *Marine Biotechnology* 9:661–8.

Strausberg, R. L., and R. P. Link. 1990. Protein-based medical adhesives. *Trends in Biotechnology* 8 (2):53–7.

Sun, C., J. M. Lucas, and J. H. Waite. 2002. Collagen-binding matrix proteins from elastomeric extraorganismic byssal fibers. *Biomacromolecules* 3:1240.

Takeuchi, Y., K. Inoue, D. Miki, S. Odo, and S. Harayama. 1999. Cultured mussel foot cells expressing byssal protein genes. *Journal of Experimental Zoology* 283 (2): 131–6.

Tatehata, H., A. Mochizuki, K. Ohkawa, M. Yamada, and H. Yamamoto. 2001. Tissue adhesive using synthetic model adhesive proteins inspired by the marine mussel. *Journal of Adhesion Science and Technology* 15:1003–103.

Tonegawa, H., Y. Kuboe, M. Amaike, A. Nishida, K. Ohkawa, and H. Yamamoto. 2004. Synthesis of enzymatically crosslinkable peptide-poly(L-lysine) conjugate and creation of bio-inspired hybrid fibers. *Macromolecular Bioscience* [Abbrev *Macromol. Biosci.*] 4:503–11.

Van Winkle, W. 1970. Effect of environmental factors on byssal thread formation. *Marine Biology* 7:143.

Villaverde, A., and M. M. Carrio. 2003. Protein aggregation in recombinant bacteria: Biological role of inclusion bodies. *Biotechnology Letters* 25:1385–95.

Waite, J. H. 1983. Evidence for a repeating 3,4-dihydroxyphenylalanine-containing and hydroxyproline-containing decapeptide in the adhesive protein of the mussel *Mytilus edulis*. *Journal of Biological Chemistry* 258 (5):2911.

Waite, J. H. 1985. Catechol oxidase in the byssus of the common mussel, *Mytilus edulis L. Journal of the Marine Biological Association of the United Kingdom* 65:359.

Waite, J. H. 1986. Mussel glue from *Mytilus californianus Conrad*: A comparative study. *Comparative Biochemistry and Physiology B* 156:491.

Waite, J. H., D. C. Hansen, and K. T. Little. 1989. The glue protein of ribbed mussels (*Geukensia demissa*): A natural adhesive with some features of collagen. *Comparative Biochemistry and Physiology B* 159:517.

Waite, J. H., and X. X. Qin. 2001. Polyphosphoprotein from the adhesive pads of *Mytilus edulis*. *Biochemistry* 40 (9):2887.

Waite, J. H., X. X. Qin, and K. J. Coyne. 1998. The peculiar collagens of mussel byssus. *Matrix Biology* 17 (2):93.

Waite, J. H., and M. L. Tanzer. 1981. Polyphenolic substance of *Mytilus edulis*—Novel adhesive containing L-DOPA and hydroxyproline. *Science* 212 (4498):1038.

Wang, Y. J., X. Zheng, L. H. Zhang, and Y. Ohta. 2004. Cloning and sequencing of the gene encoding mussel adhesive protein from *Mytilus sp* JHX-2002. *Process Biochemistry* 39 (6): 659–64.

Warner, S. C., and J. H. Waite. 1999. Expression of multiple forms of an adhesive plaque protein in an individual mussel, *Mytilus edulis*. *Marine Biology* 134 (4):729.

Weaver, J. K. 1998. Isolation, purification, and partial characterization of a mussel byssal precursor protein, *Mytilus edulis* foot protein 4. Master's Thesis. Newark, DE: University of Delaware.

Yamamoto, H., Y. Sakai, and K. Ohkawa. 2000. Synthesis and wettability characteristics of model adhesive protein sequences inspired by a marine mussel. *Biomacromolecules* 1:543–51.

Yang, M., K. Yamauchi, M. Kurokawa, and T. Asakura. 2007. Design of silk-like biomaterials inspired by mussel-adhesive protein. *Tissue Engineering* 13 (12): 2941–7.

Zhao, H., and J. H. Waite. 2005. Coating proteins: Structure and cross-linking in fp-1 from the Green Shell mussel, *Perna canaliculus*. *Biochemistry—US* 44:15915.

Zhao, H., and J. H. Waite. 2006a. Linking adhesive and structural proteins in the attachment plaque of *Mytilus californianus*. *Journal of Biological Chemistry* 281 (36):26150–8.

Zhao, H., and J. H. Waite. 2006b. Proteins in load-bearing junctions: The histidine-rich metal-binding protein of mussel byssus. *Biochemistry—US* 45 (47):14223–31.

Zhao, H., and J. H. Waite. 2006c. Linking adhesive and structural proteins in the attachment plaque of *Mytilus californianus*. *Journal of Biological Chemistry* 281 (36):26150–8.

Zuccarello, L. V. 1981. Ultrastructural and cytochemical study on the enzyme gland of the foot of a mollusc. *Tissue Cell* 13 (4):701.

4

Friction Stir Welding

Mel Schwartz

CONTENTS

Introduction and History

Friction stir welding (FSW) is a variant of friction welding that produces a weld between two (or more) workpieces by the heating and plastic material displacement caused by a rapidly rotating tool that traverses the weld joint [1]. Heating is believed to be caused by both frictional rubbing between the tool and workpiece and by visco-plastic dissipation of the deforming material at high strain rates. Like conventional friction welding processes, the FSW process is solid-state in nature. Both friction welding and FSW produce a volume of hot-worked metal along the bond line. However, FSW differs from friction welding in one aspect: the relative motion is between the workpieces that are held in compression in friction welding, whereas the relative motion in FSW is between the workpieces and the rotating tool.

Invented in 1991, the FSW process was developed and patented by The Welding Institute (TWI) in Cambridge, United Kingdom. The development of this process was a significant change from the conventional rotary motion and linear reciprocating friction welding processes. It provided a great deal of flexibility within the friction welding process group. The conventional rotary friction welding process requires at least one of the parts being joined to be rotated and has the practical limitation of joining regular-shaped components, preferably circular in cross section and limited in their length. Short tubes or round bars of the same diameter are a good example. The linear reciprocating process also requires movement of the parts being joined. This process uses a straight line back and forth motion between the two parts to generate the friction. Regularity of the parts being joined is not as necessary with this process; however, movement of the part during welding is essential. The obvious limitation of both processes is the joint design and component geometry restrictions. At least one of the parts being joined must have an axis of symmetry and be capable of being rotated or moved about that axis.

The FSW is capable of fabricating either butt or lap joints in a wide range of materials thickness and lengths. During FSW, heat is generated by rubbing a nonconsumable tool on the substrate intended for joining and by the deformation produced by passing a tool through the material being joined. The rotating tool creates volumetric heating, so as the tool progresses, a continuous joint is created. FSW, like other types of friction welds, is largely solid-state in nature. As a result, friction stir welds are not susceptible to solidification-related defects that may hinder other fusion welding processes.

FSW is a solid-state (nonmelting) joining technology that has produced structural joints superior to conventional arc welds in aluminum, steel, nickel, copper, and titanium alloys [2]. FSW produces higher strength, increased fatigue life, lower distortion, less residual stress, less sensitivity to corrosion, and essentially defect-free joints compared to arc welding. Since melting is not involved, shielding gases are not used during FSW of aluminum, copper, and NiAl bronze alloys while argon gas may be used during FSW of the higher-temperature ferrous and nickel alloys, mainly to protect the ceramic and refractory pin tools from oxidation. Simple argon environmental chambers and trailing shields are used during FSW of titanium alloys to minimize interstitial pickup and contamination. Expensive consumables and filler metals are not required. An excellent state-of-the-art review of FSW technology is provided by Mishra and Ma [2,3].

Fundamentals

Friction stir welding uses a nonconsumable welding tool to create heat locally. The rotating tool hot-works the material surrounding the weld interface to produce a continuous solid-state weld. A common tool design has the shape of a rod with a concave area (the shoulder) with a pin (or probe) that is coaxial with the axis of rotation. The workpieces are rigidly clamped and are supported by a backing plate—or anvil—that bears the load from the tool and constrains deforming material at the backside of the joint. In most cases, the pin is designed to be slightly shorter than the thickness of the weld joint to prevent contact with the backing plate and to promote complete penetration without defects. The process is illustrated in Figure 4.1 [4].

To make a linear weld in a butt joint configuration, the workpieces are positioned on the backing plate with the edges in contact. To start the process, the rotating FSW tool is plunged into the weld joint until the shoulder of the tool makes contact with the top surfaces of the workpieces. Frictional rubbing and visco-plastic dissipation cause the heated material to soften and plastically flow. The motion of the tool promotes displacement of the softened material to produce the weld. The hot-worked material is swept around the tool to produce recoalescence behind the tool. The tool shoulder provides constraint against the escape of hot-worked material, while applying a forging force to the top surface of the weld.

A particular nomenclature has been adopted to account for the asymmetry of the FSW process relative to the weld centerline as indicated in Figure 4.1;

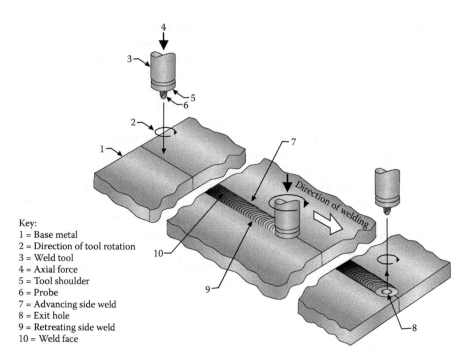

Key:
1 = Base metal
2 = Direction of tool rotation
3 = Weld tool
4 = Axial force
5 = Tool shoulder
6 = Probe
7 = Advancing side weld
8 = Exit hole
9 = Retreating side weld
10 = Weld face

FIGURE 4.1
Excerpted from the *Welding Handbook*, Vol. 3, ninth edition.

the side of the weld where the tool traverse vector is parallel to the vector of tool rotation is called the advancing side. The opposite side of the weld, where the tool traverse vector and the tool rotation vector are antiparallel, is referred to as *the retreating side*. The tool continues to rotate while traveling along the joint, completing the weld as travel progresses. When the desired length has been achieved, the tool is removed.

Why FSW?

The FSW permits the U.S. Department of Defense to meet its requirements for improved survivability, improved structural performance, enhancing capability, and affordability. The aerospace industry has been a pioneer in the application of FSW technology in the United States. FSW processing is presently used in noncritical aircraft structures. The U.S. Air Force has addressed some of its challenges in employing FSW for "critical aircraft structure." Barriers and possible solutions have been identified for technology transition opportunities in FSW processing of critical structures.

The U.S. Navy has highlighted manufacturing technologies as being essential to achieving the Littoral Combat Ship (LCS) program's strategy to reduce acquisition costs. With the planned 55 ship LCS class being a key part of the 313 ship Navy strategy, FSW for aluminum panel structures is playing a key role in reducing LCS costs. The Navy's challenge to industry is to develop low-cost FSW processing for stiffened aluminum panels and to develop a portable system that is tailored to the needs of LCS and mid-tier shipyards.

The U.S. Army has programs that show how their facilities are collaborating with industry to overcome technology barriers for rapid transition of new capabilities for the soldiers. Weld manufacturing technologies are being developed for the next generation of ground combat vehicles—Future Combat System (FCS). These vehicles must be lightweight and capable of transport by C-130 aircraft, while being fully combat capable. Manufacturing of the FCS will push the state of the art for material joining. FSW is one process that is being developed to join similar and dissimilar aluminum alloys for improved weld properties and reduced distortion. The Army has a vision to migrate FSW to depot facilities for other combat vehicles to increase knowledge and acceptance of the process within its community.

Advantages and Limitations

Since FSW is a solid-state forge welding process, it shares many of the principal advantages and disadvantages of other solid-state welding processes, particularly forge welding processes. FSW normally is done in a single pass with full penetration and with little or no joint preparation. Depending on the workpiece material and thickness, minimal distortion occurs during welding, provided proper clamping is used. The welds typically exhibit as welded mechanical properties superior to the properties of fusion welds. With the possible exception of material with high flow stress at hot-working temperatures (such as steel and titanium alloys) FSW can be done at relatively high processing speeds. Higher travel speeds can be achieved with FSW than those attained with arc welding, but FSW travel speeds may not be competitive with laser beam welding (LBW). FSW is a machine-tool process; this aspect facilitates repeatable welds to production applications with little operator input.

The FSW process provides several other advantages, including the ability to produce solid-state welds with little or no distortion of the workpieces, the avoidance of fumes or spatter, and the elimination of solidification related discontinuities, such as cracks and porosity. In addition, the process is environmentally clean. For most common applications, high-quality welds are achieved at relatively low cost using simple, energy-efficient mechanical properties. Three other significant advantages are that high travel speeds can be achieved with the process, little or no joint preparation is required, and there is no arc glare or reflected laser beams with which to contend. FSW can be used to join a variety of metals and alloys, including alloys of aluminum (Al), titanium (Ti), copper (Cu), magnesium (Mg), steel, stainless steel, and nickel (Ni). A number of joint configurations can be used, including butt, lap, corner, and T-joints. Another major advantage is that, by avoiding the creation of a molten pool that shrinks significantly on resolidification, the distortion after welding and the residual stresses are low. With regard to joint fitup, the process can accommodate a gap of up to 10% of the material thickness without impairing the quality of the resulting weld. As far as the rate of processing is concerned, for materials of 2 mm (0.07 in.) thickness welding speeds of up to 2 m/min^{-1} can be achieved, and for 5 mm (0.19 in.) thickness up to 0.75 m/min^{-1}.

One of the main limitations of FSW is that the joint is not self-supporting and must be properly restrained. If the workpiece is designed in a manner that requires support of the joint, tooling costs might be significant. However, the costs of a FSW machine and associated tooling are much the same as equipment for other solid-state forge welding processes. The initial high equipment cost typically yields a quick return on investment due to the high-volume welding capability of the process. Another limitation of the FSW process is the mechanical stability of the tool at operating temperature. During FSW, the tool is responsible for not only heating the substrate material to forging temperatures, but also providing the mechanical action of forging. Therefore, tool material must be capable of sustaining high forging loads and temperatures in contact with the deforming substrate material without either excessive wear or deformation. As a result, the bulk of the FSW applications have involved low forging temperature materials. Of these, the most important class of materials has been aluminum.

FSW Machines

Friction stir welding machine designs typically are supported by one of three different machine platforms; C-frames, gantries, or vertical structures similar to those used in boring mills. FSW typically is performed on machine tool-like equipment. The welding of thin sections with small-to-moderate forces often is done on machine tools. In heavier work, the force levels and thermal phenomena associated with the process normally require equipment specifically designed for the FSW process.

Material Suitability

A range of virtually all classes of aluminum alloys has been successfully friction stir welded. These include the 1xxx, 2xxx, 3xxx, 4xxx, 5xxx, 6xxx, and 7xxx alloys, as well as the newer Al–Li alloys. Each alloy system is metallurgically distinct. Furthermore, different alloys within the given class may have different forging characteristics. As a result, processing for each alloy may vary. However, high-integrity joints can be obtained in all classes. The process can also weld dissimilar aluminum alloys, whereas fusion welding may result in the alloying elements from the different alloys interacting to form deleterious intermetallics through precipitation during solidification from the molten weld pool. FSW can also make hybrid components by joining dissimilar materials such as aluminum and magnesium alloys. See the materials section later in this chapter.

Tool Design

The FSW tool is the most significant component of the system and usually is designed for a specific type of weld joint. Tools are manufactured from wear-resistant materials with good static and dynamic properties at welding temperature. The strength and wear resistance of the tool must be superior to the base metal used in the weldment. For example, a FSW tool made of tool steel such as H13 is commonly used to weld aluminum alloys.

The three most common variations of tool design are a single-piece tool, a two-piece retractable pin tool, and a bobbin tool. A single-piece tool is a monolithic design consisting of a pin and a shoulder, as shown in Figure 4.2. A retractable pin tool, shown in Figure 4.3, enables independent motion of the pin and shoulder in the Z-direction. This arrangement allows the pin to retract at the end of the weld, and eliminates the keyhole from the plate. A retractable pin tool allows the pin to be positioned relative to a backing plate to ensure full penetration and the shoulder can be operated in load control to ensure proper process hydrostatic pressure, all relatively independent of work-piece thickness.

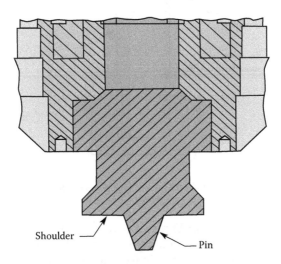

FIGURE 4.2
Single piece tool used in friction stir welding. (From AWS *Welding Handbook*, Part 2, Chapter 7, 212–58, American Welding Society, Miami, FL, 2007. With permission.)

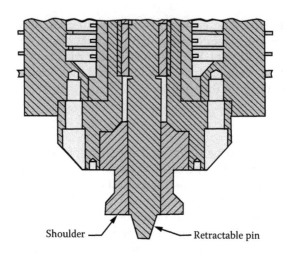

FIGURE 4.3
Retractable pin tool used in friction stir welding. (From AWS *Welding Handbook*, Part 2, Chapter 7, 212–58, American Welding Society, Miami, FL, 2007. With permission.)

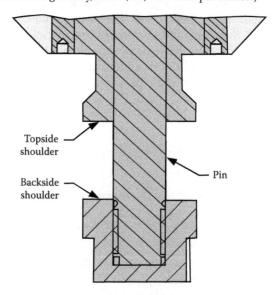

FIGURE 4.4
Bobbin-type friction stir welding tool. (From AWS *Welding Handbook*, Part 2, Chapter 7, 212–58, American Welding Society, Miami, FL, 2007. With permission.)

A bobbin tool, as shown in Figure 4.4, has a top and a backside shoulder. This tool allows for faster welding speeds and eliminates the need for a backing plate. Bobbin tooling eliminates the possibility of a root discontinuity from incomplete penetration by the pin, but bobbin tools are limited in thickness capability because of bending stresses in the pin. An issue that complicates the use of retractable pin tools and bobbin tools is the problem

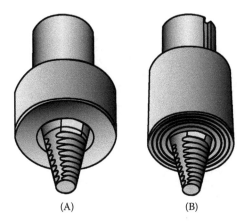

(A)   (B)

FIGURE 4.5
(A) Concave tool design and (B) scrolled design. (From AWS *Welding Handbook*, Part 2, Chapter 7, 212–58, American Welding Society, Miami, FL, 2007. With permission.)

of workpiece material extruding into the narrow clearance between the pin and shoulder and eventually binding relative motion between the tool components. These tool configurations need to be cleaned periodically to remove any extruded material.

Tool shoulders normally have either a cupped (concave) design as shown in Figure 4.5A, or a scrolled design as shown in Figure 4.5B. The concave tool design normally is used with a 1–3° tilt angle (pushing angle), while the scrolled designs normally are employed at zero tilt. With the tilted arrangement, the rear edge of the tool provides increased compressive force to aid in recoalescence of the material. The formation of the scroll is to force the material in the stir zone (SZ) to flow inward toward the pin in order to provide the compressive force [4].

In a proposed improvement of tooling for FSW, gimballed shoulders would supplant shoulders that, heretofore, have been fixedly aligned with pins. This proposal by the Marshall Space Flight Center (MSFC) researchers is especially relevant to self-reacting (SR) FSW. One consequence of the fixed alignment of the shoulders with the pin is that if the thickness of the workpiece or the slope of either the workpiece surface varies as the tool moves along the workpiece, then the leading or trailing edge(s) of one or both shoulder(s) tend to dig into the workpiece, generating excessive flashing along the weld. The proposed improvement would be a simple, relatively inexpensive means of preventing or reducing such digging. The gimballing of either or both shoulder(s) would enable the tool to better adapt to curvatures and other local variations in the slopes of workpiece surfaces, without the need for a complex, expensive shoulder-angle control system. Figure 4.6 [5] depicts a representative tool for a SR FSW incorporating the proposed improvement. (In this case, the bottom shoulder (only) would be gimballed. Optionally, both shoulders or the top shoulder (only) could be gimballed.)

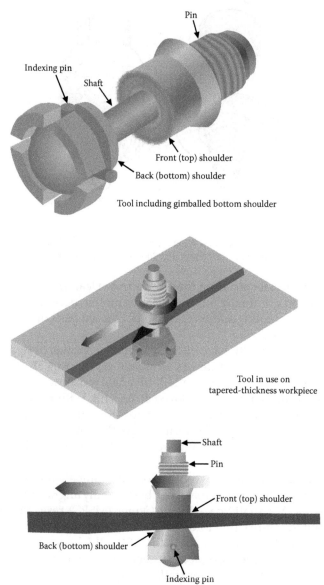

Pin

Indexing pin

Shaft

Front (top) shoulder

Back (bottom) shoulder

Tool including gimballed bottom shoulder

Tool in use on
tapered-thickness workpiece

Shaft

Pin

Front (top) shoulder

Back (bottom) shoulder

Indexing pin

Enlarged side view of tool in use on tapered-thickness workpiece

FIGURE 4.6
The back shoulder would be gimballed to accommodate local variations in the slope of the lower surface of the workpiece.

Fixturing

Fixturing for FSW usually is the most complicated and critical aspect of the process. The workpieces must be clamped to a rigid backing plate (anvil) and secured to resist the perpendicular and side forces that develop during welding. These forces tend to shift and push the workpieces apart. Fixtures are designed to restrain the workpieces and keep them from moving. A root opening (gap) of less than 10% of the material thickness can be tolerated for thicknesses up to about 13 mm (0.5 in.). The fixtures that hold the materials to the backing plate should be placed as close to the joint as possible to reduce the clamping load. Various clamping configurations are possible, but always depend on the joint configuration. Some joints are designed to allow the workpieces to provide the necessary backing, for example, a T-joint. If the vertical leg of a T-joint is of sufficient thickness, tooling would be needed only to hold it in a perpendicular position.

Tool Materials

The application of FSW on high-temperature alloys, such as steel, has not progressed as far as that for aluminum alloys due to the difficulty in finding pin tool materials able to withstand the temperatures and loads during FSW. Recent advances in the capabilities of pin tool materials have allowed for the increase of various applications as well as new materials. (See the materials section and applications on Pages 97 through 115.)

Materials that have recently been evaluated for their performance in FSW of hard metals include both refractory alloys and ceramic-based tool materials. The evaluated refractory alloys include commercially pure tungsten (CPW) and tungsten–rhenium alloy (W-25%Re). The W-25%Re alloy is one of the highest strength tungsten alloys and the W-25%Re rod used in this study was in the powder metal (P/M) condition. The examined ceramic material was polycrystalline cubic boron nitride (PCBN)—the second hardest material known to man. Materials that were welded in the above study were L80 and X70 high-strength pipe steels commonly used in the petrochemical industry. It was found in the conclusion of the evaluation and findings of the above test and the viable tool materials that:

- CPW is not a viable tool material for FSW steel due to its severe wear and deformation.
- W-25%Re pin tools performed much better than CPW tools in FSW of steels (L80 and X70) in terms of tool wear and deformation. However, there was still some wear on the W-25%Re tool after FSW L80 and X70 steels.
- The PCBN tool had no deformation issue and no visible wear. However, PCBN tools are brittle and have very high requirements on the stiffness of FSW machine and spindle run-out.

- FSW tensile properties of both L80 and X70 friction stir welds matched L80 and X70 base materials properties, respectively.
- Charpy impact results showed that L80 FSW had much lower toughness than the base material due to the large amount of martensite in the SZ. X70 FSW exhibited much higher toughness than the base material due to the formation of tough phases in the SZ.

Joint Design

The FSW can be used on a range of basic joint designs, as shown in Figure 4.7. In any joint design, fixturing must be designed to react to the forces from the FSW tool. For some applications, the joint design may allow for reaction to these forces, thus reducing fixturing requirements.

Joint Preparation

The FSW generally requires minimal joint preparation prior to welding. This requirement varies, however, depending on the base metal and the mechanical requirements of the joint. Cleaning the joint area by wiping usually is sufficient. For some applications of welds in butt joints, the heavy oxides on the joint surfaces of the workpieces must be removed, since these oxides could be swept into the joint area. For lap joints, the oxides on the overlapping surfaces must be removed.

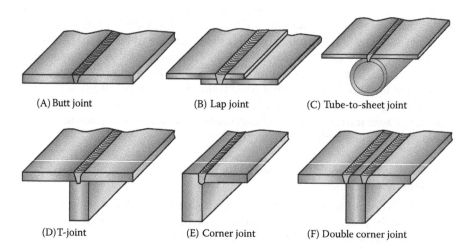

(A) Butt joint   (B) Lap joint   (C) Tube-to-sheet joint

(D) T-joint   (E) Corner joint   (F) Double corner joint

FIGURE 4.7
Basic joint designs used in friction stir welding. (From AWS *Welding Handbook*, Part 2, Chapter 7, 212–58, American Welding Society, Miami, FL. With permission.)

Materials Joining

Aluminum Alloys

Aluminum alloys possess a high strength-to-density ratio and provide good corrosion resistance in many environments. Consequently, they are often used in weight-critical structural applications in the aerospace and transportation industries. The face-centered cubic (fcc) crystal structure of these alloys provides good ductility and toughness. Many aluminum alloys have been successfully welded using FSW, including the 1xxx, 2xxx, 5xxx, 6xxx, and 7xxx series of alloys, several casting alloys, rapidly solidified aluminum–iron–silicon–vanadium (Al–Fe–Si–V) alloys, aluminum–beryllium (Al–Be) alloys, and Al-metal matrix composites. FSW has the potential to join aluminum alloys that cannot be successfully welded by fusion processes because of problems with solidification cracking and loss of properties in the HAZ (see Table 4.1 [4]).

Aluminum alloys can be welded using inexpensive and common tool materials such as tool steels. As mentioned previously, the weldability with FSW scales closely with the ease of extrusion for a given alloy. Consequently, the 1xxx alloys are more tolerant of changes in welding variables: they have a large processing window and can be welded rapidly at low forces. The 5xxx and 6xxx alloys are slightly more difficult to weld with the friction stir process, while the 2xxx and 7xxx alloys are the most difficult of the aluminum alloys. This ranking follows closely with the extrudability ranking.

Microstructures in the various regions of friction stir welds on aluminum alloys develop and evolve in accordance with the local thermomechanical cycle. In general, aluminum alloys can be strengthened by precipitation hardening during aging and/or by coldworking. The final microstructures depend on the effects of the thermomechanical cycle on the original microstructures and may develop through a variety of processes, including dissolution, coarsening and reprecipitation of precipitates as well as recrystallization, recovery, and grain coarsening. Changes in microstructure are reflected in changes in microhardness and other mechanical and corrosion properties. Because the FSW process does not melt or recast the welded material, microstructural material transformations occur during the weld's cool down—essentially taking place in the material's solid-state.

Friction stir welding may also produce significant economic advantages. The process joins aluminum alloys fairly rapidly—about 4 mm/sec (9.6 in./min)—with low heat input and without the costly shielding gases and filler materials required in fusion welding processes. The aerospace industry also uses substantial quantities of fasteners to join metallic structures—literally millions of fasteners in fabricating a large cargo or passenger aircraft. Thus, eliminating fasteners in aerospace structures by incorporating FSW joints would provide manufacturers considerable cost and weight savings. Airbus has become the first manufacturer of large civil aircraft to introduce FSW into production by incorporating this innovative technique into its A340s,

TABLE 4.1

Test Data for Tensile Properties of Base Metals and Friction Stir Welds for Common Aluminum Alloys

| | | Base Metal—Typical Properties[a] | | | Friction Stir Welds—Sample Properties | | | | | | | | |
| | | | | | Yield Strength | | | Ultimate Tensile Strength | | | % Elongation | | |
Alloy	Original Temper	Yield Strength MPa(ksi)	Ultimate Tensile Strength MPa(ksi)	% Elongation	Average MPa (ksi)	Standard Deviation MPa (ksi)	Number of Data Points	Average MPa (ksi)	Standard Deviation MPa (ksi)	Number of Data Points	Average	Standard Deviation	Number of Data Points
2014	T6	414 (60)	483 (70)	13	243 (35.2)	25 (3.6)	2	355 (51.4)	36 (5.2)	9	N/A	N/A	N/A
2024	T3	345 (50)	483 (70)	18	308 (44.7)	34 (5)	15	416 (60.4)	27 (3.9)	23	7.8	3.8	14
2195	T8	N/A	N/A	N/A	253 (36.7)	10 (1.5)	29	396 (57.5)	11 (1.6)	35	8.5	1.6	33
5083	0	145 (21)	290 (42)	22	132 (19.1)	7 (1)	4	306 (44.4)	19 (2.8)	15	22.5	0.7	2
5454	0	117 (17)	248 (36)	22	101 (14.6)	8 (1.1)	3	245 (35.5)	17 (2.4)	5	18.9	4.4	5
5454	H32/H34	207 (30)	276 (40)/	10	119 (17.3)	8 (1.1)	9	251 (36.4)	9 (1.3)	9	19.6	3.8	9
6013	T4	241 (35) 145 (21)	303 (44) 276 (40)	20	191 (27.7)	40 (5.8)	3	306 (44.4)	12 (1.8)	3	4.8	3.1	3
6013	T6	303 (44)	365 (53)	5	265 (38.5)	46 (6.6)	5	316 (45.9)	21 (3.1)	5	6.4	1.4	3
6061	T6	276 (40)	310 (45)	17	161 (23.4)	48 (6.9)	7	221 (32.1)	23 (3.3)	13	10	3.4	7
6082	T4	N/A	N/A	N/A	119 (17.2)	23 (3.3)	5	211 (30.6)	32 (4.6)	7	14.5	6.2	5
6082	T6	262 (38)	310 (45)	6	150 (21.7)	10 (1.4)	8	238 (34.6)	12 (1.7)	9	10.2	6.1	8
7050	T73/ T7451	434 (63)	496 (72)	12	444 (64.4)	63 (9.1)	6	473 (68.6)	45 (6.5)	11	6.5	0.5	6
7075	T651	503 (73)	572 (83)	11	334 (48.4)	16 (2.3)	3	439 (63.6)	30 (4.3)	7	5	2.1	3
7075	T73	393 (57)	474 (69)	7	N/A	N/A	N/A	444 (64.4)	54 (7.8)	4	N/A	N/A	N/A

Source: AWS *Welding Handbook,* Part 2, Chapter 7, 212–58, American Welding Society, Miami, FL. With permission.

Notes: The test configuration was in the transverse orientation.

N/A = Not available for this table.

Standard deviation normally applies only to Gaussian distributions and is applicable only for distributions with sample sizes greater than eight; however, standard deviations are included in this table for small sample sizes to provide some indication of the distribution about the mean.

[a] Typical properties of base metals are from *ASM Metals Handbook,* Materials Park, Ohio: ASM International. Properties of 6013 and 6075 are from Aluminum Standards and Data, 2006, Aluminum Association. Properties of 6082 T6 are from ASTM B-209, ASTM International.

and on a larger scale, into the assembly of the A350. Instead of using traditional riveted joints, Airbus has begun using FSW to join longitudinal fuselage skin joints of production A340-500 and A340-600 aircraft. Airbus noted that FSW will produce cost savings in the order of 0.9 kg for each meter of the longitudinal fuselage panel joint. Using FSW, Airbus has eliminated the need for fasteners and overlapping panels, reducing the risk of corrosion and fatigue damage. Additionally, FSW permits Airbus to join metals from different alloy families, including traditionally nonweldable alloys in thicknesses from 1 to 75 mm (0.04–2.95).

Researchers from AFRL's Metals, Ceramics, and Nondestructive Evaluation Division (Air Force Research Laboratory) [6] friction stir welded a number of aerospace alloys, including 7050-T7451, an alloy widely used in military and commercial aircraft manufacturing, to assess the effects of the process on microstructure and mechanical properties. Using optical microscopy and transmission electron microscopy (TEM) examination of the welded joint's weld-nugget region showed that FSW initially transforms the grains to fine recrystallized grains. The TEM examination also demonstrated that the FSW process redissolves the strengthening precipitates in the weld-nugget region. Furthermore, in the heat affected zone (HAZ), the FSW process preserved the initial grain size and increased both the size of the strengthening precipitates and that of the precipitate-free zone (PFZ) by a factor of 5 [6].

An analysis of residual stresses, fatigue crack closure, and fatigue fracture surfaces suggests that the decreased fatigue crack growth (FCG) resistance in the weld-nugget region is a result of an intergranular failure mechanism; in the HAZ region, residual stresses are more dominant than the microstructure and result in improved FCG resistance. The AFRL research team was the first to offer a sound theoretical explanation regarding the effects of residual stress on FCG in friction stir welds [6]. By expanding the knowledge of microstructure-property relationships, corrosion and failure modes, and life cycle benefits in friction stir welded materials, AFRL researchers are also developing databases and process specifications so that manufacturers employing these FSW tools can consistently achieve desirable and predictable properties, enabling FSW to be qualified for use in the manufacture of major structural assemblies such as reusable cryotank applications for space.

Okamoto et al. [7] evaluated the FSW of cast (A360-F) and sheet aluminum (A5052-H34) for weight reduction of structural parts, which is an important challenge in the automotive industry for improved fuel economy. Results of the study that included the effect of welding speed on the mechanical properties of the weld joints is summarized below:

1. The tensile strength of the A5052-H34 joint depends slightly on the welding speed. The joint shows lower tensile strength and elongation compared to the base metal and fails through necking at the HAZ where the lowest hardness of around 58 HV in the welded region occurs.

2. The tensile strength of A360-F joint significantly increases with high welding speed. The joint shows higher tensile strength and greater elongation than the base metal since the eutectic silicon has been dispersed in the SZ.

3. In the case of the dissimilar material joint of A5052-H34 and A360-F, the tensile strength slightly depends on the welding speed. The tensile strength depends on the strength of the HAZ in A5052 where deformation and necking occurs.

In another study and FSW evaluation of alloy 2219, LLNL (Livermore, California) and Advanced Joining Technologies Inc. (AJT; Santa Ana, California) researchers [8] successfully demonstrated the following:

1. Bobbin tool FSW produced satisfactory joints in 3.8 cm (1.5 in.) thick 2219 aluminum alloy.

2. Friction plug welding produced satisfactory closeout welds in the circumferential welds.

3. Postweld solution treatment, quenching, and artificial aging were necessary to restore the welds to near base metal strength, ductility, and toughness.

4. Satisfactory repair welds were made by FSW over the defective region in the original weld.

5. Preliminary fracture mechanics analyses indicated that the welds are fit for the intended service.

6. 6082-T6 aluminum alloy has been friction stir welded in thicknesses ranging from 1.2 to 50 mm (0.04–1.9 in.) in a single pass, to more than 75 mm (2.9 in.) when welding from both sides. Welds have also been made in pressure die cast aluminum material without any problems from pockets of entrapped high pressure gas, which would violently disrupt a molten weld pool encountering them.

Copper Alloys

Copper alloys have high thermal and electrical conductivity and are corrosion resistant in many environments. They have good ductility due to their face-centered cubic crystal structure and typically are strengthened by cold working, although some alloys can be precipitation strengthened [4]. Copper alloys can be fusion welded but the welds have problems with hot cracking and porosity. These alloys have been successfully joined used FSW. A notable application involves the fabrication of nuclear waste storage containers. Peak temperatures in the SZs have been reported to range up to 800°C (1472°F). Hence, tool materials must be chosen that will survive these temperatures without wear or fracture. Tool steels; cemented carbides; ceramics;

and tungsten, molybdenum, and nickel super alloys have been tested as tool materials for welding copper with the following results:

1. FSW tools made of tool steel are not recommended because of rapid and excessive wear
2. Carbide and ceramic stirring tools tend to fracture easily
3. The refractory metal showed wear but did not fracture
4. Nickel alloys performed best, particularly Nimonic 105

Microstructures in the SZ has been characterized by a refined grain size, probably due to classical discontinuous dynamic recrystallization. Transmission electron microscope (TEM) observations of the stirred metal region revealed a lower density of dislocations relative to the base metal. Hardness in the SZ has been shown to either increase or decrease, depending on welding conditions. Tensile testing showed that friction stir welds on copper alloys have very high ultimate tensile strengths relative to the base metal.

Magnesium Alloys

The demand for magnesium (Mg) alloys is increasing due to the useful combination of low density, tensile strength, and elastic modulus of these alloys. These alloys provide good strength-to-weight ratio and are popular choices in the aerospace and transportation industries for weight-critical applications. They are mainly alloyed with either aluminum or zinc in addition to rare earth elements and can be strengthened by precipitation hardening and/or cold working. Magnesium alloys normally are cast or produced by some type of semisolid casting process. The cast structure usually contains brittle eutectic and intermetallic phases. The hexagonal close-packed (hcp) crystal structure combined with brittle phases make the forming and shaping of Mg alloys difficult. These alloys can be fusion welded but porosity and coarse intermetallics may be present in the fusion zone. Problems resulting from the hcp structure and brittle phases also complicate the FSW of magnesium alloys. Many types of magnesium alloys have been welded using the FSW process, including AM50, AM60B, AZ31B, AZ31D, and experimental alloys. (See entries in the Bibliography for information on this). Peak temperatures of these alloys during welding have been reported to range from 325 to 450°C (617–842°F), depending on rpm and tool traverse rates. Magnesium alloys can also be welded with tools made of tool steel [4].

In a variety of reported studies [4], the SZ regions of magnesium alloys showed a refined grain size without coarse intermetallics. The grains in the SZ have been observed to show a high dislocation density, and the SZ shows a preferred texture with certain crystallographic planes called basal planes aligned with the direction of tool rotation.[a] The transition zone (TMAZ) may be recrystallized and may show some dissolution of inter-metallics. Hardness plots were either flat or showed an increase in the SZ, due to the refined grain size and higher dislocation density of the alpha phase. Typical yield strengths were reported to be approximately 60–90% of the base metal, with tensile strengths of 75–90% of the base metal and lower ductilities. The texture has been reported to have an impact on tensile properties.

Steel Alloys

Steel alloys represent one of the most common groups of engineering materials. Steels are inexpensive to produce and are used extensively by the auto and shipbuilding industries.

1018 Steels

Friction stir welds of 1018 steel made with a tungsten tool have been studied and comparisons made of tool dimensions before and after welding. These measurements indicated that most of the deformation and wear occurred during the tool plunging stage. Partial-penetration, partial-diameter, and pilot holes were used to minimize tool wear during the plunge period. Thermocouple measurements and microstructural characterization indicated that peak temperatures in the SZ reached the austenite phase field. Peak temperatures were estimated to exceed 1100°C (2012°F) and to approach 1200°C (2192°F). Friction stir welds in carbon–manganese (C–Mn) steels, such as 1018, may exhibit several microstructurally distinct regions, including the SZ (along the weld centerline), a grain-coarsened region surrounding the SZ, a grain-refined region (encompassing the grain-coarsened region), as well as an intercritical region and a subcritical region containing partially spheroidized carbides. The various regions develop in accordance with the local thermomechanical cycle.

Consistent with the slow cooling rate and limited hardenability of these steels, no evidence of martensite was found in the SZ or the HAZ. However, the SZ and HAZ normally were harder than the base metal due to a finer grain size and to the formation of fewer transformation products during cooling. Welded samples failed in regions corresponding to the base metal and demonstrated yield and ultimate tensile strengths comparable to or in excess of those of the base metal. Tensile properties of 1018 steel as well as other steels (discussed below) are presented in Table 4.2 [4].

---

[a] The basal planes are those planes normal to the C-axis of the hpc crystal on which all of the dislocations glide and consequently on which all deformation takes place.

TABLE 4.2

Test Data for Tensile Properties of Base Metals and Friction Stir Welds for Several Steels

| | | Base Metal—Typical Properties[a] | | | Friction Stir Welds—Sample Properties | | | | | | | | |
| | | | | | Yield Strength | | | Ultimate Tensile Strength | | | % Elongation | | |
Alloy	Sample Orientation	Yield Strength MPa(ksi)	Ultimate Tensile Strength MPa(ksi)	% Elongation	Average MPa (ksi)	Standard Deviation MPa (ksi)	Number of Data Points	Average MPa (ksi)	Standard Deviation MPa (ksi)	Number of Data Points	Average	Standard Deviation	Number of Data Points
101B	Transverse	310 (45)	463 (67.2)	40	332 (48.1)	21 (3)	3	476 (69.1)	21 (3)	3	22	3	3
DH-36[b]	Transverse	345 (50[a])	483 (70[a])	20[a]	439 (63.6)	24 (3.5)	7	506 (73.4)	41 (5.9)	7	5.8	2.3	7
DH-36[b]	Transverse	427 (62)	579 (84)	N/A	565 (82)	29 (4.2)	2	624 (90.5)	24 (3.5)	2	N/A	N/A	N/A
DH-36[b]	Longitudinal and all weld metal	345 (50[a])	483 (70[a])	20[a]	610 (88.5)	5 (0.7)	2	800 (116.1)	39 (5.7)	2	9.3	6	2
HSLA-65	Transverse	448 (65[a])	538 (78[a])	20[a]	470 (68.1)	45 (6.5)	4	571 (82.8)	2 (0.3)	4	24.1	6.8	4
HSLA-65	Transverse	448 (65[a])	538 (78[a])	20[a]	527 (76.5)	14 (2.1)	2	645 (93.5)	14 (2.1)	2	25.9	1.4	2
S-355	Transverse	343 (49.7)	545 (79.1)	31	345 (50.1)	10 (1.4)	5	533 (77.3)	4 (0.6)	5	18.8	0.8	5
Q&T2[c]	Transverse	1427 (207)	1731 (251)	11.3	1041 (151)	21 (3)	N/A	1234 (179)	21 (3)	N/A	2.6	3	N/A
Q&T3[d]	Transverse	1489 (216)	1765 (256)	13	1124 (163)	N/A	N/A	1213 (176)	N/A	N/A	2.8	N/A	N/A

Source: AWS Welding Handbook, Part 2, Chapter 7, 212–58. American Welding Society, Miami, FL. With permission.

Note: N/A = not available for this table.

    Standard deviation normally applies only to Gaussian distributions and is applicable only for distributions with sample sizes greater than eight. However, standard deviations are included in this table for small sample sizes to provide some indication of the distribution about the mean.

[a] Typical properties of base metals are from ASM Metals Handbook, Materials Park, Ohio: ASM International.

[b] Samples contained small discontinuities.

[c] 500 BHN quenched and tempered steel (1/4 in. thick plate).

[d] 500 BHN quenched and tempered steel (1/2 in. thick plate).

DH-36

Friction stir welds have been produced on DH-36 steel using W–Re tools by several researchers. The SZ microstructure contained grains of polygonal ferrite about 5 μm in size. No regions of martensite were observed in the SZ. The HAZ contained regions of partially spheroidized (fuzzy) pearlite but no grain-coarsened or grain-refined regions. The microhardness of the SZ averaged a Vickers hardness number of approximately 215, while that of the base metal was between 190 and 195 VHN. The weld samples showed acceptable strain-to-failure with yield and tensile strengths exceeding those of the base metal. See Table 4.2 for tensile properties.

HSLA-65

Studies of FSW on HSLA-65 steels provided similar results to those of the DH-36 tests. Corrosion testing with salt spray showed no evidence of preferential corrosion on those welds. The heat affected regions in two-pass friction stir welds on a quenched and tempered steel showed lowered hardness due to over tempering. Tensile tests showed ~70% joint efficiency fortensile strength. See Table 4.2 [4,9].

Stainless Steels

Stainless steel alloys provide a useful combination of strength and corrosion resistance. There are five major categories of stainless steels, classified according to their microstructures at room temperature. Most work on the FSW of stainless steel was conducted on alloys from the austenitic class including 304, see Table 4.3 [4], 316, and AL6-XN. This class is not strengthened by heat treatment. Austenitic stainless steels can be readily welded with fusion processes, but may have problems with hot cracking and distortion in large sections. The FSW of these materials requires the use of tools made of W, W–Re, or PCBN using higher forces and slower travel speeds relative to Al alloys. Peak SZ temperatures exceed 1000°C (1832°F). As with most high temperature materials, tool wear and deformation is a problem with metallic tools, while the service life of cubic boron nitride (CBN) tools may be limited by their low toughness.

Results of one study has revealed that the SZ of these welds was characterized by a refined grain size with equiaxed grains due to dynamic recrystallization. Microhardness results have normally showed relatively flat profiles across all of the weld zones. Tensile results have indicated increased yield strength and tensile strength with respect to the base metal, with a slight loss in elongation to failure, see Table 4.2 [4]. A few studies were completed on the FSW of duplex and ferritic stainless steels using CBN tools, with the result that the ferritic alloy showed increased SZ hardness, suggesting the formation of martensite.

Titanium Alloys

Titanium (Ti) alloys have densities intermediate to aluminum and steel alloys. They provide a good strength-to-weight ratio and can withstand higher temperatures in service than aluminum alloys. Titanium alloys also possess excellent corrosion resistance in many environments, including marine service. These alloys are classified according to their room temperature microstructure as alpha (α), alpha–beta (α–β), metastable (β), or (β) alloys. The α alloys have an hcp crystal structure, while the β alloys have a body-centered cubic (bcc) structure.

Several titanium alloys have been successfully joined using FSW, including Ti-6Al-4V, Ti-15-3, and Beta-21S. The reactivity of titanium, the high working temperatures, and large forces require the use of specialized tool materials, and wear and deformation of the tool are prevalent. Commonly used tool materials for FSW of titanium alloys include molybdenum alloy (TZM), tungsten, W–Re alloys, and W–Re containing hafnium carbide (HfC). The beta alloys are easiest to join by FSW, followed by the alpha–beta alloys.

Commercially pure (CP) titanium (alpha) appears to be the most difficult to weld because it possesses an hcp structure and higher thermal conductivity relative to the other alloy classes. Data for base metal and tensile properties of friction stir welds for three titanium alloys are presented in Table 4.4 [4]. Values are given for CP Ti, the Ti-6Al-4V (Ti-6-4) alloy, and the Ti-15-3 alloy. In Table 4.4, the average standard deviation and number of data points used were included in the friction stir weld data where available. The loss of tensile properties in titanium alloys due to welding is much less than the loss incurred when welding aluminum alloys. These differences stem from the differences in the strengthening mechanisms between titanium and aluminum alloys. In other words, high-yield strength and high-tensile strength joint efficiencies can be achieved in Ti alloys with friction stir welds. In certain cases, the joint efficiency may appear to be greater than 100%.

Ti-6-4 Alloy

Studies on the FSW of titanium alloys have included the alpha–beta alloy, Ti-6-4. Defect-free welds have been produced over a range of parameters using the tools mentioned above. The tool and workpieces were protected from surface oxidation by making the weld in an inert gas chamber with a sliding top section. In one study, peak temperatures in the SZ were reported to exceed 1045°C (1913°F) and may have ranged up to 1150°C (2102°F). The SZ in these welds contained equiaxed grains and had a refined grain size relative to the base metal. Results of microhardness tests across regions in a transverse section of a friction stir weld in Ti-6-4 do not show the pattern of microhardness; that is, the "W" shape typical of precipitation strengthened aluminum alloys with a minimum of hardness in the HAZ. Rather, the

TABLE 4.3

Test Data for Tensile Properties of Base Metals and Friction Stir Welds for 304 Stainless Steel and 2507 Duplex Stainless Steel

| | | Base Metal—Typical Properties[a] | | | Friction Stir Welds—Sample Properties | | | | | | | | |
| | | | | | Yield Strength | | | Ultimate Tensile Strength | | | % Elongation | | |
Alloy	Sample Orientation	Yield Strength MPa (ksi)	Ultimate Tensile Strength Pa (ksi)	% Elongation	Average MPa (ksi)	Standard Deviation MPa (ksi)	Number of Data Points	Average MPa (ksi)	Standard Deviation MPa (ksi)	Number of Data Points	Average	Standard Deviation	Number of Data Points
304	Transverse	689 (100)	724 (105)	28	689 (100)	N/A	N/A	738 (107)	N/A	N/A	30	N/A	N/A
304	Transverse	296 (43)	669 (97)	NA	395 (57.3)	50 (7.2)	2	707 (102.5)	40 (5.8)	2	N/A	N/A	N/A
304	Transverse	172 (25)	483 (70)	NA	340 (49.3)	32 (4.6)	3	621 (90)	21 (3)	3	N/A	N/A	N/A
304	Longitudinal and all weld metal	186 (27)	483 (70)	40	352 (51)	0	2	652 (94.5)	5 (0.7)	2	65	1.4	2
2507	Transverse	717 (104)	951 (138)	34	740 (107.4)	N/A	N/A	930 (134.9)	N/A	N/A	18	N/A	N/A

Source: AWS Welding Handbook, Part 2, Chapter 7, 212–58, American Welding Society, Miami, FL. With permission.

Notes: N/A = not available for this table.

Standard deviation normally applies only to Gaussian distributions and is applicable only for distributions with sample sizes greater than eight; however, standard deviations are included in this table for small sample sizes to provide some indication of the distribution about the mean.

[a] Typical properties of base metals are from ASM Metals Handbook, Materials Park, Ohio: ASM International.

TABLE 4.4

Test Data for Tensile Properties of Base Metals and Friction Stir Welds for Three Titanium Alloys

| | Base Metal—Typical Properties[a] | | | Friction Stir Welds—Sample Properties | | | | | | | | |
| | | | | Yield Strength | | | Ultimate Tensile Strength | | | % Elongation | | |
Alloy	Yield Strength MPa (ksi)	Ultimate Tensile Strength MPa (ksi)	% Elongation	Average MPa (ksi)	Standard Deviation MPa (ksi)	Number of Data Points	Average MPa (ksi)	Standard Deviation MPa (ksi)	Number of Data Points	Average	Standard Deviation	Number of Data Points
CP Ti (Grade 2)	275 (40)	441 (63.9)	25	N/A	N/A	N/A	430 (62.4)	14 (2)	N/A	20	2	N/A
Ti-6-4	913 (132.4)	1014 (147)	12.7	897 (130.1)	0.69 (0.1)	3	958 (138.9)	3 (0.5)	3	12.7	30.5	3
Ti-15-3	765 (111)	769 (111.5)	28	817 (118.5)	28.0 (4)	N/A	822 (119.2)	28 (4)	N/A	6.4	4	N/A

Source: *AWS Welding Handbook*, Part 2, Chapter 7, 212–58, American Welding Society, Miami, FL. With permission.

Notes: The test configuration was in the transverse orientation.

N/A = not available for this table.

Standard deviation normally applies only to Gaussian distributions and is applicable only for distributions with sample sizes greater than eight; however, standard deviations are included in this table for small sample sizes to provide some indication of the distribution about the mean.

[a] Typical properties of base metals are from ASM Metals Handbook, Materials Park, Ohio: ASM International.

hardness profile is quite flat, with the exception of an increase in hardness in regions corresponding to the HAZ. The base metal had a Vickers hardness number between 320 and 340. The Vickers hardness increased to about 370 in the HAZ and fell again to 350 in the SZ.

Ti-15V-3Cr-3Al-3Sn Alloy

Defect-free friction stir welds also were produced in the metastable beta alloy Ti-15V-3Cr-3Al-3Sn (hereafter referred to as Ti-15-3) using W–Re tools. Again, the tool and workpieces were protected from surface oxidation by welding in an inert gas and chamber. Chemical analysis of the weld metal showed no measurable pickup of oxygen during welding. Tool wear and deformation were monitored before and after each weld by inspecting the tool with an optical comparator. Results indicated no measurable tool wear, deformation, or pickup of tungsten or rhenium in the workpiece after several welds totaling about 1 m (3 ft) in length.

Microstructural characterization of the welds on annealed Ti-15-3 revealed considerable grain refinement in the SZ. No alpha phase was observed optically in the SZ, the TMAZ or the HAZ of the welds on the annealed material. Welds on the annealed Ti-15-3 exhibited high yield joint efficiencies with acceptable ductility. Properly made welds failed through the base metal away from the weld region. Little difference was observed in mechanical properties between welds produced on the annealed base metal and those given a postweld aging treatment at 635°C (1175°F) for 8 hours. The postweld aged samples exhibited the grain boundary alpha phase along the beta grain boundaries and the alpha phase dispersed throughout the grain interiors. Similar FSW success has been achieved in the beta 21S alloys.

CP Ti Alloy

Sound friction stir welds also have been produced on a CP titanium alloy using a titanium-carbide tool. The SZ contained a considerable amount of twinning and a high dislocation density. The distribution of twins was found to vary throughout the SZ, and the HAZ region was characterized by slight grain growth. Hardness results showed scatter with higher hardness in the regions with twins and higher dislocation densities.

Applications

Automotive

Prototype tailor welded blanks, in which dissimilar thicknesses of material are welded together, have been made by FSW, for example, in the inner door panels of aluminum cars. The A.O. Smith Corp. (Milwaukee, Wisconsin) has developed a prototype aluminum friction stir welded engine cradle, while

FIGURE 4.8
Two aluminum friction stir welded applications: wheel and connector. (From Friction stir welding. http://www.azom.com/details.asp?ArticleID = 1170, p. 1 of 5, 02/14/2008. With permission.)

Tower Automotive (Plymouth, Michigan) has produced the Simulform connector (see Figure 4.8) [10]. FSW has also been used to weld lightweight panels made of plastic foam sandwiched between two sheets of aluminum, for which any fusion welding technique would encounter serious problems because of the much higher temperatures involved. Foamed aluminum itself has been friction stir welded too. Other automotive examples include:

- Engine cradle
- Simulform connector for chassis members and space frames
- Bodies and floors of coaches and busses
- Suspension and auto body weldments
- Central tunnel assembly in Ford GT automobiles
- Automobile doors in Mazda Motor vehicles

Marine and Rail Transportation

- Fast ferry docking and sections of subway and rail cars

Aerospace Launch Vehicles

- First stage launch vehicle (Delta IV) containing a liquid oxygen tank, high fuel tank, and high interstage cylinder
- Upper stage of Ares I rocket [11–13]—this work was recently begun as the FSW process was applied to the Orion crew module ground test article that will serve as a production pathfinder to validate the flight vehicle production processes and tools. Ares and Orion are part of the next generation NASA project "Constellation." A new era in human space flight and space program to revisit the Moon in 2020.

The initial weld joined a 2195 aluminum–lithium (Al–Li) cone panel and an Al 2219 alloy longer on using FSW. This process will be used for all crew module welds. The weld operations occur on a Universal Weld System II (UWS II) featuring a 22-foot diameter turntable, SR friction stir weld head, and modular t-grid floor. The system provides virtually five-axis welding on fixture-mounted hardware.

Aircraft

- Circumferential aluminum fuselage stiffeners and door doublers for the Eclipse Aviation business class jet
- Airbus models A350, A340-500, and A340-600 have friction stir weldments in their fuselage and wing sections

Miscellaneous

There are numerous future potential applications undergoing evaluation prior to their utilization in production:

- C-17 Globemaster cargo aircraft floor decking.
- Main fuel tank of the space shuttle.
- Military bridge-laying vehicles (and bridges/pontoons), and waste ships.
- Joining thermoplastics such as polyethylene and plexiglass using a rotating pin that was 10.16 mm (0.4 in.) in diameter, and a rotation speed of 200 rpm, with weld travel speed of 0.8 mm/sec (2 in./min). The resulting weld strength was said to be over 90% of the strength of the base metal.

In the next millennium there is no doubt that the automotive sector will find an increasing number of uses for this process as its cost-effectiveness and ability to weld dissimilar material combinations with minimal distortion is more widely appreciated. The message for lightweight vehicle design engineers is that they can also specify alloys in welded components and structures that they had previously shied away from using because of their inherent fusion welding problems.

Variations of FSW

High-Speed FSW (HS-FSW)

The concept behind high-speed FSW (HS-FSW) is based on the premise that high spindle speeds (tens or hundreds of thousands of rpm) in FSW reduce the forces necessary to produce sound welds to a level permitting manual handheld devices. Engineers have already begun studying the effects of high rpm pin tools using a high-speed machine tool for FSW. The machine is capable of 30,000 rpm, with a travel speed up to 80 mm/sec (200 in./min). It has been used to weld 1.52 mm (0.060 in.) GR Cop-84, a copper alloy being investigated for thrust chambers. Further research is planned to investigate high-speed phenomena relative to forces, feed rates, pin-tool designs, and robotic applications [14]. The challenge for space welding technology is the constraints placed upon mass, power, and volume. Therefore, a low mass, portable device is critical to NASA's in-space, in-situ fabrication and repair according to Dr. Weija Zhow [14].

Ultrasonic Stir Welding (USW)

The ultrasonic stir welding (USW) is also a solid-state welding process. One heats the weld piece using high powered ultrasonic energy. USW is an idea based on ultrasonic assisted drilling. Without ultrasonic energy, the drill takes considerable time and force to drill through a steel plate. A load cell records the amount of force it takes to push a drill clear through the plate and is represented in a strip chart. During the drilling with no ultrasonics, the needle on the strip chart goes very high on the scale and then drops to zero when the drill pops through the steel plate. When the ultrasonic energy is applied, the drill cuts through the steel plate very quickly with significantly less force. The strip chart needle barely rises from the bottom of the scale. Without ultrasonic energy, the typical metal chips fly off the drill bit while drilling. When the ultrasonic energy is turned on, one long metallic "apple peel" is discharged, or, peels away, from the drilling process—no chips. This is because the ultrasonic energy plasticizes the steel at the interface between the drill tip and the workpiece. Therefore, heating of the metal by the drill bit and the reduction of forces when drilling has been shown and proven for drilling steel plate. And what is USW? It's primarily a very,

very small drill bit (to stir the plastic material) with a nonrotating containment plate to contain the plastic material. This process is being investigated at the MSFC for handheld welding applications. Primary data tests at MSFC have shown that one can heat metals into a plastic temperature state with ultrasonic energy and can significantly reduce plunging forces. So USW can be a way that one can take a solid-state weld process and integrate it with an off-the-shelf robot for welding. Right now a huge robust robot is required to absorb loads for FSW, but with ultrasonics, an off-the-shelf robot could perform as USW. Additionally, unlike FSW, there are no rotating shoulders producing frictional heat. This concept may be more practical than HS-FSW as an in-orbit welding and repair process because high rotational speed stability issues will be avoided.

Thermal Stir Welding (TSW)

The thermal stir welding (TSW) is an offshoot of FSW. In conventional FSW, the shoulder provides much of the frictional energy (heat) as well as the compressive or forging force. The pin, which is attached to the shoulder, "stirs" the weld joint material together. These three FSW process elements—stirring, forging, and heating—work in tandem at a desired RPM and cannot be decoupled from each other. The TSW process decouples the heating, stirring, and forging elements and allows for individual control of each. This allows for greater process control. According to Jeff Ding, aerospace welding engineer at NASA MSFC, there is one benefit of using TSW [14].

When using conventional FSW to weld .500 in thick CP titanium, the FSW pin tool must rotate between 700 and 900 rpm to generate the frictional energy required to plasticize the material. With TSW, you heat the part with a specially designed induction coil; it heats very quickly through the thickness. Once you're up to your stir temperature—it could be 1400°, it could be 1600°—whatever temperature it is that you want to stir, you move the part into your stir rod and all that stir rod does is it stirs the material together just like the little pin on a FSW tool. The stir rod protrudes through the middle of two containment plates that contain the material as it is being stirred. Containment plates are stationary—they do not rotate. The containment plates also supply the compressive load to the stirred material for microstructure consolidation. The force they compress with is also controlled independently. Thus adiabatic heating is created from the friction given off by the stir rod, but the primary source of heating is the induction coil. Welds rotating at only 200 rpm in 0.500 in thick titanium have been successfully made.

Now, what is the benefit of independent control of heating, stirring, and forging pressures? It certainly increases life of the stir rod. Since the material is already at temperature when the stirring begins, the stir rod is primarily just a mixing tool that moves plastic material within the weld zone. In FSW, the shoulder/pin assembly must provide both heat and stir functions. This

❧

reduces life of the shoulder/pin assembly at the temperatures required to join CP Ti. Other benefits of independent control are not yet known. Work is on-going with TSW whereby welding engineers are looking at the 0.500 in. thick CP Ti weld microstructures to compare those that were stirred using 100, 200, and 300 rpm to see if one notices any differences. What is suspected is that the strain rate being induced into the microstructure with the TSW process is much less than using FSW. Data shows that the FSW pin tool rotational speed to heat and stir 0.500 in. thick CP Ti must be between 700 and 900 rpm. Since CP titanium has no alloying elements, the microstructure—the grains—are free to grow very large very quickly. The high strain rate induced into the microstructure creates tears and results in wormholes. With TSW, you can rotate very slowly and put a lot less strain into the microstructure. Thus that is one reason for finding and getting good welds on 0.500 in. thick titanium. Recently an eight foot long weld was completed in 0.500 in. thick CP Ti.

The U.S. Navy is very interested in this process, because the Navy is looking at this low-cost titanium product that is processed a different way than usual titanium. Instead of costing $60 a pound, it will cost $5 a pound. The reduced cost comes from the processing of the titanium. When the titanium is processed, there are a lot of tramp elements left in the metal such as chlorine. This presents problems when welding the low-cost titanium. When it is welded with fusion weld processes, like TIG, MIG, or electron beam, a lot of oxides form, resulting in inferior weld properties. The titanium to exhibit these excellent properties has to be welded without melting; meaning, a solid-state process such as FSW or TSW must be used.

Friction Stir Spot Welding (FSSW)

This process is a potential industrial implementation as a rivet replacement technology [15]. Currently, two variations to FSSW are being used. The "plunge friction spot welding (PFSW) [16] method and the "refill" friction spot welding (RFSW) [17] method. In the PFSW process, a rotating fixed pin tool similar to that used in linear FSW is plunged and retracted through the upper and lower sheets of the lap joint to locally plasticize the metal and stir the sheets together. Even though this approach leaves a pull-out hole on the center of the spot, the strength and fatigue life is sufficient to allow application at reduced production costs on the Mazda RX-8 aluminum rear door structure (Figure 4.9). Since 2003, Mazda has produced more than 200,000 vehicles with this PFSW rear door structure. These PFSW doors provide structural stability against side impact and impart a five-star rollover protection. RFSW is a process that uses a rotating pin tool with a separate pin and shoulder actuation system that allows the plasticized material initially displaced by the pin to be captured under the shoulder during the first half of the cycle. This completely refills the joint flush to the surface. In addition to development as a rivet replacement technology for aerospace structures,

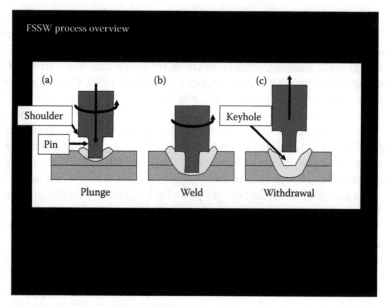

FIGURE 4.9
FSSW process overview.

RFSW is also being developed as a tacking method to hold and restrain parts during welding by linear FSW.

Friction Stir Joining (FSJ)

The friction stir joining (FSJ) of thermoplastic materials uses the controlled path extrusion characteristics of the process to join 6.35 mm (0.25 in.) thick sheets of polypropylene (PP), polycarbonate (PC), and high-density polyethylene (HDPE) materials. Recent work at Brigham Young University has shown joint efficiencies for these materials ranging from 83% for PC to 95% for HDPE and 98% for PP. These joint efficiencies compare favorably with other polymer joining methods such as ultrasonic, solvent, resistance, hot plate, and adhesive bonding. Current work at the AF Research Laboratory (Kirtland, New Mexico) is investigating the use of FSJ to join fiber-, particulate-, and nanoparticle-reinforced thermoplastic materials [3].

Friction Stir Processing (FSP)

Friction stir processing (FSP) uses the controlled path metal working characteristics of the process to perform metallurgical processing and microstructural modification of local areas on the surface of a part. Initial studies include performing microstructural modification of the cast structure of 2195 Al–Li welds to remove porosity and hot short cracks. This also improved room temperature and cryogenic strength, fatigue life, and reduced the sensitivity to intersection weld cracking by crossing welds [18].

In another study, DOE Pacific Northwest National Laboratory (PNNL; Richland, Washington) began investigating the processing by FSP of SiC powders into the surfaces of 6061 aluminum to increase wear resistance. Initial studies showed that both SiC and Al2O3 could be placed into the surface of bulk materials to create near surface-graded MMC structures. The University of Missouri–Rolla (UMR) has shown that a uniform SiC particle distribution can be achieved with appropriate tool designs and techniques, leading to significant increases in surface hardness [2].

A PNNL collaborative research program investigated increasing the wear resistance of heavy vehicle brake motors by processing TiB2 particles into the surface of class 40 gray cast iron. This resulted in a fourfold increase in the dry abrasive wear resistance. PNNL and Tribomaterials, LLC (Thousand Oaks, California) have performed subscale brake rotor pad wear tests on FSP/TiB2 cast iron rotors. These subscale brake tests have shown that FSP/TiB2 processed brake rotors have improved friction characteristics and wear resistance over baseline heavy vehicle brake friction pairs [2].

Friction stir reaction processing (FSRP) was also investigated under the above PNNL/SDSMT FSP/TiB2 program. FSRP uses the high temperatures and strain rates seen during processing to induce thermodynamically favorable in-situ chemical reactions on the surface to a depth defined by the pin tool geometry and metal flow patterns. This provides an opportunity for innovative processing methods to create new alloys on surfaces of materials and locally impart a variety of chemical, magnetic, strength, stiffness, and corrosion properties [2].

Friction Bonding (FB)

Friction bonding (FB) is a modified FSW process that allows bonding between two similar AA6061-T6 thin plates. In this regard, the process is not unlike other friction welding techniques, such as explosive welding, ultrasonic welding, roll bonding, or forge welding. In addition, the FB process is similar to FSP [19,20] in that the workpiece microstructure is locally modified. FB allows the fabrication of nuclear fuel plates for research and test reactors containing thin, monolithic fuel alloys. The FB fabrication concept is of great importance for the U.S. National Nuclear Security Administration-sponsored Global Threat Reduction Initiative. This initiative seeks to enable research and test reactors throughout the world that currently operate with highly enriched uranium (HEU $\geq$ 20%235 U) fuel to operate with low-enriched uranium (LEU < 20%235 U) fuel, which is desirable for nuclear nonproliferation reasons [21], concerns and risks. The FB process produces a surface finish that closely meets most reactor specifications and, therefore, requires minimal postprocessing surface treatment. Solutions to challenges associated with this process, in addition to the novel process itself, offer information for users of the FSW process and a potential alternative fabrication technique for laminar composite structures.

Economic Advantages and Disadvantages of FSW

Advantages

1. The process allows the welding of high-strength aluminum alloys that are not readily weldable with fusion processes. This capability allows the use of greater amounts of high-strength aluminum alloys and permits down gauging of materials with commensurate weight and cost savings.

2. FSW also can be used to fabricate wide panels of aluminum that are not available from suppliers of extruded aluminum.

3. An important consideration is that FSW normally provides better mechanical and corrosion properties for a given alloy than fusion welding.

4. Welds produced with FSW have lower values of residual stresses and less distortion, thus reducing costs associated with stress relief and straightening.

5. FSW uses no fluxes or filler metals, systemically reducing the cost of associated supplies and consumables required by other processes. Shielding gases typically are not used to weld aluminum alloys. The cost of FSW tools used for aluminum alloys is minimal, and a single tool can be used to make more than 1000 m (3280 ft) of welds.

6. Labor costs are reduced because less skill is needed to operate a FSW machine than is required for other welding processes. Most FSW machines can be run with little or no operator input, other than loading the workpieces into the machine. Fitup requirements are less stringent and joint preparation is economical because FSW requires less machining, cleaning, and etching than other processes.

7. When compared to riveting, the cost of rivets and the labor-intensive riveting process are eliminated. Startup costs and inspection costs for FSW are lower than for riveting.

8. From an energy standpoint, the FSW process is more efficient for aluminum alloys than resistance spot welding due to the low resistance of these alloys, and is more energy efficient than LBW due to the high reflectivity of aluminum. Higher travel speeds can be achieved with FSW than those attained with arc welding.

9. Cost items such as preweld and postweld machining are lower due to reduced distortion. Material costs are often lower when using FSW compared to fusion welding. This is in part due to the higher

mechanical properties in the as-welded condition, which allows the use of thinner material for a given design load.

10. Although the initial capital cost of FSW machines often is higher than machines used in other welding processes, such as arc welding, the return on investment (ROI) typically can be realized in 1 or 2 years, depending on the production volume.

Disadvantages

1. The process is still proprietary and licensing costs will continue until the patent expires.

2. The initial machine cost is high and there are a limited number of machine manufacturers.

3. The knowledge base is limited among manufacturing engineers.

4. Specifications and standards for FSW have not been established and until that occurs the process will not be more broadly accepted.

References

1. Welding terms and definitions are from American Welding Society (AWS) Committee on Definitions and Symbols, *Standard Welding Terms and Definitions*, AWS A3.0:2001, Miami, FL: American Welding Society, 2001.
2. Arbegast, W. J. Friction stir welding. *Welding Journal* 85, no. 3 (2006): 28–35.
3. Mishra, R. S., and Z. Y. Ma. Friction stir welding and processing. *Materials Science and Engineering R* 50 (2005): 1–78.
4. AWS *Welding Handbook*, Part 2, Chapter 7, 212–58, Miami, FL: American Welding Society, 2007.
5. Carter, R., and K. Lawless. Gimballed shoulders for FSW. MSFC in *NASA Tech Briefs*, 42–3, January 2008.
6. Jata, K. V., L. Semiatin, R. John, P. S. Meltzer. Friction stir welding of aerospace materials, *AFRL Technology Horizons*, 42–3, December 2005; http://www.afrl-horizond.com
7. Okamoto, K., S. Hirano, and M. Inagaki. Joining dissimilar aluminum alloys. *Welding Journal* 85, no. 4 (2006): 38–41.
8. Dalder, E., J. W. Pastrnak, J. Engel, R. S. Forrest, E. Kokko, K. McTernan, and D. Waldron. Friction stir welding of thick-walled aluminum pressure vessels. *Welding Journal* 87, no. 4 (2008): 40–4.
9. Konkol, P. J., and M. F. Mruczek. Comparison of friction stir weldments and submerged arc weldments in HSLA-65 steel. *Welding Journal* 87, no. 7 (2007): 187-s–195-s.
10. Friction stir welding. http://www.azom.com/details.asp?ArticleID = 1170, p. 1 of 5, 02/14/2008.

11. NASA's Marshall Center demonstrates first weld with tools to be used on *Ares 1. Welding Journal* 87, no. 10 (2008): 10.
12. Lockheed Martin friction stir weld currently under way on *Orion* test article. *Welding Journal* 88, no. 6 (2009): 10.
13. Iannotta, B. NASA's Michoud transferred from Lockheed to Jacobs. *Space News* (May 11, 2009): 16.
14. Ding, J., R. Carter, K. Lawless, A. Nunes, C. Russell, and M. Suits. Friction stir welding flies high at NASA. *Welding Journal* 85, no. 3 (2006): 54–60.
15. Gerlich, A., and T. North. Friction stir spot welding of aluminum alloys. *Canadian Welding Association Journal* IIW Special Edition (2006): 57–8.
16. Iwashita, T., et al. Method and apparatus for joining, U.S. Patent No. 6,601,751 B2, 2003.
17. Schilling, C., and J. dos Santos. Method and device for joining at least two adjoining work pieces by friction welding, U.S. Patent Application 0179 682, 2002.
18. Arbegast, W. J., and P. J. Hartley. Method of using friction stir welding to repair weld defects and to help avoid weld defects in intersecting welds, U.S. Patent No. 6,230,957, 2001.
19. Mishra, R. S., M. W. Mahoney, S. X. McFadden, N. A. Mara, and A. K. Mukherjee. High strain rate superplasticity in a friction stir processed 7075 Al alloy. *Scripta Materialia* 42, no. 2 (1999): 163–8.
20. Mishra, R. S., and M. W. Mahoney. Friction stir processing: A new grain refinement technique to achieve high strain rate superplasticity in commercial alloys. *Materials Science Forum* 357–359 (2001): 507–14.
21. National Nuclear Security Administration (NNSA) Web site, http://*www.nnsa. doe.gov/nuclearnp.htm#1*, accessed on April 17, 2008.

Bibliography

Burkes, D. E., N. P. Hallinan, and C. R. Clark. Nuclear fuel plate fabrication employing friction welding. *Welding Journal* 87, no. 9 (2008): 47–54.
Carter, R. Tool for two types of friction stir welding. *NASA Tech Briefs*, MSFC, Oct 1, 2006, 71–72. http://www.techbriefs.com/content/view/1055/14/-- January 21, 2008, p. 1 of 3.
Colligan, K., P. J. Konkol, J. R. Pickens, I. Ucok, and K. McTernan. Friction stir welding of thick sections 5083-H131 and 2195-T8P4 aluminum plates. *Proceedings of the 3rd International Symposium on Friction Stir Welding*, September 27–28, 2001. Kobe, Japan. Cambridge: TWI, 2002.
Defalco, J. Friction stir welding vs. fusion welding. *Welding Journal* 85, no. 3 (2006): 42–4.
Dixon, J. P., D. E. Burkes, and P. G. Medvedev. Thermal modeling of a friction bonding process. *Proceedings of the COMSOL Conference 2007*, 349–54. Boston, MA, 2007.
EWI growing capability with friction stir welder, 5. *EWI Insights*, 2004, harvey_castner@ ewi.org

Gerlich, A., G. Cingara-Avramovic, and T. H. North. The influence of processing parameters on grain size of Al 5754 friction stir welds, *ICAA Conference,* Vancouver, Canada, September 2006.

Gerlich, A., P. Su, and T. H. North. Friction stir spot welding of Mg-alloys for automotive applications, *Magnesium Technology 2005.* Edited by N. R. Neelameggham, H. I. Kaplan, and B. R. Powell. San Francisco: TMS (The Minerals, Metals & Materials Society), February 2005.

Gerlich, A., P. Su, and T. H. North. Similar and dissimilar friction stir spot welding of Mg-alloys, *International Conference of Magnesium Technology,* Montreal, November 2005.

Gerlich, A., P. Su, T. H. North, and G. J. Bendzsak. Friction welding colloquium, University of Graz, Austria, May 23, 2006.

Hallinan, N. P., and D. E. Burkes. Friction stir weld tools, methods of manufacturing such tools, and methods of thin sheet bonding using such tools. U.S. patent application, 2007.

Mendez, P. F., and T. W. Eagar. Welding processes for aeronautics. *AM&P* (May 2001): 39–43.

Nandan, R., B. Prabu, A. De, and T. Debroy. Improving reliability of heat transfer and materials flow characteristics during friction stir welding of dissimilar aluminum alloys. *Welding Journal* 86, no. 10 (2007): 313-s–322-s.

North, T. H. Friction stir spot welding of Mg alloy sheet: Research status and future needs, *First Canada-USA-China* workshop on magnesium developments and applications, October 17–19, 2005, Detroit, MI.

North, T. H., G. J. Bendzsak, A. Gerlich, and P. Su. Stir zone formation in FSW spot welds, *AWJT'2005 Conference,* Dalian, PR China, October 20–22, 2005.

North, T. H., G. J. Benzsak, P. Su, G. Cingara, and C. Maldonado. Understanding friction welding, *Proceedings of IURS 2000* in Tsinghua University, Beijing, August 15–16, 2005.

Nunes, Jr., A. C. Counterrotating-shoulder mechanism for friction stir welding, *NASA Tech Briefs* (April 2007): 58–60.

Perrett, J. Large Mg plates joined by high-force friction stir welding. *AM&P* (August 2007): 18–19. Cambridge: TWI; jonathan.perrett@twi.co.uk; http://www.twi.uk.

Su, P., A. Gerlich, and T. H. North. Friction stir welding of aluminum and magnesium sheets, *SAE World Congress,* April 2005, Detroit, MI.

TWI purchases Friction Stir Link's RoboStir System. *Welding Journal* 85, no. 3 (2006): 6.

Yang, Y. K., H. Dong, and S. Kou. Liquation tendency and liquid-film formation in friction stir spot welding. *Welding Journal* 87, no. 8 (2008): 202-s–211-s.

5

Friction Stir Welding: Aerospace Aluminum Applications

Leanna Micona and Burke Reichlinger

CONTENTS

Commercial airplane companies are constantly looking to reduce the cost of airplane manufacturing and improve airplane performance. Historically most aircraft assembly is done by mechanical joining (e.g., fasteners), fusion welding, and adhesive bonding. Improvements in cost may be achieved through improved manufacturing processes, reduced part count, raw material costs, and improved assembly rate. Performance improvements can be made by reducing the final part weight through the elimination of fasteners and overlapping splice joints. One enabling technology to be considered is friction stir welding (FSW). FSW has shown a huge potential for the integration of very complex structural elements in airframe construction that can result in significant manufacturing cost savings. This chapter will discuss the advantages of FSW for aerospace and the characteristics inherent to the process. For more specific details on FSW see Mishra and Mahoney (2007).

Friction stir welding is a solid-state joining process that utilizes friction heating resulting from pressure applied with a spinning tool to form a metallic bond. It offers advantages over fusion welding due to the higher welding rates, and the wrought microstructure and lower joining temperatures that have less impact on joint efficiency. Another advantage is that it enables the joining of alloys that both can and cannot be joined with traditional fusion welding technologies. For instance, fusion welding can be used to join 5000, 6000, and some 2000 series alloys. FSW is not only able to join these alloys but can also join 2000 and 7000 series alloys that are not typically fusion welded in structural aerospace applications.

Friction stir welding also offers several obvious ancillary cost advantages over conventional welding and fastening processes. Examples of cost advantages include not needing welding consumables (e.g., filler metal and shielding gas) and minimal safety equipment compared to that typical of fusion welding. Since FSW is a solid-state joining process that is primarily automated, there are no hazards associated with molten metal. In addition, there is a potential benefit in terms of reduced injuries associated with the ergonomic concerns of manual welding and fastening assembly. The consumable materials and safety equipment required for fusion welding is significantly reduced when FSW is applied.

Distortion caused by joint fitup, melting, and solidification of the weld is a common issue with fusion welding. When a part distorts after fusion welding it is probable that a straightening process will follow the welding prior to further processing of the part to meet engineering requirements. Since FSW occurs at a lower, solid-state temperature, distortion of the metal being joined is minimal and usually observed only in thin stock. The FSW does introduce low level residual stresses in the material but not at the levels experienced by cooling the molten metal weld bead generated by fusion welding. Lesser amounts of distortion result in less in-process rework required to meet engineering requirements. This in turn leads to reduced manufacturing cycle time and reduced cost.

The advantages discussed in the above paragraph are attractive from a manufacturing perspective. However, the thermal and microstructural effects associated with FSW require that sound joints, mechanical properties, damage tolerance, and durability properties along with corrosion be sufficiently addressed and understood prior to implementation on a commercial airplane. This can be achieved through part design, alloy selection, process control, and metallurgical understanding of the effects of FSW and postweld thermal treatments on the final detail part assembly. An integrated team approach is recommended for maximum efficiency such that manufacturing, design, and metallurgical concerns can be addressed from the onset of the project. The building blocks for implementation of FSW on a commercial airplane include (1) establishing the feasibility, (2) alloy and product form selection, (3) developing the FSW parameters and a stable process, (4) validation testing and documentation, (5) developing a statistical database for design, and finally (6) implementation.

The FSW concepts can originate in the engineering design community or during the manufacturing of a current mechanically fastened design. From a design perspective, there are many factors involved in deciding the feasibility of a FSW design that culminate into some basic questions. For an existing design, the first question would be: Can a part be manufactured more cost effectively while achieving the equivalent and/or improved performance capability as the current design? For a new application the question would be: Will the part meet the design requirements and performance goals while being more cost efficient than a traditional build concept? In some cases, the answer to these questions may be based on previously known coupon data. In other cases, validation of the answer through full-scale component testing is required. For manufacturing feasibility, major considerations are (1) the ability of the FSW tool to reach the proposed joint area, (2) fixturing, (3) raw material requirements, and (4) postthermal processing capabilities. Due to the large size of the tool, a larger working envelope is required to allow accessibility for the FSW tool than is needed for many of the fusion welding processes. Once it is determined that there is ample work space for a FSW tool, other considerations arise. When considering the workspace, the fixturing setup required to perform the welding must be able to accommodate the geometry of the work pieces being assembled. Fixturing, which is equipment and design configuration dependant, contributes to the success of the weld. Due to the high forces associated with FSW, the fixturing must be robust enough to counter the effect of these forces. In case of butt welding, the fixturing must prevent both the separation of the workpiece weld faces and lifting of the workpiece resulting from the thrust and welding forces associated with FSW. Fixturing in lap welds is often less complex and in some cases may be incorporated into the part geometry. For example, the part itself may incorporate features that act as a backing plate. Additionally, to prevent some of the aforementioned challenges, fixturing may also need to account for normal dimensional variations in the raw material products. These considerations demonstrate the necessity of an integrated design-build concept. Defects associated with improper fixturing include voids, root side defects, and backing plate penetration.

Typical FSW joints have a start and exit point that needs to be included in the geometry. Raw material requirements are driven by the final assembly dimension and additional material required to ensure the final assembly does not contain the defects associated with the start and stop of the welding process. FSW has an initial length of the joint where the process has not reached steady state and therefore is not part of the stable joint. The endpoint contains a hole from the tool exiting. These areas need to be accounted for in the material requirements and removed to produce the final part. Lastly, thermal processing facilities used for postweld treatments must be able to accommodate the entire part. This is an example where an integrated design-build approach is useful for the best chance of success.

In addition to design and manufacturing, another important piece of the integrated approach is metallurgical in nature. The FSW process has been studied extensively and has been shown to be a stable and repeatable process that can consistently produce sound joints. However, FSW does introduce heat into the metals being joined, resulting in a heat affected zone (HAZ) that contains many variations of temper, as well as a nugget in an unstable, as-quenched condition. This may not be of concern in some of the lower strength, traditionally weldable 5000 and 6000 series alloys. However, many of the large integrated structures being considered for FSW require the use of high strength 2000 and 7000 series alloys that can be prone to stress corrosion cracking (SCC) and general corrosion associated with certain tempers and/or microstructure features. Therefore, metallurgical and property perspectives need to be considered during the design process. Figure 5.1 shows a metallurgical cross section of a weld showing the various affected areas resulting from FSW.

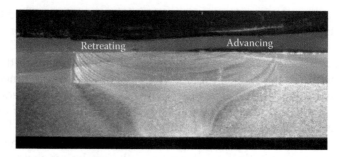

FIGURE 5.1
Cross section of a FSW joint showing the nugget, thermal mechanical affected zone, and the heat affected zone.

Properties and corrosion are controlled via alloy composition and thermal treatment. The heat generated by the FSW process produces a thermal gradient that extends from the nugget into the base metal, also known as parent material, creating a HAZ. The nugget is essentially heated to a temperature that puts the nugget material in a solution treated and quenched condition. This heat dissipates into the base metal next to the weld, putting the base metal into different stages of heat treat. There is a loss of properties in this area as a result of the FSW process due to varying degrees of aging, ranging from slight overaging to substantial dissolution of the strengthening phases associated with the alloy. It is common that a postweld age is required in order to recover most of the parent material properties in the nugget. As previously mentioned, the nugget area has essentially been resolution-treated by the FSW process and has the potential to regain most of, if not all, the base metal strength. Figure 5.2 shows a hardness profile of a typical 7000 series aluminum alloy friction stir welded followed by a postweld age matched up to the microstructure. The dips in hardness show the HAZ and the flat between the dips is the nugget area. FSW has been shown to achieve up to 90–95% of the parent material's properties for some alloys in the HAZ. However, some alloys will only see around 80% of the parent material properties due to the HAZ. This is dependent on the specific alloys, the process (including weld parameters), and thermal processing.

The interplay of the effects of FSW on the properties, the postweld age and the metallurgy of the material is shown in a study (Tuss and Reichlinger 2003) that was performed using 6.35 mm (0.25 in.) thick AA7349 extrusion in the –T6511 temper. The nominal composition for this alloy is compared to AA 7075 and shown in Table 5.1.

The extrusions were joined with a friction stir butt weld in the as-received –T6511 or peak strength temper and supplied for evaluation. The FSW produced welds parallel to the extrusion direction. The joints were naturally aged for several months prior to thermal processing and evaluation.

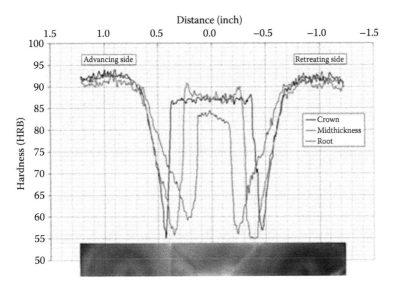

FIGURE 5.2
Hardness profile of a FSW'd Al 7xxx + postweld age (250F + 3xxF).

TABLE 5.1

Nominal Composition

Alloy	Cu	Mg	Zn	Cr	Fe[a]	Si[a]	Mn[a]	Other
7349	1.8	2.3	8.1	0.16	0.15	0.12	0.2	0.25 Ti+Zr
7075	1.6	2.5	5.6	0.2	0.5	0.4	0.3	0.25 Ti+Zr

Element ([a]max)

[a] except as noted.

The following conditions were characterized:

As-welded + natural age	AW
FSW + natural age + T6	PWA-T6
FSW + natural age + T76	PWA-T76
FSW + solution heat treatment + T6	PWST-T6
FSW + solution heat treatment + T76	PWST-T76

Hardness and conductivity profiles shown in Figure 5.3 were obtained on the weld face to measure the macroscopic response to thermal processing. The most notable features of this profile are that the nugget responds acceptably to a simple postweld age while the HAZ softens with either a –T6 or –T76 aging treatment. Tensile and fracture toughness tests were also performed. Figure 5.4 shows the orientation of the specimens relative to the weld. As can be seen by the results shown in Figure 5.5, the properties vary depending on

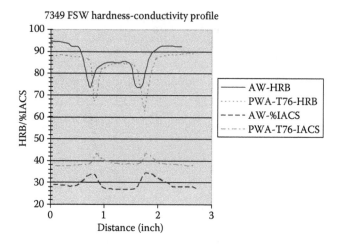

FIGURE 5.3
Hardness and conductivity profiles for FSW'd 7349-T6511 extrusions.

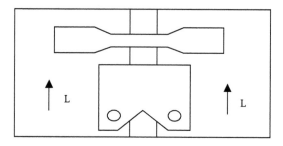

FIGURE 5.4
Test specimen location.

the postweld treatment. Comparing the properties for the joints given the postweld heat treatment, it appears that FSW + solution heat treatment + T6 is the closest to the baseline tensile and fracture toughness values.

The results of exfoliation corrosion (EXCO) testing are shown in Figure 5.6 for surface and Figure 5.7 for the cross sections of the coupons. The PWA-T6 specimen showed poor EXCO resistance generally with a bit of preferential attack in the HAZ, while the –T76 specimen had good overall corrosion behavior except for the HAZ, which was similar to the –T6 coupon indicating that a simple –T76 aging practice was not sufficient to restore the general corrosion performance of the HAZ. The PWST-T6 and PWST-T76 specimens exhibit EXCO performance typical of 7000 series alloys in –T6 and –T76 tempers across the entire coupon. Take note of the grain growth observed at the root of the nugget in the solution-treated and aged specimens. The impact of grain growth will be discussed later.

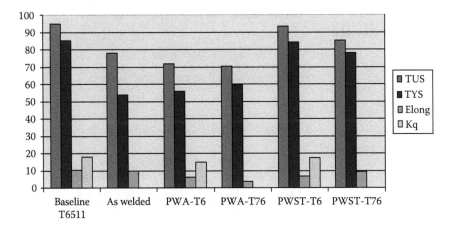

FIGURE 5.5
Tensile and fracture toughness results.

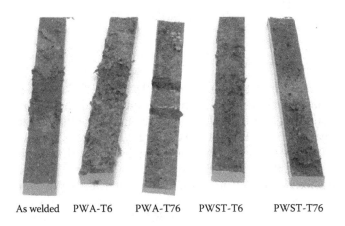

As welded   PWA-T6   PWA-T76   PWST-T6   PWST-T76

FIGURE 5.6
EXCO test results.

Now examine the results of the double cantilever beam stress corrosion crack (SCC) testing shown in Figures 5.8 and 5.9. The extent of the SCC crack growth during the test is highlighted by the dash lines to differentiate crack growth from the precracking of the specimen prior to testing. The AW coupon exhibited several cracks initiating near the chevron and extending in arcs from this point due to beam bending loads near the tip of the chevron. The PWA-T6 and PWA-T76 coupons both exhibited no SCC crack growth in the nugget of the specimen. This was somewhat unexpected, especially in the –T6 specimen, considering the relatively high SCC susceptibility documented in very high strength 7XXX alloys in peak-aged tempers. Also surprising was the fact that both the PWST-T6 and PWST-T76 specimens did exhibit significant

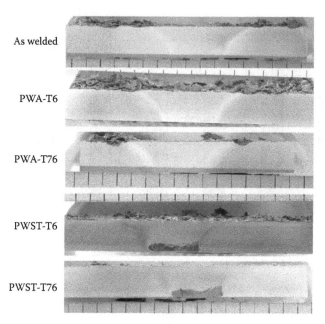

As welded

PWA-T6

PWA-T76

PWST-T6

PWST-T76

FIGURE 5.7
EXCO test results, cross sections.

SCC crack extension in the weld nugget. This time it was unexpected due to the fact that the relatively resistant –T76 temper did show significant SCC after the resolution heat treatment. Grain structure evolution during resolution heat treatment is assumed to be causing the higher SCC susceptibility.

The results of the SCC testing show that both the thermal history and grain structure of the material is important to understand in order to correctly characterize the performance of the materials being evaluated. The interface between the HAZ and the weld nugget is also a region of the weldment that is very susceptible to SCC from a microstructural standpoint and the effect of postweld thermal treatments on that structure could be completely different. Some additional thermal processing routes or external corrosion protection schemes will have to be developed in order to allow the use of high strength 7XXX aluminum alloys with an aging process only after welding. The consideration of corrosion properties is imperative in order to correctly apply FSW to structural applications in aerospace structures.

The purpose of the 7349 case study was to evaluate the affect of simple aging practices on the mechanical and corrosion performance of a friction stir welded 7000 series alloy. It should be noted that preweld temper condition and postweld thermal treatments were not optimized for the welding and associated thermal affects. The purpose of presenting this case study reinforces the need for the building blocks approach previously discussed and is not intended to implicate the alloy or FSW process.

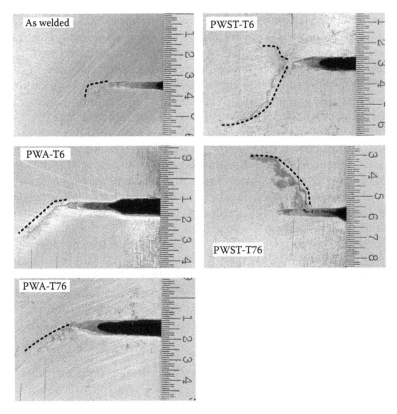

FIGURE 5.8
DCB stress corrosion results.

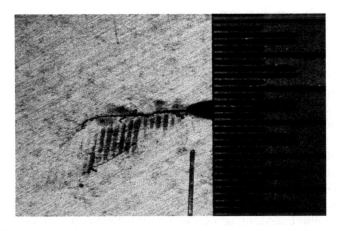

FIGURE 5.9
As-welded DCB stress corrosion.

This case study also demonstrates how one property may meet the design criteria while other properties are compromised. So when doing the design analysis, it is crucial that any loss of properties be considered with respect to the location of the weld and how property degradation affects the overall design. In other words, the designer has to determine if the location of a weld with property degradation will affect the performance of design and whether it can be tolerated. Designers can often accommodate property degradation if it is known what the degradation is. However, this may subtract from the overall cost advantage of the initial design. An example may be that the designer can thicken the affected area to accommodate for the lowest properties. This now requires purchasing thicker starting stock, leading to more machining time to thin where the extra thickness is not needed for design, adding to the overall machining costs and increased weight beyond the initial design concept. Thus, a robust, repeatable process and careful development to optimize the properties is necessary when considering FSW for any design.

Corrosion is a concern for commercial airplanes. This includes both stress corrosion and general corrosion. Stress corrosion is the more critical concern. Stress corrosion becomes through thickness cracks in the parts that require more extensive repairs. General corrosion includes pitting corrosion that can act as a crack initiation site for a fatigue critical design. Both types of corrosion can be addressed through alloy selection, temper, and finish schemes. The corrosion properties may be controlled with the processing. For instance, if FSW occurs at some point in the process after solution heat treat but prior to tempering, the stress corrosion may be mitigated. However, the thermal exposure due to the FSW will still cause over tempering by virtue of the longer time at temperature than the base metal outside the thermal gradient and the nugget that starts in the annealed condition. Thus, the HAZ may still be susceptible to pitting corrosion. If FSW is done on tempered material, stress corrosion may not be mitigated and the HAZ is likely to be more susceptible to corrosion. Aluminum is naturally not impervious to corrosion although some alloys are more resistant than others. Regardless, the traditional 2000 and 7000 series aluminums will require some finishes in order to retard the initiation of corrosion. Newly developed alloys such as those in the aluminum lithium family have shown improved corrosion but will still need finishes applied.

The benefits of FSW replace conventional weldments and fasteners, and reduce part count, material acreage, and cycle time. Figure 5.10 shows the 747 Freighter Barrier Beam. The beam at the top of the photo is the built-up design that required many fasteners installed through lap joints (overlapping material joints). Material acreage was reduced using butt welds rather than overlapping pieces for fastener installation. By using FSW on this part, the fasteners were eliminated, reducing the assembly time required for a mechanic to install each fastener. Since a mechanic was no longer required to manually install and torque the fasteners, an ergonomic benefit was captured.

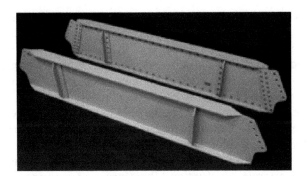

FIGURE 5.10
The 747 freighter barrier beam conventional build and FSW concept.

Statistical based design values are required for design certification, which is discussed later. Therefore, understanding, developing, and documenting a robust process is critical to implementation on commercial airplanes. Like fusion welding, FSW has key parameters that should be controlled. At a minimum, the following process parameters should be considered when developing a robust FSW process.

- Material thickness
- Force
- Spindle speed
- Travel speed
- Clamping and fixturing
- Tool design and penetration depth
- Thermal management of system
- Run on/run off tabs

Although FSW is a solid-state, it is still susceptible to defects if the process is not developed to create sound joints. Potential defects include:

- Lack of fusion
- Lazy "S" oxide strings
- Kiss bonds
- Voids, wormholes
- Inclusions: Tool breakage debris or foreign material pickup
- Root side defect

Lack of fusion, voids, and wormholes are typically indicative of incorrect welding parameters. Lazy "S" oxides strings are created from the FSW tool sweeping oxides into the nugget. These oxide strings can cause a weakened condition in the nugget that could potentially affect the joint properties. Properly cleaning the surfaces involved in the tool contact will eliminate the danger of oxide strings. Kiss bonds and root side defects are usually due to the pin of the FSW tool not penetrating deep enough for the material at the bottom of the weld to thoroughly be "stirred" together.

Throughout the previous paragraphs FSW is shown to have cost advantages over traditional manufacturing methods. However, part of the cost analysis includes the development of material design properties (allowables). The allowables are required by the Code of Federal Regulations (FAR) in order to certify aircraft structure and therefore is a necessary step for implementation. As with raw material, FSW properties must be based on enough tests from the welds and welded structure produced using an approved and appropriately documented manufacturing process to obtain a statistically based allowable. One source for the methodology for establishing allowables can be found in *Metallic Materials Properties Development and Standardization* (*MMPDS*) handbook by the U.S. Department of Defense.

When considering the overall cost, there are two approaches to address design allowables. First to consider is general multiapplication allowables. The MMPDS typically requires a significant amount of coupon level test data to statistically validate the property performance of the material. This is difficult to apply broadly to FSW since material thickness, welding parameters, tooling, product form, and alloy all impact the resulting properties across the weld joint. Using a general allowables approach to FSW would probably result in a data set that could be used across multiple applications, but such usage would be restricted to the weld parameters, joint type, and alloy system(s) used in the original testing. The advantages associated with general allowables would be broad-based usage and limited production validation testing. The obvious disadvantage is substantial up front testing. Although there is an industry interest to develop a general allowables approach, at the time of this writing, no general multiapplication FSW allowables have been published in the MMPDS.

An alternative approach to general allowables is point design. For point design, allowables are developed for specific applications. Coupon testing is still required but is focused on the design parameters for the particular application. Component testing is then used for validation of the design. The advantage to this approach is less up front coupon testing. The disadvantages would be that these allowables are not available for use for other applications unless a direct correlation to the structure can be established to the initial specific structural application and more production validation testing may be required. To date, most FSW applications have used a point design approach to these applications.

In either case careful consideration must be given to application and testing required for certification of the final part. Once allowables are documented and in place, the concept proceeds into production.

In summary, FSW may offer cost and performance improvements over traditional build concepts. Understanding the joint properties from the basic metallurgy up through the design phase is imperative. An integrated approach accounting for design, manufacturing, and metallurgy facilitates achieving the most robust process that will result in an efficient, cost-effective design-build that meets the certification requirements.

References

Mishra, R. S. and M. W. Mahoney, eds. *Friction Stir Welding and Processing.* Materials Park, OH: ASM International, 2007.

Tuss, G. D. and B. L. Reichlinger. Effect of post-weld thermal processing on corrosion and mechanical properties of 7349. Paper presented at the ASM Materials Solution Meeting, October 13–16, 2003, Pittsburgh, PA.

6

Friction Stir Welding

Arthur C. Nunes, Jr.

CONTENTS

What is Friction Stir Welding?

The friction stir welding (FSW) process joins metal by moving a rotating pin along a seam. The pin literally stirs the metal together. At the surface of the weld metal is a shoulder rotating with the pin and pressing against the workpiece surface. Without the shoulder, metal would be free to flow up the pin and the pin would plow the metal to produce a cut rather than a weld. Figure 6.1 illustrates a FSW tool embedded in a metal seam. The workpiece rests on an anvil not shown in the figure.

Solid-State Welding

FSW is a solid-state welding process. When clean surfaces of the same or not too dissimilar metals are put into contact, a metallic bond forms at the contact surface and the surfaces weld together. In fusion welding processes the surfaces to be welded are melted. The surfaces flow together and are bonded upon solidification. But melting is not required. In solid-state welding, the clean surface and the intimate contact are produced by means other than melting.

Forge welding is a form of solid-state welding. Blacksmiths have used it from ancient times. The metal is heated to soften it, making it possible for a hammer to pound down surface asperities and put the seam surfaces into intimate contact. A flux is applied to the surface to melt obstructing oxides and the surfaces to be joined are made convex so as to squeeze out the fluxed material as they are hammered together.

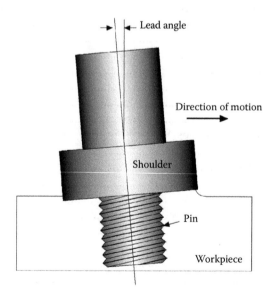

FIGURE 6.1
A friction stir welding (FSW) tool embedded in a workpiece.

The FSW process also provides clean surfaces for welding and forces the clean surfaces into contact, but it does so in a different way. Through mechanical deformation, the FSW tool generates the heat that softens the metal. A segment of seam to be welded is stretched out as it is wrapped around the tool. Stretching out the seam around the pin expands the seam to expose a large fresh area of uncontaminated, clean metal. High pressure around the pin pushes down any asperities on the expanded surface of the seam to produce full surface contact, which results in a sound solid-state weld. The high pressure around the tool is induced by a heavy *plunge force* exerted by the tool on the workpiece. The *plunge force* is balanced by a backing anvil support beneath the weld metal. If the conditions for a sound solid-state weld are not met, unsound welds with defects and low strength result.

Advantages of Friction Stir Welding

The more highly alloyed a metal is, the more it is likely to be subject to solidification problems inherent in fusion welding. Solidification from the molten pool of a fusion welding process can cause segregation of alloy constituents, coarsening of grain structure, alteration of phase structure, porosity, and cracking. The distribution of a nonmelting reinforcing phase in a metal-matrix composite is likely to be harmed by fusion melting. By using the FSW process one can avoid problems encountered in fusion welding caused by metal melting and resolidification. FSW offers the possibility of welding the hitherto unweldable, although it has its own problems as will be apparent below.

Effectively replacing a fusion welding process by a machining process may reduce skill requirements and costs and increase weld consistency and reliability. This claim does not depend upon replacing a manual with a mechanical process, for in high-quality fusion welds the heat source is normally manipulated by machine. Rather it depends upon a generally greater degree of inherent variability in fusion processes than in solid-state machining processes. For example, heat transfer through molten metal depends upon flow currents in the molten metal, which are strongly affected by surface tension gradients ("Marangoni circulations"), which in turn are sensitive to surface contaminant variations too small to be practically controlled. It is customary in gas-shielded tungsten arc welding (GTAW) to compensate by adjusting parameters on site to achieve the correct level of penetration; it would not normally be feasible to specify parameters precisely in advance. Penetration may also vary noticeably as the heat source passes near a heat sink.

Replacing a fusion welding process by FSW may reduce hazards by elimination of spatter and exposed liquid metal and high temperature heat sources, for example, electric arcs. It may reduce environmental contamination (e.g., of optics) by elimination of weld pool vapor. The deformation incorporated into a weld by the FSW process may result in stronger and/or tougher welds than obtainable with a fusion process.

Origins and Development of the Process

The Welding Institute Patent

The FSW process originated with The Welding Institute, a research and development organization, at Great Abington near Cambridge, England. The patent [1] was applied for in 1991.

The Retractable Pin Tool

The original FSW process employed solid tools with a fixed geometry. Pin length was fixed; only seams of uniform thickness could be welded. At NASA's Marshall Space Flight Center (MSFC) in Alabama, the FSW process came under investigation in the mid 1990s as a potential solution to fusion welding problems with a new alloy specified for the Space Shuttle External Tank. But tank designs required that seams with variable depths had to be welded, as well as seams around the circumferences of tanks. Welds along a tank axis could be started and stopped on entrance and exit tabs, respectively, extended beyond the tank, the tabs being removed after making the weld. This was not possible for circular seams around the tank; circumferential welds had to be closed out in some other way. A NASA-Boeing collaboration in 1997 resulted in the Retractable Pin Tool [2], a pin that could be moved in and out of the shoulder during the welding operation. The retractable pin tool could weld seams with variable depth and it could reduce the weld depth to zero so as to close out a circumferential weld.

The Self-Reacting Tool

Large tanks and space vehicles are fabricated from metal pieces. The pieces are clamped to fixtures for welding. Conventional FSW requires massive and expensive fixtures to absorb the heavy plunge force required for FSW. A way to circumvent this difficulty, the self-reacting FSW tool, now in common use, was devised at the MTS Systems Corp. [3] in 1999. The self-reacting FSW tool has two shoulders, one bearing on the weld crown, the other on the weld root, sandwiching the weld. The shoulder at the weld root is fastened to the FSW pin, which passes through the workpiece and through the crown shoulder, which can be moved along the pin to squeeze the workpiece in a *pinch force* that takes the place of the plunge force in conventional FSW. The crown and root forces essentially balance one another to leave little net plunge force to be absorbed by a fixture.

The self-reacting pin cannot be retracted during welding to close out circular welds. Circular self-reacting welds are closed out by *plug welds*, formed by pulling or pushing a rapidly rotating pin into the hole left upon extraction of the self-reacting tool. The pin seizes and is retained in the hole by a solid-state bond.

Current Status of Friction Stir Welding

As this article is being written, FSW is expanding on many fronts. The FSW process was first applied to aerospace aluminum alloys, but its range of materials has been expanded to incorporate steels, titanium alloys, copper alloys, and so on. Its range of depths has been extended to inches. Industrial applications extend to aircraft, automotive, shipbuilding, and so on. Manufacturers of FSW equipment and R&D organizations are active in developing FSW technology. This author is not aware of friction stir welders being marketed for the home workshop, but this may not be far off.

FSW technology has developed largely on an empirical basis, but a rudimentary understanding of the process has been attained.

Metal Flow in Friction Stir Welding

An understanding of metal flow in the vicinity of the FSW tool is the key to understanding weld macro- and microstructure, problem diagnosis, tool design, and process innovation. A preliminary understanding of the principal flow features and their resultant weld structures has been attained, although a great deal more remains to be learned.

Plastic Flow Stability and Instability

Suppose that within a unit volume of deforming metal there is a slip plane of slightly softer metal. An easier slip plane should take on more deformation than the other slip planes of the volume. But the extra deformation imparts extra hardening to the soft plane so as to reduce its deformation rate and bring it into harmony with the rest of the volume. This flow is stable and homogeneous.

At higher deformation rates, however, the softer plane may become hotter than its surroundings because the heat has less time to flow away from the plane of generation. At sufficiently higher temperature the flow stress of the initially soft plane does not rise, but drops with extra deformation. The initially softer plane gets hotter and hotter and softer and softer. Deformation then concentrates itself entirely on this plane. This flow is unstable and inhomogeneous.

Regions of stable and homogeneous versus unstable and inhomogeneous flow in metal processing can be mapped in coordinates of temperature and strain rate [4]. By the late 1930s it had been observed [5] that the plastic deformation by which a cutting tool produces a chip takes place in a narrow zone running from the tool cutting edge to the free surface bounding chip and workpiece. The narrowness of the shearing region implies very high shear rates and accords well with the unstable and inhomogeneous deformation mode. This narrow region of deformation is often approximated by the *shear plane model*, where the narrow deformation region is approximated by a shear plane.

The Shear Interface and the Rotating Plug Model

Note the very sharp division between light etching fine-grained metal and coarse-grained parent metal shown in the mid-sectional plan-view of a friction stir weld in the vicinity of the pin tool shown in Figure 6.2. As in metal cutting, it appears we have to do with an unstable, inhomogeneous deformation mode, appropriately modeled by a *shear surface* enclosing the FSW tool. In the case of Figure 6.2 the shear interface is circular.

A longitudinal section, Figure 6.3, of a suddenly stopped friction stir weld bead-on-plate reveals a shear interface with a funnel or vase shape enclosing the FSW pin. (Applying a welding procedure to a plate without an actual seam produces a "bead-on-plate." Beads-on-plate reveal the structural features introduced by a given welding procedure as well as an actual weld for many purposes and are much used in welding research. The observant reader will notice that the fine-grained metal produced by the tool does not descend all the way to the plate bottom. This is a partial penetration weld bead-on-plate. Figure 6.2 is not a bead-on-plate, but an actual weld; if one looks closely, one can see the weld seam extending ahead of the tool on the right.) Observations such as these present a basis for a *rotating plug model* of metal flow around the FSW pin. In the rotating plug model, a plug of metal is taken to be attached to and to rotate with the pin; the shear surface

5 mm

FIGURE 6.2
Mid-sectional plan-view of FSW macrostructure in 0.317 inch thick 2219-T87 aluminum alloy plate. The pin tool has been removed and replaced with bubble-filled mounting medium. Rotation is in the counterclockwise direction. The spindle speed was 220 RPM and the travel speed 3.5 inches per minute.

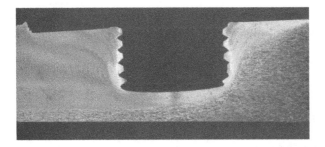

FIGURE 6.3
Longitudinal section of friction stir weld bead-on-plate. The tool has been suddenly stopped and removed. Travel is toward the right and the weld structure with its banding is seen to the left. At the upper left beyond the impression left by the tool shoulder can be seen the peaks and troughs in the weld surface constituting the tool marks. (Courtesy of John McClure, University of Texas–El Paso.)

represents the boundary between the stationary parent metal and the rotating plug.

Sticking and Slipping at the Tool–Metal Interface

If a rotating plug of weld metal sticks to the tool, then the shear stress required for slippage at the tool–metal interface must be greater than that required for shearing within the metal itself. Given a friction coefficient μ, a local pressure P, and a weld metal flow stress τ in shear, for slippage at the tool–metal interface

$$\mu P \leq \tau \tag{6.1}$$

In general one wishes to avoid slippage at the tool–metal interface because the welding mechanism depends upon flow *within* the metal to bring clean metal surfaces into contact. The high plunge forces customarily employed produce the required high local pressures P. Threads on the tool increase the effective friction coefficient μ, although this can also be done in other ways, for example, by flats machined on the tool. A pin with a square cross section can be used to make a friction stir weld. High local temperatures at deformation zones reduce flow stress τ and promote surface sticking.

Pressure is not constant but varies over the surface of the tool. Consider the flow of weld metal when the tool makes a small virtual move into the metal surface. Weld metal must be displaced to make room for the tool. Given the great differences in temperature around the tool and the sensitivity of flow stress to temperature, the displacement flow is expected to come from the hottest regions around the tool. That is, the displacement flow is presumed to be squeezed out from a thin channel along the hot shear surface. Imagining the hot flow channel to be a hollow cylinder of thickness δ enclosing the FSW pin, equilibrium of a ring element of the cylinder requires a pressure gradient along the axial direction z of

$$\frac{dP}{dz} = \frac{2\tau}{\delta}. \tag{6.2}$$

The τ is the flow stress at the boundary of the flow. In this rough approximation a linear increase in pressure with depth is obtained. But what is δ? The naturally selected value of δ would be the value minimizing the pressure gradient: too small δ requires large pressure gradients because there is so little area for the pressure to act against; too large δ requires large pressure gradients because the high boundary flow stress τ at the lowered temperatures away from the heat source offers high resistance to flow. A minimal pressure gradient is anticipated at a value of δ between too small and too large; however, the quantitative dependence of flow stress on temperature and of temperature on distance from the heat source is not known to a precision useful for a meaningful computation of δ.

The pressure reduces to zero at the edge of the shoulder, where the displaced metal emerges, and rises in accord with the required gradient to the bottom of the displacement channel. There is a region close to the edge of the shoulder where the pressure has not yet built up enough to satisfy the condition for sticking. The extent of the region of slippage at the shoulder edge depends upon the rate of pressure buildup under the shoulder.

The Wiping Mechanism of Metal Transfer

As the friction stir pin tool moves forward, weld metal continually crosses the shearing surface, entering and exiting the rotating plug. This is much like the situation when a wiping cloth picks up material from one surface and then wipes it onto another. Hence the transfer of metal into and out of the friction stir rotating plug is called the wiping mechanism.

In the wiping mechanism, metal begins to deposit on the rotating plug at the advancing side. Subsequent deposits bury previous deposits until a layer of thickness δ builds up at the retreating side. Then the forward movement combined with the rotation of the tool results in the successive abandonment of the buildup on the rotating plug. For a two-dimensional flow, conservation of metal volume requires

$$\delta \approx \frac{2V}{\Omega}. \tag{6.3}$$

The computation yields $\delta \sim 0.005$ inches for Figure 6.2. Scaling the distance from the photograph, the measured retreating side buildup appears somewhat larger, say $\delta \sim 0.010$ inches. This would be compatible with some extra metal flowing down into the rotating plug segment from above and not balanced by an equal outflow at the bottom of the segment.

The wiping mechanism of metal transfer does not per se require any translational force and is compatible with very low tool drag forces. If the pressure is distributed uniformly around the tool circumference, negligible drag force results from the pressure distribution. A circumferential variation in shear stress can still produce a drag force, however, as will be explained below.

Other mechanisms for transferring metal from the front to the back of the tool are possible. If the pin were simply a source of heat it would be possible to force metal around it by applying sufficient force.

The Mechanism of Welding

As stated above, two things are necessary for a solid-state bond. First, the pressure P must be sufficient to push down any asperities and put the surface into full contact. P must then attain the approximate pressure for making an indentation in a plastic body of shearing flow stress τ [6].

$$P \geq \left(1 + \frac{\pi}{2}\right) 2\tau. \tag{6.4}$$

Second, the contact surface must be clean. The contact surface is greatly expanded as it enters the rotating plug. In the kinematic model described above, on one side of the shear zone the radial inward velocity is $V \cos \theta$, where V is the weld speed and θ is the angle of the entry site from the direction of travel. The tangential velocity steps up from $V \sin \theta$ to $V \sin \theta + R\Omega$, where R is the radius of the shear surface and Ω is the angular velocity of the rotating plug, as an element of weld metal crosses the shear zone. The metal is thus subjected to a net lateral displacement of approximately $R\Omega \, \Delta t$ during the time $\Delta t \approx \delta/V\cos \theta$ expanded in crossing the width δ of the shear zone. The linear elongation of the seam $\Delta L/L$ can be estimated:

$$\frac{\Delta L}{L} \approx \frac{\sqrt{(R\Omega\Delta t)^2 + (V\Delta t)^2} - V\Delta t}{V\Delta t} = \sqrt{\left(\frac{R\Omega}{V}\right)^2 + 1} - 1 \approx \frac{R\Omega}{V}. \tag{6.5}$$

For Figure 6.2, $\Delta L/L \approx 78.6$; that is, 77.6/78.6 or almost 99% of the contacting area is new and clean, and 1/78.6 or somewhat over 1% is the original contaminated surface.

Flow Oscillations and Banding

A series of peaks and troughs are left on the weld crown behind the shoulder. These striations are called "tool marks," but deep within the longitudinal section there are bands in the weld metal that appear to correspond to the troughs and peaks on the crown surface. In Figure 6.3 the bands emerge out of the lower shear zone on the left and bulge out and expand into the wake of the tool.

Periodic extrusions of hot metal along the shear zone and out from under the edge of the shoulder appear to be responsible for both banding and tool marks. The periodicity is generally the same as that of the tool rotation, suggesting that the periodic extrusions are driven by tool asymmetry. The pin thread intersects the bottom of the pin at a single location on the pin circumference and creates an asymmetry. It is possible to detect small oscillations in tool forces or torque with the same periodicity and presumably due to the same cause. However, there seem to be a few cases in which the periodicity of the tool marks is not the same as that of tool rotation. If correct, this would suggest a possible alternative cause, for example self-excited oscillations generated by a stick-slip mechanism.

But why doesn't the tool leave a succession of bands imaging the shear surface contour in its wake? This is because of a secondary flow around the tool, the *ring-vortex flow component*. The FSW pin tool threads are customarily directed to force weld metal down into the workpiece when the pin tool is turning. The metal volume is conserved, so that the cylindrical downward flow along the pin surface results in a radial inward flow at the shoulder, a radial outward flow at the bottom of the pin, and a cylindrical upward flow completing the circulation away from the pin. This flow distorts the image of the shear surface. The upper portion of a band is swept back into the shear surface and obliterated. Below the obliterated portion a band is swept outward by the ring vortex flow. Hence the bands emerge as a succession of bulges on the lower part of the pin. In longitudinal section the bands appear as a series of arcs.

Figure 6.4 shows the banding in transverse section as a series of "onion rings" in an oval nucleus of recrystallized metal. Above the nucleus, where the recrystallized metal has been swept back through the shear zone into the rotating plug is a region of unrecrystallized, albeit distorted, parent metal. The parent metal is circulated into the upper weld area by the ring vortex circulation and out of the way of the captured metal emerging on the lower part of the pin.

Streamlines and Flow Traces

The qualitative features of the flow around a FSW pin tool can be easily understood if the flow is decomposed into the three components as shown schematically in Figure 6.5. The three components are viewed with respect to the tool center in the coordinates shown.

The entire weld metal volume moves back with respect to the tool with velocity -V, where V is the pin translation velocity; this is the *translation component* of the flow field.

FIGURE 6.4
Transverse section of friction stir weld bead-on-plate. Tool marks are visible on the top surface of the weld. The classic onion ring pattern is present at the weld center. Parent metal deformation in the ring vortex velocity field component surrounding the tool is visible at the weld edges. Fine-grained metal extends out to the advancing edge at the top left, where a bit of flash can be seen. Parent metal circulated in by the ring vortex field to replace metal captured in the rotating plug can be seen at the top of the weld with an increasing area toward the retreating edge on the right. Metal captured in the rotating plug at the top of the weld reappears at the bottom displacing the fine-grained metal area downward. (Courtesy of John McClure, University of Texas–El Paso.)

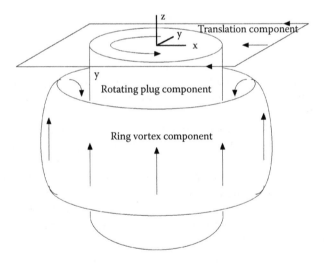

FIGURE 6.5
The flow around a FSW pin tool can be decomposed into three components (1) translation component, (2) rotating plug component, and (3) ring vortex component.

Within a cylinder of radius R the weld metal rotates with tool angular velocity Ω; outside radius R there is no motion; there is a shear plane at radius R. This is the *rotating plug component* of the flow field.

Surrounding the pin, cutting across the shear surface, sometimes inside but mostly outside the rotating plug, like a smoke ring around the pin, is the *ring vortex component* of the flow field. The ring vortex component is responsible for axial displacements along the pin. The thread on the pin and/or scrolled ridges or grooves on the tool shoulder are usually set to move metal in along the shoulder and down the pin. In this situation the radial velocity v_r at the shear surface is negative/inward toward the shoulder and positive/outward toward the pin bottom. In polar coordinates ($x = r \cos \theta$, $y = r \sin \theta$) the velocity at point r, θ in the rotating plug is

$$\frac{dr}{dt} = v_r - V \cos \theta, \tag{6.6}$$

$$\frac{d\theta}{dt} = \Omega \left(1 + \frac{V}{r\Omega} \sin \theta \right). \tag{6.7}$$

The equation for a streamline is obtained by equating dt in Equations 6.6 and 6.7:

$$dr = R \left(\frac{\dfrac{v_r}{V} - \cos \theta}{\dfrac{R\Omega}{V} + \dfrac{R}{r} \sin \theta} \right) d\theta. \tag{6.8}$$

K. Colligan [7], using steel shot tracers in aluminum alloys, has marked a number of empirical streamlines in friction stir welds that can be compared to the above model.

If the radius R of the shear surface remains constant, there is no radial velocity component v_r, and $R\Omega/V \gg 1$ R (normally the case) so that it overshadows the R/r (sin θ) term, then an element of weld metal is symmetrically buried (dr negative) and uncovered (dr positive) over the forward and rear surfaces, respectively, and the element of weld metal is not displaced laterally by the pin tool! For the most part, however, lateral displacements, as illustrated in Figure 6.6, are the norm. Lateral displacements can be estimated by Equation 6.8 by computing when (after the element enters the rotating plug at $r = R$ on the forward surface) the radial position of the element r again falls on the plug boundary R on the rear surface.

The ring vortex flow field induced by threads on the pin tool and/or scroll-work on the tool shoulder typically contributes an inward radial velocity component $-v_r$ near the shoulder that tends to hold the element in the rotating plug for a longer time and a longer displacement toward the advancing edge of the tool. Conversely, near the bottom of the classical pin tool (or near the center of the self-reacting tool) an outward radial velocity component tends to expel metal from the rotating plug prematurely so as to deflect streamlines toward the retreating side of the tool.

Close to the tool shoulder Colligan's [7] shot tracers exhibit very noticeable scatter in lateral displacement, substantially greater than is exhibited

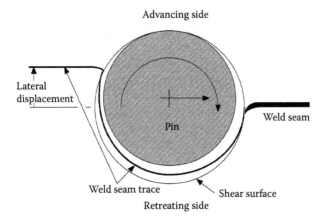

FIGURE 6.6

Weld streamlines pass through the shear surface into the rotating plug and exit at the rear of the plug. The weld seam and its traces follow a streamline, often one that enters close to the center of the pin.

at regions of the pin farther away from the shoulder. This can be attributed to the periodic extrusions of metal from the shear zone that cause banding and tool marks. If metal periodically accumulates and dissipates at the edge of the tool shoulder, and if this accumulation causes a periodic expansion and contraction of the sticking area on the shoulder, then the shear surface, which bounds the sticking region, moves in and out periodically, with greater amplitude closer to the shoulder. (The lower part of the shear surface approaches the pin surface and is less responsive to oscillations originating at the edge of the shoulder.) A tungsten wire tracer exhibits patterns similar to the steel shot tracers. The tungsten wire is broken into short segments by the action of the tool.

A Pb wire tracer that melts at shear surface temperatures has been observed redistributed as continuous wisps aligned with the band structure in the weld [8]. As the shear surface advances into the wire, the wire is continuously wrapped around the pin. A sudden contraction of the shear surface abandons a length of wrapped tracer metal, a "wisp," behind the pin. A sudden contraction of the shear surface also opens up a gap between the shear surface and the end of the wire and causes a tracer discontinuity in the weld. But the tracer wisps are sporadic. Along some weld segments the Pb tracer disappears altogether. A change in the waveform of the flow oscillations responsible for the tool marks and banding leading to a broader, more uniform dispersion of the tracer in the weld would explain the disappearance. Such a change would imply sensitivity of the waveform to small variations in some feature (temperature?) along the weld. A detailed analysis of these oscillations has not yet been undertaken.

By this point the reader should be aware that the friction stir welding tool does not rearrange the weld metal structure in a chaotic way.

Weld structures complicated at first sight can be understood by means of a simple kinematic model. But the flow pattern is not quite as simple as presented here. Outside the rotating plug flow field there is another, relatively slow induced flow field. This can be seen by looking closely at Figure 6.2 and noting the bending up of the weld seam and of the coarse parent metal grains close to the shear surface. Sometimes this secondary flow may extend over a wide region and a substantial part of the metal in the weld may flow back around the tool in this slow induced current. In this case a substantial part of the flow around the tool does not pass through the zone of high shear. The part of the flow that does not pass through the high shear zone is distorted but retains the coarse structure of the parent metal and does not show the onion ring pattern. If the weld seam lies in this flow, it is possible that insufficient exposure of fresh metal surface on the seam (Equation 6.5) will lead to reduced weld strength. Shifting the seam toward the advancing side of the pin may remedy this condition. The broad slow flow mode is probably avoidable by controlling weld parameters, but at time of writing the criteria for the flow mode have not been made clear.

Pressures and Forces

The pressures and forces on the FSW tool can, at best, only be estimated semiquantitatively as this article is being written. In part this is because of the complexity of the computations; in part, because the properties of the weld metal are only poorly known at the temperatures and deformation rates to which the FSW process subjects it. Nevertheless it is possible to construct some very simple models that can give a limited understanding of tool forces.

Torque

Approximating the geometry of the plug of metal that rotates with the tool as in Figure 6.7, the torque M is the sum of the torques upon the rotating ring elements comprising the shearing surface of the rotating plug plus the torques upon any surface of the tool that slips against the workpiece. Slippage is neglected here in the interests of simplicity

$$M \approx \int 2\pi r^2 \tau \sqrt{dr^2 + dz^2}. \tag{6.9}$$

Fluctuations in steady-state deformation mode that reduce torque can be easily accommodated by the system, but fluctuations that increase torque

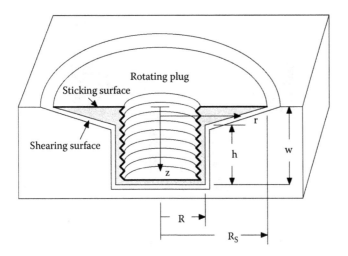

FIGURE 6.7
Simplified geometry of a hypothetical rotating plug of weld metal attached to the friction stir welding tool.

are more difficult and less likely to occur. Hence the steady-state deformation mode that actually occurs is anticipated to be the one that minimizes torque,

$$\delta \int 2\pi r^2 \tau \sqrt{dr^2 + dz^2} = 0. \tag{6.10}$$

For constant τ, an elliptic integral solution for the shape of the shear surface can be extracted. Where the plug geometry is simplified still further as in Figure 6.8 to depend upon a single parameter h, a still simpler minimization procedure estimates the shape of the shear surface:

$$\frac{dM}{dh} = 0. \tag{6.11}$$

The result of the latter minimization comes out to be:

$$\frac{h}{w} = 1 - \frac{\dfrac{R}{w}}{\sqrt{\dfrac{1}{9}\left[\left(\dfrac{R_S}{R}\right)^3 - 1\right]^2 - \left(\dfrac{R_S}{R} - 1\right)^2}}. \tag{6.12}$$

Given typical pin tool geometries, $w/R \sim 2$ and $(R_s - R)/R \sim 2$, then $h/w \sim 0.7$, and $M \sim 69.2R^3\tau$.

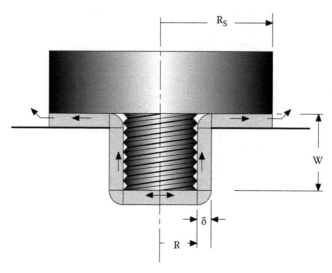

FIGURE 6.8
Extruded backflow zone approximation.

Drag

If we neglect slippage, then the entire interaction between the workpiece and tool can be taken to be that over the shearing surface. The force exerted by the tool on the weldment is

$$\vec{F} = F_x\hat{i} + F_y\hat{j} + F_z\hat{k}. \tag{6.13}$$

The F_x, F_y, and F_z are the drag force, lateral force (taken positive in the direction toward the retreating side of the tool), and plunge force, respectively, exerted by the tool on the workpiece. The $\hat{i}$, $\hat{j}$, and $\hat{k}$ are unit vectors directed, respectively, along the direction of motion, toward the retreating side of the tool, and into the weld metal. The reaction of the workpiece on the tool is, of course, equal and opposite. The pressures and shears over the shearing surface integrate to give the reaction on the tool:

$$F_x \approx -\iint \tau r\sqrt{dr^2 + dz^2}\,\sin\theta d\theta + \iint \Pr dz\cos\theta d\theta, \tag{6.14}$$

$$F_y \approx \iint \tau r\sqrt{dr^2 + dz^2}\,\cos\theta d\theta + \iint \Pr dz\sin\theta d\theta, \tag{6.15}$$

$$F_z \approx \iint \tau r dz d\theta + \iint \Pr dr d\theta. \tag{6.16}$$

The angle θ is taken from the direction of motion in the direction of rotation, so that the retreating side of the tool is at $\theta = +\pi/2$ and the advancing side at $\theta = -\pi/2$.

The accumulation of hot, newly sheared metal on the retreating side of the tool may be anticipated to raise the temperature and lower the flow stress on the retreating side of the tool somewhat above that of the advancing side. The incoming cooler metal on the forward side of the tool may be expected to reduce the temperature and raise the flow stress on the forward side of the tool. This state of affairs may be approximated over the outer edges of the shear surface by a sinusoidal stress variation. The amplitude of the circumferential temperature variation is taken proportional to the radius of the shear surface; for equivalent temperature amplitudes the temperature gradients are higher and the gradient-driven thermal conduction along the circumference larger at the smaller radii. Hence the temperature difference amplitude ought to be smaller at the smaller radii. If it is assumed that similar temperature gradients (and not temperatures) are maintained along circumferences, the temperature difference amplitude may be taken to vary proportionally with radius. For small variations in temperature and approximately proportional flow stress variations a relation for the variation of flow stress τ can be constructed:

$$\tau \approx \bar{\tau} - \Delta\tau_{lat} \sin\theta \cdot \frac{r}{R} + \Delta\tau_{fwd} \cos\theta \cdot \frac{r}{R} \tag{6.17}$$

If the circumferential variation in flow stress is the sole cause of drag and lateral force on the tool, then drag and lateral forces proportional to the torque M, and the ratio of the lateral and forward differences in shear stress to the average shear stress, $\Delta\tau_{lat}/\bar{\tau}$ and $\Delta\tau_{fwd}/\bar{\tau}$, respectively, can be extracted from Equations 6.9, 6.14, and 6.15:

$$F_x \approx \frac{M}{2R} \frac{\Delta\tau_{lat}}{\bar{\tau}}, \tag{6.18}$$

$$F_y \approx \frac{M}{2R} \frac{\Delta\tau_{fwd}}{\bar{\tau}}. \tag{6.19}$$

If a lead angle ψ is imposed on the tool, an addition to the drag is to be anticipated:

$$F_x \approx F_z \sin\psi. \tag{6.20}$$

There are many possible mechanisms by which drag and/or lateral forces could be exerted on the FSW tool. At present the work necessary to identify them and associate them with particular metals and process parameters remains to be done.

Plunge Force

There are large temperature gradients in the vicinity of the tool and significant metal flow will be restricted to channels, where high temperatures greatly reduce the flow stress of the metal. For the tool to penetrate deeper into the workpiece, metal must be squeezed out of the way of the tool. The hot metal is held back in its flow channels by resistance from the environment (Equation 6.2) as well as by internal resistance to deformation (e.g., hoop stresses in radial flow). A pressure gradient must overcome this resistance if flow is to take place and the tool is to penetrate deeper into the workpiece.

The distribution of pressure and shear force over the tool requisite for forcing metal out of the way of the penetrating tool adds up to a substantial force on the tool, which must be balanced at equilibrium by the plunge force F_z (Equation 6.16).

Many flow configurations are conceivable in the vicinity of the tool. For illustration we have taken the very simple configuration shown previously in Figure 6.8. In this configuration the extrusion zone is a uniformly thick layer of metal closely conforming to the surface of the tool. An equilibrium plunge force may be estimated from this model. The result of the computation is presented as Equation 6.21.

$$F_z \approx 2\pi R^2 \tau \left\{ \left[\frac{1}{3}\left(\frac{R_S}{R}\right)^3 - \frac{7}{3} + \frac{w}{R} \right]\frac{R}{\delta} + \left[\frac{1}{2}\left(\frac{R_S}{R}\right)^2 - \frac{w}{R} + \ln\frac{R_S}{R+\delta} + \pi - 6 \right. \right.$$

$$\left. \left. + \left[2\ln\frac{R_S}{R+\delta} + \pi - 6 \right]\frac{\delta}{R} + \frac{1}{2}\left[2\ln\frac{R_S}{R+\delta} + \pi - 6 \right]\left(\frac{\delta}{R}\right)^2 \right\}. \right. \tag{6.21}$$

The model from which Equation 6.21 is derived represents only one possibility among a number of others. Pending empirical validation it retains candidate status only.

Defects

Weld defects present a major research effort in the development of the FSW process. In general, defect formation mechanisms cannot be said to be well understood as this article is being written. The following discussion may nevertheless be of help to the reader in identifying and correcting defects.

Lack of Fill Defects: Trenches, Surface Fissures, and Wormholes

Several kinds of "lack of fill" defects, where the weld cross section exhibits gaps or breaks not filled with weld metal, are encountered in friction stir welds. The mechanisms of their formation are not well understood, but they seem to be avoidable by adjustment of weld parameters.

It is useful to distinguish between hot and cold welds. The temperature of the shear surface, if a single representative temperature may be assigned to the surface, rises to a level at which the energy fed into the surface mechanically just balances the energy conducted away as heat. At the higher RPM, more heat must be dissipated. The local temperature rises and provides a steeper temperature gradient to conduct the additional heat away. The mechanical power input is not proportional to the RPM because the shear stress (and hence torque) drops with the temperature rise. The temperature rise is small enough so that the mechanical power requirement may remain almost constant over a wide range of RPM. Nevertheless welds made at high RPM are relatively hotter and welds made at low RPM are relatively colder. In a relatively hot weld higher temperatures and softer metal extend further out from the shear surface. There is a translational weld speed effect on temperature also. Cool weld metal impinging upon the pin must be heated up to shear surface temperature; this imposes a convective power requirement in addition to a conduction loss requirement. Slower welds are hotter and faster welds are cooler.

It has been pointed out above that the shoulder of the tool prevents the upsurge of metal around the pin, which would plow and not weld in the absence of the shoulder. Nevertheless in hot welds soft metal may extend beyond the edge of the shoulder and extrude in a substantial "flash." Given an extruding flash, a counterbalancing "suckback" of hot metal may result in a trench on the opposite side of the weld.

Sometimes a fissure is seen parallel to the weld direction on the weld surface. Surface fissures are angular in contour and not rounded like trenches. They are located inside rather than on the edge of the weld surface. They may be narrow like a crack or wide, uncovering a flat surface with tool marks at the bottom of the fissure. Trenches and fissures seem to be associated with hotter welds.

Below the weld surface "wormholes," tubular gaps extended along the weld direction, sometimes open up. Wormholes form toward the bottom of the weld and are not visible on the weld surface. It is possible for plastic flow fields to generate tensile stresses, such as cause central bursts in extrusions or in forgings (Mannesmann effect). If inadequate hydrostatic pressure is maintained beneath the tool shoulder, wormholes may be expected.

Tears

What appear to be hot tears, resulting from a planar concentration of a low-melting phase, are occasionally observed near the surface close to the shoulder of friction stir welds. The tears seem to lie along the bands and so-called tool marks left by the friction stir tool. The tears tend to lie across the weld direction and could be opened up by thermal tensile stresses along the weld that arise when the hotter weld metal contracts within a colder environment. (Tensile residual stresses are well known along fusion welds.)

But what could cause a planar concentration of low-melting phase? We have called attention above to the redistribution of a Pb wire tracer as continuous wisps following the band structure of the weld [8]. The capture and planar redistribution of the low-melting phases would not be a major departure from the redistribution of the Pb wire tracer by an oscillating shear surface as described above.

Residual Oxide Defect

The trace of the joint seam exists on the weld cross section. It is not dispersed. It can be seen clearly when two contrasting metals are joined in a bimetallic weld. In conventional friction stir welds the seam trace takes on an S-shape, where the upper loop of the S extends toward the advancing side of the weld and the lower loop toward the retreating side of the weld. A trace of a marked seam is shown in Figure 6.9, a radiograph of a transverse slice of a friction stir weld in an aluminum alloy where the seam was marked by

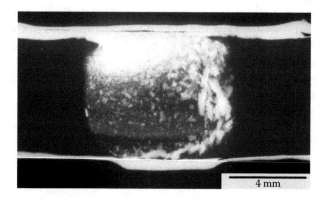

4 mm

FIGURE 6.9
Radiograph of a transverse slice of a friction stir weld in an aluminum alloy where the weld seam (and surrounding edges) was marked by a deposit of plasma sprayed nickel. The trace of the seam is deflected to the retreating side on the right over the lower part of the pin where the ring vortex radial flow component is outward and toward the advancing side of the pin on the left on the upper part of the pin just below the shoulder where the ring vortex radial flow component is inward. The marker has been torn in pieces by the stresses incurred upon entering the rotational flow field.

a deposit of plasma sprayed nickel. The lower loop of the "S" on the retreating side of the tool is prominent. The upper loop on the advancing side is narrower and is distorted by shear surface oscillation and ring vortex flow outside the rotating plug into a broad smear. The marker is torn into bits by the shear force on the marker elements that have entered the rotating flow, but are held back by the rest of the still-attached marker material embedded in the stationary environment. A few bits of marker, perhaps moved out of the flow streamlines upon which they entered the rotating flow by the tearing forces, are found in positions substantially displaced from the trace of the centerline in the weld transverse section.

With the conventional self-reacting tool, opposing thread pitches at the opposite ends of the pin produce opposing S-traces that join in the middle to form a big central loop extending toward the retreating side with end loops extending toward the advancing side of the weld. The big central loop may have a depression in its middle extended back toward the advancing side of the weld.

If the seam is contaminated, the contamination remains on the seam trace. Substantial amounts of oxide contamination may reside in and weaken the seam trace. This kind of defect can cause a serious reduction in weld strength. It has been called the Residual Oxide Defect (ROD). It may be affected through seam trace geometry, for example, spreading out and weakening the effect of detrimental material or limiting the length of seam oriented perpendicular to the direction of tension. Seam trace geometry is in turn affected by tool geometry or seam offset. Cleaning the seam surfaces prior to welding would be expected to eliminate ROD.

However, agglomerated lower melting phases have been reported in friction stir welds [9] and might also reduce weld strength. They seem to occur in hotter welds, but the mechanism of their formation is unclear. Here is a potential mechanism for segregating molten phases along the band structure in the vicinity of the tool shoulder.

Galling

Classical "galling" refers to surface damage induced by rubbing contact of relatively smooth surfaces. Stresses exerted by local seizures produce fracture and deformation at the rubbing surface. It is not clear that the surface damage sometimes encountered in friction stir welds is classical galling. The friction stir pin and shoulder are supposed to seize in the weld metal. At the edge of the shoulder, however, the pressure is anticipated to drop and approach that of a free surface, so that at some point slippage should take the place of shear. Here complex and poorly understood processes seem to take place: flow oscillations that are the origin so-called tool marks and internal banding, of redistribution of low melting phases, and perhaps of the surface cracking and deformation that have been labeled galling.

Lack of Penetration

For a strong bond to occur, the weld seam needs to undergo the conditions at the shear surface. In conventional friction stir welds it is possible that the shear surface may close up beneath the tool bottom so as to leave a thin layer of poorly bonded metal between the tool bottom and the anvil surface. The shear stress exerted by the tool through the shear surface may still cause substantial plastic deformation between anvil and tool bottom in the layer of metal often called the "ligament." The plastic deformation may produce enough bonding along the seam to make it difficult to detect discontinuities by conventional nondestructive evaluation (NDE) techniques. Pending further investigation, it appears that, if feasible, an etch revealing the presence or absence of the fine, recrystallized grains produced at the shear surface on the root surface of the weld may be able to distinguish the lack of penetration clearly.

How close does the tool bottom need to come to the anvil to avoid lack of penetration? A much-simplified model, which shows what considerations are involved, can be made. Suppose that a cylindrical tool of radius R bottoms out a distance δ above the anvil. The torque contributed by the shear surface enclosing the tool bottom is $\int_0^R 2\pi r^2 \tau dr = (2\pi R^3/3)\tau$, assuming a constant shear flow stress τ over the bottom of the tool. If the shear surface drops down by distance δ to the anvil, where the shear is τ_A at the anvil versus weld metal interface, the change in torque is $(2\pi R^3/3)(\tau_A - \tau) + 2\pi R^2 \delta \tau$. Once the lower torque configuration has been inaugurated, the torque has no reason to rise enough for a higher torque deformation mode to occur; hence, lack of penetration should not occur when the torque decreases if the shear surface drops all the way to the anvil:

$$\delta \le \frac{R}{3}\left(1 - \frac{\tau_A}{\tau}\right) \tag{6.22}$$

Equation 6.22 presents a rough idea of the clearance permissible between the bottom of the tool and the anvil. Presumably the anvil frictional shear is significantly less than the metal shear stress. For a frictionless anvil, $\tau_A = 0$ and a quarter inch diameter pin the clearance should be less than 0.042 inch according to Equation 6.22.

Allowable Parameter Window

Weld parameters have forbidden ranges outside the capability of the equipment or where weld defects of one sort or another occur. Each capability limit or defect criterion marks off regions of allowable and unallowable points in parameter space. When the unallowable regions in parameter space are superimposed, the remaining allowable space constitutes an allowable parameter window for the process.

The most critical parameters for FSW are spindle rotational speed (RPM) and weld travel speed. These are the primary parameters. Major families of defects are associated with insufficient or excessive weld temperature. For example, excessive flash formation or trenching are associated with hot welds. Excessive drag force and tool breakage is associated with cold welds. Hence, a cold and a hot isotherm in RPM-travel speed space mark out a window of allowable operation (subject to narrowing by further capability and defect criteria). The general form of such isotherms can be roughly established.

Given a cylindrical shear surface of radius R at temperature T, a length Δz loses heat to a large surrounding medium of thermal conductivity k at power rate

$$Q \approx \frac{2\pi k \Delta z}{In \dfrac{R_0}{R}} (T - T_0). \tag{6.23}$$

where T_0 and R_0 are the ambient temperature and the ambient radius, respectively. If the capture cross-sectional area of the cylinder is $2R\Delta z$ and the weld speed is V, then, in order to raise the temperature of the approaching weld metal to that of the tool, the tool needs to supply additional energy at rate

$$Q \approx \rho(2R\Delta z)VC \, (T - T_0). \tag{6.24}$$

The weld metal mass density is ρ. The weld metal specific heat is C. The mechanical power input to the tool section is

$$M\Omega \approx 2\pi R^2 \tau \Delta z \Omega \tag{6.25}$$

The mechanical power must supply the thermal conductive and convective losses plus some other losses, for example, tool conduction that has been judged small enough to neglect for a rough approximation. This power balance permits an estimate of the shear surface temperature, provided that the dependence of the flow stress τ of the weld metal on temperature can be estimated.

The same relation yields a curve for an isotherm on a plot of the principal weld parameters, tool rotational, and translational speeds:

$$V \approx \left[\frac{\pi R \tau}{\rho C (T - T_0)} \right] \left| \Omega - \frac{\pi k}{\rho C R In \dfrac{R_0}{R}} \right|. \tag{6.26}$$

If the flow stress of the weld metal depends strongly upon temperature, but only weakly upon shear rate, then the isotherms are approximately straight lines fanning out from a point on the translation velocity V-axis ($\Omega = 0$). A pair of critical isotherms tends to bound a window of acceptable operation. Low temperature problems (like tool breakage) begin to occur below some critical low temperature isotherm. High temperature problems (like excessive flash) begin to occur above some critical high temperature isotherm.

Characteristic lines isolating particular defects can be estimated theoretically if the defect is well enough understood or can be determined empirically. These lines bound regions of the parameter space where the defect occurs. When all defects have been considered, if one is fortunate, one is left with a window of parameter combinations yielding defect-free welds. The nominal weld parameters chosen for making the weld are typically taken at the center of the operation window to allow maximum leeway for variation. More parameters can be considered if necessary with multidimensional boxes as operational windows.

Here, defect is intended to signify a structural feature with a clearly detrimental effect upon weld strength or toughness. We note in passing that there are structural features that cause no appreciable detrimental effect upon weld strength or toughness, yet cause clearly visible radiographic indications called enigmas or ghosts; these structures are not considered defects. Technicians skilled in nondestructive evaluation can distinguish enigmas from defect indications.

Defect-free welds tend to be substantially stronger or tougher than defect-containing welds. But weld strength/toughness can vary in the absence of defects within the defect-free window. This could be a result of differing time–temperature-deformation histories within the defect-free window. As a rule of thumb, longer exposure to high temperatures tends to have a detrimental effect upon strength. The variation in hardness over the weld cross section can channel and concentrate slip so as to have a substantial affect on toughness and strength. Variations in the orientation of planes of weakness, for instance, the trace of a dirty weld seam, with respect to applied loading also affect toughness and strength. Hence there may be reasons to select nominal weld parameters off the center of the defect-free window or to restrict the acceptable parameter range within the defect-free window.

Tool Design

Tool design comprises two aspects: tool material selection and tool geometry. Tool material is typically selected to assure tool survival during the process, and tool geometry for its effect on weld properties, but tool material can affect weld properties and tool geometry can affect tool survival.

Fixtures, manipulators, and welding machinery of various sorts are often dealt with under the heading of "tooling." This topic is very broad and has more to do with general engineering than with the specifics of FSW. It will not be treated here, except to comment that such machinery must be designed to support the reaction forces of the weld metal upon the tool and maintain the requisite rigidity for producing the weld. The plunge force can be very large (tons), requiring appropriately massive supporting fixtures. The self-reacting tool greatly reduces the plunge force. It is still necessary to hold the FSW joint together with some degree of rigidity during welding, however, and machinery used to weld large structures will itself be large. It may be possible to reduce welding forces, perhaps by increasing tool rotation speed, sufficiently for handheld FSW; this is a matter for research and development.

Tool Materials

Typically, as the temperature of a metal rises above absolute zero, its flow stress declines gradually up to about half its melting temperature, at which point the flow stress begins to plummet. Above half the melting temperature the reduced flow stress declines still further, but less precipitously, to zero at melting temperature. The power induced by the FSW tool raises the local weld metal temperature at the shear surface close to the melting point of the weld metal. The correspondingly low flow stress of the hot weld metal adjacent to the tool limits the stresses that the weld metal can impose upon the tool.

In principle, the stresses on the tool can be reduced to arbitrarily low values by choosing suitable welding parameters. This turns out not to be so simple as raising the tool RPM, however. With conventional tooling the high temperatures extend too far out from the pin; excessive flash and lack of fill defects prevent sound welds. Substantial redesign of tooling is needed to make high RPM work, and as this is being written this is a subject of research and development.

Nevertheless, the first requirement of a tool material for FSW is not a strength requirement per se, but a requirement for strength at temperatures approaching the melting point of the weld metal. This means in general that the melting temperature of the friction stir tool should be roughly double that of the weld metal or higher. Steel tools can weld aluminum alloys. Refractory metal tools or the equivalent are needed to weld steel, titanium, or copper alloys.

With abrasive materials, for example metal matrix composites (MMCs), extensive tool wear can take place rapidly [10]. Wear requires shearing motion at the tool-weld metal interface, but there need not be slippage at the tool-weld metal interface. It is possible for hard particles gripped in a slowly moving outer flow to extend through a thin layer of metal sticking to the rapidly rotating pin and to abrade the tool metal. Toward the bottom of the tool and on the advancing side the shear surface approaches the pin closely

(Note Figure 6.2). Wear seems to concentrate initially toward the bottom of the pin. It might be possible to design a pin shape that would reduce wear by minimizing close contact between tool and shear surface. But pending an unforeseen design development, highly wear resistant tool material will be required for welding abrasive materials.

A chemical reaction between tool and workpiece materials could promote wear. Tool workpiece friction and tool thermal conductivity may affect weld structure and properties. Metals with high welding temperatures may react with the atmosphere and make gas shielding necessary as for fusion welding.

Tool Geometries

The basic dimensions of a traditional FSW tool are pin length and radius and shoulder radius.

Pin length is determined by the thickness of the seam to be welded. The pin extends from the shoulder on the seam surface to the seam bottom with allowance for enough clearance so as to avoid contact with the anvil, but not so much as to cause lack of penetration. It is customarily determined empirically.

Pin radius must be sufficient for adequate pin strength and also for stretching the seam, by factor $R\Omega/V$, so as to expose adequate clean surface for a good bond. We note in passing that for miniature FSW, where R is taken very small as for a dentist's drill, Ω needs to be made correspondingly large by the above criterion. High-speed FSW is distinct from miniature FSW in that for high-speed FSW the pin surface speed $R\Omega$ is taken much larger than normal practice. Typically a pin diameter is approximately the same as the pin length has been taken.

The tool shoulder should be just wide enough to block the escape of weld metal (flash). Greater shoulder width generates more power at the workpiece surface. In addition to wasting power, it weakens the weld joint by widening the area of metal affected by processing heat. Typically a shoulder radius 2–3 times the pin radius has been taken.

Threads on the pin increase the effective friction coefficient of the pin surface, prevent slip at the pin surface, and promote shear in the weld metal. Scrollwork on the shoulder presumably performs a similar function. Sometimes a shallow concavity is machined into the shoulder contact surface.

Practice seems to be headed toward more elaborate pin shapes. It is possible to weld with a smooth, unthreaded pin surface. FSW is also possible with many different pin shapes [11], pins with longitudinal grooves or flats, square pins, for example. The main thing is for the pin to seize in the metal and not slip so that the shearing action occurs within the weld metal and not at the metal/tool interface. Although a pin may be square, the shear zone itself must still be cylindrical, of course.

It is possible to weld with a tapered pin. Tapered pins distribute the bending stresses due to drag more evenly along the pin and so are more resistant to fracture. A taper also affects the vertical pressure distribution. For a very thin extrusion layer, $\delta/r \ll 1$, it may be estimated that a taper dr/dz increases the axial pressure gradient (Equation 6.2) needed to force metal up past the pin according to Equation 6.27:

$$\frac{dP}{dZ} \approx -\frac{2\tau}{\delta}\left[1 + \frac{1}{2}\left(\frac{dr}{dz}\right)^2\right].$$

(6.27)

According to the crude model above, for a 45° cone half-angle, $dr/dz = 1$, the pressure gradient rises by 50%. A pressure gradient increase promotes confinement of weld metal and perhaps permits some reduction of the tool shoulder. It also promotes a somewhat greater tendency to seize in the weld metal. The reduction in tool radius along the length of a tapered tool can reach a point where the stirring effect is lost; that is, $R\Omega/V$ becomes too low to maintain conditions for a sound weld. Hence tapered pins need to be truncated before coming to a point. Tool design at the time of writing is carried out by cut-and-try empirical methods, quite reasonably in view of analytical complexities, but analytical concepts can be helpful in suggesting designs worth investigating.

Although strictly a parameter rather than an aspect of tool design, it is not inappropriate to mention seam offset with respect to tool axis with other geometrical matters. The effect of the ring-vortex circulation on the trace of the weld seam in the weld cross section has been discussed above. Changing the configuration of threads on a standard pin tool can alter the shape of this trace. Seam offset can also alter the shape of the seam trace. Stretching out the seam trace may dilute the effect of seam contamination. Rotating seam segments away from a perpendicular orientation to the acting stress may reduce the effect of loading on the seam, which might be weakened by contaminants. If there is significant low speed induced circulation around the shear surface, shifting the seam toward the retreating side of the tool increases the likelihood that the seam will flow around the shear surface without crossing it into the rotating plug and without exposure to flow conditions ensuring a sound weld. Shifting the seam toward the advancing side of the tool promotes earlier entry into the rotating plug and longer dwell time therein. The dwell in the rotating plug is followed by an exit shear, opposite to the entry shear, and the slow rotation in the surrounding circulation.

Conclusion

Friction stir welding is, in a sense, a mechanization of the traditional solid-state welding process of the blacksmith. In many cases it offers more control and better quality welds than fusion processes. In some cases it enables welds not possible by fusion processes. As with any welding process, to get the best results the process needs to be understood in some detail.

Here some concepts intended to make the process more intelligible have been presented, and attention has been called to features of the process that are not well understood. Friction stir welding is a relatively new process. It is under active study. It may be anticipated that a great deal more is to be learned about the process and that more technological advancement is to come.

Acknowledgments

For reading this text and making numerous helpful suggestions I am indebted to: colleagues and FSW researchers J. Schneider and R. Carter, and Ohio State University graduate student J. Hurst. I am grateful for support received from the Materials and Processes Laboratory at the Marshall Space Flight Center both for specific support in writing this article and for the long-term opportunity to acquire the background necessary to write such an article. I am also indebted to S. Hessler of Jacobs ESTS Group (Snyder Technical Services) for her valuable IT assistance.

References

1. The Welding Institute. W. M. Thomas, E. D. Nicholas, J. C. Needham, M. G. Murch, P. Temple-Smith, and C. J. Dawes. December 6, 1991. GB Patent Application No. 9125978. U.S. Patent No. 5,460,317. Friction welding. Filed October 24, 1995.
2. NASA-Boeing. R. J. Ding and P. A. Oelgoetz. April 13, 1999. U.S. Patent No. 5,893,507. Auto-adjustable pin tool for friction stir welding. Filed August 7, 1997.
3. MTS Systems Corp. C. L. Campbell, M. L. Fullen, and M. J. Skinner. March 13, 2001. U.S. Patent No. 6,199,745. Welding head. Filed July 8, 1999.
4. Prasad, Y. V. R. K., and S. Sasidhara. *Hot Working Guide: A Compendium of Processing Maps*. Russell Township, OH: ASM International, 1997.
5. Ernst, H. "Physics of Metal Cutting." *Machining of Metals*. Detroit, MI: American Society for Metals, 1–34, 1938.
6. Hill, R. *The Mathematical Theory of Plasticity*, corrected ed., 254–61. Oxford: Clarendon Press, 1956.
7. Colligan, K. Material Flow Behavior during Friction Stir Welding of Aluminum, *Welding Journal* 78, no. 7 (1999): 229s–37s.

8. Schneider, J., R. Beshears, and A. C. Nunes, Jr. Interfacial Sticking and Slipping in the Friction Stir Welding Process, *Materials Science & Engineering A* 435–6 (2006): 297–304.
9. Cao, G., and S. Kou. Friction Stir Welding of 2219 Aluminum: Behavior of θ *(Al₂Cu)* Particles, *Welding Journal* 84, no. 1 (2005): 1s–8s.
10. Prado, R. A., L. E. Murr, D. J. Shindo, and J. C. McClure. Friction-Stir Welding: A Study of Tool Wear Variation in Aluminum Alloy 6061 + 20% *(Al₂O₃)*, in *Friction Stir Welding and Processing*. Edited by K. V. Jata, M. W. Mahoney, R. S. Mishra, S. L. Semiatin, and D. P. Field, 105–16. Warrendale, PA: The Minerals, Metals & Materials Society (TMS), 2001.
11. Elangovan, K., and V. Balasubramanian. Influences of Pin Profile and Rotational Speed of the Tool on the Formation of Friction Stir Processing Zone in AA2219 Aluminum Alloy. *Materials Science and Engineering A* 459, nos. 1–2 (2007): 7–18.

7

Nuclear Applications Using Friction Stir Welding

Douglas E. Burkes

CONTENTS

Introduction

Friction stir welding (FSW) was developed by The Welding Institute (TWI) in the United Kingdom in the early 1990s [1]. Over the past two decades, a number of developments and modifications have been made to FSW and a similar sister process, friction stir processing (FSP) [2,3]. In general, FSW involves rotating a welding tool (comprised of a shank, shoulder, and pin) at a prescribed speed and tilt and plunging the weld tool into the workpiece material until the tool intimately contacts the workpiece surface and generates friction. The friction produces intense heat that bonds the workpiece materials as the weld tool traverses the joint line. Heat generated during the

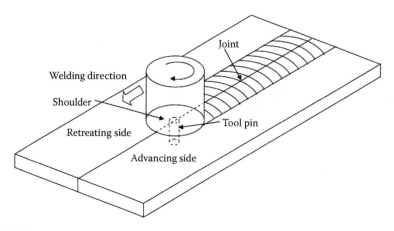

FIGURE 7.1
Schematic of the FSW process for a typical butt joint. (From Nandan, R., DebRoy, T., and Bhadeshia, H. K. D. H., *Progress in Materials Science*, 53, 980–1023, 2008. With permission.)

process due to deformation and relative motion of material around the tool pin sufficiently softens the material without reaching the solidus temperature. The leading face of the tool pin forces the plasticized material to the back of the tool while concurrently applying a substantial forging force to consolidate the weld metal, resulting in a high-integrity joint. A schematic of the FSW process is shown in Figure 7.1 [4].

Routine demonstrations of FSW to join both similar and dissimilar cast and wrought aluminum, titanium, copper, and magnesium alloys, as well as steels, have been made. FSW can be conducted in multiple joining configurations; including butt, corner, lap, T, spot, fillet, and hem joints. Many of these developments have been encouraged because FSW is a nonmelting joining technology that typically produces higher strength and ductility, increased fatigue life and toughness, lower distortion, less residual stress, less sensitivity to corrosion, and essentially discontinuity-free joints when compared to more conventional fusion welding techniques [5]. The fact that the process is solid-state eliminates most problems associated with cooling from the liquid phase, such as those often experienced with fusion welding. Similarly, porosity, solute redistribution, solidification, and liquification cracking are often not an issue with FSW. Furthermore, FSW is environmentally friendly because no fumes or spatters are generated and there is no arc glare associated with the process. This enables the FSW process to be especially suitable for components that are flat and long, such as sheets or plates, and most often on large pieces that cannot be easily heat-treated postweld to recover temper characteristics [1]. Detailed reviews of the current FSW state of the art can be found by Nandan, DebRoy, and Bhadeshia [4] and Mishra and Ma [6].

Most of the initial application of FSW was conducted in the aerospace and automotive industries [7]. More recently, the marine and defense industries have expressed greater interest in this processing technique, mainly spurred on by the ability to join more complex metals, such as titanium [8] and magnesium [9]. In addition, applications for space shuttle external fuel tank and launch vehicle welding and joining, along with the railway industry, have been investigated. The work has been conducted by all research facets, including academia, industry, and government laboratories. One industry that has not fully grasped the potential of FSW as of yet is nuclear. However, there are a number of challenging problems that FSW can help solve. This chapter discusses some of the current approaches employing FSW to solve nuclear industry-related problems; in addition, it discusses some potential areas in which FSW could be further applied within the field. Three of the most pertinent areas of FSW in the nuclear field are discussed: nuclear fuel fabrication, routine maintenance of reactor pressure vessels, and intermediate nuclear waste storage.

Nuclear Fuel Fabrication

There has recently been renewed global interest in nuclear reactors, especially in the United States. There are two perceived issues to the continued growth and development of nuclear reactors that are often discussed: (i) proliferation concerns over both highly enriched uranium (HEU) fuels, such as that used in research and test reactors; and (ii) dealing with the radioactive and toxic spent nuclear fuel upon discharge from the reactor. FSW has recently been employed to address the first issue in terms of HEU fuels to minimize proliferation concerns of general commerce including fresh and spent fuel shipment and interim storage. The second issue will be discussed later in this chapter.

Research and Test Reactors

Many high-powered research and test reactors use fuel containing HEU. The HEU is defined as a fuel containing $\geq20\%$ of the fissile isotope ^{235}U. The ability to make a readily available nuclear weapon exists with significant* amounts of material at high enrichments, should this material fall into the hands of those who intend to use the material for a purpose other than its intended application. In order to minimize proliferation of such material globally, a program has been established to convert research and test reactors that currently operate on HEU fuel into reactors that operate on low enriched uranium (LEU) fuel that contains $<20\%$ ^{235}U. As a result, the ability to manufacture a nuclear weapon from diverted material is decreased, and the complexity associated with accomplishing such a task greatly increased. This aggressive and dynamic program is the Global Threat Reduction Initiative

---

* An 20% enrichment is considered the optimum enrichment since it is difficult to breed Pu at this enrichment and the mass of material is low enough that the number of enrichment steps necessary to make a usable weapon enable such a process challenging.

(GTRI) and includes the cooperation of the majority of the countries world-wide that operate research and test reactors [10]. The reactor conversion initiative is truly a global effort, led mainly by Argentina, Australia, Belgium, Canada, Chile, China, France, Germany, Japan, Russia, South Korea, and the United States. The new fuels must behave in a manner without significant penalties in reactor or experiment performance, economics, and safety.

Typically, research reactor fuel is plate- or pin-type, and consists of dispersed uranium (U) alloy particles in a matrix clad with an aluminum (Al) alloy. A cross-sectional metallograph of a typical plate-type dispersion fuel is provided in Figure 7.2. The U can be alloyed with Al, silicon (Si), oxygen (O), or molybdenum (Mo) that increases the U density of the fuel meat (2.3 gU•cm^{-3} with Al,* 3.2 gU•cm^{-3} with O [U_3O_8], 4.8 gU•cm^{-3} for Si,[†] and 8.5 gU•cm^{-3} for Mo[‡]). Uranium density is an important consideration for converting reactors to LEU fuel because a reduction of U enrichment will require a concomitant increase in the overall uranium density. For lower-power density research and test reactors, alloying U with Si is sufficient to maintain the fuel density and decrease the amount of fuel enrichment. For many high-power density research and test reactors, this is not the case, and novel fuel concepts are being developed. The current solution to this challenge is accomplished through the use of a uranium–molybdenum alloy in a monolithic form [11]. Monolithic fuel contains a single foil to replace the multiple fuel particles that comprise a dispersion fuel plate meat, providing the highest possible uranium loading for reduced enrichment. For example, U-10 wt% Mo monolithic fuel is planned for a demonstration at 15.3 gU•cm^{-3} for qualification. In addition, implementation of a monolithic fuel will provide a smaller contact surface area with the aluminum cladding to minimize

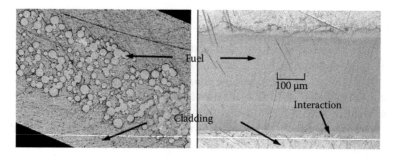

FIGURE 7.2
Typical cross-sectional metallographs showing U-Mo fuel, AA6061 cladding, and fuel–cladding chemical interaction for a dispersion fuel type (left) and monolithic fuel type (right) plate for nuclear research and test reactors.

---

* Qualified density.
[†] Qualified at 4.8 gU•cm^{-3}, although the fuel has been demonstrated at a density of 6 gU•cm^{-3}.
[‡] Planned demonstration for qualification.

interaction between the fuel and cladding. A cross-sectional metallograph of a typical plate-type fuel containing a monolithic fuel alloy is provided in Figure 7.2.

Cladding material for the fuel differs between fuel fabricators and reactor operators, depending on whether or not reprocessing of spent fuel is conducted, and also depending on the coolant chemistry of the reactor. For example, in the United States, where reprocessing is not currently conducted, precipitation-hardened Al alloys are used, mostly AA6061, while reprocessed fuel fabricated in Europe consists of solution-hardened Al alloys, such as AG3NE and AlFeNi. Furthermore, some interest in cladding research and test reactor fuel with Zircaloy alloys has been expressed to increase the melting temperature margin, since Zircaloy-2 and -4 are used in the light water power reactors; however, there is some concern over the associated weight increase with such fuels for conversion.

Fabrication of Monolithic Fuel Plates with FSW

The friction bonding (FB) process, a modified FSW process, allows bonding between two similar AA6061 thin plates as well as bonding of the AA6061 to a dissimilar U–Mo monolithic alloy sandwiched between the two AA6061 plates. The fuel foil is typically no thicker than 500 μm, while the overall fuel-plate thickness is typically no greater than 1.4 mm. Bonding of the thin sheets is accomplished without adding material, and the process must be capable of bonding without material mixing due to the fuel foil in the center of the workpiece that must not be disturbed, otherwise resulting in rejection of the fuel plate. This is a significant departure from conventional FSW that typically involves a stirring action to move material from one region to another. For example, conventional FSW tools typically have extended pins, sometimes threaded, which aid in the movement and stirring of material across the bond line or from one region of material to another. A number of unique tool designs have been investigated to achieve mixing and movement of material in conventional FSW, some examples of which are provided in Table 7.1 [4]. As previously mentioned, for the current application, the tool pin is not allowed to penetrate across the bond line. Thus, bonding must be accomplished mechanically with only minimal heat for a very short duration and is driven mainly by tool applied load [12]. A schematic of a typical tool design used for fabrication of research and test reactor fuel is provided in Figure 7.3. Some of the main features of the tool are the short pin that extends down from the shoulder, a recessed region between the shoulder and the pin that increases the stability of the process, beveled edges along the outer radius of the shoulder to aid in warp reduction, and an annular plenum that allows heat removal from the weld face via an appropriate coolant (e.g., propylene glycol).

In general, the fabrication sequence begins by machining a pocket in a piece of AA6061 sheet with the same depth as the thickness of the foil. The

TABLE 7.1

A Selection of Tools Designed by The Welding Institute Listed With Tool Specifications and Applications

Tool	Cylindrical	Whorl	MX Triflute	Flared Triflute	A-Skew	Re-Stir
Schematics						
Tool pin shape	Cylindrical with threads	Tapered with threads	Threaded, tapered with three flutes	Tri-flute with flute ends flared out	Inclined cylindrical with threads	Tapered with threads
Ratio of pin volume to cylindrical pin volume	1	0.4	0.3	0.3	1	0.4
Swept volume to pin volume ratio	1.1	1.8	2.6	2.6	Depends on pin angle	1.8
Rotary reversal	No	No	No	No	No	Yes
Application	Butt welding, fails in lap welding	Butt welding with lower welding torque	Butt welding with further lower welding torque	Lap welding with lower thinning of upper plate	Lap welding with lower thinning of upper plate	When minimum asymmetry in weld property is desired

Source: Nandan, R., DebRoy, T., and Bhadeshia, H. K. D. H., Prog. in Mater. Sci., 53, 980–1023, 2008. With permission.

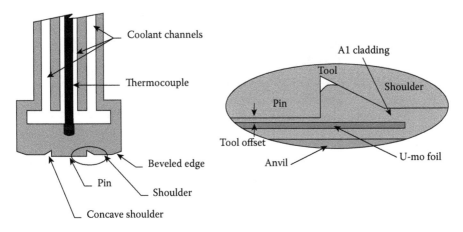

FIGURE 7.3
Schematic of an example tool design used to fabricate research and test reactor fuel plates. Note the relatively shallow pin when compared to the overall tool diameter, which is different than designs used in conventional FSW applications shown in Table 7.1.

U–Mo monolithic fuel foil is placed in the pocket, while a second AA6061 sheet is placed on the top, thereby creating the fuel assembly. The FB process is carried out with multiple passes over the surface of the fuel assembly, both top and bottom, until the entire fuel region has been subjected to the process. Regions not subjected to the process, and those that are dimensionally undesirable, are sheared from the fabricated fuel plate. This process in turn reduces the amount of residual stress, inherent because of the rigid clamping required by FSW processes, remaining in the as-fabricated fuel assembly. Photographs of typical as-fabricated fuel assemblies are provided in Figure 7.4. These particular fuel plates were used as test plates in the Advanced Test Reactor in Idaho.

A number of novel modifications to the FSW process had to be made in order to achieve the success of FB for this particular application. For example, a cooled tool design had to be incorporated, even though thin-plate geometry of high thermal conductivity aluminum was employed. Mainly, the cooled tool facilitated overheating of the weld surface that results in the generation of weld flash (shown in Figure 7.5), while maintaining sufficiently high loads (>44.5 kN) to accomplish adequate bonding between the dissimilar U–Mo and AA6061. Excessive weld flash is highly undesirable, not just for this application, but for most welding applications since it requires a significant amount of postprocess work to remove the surface irregularity that increases the time and costs associated with the fabrication process.

The FSW processes require rigid clamping to hold the workpiece in place during processing, especially in the case of joining thin sheets one on top of the other where sliding of the sheets must be prevented. Clamping in this manner will exert a high restraint on the assembly during FSW where

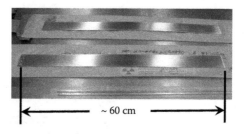

~ 60 cm

FIGURE 7.4
Photographs of typical test monolithic fuel plates fabricated by friction bonding for irradiation in the Advanced Test Reactor. Note the slight bowing of the fuel plates as a result of remnant residual stress from fabrication.

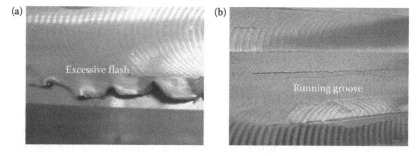

(a) Excessive flash

(b) Running groove

FIGURE 7.5
Photographs of an FB weld where process temperatures were too high, resulting in excessive weld flash (left) and in a running groove on the surface (right).

contraction of the weld nugget under the tool pin and the heat-affected zone (HAZ) under the tool shoulder is prohibited during cooling [6], ultimately resulting in the development of residual stress. In the case of the fabrication of nuclear fuel plates, heat transfer is extremely rapid in the thin assembly geometry such that residual stresses in both the longitudinal and traverse directions are significant. Residual stress in the fuel plate, particularly at the fuel–clad interface, is exacerbated by the differences in thermal properties of the dissimilar materials (Al and U–Mo). Compressive stresses introduced into the workpiece can in some circumstances be beneficial [13], while tensile stresses in the workpiece ultimately lead to crack initiation and aid in crack propagation, possibly leading to catastrophic failure. Residual stress is a concern for both fuel fabrication and fuel performance.

Crack propagation along a fuel–clad interface is an unacceptable behavior in a reactor. Such delamination would result in a significant decrease in thermal conductivity, causing the fuel temperature to increase and performance to decrease, or even worse, cause the fuel cladding to pillow, closing water coolant gaps and leading to overheating of the entire fuel assembly. Thus, residual stresses and influence on the overall structural properties of the fuel plate as a function of processing parameters must be well understood. For example, longitudinal stress increases as tool traverse speed increases due to steeper thermal gradients across the tool and reduced time for stress relaxation. Conversely, both lateral and longitudinal residual stresses decrease with an increase in the rotational speed [14]. Tool design is also an important consideration for influence of residual stress on fabrication of nuclear fuel plates. A larger tool shoulder diameter will increase the welding torque, but the shorter pin will decrease the longitudinal force exerted on the workpiece and higher residual stress [15]. Applying external tensioning during welding can induce compressive stresses that have the benefit of inhibiting or minimizing crack propagation [16]. Although delamination of the fuel–clad interface has not been observed during irradiation, it is a common observation during postirradiation examination of the fuel, as shown in Figure 7.6, after the fuel is destructively sectioned allowing the interface to delaminate.

Future Work

For FB to be a viable alternative for mass fabrication of nuclear fuel plates, the process must be demonstrated on a commercial scale. This includes the demonstration of high-product yields (>95%), high-product throughput (~hundreds of plates fabricated per day), and minimal equipment downtime. Demonstration of these factors is highly desirable and driven by two main considerations when compared to more conventionally used processes, such as hot-isostatic pressing (HIP): (i) low equipment capital cost (hundreds of thousands compared to millions) and (ii) low operator involvement. However, where HIP is already a batch process, FB must be converted to such a process to achieve throughput requirements and have very low operator involvement to keep personnel costs associated with fabrication of the fuel to a minimum. Thus far, three different modifications have been investigated to increase product throughput, each with varying degrees of success, most often at the expense of product yield.

The three modifications include a change in the process parameters, a change in the tool design, and a change in the tool raster pattern. For example, increased tool traverse rates can significantly increase the weld coverage rate, resulting in less fabrication time and overall operational costs. However, process parameters to accomplish adequate bonding and acceptable surface finish will need to be optimized for a given process coverage rate, because the parameters affect material flow, temperature, and forging action. This is not a straightforward process, since the change in one parameter requires a concomitant change in additional parameters; that is, an increased tool traverse rate will decrease temperature, requiring an increase in both process load and tool rotational rate. Such changes to multiple parameters can have

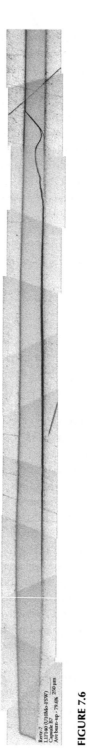

Rertz-7
U3F140 (U10Mo-FSW)
Capsule B7
Ave burn-up ~ 79.6% 250 μm

FIGURE 7.6
Partial metallograph of an irradiated fuel plate fabricated by friction bonding. Note the delamination of the fuel–clad interface that most likely occurred upon destructive sample preparation of the fuel plate as a result of residual stress relaxation.

drastic impacts on the bonding of the fuel plate, the most important factor for high-product yields.

A similar increase in the weld coverage rate can be obtained by increasing the surface area of the tool, specifically, over the tool shoulder because this portion of the tool performs most of the work while the tool is in motion. For example, given a constant tool traverse rate and bond pass length, a larger tool diameter will provide more coverage area than a smaller tool diameter, but will also increase the torque required to turn the tool. In addition, there is a trade-off between optimization of weld coverage rate and heat generation and temperature distribution across the tool surface.

Finally, the tool raster pattern can play a crucial role in both fabrication economics and successful, high-yield fabrication rates. Two types of raster patterns can be employed: a one-way raster or a true raster. A one-way raster pattern involves multiple, parallel passes over the assembly surface that run in the same direction. This pattern requires a plunge and extraction at each end of the welding pass, increasing the time required for fabrication due to the travel from the extraction end back to the plunge end for a concurrent pass, and the additional time required plunging the tool back into the workpiece to raise the bond area to the appropriate temperature. As many as nine passes may be required per side for the bonding of a fuel plate 20 cm (8 in.) wide. Thus, anywhere from 4 to 8 minutes could be added to the time it takes to fabricate each fuel plate.

A true raster pattern involves only one plunge and one extraction made per side in order to save time, but the tool and backing anvil must be equipped with adequate cooling capability to remove excessive heat from the process. Without cooling capability, multiple, extended pauses would be required to allow for heat dissipation in order to prevent tool or equipment failure. Conditions between the workpiece surface and the tool could be adversely impacted if the tool were allowed to become too hot, similar to the example provided in Figure 7.5, including decreased friction coefficients; generation of hot spots and surface melting; and uncontrolled, detrimental tool penetration into the soft surface resulting in foil impact.

Each of these process changes will have a significant impact on the final product surface finish, degree of bonding, and the degree of residual stress resident in the assembly. Techniques to measure and characterize these impacts are necessary, along with the ability to alleviate undesired assembly traits before the plates are inserted into the reactor core. Improvements to commercialize the FB process are highly desirable; that is, development of multiple-head machines that could significantly reduce the up-front capital cost and decrease the production time per plate. Some initial research on such processes and techniques exists [17], but additional development and interest in the nuclear fuel fabrication area is much needed.

Reactor Pressure Vessel Repair

Engineering components often fail during long-term service by a variety of mechanisms including: (i) fatigue under cyclic loads; (ii) creep, thermal shock, disbanding of cladding layers, or hydrogen attack during elevated temperature service; and (iii) stress corrosion cracking in corrosive or irradiated environments [18]. For nuclear reactors, failure is not an option; thus, components must either routinely be replaced or repaired. Given the interest in extending the lifetime of light water reactors in the United States before the next generation of reactors can be brought online [19], repair of components is much more desirable and practical from a cost and time savings perspective.

Current Light Water Reactor Structural Components

Most reactors operating today employ austenitic stainless steel for structural components. Helium is generated and accumulated in the inner austenitic stainless steel components of nuclear reactors based on the transmutation reaction:

$$10B + 1n = 7Li + 4He, \tag{7.1}$$

where 1n is a neutron [18]. Helium bubbles are generated by the helium migrating to grain boundaries and over time embrittle the material and result in the development of cracks. Fusion welding is one option to repair cracks that result from helium embrittlement. However, the HAZ of fusion repair welds cracked when patches were welded to the SS304 tank wall of a nuclear reactor at the Savannah River Site (South Carolina) resulting from helium embrittlement [20]. In other words, as the fusion weld repaired the initial crack, the HAZ on the outer edges of the weld released more helium resulting in the generation of cracks parallel to the repair weld. A schematic showing the typical progression of helium bubble formation, migration to the grain boundaries, coalescence, and ultimately grain boundary fracture of an irradiated stainless steel is shown in Figure 7.7 [20]. An example photomicrograph of this process is provided in Figure 7.8 [18]. Furthermore, the potential and extent of cracking will increase with helium concentration, creating additional concerns over significantly aged structural components.

The challenges associated with fusion weld repair of reactor pressure vessels has resulted in codes that require postweld heat treatment (PWHT) after welding is performed to reduce the hardness and improve the toughness in the base metal HAZ caused by welding. Obviously, PWHT in repaired components is not always possible or economical, and may exacerbate the problem of increased hardness and reduced ductility due to repeated heating

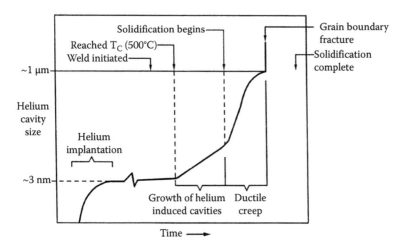

FIGURE 7.7
Typical progression from helium bubble formation to grain boundary cracking during fusion welding of an irradiated stainless steel component. (From Kanne, W. R., Jr., et al., *Weld Repair of Irradiated Materials*, Elsevier, 203–14, 1999. With permission.)

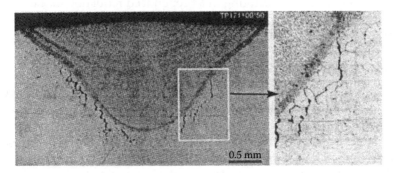

FIGURE 7.8
Crack in the HAZ of a fusion weld created by helium release of an irradiated austenitic stainless steel plate. (From Yurioka, N. and Horii, Y., *Sci. Technol. Weld. Join.*, 11(3), 255–64, 2006. With permission. http://www.ingentaconnect.com/content/maney/stwj)

[18]. For helium concentrations below 0.14 appm, grain boundary cracking can be avoided by employing a low-heat input welding method, where the arc energy is <2 kJ•mm-1 [18]. This is shown graphically in Figure 7.9 [18]. Thus, an appropriate welding method, such as the gas metal arc (GMA) overlay technique, must minimize the heat input into the base material and the amount of mixing with the base material, while providing sufficient weld penetration to develop a continuous metallurgical bond between the weld and the base metal. In this way, cracks can be sealed directly or the weld can provide a base for attachment of patches over the cracks [20,21]. Using the GMA overlay technique, weld penetration into the base metal is limited to

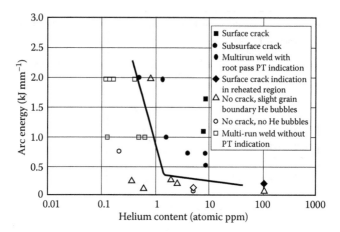

FIGURE 7.9
Helium-induced cracking of an irradiated stainless steel component as a function of welding heat input and helium content. (From Yurioka, N., and Horii, Y., *Sci. Technol. Weld. Join.* 11(3), 255–64, 2006. With permission.)

approximately 0.08 mm, resulting in a pronounced reduction in the amount of cracking, but not an elimination altogether; that is, underbead cracks are still present as shown in Figure 7.10 [20].

Potential of FSW for Reactor Vessel Repair

Based on the challenges associated with aging reactor vessel repair, FSW is adequately poised to address the problem on an industrial scale. For example, the HAZ of friction stir welds is similar to that in conventional fusion welds, but the maximum peak temperature during welding is significantly less than the solidus temperature (~1400°C for SS304). A typical temperature profile behind an FSW on ferritic stainless steel is provided in Figure 7.11 [22]. Observation of the figure reveals that the maximum process temperature was less than 1200°C and that the time taken to cool over the range 800 and 500°C (Δt_{8-5}) is about 11 seconds [22]. The Δt_{8-5} determined for this particular FSW is comparable to a manual metal arc weld with a heat input of about 1.3 kJ•mm^{-1}. As mentioned previously, grain boundary cracking can be avoided by employing a low-heat input welding method (<2 kJ•mm^{-1}).

Furthermore, the heat source for FSW is rather diffuse, meaning that steep temperature gradients across the weld tool exist, often times at 100° or more. An example of peak temperature distribution across a FSW tool for SS304L is provided in Figure 7.12 [23]. The cooling rate of the material is dependent upon the welding speed employed. Depending upon the conditions necessary to repair the structural component, the low-welding speed combined with the high-temperature gradient across the tool can lead to significantly different microstructures when compared with conventional, or even newer,

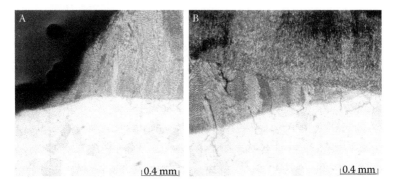

FIGURE 7.10
Metallographic sections of low-penetration overlay welds made on irradiated SS304 containing 10.4 appm He. The dark area on the top is weld metal and the light area in the lower half is base metal including the HAZ. (A) Edge location showing the absence of toe cracking in the overlay weld. (B) Underbead location of maximum cracking showing that some cracks extend from the HAZ into the weld metal. (From Kanne, W. R., Jr., et al., *Weld Repair of Irradiated Materials*, Elsevier, 203–14, 1999. With permission.)

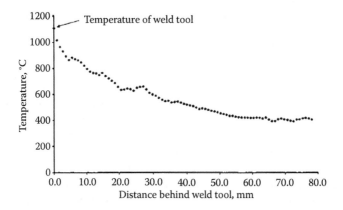

FIGURE 7.11
Temperature as a function of distance behind the tool after 250 mm of welding with a tool traverse rate of 3 mm•sec^{-1} for typical ferritic stainless steel. (From Thomas, W. M., Threadgill, P. L., and Nicholas, E. D., *Sci. Technol. Weld. Join.*, 4, 365–72, 1999. With permission.)

fusion welding processes. This can present a significant advantage in terms of minimizing any undesired phase transitions. Thomas et al. have summarized some advantages of FSW steel including: (i) parent metal chemistry is retained without any gross segregation of alloying elements; (ii) plain low carbon steel and 12% chromium alloy steel can be welded in a single pass with thicknesses of 3–12 mm; (iii) steel thicknesses up to 25 mm can be welded from two sides, as in arc welding where the double-sided weld joint is more process tolerant; (iv) since the process is solid phase, problems

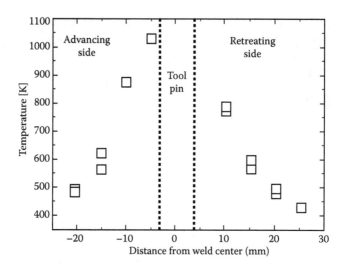

FIGURE 7.12

Peak temperature distribution along the intersection of the horizontal mid-plane and transverse plane through the FSW tool axis, for the FSW of SS304L employing a tool rotational rate of 600 rpm and a tool traverse rate of 10.2 mm•min-1. The tool shoulder and pin diameter were 25.4 and 5.6 mm, respectively. (From Cho, J. H., Boyce, D. E., and Dawson, P. R., *Materials Science and Engineering A*, 398, 146–63, 2005. With permission.)

that can occur in liquid phase, such as those with fusion welding, are eliminated; (v) the process can be carried out in all positions including vertical and overhead; (vi) the process is machine-tool technology based that can be semiautomatic or fully automated; and (vii) like most friction techniques, the process can be carried out under water [22].

The combined advantages of (iii), (v), (vi), and (vii) are especially advantageous for nuclear reactor pressure vessel repair. Underwater welding to repair damaged components in nuclear reactors is often necessary to reduce exposure to radiation [18]. The joint toughness obtained from FSW welds on particular steels is very similar to that from submerged arc welds [24], suggesting that there is no loss in material integrity of a FSW joint for repair.

Future Work

As with any developing technique, there are still some challenges that need to be addressed before FSW can be considered a mainstream option for repair of aging nuclear reactor structural components. For example, the material from which the tool is made must obviously survive much more strenuous conditions due to the increased strength of steel. It is not uncommon for conventional FSW tool materials to wear, rapidly producing debris that can frequently be found inside the weld. This has been addressed over the last decade with the development of alternative tool materials, including but not limited to, advanced ceramics (polycrystalline boron nitride, PCBN) [25] and refractory alloys based on W [26], but in some cases, such materials can increase the complexity and the cost of the relatively inexpensive FSW process. The consequences of phase transformations accompanying FSW have not been studied in sufficient depth, the requirement of which is necessary before regulatory approval will be gained [4]. The variety of steels available is much larger than for any other alloy system, requiring considerable experiments to optimize the weld for a required set of properties [4]. This is not necessarily a large concern when dealing with the repair of current nuclear reactor structural materials, since they have traditionally been fabricated from austenitic stainless steels (e.g., SS304). However, as materials science continues to evolve in the nuclear field, advanced steel alloys will most likely take the place of the traditional alloys used. Thus, it is in the best interest of FSW to demonstrate the ability to accommodate such improvements prior to the need. There are initial indications that elongation might suffer following FSW [27] that will have to be evaluated appropriately before regulatory approval is gained. This can most likely be improved through selection of appropriate process parameters and tool design. Finally, the two major drawbacks to FSW must be considered and novel approaches developed so that the technique is not at a disadvantage compared to conventional welding techniques: (i) the rigid clamping necessary to prevent the abutting joint faces from being forced apart and (ii) the run-out hole left by the tool pin at the end of the weld. The latter can be overcome with sufficient tool geometry design, run on/run off plates that take the hole from the substrate joints, or filling the hole with a taper plug or by friction hydropillar welding [22].

Fabrication of Canisters for Spent Nuclear Fuel

As mentioned in the introduction, the second perceived issue to the continued growth and development of nuclear reactors that is often discussed is how to appropriately and safely deal with the radioactive and toxic spent nuclear fuel upon discharge from the reactor. Reprocessing is most common in France, where Pu is recovered from the spent nuclear fuel, recycled into new fresh fuel in the form of UO_2–PuO_2 mixed oxide (MOX), and then returned to light water reactors. Future reprocessing schemes involve recovering not only Pu, but also other usable long-lived minor actinides Np, Am, and Cm. These can also be recycled into fresh fuel that is inserted into a fast reactor, where the long-lived constituents are transmuted to short-lived nuclides. Reprocessing significantly reduces the waste generation and concomitant heat load that must be addressed from spent nuclear fuel [28,29], but can occasionally be politically undesirable due to proliferation risks associated with separated Pu. Where reprocessing is not an option, the spent nuclear fuel must be stored in a repository, such as the proposed Yucca Mountain site in the United States. The spent nuclear fuel must reside in the repository for approximately 100,000 years before the radioactivity decreases to a safe level. As such, the storage medium for the spent nuclear fuel must be able to sustain any potential corrosion impacts and accommodate the high thermal load from spent nuclear fuel over this lengthy time period.

Copper Canisters for Spent Nuclear Fuel

Copper canisters with a nodular cast iron insert buried in a deep geological repository have been proposed as a disposition path for spent nuclear fuel, particularly in Sweden [30–32]. Copper is highly desirable for this purpose since the material satisfies the requirement on chemical resistance when immersed in groundwater in the repository, and has exceptional creep resistance given the high thermal loads of spent nuclear fuel. Proposed canisters for nuclear waste in Sweden have an outer diameter of 1050 mm and a length of 4830 mm that are capable of holding 12 boiling water reactor (BWR) or four pressurized water reactor (PWR) spent fuel assemblies [33]. A photograph of

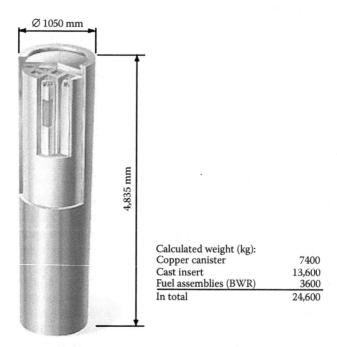

Calculated weight (kg):	
Copper canister	7400
Cast insert	13,600
Fuel assemblies (BWR)	3600
In total	24,600

FIGURE 7.13
Schematic of typical copper canister with a cast iron nodular insert for spent nuclear fuel. (From Andersson, C.-G. and Andrews, R. E., *The development of advanced welding techniques for sealing nuclear waste canisters.* Technical Report TR-01-25. Stockholm, Sweden: SKB, 2001. With permission.)

a typical spent fuel canister is provided in Figure 7.13 [34]. Over 200 canisters must be commercially fabricated per year in order to satisfy Sweden's spent nuclear fuel requirements alone.

The type of copper selected for this particular application is very close to UNS 10100 with a small addition (50 ppm) of phosphorous to further enhance the creep ductility. Copper, however, is very challenging to weld, especially given the thickness of canister and lids to be joined of 30–50 mm [33]. In particular, copper has an extremely high thermal diffusivity, anywhere from 10 to 100 times higher than that of steel and nickel alloys, making them essentially impossible to join employing conventional fusion welding techniques [13]. Thus, heat input is extremely important, and must be controlled in an opposite manner as that discussed for the nuclear reactor pressure vessel repair (i.e., high-heat input is desired for the joining of thick copper). Friction stir welding has been investigated as a potential fabrication method for these canisters since high-heat inputs can be achieved with low-welding speeds. FSW also offers the advantage of providing a very high weld quality assurance level to meet the service of the essential life requirements, can be remotely controlled, the in-process monitoring can reliably confirm if the welding conditions are being accurately maintained, and there is no risk of missing the weld interface that can occur with high-energy density welding techniques, such as electron beam welding [33]. Alternatively, FSW also has the potential for being used as a repair technique for defective fusion welds.

Initial Studies Employing FSW

Initial feasibility studies have indicated that FSW can be employed to join copper tubes of 200 mm outside diameter by 15 mm wall thickness and achieve high integrity welds with a weld cycle time of approximately 20 seconds [33]. In the case of spent nuclear fuel canisters, FSW would be employed to weld the circumferential base to the canister. Welding forces upward of 24 MN are required to accomplish such a high integrity weld. The high welding forces required along with the higher thermal diffusivity of copper has led to the development of some unique FSW concepts.

For example, the forging temperature range of copper is high (between 590 and 930°C) compared to that of aluminum alloys (340–480°C) [35]. The high-forging temperature range means that the conventional plunge procedure cannot be used; otherwise, the tool pin would either be damaged or fractured. To solve this problem, a pilot hole was drilled at the start of the weld, with the diameter being only slightly smaller than the diameter of the tool pin, in order to enable sufficient frictional heat generation. The tool must be allowed to dwell at the beginning of the weld until the temperature adjacent to the tool pin reaches 400°C given the high thermal diffusivity. As previously mentioned, tool traverse rates are very slow (on the order of 15 mm•min^{-1}) in order to minimize heat loss through the thick copper. Once steady-state welding behavior is attained, the tool traverse rate can be progressively increased. FSW has a higher heat input per unit length than EBW for comparable fabrication, but this does not appear to result in unacceptable thermal fields since heat can be lost into the canister assembly through conduction [36].

Tool material selection is of the utmost importance since the FSW tool must possess high hot strength materials and also be extremely wear resistant. In order to meet the high-welding force requirements, the machine possesses a high spindle rotation, a robust gearbox that ensures sufficient torque transfer, and a powerful screw feed system to drive the linear traversing table [32]. Water cooling is also required for both the tool and the backing anvil to keep the copper cool minimizing undesired microstructural traits and remove excess heat generated by the process. Hot work tool steel was determined as ineffective as a tool material because the fine machine features needed to accomplish material movement rapidly became filled with copper that severely restricted the stirring action. In addition, hot work tool steel softens above 540°C, well below the forging temperature range for copper, leading to rapid damage and increased replacement of the tool. W-based alloys are receiving increased attention for materials other than aluminum alloys and are ideal for this particular application [37,38].

Future Work

There are two major issues that must be addressed for fabrication of spent nuclear fuel canisters for FSW to be a viable option. Since the process must be scaled-up to accommodate the thickness of the copper and the diameter of the weld, significant improvements to the machine to accommodate necessary tool applied loads and spindle torque will increase the cost, while the rigid clamping system must provide forces even higher than those necessary for repair of nuclear reactor pressure vessels or fuel fabrication to prevent the lid separating from the canister. Initial feasibility studies have placed the cost of a dedicated FSW machine for this application at three to five million U.S. dollars [36]. Costs must be minimal in comparison to the advantages of the technique so that commercial outfits do not accept too great a risk. Furthermore, development of better tool designs that allow improved flow paths for hot softened copper are desired since they reduce the resultant traversing force required and improve the tool life, both reducing the cost of the process [36]. Nonetheless, the Swedish Nuclear Fuel and Waste Management Co. (SKB) has continued to investigate the process as a viable alternative to EBW based on FSW significant advantages [39]. Outcomes of this continued research will be of great interest to both the FSW community and the nuclear industry, especially with continued global development and implementation of commercial nuclear power.

Nondestructive Evaluation of FSW Joints

As with any welding technique, evaluation of weld quality is of particular importance, especially when addressing the stringent requirements that accompany nuclear components, whether fuel, repair of structures, or end-of-life storage. Certain conditions are inherent to the FSW process, and research is not always able to predict and define optimum operational envelopments to completely eliminate anticipated challenges. For example, when there is insufficient temperature generated by the weld tool, whether due to operational parameters or tool design; long, tunnel defects running along the weld can be developed. These defects can be either surface or subsurface with subsurface defects being of particular concern for obvious reasons. If the tool pin isn't long enough or the tool rises out of the plate during welding, then the interface at the bottom of the weld may not be disrupted and forged by the tool that can result in a lack of penetration (LOP) defect [40]. LOP is the most common defect associated with FSW and most often exists at the start of the weld where the tool pin plunges into the workpiece, or at the end where the tool is withdrawn due to inadequate frictional force. In most cases, the material at the start and the end of the process is removed, but can be impractical in certain applications. These types of defects are especially difficult to detect by low sensitivity radiography if the interface contains foil of some high-density materials (e.g., Al and U alloys). The defect can be eliminated with high confidence by either increasing the length of the pin or by controlling its relative position during welding, whichever is most practical for the fabrication application [40].

The term kissing bond is used when there is no metallurgical or physical bond between the materials being joined, but light contact between the materials still exists. This type of defect is very difficult to detect using nondestructive evaluation (NDE) methods, such as radiography or ultrasonic testing [40]. This type of bond is of particular concern for fabrication of nuclear fuel plates since an adequate bond must exist before the plates are inserted into the reactor and ensure that fission products developed during irradiation do not coalesce at the interface resulting in breakaway swelling of the overall fuel plate. Mottling effect is very common in fusion welding of Al plates, but not found with FSW.

Continued refinement of existing nondestructive (ND) techniques along with the development of new ND techniques must accompany the improvement of FSW in parallel. An example of a novel NDE technique that has been investigated for the evaluation of kissing bonds is thermography, which utilizes heat transfer rather than sound in the case of ultrasonic testing. The part is pulsed with a small amount of heat and evaluated through utilization of a high speed infrared camera to image the joined part. If there is a defect present in the joint, the defect will remain higher in temperature than the area(s) that are soundly joined together due to heat conduction. Example images obtained from a typical ultrasonic ND technique and a thermography ND

FIGURE 7.14

Example images obtained from a typical ultrasonic testing technique (right) and a flash thermography technique (left). The same sample containing a 25 μm thick SS304 foil encapsulated in AA6061 was imaged with each technique.

technique are provided in Figure 7.14. The images were each performed on a 25 μm thick SS304 foil encapsulated in AA6061. ND techniques have thus far been investigated in much less detail than the FSW process itself, with more emphasis placed on the development of optimum operated conditions. Both are needed, but significant costs in terms of time and money can be gained by quality NDE where problems can be quickly and efficiently be identified and remedied as appropriate.

Conclusions

This chapter has provided an overview of nuclear applications using FSW. Some specific examples where the process is currently being utilized, and others that are particular areas of interest, have been provided. There is a significant opportunity for FSW to expand and address some of the most challenging problems facing the nuclear industry today and in the near- to mid-future. Implementation of the FSW process in nuclear applications will only be viable if the technique can be further demonstrated as a high integrity, dependent, and cost-effective process when compared to conventional and more commonly accepted welding techniques. In addition, parallel development of NDE techniques to address the specific challenges of FSW is greatly needed, as these can be significantly different than those associated with conventional welding in comparative applications. Finally, this chapter will hopefully identify and generate opportunities for current and future FSW researchers to expand into the nuclear area to address and improve upon the applications discussed in this chapter.

Acknowledgments

Work was supported by the U.S. Department of Energy, Office of the National Nuclear Security Administration, under the U.S. Department of Energy

Idaho Operations Office Contract DE-AC07-05ID14517. Accordingly, the U.S. Government retains a nonexclusive, royalty-free license to publish or reproduce the published form of this contribution, or allow others to do so, for U.S. Government purposes.

Utilization of FSW for nuclear fuel plate fabrication has been conducted extensively at Idaho National Laboratory, and as such the author would like to graciously acknowledge those that have been involved in its continual development: Mr. N. Pat Hallinan, Mr. Michael Chapple, Mr. Curtis Clark, Mr. Jared Wight, and Mr. Gaven Knighton. Finally, the author would like to acknowledge those throughout the world that have been and continue to investigate this unique and exciting process to address problems specific to the nuclear industry, including in no small part those at TWI in the U.K., and the Swedish Nuclear Fuel and Waste Management Co. (SKB).

U.S. Department of Energy Disclaimer

References

1. Thomas, W. M., Nicholas, E., Needham, J., Murch, M., Temple-Smith, P., and Dawes, C. Friction stir butt welding (Patent Filed November 27, 1992).
2. Mishra, R. S., Mahoney, M. W., McFadden, S. X., Mara, N. A., and Mukherjee, A. K. 1999. High strain rate superplasticity in a friction stir processed 7075 Al alloy. *Scripta Materialia* 42: 163–8.
3. Mishra, R. S., and Mahoney, M. W. 2001. Friction stir processing: A new grain refinement technique to achieve high strain rate superplasticity in commercial alloys. *Materials Science Forum* 357–359: 507–14.
4. Nandan, R., DebRoy, T., and Bhadeshia, H. K. D. H. 2008. Recent advances in friction-stir welding—Process, weldment structure and properties. *Progress in Materials Science* 53: 980–1023.

5. Arbegast, W. J. 2007. Friction stir welding: After a decade of development. In *Friction stir welding and processing IV*, eds. R. S. Mishra, M. W. Mahoney, T. J. Lienert, and K. V. Jata, 3–18. Warrendale, PA: The Minerals, Metals & Materials Society (TMS).

6. Mishra, R. S., and Ma, Z. Y. 2005. Friction stir welding and processing. *Materials Science and Engineering, R.* 50 (1–2): 1–78.

7. Thomas, W. M., and Nicholas, E. D. 1997. Friction stir welding for the transportation industries. *Materials & Design* 18: 269–73.

8. Lienert, T. J., Jata, K. V., Wheeler, R., and Seetharaman, V. 2001. In *Proceedings of the Joining of Advanced and Specialty Materials III*, 160–166. Materials Park, OH: ASM International.

9. Park, S. H. C., Sato, Y. S., and Kokawa, H. 2003. Effect of micro-texture on fracture location in friction stir weld of Mg alloy AZ61 during tensile test. *Scripta Materialia* 49 (2): 161–66.

10. National Nuclear Security Administration (NNSA) website, http://www.nnsa.doe.gov/nuclearnp.htm#1 (Accessed on November 25, 2008).

11. Snelgrove, J. L., Hofman, G. L., Trybus, C. L., and Wiencek, T. C. 1996. Development of very-high-density fuels by the RERTR program. *19th International Meeting on Reduced Enrichment for Research and Test Reactors*. Seoul, Korea. October 7–10.

12. Burkes, D. E., Hallinan, N. P., Shropshire, K. L., and Wells, P. B. 2008. Effects of applied load on 6061-T6 aluminum joined employing a novel friction bonding process. *Metallurgical and Materials Transactions A* 39A: 2852–61.

13. Ohta, A., Suzuki, N., Maeda, Y., Hiraoka, K., and Nakamura, T. 1999. Superior fatigue crack growth properties in newly developed weld metal. *International Journal of Fatigue* 21: S113–8.

14. Chen, C. M., and Kovacevic, R. 2006. Parametric finite element analysis of stress evolution during friction stir welding. *Journal of Engineering Manufacture* 220: 1359–71.

15. Sorensen, C. D., and Stahl, A. L. 2007. Experimental measurements of load distributions on friction stir weld pin tools. *Metallurgical and Materials Transactions B* 38B: 451–9.

16. Staron, P., Kocak, M., and Williams, S. 2002. Residual stresses in friction stir welded Al sheets. *Applied Physics A: Materials Science & Processing* 74: S1161–2.

17. Thomas, W. M., Staines, D. J., Watts, E. R., and Norris, I. M. 2005. *The simultaneous use of two or more friction stir welding tools*. Cambridge: TWI Ltd. Available on-line at http://www.twi.co.uk/content/spwmtjan2005.pdf (Accessed on October 2008).

18. Yurioka, N. and Horii, Y. 2006. Recent developments in repair welding technologies in Japan. *Science and Technology of Welding & Joining* 11 (3): 255–64. http://www.ingentaconnect.com/content/maney/stwj

19. Hill, D. J., Sellman, M. B., and Nazar, M. K. 2007. Idaho National Laboratory/Nuclear power industry strategic plan for light water reactor research and development: An industry-government partnership to address climate change and energy security. Report INL/EXT-07-13543. Idaho Falls, ID: Idaho National Laboratory.

20. Kanne, Jr., W. R., Louthan, Jr., M. R., Rankin, D. T., and Tosten, M. H. 1999. Weld repair of irradiated materials. *Mater. Char.* 43: 203–14.

21. Franco-Ferreira, E. A., and Kanne, Jr., W. R. 1992. Remote reactor repair: Avoidance of He-induced cracking with GMAW. *Welding Journal* 71–2: 43–51.
22. Thomas, W. M., Threadgill, P. L., and Nicholas, E. D. 1999. Feasibility of friction stir welding steel. *Science and Technology of Welding & Joining* 4: 365–72. http://www.ingentaconnect.com/content/maney/stwj (Accessed on October 2008).
23. Cho, J. H., Boyce, D. E., and Dawson, P. R. 2005. Modeling strain hardening and texture evolution in friction stir welding of stainless steel. *Materials Science and Engineering A* 398: 146–63.
24. Konkol, P. J., and Murczek, M. F. 2007. Comparison of friction stir weldments and submerged arc weldments in HSLA-65 steel. *Welding Journal Research Supplement* 86: 187s–95s.
25. Park, S. H. C., Sato, Y. S., Kokawa, H., Okamoto, K., Hirano, S., and Inagaki, M. 2003. Rapid formation of the sigma phase in 304 stainless steel during friction stir welding. *Scripta Materialia* 49 (12): 1175–80.
26. Doherty, R. D., Hughes, D. A., Humphreys, F. J., Jonas, J. J., Jensen, D. J., Kassner, M. E., King, W. E., McNelley, T. R., McQueen, H. J., and Rollett, A. D. 1997. Current issues in recrystallization: A review. *Materials Science and Engineering A*. 238 (2): 219–74.
27. Sanders, D. G., Ramulu, M., Klock-McCook, E. J., Edwards, P. D., Reynolds, A. P., and Trapp, T. 2008. Characterization of superplastically formed friction stir weld in titanium 6Al-4V: Preliminary results. *Journal of Materials Engineering and Performance* 17: 187–92.
28. Burris, L., Steunenberg, R. K., and Miller, W. E. 1987. The application of electrorefining for recovery and purification of fuel discharged from the integral fast reactor. AICHE Symposium Series No. 254, 83: 135.
29. Chang, Y. I. 1989. The integral fast reactor. *Nuclear Technology* 88: 129–38.
30. Swedish Nuclear Fuel and Waste Management Co. 1998. *RD&D Programme 98.* Stockholm, Sweden: SKB.
31. Werme, L. 1998. *Design premises for canister for spent nuclear fuel.* Technical Report TR-98-08. Stockholm, Sweden: SKB.
32. Andersson, C.-G. 1998. *Test manufacturing of copper canisters with cast inserts.* Technical Report TR-98-09. Stockholm, Sweden: SKB.
33. Andersson, C.-G., and Andrews, R. E. 1999. Fabrication of containment canisters for nuclear waste by friction stir welding. *First International Symposium on Friction Stir Welding,* Thousand Oaks, CA.
34. Andersson, C.-G., and Andrews, R. E. 2001. *The development of advanced welding techniques for sealing nuclear waste canisters.* Technical Report TR-01-25. Stockholm, Sweden: SKB.
35. Bolz, R. W. 1958. Metals engineering processes. *ASME Handbook.* New York: McGraw-Hill.
36. Andersson, C.-G., Andres, R. E., Dance, B. G. I., Russell, M. J., Olden, E. J., and Sanderson, R. M. 2000. A comparison of copper canister fabrication by the electron beam and friction stir processes. *Second International Symposium on Friction Stir Welding,* Gothenburg, Sweden.
37. Densimet W + MS Metallwerk Plansee GmbH, Austria. Trade Brochure 552 DEF 8.85.

38. Burkes, D. E., Hallinan, N. P., and Clark, C. R. 2008. Nuclear fuel plate fabrication employing a friction bonding process. *Welding Journal* 87 (9): 47–54.
39. Swedish Nuclear Fuel and Waste Management Co. 2008. *RD&D Programme 2007*. Stockholm, Sweden: SKB.
40. Srivastava, S. P., Unni, T. G., Pandarkar, S. P., Mahajan, K., and Suthar, R. L. 2007. Conventional radiography: A few challenging applications. *Founder's Day* 285: 174–82.

8

Friction Stir Spot Welding

Adrian P. Gerlich and Thomas H. North

CONTENTS

Introduction to Friction Stir Welding

The friction stir welding process was developed in 1991 by The Welding Institute (TWI) for joining aluminum alloys (Thomas 1991) and subsequently the process has been used during joining of titanium, magnesium, zinc and copper alloys, and steel, in thicknesses ranging from 1 to 50 mm (Thomas, Threadgill, and Nicholas 1999; Nagasawa et al. 2000; Fukuda 2001; Cederqvist and Andrews 2004; Reynolds, Hood, and Tang 2005). The process involves plunging a rotating tool consisting of a cylindrical shoulder and protruding pin and traversing the tool across two pieces to be joined. Figure 8.1 shows a schematic of the friction stir welding process employed in producing a butt-welded section.

It has been suggested that a combination of frictional heating, plastic deformation, and viscous dissipation produces temperatures in the stir zone that approach the solidus temperature. The stir zone region of aluminum alloy friction stir seam welds comprises a fine, equiaxed grain microstructure. A thermomechanically affected zone (TMAZ) formed adjacent to the stir zone contains elongated base metal grains and a partially recrystallized grain structure as a result of the thermal and stress/strain cycle, which is

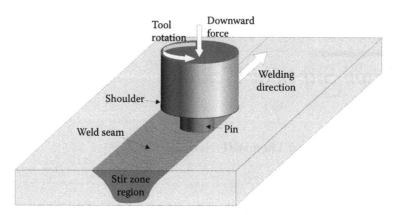

FIGURE 8.1
Schematic of friction stir seam welding.

applied during the welding operation. The travel and rotational speeds of the tool influence the joint microstructure and mechanical properties since the selection of welding parameters determines the energy input during the joining operation.

Overview of Friction Stir Spot Welding

Friction stir spot welding is a variant friction stir seam welding in which two overlapping sheets are joined without traversing the tool (Sakano et al. 2001), as shown in Figure 8.2. During the plunge stage, material in contact with the tool is heated and plasticized. Once the joining operation has been completed, the tool is withdrawn leaving a keyhole depression created by the rotating pin. The welding operation is completed in 1–5 seconds and involves very rapid heating and cooling rates. A bond is created between the overlapping sheets in the stir zone region that has been plasticized and consolidated by the tool.

The majority of published work has referred to the use of tools made from die steel (AISI-H13) when joining aluminum alloys, however, some investigators have employed tungsten-based and nickel-based superalloy tools. Ceramic coatings such as TiN are also often mentioned in conjunction with steel tools; however, no studies have been published regarding the wear performance of coated steel tools. The tool pin typically has one or a combination of threads, spiral terraces, flat sides, or flutes, which promote mixing of the upper and lower sheet materials during the welding operation (Gerlich et al. 2008a). The geometry of the tool shoulder also influences the microstructure and mechanical properties of completed joints (Pan 2008). However, the roles that tool shape and length play are only beginning to be

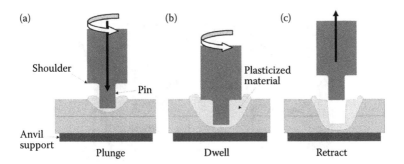

FIGURE 8.2
Schematic of the friction stir spot welding process.

understood (Badarinarayan 2008). When steel is spot-welded, polycrystalline cubic boron nitride (PCBN) is the favored tool material, since it has superior wear resistance properties compared to alternatives such as W-25Re alloy (Mishra and Mahoney 2007).

The process emerged from the need to reliably spot weld the aluminum sheet materials, since the competing resistance spot welding process is limited to between 400 and 900 welds before failure due to degradation of the copper electrode contact tips (Zhou et al. 2004). Self-piercing rivets have been employed by the automotive industry to handle this particular problem. However, consumable rivets must be driven into the overlapping sheets and this increases the cost and reduces the recyclability of the structure. Although self-piercing rivets have been widely used by Ford, for example, during the manufacture of Jaguar cars using several aluminum body designs, friction stir spot welding is being considered as a possible alternative to self-piercing rivets and clinching (Mortimer 2005).

Friction stir spot welding was first used extensively by Mazda to manufacture hoods and door panels for their aluminum alloy body RX-8 sports car (Hancock 2004) and over 100,000 of these vehicles have been produced since 2003. The joining process has also been employed by Toyota since 2004 for assembling the rear hatch of Prius hybrid vehicles. The use of aluminum instead of steel in this application produced a 6 kg weight reduction (Yamaguchi 2004).

Reduced energy consumption and capital costs of approximately 85% and 50% have been reported when resistance spot welding is replaced by friction stir spot welding (Sakano et al. 2001). High current electrical transformers, water cooling, and pressurized air supplies are not required. In addition, the problem of current shunting often encountered during resistance spot welding is avoided. This eliminates the requirement for a minimum spacing between individual spot welds.

A number of manufacturers have developed friction stir spot welding systems using C-frame designs that can be fitted to multiaxis robots, see Figure 8.3. These systems generally employ two computer-controlled servo

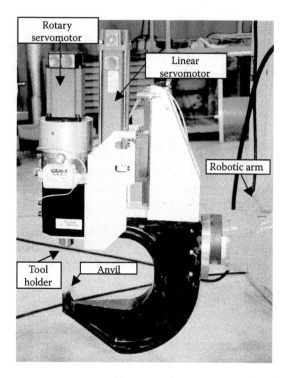

FIGURE 8.3
Detail of a friction stir spot welding gun design. (Courtesy of C. Smith, Friction Stir Link Inc.)

motors, a rotary motor that rotates the tool and provides torque, and a linear motor, which moves the tool downward and applies the axial force.

The spot welding cycle operation may be controlled either by using a displacement-controlled system or a force-controlled system. The force-controlled system was the first to be demonstrated, and was largely adapted from the feedback systems used when controlling automated resistance spot welding equipment. The force and displacement of the tool cannot be controlled simultaneously, since the sheet material is deformed and displaced upward during the welding operation (Gerlich, Su, and North 2005). The force-controlled system offers a measure of tolerance in that overloading or damage to the equipment can be avoided in the event that there is unexpected thickness variation in the parts being spot-welded.

In reality, a displacement controlled system is actually incapable of controlling the exact tool location within the sheets due to compliance in the load frame of the welding machine. The relationship between tool force and position when the two operating modes are used is schematically illustrated in Figure 8.4. Once the tool shoulder makes contact with the upper sheet surface, the tool can dwell on the surface for some time to promote bonding between the sheets. The dwell time can vary from zero to several seconds while the tool rotation continues. Both displacement and

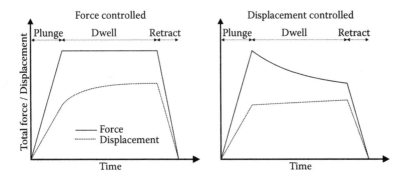

FIGURE 8.4
Comparison of the tool force and displacement outputs during friction stir spot welding using force-controlled and displacement-controlled spot welding systems.

force-controlled spot welding systems produce similar joint strengths when joints are made using the same tool and similar welding parameter settings (Tweedy et al. 2008).

While the conventional plunge and retract spot welding method described above has already been implemented in mass production, two other process variants have been investigated, see Figure 8.5. The stitch welding methodology provides increased joint strength (Okamoto, Hunt, and Hirano 2005), and completed joints can be considered as small overlap seam welds. Variations of this tool traversing methodology have also been developed, where the tool follows in a short circular or octagonal path in order to produce a larger bonded region and higher joint strength (Addison and Robelou 2004; Tweedy et al. 2008). Both the direction and geometry when traversing the tool affect final joint strength properties (Buffa, Fratini, and Piacentini 2008). Since any traversing motion requires at least one additional axis of movement by the welding robot, this increases the complexity and cost of the spot-welding equipment. A simple method of traversing the tool is shown in Figure 8.5b, where the rotating tool is mounted on a pivot and swings via a cam drive system (Hunt, Badarinarayan, and Okamoto 2006).

Over 100,000 friction stir spot welds can be made using the H13 die steel tool before tool wear becomes evident (Hinrichs et al. 2004). Conventional friction stir spot welding leaves a keyhole indentation following tool retraction. A method of refilling the keyhole has been developed, which avoids this cosmetic problem (Schilling and dos Santos 2001), see Figure 8.5c. The keyhole is refilled by allowing the plasticized material displaced by the rotating pin to flow into an annular channel formed when a rotating sleeve around the pin is retracted axially in unison with the penetrating pin. The forces between the pin and rotating sleeve are balanced, and the material is contained by a stationary outer clamp. Once a joint is made between the contacting sheets, the axial movement of the sleeve and pin is reversed, and the plasticized material is extruded downward into the keyhole. Although refilling the

(a)

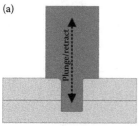

Conventional spot welding

(b)

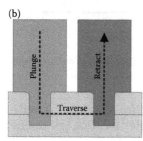
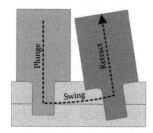

Short traverse or swing

(c)

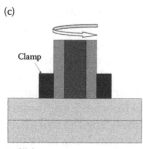
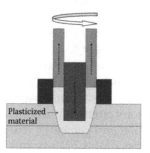
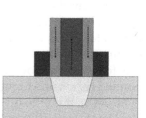

Refill friction stir spot welding

FIGURE 8.5
Schematic showing different friction stir spot welding methods: (a) conventional spot welding, (b) short traverse spot welding, and (c) refill friction stir spot welding.

keyhole takes several seconds, joint strength properties are improved and this joining technique has been suggested as a possible replacement for riveting in aerospace applications where surface quality is important (Allen and Arbegast 2005).

Spot friction weldbonding has also been proposed (Pan et al. 2006) where an adhesive is added between the sheets to increase joint strength and vibration damping. Finally, friction spot welding of an Al alloy to zinc-coated steel has also been demonstrated, where the lower steel sheet is not penetrated by the rotating pin. The temperature and pressure during welding produce a solid-state diffusion bond between the upper aluminum sheet and the zinc coating (Gendo et al. 2007).

Spot Welds in Aluminum Alloys

Friction stir spot welds made in overlapping sheets of aluminum alloy are shown in Figure 8.6. The material displaced by the pin during penetration is ejected as flash, and the bottom sheet surface in contact with the anvil has a smooth appearance. A smooth and polished anvil is needed to produce a smooth lower sheet surface. If desired, this process characteristic can be used to forge a logo into the bottom side of the spot weld (Pan et al. 2004).

Figure 8.7 shows the different regions found in friction stir welded 7075 aluminum alloy. Three microstructural regions are produced when aluminum alloys are spot welded. The heat and deformation imposed by the tool creates a stir zone around the tool periphery comprising fine

FIGURE 8.6
Friction stir spot welded aluminum alloy sheets.

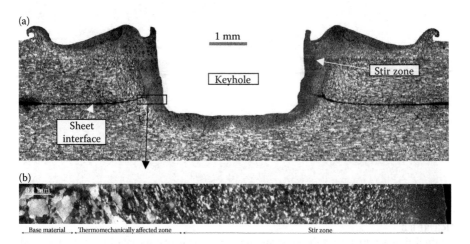

FIGURE 8.7
Polarized light optical micrographs showing the microstructural features found in friction stir spot welded 7075 aluminum alloy. (a) Transverse section showing the keyhole, stir zone, and the interface between the contacting sheets. (b) Microstructures in the stir zone, thermomechanically affected zone, and base material.

equiaxed grains resulting from dynamic recrystallization. The TMAZ is produced at the extremity of the stir zone and contains grains, which although heated and severely deformed, are not completely recrystallized, see Figure 8.7b.

When the pin plunges into the contacting sheets, the material below the pin is compressed, and extruded upward around the pin periphery. Material flow during friction stir spot welding has been investigated by incorporating tracer particles and by examining dissimilar aluminum alloy spot welds (Su et al. 2005b). Bonding between the contacting sheets occurs prior to penetration into the lower sheet since a layer of dynamically recrystallized material forms beneath the tip of the rotating pin. Once the rotating pin begins to penetrate into the lower sheet, a bond is established across the diameter of the tool pin and intermixing between the upper and lower sheets begins, see Figure 8.8.

The tool shoulder contacts material expelled during pin penetration and applies compressive loading, which promotes joint formation. Both tool geometry and welding parameter selection have a drastic influence on the stir zone microstructure. For example, incorporating a dwell time within the

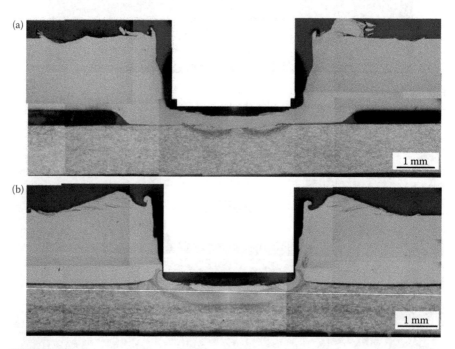

FIGURE 8.8
Controlled penetration tests showing: (a) initial bonding between contacting Al 5754 and Al 6111 sheets and (b) formation of an Al 5754 layer immediately below the keyhole and a stir zone comprising Al 5754 and Al 6111 material. Al 5754 is the upper sheet and Al 6111 is the lower sheet in the dissimilar sandwich. (After Su, P., Gerlich, A., North, T. H., and Bendzsak, G. J., *Sci. Technol. Weld. Join.*, 11(1), 61–71, 2005. With permission.)

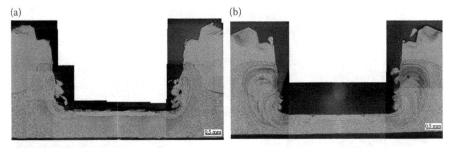

FIGURE 8.9
Stir zone profiles in dissimilar Al 5754/Al 6111 spot welds made with and without a dwell time: (a) when no dwell time is applied and (b) when a dwell time of 2 seconds is applied. The rotational speed of the tool was 2500 rpm and the plunge rate was 2.5 mm/s. The upper sheet is Al 5745 and the lower sheet is Al 6111. (After Su, P., Gerlich, A., North, T. H., and Bendzsak, G. J., *SAE Technical Series*: 2006-01-0971, 2006. With permission.)

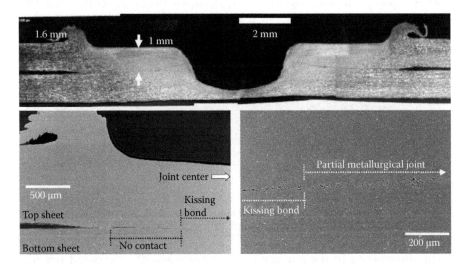

FIGURE 8.10
(a) Optical micrograph showing the microstructural features in a friction stir spot weld made in Al 6111 aluminum alloy sheets. (b) SEM micrograph showing the sheet interface under tool shoulder (see region A). (c) SEM micrograph showing partial metallurgical bonding in region B. (After Mitlin, D., Radmilovic, V., Pan, T., Chen, J., Feng, Z., and Santella, M. L., *Mater. Sci. Eng. A*, 441, 79–96, 2006. With permission.)

welding cycle dramatically increases the dimensions of the stir zone region formed adjacent to the periphery of the rotating pin, see Figure 8.9.

Mitlin et al. (2006) noted the formation of a region adjacent to the stir zone extremity where partial bonding occurred in Al 6111 alloy friction stir spot welds, see Figure 8.10. Compressive loading by the tool shoulder produced a hot-pressure bond around the periphery of the stir zone.

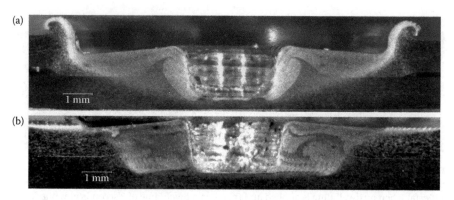

FIGURE 8.11
Optical micrographs showing the microstructure features of dissimilar friction stir spot welds where Al 7075 is the upper sheet and Al 2024 is the lower sheet. (a) A conventional spot weld, and (b) a short traverse spot weld made using an octagonal traversing path. (After Tweedy, B., Widener, C., Merry, J., Brown, J., and Burford, D., *SAE Technical Series*: 2008-01-1135, 2008. With permission.)

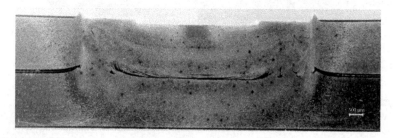

FIGURE 8.12
Optical micrograph showing the microstructure of a refill friction stir spot weld made in Alclad 2024 alloy sheet. (After Tier, M. D., dos Santos, J., Rosendo, T., Mazzaferro, J. A., and da Silva, A. A. M., *SAE Technical Series*: 2008-01-2287, 2008. With permission.)

The dimensions of the stir zone region are markedly increased when a short traversing motion is incorporated in the spot welding cycle (Tweedy et al. 2008). The stir zone dimensions are uniform around the periphery of the keyhole during conventional friction stir spot welding, see Figure 8.11a. When a single traverse direction is applied, the bonded region adjacent to the keyhole is similar to that in a conventional spot weld and may result in premature failure. For this reason, two dimensional traversing paths have been investigated where the rotating tool returns to the center and produces a bonded region, which has a more uniform and symmetrical bonded region adjacent to the keyhole periphery, see Figure 8.11b.

The microstructural features produced in spot welds made using refill friction stir spot welding are shown in Figure 8.12. Following filling of the keyhole, the locations of the rotating pin and sleeve are apparent from impressions produced on the upper surface of the welded joint. The bonded area between the two sheets is much larger than that formed by a conventional spot welding operation where the pin creates a central keyhole. Since Alclad 2024 base material has a corrosion resistant coating of pure aluminum on its surface, this layer serves as a tracer, which indicates material flow during the joining operation. The layer of aluminum apparent at the base of the weld in Figure 8.13 results from the formation of a dynamically quiescent region beneath the tip of the rotating pin (Bendzsak, North, and Li 1997; Su et al. 2005b).

Dissolution of Precipitates and Microhardness

Heat-treatable Al alloys contain second-phase particles, which are precipitated at elevated temperatures during ageing. This increases the yield strength of the Al alloys since dislocations are forced to bypass and cut through particles during plastic deformation (Orowan 1947). Following friction stir welding the heat affected zone (HAZ) located between the TMAZ region and the as-received base material has decreased yield strength, since second-phase particles are coarsened (Sato et al. 1999; Lippold and Ditzel 2003). In this connection, the HAZ region cannot be readily observed using optical microscopy and microhardness testing is required to identify this region.

The particles that are precipitated during ageing are not stable above a critical temperature, and they dissolve in the stir zone during friction stir spot welding. For example, based on the stir zone temperatures measured during spot welding, the majority of the $MgZn_2$ precipitates in the Al 7075 are dissolved, while the average dimensions of Al_2CuMg precipitates in Al 2024 are

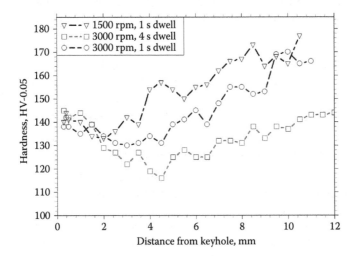

FIGURE 8.13
Influence of tool rotational speed and dwell time on the width of the softened region produced during friction stir spot welding of Al 7075-T6 base material using a plunge rate of 2.5 mm/s. (After Gerlich, A., Yamamoto, M., Shibayanagi, T., and North, T. H., *SAE Technical Series*: 2008-01-0146, 2008. With permission.)

significantly reduced (Gerlich, Cingara, and North 2006; Gerlich et al. 2007a). Although reprecipitation occurs during cooling following spot welding and at room temperature (Rhodes et al. 1997; Su et al. 2003), post-weld heat treatment can be used to reestablish a fully aged microstructure in the stir zone (Heinz and Skrotzki 2002).

The hardness increases when reprecipitation occurs within the stir zone and a microhardness profile such as that in Figure 8.13 is produced. The hardness increases in the location closest to the keyhole, and a minimum is observed outside of the stir zone. Fujimoto et al. and Wang and Lee have also shown that the minimum in hardness occurs at the location 5–10 mm from the extremity of the stir zone in friction stir spot welds made between overlapping sheets of Al 6061 alloy (Wang and Lee 2007; Fujimoto et al. 2008). Although a model has been developed predicting reprecipitation in Al-alloy friction stir seam welds (Kamp et al. 2006) one is not available for friction stir spot welding.

Stir Zone Grain Size

It is well known that the strength of an alloy increases with decreasing grain size according to the Hall–Petch relationship (Petch 1953). A characteristic feature of friction stir welding is that the stir zone has a fine equiaxed grain microstructure that is quite different from the as-cast structures typically formed during arc welding or resistance spot welding. Although there is some potential to increase the strength of joints by refining the grain size in the stir zone, the Hall–Petch coefficient for aluminum alloys is low (Embury 1971) and the mechanisms promoting grain refinement are a subject of fierce debate. It is well known that the grain size produced during deformation of aluminum alloys is a function of the strain rate and temperature (McQueen and Ryan 2002). Finally, although very fine stir zone grain sizes can be achieved during friction stir spot welding of Al alloys (Gerlich, Yamamoto, and North 2008b), the potential strengthening resulting from grain size refinement may be limited since grain growth can occur following tool withdrawal from the workpiece (Gerlich, Yamamoto, and North 2007b).

Mechanical Properties of Friction Stir Spot Welds

The mechanical properties of aluminum alloy spot welds have been examined during both overlap shear testing and cross-tension testing methods (see Figure 8.14). Overlap shear testing imposes shear loading at the joint interface, and the tendency for the sheets to peel apart is reduced by incorporating alignment spacers at the gripped ends of the test specimen. Cross-tension fracture testing applies a tensile load normal to the spot weld using a fixture to grip the ends in a mechanical testing machine. In addition to these techniques, peel testing is frequently applied during qualification testing of resistance spot welds where nugget pull-out failure is the key objective (see SAE Standard AMSW6858A). Since conventional friction stir spot welds have

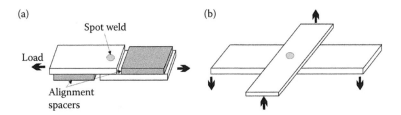

FIGURE 8.14
(a) Overlap shear and (b) cross-tension fracture test specimen configurations.

a keyhole indentation centerline of the welded joint, peel tests conducted to examine nugget pull-out are not meaningful, although some investigators have employed the forces applied during peel testing as the indicator of joint strength (Blundell et al. 2006).

It is difficult to compare the failure properties of friction stir spot welds and resistance spot welds since the size and geometry of the weld zones differ markedly. Since no published specification regarding the recommended joint strengths of friction stir spot welds exist at the present time, some investigators have referred to the Society of Automotive Engineers (SAE Standard AMSW6858A), the American Welding Society (AWS D8.1M), or Japanese Industrial Standard (JIS Z 3140) standards for resistance spot welds. These standards take into account the sheet thickness and base material strength, and recommend a minimum failure load during mechanical testing. It has been confirmed that conventional friction stir spot welds can meet these standards once the optimum welding parameter settings have been established for a particular tool geometry and material thickness (Tran et al. 2007; Arul et al. 2008). For example, the overlap shear fracture strengths of friction stir spot welds made in 1 mm thick Al 6000 series alloy sheets ranges from 2 to 4 kN (Pan 2008). When a short linear traversing movement is incorporated into the welding cycle (Okamoto, Hunt, and Hirano 2005) the weld strength can be increased by more than 30%. Application of a two dimensional traversing path (Tweedy et al. 2008) or the use of refill friction stir spot welding also produces higher joint strength properties than conventional friction stir spot welding (Allen and Arbegast 2005).

Sakano et al. (2001) examined the relationship between processing parameters and the tensile shear fracture loads in spot welds made in aluminum alloy Al 6111 base material. The tensile failure loads exceeded the mechanical properties of resistance spot welded joints having comparable weld diameters when a high welding force in combination with high tool rotation speed and short welding time settings was applied. Joint strength was degraded when the tool rotation speed increased beyond a critical value. In contrast, the highest joint strength in spot welds made in Al 5052 alloy were attained when lower tool rotation speeds were used (Freeney et al. 2006) and increasing the cycle time leads to improved mechanical strength in short traverse spot welds (Hunt, Badarinarayan, and Okamoto 2006). These inconsistent

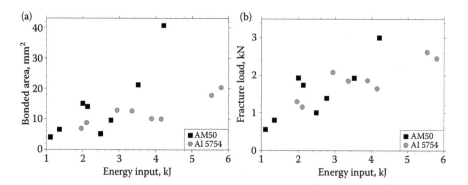

FIGURE 8.15
Relation between energy input and (a) bonded area, (b) overlap shear failure load in friction stir spot welds made in 1.5 mm thick Al 5754 and AM50 alloy sheets. (After Su, P., Gerlich, A., North, T. H., and Bendzsak, G. J., *SAE Technical Series*: 2006-01-0971, 2006. With permission.)

correlations between processing conditions and mechanical strength support the notion that there are many interacting parameters controlling joint strength (Su et al. 2006a).

The failure load properties during overlap shear testing are determined by the energy input during friction stir spot welding and the bonded areas found when examining broken mechanical test samples. If one measures the force and torque at the tool in a series of discrete samples or steps, the energy input during friction stir spot welding can be expressed as (Su, Gerlich, and North 2005a):

$$Q_{applied} = \sum_{n=1}^{n=N} Force(n)\ (x_n - x_{n-1}) + \sum_{n=1}^{n=N} Torque(n)\ \omega(n)\ \Delta t, \qquad (8.1)$$

where x_n is the penetration depth at sample (n), ω is the angular velocity (rad/s), n is the sample number, N is the final sample, and Δt is the sampling time. A strong positive correlation was found between the energy input, the bonded area, and fracture load that are produced during friction stir spot welding of both Al-alloy and Mg-alloy spot welds (see Figure 8.15). On this basis, ultrasonic nondestructive testing may be used to determine the bonded area between the sheets in order to provide quality control (Pan 2008). The fracture load during spot welding has also been related to the peak temperature produced during the welding operation (Arul et al. 2008), which is expected since almost all of the energy input during welding is converted to heat.

Using calorimetry, Su et al. (2006a, 2006b) confirmed that only 4% of the total mechanical energy generated during tool rotation was required for stir zone formation with the remainder of the energy dissipated into the sheets being welded, the anvil supports, tool assembly, and the atmosphere. In

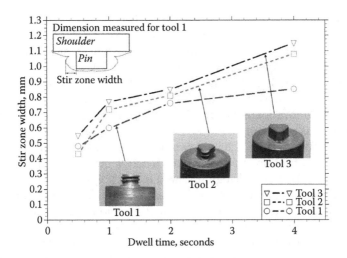

FIGURE 8.16

Stir zone width changes when the dwell time is varied during friction stir spot welding of AM60 sheet using different tool geometries. During spot welding using a tool rotation speed of 3000 rpm. (After Gerlich, A., Yamamoto, M., Shibayanagi, T., and North, T. H., *SAE Technical Series*: 2008-01-0146, 2008. With permission.)

addition, >99% of the energy generated results from the torque (Su, Gerlich, North 2005a). The estimated power density during friction stir spot welding was 10^{10} W/m^3 based on the stir zone volume measured from metallographic cross-sections and the total energy input during welding. Su et al. (2006a, 2006b) also found that the tool shoulder accounted for 48–65% of the total heat input during friction stir spot welding with the remainder contributed by the rotating pin. These values were much less than the tool shoulder energy contributions that have been estimated (83%) during the examination of friction stir seam welds (Schmidt and Hattel 2008).

The failure load during mechanical testing is directly related to the energy input during friction stir spot welding, since the stir zone dimensions and bonded area increase when the energy input increases. Although incorporating threads on the rotating tool has a minimal influence on the energy generated during friction stir spot welding (Su et al. 2006a), tool geometry does influence stir zone dimensions when a dwell time is applied (Gerlich et al. 2008a). For example, incorporating flat sides on the rotating pin significantly increased the stir zone width, see Figure 8.16.

Fracture Modes and Fatigue

Although there is a strong positive correlation between energy input, bonded area, weld temperature, and failure load properties, there is considerable scatter in plotted results. This occurs since the fracture mode changes with processing parameters. Two primary failure modes are observed during the overlap shear testing of conventional friction stir spot welds as follows (Lathabai et al. 2005; Tozaki, Uematsu, and Tokaji 2006): (i) shearing across the stir zone material and (ii) circumferential failure around the stir zone through the upper or lower sheet (Lin et al. 2004; Arul et al. 2005), see Figure 8.17. Arul et al. (2008) have suggested that the fracture areas resulting from both failure modes should be taken into account, and indicated that the highest fracture loads occur when an optimum ratio between the two fracture areas is achieved.

Lin, Pan, and Pan (2005a, 2005b) demonstrated that the overlap shear fracture strength of friction stir spot welds depends markedly on the geometry of the tool and the bonded region. There is much variability in the fracture strength properties of the spot welds made using different welding parameter settings, since this affects the stir zone dimensions, thinning of the upper sheet, and the formation of discontinuities in the joint (Su et al. 2006a). Lin et al. (2004) examined friction stir spot welded joints produced in Al 6111 alloy and proposed that the failure load depended on the geometry of the bonded region as well as on thinning of the upper sheet. Both vary considerably with processing parameters.

The orientation of the unbonded interface between the contacting sheets markedly influences the fracture load during overlap shear testing, see Figure 8.17. This region is often referred to as a hook (Okamoto, Hunt, and Hirano 2005). For example, Figure 8.18 shows transverse cross-sections through friction stir spot welds made in Al 6111 alloy sheets. When the tool rotational speed increased from 1500 to 2250 rpm it was found that the unbonded region was oriented away from the stir zone (Arul et al. 2008). Finally, when fracture occurs through the upper sheet, the use of a concave shoulder will promote higher fracture loads by increasing the thickness of material between the unbonded region and the upper sheet surface (Arul et al. 2005).

The fatigue properties during tensile shear testing of friction stir spot welded Al 6111 alloy were comparable with those of resistance spot welded joints (Lin, Pan, and Pan 2006). The influence of tool shoulder geometry on the fatigue lives and fracture modes of Al 6111 friction stir spot welds have also been studied extensively by Lin, Pan, and Pan (2005a, 2005b, 2006). The fatigue

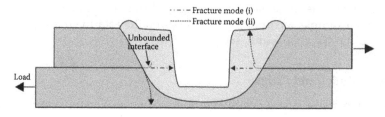

FIGURE 8.17
Schematic showing the different failure modes found during overlap shear testing of friction stir spot welds.

(a) (b)

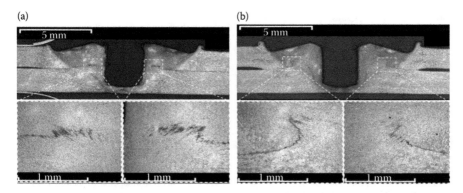

FIGURE 8.18
Optical micrographs shows transverse sections through friction stir spot welds made in Al 6111 alloy sheets. The unbonded regions formed at the extremity of the stir zone are indicated in spot welds made using tool rotation speeds of: (a) 1500 and (b) 2250 rpm. (After Arul, S. G., Miller, S. F., Kruger, G. H., Pan, T.-Y., Mallick, P. K., and Shih, A. J., *Sci. Technol. Weld. Joining*, 13(7), 629–37, 2008. With permission.)

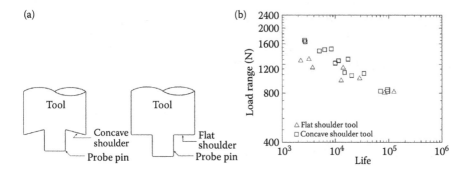

FIGURE 8.19
(a) Comparison of concave versus flat tool shoulder geometries when investigating the fatigue strength of friction stir spot welds made in an Al 6111 alloy sheet. (b) Fatigue strength properties produced using both tool shoulder designs. (After Lin, P.-C., Pan, J., and Pan, T., *International Journal of Fatigue*, 30, 74–89, 2008; Lin, P.-C., Pan, J., and Pan, T., *Int. J. Fatigue*, 30, 90–105, 2008. With permission.)

lives of Al 6111 spot welds are shown in Figure 8.19. The presence of a concave shoulder on the tool produced joints having superior low cycle fatigue performance. However, the high cycle fatigue properties were similar for both concave and flat shoulder geometries (Lin, Pan, and Pan 2008a, 2008b).

The fracture path followed during cyclic loading takes different paths depending on the loading that is applied (Lin, Pan, and Pan 2008a, 2008b), see Figure 8.20. When low loads are applied, the fracture in high-cycle fatigue tests propagated through the bottom of the sheet near the stir zone, and through the upper sheet beyond the periphery of the impression made

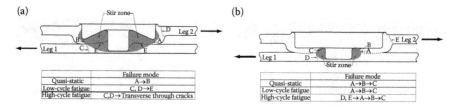

FIGURE 8.20

Comparison of fracture modes produced during fatigue testing of friction stir spot welds made in an Al 6111 alloy sheet: (a) using a concave tool shoulder and (b) using a flat tool shoulder geometry. The loading direction on Leg 1 and 2 is indicated by the arrows. (After Lin, P.-C., Pan, J., and Pan, T., *International Journal of Fatigue*, 30, 74–89, 2008; Lin, P.-C., Pan, J., and Pan, T., *Int. J. Fatigue*, 30, 90–105, 2008. With permission.)

by the tool shoulder. When high loads were applied, the low-cycle fatigue fracture propagated through the stir zone in spot welds were made using a tool with a concave shoulder, and through the thinned region of the upper sheet in spot welds were made using the tool having the flat shoulder (see Figure 8.20). These test results emphasize that a number of factors determine the fracture mode in spot welded joints made between contacting sheets. The geometry of the unbonded regions is particularly important, since this affects the stress concentration during loading. Methods for characterizing these factors based on the geometries observed in spot-welded joints have been suggested (Lin, Pan, and Pan 2008a).

Friction Stir Spot Welding of Magnesium Alloys

Arc welding of Mg-alloys is problematic since fusion welded deposits are prone to hydrogen porosity formation and liquation cracking; preheating temperatures as high as 400°C are also required in highly restrained joints (Avedesian and Baker 1999; Davis 2002). Acceptable joint strengths can be achieved during resistance spot welding of AZ31 Mg-alloy, even though liquation cracking may be observed (Sun et al. 2007). Liquation cracking has also been found in the TMAZ regions of AZ31, AM60, and AZ91 friction stir spot welds while liquid penetration induced (LPI) cracking has been found in the stir zone close to the extremity in AZ91 friction stir spot welds (Yamamoto et al. 2007a). Although liquation cracking may occur in spot welded Mg-alloys, it may be limited and does not necessarily have a severe influence on the fracture strength that can be achieved since Figures 8.15 and 8.21 indicate that 3 kN and 2 kN loads are obtained in AM50 and AZ31 magnesium alloy spot welds.

In the case of AZ91 friction stir spot welding, much of the stir zone may be removed when the rotating tool is extracted at the termination of the

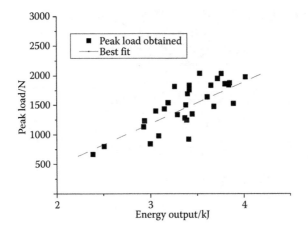

FIGURE 8.21
Relation between energy input overlap shear failure load in friction stir spot welds made in 1.5 mm thick AZ31 alloy sheets. (After North, T. H., Unpublished research, Department of Materials Science and Engineering, Toronto, Ontario, Canada: University of Toronto, 2009. With permission.)

friction stir spot welding operation (Yamamoto et al. 2007b). This has been attributed to liquid penetration induced cracking, and is characterized by the following steps: (i) eutectic melting in the TMAZ region, (ii) penetration of melted films from the TMAZ into the stir zone, and (iii) crack propagation along the grain boundary due to the presence of liquid films and the application of shear loads imposed by the rotating tool during the welding process (Su et al. 2007). Yamamoto et al. (2007b) suggested that rapid heating during the tool penetration stage in spot welding created undissolved $Mg_{17}Al_{12}$ particles and facilitated eutectic melting when the stir zone temperature reached 437°C, the (α-Mg + $Mg_{17}Al_{12}$) eutectic melting temperature in the binary Mg-Al equilibrium phase diagram. During AZ31 and AM60 friction stir spot welding cracking is only apparent very early in the dwell period and crack-free joints are produced when the dwell time is extended to 4 seconds (Yamamoto et al. 2007a).

Friction Stir Spot Welding of Steels

The demand for weight reduction in the automotive industry to improve fuel economy and roll-over standards has led to the increased use of advanced high strength steels having yield strengths exceeding 1000 MPa. Traditional approaches when welding these higher strength steels are

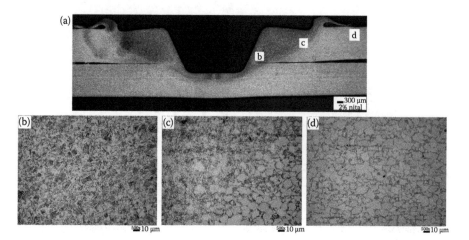

FIGURE 8.22
Optical micrographs showing: (a) the transverse section through friction stir spot welds made in DP600 sheets, and detailed microstructures in the (b) stir zone, (c) TMAZ, and (d) base material. (After Feng, Z., Santella, M. L., David, S. A., Steel, R. J., Packer, S. M., and Pan, T., *SAE Technical Series*: 2005-01-1248, 2005. With permission.)

remarkably challenging in order to avoid cracking or embrittlement (Gould and Peterson, 2005). Friction stir spot welds made in steels have mechanical strengths comparable to those produced using resistance spot welding (Feng et al. 2005; Khan et al. 2007). Friction stir spot welding of steel sheets is readily accomplished using PCBN, Si_3Ni_4, or tungsten–rhenium alloy tool materials since these materials retain their strength at high temperature, and allow high axial loads to be applied without exhibiting serious tool wear problems. For example, it has been reported that PCBN ceramic tools survive for at least 100 spot welds during friction stir spot welding of dual phase and martensitic steels without noticeable wear problems (Feng et al. 2005).

It has been shown that friction stir spot welding of advanced high strength steels can be successfully accomplished in dual phase DP600 (Feng et al. 2005; Khan et al. 2007), DP800 (Kyffin et al. 2006; Miles et al. 2007), DP980 (Miles et al. 2007) grades, as well as in martensitic hot-stamped boron-containing steels (HSBS; Hovanski, Santella, and Grant 2007) and M190 grade steels (Feng et al. 2005). The microstructures produced in friction stir spot weld DP600 sheets consist of bainite and acicular ferrite, while the base material consists of ferrite plus islands of martensite, see Figure 8.22. Friction stir spot welding of zinc-coated and mild steels is also feasible, and produces acceptable joint strengths (Pan 2008). Dissimilar spot welding of steel to aluminum alloy is also possible, and joints typically fracture through the aluminum sheet and contain Fe–Al type intermetallics at the interface (Figner et al. 2009).

Conclusions

Considerable progress has been made in terms of understanding the friction stir spot welding process. The welding process has been implemented in the automotive and high-speed rail industries. Niche applications in these industries have arisen from the need for an alternative joining methodology to resistance spot welding in the case of aluminum alloys. The potential for joining magnesium alloys, steel, and dissimilar combinations have also been investigated. The mechanical properties of friction stir spot welds are similar to those of resistance spot welds and the joining process can provide reduced costs. However, a number of key issues remain with regard to improving the cycle time during spot welding and the cosmetic appearance of the joints. In addition, a number of other joint applications have yet to be investigated in other alloy systems including stainless steels, nickel based alloys or super-alloys, titanium alloys, and other dissimilar combinations. A great deal of work also remains to be done vis-à-vis the corrosion properties of steel and Mg-alloy friction stir spot welded joints.

References

Addison, A. C., and Robelou, A. J. 2004. Friction stir spot welding: Principal parameters and their effects. *Proceedings of the 5th International Symposium on friction stir welding*. Metz, France: The Welding Institute.

Allen, C. D., and Arbegast, J. A., 2005. Evaluation of friction stir spot welds in aluminum alloys. *SAE Technical Series*: 2005-01-1252.

Arul, S. G., Miller, S. F., Kruger, G. H., Pan, T.-Y., Mallick, P. K., and Shih, A. J. 2008. Experimental study of joint performance in spot friction welding of 6111-T4 aluminium alloy. *Science and Technology of Welding and Joining* 13 (7): 629–37.

Arul, S., Pan, T., Lin, P., Pan, J., Feng, Z., and Santella, M. 2005. Microstructures and failure mechanisms of spot friction welds in lap-shear specimens of aluminum 5754 sheets. *SAE Technical Series*: 2005-01-1256.

Avedesian, M., and Baker, H. 1999. *Magnesium and magnesium alloys*, 106–18. Materials Park, OH: ASM International.

Badarinarayan, H. 2008. Effect of tool geometry and pin length on failure mode and static strength of friction stir spot welds. *SAE Technical Series*: 2008-01-0147.

Bendzsak, G. J., North, T. H., and Li, Z. 1997. Numerical model for steady-state flow in friction welding. *Acta Metallurgica* 45 (4): 1735–45.

Blundell, N., Han, L., Hewitt, R., and Young, K. 2006. The influence of pain bake cycles on the mechanical properties of spot friction welded aluminum alloys. *SAE Technical Series*: 2006-01-0968.

Buffa, G., Fratini, L., and Piacentini, M. 2008. On the influence of tool path in friction stir spot welding of aluminum alloys. *Journal of Materials Processing Technology* 208: 309–17.

Cederqvist, L., and Andrews, R. E., 2004. Friction stir welding of copper canisters—A weld that lasts for 100,000 years. *Materials Solutions '03: Joining of Advanced and Specialty Materials VI*: 35–39.

Davis, J. R. 2002. *Metals handbook*. 9th ed., Vol. 6, 429–35. Materials Park, OH: ASM International.

Embury, J. D. 1971. *Strengthening methods in crystals*. eds. A. Kelly and R. B. Nicholson. New York: Wiley.

Feng, Z., Santella, M. L., David, S. A., Steel, R. J., Packer, S. M., and Pan, T. 2005. Friction stir spot welding of advanced high-strength steels—A feasibility study. *SAE Technical Series*: 2005-01-1248.

Figner, G., Vallant, R., Weinberger, T., Schröttner, H., Pasic, H., and Enzinger, N. 2009. Friction stir spot welds between aluminium and steel automotive sheets: Influence of welding parameters on mechanical properties and microstructure. *Welding in the World* 53 (1/2): R13—23.

Freeney, T., Sharma, S., and Mishra, R. 2006. Effect of welding parameters on properties of 5052 Al friction stir spot welds. *SAE Technical Series*: 2006-01-0969.

Fujimoto, M., Koga, S., Abe, N., Sato, Y. S., and Kokawa, H. 2008. Microstructural analysis of stir zone of Al alloy produced by friction stir spot welding. *Science and Technology of Welding and Joining* 13 (7): 663–70.

Fukuda, T. 2001. Friction stir welding (FSW) process. *Welding International* 15 (8): 611–5.

Gendo, T., Nishiguchi, K., Asakawa, M., and Tanioka, S. 2007. Spot friction welding of aluminum to steel. *SAE Technical Series*: 2007-01-1703.

Gerlich, A., Cingara, G. A., and North, T. H. 2006. Stir zone microstructure and strain rate during Al 7075-T6 friction stir spot welding. *Metallurgical and Materials Transactions A* 37A: 2773–86.

Gerlich, A., Su, P., and North, T. H. 2005. Tool penetration during friction stir spot welding of Al and Mg alloys. *Journal of Materials Science* 40 (24): 6473–81.

Gerlich, A., Su, P., Yamamoto, M., and North, T. H. 2007a. Effect of welding parameters on the strain rate and microstructure of friction stir spot welded 2024 aluminium alloy. *Journal of Materials Science*: DOI: 10.1007/s10853-006-1103-7.

Gerlich, A., Yamamoto, M., and North, T. H. 2007b. Strain rates and grain growth in Al 5754 and Al 6061 friction stir spot welds. *Metallurgical and Materials Transactions A*: DOI: 10.1007/s11661-007-9155-0.

Gerlich, A., Yamamoto, M., and North, T. H. 2008b. Local melting and tool slippage during friction stir spot welding of Al-alloys. *Journal of Materials Science* 43 (1): 2–11.

Gerlich, A., Yamamoto, M., Shibayanagi, T., and North, T. H. 2008a. Selection of welding parameters during friction stir spot welding. *SAE Technical Series*: 2008-01-0146.

Gould, J., and Peterson, W. 2005. Advanced materials require advanced knowledge. *Fabricator* 35 (8): 34–5.

Hancock, R. 2004. Friction welding of aluminum cuts energy costs by 99%. *Welding Journal* 83 (2): 40.

Heinz, B., and Skrotzki, B. 2002. Characterization of a friction-stir-welded aluminum alloy 6013. *Metallurgical Transactions B* 33B: 489–98.

Hinrichs, J. F., Smith, C. B., Orsini, B. F., DeGeorge, R. J., Smale, B. J., and Ruehl, P. C. 2004. Friction stir welding for the 21st century automotive industry. *Proceedings of the 5th International Symposium on Friction Stir Welding*, Metz, France: The Welding Institute.

Hovanski, Y., Santella, M. L., and Grant, G. J. 2007. Friction stir spot welding of hot-stamped boron steel. *Scripta Materialia* 57: 873–6.

Hunt, F., Badarinarayan, H., and Okamoto, K. 2006. Design of experiments for friction stir stitch welding of aluminum alloy 6022-T4. *SAE Technical Series*: 2006-01-0970.

Kamp, N., Sullivan, A., Tomasi, R., and Robson, J. D. 2006. Modelling of heterogeneous precipitate distribution evolution during friction stir welding process. *Acta Materialia* 54 (8): 2003–14.

Khan, I., Kuntz, M. L., Su, P., Gerlich, A., North, T. H., and Zhou, Y. 2007. Resistance and friction stir spot welding of DP600: A comparative study. *Science and Technology of Welding and Joining* 12 (2): 175–82.

Kyffin, W. J., Threadgill, P. L., Lavlani, H., and Wynne, B. P. 2006. Progress in FSSW of DP800 high strength automotive steel. *Proceedings of the 6th International Symposium on Friction Stir Welding*, October 10-13, Saint-Sauveur, Quebec, Canada: The Welding Institute.

Lathabai, S., Painter, M., Cantin, G., and Tyagi, V. 2005. Friction stir spot welding of automotive lightweight alloys. *Proceedings of the 7th International Conference on Trends in Welding Research*, 207–12. Materials Park, OH: ASM International.

Lin, P., Lin, S., Pan, J., Pan, T.-Y., Nicholson, J., and Garma, M. 2004. Microstructures and failure mechanisms of spot friction welds in lap-shear specimens of aluminum 6111-T4 sheets. *SAE Technical Series*: 2004-01-1330.

Lin, P.-C., Pan, J., and Pan, T.-Y. 2005a. Investigations of fatigue lives of spot friction welds in lap-shear specimens of aluminum 6111-T4 based on fracture mechanics. *SAE Technical Series*: 2005-01-1250.

Lin, P.-C., Pan, J., and Pan, T.-Y. 2005b. Fracture and fatigue mechanisms of spot friction welds in lap-shear specimens of aluminum 6111-T4 sheets. *SAE Technical Series*: 2005-01-1247.

Lin, P.-C., Pan, J., and Pan, T.-Y. 2006. Fatigue failures of spot friction welds in aluminum 6111-T4 sheets under cyclic loading conditions. *SAE Technical Series*: 2006-01-1207.

Lin, P.-C., Pan, J., and Pan, T. 2008a. Failure modes and fatigue life estimations of spot friction welds in lap-shear specimens of aluminum 6111-T4 sheets. Part 1: Welds made by a concave tool. *International Journal of Fatigue* 30: 74–89.

Lin, P.-C., Pan, J., and Pan, T. 2008b. Failure modes and fatigue life estimations of spot friction welds in lap-shear specimens of aluminum 6111-T4 sheets. Part 2: Welds made by a flat tool. *International Journal of Fatigue* 30: 90–105.

Lippold, J. C., and Ditzel, P. J. 2003. Microstructure and properties of aluminum friction stir welds. *Materials Science Forum* 426–432: 4597–4602.

McQueen, H. J., and Ryan, N. D. 2002. Constitutive analysis in hot working. *Materials Science and Engineering A* 322: 43–63.

Miles, M. P., Sederstrom, J., Kohkonen, K., Steel, R., Packer, S., Pan, T., Schwartz, W. J., and Jiang, C. 2007. Spot friction welding of high strength automotive steel sheets. *MS&T '06: Materials Science and Technology 2006 Conference and Exhibition*, October 15–19, Duke Energy Center, Cincinnati, OH.

Mishra, R. S., and Mahoney, M. W. 2007. *Friction stir welding and processing*. Materials Park, OH: ASM International.

Mitlin, D., Radmilovic, V., Pan, T., Chen, J., Feng, Z., and Santella, M. L. 2006. Structure–properties relations in spot friction welded (also known as friction stir spot welded) 6111 aluminum. *Materials Science and Engineering A* 441: 79–96.

Mortimer, J. 2005. Jaguar "roadmap" rethinks self-piercing technology. *Industrial Robot: An International Journal* 32 (3): 209–13.

Nagasawa, T., Otsuka, M., Yokota, T., and Ueki, T. 2000. Structure and mechanical properties of friction stir weld joints of magnesium alloy AZ31. *Magnesium technology 2000*, 383–7. Warrendale, PA: The Minerals, Metals & Materials Society.

North, T. H. 2009. Unpublished research, Department of Materials Science and Engineering, Toronto, Ontario, Canada: University of Toronto.

Okamoto, K., Hunt, F., and Hirano, S. 2005. Development of friction stir welding and machine for aluminum sheet metal assembly. *SAE Technical Series*: 2005-01-1254.

Orowan, E. 1947. Discussion in October 1947 Symposium, 451-453. London: Institute of Metals.

Pan, T.-Y. 2008. Friction stir spot welding (FSSW)—A literature review. *SAE Technical Series*: 2007-01-1702.

Pan, T.-Y., Joaquin, A., Wilkosz, D. E., Reatherford, L., Nicholson, J. M., Feng, Z., and Santella, M. L. 2004. Spot friction welding for sheet aluminum joining. *Proceedings of the 5th International Symposium on friction stir welding*. Metz, France: The Welding Institute.

Pan, T.-Y., Shwartz, W. J., Lazarz, K. A., and Santella, M. L. 2006. Spot friction weld-bonding for sheet Al joining. *Proceedings of the 6th International Symposium on Friction Stir Welding*, October 10–13, Saint-Sauveur, Quebec, Canada: The Welding Institute.

Petch, N. J. 1953. The cleavage strength of polycrystals. *Journal of the Iron Steel Institute* 174: 25–8.

Reynolds, A. P., Hood, E., and Tang, W. 2005. Texture in friction stir welds of Timetal 21S. *Scripta Materialia* 52 (6): 491–4.

Rhodes, C. G., Mahoney, M. W., Bingel, W. H., Spurling, R. A., and Bampton, C. C. 1997. Properties of friction-stir-welded 7075 T651 aluminum. *Scripta Materialia* 36 (1): 69–75.

Sakano, R., Murakami, K., Yamashita, K., Hyoie, T., Fujimoto, M., Inuzuka, M., Nagao, Y., et al., 2001. Development of spot FSW robot system for automobile body members, 645–50. *Proceedings of the 7th International Symposium of the Japanese Welding Society*, Kobe, Japan.

Sato, Y. S., Kokawa, H., Enomoto, M., and Jogan, S. 1999. Microstructural evolution of 6063 aluminum during friction-stir welding. *Metallurgical Transactions A* 30A: 2429–37.

Schilling, C., and dos Santos, J. 2001. Method and device for linking at least adjoining workpieces by friction welding. Patent No. W01/36144 A1.

Schmidt, H. B., and Hattel, J. H. 2008. Thermal modelling of friction stir welding. *Scripta Materialia* 58: 332–7.

Su, J.-Q., Nelson, T. W., Mishra, R., and Mahoney, M. 2003. Microstructural investigation of friction stir welded 7050-T651 aluminium. *Acta Materialia* 51: 713–29.

Su, P., Gerlich, A., and North, T. H. 2005a. Friction stir spot welding of aluminum and magnesium alloy sheets. *SAE Technical Series*: 2005-01-1255.

Su, P., Gerlich, A., North, T. H., and Bendzsak, G. J. 2005b. Material flow during friction stir spot welding. *Science and Technology of Welding & Joining* 11 (1): 61–71.

Su, P., Gerlich, A., North, T. H., and Bendzsak, G. J. 2006a. Energy generation and stir zone dimensions in friction stir spot welds. *SAE Technical Series*: 2006-01-0971.

Su, P., Gerlich, A., North, T. H., and Bendzsak, G. J. 2006b. Energy utilization and generation during friction stir spot welding. *Science and Technology of Welding and Joining* 11 (2): 163–9.

Su, P., Gerlich, A., Yamamoto, M., and North, T. H. 2007. Formation and retention of local melted films in AZ91 friction stir spot welds. *Journal of Materials Science*: DOI: 10.1007/s10853-007-2061-4.

Sun, D. Q., Lang, B., Sun, D. X., and Li, J. B. 2007. Microstructures and mechanical properties of resistance spot welded magnesium alloy joints. *Materials Science and Engineering A* 460–461: 494–8.

Thomas, W. M. 1991. Friction stir butt welding. International Patent Application No. PCT/GB92. U.S. Patent Application No. 9125978.8.

Thomas, W. M., Threadgill, P. L., and Nicholas, E. D. 1999. *Science and Technology of Welding & Joining* 4 (6): 365–72.

Tier, M. D., dos Santos, J., Rosendo, T., Mazzaferro, J. A., and da Silva, A. A. M. 2008. The influence of weld microstructure on mechanical properties of alclad AA2024-T3 friction spot welded. *SAE Technical Series*: 2008-01-2287.

Tozaki, Y., Uematsu, Y., and Tokaji, K. 2006. Effect of welding condition on tensile strength of dissimilar FS spot welds between different Al alloys. *Proceedings of the 6th International Symposium on Friction Stir Welding,* October 10–13, Saint-Sauveur, Quebec, Canada: The Welding Institute.

Tran, V.-X., Lin, P.-C., Pan, J., Pan, T.-Y., and Tyan, T. 2007. Failure loads of spot friction welds in aluminum 6111-T4 sheets under quasi-static and dynamic loading conditions. *SAE Technical Series*: 2007-01-0983.

Tweedy, B., Widener, C., Merry, J., Brown, J., and Burford, D. 2008. Factors affecting the properties of swept friction stir spot welds. *SAE Technical Series*: 2008-01-1135.

Wang, D.-A., and Lee, S.-C. 2007. Microstructures and failure mechanisms of friction stir spot welds of aluminum 6061-T6 sheets. *Journal of Materials Processing Technology* 186: 291–7.

Yamaguchi, J. 2004. Toyota Prius: AEI best engineered vehicle 2004. *SAE Automotive Engineering International* 3: 58–68.

Yamamoto, M., Gerlich, A., North, T. H., and Shinozaki, K. 2007a. Cracking in the stir zones of Mg-alloy friction stir spot welds. *Journal of Materials Science*: DOI: 10.1007/s10853-007-1662-2.

Yamamoto, M., Gerlich, A., North, T. H., and Shinozaki, K. 2007b. Mechanism of cracking in AZ91 friction stir spot welds. *Science and Technology of Welding & Joining* 12 (3): 208–16.

Zhou, Y., Fukumoto, S., Peng, J., Ji, C. T., and Brown, L. 2004. Experimental simulation of surface pitting of degraded electrodes in resistance spot welding of aluminium alloys. *Materials Science and Technology* 20: 1226–32.

9

Properties of Materials from Immersed Friction Stir Welding

Thomas S. Bloodworth III, George E. Cook, and Alvin M. Strauss

CONTENTS

Introduction

Friction stir welding (FSW), since its invention by TWI in 1991, has grown rapidly into a preferred method of welding many lighter materials such as magnesium and aluminum alloys [1,2]. In this innovative process, a pin and shoulder configuration are rotated and plunged into a work piece at the joint line, the pin then traverses the joint forming the bond (see Figure 9.1). The shoulder of the tool above the pin inputs the majority of the frictional heat as well as provides the forging force necessary to contain and compact the weld material. The material is plastically deformed, extruded, and stirred around the pin depositing behind it to form a solid-state bond. This solid-state process avoids many of the disadvantages of other fusion welding processes such as solidification cracking due to phase change and high distortion requiring post weld processing [2]. Since welding occurs at approximately 80% of the solidus temperature, members can be joined at any orientation including upside-down arrangements. Welding alloys with high quench rate dependence such as 2XXX, 6XXX, and 7XXX series aluminum alloys once deemed unweldable by most arc welding techniques are now easily joined using the solid state FSW process [4]. Additional advantages of FSW include the lack of shielding gases and high repeatability of the process by automation. The repeatability is attributed to the fact that FS welders do not need extensive on-site training required for other fusion or arc welding work.

Immersed friction stir welding (IFSW) is the FSW process performed while the coupon is entirely immersed in a fluid such as water. The higher quench rate of the water inhibits the growth of large grains and therefore increases the hardness of the welded zone. This was first observed by Hoffman and Vecchio who determined that immersed friction stir processing (FSP) of aluminum produced ultrafine grains (<200 nm) by severe plastic deformation [5,6]. The ultrafine grains produced in their experiments were an order of magnitude smaller than the grains produced by traditional FSP done in air. The advantages of IFSW were also recently used to weld 304L (stainless steel) using a poly crystalline cubic boron nitride (PCBN) tool in order to quantify corrosion resistance and microstructure [7].

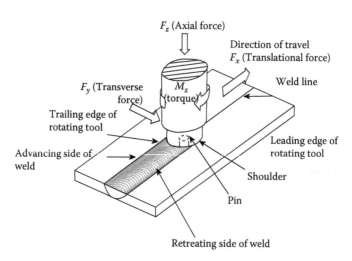

FIGURE 9.1
Schematic of friction stir welding. (From Crawford, R. M. S. Thesis, Nashville, TN: Vanderbilt University, 2005.)

This drastic decrease in grain size was expected to increase the hardness of the weld possibly having comparable, or in some cases superior, properties to conventional FSWs. The comparison of forces, torques, and other properties of the two techniques quantified the differences that were ultimately used as a gauge as to the feasibility of this process technique. The increased quench rate of water requires a necessary increase in power output by the welding machine creating different ideal welding conditions.

Experimental Setup 1

All conventional and immersed FSW experiments were conducted using a Milwaukee #2K Universal Milling Machine modified with a Kearney and Treker Heavy Duty Vertical Head Attachment at Vanderbilt's Welding Automation Laboratory [19]. Welding coupons were 8 inches long by 3 inches wide full penetration butt joints. The tool used a shoulder to pin a diameter ratio of 2.5. The tool profile is discussed in the next section and exact dimensions of the tool are shown in Figure 9.2.

Trivex

The pin was a Trivex design by TWI [8] noted for its decrease in welding forces with a static tool pin diameter of ¼ inch (6.35 mm). The pin cross section was developed by TWI as an equilateral triangle with sides given a specified radius. As the radius of the sides goes to infinity the cross section goes to a geometric equilateral triangle, the radius used for this study was centered at one point of the triangle and arcs using the other two points. This configuration gives the tool pin a tool area to swept area ratio of approximately 68%. Experimental results by TWI show that the Trivex tool welds were comparable to those of the more complex Triflute or Triflute-MX designs [8,9]. Tool profiles for Trivex and Triflute tools are given in Figure 9.3. The shoulder diameter was 5/8 inches (15.875 mm) and featureless. The tool and pin were machined from 01 tool steel and heat treated. During the experiment no visible wear or deformity was observed on the tool pin or shoulder. A tool lead angle and plunge depth were held constant at 1° and 0.009 inches (0.229 mm), respectively. This plunge was determined to produce the 80% shoulder contact condition desired.

A Kistler rotating cutting force dynamometer (RCD) Type 9123C was used to measure translational force (F_x), lateral force (F_y), axial force (F_z), and tool moment (M_z). The dynamometer was rated to measure up to 20 kN of axial force and 200 Nm of torque. Experimental force measurements for both IFSW and FSW were found to be well below the limits. The welding machine was fitted for position control using string potentiometers for translational and lateral location tracking.

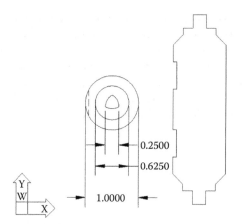

FIGURE 9.2
Experimental tool (all dimensions given in inches).

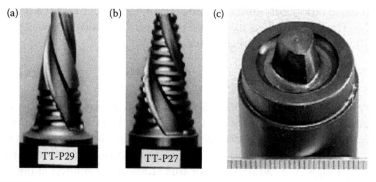

FIGURE 9.3
Tool profiles for (a) and (b) Triflute and (c) Trivex tool pins. (From Dubourg, L., and P. Dacheux. *6th International Symposium on Friction Stir Welding*. St. Sauveur, Canada, 2006.)

The effort implemented to ensure a constant plunge depth was critical, as small changes in vertical position cause significant changes in weld quality as well as excess flash [10]. The vertical axis was instrumented with a magnetic position transducer with quadrature output leading to position resolution on the order of <0.0005 inches (0.0127 mm). All welds in this experiment were kept at a plunge depth of 0.009 inches.

Thermocouple Implantation

Welds were implanted with thermocouples to determine characteristic temperatures and quench rates into the medium, whether it was air or water. It had been previously observed that the welding temperature at a lateral location was not greatly affected by the traversal distance [11]. Multiple thermocouples were imbedded to ensure an accurate temperature reading. Four equally spaced thermocouples were placed into each weld at a depth of 1/8 inch, or half the thickness of the ¼ inch coupons, and a distance of 5/16 inch from the weld center line. This corresponded to the lateral position of the shoulder edge during welding. This layout is shown in Figure 9.4.

The heat input into the water was important to verify certain process trends. It also served to quantify the power increase required to successfully produce IFSWs. A lower limit for heat input was computed using the change in water temperature before and after welding. Heat input was measured for IFSW using enthalpy change.

$$\Delta H = mc_p \Delta T.$$

Heat input is simply the change in enthalpy of a substance in a constant state where m is the mass of water, c_p is the specific heat at constant pressure, and ΔT is the change in water temperature before and after welding. This was ideal since the variation in water temperature was relative and not absolute and thus the water was not needed to return to room temperature prior to welding again. The specific heat, c_p, of water is 4189 J/kg K. Water for the experiment was initially room temperature (~298 K) and kept at a constant volume of about 3 L for all immersed welds. It should be noted that this method does not assume a loss of heat due to conduction through the backing plate or air convection at the surface. All that is measured is the amount of heat input into the water through the heating due to welding.

An additional heat input equation is given by Nunes [12], which gives the heat input during FSW. The heat input, ΔH, in energy per unit distance traveled is

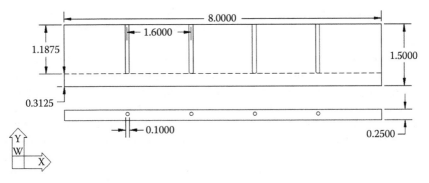

FIGURE 9.4
Thermocouple dimensions (all units in inches).

$$\Delta H = P/v.$$

Where v is the travel speed (m/s), and P is the power (J/s).

$$P = M_z \omega.$$

Where M_z is the torque required to weld in Nm and ω is the rotational speed in Hertz. This simple heat input and power equation predicted accurate trends and were used to calculate lost terms in the lower limit heat input given above. The power estimation predicted an approximately constant power output for a range of spindle speeds. This was modeled and experimentally verified by Crawford et al. where for a substantial increase in rotational speed a substantial decrease in torque followed [3,10].

Experimental Setup 2

The second setup for IFSW involved many of the same instruments and values as the first setup with a different tool pin configuration and modified weld matrix. Welding was performed on the same welding machine and dynamometer [19]. This was due to the expectation of similar forces and comparable parameter matrices as the prior setup. The tooling used was a more conventional threaded cylinder and had a flat featureless shoulder. It is noted by many as the most common weld tool pin for academic use and research due to its simplicity and ease of production. Its advantages include imposing a downward flow that is complimentary to the rotary flow around the pin. Together, the two flows increase material stirring leading to a further breakdown of the oxide layer and greater root fill and mixing, producing good welds. The tool pin and shoulder diameters were 0.25 inches and 0.625 inches respectively, with a thread pitch of 20 threads per inch. The pin length of the nonconsumable heat-treated 01 steel tool was 0.235 inches.

Disadvantages of the threaded cylinder tool design include the lack of a significant dynamic volume greater than that of the static tool volume during welding that decreases mixing when compared to the Trivex profile. Secondly, the threaded cylinder produces higher traversing, moment, and forge (F_x, F_z, M_z) forces compared to Experiment 1's tool. The Trivex design was for this purpose, however, the threaded cylinder's increased forces led to a higher work envelope and fewer defects than a similar Trivex matrix. It is for this reason that the work envelope for Experiment 2 includes rotation speeds (RS) of 2000–3000 rpm and travel speeds of 10–16 ipm (inches per minute). Tool tilt angle and plunge depth were kept constant for all Experiment 2 welds at 2° and 0.004 inches (0.102 mm).

FIGURE 9.5
Modified backing plate for IFSW.

Tank Construction

The backing plate, or backing anvil, was modified to contain approximately 3 liters of water for the experiments (see Figure 9.5). The internal dimensions for the containment tank were 12 inches wide by 29.75 inches long. Water was placed in the tank to a level of approximately ½ inch deep using a graduated cylinder to ensure volume. This gave a total volume of approximately 178.5 cubic inches or 2.925 L.

Results

Welds for the conventional FSW and IFSW matrices were run for each parameter set at least once and in most cases two or three welds were done in order to verify trends and build confident and competent data. The welding matrix for Experiment 1's conventional and IFSW are given in Table 9.1 and cover a broad range of travel speeds and rotational speeds.

Weld Pitch

Weld pitch, W_P, is often used to characterize a welding envelope and determine defect trends due to hot or cold welding [13]. W_p is simply the ratio of the rotational speed to the travel speed and has units of rev/inch (rev/mm). The first experimental matrix uses a range of weld pitches from $1000/14 = 71.4$ to $2000/5 = 400$ rev/inch. The weld pitch can give a general trend for the expected quality of the FS weld. A low weld pitch indicates that the travel

TABLE 9.1

Experimental Matrix for IFSW and FSW
($x = 1$ or More Welds Performed)

	1000 rpm	1500 rpm	2000 rpm
5 ipm	x	x	x
8 ipm	x	x	x
11 ipm	x	x	x
14 ipm	x	x	x

speed is relatively high in comparison to the rotational speed. This leads to a weld with a low heat input and poor mixing. Such welds can be expected to form worm holes at the base of the pin where temperatures are lowest and mixing is poor. The high end of the weld pitch spectrum indicates that the travel speed is relatively low compared to the rotational speed. This can lead to discontinuities discussed by W. Arbegast related to the overheating of the weld zone such as excess flash, expulsion, or surface galling [14].

An optimum weld pitch is not universal. It is dependent on many factors including the welded alloy, welding tool, and other parameters. Variation in heat dissipation due only to a change in welding machine can alter the optimum pitch parameters. Also, a weld pitch is not deterministic in even its own matrix. That is to say that similar weld pitches with differing parameters may not lead to optimal welds. For example, a weld at 2000 rpm and 10 ipm may have produced a good weld, however, a weld at 3000 rpm and 15 ipm produced a worm hole even though the weld pitch for both are 200 rev/inch. It can not be stressed enough that the weld pitch parameter is for general envelope trends and should not be construed to verify specific weld characteristics.

Axial Force

Axial force for both experimental setups was measured using the aforementioned Kistler Dynamometer. Since welding was position controlled, trends were experimentally verified to determine force dependence on processing parameters. Data for conventional welding as well as IFSW showed expected trend lines in which an increased rotation speed/decreased force relation was evident. This trend was also observed by previous research and was further validated for both processes here [3,10,13]. Axial force was expected to behave inversely proportional to weld pitch as a general trend. The limit to this axial force drop as a function of travel speed has not been met. The data for both IFSW and conventional FSW indicated that axial force was independent of either process. This was evident in Figure 9.6 for Experiment 1. The welding machine was called on to maintain an axial load in excess of 7000 N. The experimental setup at Vanderbilt Welding Automation Laboratory is

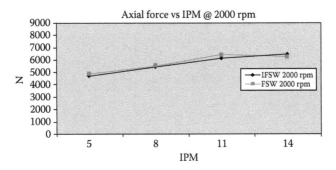

FIGURE 9.6
Axial force (N) vs. travel speed (ipm) at 2000 rpm.

currently set to maintain greater than 12 kN of axial force. Similar axial force data was generated for 1000 and 1500 rpm showing an identical independence on welding conditions.

A negligible change in axial force for the two processes is advantageous since prior hypotheses assumed an increase due to quench rate dependence of the alloy. The axial force was found to be independent of the process at all parameter values for this data set. It has been observed that axial force is a quality indicator for friction stir welds [3,10,13,14]. An insufficient axial force indicates lack of shoulder pressure and can indicate lack of containment of surface flash and lead to worm holes or other lack of fill defects.

Torque

Torque values were recorded in order to quantify the power generation requirements for IFSW over conventional FSW in both experiments. It was expected that the required torque would increase, as much of the frictional heating for the process would dissipate, by heating the water. The quantified increase in power output required for optimal IFSW conditions is an indicator of feasibility of the process in industry. Torque values recorded for 1500 rpm are given for both processes in Figure 9.7 for Experiment 1. A similar torque curve was generated for 1000 and 2000 rpm showing an increased rpm/increased torque trend.

A clear increase in torque was visible from 5 to 14 ipm for IFSW over FSW. It was clear from Figure 9.7 that the welding machine produced welds requiring 10–25 Nm. An increase of 2–5 Nm was required and was found to be a highly parameter independent change in torque. The increased weld pitch/decreased torque relationship was observed for both processes [3,8,10,13–15]. This trend had been observed in even greater weld pitches and the limit to this trend has not yet been identified. The torque increase requirement was less than 25% for both experimental setups, which is good for industrial

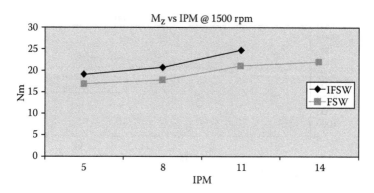

FIGURE 9.7
Torque output (Nm) vs. travel speed (ipm) at 1500 rpm.

applications because it would determine only a marginal increase in power output. This will be discussed in the next section.

Power

The increase in power output was found to be consistent with the increase in torque and rotational speed as evidenced by the power equation given earlier. Power increased linearly as a function of travel speed that indicates a travel speed to moment relationship observed by other authors [3,10,11,15]. From Figure 9.8 it can be observed that the welding machine outputs between 1–5 kW for FSW and IFSW. The observed increase due to the IFSW process is marginal at approximately 0.5 kW or 15–20% for both experiments. Similar power curves were generated for 1000 and 1500 rpm showing an increased rpm/increased power output trend.

Power is experimentally determined by the previous equation:

$$P = M_Z \omega.$$

This form of the power equation is actually a simplified form that assumes that the power input due to tool travel is negligible. This assumption is correct for ¼ inch AA6061 in which the tool travel term was found to be less than 5% of the total power input. Including both rotation and translation terms, the power equation would be:

$$P = M_Z \omega + F_X v_t.$$

Where F_x is the traversing force and v_t is the linear travel speed (TS). The remainder of the power input was due to the tool rotation causing frictional heating as well as plastic deformation of the welded material. Additional heat loss up the tool shank was also accounted for.

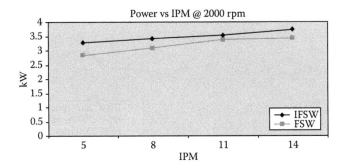

FIGURE 9.8
Power output (kW) vs. Travel speed (ipm) at 2000 rpm.

Power, torque, and axial force all have important factors in producing optimal weld parameters. Optimal welds were determined by two factors that included ideal weld cross sections and ultimate tensile strength (UTS). An increase in fatigue properties was also an advantage that has been touted for FSW over other fusion welding techniques [13,15]. The increase in tensile properties was also beneficial as a greater weld/base material UTS ratio exists due to the solid state process.

Tensile testing was performed on welds in order to judge optimal parameters and for microhardness comparison. Hoffman and Vecchio observed an order of magnitude decrease in the weld nugget grain size over conventional FSP [5,6]. Microhardness tests performed on IFSWs showed an increase in local weld hardness over standard FSW. Cross sections also indicated a consistent root flaw size for single pass FS welds. Microscopy indicated that both optimal conditions retained the same or better root properties. Although porosity was evident in optical microscopy of the weld zone, tensile properties matched or exceeded those of friction stir welds done in air. Additional discussion of porosity as a function of water depth as it relates to FSW and arc welding processes will be summarized at the conclusion of the chapter.

Materials Testing

Materials testing included microhardness analysis, cross sectioning, and tensile testing of all weld coupons for both experimental setups. Some parameters were not run for the immersed matrix in Experiment 2 as it was determined that the rotational speed was not great enough to produce welds at 14 ipm with the exception of 2000 rpm for the Trivex tool. The wormhole defect was prevalent in these welds so further welds of lower RS at 14 ipm were not run as it was assumed that the wormhole would only increase in size for those sets.

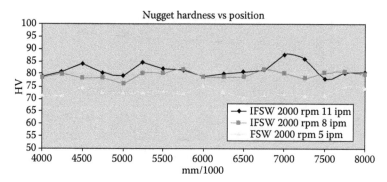

FIGURE 9.9
Hardness (HV) vs. Weld Nugget location (mm).

The increase in quench rate due to IFSW causes the grains to quench quickly and solidify from its plastic state without excessive grain grown leading to a harder weld nugget. Hardness testing indicated a 10% increase in weld zone hardness. Weld zone hardness test results showed an average weld nugget hardness of 73 for conventional FSW and 81 for IFSW (see Figure 9.9). All hardness tests were performed only on the highest tensile test welds for either process. These included welds that when cross sectioned showed no evidence of defects including worm holes or excess flash. The observed increase in nugget hardness was assumed to create a decrease in fatigue properties. An increase in hardness may affect fatigue properties.

Cross sections for all welds were polished and etched using Boss's reagent at 10:1 ratio of water to hydrofluoric acid (HF) for 15–20 seconds. Weld zone cross sections showed an expectedly porous heat affected zone and joint root flaw for IFSW when compared to conventional FSW (see Figure 9.10).

Tensile Testing

Experimental Setup 1

Tensile tests were conducted to determine the optimum weld parameters for both FSW and IFSW for the tool and plunge used in this study. Tests were conducted according to the ASM standard for materials testing. Ultimate tensile strength (UTS) was the criterion for ranking weld quality. Optimal welds were those whose weld to base metal UTS ratio was 80–95%. For the weld matrix given above for Experiment 2, the optimal weld conditions for FSW were 2000 rpm at 11 ipm while the IFSW required 2000 rpm at 8 ipm, a required decrease of 3 ipm.

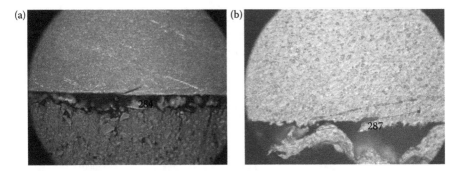

FIGURE 9.10
(a) Conventional FSW root flaw (10×) and (b) Immersed FSW root flaw (50×).

Experimental Setup 2

Experiment 2 used the standard threaded cylinder tool pin and produced welds in excess of 90–95% of the parent material strength for both FSW and IFSWs. Welds performed in water produced higher UTS when compared to conventional FS welds. The tensile strength of the optimal conventional FS welds, experimentally and found in literature, was approximately 281.5 MPa compared to the UTS of the optimal immersed weld in excess of 296.1 MPa, an increase in UTS of 5.2% [15]. This is in agreement with previous findings by Fratini et al. indicating a 5% increase in tensile properties by running water in situ over the weld surface as a form of heat treatment [16]. Optimal parameters varied as expected for IFSW and FSW. Optimal FSW parameters for Experiment 2 (threaded cylinder) were 2000 rpm at 16 ipm. The immersed parameters were optimized at 2000 rpm at 10 ipm.

As previously observed this is due to the power increase required to form the bond while heat is also subsequently lost in the form of heating water. Water has a heat capacity of roughly four times that of room temperature air, thus, it requires four times the heat input to heat an equal mass of water than that of air. It is therefore obvious that a decrease in travel speed is required in order to increase the heat input into IFSWs.

For a constant weld travel speed (TS) it was observed that the weld quality increased with increasing rotational speed (RS). This was observed most conspicuously in FSW while IFSW seem to have indicated a logarithmic trend of UTS with respect to RS for the matrix run (see Figure 9.11). For each TS run (5, 8, 11 ipm) in the IFSW matrix the trend for tensile strength versus RS remained a logarithmic function of RS. Figure 9.9 illustrates the logarithmic relationship between weld UTS versus RS at a constant TS. Results for a constant rotational speed indicated independent UTS with increased TS. The range of RS for the study was found to be large in comparison to TS and therefore no trend was identified for either FSW or IFSW.

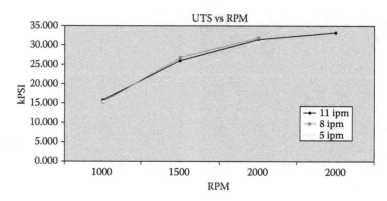

FIGURE 9.11
UTS vs. RS at a constant TS (IFSW); WA = 1000 rpm, WB = 1500 rpm, WC = 2000 rpm.

Summary and Conclusions

Experimental data for immersed friction stir welds parameterize and quantify forces and process properties unique to this variation of FSW. For the weld matrix given, a database was compiled of process quantities such as characteristic force, torque, weld zone properties, and materials properties such as butt joint tensile strength. Upon comparison to the conventional FSW matrix the parametric increase in properties were observed. These include the requisite increase in welding machine power output due to the torque increase required. Also of great interest was the observation of an independent axial force to process medium. Axial force (F_z) had no observable change for identical parameters for the FSW and IFSW process.

Generally, the properties for IFSW required a marginal increase in value for identical parametric conditions. Evidence for this was in the quantification of torque and power that required increases of approximately 20% for producing similar welds. This marginal increase in process parameters for IFSW is believed to be due to the quench rate of the water defined by the lower bound heat input equation. The production of ideal IFSWs is restricted by the lack of mixing due to reduced local heating at identical welding parameters as FSW. The Trivex tool pin was noted by TWI for its reduction in forces, however, the decrease in forces and lack of a sufficient downward flow limited the envelope of acceptable weld parameters for IFSW for the matrices investigated.

Porosity

Investigations into the effect of water depth on the porosity of ferroalloys (iron-based alloys) welded by traditional arc processes were made by Rowe et al. and others [17,18]. Porosity was found to be highly dependent on the water depth. This is commonly attributed to hydrogen gas as well as iron oxidation at high pressure. Water pressure increases at a rate of 1 atmosphere for every 10 m (33 ft); hydrogen and oxygen/chemical oxidation, increases proportionally. In arc welding processes, porosity is mitigated by the introduction of coatings that lower oxidation such as calcium carbonate or titanium that is especially noted as a strong deoxidant, and many other formulations. Since porosity is seen to increase dramatically as a function of depth for arc welding, then similar studies must be investigated to define trends in FSW as well as fixes as they are needed for unique process characteristics. Porosity in ferro-alloys was observed to exceed 5% in conventional arc wet welds performed at greater than 15 ft of water [18]. AWS standards for wet welds (D3.6 Class B) specify a maximum allowable porosity of 5% as seen by metallographic cross sections. Although no such standard exists yet for FSW one can infer from the advantages of FSW that low porosity can be expected. Porosity is assumed to be the product of the oxidation of the fresh weld material as it is drawn to the surface by the mixing process. The pure aluminum quickly bonds to oxygen drawn from water molecules and hydrogen gas left over from the dissociation of the water that creates porosity in the aluminum oxide. Due to the solid state nature of FSW it is expected that the porosity dependence on depth would be mitigated although future research is needed in this area to quantify characteristics and process standards. Pressure at a depth of water would allow high porosity to develop more prevalently when the weld is in liquid phase. This is related to the intermolecular forces between the weld alloy molecules themselves. A higher temperature seen during arc welding leads to weaker bonds between alloy molecules that make them more susceptible to oxidation than the lower temperature solid state process.

Acknowledgments

This work was completed in part due to the financial and intellectual support from the following individuals and organizations. University of Tennessee Space Institute (UTSI) donated their time and facilities for cross sectioning preparation, optical microscopy, and microhardness testing. Robin Midgett donated his expertise in the use of tensile testing on an Instron Universal Tester. A NASA GSRP fellowship grant and Tennessee Space Grant Consortium gave needed funding that allowed these experiments to be run and was instrumental to the project. Dr. Arthur Nunes from Marshall Space Flight Center (MSFC) gave his valuable expertise during personal communications.

References

1. Thomas, W. M., Nicholas, E. D., Needham, J. C., Murch, M. G., Temple-Smith, and P., Dawes, C. J. 1991. Friction stir butt welding. International Patent Application # PCT/GB92/02203 and GB Patent Application #9125978.8.
2. Dubourg, L., and P. Dacheux. "Design and properties of FSW tools: A literature review." Paper 62. *6th Symposium of Friction Stir Welding*. St. Sauveur, Canada, 2006.
3. Crawford, R. "Parametric quantification of friction stir welding." M.S. Thesis, Nashville, TN: Vanderbilt University, 2005.
4. Shi Qing, Y., T. Dickerson, and R. Shercliff Hugh. "Thermo-mechanical analyses of welding aluminium alloy with TIG and friction stir welding." *6th International Conference: Trends in Welding Research*, 247–52. April 15–19. Pine Mountain, GA: ASM International, 2002.
5. Hofmann, D. C., and K. S. Vecchio. "Submerged friction stir processing (SFSP): An improved method for creating ultra-fine-grained bulk materials." *Materials Science and Engineering A* 402 (2005): 234–41.
6. Hofmann, D., and Vecchio, K. "Thermal history analysis of friction stir processed and submerged friction stir processed aluminum." *Materials Science and Engineering A*. (2007). 465:165–175.
7. Clark, T. D. "An analysis of microstructure and corrosion resistance in underwater friction stir welded 304L stainless steel." M.S. Thesis, Provo, UT: Brigham Young University, 2005.
8. Colegrove, P. A., and H. R. Shercliff. "Development of Trivex friction stir welding tool. Part 1—Two dimensional flow modeling and experimental validation." *Science and Technology of Welding & Joining* 9 (2004): 345–51.
9. Dubourg, L., and P. Dacheux. "Design and properties of FSW tools: A literature review." Paper 02. *6th International Symposium on Friction Stir Welding*. St. Sauveur, Canada, 2006.
10. Crawford, R., T. Bloodworth, G. Cook, and A. Strauss. "High speed friction stir welding process modeling." Paper 106. *6th International Symposium on Friction Stir Welding*. St. Sauveur, Canada, 2006.
11. Mitchell, J. E. "The experimental thermo-mechanics of friction stir welding." M.S. Thesis. Nashville, TN: Vanderbilt University, 2002.
12. Schneider, J. A., R. Beshears, and A. C. Nunes. "Interfacial sticking and slipping in the friction stir welding process." *Materials Science and Engineering A* 435–436 (2006): 297–304.
13. Khaled, T. "An outsider looks at friction stir welding." FAA Report #: ANM-112N-05-06. 2005.
14. Arbegast, W. J. "A flow-partitioned deformation zone model for defect formation during friction stir welding." *Scripta Materialia* 58 (2008): 372–6.
15. Mishra, R. S., and M. W. Mahoney. "Friction stir welding and processing." Materials Park, OH: ASM International, 2007.
16. Fratini, L., G. Buffa, and R. Shivpuri. "In-process heat treatments to improve FS-welding butt joints." *International Journal of Advanced Manufacturing Technology* 43 (2008): 664–70.

17. Rowe, M., S. Liu, and T. Reynolds. "The effect of ferro-alloy additions and depth on the quality of underwater wet welds." *The Welding Journal*, August 2008.
18. Suga, Y., and A. Hasui. "On formation of porosity in underwater wet weld metal (the first report)." *Transactions of the Japan Welding Society* 17, no. 1 (1986): 58–64.
19. Fleming, P. "Weld control user's manual." http://research.vuse.vanderbilt.edu/vuwal/index.shtml. Nashville, TN: Vanderbilt University, 2007.

Appendix

Raw tensile data for Experiment 1 and 2 for various weld parameters are given below in tabular or chart format.

A1–A6: Control force data for Experiment 1 welds

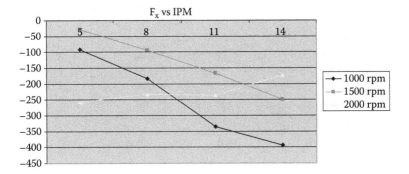

FIGURE A9.1
Experiment 1 control (air) F_x (N) vs. TS (ipm); negative force opposes tool travel direction.

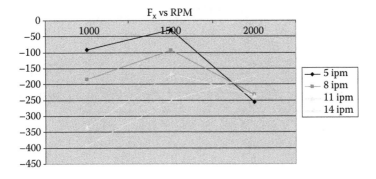

FIGURE A9.2
Experiment 1 control (air) F_x (N) vs. RS (rpm); negative force opposes tool travel direction.

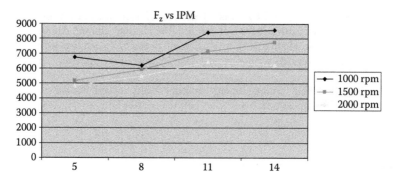

FIGURE A9.3
Experiment 1 control (air) F_z (N) vs. TS (ipm).

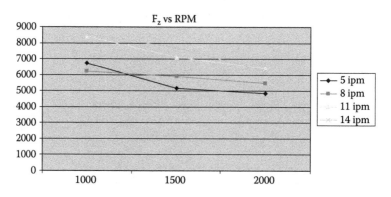

FIGURE A9.4
Experiment 1 control (air) F_z (N) vs. RS (rpm).

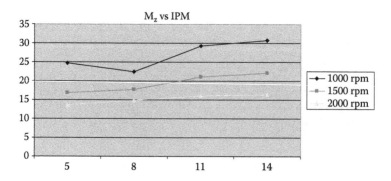

FIGURE A9.5
Experiment 1 control (air) M_z (Nm) vs. TS (ipm).

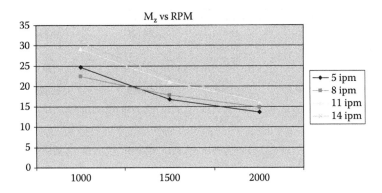

FIGURE A9.6
Experiment 1 control (air) M_z (Nm) vs. RS (rpm).

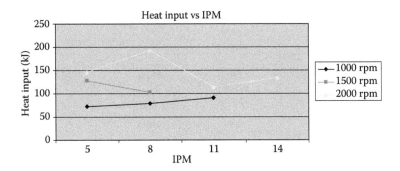

FIGURE A9.7
Experiment 1 IFSW heat input (kJ) vs. TS (ipm).

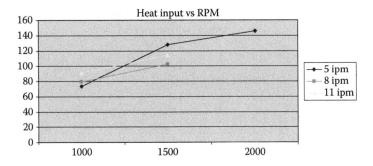

FIGURE A9.8
Experiment 1 IFSW heat input (kJ) vs. RS (rpm).

Specimen ID	Specimen #	UTS (ksi)	Average UTS (ksi)	Percent UTS of Parent
WC1	1	29.046	29.125	64.896
3000–15	2	30.250		
(void defect)	3	28.079		
WC3	4	40.368	40.987	91.326
2000–10	5	40.250		
	6	42.344		
WC4	7	43.038	42.752	95.258
2000–10	8	42.960		
	9	42.257		
WC5	10	41.671	41.485	92.435
2200–10	11	41.384		
	12	41.399		
BW2	13	43.150	42.721	95.190
2000–16	14	42.679		
	15	42.335		
BW3	16	42.803	42.126	93.863
2000–16	17	40.869		
	18	42.705		
Parent Material	1	44.7	44.88	
Certified Test	2	43		
Report	3	47.1		
	4	44.5		
	5	45.1		

FIGURE A9.9
Experiment 2 tensile data for IFSW (WC) and FSW (BW).

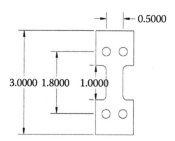

FIGURE A9.10
Schematic of a tensile specimen for both experiments.

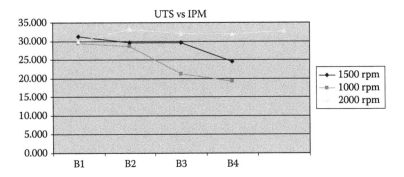

FIGURE A9.11
Experiment 1 control (air) UTS (ksi) vs. TS (ipm) where B1–B4 denote 5, 8, 11, and 14 (ipm) respectively.

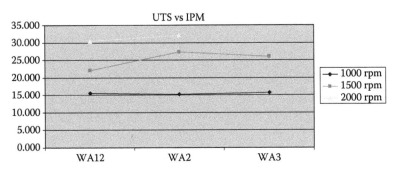

FIGURE A9.12
Experiment 1 IFSW UTS vs. TS (ipm) where WA12–WA3 denote 5, 8, and 11 (ipm) respectively.

10

Friction Stir Welding and Related Processes

Christopher B. Smith and John F. Hinrichs

CONTENTS

Introduction

Friction stir welding (FSW) and related processes (or process variants) are emerging technologies that can be used to generate high quality welds or process various materials. Friction Stir Welding was invented in 1991, by Wayne Thomas at The Welding Institute in Cambridge, England [1]. Since then, significant research and development projects have been initiated throughout the world to develop production capable processes for numerous applications. The first production implementation involved fabrication of aluminum paneling for the marine industry at Marine Aluminum in Haugesund, Norway in 1995. Since then, the process has expanded to worldwide use, with a majority of the fabrication based in Europe. Although there is nothing published worldwide on the use of FSW, based on information provided by multiple FSW fabricators, it is estimated that currently between 300 and 400 miles of friction stir weld is produced annually as of 2008.

Since its inception there has been a tremendous interest in FSW and its related processes, due to the potential advantages over traditional joining processes. This has led to numerous development efforts throughout the world and in numerous industries. Due to this level of interest, it is not practical to review FSW and the related processes in any significant detail. Rather, this chapter will describe FSW and the related processes, their characteristics, their advantages and challenges, areas of applicability, and equipment solutions at a general overview level. The intent is for an engineer or technician to develop a basic understanding of FSW and the related processes, as well as where and how the processes can be applied.

Process Description

FSW is a rather simple process as shown in Figure 10.1. A diagram of the process is shown above and a photograph of an example friction stir weld in aluminum is shown below. The process involves multiple steps including:

1. The friction stir machine first starting rotation of a specially design tool (referred to as the friction stir tool).
2. The machine plunging the pin of the friction stir tool into the material, heating the material to a temperature near its melting point where it is soft and can be readily plasticized. The tool is plunged into a weld joint (the faying surfaces that are to be joined) until the

(a)

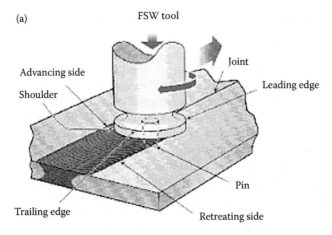

(b)

FIGURE 10.1
FSW process description (a) and example weld (b).

shoulder of the friction stir tool contacts the top surface. This location is referred to as the weld start.

3. Once the shoulder of the friction stir tool contacts the surface, the machine starts traversing the friction stir tool (while rotating) along the joint line or along a path that is to be processed. During this time, the friction stir tool is generating heat via friction and mechanical working of the material. In addition, the rotation action of the tool and the geometry of the pin and shoulder allow for movement of plasticized material from one side of the friction stir joint to the other. During this portion of the process, the shoulder of the friction stir tool performs the role of encapsulating material, ensuring that the weld or processed material is trapped below the shoulder.

4. Once the machine traverses the friction stir tool to the end of the desired path, the friction stir tool is retracted from the material, generally leaving an exit hole where the pin was last located.

FSW and its related processes can be used to join or process numerous materials. Since the inception of FSW, the majority of development and production applications have involved joining of various aluminum alloys. This is due to significant technical and cost advantages of FSW versus traditional joining processes for aluminum. The technology has also been demonstrated to weld or process many other metals, including magnesium, lead, zinc, copper, bronze, titanium, and steel, to name a few. At this time, there are also a small number of copper and steel production applications in existence. As the melting point of the material to be joined or processed increases, the process generally becomes more costly to implement, due to increased cost and wear of materials suitable for the friction stir tool.

Advantages and Disadvantages

FSW and the related phase processes are solid; that is, no melting occurs. This allows for significant advantages over many of the fusion welding technologies. Some advantages include:

1. Low operating cost due to lack of consumables. The friction stir tool life is generally long and there is no requirement for filler wire or shielding gas in most applications.

2. Very low distortion. This can provide dramatic cost benefits, especially on large parts, where straightening and other distortion remediation operations can be avoided.

3. The process is very repeatable and consistent. The process is not sensitive to disturbances that wreak havoc with fusion welding processes. Such disturbances include contamination caused by lubricants, moisture, or oxides on surfaces, and so on.

4. Minimal affect on material properties. Due to the low heat input and lack of existence of melting, the degradation of material properties associated with other joining processes is reduced. In some materials and alloys, especially nonheat treated materials, the friction stir joint is stronger than the parent material.

5. The process is environmentally friendly. The process is energy efficient and does not generate harmful fumes.

6. Friction stir welding has been proven to be capable of joining almost any metal, but is especially beneficial for materials that are more difficult to weld such as aluminum, copper, lead, and so on, which have a lower melting point.

7. In many applications, special precautions or joint preparations are not required to create quality welds, unlike many fusion welding processes.

As with any process, there are also challenges or disadvantages, including:

1. The process requires very high levels of force to hold the friction stir tool within the material. Actual required forces depend on many factors (material, alloy, thickness, etc.) but can range from several hundred pounds to tens of thousands of pounds. This poses significant challenges for typical off-the-shelf equipment (e.g., Computer Numerically Controlled (CNC) machines, robots, spindles, etc.). Because of the high levels of force required, the process must be performed by an automated machine.

2. In most cases, the process requires a flat or nearly flat surface in at least one plane. This can cause challenges for implementation on highly contoured parts.

3. Due to the high forces, very robust fixturing is required. Any fixture must support the loads of the process, but also must resist joint parting forces. For example, the operation of plunging the pin through a butt joint necessarily causes forces that tend to separate the joint line. The fixture must counteract these forces. These issues can partially be mitigated by self-clamping joint geometries.

4. The process leaves an exit hole (see Figure 10.1). In reality, this exit hole is no worse than a weld start or stop using traditional fusion welding processes. In addition, there are numerous workarounds. In most applications there is a location where the exit hole can be placed where it will not have a deleterious effect on the final part and in many other situations the exit hole can be placed in a location where there is offal. In these cases the offal is used for quality control purposes. There is also an equipment solution, referred to as a retractable pin tool, where the pin of the friction stir tool can retract

up into the shoulder, which can be used to eliminate the exit hole. See the section on special equipment implementations for a more detailed discussion.

5. Welding on hollow sections requires access for an internal mandrel or the component(s) to be welded or processed to be self-supporting, unless special friction stir tools are used (see the section for self-reacting tools).

6. Friction stir welding and its related processes are more suitable to lower melting point materials. It is believed that any metal can be friction stir welded. However, due to the high forces required for the friction stir processes, the friction stir tool material must retain sufficient strength and toughness at temperatures near the melting point of the material. This limits choices for the tool material and significantly increases friction stir tool costs for higher melting point materials, such as steel and titanium.

Related Processes or Process Variants

There have been two significant variants of the process that have been further developed since the original invention. Friction stir welding was originally invented with the assumptions that it would be used to join multiple sections and there was a requirement for a translation of the friction stir tool along a joint line. The first variant is friction stir spot welding (FSSW) and the second variant is friction stir processing (FSP).

1. Friction stir spot welding (FSSW): This variant of the process essentially eliminates the traverse portion of the process. FSSW is a spot process, similar to resistance spot welding (RSW) or riveting [2,3]. A typical example of FSSW is shown in Figure 10.2. The FSSW machine simply plunges the friction stir tool into the material and then retracts, without traversing the friction stir tool. The FSSW process has been demonstrated to have superior static and fatigue properties versus resistance spot welding, especially when joining aluminum. The FSSW process can readily be used to supplant RSW in aluminum, which is primarily used in the automotive industry for body panel joining [4,5,6]. FSSW has been demonstrated to be capable of producing welds in less than one second in one millimeter material, which is similar to cycle times for RSW. In addition, the FSSW process has been shown to have significant advantages over RSW in aluminum including; far superior consumable life, less peripheral equipment (no water cooling, etc.), no shunting problems, less sensitivity to contaminants (oxide, oils, etc.), and no need for costly electrical equipment to manage power factor corrections for the very high currents required for RSW.

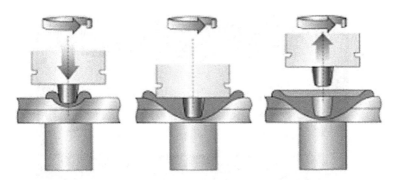

FIGURE 10.2
Friction stir spot welding.

2. Friction stir processing: FSP is a variant of the friction stir technology where there is no weld joint, but rather the friction stir tool is traversed through parent material to process the material. FSP is similar or equivalent to bead-on-plate when related to traditional welding processes. FSP results in a very fine grain structure, locally, and can be used to modify the material properties in the FSP region. With castings this can be used to eliminate porosity, as well as improve material properties [7]. Two examples are shown in Figure 10.3, where nickel aluminum bronze has been friction stir processed to locally remove porosity (left) and the same is shown for an aluminum casting on the right. In wrought material or thin sheet, FSP can be used to improve material properties, to enhance high temperature elongation, and to allow or enable superplasticity (elongation > 250%) in certain alloys [8]. In addition, FSP can be used to enhance room temperature ductility and to enable or allow for improved formability [9,10]. This is especially true for materials that are traditionally difficult to form, such as aluminum, as is shown in Figure 10.4.

Joint Configurations

Similar to other welding processes, there are a multitude of joint configurations to which FSW can be applied. The joint configurations that are generally accepted as being feasible for FSW are shown in Figure 10.5. A few notes follow with regard to some of the joint configurations:

1. Full penetration butt weld: This is defined as a joint where both materials that are to be joined are of equal thickness and the welded area extends 100% of the way through the thickness (full penetration). This particular joint design generally requires no preparation

(a)

(b)

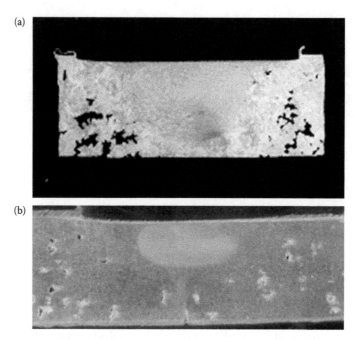

FIGURE 10.3
Friction stir processing of nickel aluminum bronze and aluminum casting.

FIGURE 10.4
FSP of aluminum to allow for improved formability.

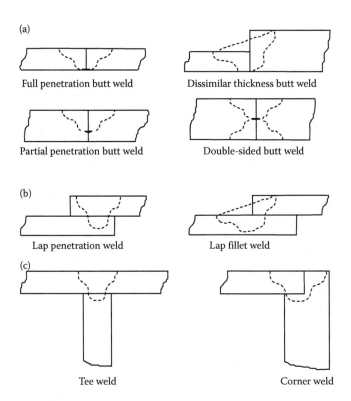

FIGURE 10.5
Friction stir welding joint configurations.

(e.g., chamfers) prior to welding, unlike other processes. To date, the butt weld is the most common production joint using the FSW technology.

2. Partial penetration butt weld: This is similar to the full penetration butt weld, except for the fact that the depth of the weld is less than 100% of the thickness of the material.

3. Lap penetration weld: As shown in Figure 10.5, a lap penetration joint is defined as a joint where one part is lapped onto the top of a second part. The friction stir tool penetrates through both surfaces and mixes material between both to generate a joint.

4. Lap fillet joint and dissimilar thickness butt joint: Both of these joint designs are similar with respect to the friction stir process. They both require the FSW to be tilted sideways. By tilting the tool sideways, material is forged from the upper plate toward the lower plate to create a smooth transition between the upper and lower plates. For the lap fillet joint, it should be noted that the resulting weld thickness (distance from original interface to weld surface) is less than

the material thickness (see Figure 10.5). Thus, this joint cannot create fillet weld joints with proper theoretical throat or actual throat and should be used with care in structural applications.

5. Tee joint: A tee joint (similar setup to a tee-fillet configuration) can be accomplished with FSW by approaching the joint from the opposite side of the flange and welding into the web, similar to a lap penetration weld. In normal processing conditions, the joint coverage is less than the thickness of the web as is shown in Figure 10.5. However, recent developments have shown that it is possible to generate 100% joint coverage and create a small fillet on both sides of the web on the back side [11,12]. This new technique requires access to the back side of the part for fixturing purposes. Lastly, the traditional tee-fillet approach cannot be accomplished with FSW, unless a strip of filler material is placed in the corner to allow for a planar welding surface or a special tool with a stationary shoulder is used.

6. Corner: A butt weld or a combination of a butt and lap weld where one plate is brought into contact with another plate at their edges. One plate is oriented horizontally while the other is oriented vertically.

7. Others: There are other joint configurations that can be welded. These may include, but not be limited to, the ones listed below.

    a. Self-locking: An irregular joint configuration where the material from one side of the joint locks into the other side of the joint. A tongue and groove configuration is one such example. This concept can be used to reduce fixturing requirements, especially for areas where a butt weld would otherwise be used.

    b. Various modifications or combinations of the joints shown.

Terms and Definitions

As with any new welding or processing technique, terms and definitions that are typically used for other manufacturing processes may or may not apply. For consistency, as many terms that can be common are typically made so. Terms similar to those indicated in the American Welding Society (AWS) 3.0: Welding Terms and Definitions are used as much as possible. However, there are many characteristics of a new process that cannot be described by terms used for existing technologies. In addition, with a new technology there are often multiple terms used to describe the same characteristics or features where there are known multiple terms to describe the same characteristics or features, both terms are used.

Weld or Processed Region Terms

As with other joining processing, the welded region has specific zones that are used to describe the weld. Since FSW and FSP are essentially the same process, except for a lack of a joint in FSP, the zones or characteristics of the welded or processed regions are similar. The characteristics of a friction stir spot weld are quite similar, but there are some basic differences.

Friction Stir Weld and Friction Stir Processing Region Terms

The terms that are generally used to describe a friction stir weld are detailed in this section. The first list refers to regions within the weld or processed region, after completion of the process [13,14]. The regions and related terms are shown in Figure 10.6. Figure 10.6 is a cross-section of a typical weld (lap penetration weld in this case).

1. Stir zone or nugget: The stir zone is the area of the weld or processed region, generally described by the location through which the friction stir tool pin has traversed. This area is also below the friction stir tool shoulder. The area is typically somewhat larger than the area of the pin and has a fine grain structure. The material is typically fully recrystallized.

2. Thermomechanically affected zone (TMAZ): An area outside the stir zone where the material has been plastically deformed, but not mixed, and possibly has undergone changes due to elevated temperatures experienced by the material. This zone can be nonexistent or nearly so, depending on the material that is being welded or processed.

3. Heat affected zone (HAZ): An area outside the TMAZ where the material has not been plastically deformed, but has possibly under-

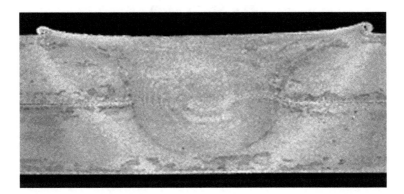

FIGURE 10.6
Friction stir weld cross-section and regions.

gone microstructural and material property changes due to the elevated temperatures experienced during welding or processing.

A second set of terms is used to describe the regions surrounding the weld or processed region. These terms are also related to the friction stir tool [13,14]. These terms are shown in Figure 10.7 and include:

1. Advancing side: This is the side of the weld or processed region where the rotation of the friction stir tool and the travel direction are parallel or where the net velocity of the friction stir tool is highest. It is dependent on the rotation direction. If we refer to Figure 10.7, where the tool is shown to be rotating clockwise, the advancing side is on the left side of the weld or processed region as viewed toward the travel direction or behind the friction stir tool. The advancing side of the weld or processed region typically exhibits the most mixing and is typically the location of most internal defects.

2. Retreating side: This is the side of the weld or processed region where the rotation of the friction stir tool and the travel direction are antiparallel. If we refer to Figure 10.7, where the tool is shown to be rotating clockwise, the retreating side is on the right hand side of the weld or processed region as viewed toward the travel direction or behind the friction stir tool. This retreating side of the weld or processed region typically exhibits little mixing, but rather more extrusion or forging of material.

3. Leading edge: This is the area in the front of the friction stir tool during welding or processing. It has unprocessed material, which is being drawn into the welded or processed region.

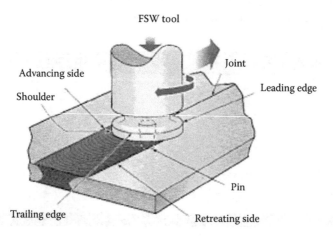

FIGURE 10.7
Friction stir regions related to friction stir tool. (Courtesy of TWI, Ltd.)

4. Trailing edge: This is the area behind the friction stir tool during welding or processing. It contains the welded or processed material in its final state.

Friction Stir Spot Weld Region Terms

As noted, FSSW is one variant of the FSW process. It is similar to FSW, with the exception that the traverse portion of the processes is eliminated. Thus, any terms that reference the traverse motion of the friction stir tool are not applicable. This includes the travel speed, advancing side, retreating side, and so on. However, most of the weld zones also apply to FSSW. The regions and related terms that are specific to a friction stir spot weld are shown in Figure 10.8. The specific terms related to FSSW include the exit hole and the flash ring. The exit hole is similar to the exit hole left after friction stir weld at the end of the weld where the friction stir tool is retracted from the material. It should be noted that this exit hole can be eliminated with the refill FSSW process described in the section on special implementations [15]. The flash ring is the material that is extruded to the outside of the friction stir tool shoulder (See Section 2.2) during welding. Since material cannot be destroyed, the material that is displaced from the friction stir tool pin (See Section 2.2) region must be displaced somewhere.

Friction Stir Tool Terms

Friction stir welding and processing requires a specially shaped friction stir tool. When rotated, the tool helps generate heat through friction and mechanical work of the material. Furthermore, the tool serves the critical functions of mixing, displacing, and consolidating the material behind the tool. The friction stir tool has several generic features and geometrical dimensions that are described below. If we refer to Figure 10.9, there are generally two basic configurations of friction stir tools. The first is shown as Configuration A (on the left in Figure 10.9) and typically inserts directly into a spindle or collet holder. It is typically at least several inches long, with the upper portion containing critical geometry to appropriately allow the tool to be seated

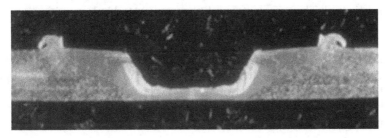

FIGURE 10.8
Friction stir spot weld regions or zones.

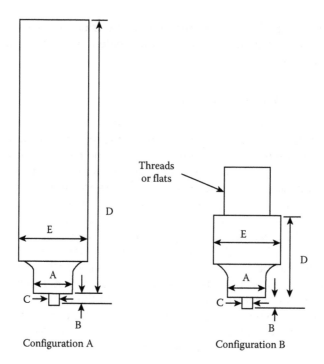

FIGURE 10.9
Friction stir tool descriptions and geometry.

into the tool holder with minimal run-out. In addition, this tool configuration has other features (e.g., flats) to enable the tool to be secured into the spindle. The second configuration, shown on the right, is a smaller version and can be more appropriately described as an insert. For Configuration B, there typically is a specially designed tool holder that will mate with the friction stir machine's spindle. The advantage of this configuration is that it minimizes the use of friction stir tool material, which can be costly and difficult to machine. In addition, this configuration can allow for quicker tool changing procedures.

There are several basic features and terms that are used to describe the friction stir tool. These features and terms are listed below, with discussion of design variants to follow.

1. Friction stir tool: The friction stir tool is the object with specific geometry that the friction stir machine plunges into and traverses through the material to be welded or processed. The friction stir tool is rotated by the friction stir machine and performs the functions of heating, mixing, and consolidating the material that is welded or processed.
2. Shoulder: The shoulder is the portion of the friction stir tool that contacts and moves along the top surface of the weld joint (see

Figure 10.1). The shoulder may have various geometrical shapes or configurations.

3. Shoulder diameter: The shoulder diameter is the diametrical dimension of the friction stir tool shoulder. This is Dimension A in Figure 10.9.

4. Pin or probe: The pin is the part of the friction stir tool that extends below the shoulder of the friction stir tool. The pin may have various geometrical shapes or configurations.

5. Pin or probe length: The pin length is the linear distance in the axial direction of the friction stir tool from the most distal portion of the shoulder to the most distal portion of the pin. This is Dimension B in Figure 10.9.

6. Pin or probe diameter: The pin diameter is the diametrical dimension of the pin of the friction stir tool. The diameter may vary as a function of axial position along the length of the pin. This is Dimension C in Figure 10.9.

7. Tool length: Most often, the tool length is the linear dimension from the critical mating point of the friction stir tool to the shoulder. In some instances, the tool length is the dimension to the most distal point on the friction stir tool pin. This is Dimension D in Figure 10.9.

8. Shank diameter: The shank diameter is the diameter of the friction stir tool at the contact point or critical mating point between the friction stir tool and the tool holder or spindle. This is Dimension E in Figure 10.9. The diameter may vary or be tapered in some implementations.

9. Tool coating: The tool coating is the type of material used to coat the tool and provides some benefit in certain circumstances. Coatings are typically used for improved wear resistance or modification of the friction coefficient.

10. Tool material: The tool material is the material from which the friction stir tool is fabricated.

11. Tool hardness: For certain tool materials, heat treatment is employed to harden the tool. This is the specified hardness and should be indicated on the friction stir tool drawing.

12. Tool number: The label or identification process used to designate the friction stir tool. The tool number, placement, and identification method should be indicated on the tool drawing.

Welding or Processing Parameters

Like most other welding or joining techniques, FSW and its related processes have processing parameters or conditions that are unique to the process. The process parameters can be separated between those during welding or processing and those at the start and the end of the process. The significant processing parameters during welding or processing include:

1. Travel speed or welding speed: The linear speed at which the friction stir tool is traversed through the material by the friction stir machine, typically expressed in inches/minute, meters/minute, or millimeters/second. This parameter does not apply to FSSW.

2. Rotation speed: The rotational speed of the friction stir tool as rotated by the friction stir machine, typically expressed in revolutions per minute (RPM).

3. Rotation direction: The direction of rotation of the friction stir tool as viewed from the top of the tool looking down on the process, expressed as clockwise (–) or counterclockwise (+).

4. Travel angle or tilt angle: The angle of rotation about the axis perpendicular to the welding direction. The axis can be calculated by taking a cross product of the travel direction and the axial direction of the friction stir tool at any instance in time. A tilt backward is defined as positive. The travel angle is shown in Figure 10.10.

5. Work angle: The angle of rotation about the axis parallel to the welding direction. As viewed from the trailing edge of the tool, a positive work angle would be such that the tool is tilted to the right. This angle is generally zero for most joint configurations, with the exceptions being the lap fillet weld or dissimilar thickness butt weld (refer to Figure 10.5). The work angle is shown in Figure 10.10.

6. Vertical position: The axial position of the friction stir tool with respect to the surface of the material. The zero position is typically defined as the position of the friction stir machine where the pin of the friction stir tool contacts the surface to be welded or processed.

7. Thrust or vertical force: This is defined as the force in the axial direction of the friction stir tool during welding or processing.

It should be noted that the friction stir tool design is a critical parameter in creating quality welds or processed regions. It can be considered a parameter. Tool features are defined in pre-going areas and tool design variants are discussed later.

As with many welding processes, there are additional parameters that are used at the start and stop of the weld. The parameters specifically related to the start and stop include:

1. Pounce position: This is the position of the friction stir tool above the material at the start of the weld or processing. The friction stir machine typically moves the friction stir tool relatively quickly to this position from some previous position before starting the weld.

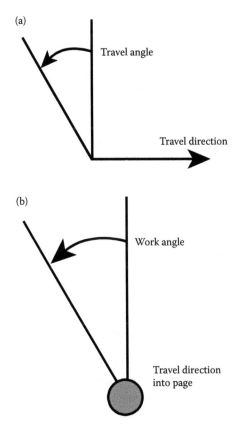

FIGURE 10.10
Work and travel angle description.

2. Plunge depth: This is the distance the friction stir tool plunges into the material from the pounce position or from the surface of the material at the start of the weld.

3. Plunge speed: This is the speed at which the friction stir machine plunges the friction stir tool into the material at the start of the weld or processing. This is typically relatively slow, so as to reduce forces on the friction stir tool pin and to allow sufficient time for the material to be heated to a temperature where it can be readily plasticized. For FSSW, this speed is typically faster than for FSW or processing so as to minimize the cycle time of the welding process.

4. Start rotation speed: This is the rotation speed of the friction stir tool during the plunging operation. It may or may not be different than the rotation speed during welding or processing.

5. Start dwell time: In certain circumstances, the friction stir tool will be paused at the bottom of the plunge motion before the machine

starts traversing the friction stir tool through the material. This is referred to as the start dwell time.

6. End dwell time: In certain circumstances, the friction stir tool will be paused at the end of the welding or processing path before the machine retracts the friction stir tool from the material. This is referred to as the end dwell time.

7. Retract speed: This is the speed at which the machine exits the friction stir tool from the material at the end of the weld. It is typically much faster than the plunge speed.

Welding Setup Conditions

As with any welding or joining process, it is possible to have setup conditions that are not ideal or are error conditions. If not properly controlled, these conditions can lead to undesirable or poor welding results. In the extreme, when not properly controlled, these conditions can lead to welding defects. The following is a list of setup conditions that should be properly controlled:

1. Gap: In a butt weld, this is the horizontal distance between the two faying surfaces. This is shown in the upper portion of Figure 10.11. In a lap penetration or fillet weld it is the vertical distance between the welds, as shown in the lower part of Figure 10.11. For a butt joint, the FSW process is generally considered to be fairly tolerant to this condition. This condition can occur when part quality is deficient (e.g., edge of parts not straight) or when fixturing does not sufficiently hold the joint together.

2. Mismatch or offset: In a butt weld, this is the difference in thickness of each of the faying surfaces or the difference in vertical position.

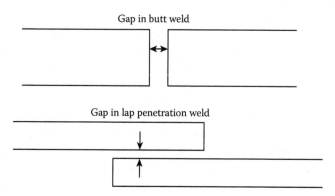

FIGURE 10.11
Gap condition.

This is shown in Figure 10.12. This condition is generally caused by poor part quality.

3. Overlap: In a butt weld, this condition occurs when one side of the faying surface is lapped onto the other faying surface. This is shown in Figure 10.13. This condition most often occurs with parts that are not flat, but is exacerbated with parts with beveled or sharp edges (poor shearing) and with poor fixturing.

4. Off-seam: This condition can be described as improper alignment of the friction stir tool with the joint line, as shown in Figure 10.14. The process is tolerant to misalignment of the joint such that the joint is on the advancing side of the tool, but relatively intolerant to misalignment to the retreating side of the tool [16]. The distance by which the tool can be off-seam is generally correlated to the size of the friction stir tool pin diameter.

For each of these setup conditions, the production control plan will indicate the acceptable limits. If the condition is above the suggested limit, the part will be scrapped or reworked with special processing conditions. A weld qualification process is typically performed to determine the acceptable limits of these conditions.

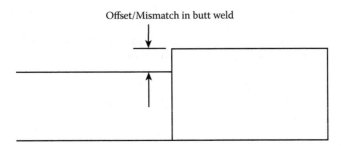

FIGURE 10.12
Mismatch or offset condition.

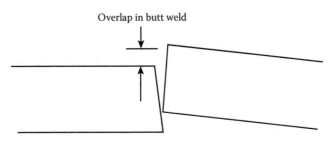

FIGURE 10.13
Overlap condition.

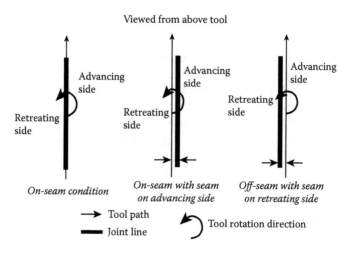

FIGURE 10.14
Off-seam condition.

Welding or Processing Measurement and Defects

The various regions of the weld or processed areas were described in a previous section. The regions have various measurements or metrics that are used to describe the shape of the weld. Depending on the actual value, these measurements, above or below certain limits, can be considered to be defects. The defects are described in detail afterward. The following list details the various weld or processed region measurements.

1. Penetration depth: The distance from the lowest portion of the top of the weld to the lowest portion of the stir zone. This is Dimension B in Figure 10.15.

2. Penetration ligament: The distance from the bottom of the material to the lowest portion of the stir zone in a butt weld. This is Dimension C in the upper portion of Figure 10.15.

3. Weld width: The width of the weld at the upper surface of the weld (surface on which the tool shoulder was located). This is Dimension A in Figure 10.15.

4. Weld interface width: In a lap penetration weld, this is the distance from one side of the stir zone to the other at the height of the original joint interface. This is Dimension C in the lower portion of Figure 10.15.

5. Flash height: Flash is an expulsion of material that is not contained underneath the tool shoulder. Flash height is defined as the maximum height of the flash. The flash height is shown as Dimension A in Figure 10.16.

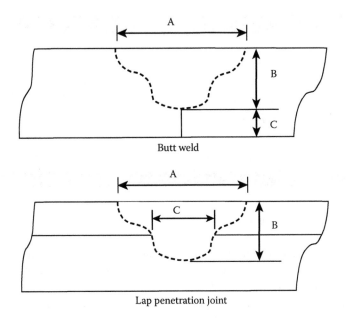

FIGURE 10.15
Basic weld measurements.

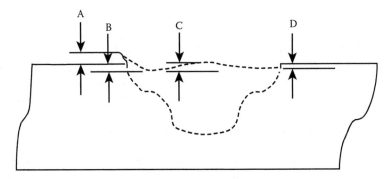

FIGURE 10.16
Surface feature measurements.

6. Undercut: A notching effect at the edge of the weld, generally noted as a sharp corner at the edge of the weld. The measurement for undercut is shown as Dimension D in Figure 10.16.

7. Underfill/concavity/thinning: A condition where the surface of the resultant weld is depressed below the surface of the original material and/or the overall thickness of the material after welding is less than the original material thickness. This is similar to undercut, but is a more gradual condition. This condition is not necessarily unacceptable and some may be required. Generally as the thinning increases,

FIGURE 10.17
Interface deformation in a lap weld.

the tensile strength of the weld decreases. This condition should not be confused with undercut. The amount of thinning is measured as Dimension B in Figure 10.16.

8. Crown/convexity: A condition where the weld appears convex near the center of the weld. The amount of convexity is defined as the distance from the lowest portion of the weld surface to the highest. This is shown as Dimension C in Figure 10.16. If there is no crown in the weld then this value is zero.

9. Interface deformation/hook/thinning (lap penetration weld): A condition with lap penetration welds where the original material interface is plastically deformed in the TMAZ. Figure 10.17 shows a cross-section of a weld with such a condition. This condition effectively thins the upper sheet. In a lap shear tensile test this will reduce the strength of the weld. It is controlled by FSW tool design and weld parameters. Lastly, it is generally more severe on the retreating side of the weld. This condition is generally not acceptable and should be limited to less than 10% of the material thickness, but it can be beneficial under certain conditions. One such condition is if the interface deformation folds over such that it hooks the upper sheet onto the lower sheet. This condition is not uncommon in a friction stir spot weld.

Like all other joining processes, it is possible for defects to occur. Some of the previously mentioned measurements, if too large or too small, can be considered to be defects. The following lists other potential defects and their general potential effects:

1. Lack of penetration/kissing bond: An avoidable condition where full weld penetration butt weld is not achieved where there is inadequate bonding, caused by inadequate tool pin length or insufficient plunge

depth. If significant, then the original joint line can be seen on the bottom side of the weld. If minor, this condition is difficult to observe even in a cross-section, but can be detected through root bend testing and tensile testing where failures will emanate from the root of the weld. This condition generally reduces the strength of the weld.

2. Remnant oxides or lazy-s defect: Remnant oxide traces are remnants of the oxide layers from the original joint area that are not fully mixed in the stir zone. They can be seen upon viewing a weld cross-section under magnification. Generally, they always can be found, if enough effort is spent looking for them. However, depending on the severity and application, they are not necessarily unacceptable. If significant, they can reduce static and fatigue properties of the welds. They can be indicative of less than optimal weld parameters, heavy original oxides (old material), or off-seam conditions. Depending on the joint configuration, they will appear differently.

    i. Butt weld: In butt welds, remnant oxide traces often appear as s-shaped structures and sometimes are referred to as lazy-s defects. See the upper photograph of Figure 10.18 for an example. The particular remnant oxide trace shown in Figure 10.18 extends from near the top of the weld to the bottom of the weld.

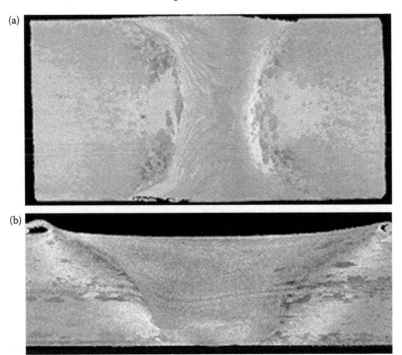

FIGURE 10.18
Remnant oxide traces in a butt weld (a) and lap weld (b).

ii. Lap weld: In lap welds, remnant oxide traces extend from the original interface into the weld. They can extend from one side through to the other, but often extend only partly through the weld. They also may deviate in a vertical direction from the original interface. See the lower portion of Figure 10.18 for an example.

3. Off-seam conditions: This is a condition where the FSW tool is not centered with respect to the seam for a butt weld condition. Depending on the direction (see Figure 10.14) and amount the friction stir tool is off-seam, the ramifications vary. If the seam runs through the retreating side of the weld, then significant reductions in strength can be observed. This can be an unacceptable condition. If the seam runs through the advancing side of the tool, the resulting loss in strength is less significant.

4. Voids/wormholes: Voids are discontinuities or holes in the stir zone and in most cases are considered to be unacceptable. These discontinuities are, at times, referred to as porosity. However, since porosity is a cluster of voids related to the generation of gases, and FSW does not create any gas, the term porosity is not applicable. These typically occur on the advancing side of the weld. They generally appear in two forms

   i. Internal void: A void that is internal to the weld and not visible from the surface. Large internal voids possibly can be noticed by viewing the end of the weld or a simple saw cut. Smaller voids are only viewable upon observing a cross-section. They can also occur in various sizes and quantities. They are typically caused by incorrect weld parameters, tool designs, insufficient vertical force, or gaps. An example is shown on the left side of Figure 10.19.

   ii. Surface void: A void that is visible from the surface of the weld. It usually appears as a trench and results from insufficient vertical force, incorrect weld parameters or tool designs, or gaps. An example is shown on the right side of Figure 10.19.

5. Undercut: Undercut (described previously) should generally be avoided, as it affects tensile strength and fatigue properties. It is generally caused by material mismatch, overlap, or weld depth being too great. The latter two conditions will cause more negative effects than material mismatch conditions.

6. UnderFill: Underfill (described above) may or may not be a defective condition and will depend upon the amount. In some case settings, underfill will naturally occur, especially as the travel angle is increased or there are larger gaps. In some instances, a weld land is included in the part geometry to minimize any potential effect of this condition.

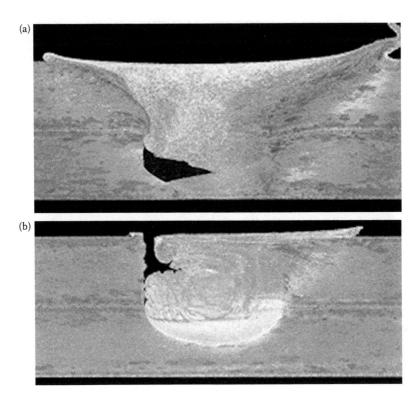

FIGURE 10.19
Surface and internal voids.

7. Flash: Flash is generally not a defect, but is aesthetically displeasing. Its effect on mechanical properties tends to be negligible, unless it is also associated with significant undercut or significant underfill. Flash can always be removed with a post welding process such as grinding. Flash can result from too much vertical force, incorrect welding parameters, mismatched material thickness, or increased material thickness. A picture of typical flash is shown in Figure 10.20.

8. Blistering or scalloping: This is a range of conditions on the top weld surface. This condition can be described as a flaky or a blistered appearance. It can be caused by incorrect parameters, damaged tool shoulder, or insufficient force. This condition generally has a negligible effect.

9. Fuzz: This is a condition where the surface of the weld has fine hair-like or fibrous features. It is typically caused by slight damage to the friction stir tool shoulder. It has negligible effect on weld properties.

FIGURE 10.20
Flash.

It should be noted that the effects of these various defects can range from negligible to significantly detrimental. Thus for a defect to be considered important or critical, it must have some sort of negative effect. This can be application dependent. Production control plans should indicate the acceptable limits of each of these potential defects.

Friction Stir Tool Material and Designs

A variety of tool materials and a large number of tool designs have been demonstrated to generate quality friction stir welds or processed regions. Given the general nature of this chapter, an exhaustive discussion of all friction stir tool designs and potential tool materials is not practical. For a more detailed discussion on any one design and/or tool material, there are numerous references that can be reviewed.

From the time of its invention in 1991 at The Welding Institute (TWI), through the first several years afterward, the majority of FSW development investigations were performed on aluminum. Early investigations concentrated on tool design and weld parameter development. A significant portion of the work was performed at TWI and by the first FSW licensees. In 1994, TWI created and started a group sponsored project for which many of the first FSW licensees provided funding. This project is generally credited with generating the first viable tool design concept and is often referred to as the 5651 tool (after the project number) [17]. This tool design is shown in Figure 10.21. The tool can be described as having a cylindrical pin with

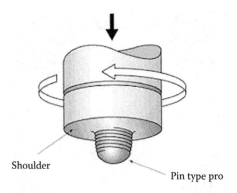

Shoulder

Pin type pro

FIGURE 10.21
The welding institute group sponsored project 5651 FSW tool design. (From Dawes, C. J., et al. TWI Member Report 5651/35/95, November 1995. With permission.)

threads and a round bottom. The shoulder was concave by a few degrees, had a diameter 2–3 times the diameter of the pin, and was smooth. The tool material was H13 Tool Steel heat treated to a moderate level of hardness.

Friction Stir Tool Design Advancements

There have been numerous advancements to the design of the friction stir tool since the early stages. The advancements have been classified below, with the benefits indicated. Many of the improvements have been related to increasing the swept volume of the friction stir tool versus the actual volume [18], and to improve the ability of the tool to mix material. The advancements include:

1. Conical pins: This reduces the stress levels within the friction stir tool pin and improves tool life. This innovation reduces thrust loads and also reduces the force required to traverse the tool through the material.

2. Flat bottom: This was a relatively early advancement that increased mixing at the bottom of the friction stir tool. This allows for a more stable process and less chance for lack of penetration at the root of the weld.

3. Flats on pins: Existing in varying numbers (2, 3, or 4), flats on the pins increase the swept volume to actual volume ratio. As noted, this tends to increase the ability for the tool to mix material and reduces traverse and thrust forces. A typical tool with a conical pin and flats is shown in Figure 10.22.

4. TriFlute: This is a pin design innovation developed by TWI that includes three flutes on the pin. This concept is similar to, and has similar benefits to, flats. The main difference is that the flutes are grooves with a round shape. This concept is shown in Figure 10.23.

FIGURE 10.22
FSW tool design with conical shaped pin, threads, three flats, and a flat bottom.

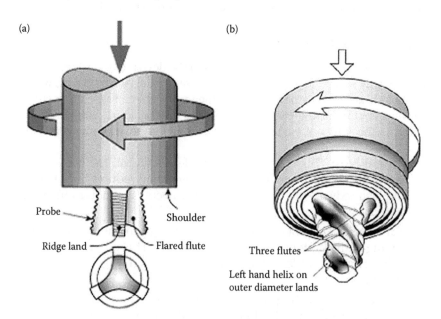

FIGURE 10.23
Examples of TriFlute friction stir tool design. (Courtesy of TWI.)

5. Whorl: This is a pin design innovation developed by TWI. This design involves the pin having square grooves versus the threaded shapes shown previously. This allows for additional tool surface area forcing or pumping material downward. This design acts more like an auger. An example is shown in the left side of Figure 10.24. This

(a)   (b)

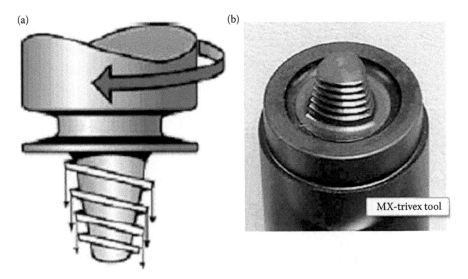

MX-trivex tool

FIGURE 10.24
Whorl (a) and Trivex (b) friction stir tool designs. (Courtesy of TWI.)

pin design concept can be in several formats, including as shown in Figure 10.24, cylindrical pins, various pitches, and so on.

6. Trivex: This is another innovation developed by TWI. The concept relates to the pin design, whereby the pin has three sides, and convex in shape. The three sides can have threads, grooves, or other features. An example is shown in the right side of Figure 10.24.

7. Skew-Stir: This is an innovation developed by TWI whereby the pin or whole tool is tilted on an angle with respect to the axis of rotation. This allows the pin of the friction stir tool to sweep through a larger volume of material. However, it can cause highly oscillatory loading and vibration. This concept is shown in Figure 10.25.

8. Whisks: These are tools with multiple pins and have been shown to have benefits for lap penetration welds and other soft materials (e.g., lead).

9. Lap weld designs: Lap penetration welds require different tool designs to limit the interface deformation or thinning defect. The lap penetration joint requires material mixing to favor the horizontal direction rather than the vertical direction. There are numerous designs that can be used for this purpose.

10. Stepped spiral: The stepped spiral is primarily used on the pin and is a step that spirals around the pin. This design concept requires a conical shaped pin. This tool design has been demonstrated to be especially useful for processing applications, but not as beneficial for welding, when a joint is present. This tool tends to increase traverse

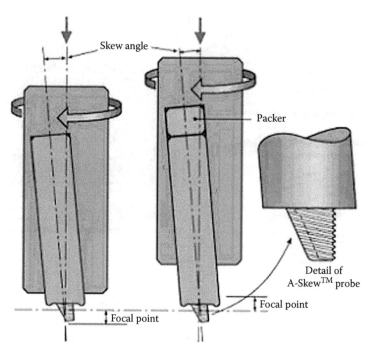

FIGURE 10.25
Skew-stir design. (Courtesy of TWI.)

and thrust forces. A typical stepped spiral tool design is shown in Figure 10.26.

11. Scroll shoulder: This is a tool design where there exists scroll like features on the surface of the shoulder [1]. An example is depicted in Figure 10.27. This design concept can be used to draw material in toward the pin. It has been shown to improve process stability (allowing for a wider range of parameters) and enables welding with a zero degree travel or tilt angle. There can be multiple scrolls, various pitches, various heights, and so on. In addition, the shoulder surface can be tapered [19].

12. Other shoulder features: It is possible to have other shoulder features, such as flaps, stepped spirals, dimples, and so on [1]. Each may provide benefits over other designs.

13. Other design concepts: There are infinite potential designs for the friction stir tool. The above list is not exhaustive and may not include specific advancements that are proprietary or internal to an organization, but is intended to provide guidance on what concepts are possible. Some designs are also patented.

FIGURE 10.26
Stepped spiral design.

FIGURE 10.27
Scroll shoulder.

Friction Stir Tool Materials

As noted, early developments of the FSW technology concentrated on aluminum. The tool material used for this work was mostly various tool steels. Since then, there has been significant interest in tool materials and expanding the friction stir technology into welding and processing of materials other than aluminum. Welding or processing of other materials can require the use of different tool materials, especially higher melting point materials. The tool material must have a melting point above the material that is to be welded or processed and must retain sufficient strength and toughness at these temperatures. Thus, a material that melts at a significantly higher temperature requires a tool material that may be different than that required for a lower melting point material. A good example is welding or processing of steel versus aluminum. For welding or processing of aluminum, the use of a tool steel for the friction stir tool is feasible. However, the use of tool steel is not feasible for welding or processing steel, as the tool itself will be deformed and destroyed quickly. The following lists a selection of tool materials and their potential use.

1. H13 tool steel: This is a low-cost material that is widely used for welding or processing of aluminum and lower melting point materials, such as magnesium, lead, zinc, and so on.

2. Tungsten carbide: This is a low to moderate cost material that can be used for welding aluminum, other lower melting point materials, and moderately higher melting materials (e.g., copper). This material can have better wear properties, but is more brittle than other materials.

3. Densimet (tungsten): This is a moderate cost tool material. Tungsten is a higher melting point material with high thermal conductivity. It can be machined but it is relatively soft, which limits the ability to include intricate features on the pin and shoulder. It can be used to successfully weld or process copper, bronze, and aluminum.

4. Tungsten–Rhenium: This is a costly material with better strength than pure tungsten. It has been used to weld steel with some success and titanium with limited success.

5. Tungsten–Lanthanum: This is a more moderate cost material that has been demonstrated to be capable of welding titanium and some steels.

6. PolyCrystalline boron nitride (PCBN): This is costly material that is manufactured at extreme pressures and temperatures. It is primarily used for welding and processing of steel. To date, it has been demonstrated to have the longest tool life when welding various types of steel material. However, due to its brittleness, it is sensitive to vibration and requires a low run-out spindle. It is available in multiple different grades. There is research and development being performed in developing different grades that may allow for improved FSW and processing capability.

7. Nickel and cobalt-based alloys: These high temperature alloys have high strength and good ductility, as well as good creep and corrosion resistance. Several alloys have been investigated for suitability for FSW. Of special note are:

   a. Nimonic 105: This alloy has been demonstrated to successfully weld thick section copper [20].

   b. MP159 steel: This is a moderate cost material, which has better high temperature fatigue properties than H13 tool steel, but reacts with aluminum at high temperatures. This can add difficulty during welding or processing. Coatings are required with its use, or otherwise, tool life is insufficient.

8. Others: The above list is not exhaustive. As such, there are other materials that can be used for welding and processing, with some being proprietary or patented.

Friction Stir Equipment

Like most other processes, equipment to perform FSW and the related processes are engineered in a variety of configurations. The requirements of the application will dictate the eventual type of equipment and its capabilities. Every manufacturing process and application have different technical requirements in four categories. These categories include payload/force, intelligence, accuracy/stiffness, and flexibility [13,21]. For the machine to be successful it must have capabilities that meet or exceed the requirements of the application. Similar to most other manufacturing processes, FSW and the related processes have requirements that are dependent on the application. It is noted that FSW and processing cannot be performed manually due to the significant forces that are required. The following sections will discuss the various types of equipment categories and their general capabilities.

Friction Stir Welding and Processing Equipment

Friction stir welding and processing require a very high level of force compared to other manufacturing processes. This requirement results in a situation where there is limited standard off-the-shelf equipment available to perform FSW. Thus, any FSW and FSP equipment tends to be customized and highly engineered. This necessarily causes the equipment to be expensive. However, more recent developments have shown that standard industrial robots (lower cost, off-the-shelf-equipment) can perform FSW and FSP. However, these solutions are not feasible for all applications [22–24].

Two-Axis Welding Equipment

The first production FSW machines and the first development machines were two-axis machines. These are still the most common type of machines today. These machines have a vertical axis and traverse axis. The vertical axis is used for plunging the friction stir tool into the material and maintaining the force during welding. The traverse axis is used to force the friction stir tool through the material and along the joint line. Two examples of typical machines of this type are shown in Figure 10.28. Both machines are panel welding machines, used for fabrication of extruded paneling. The machine on the left has the vertical axis, spindle, and clamping all mounted on the long beam. This machine was the first production FSW machine and is located at Hydro Aluminum, and was engineered and fabricated by ESAB AB [25]. The

(a)

(b)

FIGURE 10.28
Beam and C-frame two-axis panel welding machines. (Courtesy of ESAB AB and Friction Stir Link, Inc.)

machine on the right is a C-frame style machine, where the C-frame travels, engineered and fabricated by Friction Stir Link, Inc. [26]. The C-frame style machine is generally lower in cost, but does not readily allow for significant amounts of clamping near the joint line. Both of these machines have limited flexibility (can only manage long linear welds). There are many suppliers of this type of machine. In addition, they are engineered in all sizes and force capabilities depending on the application. Smaller versions are often used for research and development and qualification programs, while larger ones are typically used for production.

Three-Axis Welding Equipment

Three-axis machines are similar to the two-axis machines, except with the addition of a second traverse axis that is perpendicular to the first traverse axis. This can allow for motion in both the X and Y planes. These type of machines have two significant applications. The first is with long panel welding machines, similar to what is shown in Figure 10.28. An additional traverse axis can allow for the machine to readily adapt to different extrusion widths. In this case, the second traverse axis tends to have a relatively small range of motion. The second approach is with a shorter first traverse axis and a larger range second traverse axis, similar to a milling machine. This can allow for welding or processing of parts that have weld paths that are rectangular, circular, and so on, in nature. However, the machine must be operated with a zero degree travel angle. Operation at a zero degree travel angle is feasible, but significant attention must be paid to developing a process that is stable enough to manage the variation in the parts that will be encountered. The use of a zero degree travel angle tends to generate a less stable process and generally requires friction stir tools to have shoulder features.

As noted, these machines can be similar to CNC milling machines. In fact, it is possible to use a CNC milling machine for FSW and processing. However, extreme care must be taken, as the forces that are required for FSW are generally significantly in excess of those required for milling. In addition, many CNC machines are not designed to manage the level of the heat that is generated with the friction stir processes. Lastly, with standard CNC machines it is difficult to implement any sensing feedback. Thus, the ability to add intelligence to the machine, such as force control, can be very difficult as compared to a custom or robotic machine.

Multiaxis Welding Equipment

A multi-axis friction stir machine is any machine with five or more motion axes. Five or more axes (or degrees of freedom) allow for the friction tool to be oriented at any posture with respect to the world or part, as long as the range of motion of the individual axes permits. There are two major architectures for these machines. The first is a custom engineered gantry solution and the second is a robotic solution.

Custom Equipment

Multi-axis machines are engineered in many configurations. The most common configuration is a gantry type machine, where the three major axes provide linear motion in the horizontal plane (two axes) and third linear axis provides vertical motion. The fourth and fifth axes provide rotary motion and are perpendicular (orthogonal) to one another as well as the spindle axis. This allows the machine to approach the part with any orientation or posture. Thus a range of both work and travel angles can be used. Furthermore, welding or processing on three-dimensional paths can be accomplished. The gantry style is a common architecture, due to its inherent stiffness and ability to apply high forces efficiently. One such example machine is shown in Figure 10.29, engineered and fabricated by MTS Corporation. Other suppliers include, ESAB, Friction Stir Link, General Tool, Hitachi, and Transformation Technologies. These machines are typically custom engineered for the application, in that they are designed for the size and shape of the part that is to be welded or processed. They tend to be the most costly type of equipment.

As noted, there are different forms of custom multiaxis friction stir equipment. They tend to be designed specifically for the application, so there are many forms. A good example is a machine designed for the welding of tanks

FIGURE 10.29
Custom multiaxis friction stir welding machine. (Courtesy of MTS Corporation.)

or other round parts. In these cases, the machine may hold and move the part on an axis separated from the machine motion axes.

Robotic Welding Equipment

Robotic machines used for FSW and processing are the newest types of FSW machines [22–24]. They have six axes of motion and employ a standard industrial robot. They are desirable because they are much lower in cost, are very flexible, and can have significant added intelligence. They are lower in cost because they employ a common machine (industrial robot) that has been previously engineered and are fabricated in large quantities. In addition, they are flexible, allowing for multiple parts to be processed at the same time. This can improve productivity significantly, reducing operating costs. Lastly, with peripheral sensing equipment and/or software they can be made to have significant decision-making and adjustment capability (intelligence).

Until the early 2000s, it was impractical to perform FSW and processing with a robot. The main reason was that robots lacked the force capability until that time. However, in the early 2000s industrial robot companies (ABB, Kuka, Fanuc, etc.) began to produce robots with enough force capability. It should also be noted that FSW requires a significant level of intelligence capability as well. Thus, a robot with an open architecture is required. To date, only two robots have both the force capability and open architecture to allow friction stir welding and processing to be performed. These are ABB and Kuka. The first production capable FSW system was developed by Friction Stir Link, Inc. and is shown in Figure 10.30. This system has been used for production friction stir welding since 2004.

It should be noted that these robots do not possess the stiffness of many custom friction stir machines. Although stiffness is not absolutely required, any lack of stiffness must be managed. Any robot application software for friction stir welding and processing must have means of managing the relative lack of stiffness of the robot. In addition, these machines still do not possess the force capability of some friction stir machines. Thus, they can be limited in the material or thickness that can be welded or processed versus a custom machine. However, they can readily be used for many aluminum welding applications.

Friction Stir Spot Welding Equipment

Friction stir spot welding (FSSW) only requires the friction stir tool to be plunged into the material and retracted. Thus a spot welding machine does not need to be capable of traversing the tool, except for special implementations. Friction stir spot welding still requires a high level of force, but there are several other spot joining processes that have a similar characteristic. These include resistance spot welding and riveting. Thus, equipment that performs FSSW can be somewhat similar to equipment used for resistance

FIGURE 10.30
Robotic friction stir welding machine.

spot welding and riveting. In the case of FSSW, the equipment must have a rotary axis to rotate the friction stir tool and a vertical axis to plunge the tool into the material. The friction stir process requires relatively precise control of the vertical position, so electrical control is more desirable, versus the pneumatic or hydraulic control that is often used for resistance spot welding and riveting, respectively. As with resistance spot welding, a significant application for FSSW is within the automotive industry on the end of a robot arm. Two examples are shown in Figure 10.31, where FSSW units from Friction Stir Link (left) and Kawaski (right) are shown.

FSSW machines are also engineered and fabricated in different formats. These are pedestal units and benchtop units. These types of machines can be used where a robot or person places the part to be welded under the machine. Examples of these machines are shown in Figure 10.32.

Friction Stir Control Modes and Monitoring

Friction stir welding and the related processes can be performed using several basic control strategies. The control strategies are referred to as force control, position control, and hybrid control. These control strategies are

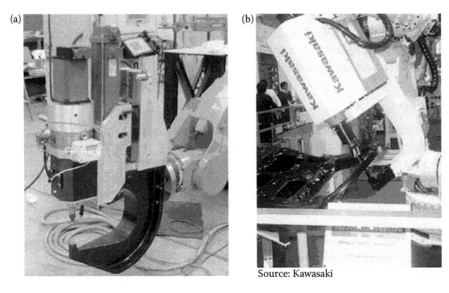

Source: Kawasaki

FIGURE 10.31
Robotic FSSW equipment Friction Stir Link, Inc. (a) and Kawasaki (b). (Courtesy of Kawasaki.)

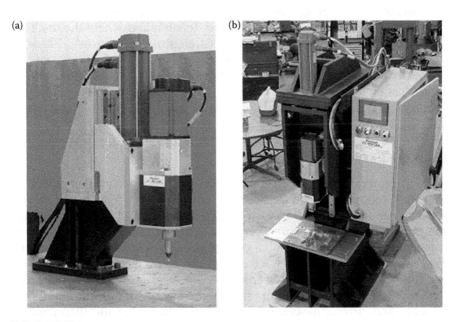

FIGURE 10.32
Benchtop and pedestal FSSW machine.

generally applied to the vertical position of friction stir tool. Friction stir welding, FSSW, and friction stir processing can all be performed with these control strategies. However, the exact implementations can be somewhat different based on the machine and process.

Position Control

Position control is the simplest of all control strategies. With this strategy, a predetermined vertical position or motion profile is programmed into the friction stir machine. The friction stir machine then relies on internal motion control equipment (e.g., servo drives) to cause the machine to maintain the programmed vertical position or follow the desired motion profile. For example, during welding or processing, the machine controls the vertical position of the friction stir tool to a predetermined or set value. In response to disturbances (e.g., thicker material), the machine will attempt to maintain the same position.

Position control can be effective, but needs to be used with certain cautionary notes.

1. In situations where gaps exist, the resulting weld quality may be reduced, especially when the gap gets large.
2. In situations where material thickness increases or decreases, weld quality may not be consistent. In the case where the thickness increases, weld flash will be created. In situations where the thickness decreases, the weld quality (strength) may decrease. These tendencies are dependent on the process, material thickness, and friction stir tool design and should be investigated in a process qualification program.
3. In situations where there is mismatch, similar results to material thickness variation will occur.

Force Control

Early prototype and production friction stir equipment demonstrated that force control could be an effective control strategy for the friction stir welding process [22,27,28]. In each case, a digital control loop is closed on the thrust force (in axial direction of tool) to maintain the thrust force at a predetermined level. The force control strategy requires some means of measuring the actual thrust force during the welding process. This can be accomplished by numerous means. A few measurement schemes include load cells, strain gauges, pressure measurement on hydraulically controlled systems, and current measurement (converted to torque and then applied force) for electric motors, to name a few. A proportional, integral, and derivative (PID) digital control loop is then closed on the measured force with controller gains tuned appropriately [29], as shown in Figure 10.33. Once tuned appropriately, the PID digital control loop applies adjustments to the

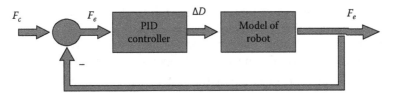

FIGURE 10.33
PID force control.

commanded position of the machine or the pressure (hydraulic systems) to maintain the predetermined level of force, while reacting to disturbances with appropriate reaction time, without oscillation.

Force control can be beneficial. It can allow for the machine to automatically adjust to height or material thickness variations of parts in a production mode. It can also allow for developing an understanding of the range of forces over which successful welds can be generated. However, force control has several drawbacks that must be managed, otherwise unacceptable quality can result. The first drawback is caused by the force profile versus vertical position for friction stir welding and processing. The force versus vertical position profile has an unstable region. At the start of the plunge operation, the force versus position profile generally increases as the depth of the pin increases until the shoulder is fully engaged. The maximum thrust force occurs as the shoulder engages the surface of the material. However, once the shoulder is plunged below the top surface of the part, the force versus vertical position profile decreases with increasing depth. This results in a positive feedback condition where, if the tool gets too deep, the force control system will drive the friction stir tool deeper and deeper into the material. The second drawback is that the force required in certain production conditions is not the same as that required for nominal conditions. Conditions where the required thrust force level changes include the existence of gaps, mismatch, and thermal build-up (higher temperature). Each of these conditions can exacerbate the unstable mode previously discussed. Given these conditions, a pure force control strategy is rarely implemented into production, unless there is precise or good control of inbound parts and alternative control strategies are put in place to avoid the friction stir tool over plunging. These alternative control strategies are discussed in the next section.

Hybrid Control

Due to the drawbacks noted for both position control and force control, many production implementations use some sort of hybrid control strategy. Some examples include:

1. A master-servant control strategy where the position control is the master and force control is the servant. In this situation, the force

control only operates within certain limits of position. Thus, upon the condition that the actual position is within certain predefined high and low limits, the machine software will activate the force control, otherwise the machine will operate in position control.

2. A force control strategy where the system is prevented from welding or processing too deep by an additional mechanical and/or electrical system, such as hard-stops, rollers, height sensors, and so on, or a combination thereof.

3. A force or position control system where an operator is present and observing the weld in real time, sometimes through a camera. In this case, the operator acts as the master and modifies commanded position or force, depending on welding results.

Force and other Data Monitoring

Almost all friction stir welding machines have some means of measuring thrust force and other important processing variables. Other processing variables include rotary torque, traverse force, rotary speed, travel speed, and so on. These variables are often displayed on the operator interface panel. The operator views the feedback during welding or processing to ensure the process is within specified limits. This is referred to as process or data monitoring. The process monitoring software does not control the process based on the feedback, except for cases when the values exceed high or low limits. On many machines there is a warning and abort limit. In the cases when a warning limit is exceeded, but not an abort limit, the machine will keep welding or processing while creating some audible or visible alarm. In the case where the abort limit is exceeded, the process will be halted or aborted.

For record keeping purposes, these data can be stored for future investigation. This is especially helpful for qualification programs. In production, data storage can become problematic, especially for high sample rates or long welds. Physical storage and analysis of the data can be burdensome, unless care is taken.

Special Equipment Implementations

There are several special implementations of the friction stir technologies that can be used or provide benefit in certain circumstances and applications. These implementations generally require some additional equipment.

Self-Reacting FSW

With the standard FSW process, the FSW tool has a pin and a shoulder. That standard welding process requires the back side of the FSW joint to be fully supported with fixturing or material from the part. However, it is possible to implement a second shoulder on the lower side of the friction stir tool to eliminate the need to support the part on the back side. This friction stir tool configuration is referred to as a self-reacting friction stir welding (SRFSW) tool or sometimes as a bobbin tool, since the tool is similar to a bobbin [1,30,31]. An example of a self-reacting tool is shown in Figure 10.34. The SRFSW tool can come in two basic forms; a fixed gap version where the distance between the two shoulders is fixed and an adjustable gap version where there is the capability to adjust the gap between the shoulder in real-time. The gap can be adjusted by adding actuation to the second shoulder. The first implementation (fixed gap) normally requires additional tool shoulder features to be effective [30], unless the machine and parts are very precisely controlled. The adjustable gap SRFSW tool requires significantly more complex and costly equipment.

The SRFSW tool is beneficial in certain applications. One excellent application is fabrication of tubular or hollow structures. For example, two long C-channels can be welded together to form a rectangular tube. This eliminates the need for an internal mandrel that would be required with the standard FSW process. These internal mandrels can be costly and cumbersome. The SRFSW process can also eliminate the need for a vertical axis, if all of the welding is at the same relative height.

Like any other technology, there are also challenges or disadvantages. The SRFSW process eliminates the thrust force requirement (both shoulders react to equal and opposite forces), but it significantly increases transverse

(a)

(b)

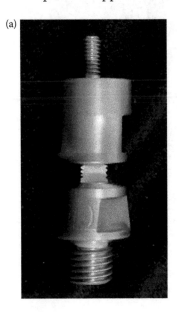

FIGURE 10.34
Self-reacting FSW tool and SRFSW tool welding. (Courtesy of MTS Corporation.)

loads (perpendicular to weld direction). In addition, the concept is generally more sensitive to vibration or lack of stiffness in the machine. Furthermore, unless special features are included on the shoulder, the process can be very sensitive. Lastly, there is the risk of material extruding between the pin and shoulder, causing the tool to jam.

Retractable Pin Friction Stir Welding

Similar to the adjustable gap SRFSW tool concept, it is also possible to have an adjustable length pin. This is referred to as the retractable pin friction stir welding tool [13,32–34]. An example system, engineered and fabricated by MTS Systems Corporation, is shown in Figure 10.35. This concept can be used to eliminate the exit hole. It is most applicable to butt welds or friction stir processing. However, the pin must be retracted over some distance, to avoid leaving unwelded regions near the end of the weld or processed region. Thus, for a butt weld, one potential drawback is an unwelded region (or lack of penetration) near the end of the weld joint, unless the machine traverses the tool through a previously welded region or into a region where

FIGURE 10.35
Retractable pin tool system. (Courtesy of MTS Corporation.)

there is no joint. This technology is most applicable to circumferential welds on round parts, where the weld path can overlap (greater than a 360° path) and the pin can be retracted after one complete revolution. Like the adjustable self-reacting FSW tool, this concept adds complexity and significant cost. In addition, there are precautions that must be taken to prevent the pin and shoulder from jamming from extrusion of material between the two surfaces. Thus, it is most applicable where there is no other alternative or ability to place the exit hole in a benign location.

Independent Pin and Shoulder Rotation or Stationary Shoulder

It is feasible for the pin to rotate independently from the shoulder or even to have a stationary shoulder. This concept has been demonstrated to be beneficial for materials with low thermal conductivity, such as thermoplastics and titanium [35–37]. In these cases, the shoulder is stationary and can be considered to be a pad. Otherwise there has been little published work on the merits of independent rotation of pin and shoulder at different speeds. However, in theory this could also provide benefit for materials that are sensitive to high-heat input/rotation speeds.

Preheating

Preheating of the material ahead of the friction stir tool has been investigated [37–39]. The objective of the preheating is to reduce the forces required to weld or process material, as well as increase tool life. In theory, preheating should provide more benefit for higher melting point alloys, such as steel and titanium [37]. Preheating has also been investigated for lower melting point alloys such as aluminum and magnesium [38,39]. Several preheating mechanisms have been considered, including induction and laser. In all cases, the preheating requires extra equipment, cost, and adds more process variables. In addition, preheating can require different friction stir parameters and tool designs, due to the existence of locally softened material near the friction stir region. In some of these studies preheating has been shown to be beneficial [37].

Refill Friction Stir Spot Welding

Refill FSSW is a method by which the exit hole is filled during the FSSW process [15]. It employs a similar technology to the retractable pin tool process. In this case, after the pin plunges into the lower sheet, it is retracted into the shoulder, as the shoulder is forced into the upper sheet. This concept provides more of an aesthetic benefit than structural benefit. The process eliminates the exit hole and reduces the flash ring that is inherent with the standard FSSW process, but provides minimal structural benefit. The standard FSSW process has been shown to be capable of generating weld failures

1500 μm

FIGURE 10.36
Refill friction stir spot weld cross-section.

around the periphery of the stir zone, demonstrating that the exit hole is not the weakest point in a friction stir spot weld [2,3]. Thus, the strengths of welds from each process are about the same. A cross-section of a refill friction stir spot weld is shown in Figure 10.36. This solution increases cycle time, in addition to adding cost and complexity to the system, similar to the retractable pin tool solution.

Stitch or Swing Friction Stir Spot Welding

Another variant of friction stir spot welding is to include a small translation motion. This is referred to as stitch welding and is equivalent to a short weld using the standard friction stir welding process. The length of the stitch weld is typically less than 10 mm. In one such implementation, developed by Hitachi, the translation is accomplished with a swinging motion [40]. This is referred to as a swing-stir system and is shown in Figure 10.37. The addition of the translation allows for a larger weld zone. Thus, it should allow for higher strength welds. The disadvantage of this concept is more complex equipment and fixturing, as the standard FSSW technique allows for simple clamping techniques.

Friction Stir Parameters and Properties

Friction stir welding or processing parameters and resulting mechanical properties vary significantly depending on the material, alloy, and thickness. Given that there have been numerous studies conducted to developed welding or processing parameters and to understand resulting material properties, an exhaustive discussion of details and results for all materials, alloys, and thicknesses is not feasible. Rather, a general discussion of trends and special conditions for various alloy and material families is presented. The materials for which the most development and production implementation has occurred will be discussed, with a few notes on other materials to follow.

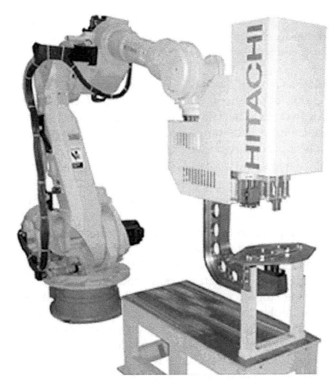

FIGURE 10.37
Hitachi swing-stir system. (Courtesy of Hitachi America, Ltd.)

Aluminum

As noted, the majority of the friction stir welding and processing research and initial production applications have involved aluminum. There are several reasons for this, including:

1. The relative ease by which aluminum is friction stir welded.
2. Conversely, the relative difficulty to weld aluminum with traditional processes. Some aluminum alloys are nearly impossible to weld with traditional processes.
3. Significant benefits can be realized, such as improved mechanical properties, lower distortion, and lower operating costs.
4. Aluminum is a widely used material in many industries, so there is a large potential market.

The welding parameters or conditions and resulting properties for aluminum vary significantly among the alloy families. In general, the ease by which a material is friction stir welded or processed is related to its extrudability. That is, a material that is more difficult to extrude is also more difficult to friction stir weld. A material that is more difficult to friction stir weld tends to require higher forces and has a more limited weld or processing parameter envelope. Some trends versus the families are listed.

1. 1xxx alloys: These alloys are near pure aluminum, with relatively few welding applications. The base material has low mechanical properties versus other alloys. They can be friction stir welded or processed with a wide range of parameters and they require very low levels of processing force.

2. 2xxx alloys: These are moderate to high cost, heat treated alloys that are commonly used in the aerospace industry. They tend to have moderately high mechanical properties. These 2xxx alloys are one of the four alloy groups, including 5xxx, 6xxx, and 7xxx, which are widely friction stir welded and used throughout industry. They are moderately difficult to friction stir weld, having a moderate range of applicable parameters/conditions. The range of applicable parameters for 2xxx is larger than the range for 5xxx or 7xxx, but smaller than the range for 6xxx. Given these are heat treated alloys (and are commonly implemented in that condition), friction stir welding will result in some reduction in material properties, as the material is over-aged as a result of the heat from friction stir welding. Two commonly used alloys, 2024 and 2219 will have post weld strengths of 45–50 kilopounds per square inch (ksi), retaining 70–80% of the base material properties. Postweld yield strengths of these alloys tend to be around 30 ksi. After welding, these alloys will naturally age, resulting in minor increases in weld strength versus time. Aluminum lithium alloys (e.g., 2195) have also been investigated, but more significant reductions in strength have been observed after friction stir welding [13].

3. 3xxx alloys: These alloys are commonly used for deep drawing applications, such as pots and pans. There are some welding applications. The base material has relatively low mechanical properties versus most other alloys, but higher than 1xxx alloys. They can be friction stir welded or processed with a wide range of parameters and they require low levels of processing force.

4. 5xxx alloys: These are relatively low-cost, work-hardened alloys that are commonly used in the marine industry, other transportation industries, and general industry. They are mostly used in plate or sheet form, but they are also available in simple shaped extrusions. They have the best combination of strength and ductility, while having good corrosion resistance. They are commonly used in the marine industry and other industries that require fabrications because the post weld material properties (with traditional processes) are near the parent material. Because they are difficult to extrude, they are

quite difficult to friction stir weld or process, with a limited range of processing conditions. They are sensitive to high rotation speeds, require a very high level of welding or processing force, and some 5xxx alloys present wear challenges for friction stir welding tools. Because they are not heat treated, the post FSW properties often meet or exceed the base material, unless the base material is highly work hardened. These alloys represent the second largest market for FSW and FSP to the 6xxx alloys.

5. 6xxx alloys: These are low-cost, heat-treated alloys that are readily extruded. They are commonly used throughout industry for structural applications, except where high strength and weight are critical. They possess moderate strength, but lower ductility. These alloys currently represent the largest market for FSW. FSW is commonly used to fabricate large paneling from individual extrusions. Given their relative ease to extrude, they are friction stir welded with ease. Since they are also heat treated, in their most commonly used form, FSW results in some reduction in mechanical properties. The relative reduction in mechanical properties is dependent on the initial heat treatment temper. They will also naturally age after welding. For example, the commonly used 6061-T6 typically has ultimate strengths of 28–30 ksi after welding (70–80% of the base material), but after several days, a 2–3 ksi increase in strength can be observed. The yield strength of friction stir welds of this alloy is typically in the low 20 ksi range. Post weld aging can be used to bring the weld properties back to very near the parent material properties.

6. 7xxx alloys: These are heat-treated alloys, possessing very high strength and low ductility. They are commonly used in the aerospace industry and are relatively high in cost. They are difficult to extrude and thus difficult to friction stir weld. The 7xxx alloys have the narrowest range of FSW or FSP parameters of all of the alloys and require the greatest level of force. The level of friction stir difficulty is similar to that of the 5xxx. Because they are heat-treated in their most common forms, they also experience a reduction in mechanical properties after FSW or FSP. However, post weld strengths can be 55–65 ksi or higher, depending on the alloy. They will also naturally age significantly after welding and do so over a long period of time. To overcome the extended natural aging cycle, post weld artificial aging is often considered for these alloys to stabilize them. In addition, the welds are also less corrosion resistant than the base material. The post weld artificial aging process also helps manage this corrosion issue. These alloys are extremely difficult to impossible to join with traditional arc welding processes, so they represent a good potential market for friction stir welding and processing.

7. Other alloys (4xxx and 8xxx): There is little experience with these alloys and relatively little use in the marketplace.

8. Cast alloys: Many of the cast alloys have chemistries similar to the wrought alloys (2xxx, 5xxx, 6xxx, etc.). Thus, they tend to weld with similar conditions as the wrought alloys with a similar chemistry. Castings tend to have some level of porosity and lower ductility than wrought alloys. There has been interest in using friction stir processing to locally improve castings to eliminate porosity, improve ductility, and improve the cast material in areas where sealing is required. For friction stir welding, the cast alloys represent a fairly small market versus welding or processing of the wrought alloys (plate or extrusion) for a couple of reasons. First, castings are more typically joined with fasteners, so they are not designed for welding. Thus there is less historical use or experience with welding, which results in hesitancy to use welding for joining. Secondly, castings are often complex in shape and/or hollow. Both of these conditions reduce the opportunity to use friction stir welding. However, the lack of friction stir welding applications may be supplanted by additional opportunities for friction stir processing to locally improve the cast material.

9. Metal matrix composites: These are aluminum alloys with reinforcing particles. Their use is not widespread, due to difficulty in processing. There has been interest in friction stir welding and processing these alloys. However, tool wear has been a significant issue with no known publications indicating this issue has been overcome.

10. Various combinations: It is possible to friction stir weld dissimilar aluminum alloys. In general, the conditions or parameters that are used to weld the more difficult alloy will be applicable.

Magnesium

Magnesium alloys have experienced a moderate level of friction stir research and development activity. Magnesium is a lightweight material with less density than aluminum. Magnesium use is a growing market, especially in the automotive industry. It is primarily used in cast form with some use of extrusions as well. It is rarely used in sheet form, since it is difficult and very costly to roll with current magnesium processing technology. It is challenging to join with traditional processes and there is concern over its flammability. Thus, magnesium represents a potential market for friction stir welding. However, this market is not as large as it would otherwise be, since a large percentage of magnesium is in cast form. As noted, castings are often hollow and complex in shape, making application of FSW more difficult. Most magnesium alloys are relatively easy to friction stir weld, similar to the 6xxx aluminum alloys. Magnesium also has a similar melting point as aluminum. The parameter envelope may be somewhat more restricted than 6xxx aluminum and welding forces somewhat higher. Magnesium has a tendency to have a heavier oxide, so some attention needs to be given to this condition.

Copper

Copper has also experienced a moderate level of research and development activity to understand its friction stir weldability. There are some production applications, as well. Applications include heat exchangers, heat sinks, and large corrosion resistant containers [20]. Copper is difficult to join with traditional welding processes, so it represents a good potential market for friction stir welding. It has a moderately high melting point (near 1100°C) and has very high thermal conductivity. These two issues make it somewhat more challenging to friction stir weld than aluminum. Friction stir tool material selection is not as expansive as that for aluminum. However, it is possible to weld copper with tool steels due to the softness and low strength of copper. Copper has been welded in thicknesses up to two inches, but precise temperature control has been found to be required to obtain consistent results [20]. Also, copper oxidizes more readily if welded with parameters that result in temperatures that are too high. Copper can be welded with a large range of parameters, with forces being somewhat higher than 6xxx aluminum.

Bronze

Bronze has gained a moderate level of development activity, but mostly for application of friction stir processing. Nickel aluminum bronze alloys have been the subject of a number of investigations [7,35]. This particular alloy is used for ship propellers and has excellent corrosion resistance. These ship propellers are cast, but can have porosity. This porosity can result in significant rework and repair costs. Friction stir processing has been investigated for its ability to eliminate the porosity and resulting rework and repair (refer back to Figure 10.3). Studies have shown that a significant local improvement in material properties can be realized by use of FSP on these castings [7]. Because the major alloying constituent in bronze is copper, it tends to weld or process similarly to copper. That is, it has a relatively large processing envelope and requires forces similar to or somewhat higher than aluminum. However, friction stir tool wear is of more concern with nickel aluminum bronze than with copper.

Titanium

Titanium is another material that has received a moderate level of research and development activity. Titanium is costly, but has the highest strength to weight ratio of any common metal. It is widely used throughout the aerospace and defense industries, where weight is critical. It also has good corrosion resistance. Titanium is possible to join with traditional fusion welding processes, but not without challenges. Titanium has low thermal conductivity, has a high melting point, and is very reactive at high temperatures. Lastly, some alloys have variable microstructure versus temperature (e.g., Ti-6Al-4V), especially near FSW temperatures. Given these conditions, titanium is much more difficult to friction stir weld than aluminum. However, it is possible to friction stir weld [13,36,37,41–43]. Because of the aforementioned characteristics, it is difficult to develop a stable or robust friction stir welding process. The low thermal conductivity creates a situation where standard shaped friction stir tools generate too much heat on the surface. Thus, small shoulders versus pins must be used. A smaller shoulder versus pin diameter decreases process stability. In addition, the high melting point and reactivity limits the tool material choices. These limited tool materials tend to also be difficult to machine. This limits the ability to include features on the pin that are normally used to aid mixing. In addition, the reactivity of titanium creates a requirement for shielding gas. Successful laboratory level implementations have been demonstrated with tungsten-based tool materials, using tools with small shoulder to pin diameter ratios [42,43] or stationary shoulders [36,37], relatively featureless pins, shielding gas, and fairly precise machine control. Typical rotation speeds used to weld titanium are very slow (a few hundred RPM or less), travel speeds are slow, and welding forces are significantly higher than for aluminum.

Steel

After aluminum, various steel alloys have received the most research and development activity in friction stir welding. Steel is used throughout the manufacturing industry, so welding of steel represents a potential large market for friction stir welding. However, many steel alloys are readily joined with traditional fusion welding processes. Because steel melts at a high temperature, choice of tool materials is limited and potentially applicable materials are very costly. To-date friction stir tool wear is a limiting factor. However, steady improvements in friction stir tool life are being made [44]. Use of polycrystalline boron nitride (PCBN) and tungsten–rhenium as the tool material have provided for the best results. Given the current state-of-the-art, friction stir welding or processing of steel is limited to high value applications where a high consumable cost can be justified, where weld quality is of utmost importance, or where lack of distortion is critical. One of the first production applications for friction stir welding or processing of steel that has been implemented is for improving knife blades [45].

Like aluminum, there are numerous steel alloys. Similarly, there is a range of conditions and result mechanical properties versus the alloy that is welded. Although there is a range of parameters that can be used to weld or process steel, most successful friction stir welding of steel has been performed with lower rotation speeds (<1000 RPM) and travel speeds less than 10 inches per minute. In addition, the forces required to weld or process steel are significantly higher than for aluminum. Lastly, machine requirements, (e.g., spindle run-out with use of PCBN for tool material), can be considerably more stringent than for aluminum.

Other Materials and Dissimilar Joints

There are other materials that have successfully been friction stir welded or processed. Some examples include nickel-based alloys, molybdenum, lead, zinc, thermoplastics, and so on. As noted previously, the relative ease or difficulty of welding is primarily related to the melting point of the material. However, low thermal conductivity and reactivity of the material can cause additional challenges.

It is also possible to join dissimilar materials. There have been investigations into welding of aluminum to copper, aluminum to magnesium, and aluminum to steel, to name a few [46–48]. Joining of dissimilar materials can be very challenging. There are a couple limiting factors. The first limiting factor relates to the fact that FSW and processing operates near the melting temperature of the material that is to be welded or processed. If there is a significant difference in melting temperature between the two materials, welding or processing when favoring the higher melting temperature material will cause local temperatures to be above the melting point of the lower melting point material. This causes melting of the lower melting point material. This places limitations on how the materials are oriented, the joint configurations, and where the friction stir tool path is located. The weld path must generally favor the lower melting point material. A second potential issue is that some of the dissimilar combinations will alloy with each other. This means there is a phase diagram that needs to be investigated. In some cases there are lower melting point eutectics, created below the welding temperature of either material, or intermetallics that are created. These issues limit the ability to join the materials. Aluminum and magnesium are one such combination where this condition exists.

Quality Control, Testing, and Specifications

Quality Control and Testing

As with any manufacturing process, there are testing methods that are used to initially qualify the process prior to production. There are also testing methods that are implemented during production, which may or may not be the same as during qualification. The testing methods and frequency for production applications are dictated by quality control plans. It is often the case that more testing is performed to qualify the process than what is performed to ensure a consistent and quality process during production. These testing methods consist of destructive and nondestructive methods. There are multiple types of both destructive and nondestructive testing, with varying types and amounts of use throughout the industry.

Destructive Testing

By definition, destructive testing includes a variety of test methods for which the weld or processed region will be destroyed after testing. This testing is often used more thoroughly during the qualification process and less so in production. Some methods are not practical for production or require reduced frequency for reasons including: (1) the test method requires too much time or effort, or (2) inability to destroy the actual part. The second issue can be overcome by either performing the process on test coupons before and/or after welding on the product or allowing for extra material that can be removed after completion of the part. For FSW, there is often extra material in cases where starts and stops of welds are removed. Destructive testing can be performed in these areas of extra material.

A list of common destructive testing methods is shown below with a general discussion of each method. The list is not intended to be exhaustive, but rather to provide a list of the more typical test methods and their applicability. There are specifications that can be reviewed to provide detailed information on the test method. In addition, there are numerous FSW and processing references that can be reviewed to gain an understanding of the destructive test results versus material, alloy, thickness, joint configuration, and so on.

1. Tensile testing: Tensile testing is used significantly during qualification and often during production to determine the mechanical properties (ultimate strength, yield strength, and elongation) of a weld or processed region. In production, the testing is performed in offal (an area that is to be removed prior to shipment) or in test coupons that are created before, after, or in between the welding of the parts. Tensile testing will at least determine the ultimate strength. If desired, the yield strength and elongation can also be determined. This testing is often performed to some specification, such as the AWS D1.2, AWS B4.0, and the American Society for Testing of Materials (ASTM) E8 [51], among others. Tensile testing can be used to detect internal voids, lack of penetration, heavy remnant oxides, and other general weld quality issues.

2. Bend testing: Bend testing is used significantly during qualification and during production to test the ductility of a weld, primarily for butt welds. In production, the testing is performed in offal (an area

that is to be removed prior to shipment) or in test coupons that are created before, after, or in between the welding of the parts. There are two basics types of bend testing for butt welds. These include wrap-around and guided bend tests [49,50]. Each of the methods can be performed in three different modes to test various locations of the weld. These modes include root bends, face bends, and side bends. Each of these modes puts the different locations of the weld in tension. For example, the root bend test places the root of the weld in tension. This test method is typically performed per AWS D1.2 or AWS B4.0 standards. This test method can detect lack of penetration or heavy remnant oxides. For other joint configurations, a variety of other special bend tests can be considered. Examples include a hammer bend or peel test on a lap weld or a three point bend test on a tee-weld, and so on.

3. Cross-sectioning: Cross-sectioning is used to visualize the internal quality of the weld. It is performed by first removing a small section of the weld. The small section is often mounted into a button, such that the face of the sample is perpendicular to the weld direction. The sample is then polished to attain a smooth surface. Lastly, the sample is etched to reveal the various microstructures within the weld or processed region and the HAZ. Etchants that are used for other welding processes are often applicable for friction stir welding. The etchants vary depending on the material [52]. However, due to the lack of heat and lack of introduction of filler material (lack of alloy difference), some standard etchants do not perform as effectively for friction stir welded or processed regions.

4. Fatigue testing: Fatigue testing is often used for initial qualification. Fatigue testing subjects the weld to cyclic loading. In any one test, the weld sample is subject to a cyclic load profile with predetermined maximum and minimum loads and frequency. Depending on the maximum load level, fatigue testing can be very time intensive. Given fatigue testing is a time intensive test method, it is rarely used for production quality control and is typically only performed on a minimal number of weld conditions during qualification. Fatigue testing is typically performed to some standard such as ASTM E 466.

5. Fatigue crack growth rate testing: This testing is occasionally included in initial qualification programs, primarily where welds experience a cyclic loading environment and the component is relatively critical, such as aerospace structures. The goal is to determine the rate at which any crack will grow versus the parent material or some other joining process.

6. Corrosion testing: This testing is occasionally performed in qualification programs and can be done in several forms (stress-corrosion testing, salt spray, etc.). The objective is to determine the sensitivity of the weld or HAZ to corrosive environments. It is typically used for applications where the component has a significant life expectancy or subjected to corrosive environments, such as aerospace structures or marine structures. The testing is usually performed to some standard, such as ASTM B117.

7. Others: There are other destructive test methods that can be used. But these tend to be application specific and may include ballistics testing, shock testing, leak testing, and so on.

At the end of any qualification program, with destructive testing complete, a weld procedure specification (WPS) and procedure qualification record (PQR) are completed. The WPS records the nominal welding parameters and conditions at which the weld or processing was performed. The PQR records the destructive and nondestructive testing results. The PQR is the record indicating the process is qualified and meets acceptable standards. The format of these documents is typically organization specific, but examples for other processes can be seen in AWS welding standards. These documents can be modified for FSW and processing.

Nondestructive Testing

Similar to destructive testing, there are a variety of nondestructive testing methods that can be implemented to test the quality of friction stir welds and processed regions. Nondestructive testing is most often implemented where it is difficult to employ destructive test methods (where the welded component has a high value or where there is no offal) or where the criticality of the weld is relatively high. These methods are most often performed in the aerospace, nuclear, and defense industries, to name a few. However, some low-cost methods are used in other industries such as the automotive industry. Nondestructive test methods have been an area of significant research, since some friction stir defects are difficult to detect. Like the destructive testing procedures, they are usually performed to some specification. In addition, they are typically required to be performed by certified personnel, due to their relative difficulty versus destructive test methods. There are also three levels to which the personnel can be certified, which dictates the actual roles that can be performed during the testing process. The personnel are typically certified to American Society of Non-Destructive Testing (ASNT) standards. A list of the most common nondestructive test methods includes:

1. Visual inspection: Almost all industries use some sort of visual inspection. Even highly automated applications require some periodic visual inspection. For FSW, the goal of the visual inspection is

to determine if there are any aesthetic issues (flash) or defects that can be visually detected (e.g., surface void). Both the face and root sides of the welds may be inspected.

2. Dye penetrant inspection: This inspection method is generally used to determine if there are any cracks or lack of penetration, especially in a butt weld. It is a relatively common nondestructive inspection method and is used throughout industry. It involves first cleaning the surface that is to be inspected. Secondly, a dye is sprayed on the material and lastly, a developer is applied. Cracks or unbonded regions will tend to be highlighted by the dye. After a certain amount of time, the developer is cleaned off to reveal the original surface. For FSW, this test method can yield false positive readings, especially on the root side. The high pressures of the FSW can fracture the original oxide surface. To avoid this issue, the material can be subjected to a grinding operation. However, grinding can potentially smear defects. Etching procedures have also been considered to solve this issue.

3. Radiography: Radiography is used to sense relative differences in density. Thus, it can be used to detect internal defects such as internal voids [53]. However, radiography has extreme difficulty in detecting lack of penetration (especially small amounts) or heavy remnant oxides. There are both fixed and portable systems that can be used. This testing is usually performed to some standard, such as ASTM E1742.

4. Ultrasonics: Ultrasonic inspection (UT) uses cyclic sound pressure at frequencies that are not audible. The sound will reflect off internal surfaces or features differently, providing information about the internal structure of a weld or component. This technique can be used to detect flaws and the thickness of the material. Ultrasound requires a probe that contains a transmitter and receiver. With the transmitter, a couplant, typically a gel, is required to transfer the sound energy from the transmitter into the part that is to be inspected. Ultrasonic testing equipment is available in various forms, from single transmitters (conventional UT) to phased array systems (with multiple transmission and receiving units). Conventional UT systems have been found to have fairly limited capability to detect flaws in friction stir welds, but the phase array systems have been demonstrated to be more capable [54]. Phased array systems have demonstrated the ability to detect a lack of penetration and internal voids. Ultrasonic inspection typically requires a set of calibration blocks (with and without defects) to allow for accurate detection of flaws. The inspection process is more applicable to thicker material (>6 mm), with reduced capability when material thickness is less than 6 mm. The phase array type of equipment has been used for

various FSW aerospace applications (e.g., Delta Rocket Tank, Eclipse Business Jets, etc.).

5. Eddy current: The eddy current inspection process can also be used to detect flaws in friction stir welds [54]. The technology uses electromagnetic induction. Magnetic fields are created within the material, but near the surface. These fields are affected by areas of varied resistance. These areas of varying resistance can be detected and correlated with defects within the material or weld. Defects that can be detected include lack of penetration (equivalent to a crack), internal voids, or heavy remnant oxides in lap welds. Similar to UI, there are probes with arrays of coils for application over a larger surface. Eddy current inspection only works on conductive materials, requires calibration blocks, and depth of detection is limited to approximately 6 mm. However, the technique is fast and does not require a couplant [55]. As with UI, it has primarily been investigated for FSW applicability in aerospace applications.

Specifications

Most mature manufacturing processes have standards or specifications related to their use. Because friction stir welding and processing is a nascent technology there are limited standards and specifications at the current time. However, there are several that are in process. To date, the following specifications or standards exist or are being developed:

1. "NASA PRC-0014—Process Specification for Friction Stir Welding of Flight Hardware": Published in October 2002, this is one of the first publically available specifications. This document is publically available at http://ams-02project.jsc.nasa.gov/life_cycle/cdr051003/1.4-process_specificatio_prc0014.pdf

2. ISO 25239—Friction Stir Welding—Aluminum: This specification has been under development for several years with final approval and issuance expected in 2010.

3. AWS D17.3/D17.3M: 200x Specification of Friction Stir Welding of Aluminum Alloys for Aerospace Hardware: This specification was published in January 2010.

4. AWS D8H: 20xx Specification for Automotive Weld Quality—Friction Stir Welding: This specification is in the early stages of development.

5. Others: There are other completed specifications that are being used. However, most of these other specifications have been created by organizations performing friction stir welding. They have been developed internally within the organizations and are considered proprietary in most cases.

Summary and Future

Friction stir welding and its related processes are relatively new, with friction stir welding being invented less than 20 years ago and the related processes even newer. For a welding process, FSW has seen a rapid increase in its use, especially for aluminum. Aluminum joining applications represent a significant and growing market, due to the relative difficulty in joining aluminum with traditional welding processes and the advantages that FSW can provide. For the foreseeable future, aluminum will likely be the predominant area of expansion of friction stir welding for a number of reasons:

1. Current use is still a fraction of the potential market.
2. The market itself is very large, as aluminum is the most widely used metal, only behind steel.
3. The benefits of FSW are dramatic for aluminum.

For other reasons, the general market for friction stir welding and related process will also likely increase.

1. Knowledge and awareness of the technology will increase. Although increasing, there is still relatively little expertise in friction stir welding or inclusion in engineering degree programs.
2. As the benefits of the process are more widely known and demonstrated, products will be increasingly designed for friction stir welding. To-date, very few products are designed for FSW. Rather, they are designed for traditional joining processes, such as riveting or fusion welding, many of which do not readily lend themselves to direct use of friction stir welding. As an example, many products are designed with tee-fillet joints for fusion welding processes. This joint configuration is widely used throughout the fabrication industry and generally not feasible for use with FSW. Avoidance of this joint configuration, or conversely, design for FSW will allow for access to a much larger market.
3. As with other technologies, machine or equipment cost will reduce over time, allowing for the process to be justified in more applications.
4. There will likely continue to be friction stir tool material and equipment improvements that will allow FSW and its related processes to be more readily used in steel and titanium.
5. Increasing creation and availability of specifications. This will aid in improving the knowledge base as well as the comfort level/risk reduction of implementation of friction stir welding and related processes.

Acknowledgments

The authors appreciate the support of those who have influenced and contributed to the friction stir welding and processing developments that have enable our ability to create this chapter. This first includes Wayne Thomas, at The Welding Institute, who invented friction stir welding in 1991 and those at TWI who were responsible for the critical early FSW developments. The authors would also like to thank those who funded and contributed to the author's first FSW efforts at A.O. Smith Automotive Products Company, in the middle to late 1990's, leading to the first automotive FSW application in North America and the first industrial robot based friction stir welding systems. In the early 2000's, the authors established Friction Stir Link, Inc. (FSL). The success and knowledge developed at FSL has been influenced by many contributors, including, but not limited to FSL engineers, technicians, and production associates, as well as FSL ownership and customers who have provided the funding for the various projects which contributed to our knowledge base. Last but not least, the numerous other investigators of friction stir welding throughout the world are appreciated for their own contributions to the community.

References

1. Thomas, W. M., et al. "Friction Stir Butt Welding," U.S. Patent #5,460,317, December 6, 1991.
2. Hinrichs, J., et al. "Friction Stir Welding for the 21st Century Automotive Industry," *Proceedings of the 5th International Friction Stir Welding Symposium*, Metz, France, September 2004.
3. Sakano, R., et al. "Development of Spot FSW Robot System for Automobile Body Members," *Proceedings of the 3rd International Symposium of FSW*, Kobe, Japan, September 2001.
4. Iwashita, T. "Method and Apparatus for Joint," U.S. Patent 6602751 B2, August 5, 2003.
5. Kashiki, H., et al. "Friction Stir Joining Apparatus," U.S. Patent Application US2003/0029903 A1, February 2003.
6. Strombeck, A. V., et al. "Robotic Friction Stir Welding—Tool Technology and Applications," *Proceedings of the 2nd International Symposium on FSW*, Gothenburg, Sweden, June 2000.
7. Mahoney, M. W., et al. "Microstructural Modification and Resultant Properties of Friction Stir Process Cast NiAl Bronze," *Material Science Forum* 426–432, Part 4 (2003): 2891–6.
8. Mishra, R. S., and M. W. Mahoney. "Friction Stir Processing: A New Grain Refinement Technique to Achieve High Strain Rate Superplasticity in Commercial Alloys," *Material Science Forum* 357–359 (2001): 507–14.

9. Mahoney, M., and W. Bingle. "Thick Section Bending Patent," U.S. Patent #6,866,180, March 2005.

10. Mahoney, M., et al. "Thick Plate Bending of Friction Stir Processed Aluminum Alloys," *TMS Proceedings: Friction Stir Welding and Processing III* (2005): 131–7.

11. Smith, C. B., et al. "Fabricated Shapes Using a Friction Stir Welding/Forging Process," *Proceedings of the 7th International Symposium on Friction Stir Welding*, Awaji Island, Japan, May 2007.

12. Dracup, B., and W. Arbegast. "Friction Stir Welding as a Rivet Replacement Technology," U.S. Patent #6986452, January 2006.

13. Mishra, R. S., and M. W. Mahoney. *Friction Stir Welding and Processing*, Materials Park, OH: ASM International, March 2007.

14. Threadgill, P. L. "Terminology in Friction Stir Welding," *Science and Technology of Welding & Joining* 12, no. 4 (2007): 357–60.

15. Patnaik, A., et al. "Static Properties of Refill Friction Spot Welding of Skin Stiffened Compression Panels," *Society of Automotive Engineering*, April 2006.

16. Smith, C., et al. "Development and Qualification of a Production Capable FSW Process for 25 mm Deep Robotic Friction Stir Welds," *Proceedings of the 6th International Symposium on Friction Stir Welding*, St. Sauvier, QC, Canada, October 2006.

17. Dawes, C. J., et al. "Development of the New Friction Stir Technique for Welding Aluminum—Phase II," TWI Member Report 5651/35/95, November 1995.

18. Thomas, W. M., et al. "Friction Stir Welding Tools and Developments," *FSW Seminar, Instituto* de Soldadura e Qualidade, Porto, Portugal, December 2002.

19. Colligan, K. "Tapered Friction Stir Welding Tool," U.S. Patent 6,669,075, December 2003.

20. Lederkvist, L. "FSW to Seal 50 mm Thick Copper Canisters—A Weld that Lasts for 100,000 Years," *Proceedings of the 5th International Symposium on Friction Stir Welding*, Metz, France, September 2004.

21. Smith, C. B. "Friction Stir Equipment Solutions," *TWI FSW Workshop*, Sheffield, UK, November 2007.

22. Smith, C. B. "Robotic Friction Stir Welding Using a Standard Industrial Robot," *Proceedings of the 2nd International Friction Stir Welding Symposium*, Gothenburg, Sweden, June 2000.

23. Voellner, G., et al. "Three-Dimensional Friction Stir Welding using a Modified High Payload Robot," *Proceedings of the 6th International Symposium on Friction Stir Welding*, St. Savieur, QC, Canada, October 2006.

24. Soron, M. "Friction Stir Welding of High-Strength Aluminum Alloys Using an Industrial Robot System: A Feasibility Study," *Proceedings of the 7th International Symposium on Friction Stir Welding*, Awaji Island, Japan, May 2008.

25. Midling, O., et al. "Industrialization of the Friction Stir Welding Technology in Panels Production for the Maritime Sector," The First International Symposium on Friction Stir Welding, Thousand Oaks, CA, 1999.

26. "Friction Stir Welding (FSW) System Integration Services", http://www.frictionstirlink.com, December 2008

27. Cook, G., et al. "Robotic Friction Stir Welding," *Industrial Robot* 33, no. 1 (2004): 55–63.

28. Loftus, Z., et al. "Friction Stir Weld Tooling Development for Application on the 2195 Al-Li-Cu Space Transportation System External Tank," *Proceedings of the 5th International Conference on Trends in Welding Research*, 590, Pine Mountain, GA, June 1–5, 1998.

29. Bollinger, J., and N. Duffie. *Computer Control of Machines and Processes*, 1st ed. Reading, MA: Addison-Wesley Longman, 1988.

30. Colligan, K. J., and J. R. Pickens. "Friction Stir Welding of Aluminum Using a Tapered Shoulder Tool," *Friction Stir Welding and Processing III*. Edited by K. V. Jata, et al., 161–70. Warrendale, PA: The Minerals, Metals & Materials Society, 2005.

31. Stol, I. "Multi-Shoulder Fixed Bobbin Tools for Simultaneous Friction Stir Welding of Multiple Parallel Walls between Parts," U.S. #7,198,189, April 2007.

32. Ding, R. J., and P. A. Oelgetz. "Autoadjustable Pin Tool for Friction Stir Welding," U.S. Patent 5,893,507, April 1999.

33. "Friction Stir Welding Tool with Real-Time Adaptive Control," *NASA Tech Briefs*, (February 1997): 70–1.

34. Ding, R. J., and P. A. Oelgetz. "Mechanical Property Analysis in the Retracted Pin-Tool Region of Friction Stir Welded Aluminum Lithium Alloys," *Proceeding of the First International Symposium on Friction Stir Welding*, Thousand Oaks, CA, June 1999.

35. Arbegast, W. "Friction Stir Welding after a Decade of Development," *Welding Journal* 85, no. 3 (March 2006), 28–35.

36. Norris, I., et al. "Friction Stir Welding—Process Variants and Recent Industrial Developments," *10th International Aachen Welding Conference*, Eurogress, Aachen, October 2007.

37. Walker, B. "Robust, Low Cost Stir-Welding of Titanium Alloys," http://www. virtualacquisitionshowcase.com/docs/2008/Keystone-Brief.pdf, 2008.

38. Kohn, G., et al. "Laser-Assisted Friction Stir Welding," *Welding Journal* February (2002): 45–8.

39. Able, N., and F. Pfefferkorn. "Laser-Assisted Friction Stir Lap Welding of Aluminum," *National Heat Transfer Conference*, San Francisco, CA, HT2005-72829, 2005.

40. Okamoto, K., et al. "Development of Friction Stir Welding Technique and Machine for Aluminum Sheet Metal Assembly," Paper 2005-01-1254, *2005 SAE World Congress, Society of Automotive Engineers*, Detroit, MI, 2005.

41. Jata, K., and A. Reynolds, *Metallic Materials with High Structural Efficiency*, Part 6, Vol. 146. The Netherlands: Springer, 2004.

42. Bernath, J. "FSW of Titanium 6Al-4V Structural Components," *Proceedings of the 6th International FSW Symposium*, St. Sauvier, Quebec, October 2006.

43. Jones, R., and Z. Loftus. "FSW of 5 Millimeter Titanium 6Al-4V," *Proceedings of the 6th International FSW Symposium*, St. Sauveur, Qc, October 2006.

44. Nelson, T., and C. Sorensen. "Advances in PCBN Tooling for Friction Stirring of High Temperature Alloys," *Proceedings of the 6th International FSW Symposium*, St. Sauveur, Qc, October 2006.

45. "Friction Forged Technology: Engineering a Super Blade," http://www. diamondbladeknives.com, December 2008.

46. Savalainen, K., et al. "A Preliminary Study on FSW of Dissimilar Metal Joints of Copper and Aluminum," *Proceedings of the 6th International FSW Symposium*, St. Sauveur, Qc, October 2006.

47. Okamoto, K., et al. "Friction Stir Welding of Magnesium for Automotive Applications," *SAE Transactions* 114, no. 5 (2005): 392–6.
48. "A Friction Heat First for Mazda, Friction Stir Welding Joins Aluminum and Steel on MX-5," *Aluminum Now Online, Aluminum Association* 7, no. 6 (November–December 2005).
49. American Welding Society Standard, *AWS D1.2:2003—Structural Welding Code—Aluminum*, 4th ed. Miami, FL: American Welding Society, February 2003.
50. American Welding Society Standard, *AWS B4.0:2007—Standard Methods for Mechanical Testing of Welds*, 7th ed. Miami, FL: American Welding Society, May 2007.
51. American Society for Testing of Materials Standard, *ASTM E-08—Standard Test Methods for Tension Testing of Metallic Materials*, 6th ed. West Conshohocken, PA: American Society for Testing of Materials, April 2004.
52. American Society for Metals, *Metals Handbook, Metallography, Structures, and Phase Diagrams: Eighth Edition*, Vol. 8. Materials Park, OH: American Society for Metals, 1973.
53. Kinchen, D. G., and E. Aldahir. "NDE of Friction Stir Welds in Aerospace Applications," *American Welding Society—Inspection Trends*, July 2002.
54. Lamarre, A. "Eddy Current Array and Ultrasonic Phased Array Technologies as Reliable Tools for FSW Inspection," *Proceedings of the 6th International Symposium on Friction Stir Welding*, St. Sauveur, Qc, October 2006.
55. Rao, B. P. C. "Eddy Current Non-Destructive Testing," http://www.geocities.com/raobpc/EC-Def.html, December 2008.

11

Laser and Laser Processing

Mel Schwartz

CONTENTS

Introduction

Laser beam technology promises both quality and cost advantages compared with the current arc welding processes. The laser beam is a monochromatic light; that is, it radiates with a characteristic wavelength of the respective beam source. Depending upon the laser-active medium, the laser beam can be ultraviolet (UV), green, blue, red, or infrared. Depending upon the structure and kind of the beam source, the laser can radiate steadily or emit extremely short pulses. Currently, the following laser beam welding (LBW) devices are used in shipyards, for example, the CO_2 laser with power performance $P = 40$ kW and wavelength $\lambda = 10.6$ μm; the Nd:YAG laser with $P = 10$ kW, $\lambda = 1.064$ μm, and $\eta = 3$–10%; and the fiber laser with $P = 20$ kw, $\lambda = 1.070$ μm, and $\eta = 30$%.

Depending on the laser type, the light beam is delivered, without loss, to the weld site using a mirror-and-lens system or light cables. (10 mm diameter fiber-optic cable or beam fiber cable). Laser beam delivery to the weld area with a beam cable depends on the wavelength of the laser beam. For example, the beam delivery of the Nd:YAG lasers (short wavelengths) uses a beam cable, and the beam delivery of the CO_2 lasers uses mirror systems.

The absorption and reflection characteristics of the obtained beam depend on the material and the wavelength of the beam. For example, the Nd:YAG laser beam ($\lambda = 1.06$ μm) is absorbed more than the CO_2 laser beam and is reflected from materials less than the CO_2 laser beams. There are different absorption characteristics for different materials; for example, absorption of the laser beam by aluminum is less than by iron and steel.

Welding: Laser Types

With fiber lasers, new welded structure designs can be realized that were not possible before. With arc welding, heat distortion and deformation often appear in the panels because of the high-heat input. Laser welding of the same parts produces minimal heat distortion. The laser permits welding of pipes with a number of joint configurations, especially in underwater applications. High-power fiber lasers with high-beam quality and high efficiency are available for material processing, particularly for welding and cutting. High-power fiber lasers can weld a thick sheet [about 5 m/min (15 ft/min)] and on one side up to a wall thickness of 15 mm (0.59 in.), and they can be welded bilaterally up to 30 mm (1.18 in.) economically with acceptable quality [1–5].

Components of Laser Weld System

The majority of industrial welding applications use CO_2 and solid-state lasers such as the thin disk laser. The choice between a CO_2 and a solid-state laser needs to be made on a case-by-case basis and requires knowledgeable partners in the analysis of the entire system layout and its impact on cost, flexibility, and performance. Consider the thin disk laser with its versatile fiber-optic beam delivery. The laser medium of a thin disk laser is a synthetic yttrium aluminum garnet (YAG) crystal. Embedded in the crystal lattice are the ions of a laser-active medium such as ytterbium (Yb). The higher beam quality attainable with the diode-pumped thin disk laser is primarily a function of the ability to dissipate heat from the lasing medium. The Yb:YAG disk [roughly 14 mm (0.55 in.) diameter by 0.15 mm (0.059 in) thick] is mounted to a water-cooled block (heat sink). Since the disk is face surface mounted to the block and because the disk is very thin (i.e., high area to lasing medium volume ratio), cooling is extremely efficient and results in a nearly negligible thermal gradient. The beam quality of a laser affects four parameters: focus diameter, depth of focus, working distance, and raw beam diameter. The effects of good focusability include:

- Small focus spot diameter: The smaller the focus spot, the higher the power density on the workpiece. There is a threshold power density for deep penetration welding. Deep penetration welding allows greater penetration and speed than conventional heat conduction welding. Therefore, it is important to reach at least this power density for most applications.

- Greater working distance: The working distance is the distance between the processing optics and the workpiece. The beam quality of the thin disk laser enables these power densities even at a working distance of 50 cm (1.9 in.), and thereby enables remote welding where the laser beam is steered onto the workpiece remotely, using a 2-D scanner system.

- Greater depth of focus: A greater depth of focus enables accurate processing of thick sheets while expanding the tolerance limits for adjustment of the working distance.

- Small optics diameter: A small raw beam diameter enables more compact design of the processing optics.

Finding the optimum focus spot for each individual process is crucial for obtaining good laser processing results. The beam quality of the disk laser combined with the modularity of the fiber delivery and processing optics make it possible to select the appropriate beam quality at the workpiece and, with it, the appropriate focus spot diameter for particular tasks [6].

Focusing Optic and Fiber Delivery

The beam emitted from a thin disk laser has a wavelength of around 1 μm, putting it within the near-infrared spectral band. As a result, optical components constructed of glass and fiber-optic laser cables can be used for beam delivery. Fiber-optic laser cables can carry multi-kW cw laser power over several hundreds of meters without notable loss, enabling the use of thin disk lasers in a variety of locations. The laser beam is focused into the cable and collimated after it exits the fiber so that it is parallel again. Focusing optics are then used to focus the beam at the desired point on the workpiece.

Modern laser systems, such as the thin disk laser, can provide uptime of around 99% or more, even at power levels of 8 kW for solid-state lasers and up to 29 kW for CO_2 lasers. These modern laser sources have design concepts that allow easy exchange of components in the field. Beam parameters are standardized and closed-loop controlled so that a process developed on a specific laser system can be transferred to a different laser system. Due to their flexible beam delivery and integration into other systems, thin disk lasers are often used in robot-aided manufacturing processes. There are many different options available for the focusing optics. Modern focusing optics can be equipped with process control (sensors and cameras), closed feed-back loops that ensure the automatic laser power control, wire feeders, even scanner optics for remote welding applications.

System Integration

There are many choices for laser welding system setups. Choices include a robot-based laser welding cell with a fiber delivered thin disk laser, gantry-based systems, systems optimized for linear welds or circular welds, and even scanner welding systems for remote welding. Often the laser is at the center of attention. However, the modern laser sources hardly ever pose the greatest financial or reliability risk. The future and part fitup tolerances, beam delivery fiber management, the optical system, and the programming of the motion system are often more important factors in determining the success or failure of a laser welding application. The economical success of a new laser welding application is clearly linked to the choice of the right system integrator. It is beneficial if the integrator has access to the most modern components in the following fields:

- Laser: Access to multi-kW high power solid-state laser and modern fast axial flow CO_2 laser technology
- Optics: 2-D and 3-D scanner optics, welding optics with integrated process control and optional wire feeder, specialty optics for optimized weld geometries
- Beam delivery: Fiber delivery cables that enable exchange in the field in the case of required maintenance. Fast plug-and-play technology to switch fibers between workstations and lasers. Fiber delivery cables and beam switches that have proven performance at power levels greater than 4 kW.
- System components and motion systems: Robot cells, 5-axis gantry welding systems, linear and rotary welding systems, off-line programming, teach-in concepts, fixture design capabilities

- Process competency: application laboratory with different lasers and delivery systems, access to a metallurgical laboratory, online process control equipment, design capabilities, and training classes

Temporal Pulse Shaping

Encapsulation of electronic and optoelectronic components or assemblies in hermetic enclosures or packages is performed to improve the reliability of these components by providing mechanical support, shielding from electromagnetic interference, better thermal management, and environmental protection of the environmentally sensitive components. Hermetically sealed electronic packages are widely used in the electronics, communications, automotive, computer, and medical industries.

Temporal pulse shaping can be used to affect the solidification time, the dendrite arm spacing, the size of the mushy zone, the strain rate, and the severity of interdendritic and intergranular microsegregation of alloying elements during weld metal solidification. These in turn influence the resultant solidification cracking susceptability. Crack-free pulsed laser seam welds could be made in 6061-T6 aluminum alloy by using a limited range of intermediate ramp-down gradients where both the positive and the detrimental effects of laser pulse ramp-down rate were at an optimum level [7].

Comparison of LBW with Other Technologies

The advantages of LBW in relation to conventional welding methods can be summarized as follows.

Materials Advantages

- Minimum, concentrated, point-shaped heat entry; hence, deep and slim zones of fusion with narrow heat-affected zone (HAZ); (Figure 11.1)
- High melting metals are weldable due to the high power density.
- Reduction of microcracks
- Liquation by rapid cooling
- Suitability for joining numerous materials with different characteristics as well as material combinations
- Highly reactive as well as very pure materials are weldable under inert gas or vacuum
- Smooth joint surfaces; therefore, finishing is most unnecessary
- Decreased chemical changes such as small burn-up of alloying elements
- Fine-grained, generally detect-free joints with more favorable mechanical and technological characteristics can be made

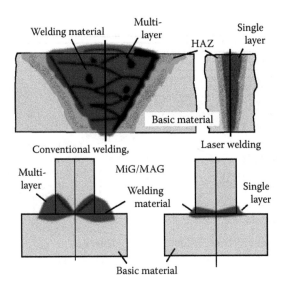

FIGURE 11.1
Comparison of joints welded with gas metal arc and laser beam welding.

Construction Advantages

- High strength of the weld material and the HAZ (approximately the same strength of the base metal)
- Small residual stresses; small thermal distortion
- Accurately controllable with reproducible weld depths
- Production of narrow welded joints and small spot welds
- Production of the current conducting connections of metals
- Production of connecting kinds that is not possible or is uneconomical with conventional welding methods such as welding below-deck sheet metals of ships and marine structures
- Manufacturing of sheet metals with thicknesses up to 15 mm (0.059 in.) faster and more economically than by welding methods with multilayer technology
- Smaller transition radius between the workpiece and the welded joint, and thus, more favorable fatigue resistance
- Suitability for welding of heat-sensitive construction units
- Welding by transparent materials is possible

Manufacturing Advantages

- Because of noncontact welding, tool wear does not occur
- Higher welding speeds offer much shorter welding times than with conventional welding methods (Figure 11.2)

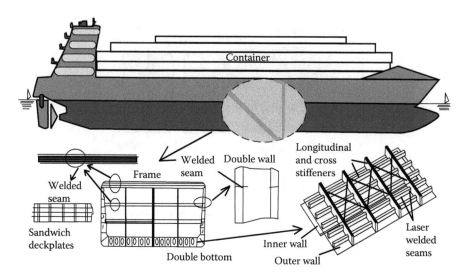

FIGURE 11.2
Schematic presentation of the shipbuilding units that are suitable for LBW (not to scale).

- Higher degree of automation and mechanization is possible
- Possibility for the combination of welding, cutting, and machining processes
- Higher reliability and reproducibility than with conventional methods
- Suitability of process control for quality increase is available in real time
- New and economical connection techniques in shipbuilding such as welding in ships with double walls
- Application in underwater welding under hyperbaric chamber gas conditions
- Property accessability by large work distance and/or by inlet and by focusing of the beam over a long distance (fiber cables) in hard to access places
- Work in various atmospheres such as air or inert gas depending upon requirements (e.g., material ones)
- Larger flexibility using movable optics or fiber cables
- Supply of several work terminals over a multiplexer with fiber optics is possible (time-sharing operation)
- The parts to be joined require only a small clamping force
- Short warming-up and cooling times make short cycle times possible
- Little or no pollution of the environment

Disadvantages

The following are some of the disadvantages of LBW. The relatively narrow welds can lead to unacceptable weld joint failures, such as joints failing or burning notches. A further disadvantage in the application of laser welding at the thick sheet metal connections is the occurrence of cracks and pores. With increasing joint depth and plate thicknesses, the frequency of the micropores increases due to the fast cooling and narrowed deep welding caverns. Degassing from the melting pool is hindered. Further, in LBW a multiplicity of factors such as laser beam characteristic values, welding process parameters, and material properties are present, which affect the reliability and the reproducibility of the welding results.

High-Power Fiber Laser

Fiber lasers combine the advantages of diode-pumped, solid-state lasers (Nd-YAG lasers) in an outstanding way with those from transverse electromagnetic waves. Fiber lasers offer high-power outputs with distinguished jet quality at the same time. The state of the art constantly changes with innovations in technology and with higher speeds. The laser welding machines are becoming more efficient and less expensive. Thus, LBW and cutting of the thick sheet metals in shipbuilding and marine engineering is becoming more attractive. The newest fiber lasers are able to do jobs that previously required a CO_2 laser.

Very good results have been obtained regarding flexibility and reliability using a 17 kW fiber laser [5]. The most important advantages of the fiber lasers compared with CO_2 and Nd:YAG lasers can be summarized as follows [1–5]: high output powers, excellent beam quality, electrical efficiencies above 25%, high flexibility, maintenance-free operation, simple operation, truly mobile system for service under field conditions, small size and weight, complete building method, and low purchase prices. In the near future, it is expected that the conventional welding methods used in shipyards will be replaced with Nd:YAG and Yb laser procedures due to their low initial costs, mobility characteristics, and higher power performance by which the thick sheet metals can be welded.

LBW: Repeatability

Over the last two decades LBW has become established as an economically sound and high-quality joining process ideal for many industrial applications. Welding with laser technology offers a number of advantages over conventional welding methods. It improves efficiency, simplifies handling, provides high quality, and laser-welded parts require less refinishing. These advantages are a result of focused energy input, narrow heat-affected zone (HAZ), and minimal distortion of the workpiece. Laser welding also offers higher productivity due to high welding speeds, process reliability, and good accessability. Repeatability is also an important strength of laser welding. The laser welding process is almost always highly automated and often accompanied with an online process control. This control gives highly repeatable results and makes manufacturers less dependent on workers' experience.

Slab Lasers

Rofin-Sinar Laser GmbH recently delivered a DC 050 to Mercedes-Benz in Stuttgart-Hedelfingen, Germany. The company provided the 5000th CO_2 slab laser machine it had produced to the automobile manufacturer. The laser will be used in an existing two-station LBW machine from the German mechanical engineering company Arnold to weld planet carriers. Daimler has been welding the planet carriers since 2002. The laser processes the planet carrier in the all-wheel component of the automatic gearbox 7G-TRONIC. This all-wheel component has helped Daimler increase traction while reducing consumption of fuel from 1 to 0.4 L per 100 km traveled.

Hybrid Laser Systems

The combination of laser light and an arc in one welding process has been known since the 1970s; however, the process has not been extensively developed [8,9]. This technology was taken up again recently in an effort to combine the benefits of the arc with those of the laser in one hybrid welding process [10,11]. The combination of laser welding with any other welding process is called a hybrid welding process. This means that both a laser beam and an arc simultaneously act on the welding zone and each affects and complements the other. Here, one's imagination knows no bounds. As an example, recent process-technological examinations of CO_2 LBW with welding wire in combination with the gas metal arc welding (GMAW) process have been carried out [12].

Laser-hybrid welding not only requires a high-power laser but also high beam quality in order to achieve the so-called deep-weld effect [13]. Figure 11.3 shows the principle of laser-hybrid welding [14]. The laser beam provides additional heat to the weld metal in the upper weld region in addition to the arc. In contrast to a series-connected arrangement, hybrid welding is the combination of both welding processes in one process zone. The resultant mutual influence of the processes can have different intensities and characteristics depending on the arc and laser process used and on the process parameters applied. Compared to the individual processes, the welding depth and the welding speed are increased together with the realization of the laser process and the arc heat. The metal vapor evaporating from the cavity reacts on the arc plasma. The absorption of the Nd:YAG laser radiation in the working plasma remains negligibly low. Depending on the selected ratio of the powers, either the laser or the arc character will prevail.

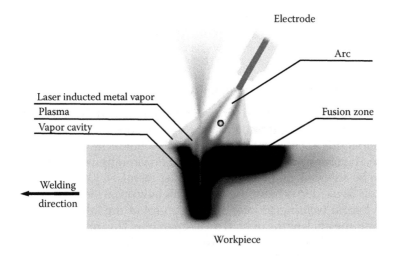

Electrode

Arc

Laser inducted metal vapor

Plasma

Fusion zone

Vapor cavity

Welding direction

Workpiece

FIGURE 11.3
Schematic representation of laser hybrid welding (From Staufer, H., *Welding Journal*, 86(10), 36–40, 2007. With permission.)

Synergies by Laser-Hybrid

The following benefits can be achieved by combining the arc and the laser beam. Compared to the laser welding process, the laser-hybrid process features the following advantages: high root opening bridging ability in the case of a gap existing for a short period, wider and deeper penetration, significantly wider range of applications, lower investment costs by saving laser power, and increased toughness. Laser-hybrid advantages compared to GMAW: higher welding speeds, deeper penetration at higher speeds, lower heat input, higher strength, and narrower joint. By combining the laser beam and the arc, a larger molten pool is formed compared to the LBW process. Consequently, components with larger root openings can be welded. The arc welding process is characterized by a low-cost energy source, good root opening bridging ability, and a microstructure that can be influenced due to the filler material added. The laser beam process features a deep penetration depth, a high welding speed, a low thermal load, and a narrow joint. In the case of metal workpieces, the laser light produces the so-called deep-weld effect as of a determined beam density, so that components with an increased wall thickness can be welded providing the laser power is sufficient.

Laser-hybrid welding allows, therefore, higher welding speeds, process stabilization due to the interaction between arc and laser light, and neutralization of tolerances. The smaller molten pool compared to the GMAW process results in lower heat input and thus in a smaller HAZ, which reduces distortion and consequently the subsequent straightening work. In the event that two separated molten pools are available, the laser-beam-welded area—especially in the case of steel—is tempered due to the subsequent heat input via arc, and furthermore, it is possible to reduce hardness peaks. The higher welding speeds allow reducing fabrication times and hence fabrication costs [14].

The Laser Tandem Hybrid Welding Process

The leading laser beam is used for welding the root, and the trailing tandem process is for increasing the ability to bridge root openings and throat thickness. One significant aspect of the process as a whole is its high flexibility. For example, the user can select three different power outputs, the depth of the root can be adjusted, and two different filler metals, with various different wires, can be used to achieve the desired metallurgical effects [14]. The laser tandem hybrid welding process allows control of the laser power and the power and the arc lengths of both arcs separately. This results in the perfect drop detachment, stable arcs, and fewer spatters. High deposition efficiencies and high welding speeds can be achieved. Furthermore, laser single wire welding with one arc only is also possible. As you can imagine, it may be useful in the case of large root openings if the two electrodes are not guided exactly one after the other, but if there is a lateral displacement of the electrodes from the weld direction. A major advantage of combining processes in this way is the fact that the arc processes generated as the filler metal melts off does not act on the workpiece and the melt at just one arc root, but is distributed across separate arc roots. With the laser tandem hybrid process, it is possible to increase not only the welding speed, but also the ability to bridge root openings compared to the conventional laser single wire welding.

Hybrid Laser MAG Welding System

In a report by C. M. Allen, TWI Ltd. (Cambridge, UK), a new hybrid laser arc welding technique combines deep penetration keyhole laser welding with an arc welding process in a single process zone. The report describes hybrid laser MAG welding procedures and weld properties in 4, 6, and 8 mm (0.15, 0.23, and 0.31 in.) thick C-Mn steels. The report further describes the development of welding procedures with lower power output CO_2 and Nd:YAG lasers, for a selection of DH36, D36, and S275 grade steels. The hybrid process offers the benefits of the separate processes, and overcomes some of their respective drawbacks, such as the lower tolerance to joint fit-up of laser welding, or the higher heat input of arc welding, which increases distortion and subsequent rework costs. These costs have been estimated to be up to 15–30% of the total labor cost for new construction. For hybrid welding of steel plates, high power output CO_2 lasers are typical.

Another hybrid system has been reported for a robotic hybrid laser gas metal arc (GMA) pipe welding system. This is the first qualification of hybrid laser GMA welding by the American Bureau of Shipping in the United States, the first demonstration of hybrid welding in a U.S. shipyard, the first production components hybrid welded in a U.S. shipyard, and first hybrid welded components installed on a U.S. ship. The hybrid process was developed for this application and qualified by the American Bureau of Shipping for a wide range of pipe schedules. A system to realize this application was specified, designed, built, and implemented, and subjected to a 7 month evaluation and numerous production pipe spools were manufactured using the system as well.

Finally, Figure 11.4 [15] depicts a hybrid laser welding technology that combines the best attributes of laser welding with those of conventional GMAW. The laser welding component offers deep weld penetration with very low heat input and small HAZ. The addition of the GMAW component vastly expands the tolerance of joint root openings, surface conditions, and impurities; improves root opening filling and contouring; and enhances control of the weld metallurgy. This new system includes the following:

- Active joint tracking and measurement
- In-process weld monitoring
- Real-time, closed-loop control of all critical welding parameters
- Automated weld surface inspection, measurement, and quality attribute checking
- Automatic weld flaw identification, flagging, and/or repair
- Complete weld process and quality documentation and reporting

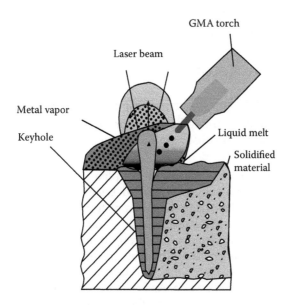

FIGURE 11.4
Hybrid laser welding combines the attributes of GMAW and LBW (From Defalco, J., *Welding.*, 86(10), 47–51, 2007. With permission.)

The control system monitors and adapts the welding process faster than a human operator could, allowing travel speeds that would not be possible with conventional control systems. This technology is available in a 2-D gantry, 3-D robotic, or custom mechanized solution [15]. This system has welded beams from ordinary and high strength plate at rates that are 5–10 times that of conventional welding processes. Additionally, the finished beams are produced to tolerances that are 1/4–1/10 of the allowed tolerances for hot rolled beams. Additionally, tests have shown that the laser welds can achieve fatigue lives that are 2–3 times as long as conventional GMAW and submerged arc (SAW) processes.

In comparison to other high production welding processes, such as the tandem GMAW and rapid arc welding, the hybrid laser arc welding (HLAW) has demonstrated these benefits:

- Higher welding speed
- Deeper weld penetration and wider weld root
- Lower heat input, resulting in fewer heat-induced deformations
- Better mechanical properties
- Butt-joint welding without ceramic backing
- Very smooth transition between weld and base material on the root side
- Greater stability of the welding process
- Better fatigue performance
- Better weld toughness

The process has been applied to:

Shipyards: In ship applications, the combinations of laser/GMAW capabilities has enabled the reduction of beam structure weight 20 –50% and is projected to substantially reduce structural fabrication and assembly costs [15].

Bridges: Bridge builders regularly use plate-fabricated beams where size precludes the use of hot-rolled beams. Some engineers also specify high-yield steels in these beams to reduce structural weight. Laser welding processes facilitate the production of these beams at lower costs and higher overall throughput. Additionally, by using new, affordable, high-strength steels to reduce weight, laser welding enables smaller plate fabricated beams to be produced competitively with hot-rolled beams. As pricing for ordinary steel continues to escalate, processes that can reduce total steel weight will provide cost advantages for producers and consumers of structural steel. Another application for HLAW in bridge fabrication is the use of sandwich panel bridge decks. Steel sandwich panels have the potential to be highly competitive in cost and weight with conventional bridge deck systems currently on the market [15].

Railcars: The railcar industry shares many of the same challenges faced by shipyards. Railcars are largely constructed of a steel subframe, a steel exoskeleton, and steel side panels. Side panel buckling is a persistent and noticeable problem in many designs. One railcar fabricator recently studied the possibility of using HLAW in its side panel assembly area. It hoped to increase the throughput in this area, eliminate side panel buckling, and reduce labor costs [15]. The manufacturer's current SAW process produced unacceptable levels of distortion. The potential for replacing the SAW butt-joint welding process with a hybrid laser welding process was investigated. It was estimated that switching to HLAW would reduce operational times by more than 20%. The additional capacity possible with the new system would produce an additional 1000 railcars per year. Coupled with the labor savings, this would allow a return on investment in just over 1 year.

Pipelines: In an attempt to control the high cost of pipeline construction and enhance safety though implementation of innovative technology, pipeline owners and operators have been pursuing a number of next-generation pipeline technologies including higher strength pipeline steels and fittings, multiwire mechanized/automated welding, ultrasonic inspection, advanced coating systems, and alternative integrity-validation processes. The current state-of-the-art welding processes for onshore pipelines involve mechanized GMAW. However, it is highly likely that the next generation of automated pipeline welding equipment may be built around HLAW [15].

Applications

Shipbuilding and Marine

Currently, many ships are prefabricated as a number of individual sectional and modular construction units, and then assembled on the slipway partially in the water (Figure 11.2). For the manufacturing of sectional and modular constructions of marine structures, the application of LBW technology proves to be very favorable due to combined welding and cutting automation with the laser beam technique. Thus, productivity could be greatly improved with this technology compared with the arc welding method.

In shipbuilding, using fiber laser technology, the components can be joined together without welding edge preparation and pre-or postweld heat treatments. The resulting joints are more economical and are of better quality than arc welded joints. The laser produces a narrow weld joint with a small

HAZ compared with arc welding, plus there are no weld defects caused by arc blow or electrode wear, which can occur during conventional arc welding methods.

The high-quality steels and light metals alloyed with Al, Mg, and Ti, which present various problems when welded with the conventional methods, can be easily welded with the laser beam technique. Thus, the unladen weight of ships is decreased, and production and operating costs as well as production times are reduced. For the manufacturing and/or building of sectional and modular constructions of marine structures, the application of laser technology proves favorable, because LBW and cutting can be combined and automated very well with each other. However, for an economic evaluation of the laser beam welded constructions, knowledge is necessary for the correlation among welding parameters, joining geometry, and fatigue strength characteristics of the materials used. The advancement of the high-power laser technology for welding of construction units in the I-butt joint connection with joint columns can give new advantages to the welding technique. For example, the application of LBW in ships, plants, and other steel buildings as well as in the offshore technology can spread rapidly. The laser beam connecting methods make newer construction techniques for ships and other marine structures possible [1–5].

Disk lasers have been used for hybrid welding of thick I-sheet steel. The important factor in selecting the 10 kW disk laser is the absolute control over output power. Reproducible process results that do not depend on ambient conditions are crucial to subsequent production. Another benefit of the system is beam guidance via a laser light cable because the disk laser's setup does not depend on the location of the processing site, an aid to production versatility.

PSA Peugeot Citroën used three TRUMPF TruDisk 6002 and one TruDisk 4002 lasers to weld the doors and body reinforcements of its new model that was launched in the summer of 2009. The disk lasers, which range in power from 4 to 6 kW, lay a full-length weld joint rather than individual spot welds. "This enables the Peugeot 3008 to achieve high body stiffness," said Jean-Charles Schmitt, product and process laser manager at PSA. In addition, use of the disk laser allowed the company's engineers to develop new design options. "Compared to conventional spot welds, full-length weld seams in car body manufacturing require less sheet folding and thickness due to a specific laser design, and thus saves about 5 kg of weight per vehicle," Schniitt said. The company could also increase the size of the rear quarter windows, allowing better visibility for the driver and passengers.

Aircraft Blades

A laser welding robot system that can quickly and accurately repair a damaged aircraft turbine blade or a worn injection mold has been reportedly been developed at the Fraunhofer Institute for Material and Beam Technology,

Germany [16]. To begin the process, a flexible robot arm accurately focuses the laser beam on the damaged section of the component. The energy from the laser beam causes the surface to melt as the laser scans the area dot by dot, producing microscopic puddles no wider than a few tenths of a millimeter each. At the same time, powder is blown onto the surface by a stream of gas, and it bonds with the melt. Because the powder granules have diameters in the micrometer range, they are completely melted by the laser beam and rapidly form a tight bond with the base material. Powder materials include titanium, nickel, cobalt, hard metals, or even ceramics.

Laser beam deposition welding has been available for several years. However, the new system enables surfaces to be processed more accurately than before, thanks to the fiber laser. This laser is able to deposit material with unprecedented accuracy, yet without putting any strain on the component. This makes it possible to produce metal structures at a resolution of as little as 100 micrometers.

Thermoplastics

Unlike laser welding of metals that focuses light on a joint edge, thermoplastics use a through-transmission technique. Here, laser light passes through a part that is transparent to the laser wavelength (typically near infrared) and is absorbed by a part filled with carbon black or other colorant. The absorbent part melts and conducts heat to the transmissive part, welding the two together.

There are two kinds of through-transmission laser welding: collapse and contained. In the collapse mode, laser light illuminates and welds the entire joint width. Clamp force applied to the parts squeezes molten material from the immediate joint area as flash. No particulate forms in most cases, making the technique suitable for high cleanliness applications. Clamp force must be carefully controlled. Too little and the parts may not fully touch one another during welding, which can hurt quality. Too much clamping force pushes excessive material from the joint, causing a thinning of the bond line and poor molecular orientation, both of which lower weld strength. The collapse mode process tolerates air gaps to about 100 μm or more. The process better compensates for surface inconsistencies than contained welding, though not as well for part wrap.

Contained welding shines laser light on an area smaller than the total wall section of a weld. Molten material stays within the part edges, hence the name. The contained approach eliminates both flash and particulates, though gaps exceeding about 75 μm compromise joint strength. A minimum amount of clamp force holds parts for welding, the upper limit of which is set by part structural strength. Clamp force, therefore, is not considered a process variable. Because contained welding does not produce flow-induced molecular orientation. It is theoretically capable of weld joints that are nearly as strong as the parent material. There are four ways to introduce laser light into a thermoplastic assembly: contour, mask, simultaneous, and quasi simultaneous.

1. Contour welding trains a single point of laser light on the workpiece and is limited to contained welding mode. Systems move the beam, the part, or both. They are energy efficient and highly programmable, have low tooling costs and quick setup, but relatively long cycle times.

2. Curtain mask systems drag a line of high-intensity laser light—from banks of diode lasers—across a glass mask. A coating on the glass shades the workpiece from the laser light except where it is needed for welding. The method works only for near-planar joints and contained-node welding.

3. Simultaneous welding systems use an output lens shaped specifically for the part to be welded. Multiple diode lasers deliver light (through bundles of optical fiber) to the lens. Such systems are typically designed for collapse-mode welding, though they can accommodate 3-D geometries in contained-welding mode.

4. A quasi-simultaneous system uses a pair of servocontrolled mirrors (galvo head) to steer a single laser beam on several trips around a joint. High heat loss associated with the technique requires extra laser power to overcome. Quasi-simultaneous systems typically work in collapse mode and joints must be nearly planar. Galvo heads limit beam travel to their field of illumination, typically about 200 mm (7.8 in.) diameter. Cycle times can be about equal to that for mask welding, but are longer than simultaneous welding.

A tube of glass reinforced polyamide carries oil from the filter to the six-speed, direct-shift gearbox control unit of Volkswagen's dual-clutch gearbox. To produce the complex tube and to avoid demolding problems, a German automotive supplier IBS Filtran (Deutschland) combines injection-molding with laser-transmission welding to join separately molded parts without risk of producing particles that might contaminate the oil. Zytel 70G30 HSLR type reinforced with 30% glass fiber, makes weight-saving, thin-wall moldings possible, but still has very high strength. Even when the temperature and the vacuum in the system reach peak valves, and vibration and impacts create high acceleration forces at the same time, deformation remains minimal. As a result, the upstream filter always works optimally and there is no risk of collision with gearbox parts that are very close.

Blow molding was rejected as a potential process and the solution decided by IBS Filtran was a tube with offset open areas. One part made of a laser-absorbent variant of Dupont's Zytel 70G30 HSLR molded in a tool with a single parting line. Onto the open areas two injection-molded tops of the same polyamide but a laser-transparent grade was laser welded. As the welding does not involve oil-contact surfaces, there is no risk of plastic particles coming loose and entering the oil stream.

Pipelines

So far, LBW has been successfully applied to the stationary production of (longitudinally welded) pipes [17,18]. For girth welding, different laser welding techniques have been suggested. Among the requirements associated with such systems for welding of land pipelines (for quality as well as for economic reasons) are the following:

- The welds should only be made from the outside.
- The beam source has to withstand vibrations and adverse environmental conditions.
- The beam has to be guided over a distance of approximately 30 m (9.8 ft) to the welding head.
- The power requirements may not exceed 160 kVA, because larger generator sets are difficult to handle on the construction site.
- The weld quality has to comply with the requirements.
- The welding speed has to be higher than for orbital GMAW.
- The investment costs have to be economically justifiable.

Recent developments in laser physics have made fiber lasers with beam powers exceeding 15 kW available to the market. This type of laser belongs to the solid-state laser group. Its laser active medium consists of an Yb doped glass fiber that is optically pumped by means of diodes. The wavelength is 1.07 µm; its wall-plug efficiency may reach up to 30%. To improve both quality and cost in the construction of land pipelines, a high-power fiber laser welding system has been designed. Due to the unique features, this system has the potential to bring laser welding of pipelines for the first time into the field. For prequalification, the laser system and the welding process have been developed and tested under laboratory conditions. It was established that the process is, in principle, capable of fulfilling all associated requirements, and especially that pipeline segments of X70 [thickness 12 mm (0.4 in.)] could be welded in one layer at a welding speed of 2.3 m/min (6.9 ft/min) [19].

Remote Laser Welding

Remote laser welding is quickly becoming an accepted method for LBW. This method differs from conventional laser welding in that the beam is steered to the work surface using small lightweight mirrors instead of a conventional focusing head, and either moving the part or positioning the focusing head with a robot. Much faster positioning speeds can be realized utilizing remote welding technology. Positioning speeds of up to 20 m/s (800 in./s) can be achieved with today's technology.

There are a few design considerations that welding engineers need to be aware of for proper design of parts and assemblies that are going to be remote laser welded. An informed engineer can design assemblies and parts so that manufacturing can fully use the benefits of remote laser welding. One consideration is how the focused laser beam is positioned onto the work surface. Second is how the weld profile changes as the beam is positioned around the work surface. A third consideration is the type or shape of the laser weld required to get the most strength with the least number of welds.

Process Description

The key technology behind remote laser welding is using a scan head to rapidly position and focus the laser beam on the work surface. The laser source, lens, and X- and Y-mirrors comprise the elements of the scan head where the beam is first sent through a set of lenses where the first lens is moved along the optical axis to change the focal position of the laser beam. The beam is then steered with two mirrors. One mirror steers the beam along the x-axis; the other mirror steers the beam along the y-axis. This allows for complete access to the entire work surface without having to move the work surface or the scan head. The only moving parts are the two lightweight mirrors and the first lens. For more details of the scan head, focusing lens, laser stationary see Klingbeil [20] and Cann [21].

Part Design and Process Development

As stated earlier, the beam is directed from the center of the field. When a part is presented to the remote laser welding system, there needs to be a line of sight to all welding areas. For instance, if the joint is located below an edge that cannot be seen from the center of the field, it would be necessary to move the part so that line of sight is reestablished. If a part is redesigned, the manufacturer can take more advantage of the benefits offered by remote laser welding. Tooling design is usually much easier for stationary parts, and process repeatability is usually much higher the less the part has to be moved. Every time the part is moved, you increase the number of opportunities for error. There is a great deal of flexibility in the remote laser welding process. Because of this flexibility, process development should be done for each family of parts that is going to be manufactured. There are several items that should be investigated before finalizing the process. The programmed weld traverse, weld penetration, weld sequence, and number of welds will affect the final part strength and fatigue life.

Applications

Remote laser welding is finding new applications every day. It is easy to imagine how remote laser welding can be used in the automobile industry. Many assemblies and subassemblies could benefit from this process. Applications where a long reach is required, such as internal components in a muffler, are a good example. GMAW is often used for these applications because the weld gun is small and can be easily manipulated inside these components with a robot. Conventional LBW may not work well because of the small opening and long distance that the head would have to reach into the assembly. Remote laser welding is capable of keeping the physical components outside of the assembly and only the laser beam is directed to the weld locations inside the assembly. Another benefit beyond the long focal lengths possible with remote laser welding is that both the welding head and assembly can be stationary. This reduces the amount of tooling and programming necessary to complete the welding of the assembly.

Remote laser welding also lends itself well to spot welding. Applications where multiple spot welds need to be placed in a very short time are prime examples. One example would be spot welding plate and ring assemblies for oil filters. These assemblies are usually resistance welded. Resistance welding can take a significantly longer time to weld the assembly because of having to reposition either the part or the resistance welding machine. Remote laser welding is able to complete six spot welds in slightly less than 0.3 seconds. Repositioning of the laser beam is an insignificant portion of the cycle time. An added advantage to using remote laser welding is that the parts can be processed while in motion. Remote laser welding also has the typical advantages that are associated with LBW. Some of these are that it is a noncontact process, it does not require a filler material, it is a low heat input process. These advantages along with the added benefits of the remote laser welding process should help the automotive market reduce cost and increase production, if it is applied properly.

3-D Laser Combination Head

Laser manufacturing with a combi-head is a new, highly flexible solution based on old ideas [22–26]. Recently, these ideas have been taken up again by several groups [27–29] who were motivated by the availability of laser sources with simultaneously high power and high beam quality and by increasing market demands on flexible production. In a multifunctional laser cell, the combi-head is able to perform 3-D cutting and welding tasks in an arbitrary sequence without retooling. The part handling, positioning, and clamping steps are omitted. This results in shorter and more flexible process chains and with them reduced production time and costs as well as improved manufacturing accuracy [30,31]. Moreover, the combi-head opens up efficient solutions for innovative products from sheet metal with a wide range of variants. The main technical features of the laser combi-head can be summarized as follows:

- Quick, software-controlled switching between cutting and welding in arbitrary sequences
- One machine, one setting, one tool for cutting (N_2 and O_2) and welding processes
- Constant tool center point (TCP)
- Rotationally symmetrical, coaxial, and slim head design with the autonomous nozzle, integrated capacitive clearance sensor, and effective crossjet
- Gas functions decoupled from the optical system
- No principal limits regarding beam power and gas pressure

These features can lead to the following benefits for production when manufacturing sheet component assemblies:

- Short, integrated process chains for high productivity
- High flexibility and cost-efficient production of options and variants
- Savings in handling, positioning, and clamping operations
- Savings in machine investment and floor space
- Increased utilization of the machine
- Easy reconfiguration of the system for new products
- Short tolerance chain improving accuracy of parts and components
- Good accessibility of narrow workpiece areas
- 1-D, 2-D, and 3-D applications
- Suitable for lens as well as mirror optics
- Suitable for all laser types with appropriate beam quality [32–33].

Joint Designs for LBW

The key for the various joint designs [34] in LBW is to bring the components in direct metal contact with relatively high precision. The laser beam is focused at the contact line. In deep penetration welding, both components are molten and subsequently fuse together while resolidifying. The most common joint types for LBW are butt, lap, fillet, and flange. Butt joints allow for complete penetration welding of the workpieces with single-sided access. Butt joints give high strength because the entire welded cross section carries the load. Additionally, butt joints require the laser focus position relative to the joint to be accurately controlled. Edges have to be sharp and straight to ensure good part fitup. Typically the edges are prepared with precision shears, laser beam cutting (LBC) machines, or mechanical machining in order to ensure optimum welding conditions.

In lap joints, two or more layers of components are stacked. By deep penetration welding, the laser beam melts through the layers. Subsequent resolidification fuses the components together at the interface. Compared with butt joints, lap have reduced weld strength due to the smaller joining area. Compared to butt joints, lap joints allow for significantly wider focus positioning tolerances and, therefore, is a preferable technique to compensate for part tolerances. Clamping along the weld interface ensures contact between the components to be welded.

Fillet welding is effective for welding T-joints. Welding from one side, the fillet weld gives a symmetrical joint across the joining area. It offers good weld strength because the cross section equals the material thickness. Comparable to butt joints, focus positioning, and edge preparation are critical. In flange joints the laser beam is aimed at the contact line of the two components. By multiple reflections at the interface of the joint, almost all the laser power reaches the contact line and is utilized for welding. Flange welds result in a narrow joint, with high penetration depth resulting in a high depth-to-width ratio. The weld-focusing effect at the joint interface makes flange welds relatively insensitive to lateral focus positioning. Each joint has unique benefits and requirements for successful welding. Part tolerances, edge preparation, and clamping techniques are critical for laser welding. The following section gives design guidelines and best practices for designing laser beam welds for producing high-quality, cost-effective parts [34].

Table 11.1 [4] summarizes guidelines for weld joint preparation and cleanliness, along with methods on how to achieve these conditions. In general, the surfaces shall be clean and dry. Contamination may vaporize during

TABLE 11.1

Surface and Edge Condition Guidelines for LBW

Guidelines	Remarks
Clean surface: no dirt, rust, oxides	Ultrasonic bath or washer
No grease, heavy oil, heavy oil-based lubricants	• Ultrasonic bath or washer • Light protective oil film is acceptable • Water-based lubricants are usually acceptable for steel
No water or water-based lubricants for Al	Will cause porosity
No coatings: paint, chrome, zinc, phosphate	• Remove mechanically • Coated surfaces may be welded with special techniques
No case hardened layer	Remove mechanically
No sand-blasted surface	Sand residue may cause porosity
No grinding residues	May cause porosity

Source: Verhaeghe, G., *Welding J.*, 84(8), 56–60, 2005; Cann, J., *Welding J.*, 84(8), 34–7, 2005. With permission.

the welding process, but the risk remains that residues will adversely affect weld properties. Coated surfaces, especially zinc-coated, can be laser welded if special design rules allowing for ventilation of evaporating zinc are observed [34].

Newcomer for Material Welding, Cutting, and Medicine: The Fiber Laser

Fiber lasers (not to be confused with fiber-delivered lasers where the fiber is merely an optical delivery mechanism) are solid-state lasers in which an optical fiber doped with low levels of a rare earth element is the lasing medium. Laser diodes are used to stimulate the lasing medium to emit photons, an action known as pumping, at a wavelength specific to the rare earth element used as the doping element. Ytterbium is generally used for the high-power fiber lasers currently available for material processing, and it emits a wavelength approximately the same as Nd:YAG lasers (i.e., between 1060 and 1085 microns). The doped fiber is surrounded by a low refractive index material that acts as a waveguide for the pumping light and ensures optimum transfer of this energy to the lasing medium. Diffraction gratings are used as rear mirror and output coupler, to form the laser resonator, creating a long thin laser, which due to the flexibility of the optical fiber (which is simply coiled up) can be very compact. Although it is possible to use the lasing fiber as the beam delivery fiber, with the appropriate beam shaping and focusing optics at its end decoupling of the beam delivery fiber from the lasing fiber is preferred for lasers used for material processing to protect from unwanted back-reflections from the workpiece surface.

To date, 700 W single-mode fiber laser modules are commercially available, with prototype single-mode Yb-fiber lasers up to 1000 W of output power already being assessed under laboratory conditions [35,36]. The manufacturing route currently preferred for achieving multikilowatt output powers suitable for deep penetration keyhole welding of metals is by combining the outputs from a series of commercially available single-mode units into a single fiber output for the 7 kW fiber laser [35]. Fiber laser technology seems to allow, for the first time, the manufacture of scalable lasers, in a compact form, with no obvious limit to the power available. Today, the output power of a fiber laser exceeds that of commercially available Nd:YAG laser technology, while also offering a better beam quality. In fact, fiber laser power and beam quality are approaching, and in certain cases exceeding, those of CO_2 lasers. For instance, a 17 kW Yb fiber laser with a beam parameter product (BPP) of around 12 mm.mrad was recently installed in Europe and a 5 kW system with a BPP of 2 mm.mrad is now available (see Table 11.2) [34].

Welding trials at various laboratories using the latest fiber laser technology confirm that this new type of laser source should now be considered as an alternative to CO_2 or Nd:YAG lasers for the welding of materials such as steel or aluminum. The initial cutting results also show considerable

TABLE 11.2

Paser Source Comparison

	CO$_2$	Lamp-Pumped Nd:YAG	Diode-Pumped Nd:YAG	Yb Fiber (Multimode)	Thin Disc Yb:YAG
Lasing medium	Gas mixture	Crystalline rod	Crystalline rod	Doped fiber	Crystalline disc
Wavelength, micron	10.6	1.06	1.06	1.07	1.03
Beam transmission	Mirror, lens	Fiber, lens	Fiber, lens	Fiber, lens	Fiber, lens
Typical delivery fiber, Micron	—	600	400	100–200	150–200
Output powers[a] kW	Up to 15 kW	Up to 4 kW	Up to 6 kW	Up to 20 kW	Up to 4 kW
Typical beam quality[b], mm.mrad	3.7 / 3.7	25 / 12	12 / <12	12 / 1.8	7 / 4
Maintenance interval, kh	2	0.8–1	2–5	100[c]	2–5
Power efficiency %	5–8	3–5	10–20	20–30	10–20
Approximate cost per kW, k$	60	130–150	150–180	130–150	130–150
Footprint of laser source	large	medium	medium	small	medium
Laser mobility	low	low	low	high	low

Source: Zefferer, H. and Morris, T., *Welding J.*, 84(8), 23–8, 2005. With permission.

[a] Commercially available at the time of writing.

[b] The top figures are for the maximum available output powers, the bottom figures for the same type of laser but configured for optimum operation at 1 kW.

[c] Manufacturer's claim.

promise for the new technology. Moreover, its compact design, easy set-up, and lower power and cooling requirements compared with existing laser technology, make it ideal for on-site welding, for pipeline welding or ship-building, or for remote repair, as an example. As the fast-paced development continues to push up the power levels and improve the beam quality, the range of industrial applications for this new type of laser will undoubtedly expand. It remains to be seen, however, if this competition between laser technologies will also result in more affordable lasers [37,38].

In some recent studies Miyamoto et al. conducted experiments using a single-mode fiber laser for high-speed microkeyhole laser welding [39]. Paleocrassas and Tu [40] explored the use of a 300 W single-mode, Yb fiber-optic laser for low-speed keyhole welding of heat-treatable aluminum (7075-T6) for potential applications of crack repair. The characteristic that sets fiber lasers apart from Nd:YAG lasers is its excellent beam quality, with an M2 value of approximately 1.04. This allows the beam to be focused on much smaller spots and, therefore, it is possible to achieve higher power densities. In addition to the excellent beam quality, this laser has a wavelength of approximately 1075 nm (near infrared spectrum), which is relatively short compared to CO_2 lasers. This allows good beam absorption that is needed when welding highly reflective materials like aluminum.

Therefore, a fiber laser combines the advantages of CO_2 and Nd:YAG lasers and the optimum welding speed is between 2 and 3 mm/s (592 and 394 ft/min), focused slightly into the workpiece (no deeper than 0.75 mm (0.29 in.), for a maximum penetration of 1.02 mm (0.04 in.) [40]. This investigation showed that there is great potential for welding heat-treatable aluminum alloys with fiber-optic lasers. The biggest obstacle so far has been aluminum alloys' high reflectivity. If the workpiece is positioned perpendicular to the beam, it reflects some of it back into the collimator and damages it. One of the solutions to this problem is to fit an optical isolator onto the collimator, which will divert the reflected beam away from it. Although the beam quality may be sacrificed slightly, some of the problems that occur with the current (tilted) setup will be eliminated. Currently, any variation perpendicular to the weld direction will change the focus of the beam with respect to the workpiece. Other areas in which research needs to be pursued in the future include vapor/plasma effects, shielding gas flow rate control, different joint configuration (leading up to crack repair), microstructure examination, and defect suppression [40].

Another fiber laser application has been reportedly developed by the Fraunhofer Institute, Germany [41]. It is a fiber laser system that is capable of hardening, cutting, and even welding. In the last few years, fiber lasers have been developed that can generate light with an output of several kilowatts in fibers with a thickness of only 50 microns. The fibers are as flexible as a cable, allowing them to get close to components with a complex geometry. Additionally, fiber lasers generate light with a wavelength of around one micron, a good wavelength for absorption by metals such as steel and aluminum. The fibers generate very uniform light and a very small focal spot, operating more rapidly and with greater precision. Moreover, the fiber laser has an energy conversion efficiency of 20%, compared to the 6–10% achieved by CO_2 lasers.

In another application, IPG Photonics Corp. in Germany achieved record high-quality, deep penetration welds on thick stainless steel plates at high speeds. Using the company's 20 kW continuous wave, 1070 nm commercial fiber laser, it was possible to weld 2.54 cm (1 in.) thick stainless steel at a speed of 0.85 m/min (2.82 ft/min) and 19.05 mm (0.75 in.) samples at a speed of 2 m/min (6.6 ft/min) using a 200 micron fiber cable with a 420 micron focus spot size. In other experiments, IPG was also able to produce high-quality welds on 5.08 cm (2 in.) thick 304 stainless steel samples by applying laser beams from both sides with a penetration depth of 54–56 mm per pass.

There are other exciting and growing markets of biotechnology and semi-conductor manufacturing that demand the very latest in state-of-the-art fiber-optic laser technology. In biotechnology, for example, there is a drive for effective new drug discovery using techniques such as confocal laser scanning microscopy to understand the biology of living cells, plus high throughput screening to enable millions of simultaneous experiments to be carried out to test the reaction of human cells to new drug compounds. In clinical medicine there has been high demand for blood screening (hematology) using a technique called flow cytometry due to the worldwide epidemic of HIV/AIDS and other contagious viral diseases such as hepatitis.

Cutting, Machining, and Drilling Processes

Cutting

The LBC process is used mainly in the transportation equipment manufacturing industry, the construction industry, and other sectors of the metal fabricating industry. During the process, a laser beam is focused on the base metal, while an assist gas is used to eject the molten metal along the kerf. The laser beam is physically produced by an electrical discharge that is induced in a cavity filled with a gaseous medium composed mainly of nitrogen, carbon dioxide, and helium, which produces a consistent monochromatic light. This process is used primarily for cutting metal sheets when extreme precision is required and where a large number of pieces must be cut at high speed.

While laser cutting technology requires maintaining control over many parameters and major investments in machinery, the cutting speed and quality that are its hallmarks make LBC a preferred process. A new laser technology has increased laser cutting speeds even more. According to observations made since it was first marketed, this technology, which is sold under the name Bifocal, increases cutting speed for metals that use nitrogen as an assist gas (e.g., stainless steel and aluminum alloys) by 20–40%. The

Bifocal effect is created using a mirror or a replaceable lens. The technology focuses the laser beam at the base of the workpiece, which prevents the formation of burrs. The second focus point for the new optics is located on the same axis as the first, but is directly on the surface of the workpiece. It is the second focus point that speeds up cutting. It concentrates sufficient energy at the surface to initiate penetration of the beam. In general, increases in cutting speed of 20–40% have been observed, depending on the type of laser source, the type of alloy, and the thickness of the metal sheet. In addition, by focusing the energy and the beam on two points, the technology can cut metal sheets that are 10–20% thicker.

In the case of carbon steel, the fact that the laser beam used with this technology is narrower at the base of the material (first focus point) than that used by traditional technologies also reduces surface oxidation. With traditional cutting technologies, slag is a by-product that results from a more diffuse beam. Because it is applied uniformly to a larger surface of the material, the beam used with a standard focus acts more slowly, which causes impurities that are contained in the molten metal, or dross, to rise to the surface. The absence of slag and burrs results in a much cleaner cut, which facilitates the subsequent processing of cut workpieces. For example, paint adheres better to a workpiece that is free of slag and burrs. As a result, the new technology contributes to increased productivity during subsequent steps of the fabrication process. The faster cutting speed also generates savings in terms of energy and the volume of assist gas, such as nitrogen or oxygen.

Implementing the technology in a plant is easy. In fact, the optics (lens or mirror) can replace the standard optics that are installed on most CO_2 laser cutting machines that use nitrogen or oxygen as an assist gas. However, the geometry of each optics (lens or mirror) must be optimized for the specific anticipated use of the machine. The optics on the cutting machine are installed during testing while the parameters of the laser cutting machine that is equipped with the new technology are optimized. This step is carried out on site, with the collaboration of the cutting operators.

Anyone who works in the field of manufacturing production can appreciate the importance of new technological progress. Steel cutting is one of the more common operations in the manufacturing industry. Any productivity increases at this stage have a significant impact on manufacturing productivity as a whole. By increasing cutting speed and the capacity of cutting machines, the laser beam also significantly improves the quality of cuts, which results in a substantial increase in productivity in a number of areas.

Machining

The cutting, drilling, and shaping of glass, quartz, and related materials have traditionally been achieved by skilled individuals using cleaving, grinding, polishing, or hot-blowing techniques. Precision laser-based machining systems are now providing a fast, accurate, and reproducible alternative that allows glass and glass-like materials to be processed as an engineering material in a production environment.

Silicon dioxide, in its many familiar forms of glass, silica, and quartz, represents one of the most useful materials known to mankind. It provides a unique combination of high tensile strength, excellent chemical resistance, low thermal conductivity, high temperature compatibility, excellent electrical insulation, and high optical transparency at a relatively low cost, all of which make it the material of choice in applications from windows to test tubes, optical fibers to attic insulation, and headlamp bulbs to furnace baffles.

Long-established methods exist for producing the raw material in the basic geometric forms of rods (including fibers and filaments), tubes (including capillaries), sheets, and balls. However, if an application requires a shape different from or more complex than these basic forms, the available processing techniques become a limiting factor. Indeed, the methods for processing the raw product into usable components have seen very little technological development for decades and can often be traced back hundreds if not thousands of years. Even simply cutting these materials remains a difficult process and typically is avoided by using a "cleaving" technique in which a stress raising scratch is made on the surface of the material, which is then stressed to the point at which a fracture propagates through it causing a break. This technique is generally limited to nominally straight lines, has a less-than-perfect yield, does not operate to conventional engineering tolerances, leaves sharp edges that often need further processing, and generates shards of sharp and potentially hazardous material. The alternative—machining the part with a diamond saw—is slow, expensive, and still limited in the geometries that can be achieved. Some shapes can be created by grinding and polishing but these are laborious processes with significant consumable costs and in any case are generally limited to flat and smoothly curved (preferably spherical) surfaces. For nonspherical (aspheric) surfaces significantly more complex techniques are required.

Advances in laser technology have led to the development of high-throughput, automated machines for precision cutting and shaping of components. These systems are operator-independent and provide consistent high-quality results on a range of materials from the optical fiber used in optical communications networks to quartz used in gas-discharge and tungsten–halogen lamps.

Modern manufacturing environments are often highly automated and demand high levels of quality, consistency, yield, and throughput. As the demand for a given product grows, tooling must evolve to enable increased production volumes while maintaining a quality as good or better than that which was previously achievable only by the most experienced craftsman. The need for advanced production techniques for glass-like materials has fueled development of a range of laser-based manufacturing tools. Using a laser to precisely cut and shape glass components has several advantages over conventional processing methods. The laser beam cuts through thin glass in a fraction of a second, can be directed with extremely high precision, and can be controlled to remove material in a predictable and reproducible manner. At its simplest level the laser can cut accurate lengths of material, leaving a clean, flat face with a high-quality surface finish and robust, slightly rounded (burr free) corners. The laser can also create a range of edge profiles—straight, angled, rounded, pointed, or tapered—and it is possible to cut flats or combine these shapes in a number of directions. In fact the variations in the profiles that can be produced are almost limitless. In addition, the noncontact laser process will not leave unwanted cracks, chip outs, or scratches that can act as failure initiators in glass-like materials. The laser process can also, therefore, benefit the in-use reliability and lifetime of the final component.

Applications

Many applications can benefit from precision-shaped quartz components. Both rotationally symmetric and asymmetric forms can be made. In these types of applications the laser system can produce, in a few seconds, parts that would otherwise take 10–30 minutes to produce using conventional polishing techniques. The automated system is a stand-alone machine that can routinely operate to a precision of netter than 0.5 μm. As is generally the case with laser machining, the noncontact nature of the process also permits a range of on-line diagnostics, data-logging, and quality control to be incorporated within the system to endure product quality and traceability. Several examples serve to illustrate the range of applications of this technique.

Lens produced on the ends of optical fibers, for instance, have a host of uses based around their ability to improve and control the coupling of light into and out of the fiber. These lenses are used extensively in fiber-optic telecommunications, but also provide efficient and flexible optical probes in other applications like biomedical gel scanning, silicon-wafer fluorescence spectroscopy, and on-die VCSEL (vertical cavity laser) characterization.

The quartz vacuum chamber supports provide mechanical support for components and samples within a range of treatment chambers, including ion-implantation, X ray crystallography, and X ray photoelectron spectroscopy. These supports must be vacuum-compatible with very low thermal conductivity, and the material must be capable of withstanding the considerable temperature difference between the sample under test and the enclosure. Laser machining of the part allows the mechanical strength to be achieved while minimizing physical contact with the sample being held, thereby further reducing heat transfer as well as minimizing contamination of the sample or the measurement data.

The end of a quartz lamp cut by the laser is clean, smooth, and has gently rounded robust corners with no need for postprocess flaming. Moreover, these parts are directly cut to precise length tolerances, eliminating the need for grinding, polishing, and cleaning steps to achieve tolerance. Laser cutting also allows the cut ends to be angled or chamfered to facilitate subsequent assembly processes. The precise cutting and milling of the features in capillaries and flat plates, and the ability to laser weld these parts together is finding increasing use in "lab-on-chip" devices. In applications like this, the laser machining system is typically fully integrated with the production line and parts handling is automatic.

Both lab-on-chip and planar optoelectronic devices require the fiber end to be cut at angles of about 45° and rely on achieving total internal reflection off the inside of the angled face to couple light either in or out of the side of the fiber. The laser-cutting process can be applied to any requirement in which end-face angles from 0° to greater than 50° are required, and can also tailor the optical performance by using curved surfaces. The typical laser processing time for features such as this is less than 1 second.

The ability to reliably and efficiently process glass and quartz materials into new and complex geometries is opening new and exciting applications for what is a well-characterized and inexpensive material. Moreover, the capabilities offered by laser machining are bringing routine production to an area previously characterized by craftsman skill.

Researchers at Purdue University, West Lafayette, Indiana, are perfecting a technique to manufacture parts that have complex shapes and precision internal features by depositing layers of powder materials, melting the powder with a direct-diode laser producing coatings in a one-step process that is less expensive and faster than conventional technologies, and immediately machining each layer. The method can be used to create parts made of advanced materials such as ceramics, which are difficult to manufacture and cannot be machined without first using a laser to soften the material according to Yung Shin, professor of mechanical engineering and director of Purdue's Center for Laser-Based Manufacturing.

Because the technique enables parts to be formed one layer at a time, it promises new industrial applications for manufacturing parts containing myriad internal features. Although the basic laser deposition technique is not new, the researchers have increased its precision by adding the machining step. The method is about 20 times more accurate by adding the ability to machine the part while it is being formed. They have developed a facility that can actually deposit the powder, heat it with the laser, and machine it at the same time. This new technique could be ideal for manufacturing certain kinds of ceramic components that are not produced in large enough quantities to justify the expense of designing costly dies. Components made in small lot sizes might be produced far more economically by machining instead of using dies. But there has been no practical way to machine the brittle ceramic materials economically with the high precision needed for many applications. The technique is potentially practical for industry because it does not require expensive clean room environments.

Micromachining

SiC Mirrors

Laser micromachining offers the ability to machine products with significant automation while also requiring less postprocessing and enabling machining that is impossible with conventional equipment. Mound Laser & Photonics Center (MPLC, Miamisburg, Ohio) has developed new techniques to use laser micromachining to produce objects such as mirrors and 3-D parts for medical devices.

The U.S. Government Missile Defense Agency (MDA) and other government entities are interested in replacing beryllium components, such as mirrors, with other materials since beryllium is hazardous to work with and that dramatically increases costs. Silicon carbide (SiC) is a possible candidate to replace beryllium, but requires new processing technologies. MLPC has developed laser-micromachining techniques for fabricating a mirror shape out of SiC.

The MLPC used an 8 W laser system that emits laser pulses only 10 picoseconds in duration. The time between pulses is adjustable down to 20 nanoseconds of separation. The laser system can machine features on the order of a few microns over a 305×305 mm (12×12 in.) area. Essentially, the laser operates like an adjustable drill bit, capable of being 5 microns in diameter, up to a few 10s of microns in diameter. In comparison, traditional micromachining uses tool bits that are one or two thousandths of an inch or millimeter, and they tend to break, dull, and have various other problems. Lasers do not dull and need very little maintenance. The techniques being developed by MLPC are capable of creating 3-D structures that could be used in a variety of industries. Automation using the micromachining technology could change many fields, including the polishing of mirrors used in missile defense applications.

The first application for this technology was the production of mirrors for the MDA and other defense clients. However, there are potential customers, especially in the medical device industry, for very small 3-D structures made by laser micromachining where accurate manufacturing is essential. Commercial uses for laser micromachining focus on the technique's ability to produce very small objects very efficiently and in a highly automated fashion. This enables the manufacture of much smaller devices.

Glass

The vast majority of microdevices are built out of silicon using photolithographic processes. This activity is supported by a large industry, which provides the necessary materials and tools. Yet, while the return from investment in silicon technology has been enormous, it is clear that for some applications silicon is not an appropriate base material—and that for a significant subset glass is better. Many designers have contemplated glass, but fabricating microdevices out of it is not easy. Glass producers and machine tool manufacturers have not developed glass micromachining capabilities. With few exceptions (optical fibers, for example) the glass-based devices people use today are produced with 50-year-old technology, using tools designed to fabricate much larger components. Consequently, even when glass is the most desirable material for a given application, the microdevice designer is generally forced to settle for another material.

This situation is changing. For the last decade, numerous research groups have been working to develop processes to micromachine fused silica (a high-end, ultra pure glass). Translume, Inc. (Ann Arbor, Michigan) has developed a manufacturing platform that relies on ferrosecond lasers to micromachine fused silica substrates. This workstation has a capability to machine complex geometric contours and shapes in three dimensions. This computer controlled tool is able to ablate glass, to induce local permanent index of refraction changes in glass, and to locally change the chemical reactivity of glass. Typically, the company uses index change ability to create waveguides in the glass substrate, and relies on the other processes to shape the boundaries of the glass substrate. In addition, often microfluidic features have been created using the same tool. Initially this novel manufacturing capability was used to manufacture glass-based components for the telecommunication market. However, the company is now using its fabrication capability to manufacture small instruments and sensors for customers in the biomedical, aerospace, and defense industries. Yet before the manufacturing process finds wide acceptance, it had to overcome two psychological barriers.

Fused silica is a well-known material, which has been used for decades. Ironically, it is not considered a high-tech material, except when deposited on top of a silicon wafer. Yet, fused silica offers a set of characteristics that compare favorably with silicon. It is transparent from the deep UV to the mid-infrared. It is comparable with all industrial and biological fluids (except hydrofluoric acid). And fused silica offers excellent thermal stability (its expansion coefficient is similar to that of Invar). Lesser known, or even counterintuitive, is the fact that fused silica is a good material for manufacturing mechanical pieces that must flex or move (known as MEMS in the silicon world). Glass is the epitome of a breakable material, or what material scientists call a brittle material; and thus one naturally assumes it can't be used to manufacture MEMS. At the macroscopic level this is indeed true, but at the microscopic level fused silica is rather elastic and can be used successfully to manufacture glass MEMS (GMEMS).

The other mental barrier is related to the linear nature of the manufacturing process. The translume [42] process, starts with a short laser pulse, the glass is turned instantly to plasma but since the pulse is so short the glass almost immediately resolidifies. With proper laser parameters, one can control the local nanostructure of the resolidified material and modify its physical properties. This creates an ability to fabricate devices using a maskless approach as a key economic advantage for some high value markets such as the aerospace industry and the military. The lack of hard tooling for a given device has important cost-saving implications for both prototype development and long-term manufacturing.

The ability to use a single manufacturing step to define both optical and mechanical features dramatically simplifies device fabrication and eliminates alignment issues associated with sequential fabrication processes. Activities in glass micromachining have historically been extremely limited. However, the recent development of manufacturing processes based on femtosecond lasers has created commercial opportunities. In the next few years glass-based microinstruments and glass-based sensors will play a much more important role than is generally envisioned today and that engineers and instrument designers may gain a competitive advantage using glass as an alternative to metal and silicon.

Drilling

There are new techniques as well as equipment that are creating new breakthroughs with laser drilling. Japan's Mitsui Seiki has developed a drilling machine for aero-engine cooling holes that combines a laser with the massive, precise construction of a high-grade mechanical tool. The company says it developed the drilling machine, in cooperation with component makers of aero-engines that were seeking faster, more precise laser drilling. The goal was a Nd:YAG laser that could produce 0.001 mm (0.00004 in.) of the specified point on the metal piece to be worked on. Cooling holes in components such as turbine blades are made with lasers to avoid mechanical effects on the metal from traditional drilling, but Mitsui Seiki says it has improved the technology by combining the laser with the precision that is built into the highest grade of machine tools.

Another unique laser machining process for drilling is a five-axis Lasemill that will mostly burn precision and complex holes in aerospace and industrial gas turbines. This five-axis, laser-machining process forms complex geometries into gas-turbine blades and vanes with a level of accuracy previously possible only with electrodischarge machining (EDM). Winbro Group Technology [43] in the UK, developers of Lasemill, says laser machining costs less and more quickly creates complex features. The laser-machining process can also manufacture and repair hard-metal parts even if they are ceramic coated. The main application of the machine is expected to be in the precision drilling of holes for air cooling in stainless steel, titanium, and nickel components. This is a critical process because the aim is to create as many holes as possible in the part without jeopardizing its structural integrity. Furthermore, drilling usually takes place in final production so each part already has a high value. For example, the cost of scrapping a single part in one project was an estimated $20,000.

Machining speed is as critical as accuracy. Any delay in delivering engines for new aircraft can bring financial penalties. Similarly, component repairs must take place quickly because there's lost revenue for each day a plane spends on the ground. The developed plans are to supply the Lasemill complete with the five-axis laser-machining center, jigs and fixtures, and Delcam NC software. Delcam was chosen because it can model an array of holes, develop machining paths for the laser, and handle data from various CAD systems.

New technology [44] provides a method of laser drilling any component under the control of real-time feedback to provide for predictable, calibrated fluid flow through one or an array of holes formed in the component. The technique can improve the flow control through a machined part by an order of magnitude or more over conventional open-loop laser drilling technologies, thus increasing the predicted lifetime of component parts. The method can be applied to any component or device that requires controlled flow of any fluid. The technology [45] is accurate to a fraction of 1% of the resulting flow, and can be automated for high-volume production. It is applicable to one or many holes through a single workpiece, and is valid for various fluid types (gas, liquid, plasma) at a wide range of temperatures and pressures.

In laser-drilling ceramic composites for applications such as cooling holes— including holes drilled at highly acute angles—this technology [46] creates a smooth bore that seals the composite's reinforcing fibers. The seal created by the laser also preserves the surface oxidation barrier of the composite where mechanical and even laser drilling would compromise the ceramic's integrity. This tested technology overcomes the difficulties posed by conventional laser drilling with techniques that recast the composite into a glassy-walled hole, thus integrating reinforcement protection and an oxidation barrier into the drilled material.

Primary applications of the technology [45,47] include drilling cooling holes in critical parts subject to high temperature and harsh environments, to provide an array of holes for atomizer or injector applications, and where knowing and controlling the precise flow of eventual coolant (gaseous or liquid) is of paramount importance. The technique helps eliminate the waste that results when finished parts are tested and their fluid flow does not perform in accordance with requirements. As an example [47], in engineering practice, an aircraft engine with a variation of $+/-10-15\%$ in terms of accuracy of coolant air flow requires that the engine be designed against the worst case. However, a coolant air flow accuracy of $+/-2\%$ could conceivably provide $20°-50°$ less variation in maximum temperature, resulting in an approximate two times increase in component parts lifetime.

Ceramics can be molded, formed, spray coated, and machined into many useful shapes and sizes, and can also be laminated, plated, and coated. Every year, more devices are made using ceramic substrates because of these useful physical properties. These devices often need to be cut into shapes and frequently require holes, or vias, drilled into or through the substrate. These holes must be of high quality and accurately located, and the technology used to drill the holes must be reasonably priced. Laser drilling technology can perform many tasks that are either not possible or more expensive using other technologies.

Several common laser-types are used in precision laser micromachining [48–50]. Typically, lasers with output from the infrared region to the UV region of the spectrum can be used or ceramic processing since most ceramics couple nicely with many color photons. There is a difference, however, in just how the processing occurs and in the end result.

Wavelength

In general terms, the longer the laser wavelength, the faster it will process ceramics. Shorter wavelength lasers can attain smaller hole sizes and form higher quality holes with less debris or residue, but usually at a sacrifice in processing speed (and therefore probably cost). Most ceramic processing is done with CO_2 lasers that have a laser emission around 10 microns in wavelength. These lasers are fast, inexpensive and highly reliable. Minimum hole sizes of about 50 microns are possible. The primary photon/material interaction is thermal, so the processing takes place by locally superheating the material and ejecting it in a vapor form. However, CO_2 processing is frequently accompanied by a HAZ or a melt zone located directly adjacent to the hole walls. Glazing, porosity, microcracking, and scaling are some of the detrimental changes that can occur in the material within this zone. Using a shorter pulse length and/or a shorter laser wavelength can minimize these effects.

The neodymium:yttrium aluminum garnet (Nd:YAG) lasers operate at a fundamental wavelength of around 1 micron and can be frequency shifted to give higher harmonic outputs well into the UV range. UV light at the tripled wavelength of 355 nanometers (nm) is usually used in high speed drilling applications. These lasers have a very small beam size and a high pulse repetition rate, and are typically used with galvanometers that steer and position the beam. Hole drilling is accomplished by trepanning. On the far UV side is the excimer laser, usually used at a wavelength of 248 nm. It is possible to drill holes as small as a few microns with this laser, but the drilling time is fairly high, especially for thicker materials. The hole is typically very clean using excimer laser processing. Another virtue of excimer lasers is that the beam is delivered to the target not at the focal point of the lens, but at the image plane. A mask is placed into the beam somewhere after the laser, and this mask is imaged onto the work surface at the image plane. If the mask contains many holes that fit into the usable area of the beam, all of these holes can be drilled simultaneously if the holes are packed densely enough and are small enough in size.

CO_2 Processing

When using the CO_2 laser, it is best to control both the pulse width (and thereby the energy) and pulse spacing or repetition rate. If these parameters are not controlled and optimized, the material will crack. Long pulses (>100 microseconds) produce relatively larger holes. Shorter pulses are necessary to obtain hole exit diameters of less than 100 microns. If possible, the laser should also be programmed to fire in short bursts of about 12 pulses at 1000 Hz, with a few milliseconds pause between bursts to allow the molten material to be ejected. This simulates a long pause/low repetition rate condition that produces less thermal stress, but the resulting heat penetration depth will be much less in this case, resulting in smaller hole diameters (35–80 microns).

Airflow also plays a large role in CO_2 processing. Because of the thermal nature of the material interaction with this laser, a few microns of molten material are created with even a microsecond pulse. Airflow should be optimized for each application to efficiently remove this molten material. If the airflow is not strong enough, a "crust" of glazed ceramic, which appears to be formed by a flow of molten matter off of the hole walls, blocks the exit. If the airflow is too high, it may create turbulence in the hole or cool the molten material too fast, resulting in poor hole quality.

With excimer laser drilling, the laser intensity usually far exceeds the air and vapor breakdown threshold. This breakdown is accompanied by a strong shock wave and pressure gradients in the plume of ejected material, and both of these factors contribute to ejected and ablated material removal from the hole. Looking at the smooth bottom of an excimer drilled hole, one can see that bond breaking accompanied by some melting. The molten material may become highly reflective to the beam, resulting in undercutting of the walls and reduced taper. For a particular application, this effect could either be desirable or undesirable, depending on whether a lot of hole taper is required. Sometimes it is even necessary to use a two-step process or even two lasers to achieve desired results, as shown in the examples given later.

Thick Ceramics

Thick ceramics are usually defined as ceramics over 1 mm in total thickness. In reality, because the aspect ratio is extremely important, even a 100 micron piece of ceramic is "thick" if one is trying to reproducibly drill 1 micron diameter holes. However, most holes drilled in ceramics are 100 microns or larger in exit diameter, hence the "over 1 mm" definition. For materials where the aspect ratio is over about 10–1, the laser light gets channeled into the hole, and the hole itself acts as a waveguide with internal wall reflections, helping to keep the hole relatively straight and of somewhat consistent diameter through the exit.

Thin Ceramics

Thin ceramics are those between about 10 microns and 1 mm in thickness and are usually characterized by having holes with diameters approximately equal to the part thickness, or an aspect ratio of one or some other low number. Under these conditions, the drilled hole taper is present and can actually be controlled to some extent.

Ceramic Films

Lasers have also been used on very thin ceramic films, both freestanding and laminated or sprayed onto some other substrate. In general, films are quite easily drilled and patterned. One interesting note about thin films is that at a certain point, the material may be so thin that it no longer displays bulk properties. Frequently, thin films can be processed at very low laser fluences—in many cases lower than the ablation threshold of the bulk material at that laser wavelength.

Green ceramic tape can be drilled and cut quite easily with several different lasers. Low-temperature co-fired ceramics (LTCCs) are used extensively in the microelectronics industry, and many of the devices are manufactured using lasers to drill holes, as well as to pattern the final part shape.

Another material that reacts to laser light very much like ceramics and also shows other ceramic-like qualities is thin-film chemical vapor deposited (CVD) diamond. Lasers have been used on both freestanding CVD films and films grown onto different substrates. One difference between diamond films and ceramics, however, is that the CO_2 laser is not a useful tool for processing a diamond because the thermal interaction degrades the carbon structure from a tetrahedral bonding configuration and turns the diamond into graphite or other carbon forms.

Finally, lasers can be used to remove thin films of other materials from ceramic substrates. Because bulk ceramic is fairly robust and many thin films are easily removed at low laser fluences, patterning of films on hard substrates, including ceramics, is straightforward. Gold and other metal films, indium tin oxide (ITO) conductive films, dielectric films and many organic films have all been patterned using lasers.

Future Applications

In the future as ceramics continue to be used in new applications, lasers provide excellent tools for cutting, drilling, marking, or patterning them for these applications. With the appropriate setup and laser, holes can be drilled through very thick material with minimal taper and very small diameter. If taper is desirable, it is also possible to use setups that will allow the shaping of holes within a fairly large range. Ceramics are currently being used in application areas such as filters, ink jet nozzles, microwave substrates, microelectronic devices, wafer probes, and electrostatic wafer chucks. Design rules are demanding smaller hole diameters, and in many cases, exotic hole shapes. Lasers are playing a big role in the processing of ceramic substrates for these devices, and they will continue to play an even larger role in the future.

Shot Peening

Laser shock peening, or simply laser peening, is an emerging surface treatment technology that is an extension of conventional shot peening. The process involves the application of a high-intensity laser beam, often a high-energy pulsed neodymium (Nd) glass laser, to impart compressive forces to a component surface [51]. The surface of the component is covered with an insulation/absorption layer—typically wet or dry paint, metal, or plastic tape—followed by a transparent layer that acts as a tamper to confine the pressure generated during the process. The tampering layer most often utilized is flowing water. When the laser beam strikes the surface, the beam is absorbed and immediately forms a high-intensity plasma. The tamper layer confines the plasma against the surface and consequently causes a pressure buildup of approximately 6.9 GPa (1 million lb.in.2) [52]. This rapid rise to high pressure causes a shock wave to propagate into the material causing plastic deformation beneath the surface, thus plastically straining the material. The laser intensity is optimized to create a shock that is above the yield strength of the material to induce permanent residual compressive stresses. Deeper residual stresses can be obtained by the use of successive shocks to drive the stress deeper and deeper provided that the material limits at the surface are not exceeded. The principal processing parameters that influence the residual stress distribution are the laser's power, spot size on target, duration of the shock, number of shocks, and type of confining medium and absorptive layer [53].

Laser peening offers advantages over conventional surface treatments. The process imparts deeper compressive residual stresses, 8 to 10 times deeper in comparison to other conventional surface-treatment technologies such as shot peening [54]. The process does not damage the surface of the peened component and is not a thermal treatment. The surface finish of the peened component remains virtually unaltered and the temperature of the peened component rises to a maximum temperature around 149°C (300°F) for a short time.

Applications

The potential applications for the laser peening process are driven by the need to enhance fatigue strength, enhance stress corrosion cracking (SCC) resistance, and to use residual stress to form shapes such as aircraft wing skins. The initial applications for laser peening have predominately been in the aerospace industry. The need for a long service lifetime, cyclic loading, and environmental conditions encountered during service make surface modifications such as laser peening an essential part of materials technology for aerospace applications. These applications are typically aircraft engine parts including turbine blades, aircraft structural components such as landing gear, and other components with notches, holes, and corners that are prone to fatigue failure. Laser peening has also been used for enhancing the fatigue life of automotive components including camshafts, crankshafts, rocker arms, and gears [55,56].

Laser peening technology to improve the fatigue strength of titanium last-row blades on advanced steam turbines has recently been developed by Curtiss-Wright Corp., Livermore, CA, for Siemens Power Generation, Houston, TX (*www.powergeneration.siemens.com*). The last-row blades extract energy from the steam to drive electrical generators. Although laser peening technology has been strengthening critical titanium components in commercial and military turbine engines for several years, this represents its first production application in power generation steam turbines.

Curtiss-Wright's laser peening technology is based on a neodymium glass laser technology originally developed in conjunction with the LLNL (Lawrence Livermore National Laboratory). The laser beam, with a peak power output of 1000 megawatts, is pulsed and directed at the surface of metal parts to be treated. One million pounds per square inch pressure waves compress the metal and leave behind a protective residual compressive stress layer beneath the surface. This compressive stress acts to increase the component's resistance to fatigue, fretting fatigue, and SCC.

The biggest barrier to wider application of laser peening has been the relatively high cost and, to a lesser extent, the slow throughput of the process [57]. However, advances in processing technology with higher processing speeds and repetition rate, better beam uniformity, and optical delivery hardware, as well as better understanding and realization of the technology by engineering organizations, has broadened the application of laser peening. Laser peening has been identified as a solution to a critical SCC problem for canisters filled with nuclear reactor waste for the U.S. Department of Energy's Yucca Mountain Nuclear Waste Disposal Program [52]. Other research efforts include studying the effects of laser peening on as-received and welded A-656 grade structural steel piping for upstream offshore oil and gas transport [51]. Opportunities also are ongoing for medical applications including treatment of orthopedic implants to improve fatigue performance of hip and joint replacements [58].

Test Results

Laser shock peening is an emerging surface treatment technology for metals that imparts deep compressive residual stresses into surfaces. The application of compressive residual stresses to metal surfaces can significantly improve fatigue life of as-welded and as-machined specimens. Preliminary research shows that laser peening is an effective technology for enhancing the fatigue life of U. S. Navy alloys, aluminum 5059-H116, titanium Ti-5111, and high-nickel alloy MP35N. The results of four-point bend fatigue testing showed a 3.5 times improvement in the fatigue life of welded aluminum alloy 5059-H116. A significant fatigue life improvement was also seen in the four-point bend testing of as-machined MP35N. Future work would include Ti-5111 titanium alloy and Al-6XN stainless steel weldments since residual stress measurements indicate that Ti-5111 and Al-6XN stainless steel are favorable to the laser peening process.

Coatings

Recent developments have shown a new technique that enables coating technicians to measure light absorption and scattering losses of less than one part per million in real time, as the coating process is underway inside a vacuum chamber. By providing technicians a means to optimize materials processing and coating parameters on a routine basis, the new method enables very low absorption (VLA) optical thin film coatings to be produced in less time and at reduced cost than previous methods.

Highly reflective multilayer optical coatings are essential to high-power laser systems, such as the airborne laser. Because these lasers are so powerful, even small amounts of incident laser energy absorption can result in damage to the optical coatings. Since damaged steering optics produce distorted laser beams, VLA coatings—as well as accurate and timely nondestructive VLA measurements—are necessary to efficiently and cost effectively optimize the optical thin film deposition process.

Coating technicians can now quickly and routinely optimize materials processing and coating parameters in order to minimize coating absorption. This new method causes light that is undergoing total internal reflection at the surface being coated to be evanescently coupled into a waveguide within the thin film. Absorption or scattering losses to the waveguide reduce the total internal reflection of the light. Therefore, measuring the reduction in internal reflection can determine the extinction coefficient of the thin film. The measurements collected for this effort, which represent the most sensitive ever achieved in vacuum conditions, could improve instrumental for ongoing development of high-energy lasers.

Cladding

Precision repairs and rapid design changes of molds, tools, and high-value components are possible with direct metal deposition using modern laser technology. Repair and rapid geometrical modifications of high-value tools and components are demanding challenges of modern manufacturing technology. Advanced laser cladding techniques with fiber lasers offer outstanding possibilities for applications in aircraft maintenance as well as mold and die industry. The reason the interest in this technology is increasing is its features. High-power fiber lasers open a completely new dimension in material deposition. The beam quality is extremely high, and this results in both small laser focus diameters (10–100 μm) and very long focal lengths. In addition, the current system technology, laser optics, powder feeders and nozzles, as well as the CAD/CAM software, permit an easy and efficient integration of the laser process into manufacturing systems.

The laser beam generates a localized melting bath on the workpiece surface. The filler material is fed as powder or wire and is heated when moved through the laser beam. However, it only melts once it is in the melting bath. The formation of a metallurgical bond requires a slight melting of the base metal, which happens through thermal conduction. Thermal conduction into the cold substrate facilitates rapid solidification of the molten filler material, which results in the formation of the deposited material. The width of these tracks can typically be varied between 0.2 and 6 mm (0.007 and 0.23 in.). The height depends on the application and is between 0.1 and 2 mm (0.003 and 0.05 in.). Tracks can overlap and coat entire areas. Multiple layers can be deposited to form 3-D structures. The characteristic deposition rate is 0.1–1.5 kg/h. A higher deposition rate leads to a reduced precision of the depositions.

As beam sources, currently high-power diode lasers, traditional Nd:YAG slab lasers, and the new Yb:YAG fiber lasers are used in the industry. The typical power range lies between 1 and 6 kW. Lasers with higher output powers are available, but for the most part not cost effective for cladding applications. The fiber laser, a special type of the solid-state laser, represents a new generation of high-power lasers for materials processing. The Yb-doped core of a YAG glass fiber is the active medium in this laser. The beam quality is increased about four times compared to conventional Nd:YAG slab lasers. This results in an extremely improved ability to focus the laser beam, which can be as small as 100–10 μm at lengthy working distances. This way, the user has a wider window of process options: long and slight cladding optics and powder nozzles for a better powder efficiency, better accessability of complex welding positions, and better 3-D operation capability. Compared to the traditional Nd:YAG slab laser, the fiber laser is more compact, smaller, the efficiency is much higher (wall plug efficiency >30% [59,60], compared to 5–15%), and the investment costs are less by a factor of 2–2.5.

For applications that require less accuracy, the high-power diode laser is a cost-effective alternative. This type of laser is available in the power range of up to 6 kW. Compared to other beam sources, it has the highest available power efficiency of 35–50%. The equipment costs are comparatively low. Since diode lasers are very compact, they can be directly integrated into a machine tool or robot system without the need to transport the laser light via fibers. The low beam quality of this laser type limits the minimum dimension of the laser focus. Thus, the diode lasers are not suitable for high-precision claddings and microprocessing.

Scanning

In the past, one of the world's leading manufacturers of fuel tanks used coordinate measuring machines (CMMs) to inspect the first articles. However, the geometry of the tanks is so complex that it was difficult to fully inspect the surface one point at a time [61]. Inergy Automotive Systems has achieved significant improvements in quality by switching to laser scanning, which "paints" the surface of the tank with a laser and then uses a sensor to capture all the points in the laser's path. The new approach generates a solid model of the as-built part that can be compared to the design intent to highlight the full extent of any differences between the two. As a result, Inergy can make required tooling changes with a higher level of confidence and ensure that production parts meet customer specifications.

An automotive fuel tank is one of the most difficult parts of a vehicle to inspect because of its complicated geometry. When designing the vehicle, fuel tanks are often one of the last parts to be designed because their functionality does not depend on their shape. So, after the other parts have all been designed, the fuel tank shape is designed around the other components to fit the available space. This explains why fuel tanks have few uniform features and instead incorporate many different contours that typically match whatever space is available around the other vehicle components. However, after the fuel tank has been designed, each of its many geometrical features must be positioned perfectly in order to avoid interfering with other components while at the same time ensuring that the tank provides the full stated capacity. It's critical to try out the mold, accurately measure the geometry of the tanks that it produces, and make corrections when needed to the mold to ensure that the final tank geometry matches the customer's design.

Operative System

Laser scanning systems work by projecting a line of laser light onto surfaces while cameras continuously triangulate the changing distance and profile of the laser line as it sweeps along, enabling the object to be accurately replicated. The laser probe computer translates the video image of the line into 3-D coordinates, providing real-time data renderings that give the operator immediate feedback on areas that might have been missed. Laser scanners are able to quickly measure large parts while generating far greater numbers of data points than probes within the need for templates or fixtures. Since there is no contact tip on a laser scanner that must physically touch the object, the problems of depressing soft objects, measuring small details, and capturing complex free-form surfaces are eliminated.

Instead of collecting points one by one, the laser scanner picks up tens of thousands of points every second. This means that reverse engineering of the most complicated parts can often be accomplished in an hour or two. Laser scanning can reverse engineer parts that are so complex that they would be practically impossible to do one point at a time. Finally, the software provided with the scanner greatly simplifies the process of moving from point cloud to CAD model, making it possible in minimal time to generate a CAD model of the scanned part that faithfully duplicates the original part. Special software can be used to compare original design geometry to the actual physical part, generating an overall graduated color error plot that shows at a glance where and by how much surfaces deviate from the original design. This goes far beyond the dimensional checks that can be performed with touch probes on CMMs.

As a result engineers can now see the complete surfaces of a part rather than just a few points. This makes it possible to visualize the entire extent of the discrepancy so one can be much more precise in tooling changes. One also can save money on the first article inspection because the cost of the service bureau is considerably less than what it cost to do the job internally on a CMM.

Smart laser sensors are used in tire production to monitor such metrics as thickness profile, splice width radial runout, and sidewall integrity. Tire makers inspect the product in process and when it's finished. For both jobs, laser measurement sensors have become the tool of choice over contact/ mechanical followers and capacitive sensors. Recent laser measuring systems that employ high-speed digital data communications nullify electrical noise and eliminate separate analog-to-digital converters. This makes 100% inspection feasible, which lowers scrap rates.

A big advantage of laser sensors is that they work without getting close to or touching a tire. Capacitance sensors, in contrast, must get extremely close and perpendicular to a tire surface to be measured. This requires complex, multiaxis positioning mechanisms and frequent sensor calibration because material properties change during the manufacturing process and capacitive sensors are sensitive to those changes. Also, capacitance sensors don't work reliably on certain materials such as high silica rubber. Not to mention, tires can hit and damage improperly positioned sensors. Capacitance sensors measure over a relatively large surface area and can't image grooves, lettering, and bar codes. They must instead be positioned over a clear path on tire sidewalls.

Contact sensors such as LVDTs (linear variable differential transformers) have similar shortcomings. Contact pressure deforms the rubber, resulting in erroneous data. In-process measurements, such as extrusion profiling, are impractical with contact sensors because tire surfaces tend to be hot and gummy at this stage of production. Properly designed laser sensors need no recalibration, are insensitive to changes in material properties and surface condition, such as color, finish, or the presence of bead lubricant. There are two basic types of laser-measurement sensors; point triangulation and line.

Point triangulation sensors, as the name implies, take data at a single point, effectively acting as a noncontact LVDT. Such lasers employ a low-powered laser that projects a spot about 0.1–0.3 mm (0.003–0.011 in.) in diameter on the surface to be measured. Point triangulation sensors support high data rates (16–32 kHz) and resolutions to 25 µm.

Laser line sensors operate much the same way as point triangulation types. A key difference is that the laser beam is optically expanded in one dimension to form a line of laser light on the surface to be measured. Laser line sensors take data over multiple points on the surface. Most laser line sensors output at a slower rate than point sensors—typically 10–60 frames/sec. for each line of data—though some are capable of 1-kHz frame rates or higher. PSD-based laser sensors (position-sensitive detector) take 16,000–32,000 data points/rev at 60 rpm, far greater than possible with a line sensor. As such, PSD sensors can measure bulges and other deformities to better than ±0.025 mm (0.001 in.).

LENS (Laser Engineered Net Shaping)

LENS [62] advances additive manufacturing and repair and the LENS technology is transitioning into a process accepted by a growing number of commercial, aerospace, and U.S. Department of Defense customers. LENS may be characterized as a "disruptive additive process" that may be utilized for a variety of repairs and free-form fabrications. Disruptive, not in the negative sense of the word, but disruptive in the fact that this technology challenges one to think outside of the box because of the unique capabilities it possesses. No other additive process combines excellent material properties with near net shape, direct from CAD, part building and repair quite like this process. Applications include the repair of worn components, performing near net shape free-form builds directly from CAD files, and the cladding of materials.

Figure 11.5 shows the typical process layout. The deposition substrate or "target" is aligned to the desired start point of the deposit. The powder feeder(s) feeds the powder delivery nozzle assembly, which creates a powder stream that converges at the point of the deposit. Next, the laser provides a focused beam that is delivered to the point of deposit. The focused laser beam melts the surface of the target and generates a small molten pool of base material. Powder that is being delivered to this same spot is absorbed into the melt pool, thus generating a deposit that may range from 0.127 to 1.023 mm (0.005–0.040 in.) thick and 1.023–4.06 mm (0.040–0.160 in.) wide.

All LENS deposits are metallurgically bonded and exhibit HAZ and dilution zones ranging from 0.127 to 0.635 mm (0.005–0.025 in.) thick. Mechanical properties and the quality of the deposits are typically better than castings and approach properties of wrought products. In some cases, like titanium, the deposits may actually exceed typical handbook values.Flexibility is a key ingredient guiding this technology. LENS systems are typically coupled with lamp-pumped Nd:YAG lasers or more recently the new fiber lasers. Both lasers have wavelengths that are ~1 micron long. The optical absorption of these laser beams is much higher for the Nd:YAG and fiber laser beams than that of the CO_2 laser beam, whose wavelength is 10 microns. Typically, the Nd:YAG and fiber laser beams require only one-half the wattage of a CO_2 laser to achieve the same deposition rates. The Nd:YAG and fiber laser beams

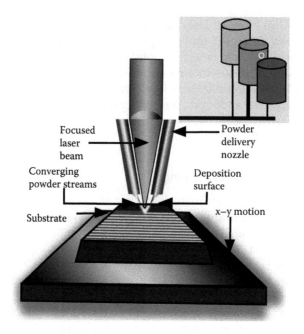

FIGURE 11.5

Basic layout and flow paths for a typical LENS system (From Mudge, R. P. and Wald, N. R., *Welding J.*, 86(1), 44–8, 2007. With permission.)

may also be delivered using fiber optics where the CO_2 beam must be delivered via reflective mirrors. This flexibility opens up many more potential applications.

The final considerations for the laser power sources are the floor space required and overall wall plug efficiencies to deliver power to the workpiece. The CO_2 laser requires the most floor space and the most energy. The lamp pumped Nd:YAG requires less floor space and less energy, while the fiber laser requires the least floor space and the least energy, but arguably it provides the highest quality laser beam.

Typical Free-Form Applications

The process may be used to deposit free-forms of near net shape metal components that are nearly 100% dense with mechanical properties comparable to wrought materials directly from processed CAD files. Free-forms may be thin wall [minimum 1.52 mm (0.060 in.)] thick or solid to any thickness. In any event, it is necessary to overbuild or add some "protect" material that is typically machined off to achieve the final desired component. This capability may be employed to make the complete part or to add special features to a simplified casting or forging.

Cladding and Composite Manufacturing Applications

Cladding is actually a form of repair buildup applied to the manufacturing of new components. The concept of composite manufacturing has been employed for many years. Stellite products have been available for many years and have been applied via a variety of conventional processes. The LENS process takes the depositing of Stellite to the next level. Very consistent high-quality thin deposits can be made with little dilution. Overbuilding is kept to a minimum, which reduces finishing costs. Hardfacing and cladding using CO_2 lasers are highly successful. Combining the LENS system with the new fiber lasers improves on this success. The fiber laser, with the shorter wavelength laser beam, can achieve equivalent deposition rates with approximately 50% of the wattage required by a CO_2 laser. The net result is similar to production rates with less heat and less stress conveyed into the part being cladded. The surface finish of the cladding may be left as-deposited or ground-to-finish dimension.

Future of Process: LENS

LENS is a maturing "disruptive additive technology" that provides new capabilities for creative repairs on components that may have previously been considered unrepairable. Its small HAZ and dilution zones will lead into more cladding and composite manufacturing applications. The process also has the ability to change a CAD file for free-form production revisions versus remanufacturing hard tooling. This is not only cost effective, but will save months of development time on some projects. This freeform ability may revolutionize existing manufacturing processes by employing the concept of simplifying castings and forgings, and then applying special features utilizing this process. We must think outside the box in all of the above-described scenarios.

Laser Brazing

Furnace brazing can braze much more product in a single operation, but laser brazing, especially with the diode laser (LD laser) that is useful for its short wavelength and high absorption by metals, produces less thermal damage and erosion to the base metal and allows for better local heating and controllability. Development of a technology for tandem laser brazing superalloys was reported by Osaka University (Saida, Song, and Nishimoto) [63]. Tandem laser brazing with main and preheating laser beams was used on 1 mm (0.04 in.) thick plates of nickel-based, heat-resistant superalloy Inconel 600 and iron-based heat-resistant Superalloy A-286. Thin Inconel tubes of 5 mm (0.2 in.) wall thickness and thin SUS 304 tubes were also tested, in bundles, with various clearances. The precious filler metals that were tested for all samples were Au-18 Ni, Ag-10Pd, and Ag-21Cu-25Pd.

A 50 W laser was used to preheat the base metal, which was on both parts gently topped with F10SU flux beforehand, while a substantially more powerful 2000 W laser was used to melt the filler metals. Spot water cooling was employed. The filler metal temperature was measured with a thermocouple placed underneath it, embedded inside the base metal. It was determined that preheating resulted in superior wetting and limited erosion, while increasing the main laser power increased erosion depth. Sound butt joints were also obtained regardless of whether from wide or narrow joint clearances within a certain reasonable range, and the maximum tensile strength of the butt joint was 70% that of the base metal at the 0.1 mm (0.004 in.) joint clearance.

Tandem laser brazing of thin tubes of Inconel 600 and SUS 304 stainless steel produced good wetting and sound fillets. The fracture strength was measured to be 300–550 MPa (43–80 ksi). Only the Ag-pd fractured in the braze layer instead of the base metal.

Nanolasers

Memory capacity has undergone impressive advances in recent years, but could take a quantum leap thanks to advanced nanolasers now under development in laboratories such as at the University of California in Riverside [64]. Led by associate professor of engineering Sakhrat Khizroev, the research team has explored using tiny lasers that could lead to the development of hard drives able to cram 10 terabits into a one-half inch square footprint. That is 50 times the data density of today's magnetic storage technology. Thus far, the nanolaser can concentrate light as small as 30 nanometers and focus 250 nanowatts of power. The researchers hope to improve the laser to produce light beams as small as 5–10 nanometers. To achieve this goal, the scientists need to refine the precision of the focused gallium ion beams used for their fabrication. Khizroev's lab adapted the ion beam technology, commonly used for diagnostics in semiconductor manufacture, to cut the components of their laser.

Direct Metal Laser Sintering (DMLS)

Laser sintering is transforming into an accepted process for fulfilling rapid manufacturing strategies. The equipment is improving steadily, creating parts with end-user materials, smoother surfaces, greater precision, and higher durability, which are all requirements for making production-grade components. The number of useable plastics materials has increased, and the introduction of direct metal laser sintering (DMLS) has opened up new application areas for the processing of metal materials. The DMLS [65] is an additive technology that can turn 3-D CAD designs with complex geometries into production-quality metal parts in hours. Today, available plastic laser sintering materials include fine polyamides, aluminum-, glass-, or carbon-fiber-filled polyamides, a polystyrene suitable for use in plaster, shell, and vacuum casting operations, and a high-temperature PEEK polymer. Among the metal DMLS offerings are maraging and stainless steels, cobalt chrome alloys, Inconel, and titanium.

This additive process works by sintering very fine layers of metal powder layer-by-layer from the bottom up until the build is complete. The process begins by sintering a first layer of 20 microns powder onto a steel platform. The platform then lowers by 20 microns, a fresh layer of powder is swept over the previously sintered layer, and the next layer is sintered on top of the previously built one. A powerful 200 W Yb-fiber laser is precisely controlled in the X and Y coordinates allowing for exceptional tolerances to be held ±0.025 mm (0.001 in.). The latest technology takes advantage of a dual spot laser allowing feature sizes as small as 0.203 mm (0.008 in.) to be built. With a machine build envelop of 250 × 259 × 215 mm (9.8 × 10 × 8.4 in.), many medium to small parts and inserts are able to be constructed in hours and days versus days and weeks using traditional processes. Once started, the machine builds unattended, 24 hours a day. Parts and inserts that come out of the machine typically will go through a series of post steps including support removal, shot peening, and so on.

Advantages of Laser Sintering

Because laser sintering differs significantly from traditional manufacturing methods, it offers attractive advantages to product developers. Of course, one advantage is that it's rapid, but there are other, more important reasons that manufacturers would select laser sintering in place of traditional processes:

- Creating complex geometries and integrating parts: Parts that are expensive, difficult, or even impossible to manufacture using traditional cutting or molding processes are easy to grow layer by layer. Laser sintering can make nearly any shape, even interlocking parts such as a hinge, chain link, or braids. Designers and manufacturers are discovering that manufacturing with laser sintering eliminates secondary processes such as drilling holes: parts can be created, holes and all. In addition, laser sintering can enable the integration of several individual parts into a single complex one, reducing assembly times and manufacturing costs.
- *Enabling mass customization:* Do you need to make hundreds of different versions of the same product—maybe with a different logo or a slightly altered geometry? No problem. A tweak of the 3-D CAD model is all that's required to manufacture each variation of the original design, frequently side by side in the same batch.
- *Producing low volumes at low costs:* Sometimes the demand for products is at an awkward level—too many to machine cheaply, and not

enough to support the additional cost of tooling or a mold. Laser sintering will never replace traditional processes for all manufacturing tasks, but it cost effectively fills an important niche in low volume production that other methods can not.

- *Providing mold making tools:* DMLS has become an adjunct manufacturing process for tooling and mold making. For example, suppliers have found that by laser sintering mold inserts, they are able to eliminate out-of-house processes such as EDM or five-axis machining—reducing production times.

Such capabilities benefit designers and manufacturers in many industries and have made laser sintering ideal for a wide array of applications.

Applications for Laser Sintering

Applications are wide ranging and include inserts for plastic injection molding and die casting, as well as direct parts for a variety of applications and industries including aerospace, automotive, medical, electronics and many others. Complex parts have been fabricated from 17-4 PH stainless steel, superalloy CoCr, and more recently titanium and Inconel 718. The titanium material is Ti64 (a pre-alloyed Ti6Al4Valloy), which has excellent mechanical properties and corrosion resistance, low specific weight, and biocompatibility. Parts built from this alloy can be machined, spark eroded, welded, microshot peened, polished, and coated as needed. Typical uses include dental, orthopedic, and airframe applications.

In medical applications, dental labs laser sinter cobalt-chrome, mass customized dental prostheses such as crowns and bridges, hundreds at a time—an enormous improvement over previous manufacturing processes. With DLMS, medical design companies create prototypes of customized surgical instruments for spinal surgery and implant prototypes for spinal and knee surgery. Laser sintered plastics are gaining use for manufacturing perfect-fit prosthetic limbs. In addition, medical product developers have been able to integrate parts on existing designs by switching to plastics laser sintering. In one instance, a company that manufactures centrifuges was able to reduce the part count on a washing rotor from 32 parts to 3, with 2 of the remaining parts being laser sintered.

In aerospace, plastics laser sintering creates actual parts such as cylinder head temperature controllers, cooling shrouds, heat exchangers, fuel tanks, and air intakes for the engines of unmanned aerial vehicles. Laser sintering also produces flame retardant polyamide air plenums for luxury commercial planes, slashing production lead times and completely eliminating tooling costs.Recently, mold makers have discovered another, powerful advantage to creating molds using laser sintering: the ability to optimize cooling and heating channels for maximum heat transfer (conformal cooling). With traditional processes, a mold or a cooling insert is machined to its overall geometry and then, in a separate operation, straight-line channels are drilled into it to pass near crucial features. Because they are straight, these channels can't always reach all hot spots effectively.

With laser sintering, mold makers create the mold or mold insert geometry and the cooling channels in a single operation. This conformal cooling eliminates the need for straight (drilled) channels, allowing designers to create cooling channels that conform to the part geometry and manage the flow of heat more efficiently. Because cooling normally can take up as much as 70% of production cycles, the resulting reduced cooling time dramatically increases workflow.

A Look at the Future

What happens from here? If the past is any indication, customers will often lead the way, as their own prior successes and the clear advantages of laser sintering motivate them to explore other applications currently created with traditional processes. An even wider selection of market-ready plastics and metals that exhibit properties very similar to their traditional counterparts in manufacturing are being evaluated. For instance, new metal alloys, including nickel alloy variants and other high-temperature superalloys. The new materials will be tailored specifically for rapid manufacturing applications in the medical and aerospace industries, among others.

Recently, a group of rapid manufacturing equipment suppliers, joining such companies as Boeing and Siemens, created a partnership with the University of Paderborn in Germany to found a Direct Manufacturing Research Center (DMRC). Among the goals of the DMRC are to make additive manufacturing methods such as laser sintering more reliable, to establish industry requirements for materials and training, to develop standards, and to make the additive manufacturing process faster still. These activities will help foster the kind of quality certification that industry has found essential for aerospace and medical manufacturing.

Laser Soldering

Selective soldering [66] is seeing a rise in popularity coinciding with an uptake in mixed surface mount and thru-hole assemblies. Choosing a selective soldering system that meets your needs requires understanding of the entire production process. Selective soldering has found wide application as the use of SMT (surface mount technology) devices increase. Selective soldering basically is a choice of process more than of machine; the choice of system comes as a consequence. The selective soldering technological solutions available to EMS (Electronic Manufacturing Services) providers looking to avoid manual soldering are somewhat limited: wave soldering with masked carriers, dipping, or selective soldering (mini wave, hot iron, and laser) with robot systems. Every selective soldering procedure is unique and needs specific management for operation, including dedicated tools. To choose a system that meets one's needs, a whole vision about the assembly to be soldered is required.

Selective soldering is an operation that occurs after other thermal phases, as it is generally performed postreflow on mixed assemblies. Therefore, it is important to foresee any possible board distortion. A planarity sensor indicates suitable adjustments to maintain soldering device synchronized with the soldering plane. When dealing with process considerations, productivity can't be left as a separate part of the equation. Productivity must be considered, not only as number of soldering actions per unit of time, but also as production changeover speed with minimal maintenance and irrelevant stoppages. The most important element for the final assembly is the surface finishing of components, pads and pins.

The tin/lead alloys can be defined today without any secret, but the lead-free alloys still present some unknown behavior aspects. Although the group's main characteristics are known, certain aspects remain unpredictable. Alloys similar in composition, but made by different manufacturers, often present different behaviors in the same working conditions. These diverge completely when operators change the working variables. Every solder wire manufacturer also requires a specific flux. These elements are enough to make a difference in production, evident when considering residues and postsolder cleaning.

Soldering System

The most suitable selective soldering system for a majority of worldwide users in production is a robot soldering machine. Robot systems help a gripper bring the board over the mini wave pot, or to move a hot iron together with a wire feeder unit, or point a laser beam for reflow. Availability of a movable axes system is fundamental to accuracy and repeatability. Laser systems' wire dispensers host large reels with various solder wire diameters accommodated. Some systems include a sensor at the dispenser exit. A high-speed reading pyrometer on a small area makes the whole process easier, due to its double function as control instruments and acquiring instrument.

The software links all subsystems in the soldering machine and must be simple in terms of programming and use for the operator. Good comprehension of the process and available system capabilities is part of the technician's knowledge. Besides, knowledge of soldering basic elements and a sensibility about the process makes learning and using a system easier.

Among the several advantages of laser soldering, there's accurate and repeatable thermal transfer. Since this is a contactless soldering technique, it realizes a joint without stressing the board or components, without wearing out the bits—unlike hot iron robots—or producing waste. This prevents any board contamination. By changing the size of the laser spot, it is possible to work on different pads without changing the soldering nozzle, as on mini wave soldering machines.

Consistent quality targets soldering of lead-free alloys without using nitrogen. It also works with traditional wire; users simply change the solder alloy and corresponding thermal profile. During the soldering phases, interaction between metallic and chemical components influences soldering performance. Inside the process, temperature is the independent variable, but it is also the one having the highest impact on wettability. The aim of flux is to take away oxidation from surfaces to be soldered in order to make their wettability easier with the melted alloy. The flux must be able to deoxidate pads and terminals at soldering temperature without decomposing. When considering the influence of temperature on flux, we must focus our attention on three parameters: evaporation of solvent and of volatile substances, speed of distribution and viscosity of solid substances under transformation, and decomposition of organic material.

With the help of system software advances, laser selective soldering systems are able to determine the necessary data—thermal profile, laser spot size, volume of tin required—for every solder joint. Since each joint has its own thermal mass, changed by varied paths on the printed circuit board (PCB), it has its own thermal profile. The soldering profile is characterized by the trend of temperature according to time, and is obtained by changing the involved powers in the three phases and the corresponding time of application.

The bead is in the operating center of the soldering system, since it contains the laser optics, wire dispenser, camera, pyrometer, and planarity sensor necessary for controlled operation. In laser selective soldering systems the laser unit typically is located at the back of the machine. The laser is routed through the optical fiber and focused by the motorized optic group in the required size and such to meet the pad size. The whole head rotates a little more than 180°; rotation allows changes in working angle and rotating during the soldering process, helping improve wettability for a large-sized pad. The growth of laser soldering is due to its capability of exploiting the high energy enclosed in the ray and localized only on the joint, a characteristic that avoids involving the substrate and nearly components, even with the higher working temperatures required for lead-free alloys.

Noncontact Laser Soldering

A noncontact laser soldering system for joining arrays of silicon solar cells has reportedly been developed by researchers at the Fraunhofer Institute for Laser Technology, Germany [67]. To melt the solder, a laser beam is passed over the solder-coated stringer. An infrared heat camera detects the temperature of the silicon and of the metal strip from real-time measurements of their emitted radiant heat. If the temperature is too high or too low, a feedback control circuit automatically adapts the laser output within milliseconds, enabling every joint to be effectively soldered. The current goal is to develop a faster, more reliable method of connecting solar cells by means of laser welding. This means applying more heat than needed for soldering, but only for a very short time. However, the conductive metallic coating on the silicon has a thickness of only 10 microns. Therefore, the laser beam has to be modulated in such a way that the stringer will melt while leaving the coating intact.

Microscalpel

Scientists at the University of Texas at Austin have developed a laser "microscalpel" that destroys a single cell while leaving nearby cells intact, which could improve the precision of surgeries for cancer, epilepsy, and other diseases. The device uses femtosecond lasers, which produce extremely brief, high-energy light pulses that sear a targeted cell so quickly and accurately, the laser's heat has no time to escape and damage nearby healthy cells. One can remove a cell with high precision in 3-D without damaging the cells above and below it, claims mechanical engineering Assistant Professor Adela Ben-Yalar, who developed the scalpel. Ben-Yalar's laboratory created a microscope system that uses a tiny, flexible probe to focus laser light pulses up to 250 microns deep inside tissue, to a spot size smaller than human cells. The medical community envisions the lasers being used to accurately destroy many types of unhealthy material. These include small tumors of the vocal cord, cancer cells left behind after the removal of solid tumors, individual cancer cells scattered throughout the brain or other tissue, and plaque in arteries [68].

Marking

Laser printers often have a hard time printing on curved surfaces. Sometimes letters get fuzzy, elongated, or foreshortened because the printing head cannot take into account the surface topology. And even flat surfaces present problems as the laser's spot size varies from the edge of the marking area to the center. To account for these issues, engineers have developed CO_2 laser markers. It can determine and adjust to the correct focal length for any surface within its 42 mm (1.6 in.) range. This eliminates the need for complex indexing mechanisms that change the height of the marking unit as the distance to the surface changes. It also has an X, Y, and Z scanner that controls the marking laser in three dimensions by adjusting its focal point. This lets the marker put clear letters and symbols on stepped surfaces, cylinders, and cones, even if these shapes are moving. The controls also have a warm-up feature that ensures the 30 W laser has reached its stable operating power before beginning to mark, so there are no variations in lettering from the beginning of the process through to the end.

Photolithography Without UV Light

Using photolithography to create the tiny patterns needed to fabricate computer chips and other applications of nanotechnology requires the use of UV light, which is difficult and expensive to work with. A University of Maryland College of Chemical and Life Sciences research team led by John Fourkas, Professor of Chemistry and Biochemistry, has developed a lithographic technique called RAPID (Resolution Augmentation through Photo-Induced Deactivation), which makes it possible to create small features without using UV light [69].

According to Fourkas, the process is similar in principle to current dental cavity filling, where a liquid is squirted into a cavity and a blue light is used to harden the liquid. The RAPID technique uses two pulsed laser light sources of the same color. The laser used to harden the material produces short light bursts, while the second laser stays on constantly. The second laser beam also passes through a special optic that allows for sculpting the hardened features into the desired shape. Fourkas said, "The RAPID lithography technique we have developed enables us to create patterns twenty times smaller than the wavelength of light employed, which means that it streamlines the nanofabrication process" [69].

Laser Repair of Composites

A joint investigation and agreement has been initiated between GKN Aerospace (Isle of Wight, UK) and SLCR Lasertechnik (Düren, Germany) in the use of automated lasers in composite aircraft structure repair. Intended to replace the manual grinding currently required to prepare damaged surfaces for bonding, the laser technique reportedly will apply no force or vibration to the structure and, therefore, will have no detrimental impact on the composite component's strength or integrity.

The process uses a CO_2 laser generator the size of a small, three-drawer filing cabinet, with a camera mounted on the top. It is linked to a receiver camera mounted on a small robot of the appropriate size for each task. The robot contains a laser gun with the laser optics, in a package small enough to permit accurate movement over the part surface. The laser evaporates the organic resin one ply at a time, leaving the damaged fiber behind, until good structure is reached. It should be noted that part of the development effort will involve refining the process for clearing away the fibers left behind. Damaged areas prepped with the laser can be repaired using conventional methods, achieving comparable strength in one-third of the time required for hand grinding, with a potential 60% cost reduction. The technique is reportedly useful in the workshop or while the part is still mounted on the aircraft and can be applied to both monolithic laminates and honeycomb structures.

As composite materials increasingly dominate the future airframes of aircraft as well as composite fan blades, their effective and swift repair becomes even more critical. It is estimated that this equipment will be developed within 2 years.

References

1. Trumpf Group. 2006. *Laser Tools*, in German. Germany: Vogel Buchverlag, D.-Würzburg.
2. Sumpf, A., and Jasnau, U. 2005. Usage of fiber lasers as mobile beam source in the shipbuilding industry, I. Intern, Fraunhofer Workshop, *Fiberlaser*, Dresden, Germany.
3. Jasnau, U., and Schmid, C. 2006. Fiber laser cutting of thick metal sheets. 2. Intern, Fraunhofer Workshop, *Fiberlaser*, Dresden, Germany.
4. Verhaeghe, G. 2005. The fiber laser—A newcomer for material welding and cutting, *Welding Journal* 84 (8): 56–60.
5. Thomy, C., Vollertsen, F., and Seefeld, T. 2005. Welding with high-power fiber laser. In German. *Laser Journal* No. 3: 28–31.
6. Schluetter, H. 2007. Laser beam welding: Benefits, strategies, and applications, *Welding Journal* 86 (5): 37–9.
7. Zhang, J., Weckman, D. C., and Zhou, Y. 2008. Effects of temporal pulse shaping on cracking susceptibility of 6061-T6 aluminum Nd:YAG laser welds, *Welding Journal* 87 (1): 18s–30s.
8. Matsuda, J., Utsumi, A., Katsumura, M., Hamasaki, M., and Nagata, S. 1988. TIG and MIG arc augmented laser welding of thick mild steel plate, *Joining & Materials*.
9. Steen, W. M., et al. 1978. Arc-augmented laser welding. *4th International Conference on Advances in Welding Processes,* Paper No. 17, 257–65.
10. Cui, H. 1991. Untersuchung der Wechselwirkungen zwischen Schweisslichtbogen und fokussiertem Laserstrahl und der Anwendungsöglichkeiten kombinierter Laser-Lichtbogentechnik, TU Braunschweig, Dissertation.
11. Maier, C., Beersiek, J., and Neuenhahn, K. 1995. Kombiniertes Lichtbogen-Laserstrahl-Schweißverfahren. On-line-Prozessüberwachung. DVS 170 S. 45–51.
12. Haberling, C. 1994. Prozesstechnische Untersuchungen des CO_2-Laserstrahlschweissens mit Zusatzdraht und in Kombination mit dem MIG-Schweissverfahren. Diplomarbeit, RWTH, Lehestuhl für Lasertechnik.
13. Dausinger, F. 1995. Hohe Prozesssicherheit beim Aluminiumschweissen mit Nd:YAG-Lasern. Bleche und Profile 42 Nr. 9, S544–7.
14. Staufer, H. 2007. Laser hybrid welding in the automotive industry, *Welding Journal* 86 (10): 36–40.
15. Defalco, J. 2007. Practical applications for hybrid laser welding, *Welding Journal* 86 (10): 47–51.
16. Nowotny, S. Laser robot welding system fixes microcracks in turbines, Fraunhofer Institute, Dresden, Germany, *AM&P* (February 2007): 15.

17. Vietz, E. 2004. Welding methods used in worldwide pipeline construction, tuned to the quality of the pipe—Yesterday, today, and tomorrow. *Proceedings 18, Oldenburger Rohrleitungsforum 2004*, 14–29. Rohrleitungen im Jahr der Technik, Vulkan-Verlag, Essen.

18. Kohn, H., Thomy, C., Grupp, M., and Vollertsen, F. 2004. New developments in laser welding of pipe. *Proceedings 18, Oldenburger Rohrleitungsforum 2004*, 50–74. Rohrleitungen im Jahr der Technik, Vulkan-Verlag, Essen.

19. Thomy, C., Seefeld, T., Vollertsen, F., and Vietz, E. 2006. *Welding Journal* 85 (7): 30–3.

20. Klingbeil, K. 2006. What you need to know about remote laser welding, *Welding Journal* 85 (8): 44–6.

21. Cann, J. 2005. A look at remote laser beam welding, *Welding Journal* 84 (8): 34–7.

22. JP 60 108 191 A, Japanese patent application: Working head of laser working device, publ. 1985.

23. GB 2 163 692 A, UL patent application: Laser apparatus, publ. 1986.

24. Felleisen, R., and Kessler, B. 1994. Higher productivity by combining CO_2 laser beam welding and cutting in one processing head. *Proceedings of ECLAT '94, 5th European Conference on Laser Treatment of Materials*, 468–74. Bremen, Germany.

25. Geiger, M., Neubauer, N., and Hoffmann, P. 1994. Intelligent processing head for CO_2 laser material processing, *Production Engineering*, 1 (2): 93–8.

26. Kaplan, A. F., Zimmermann, J., and Schuöcker, D. 1997. Combined laser welding, cutting and scribing. *Proceedings of LANE '97, Laser Assisted Net Shape Engineering* 2, 757–66. Erlangen, Germany.

27. EP 0 741 627 B1, European patent: Nozzle assembly for laser beam cutting, publ. 1997.

28. Appendino, D. 2003. Saldatura laser flessibile di Tailored Blanks, *Proceedings of Expo Laser 2003*, Ancona, Italy.

29. NN: Cutting and welding with one processing optic. *TRUMPF EXPRESS* (July 2003)" 13.

30. Schneider, F., Wolf, N., and Petring, D. 2003. Cutting and welding with the "Autonomous Nozzle," 68. *Annual Report 2003*, Fraunhofer ILT, Aachen, Germany.

31. Schneider, F., and Petring, D. 2004. Cutting and welding with a combined processing head without retooling. *Annual Report 2004*, 64. Fraunhofer ILT, Aachen, Germany.

32. Schneider, F., Petring, D., and Poprawe, R. 2005. Integrated laser processing—Cutting and welding with a combined processing head. *Proceedings of LIM 2005, 3rd International WLT-Conference on Lasers in Manufacturing*, 133–5. Munich, Germany.

33. Petring, D. 2005. One head does it all, *Welding Journal* 84 (8): 49–51.

34. Zefferer, H., and Morris, T. 2005. Guidelines for laser welding of sheet metal, *Welding Journal* 84 (8): 23–8.

35. Verhaeghe, G. 2005. The fiber laser—A newcomer for material welding and cutting, *Welding Journal* 84 (8): 56–60.

36. Woods, S. 2003. Fiber lasers—The new high power, high quality, high efficiency source. AILU magazine, *The Industrial Laser User* Issue 33 (December): 32–3.

37. Hilton, P. A. 1998. Fiber optic beam delivery for high-power CW Nd:YAG lasers. Confidential TWI Report 88277/49/98.

38. Shiner, B. 2004. kW fiber lasers for material processing markets. AILU magazine, *The Industrial Laser User* Issue 35 (June): 23.
39. Miyamoto, I., Park, S.-J., and Ooie, T. 2003. Ultrafine-keyhole welding process using single-mode fiber laser, *Proceedings of ICALEO*, 203–12.
40. Paleocrassas, A. G., and Tu, J. F. 2007. Low-speed laser welding of aluminum alloy 7075-T6 using a 300-W, single-mode, Ytterbium fiber laser, *Welding Journal* 86 (6): 179s–86s.
41. Himmer, T. 2007. Fiber laser system is capable of hardening, cutting, welding, Fraunhofer Institute, Dresden, Germany, *AM&P* (July): 16.
42. Bado, P. 2007. Laser-based glass micromachining, *Small Times* (January/February): 18–9.
43. Available at online http://www.winbrogroup.com
44. Available at online http://link.abpi.net/l.php?20070807A3
45. Available at online http://www.techbriefs.com/techsearch/tow/200709b.html, *NASA Tech Briefs* (September 2007): 53.
46. Available at online http://link.abpi.net/l.php?20070807A4
47. yet2.com and mhtml:file://C:\Documents&Settings\Owner\My Documents\GE Precision Feedback-Co., page 1 of 4, 01/23/2008; Precision feedback-controlled laser drilling for fluid flow apertures.
48. Schaeffer, R., and Angell, J. 1996. Laser processing of ceramics and CVD diamond film, *Proceedings from Advances in the Application of Ceramics in Manufacturing.* Newton, MA: Society of Manufacturing Engineers, October.
49. Schaeffer, R. 1998. An overview of laser microvia drilling, *Future Circuits International*, Issue #3 (June): 159.
50. Schaeffer, R. 1998. A status report on laser micromachining, *Industrial Laser Review* (September): 29.
51. Hackel, L. A. 2004. High throughput production laser peening: A very successful technology transfer, *Aeromat 2004*, Seattle, WA: Lawrence Livermore National Laboratories.
52. Hackel, L. A., Chen, H., Hill, M. R., Rankin, J., Dane, C. B., and Harris, F. 2004. Laser peening—A technology to improve lifetime, reliability and safety while reducing maintenance costs. White paper for NSWCCD.
53. Lu, J., Peyre, P., Oman, C. N., Benamar, A., and Flavenot, J. F. 1994. Residual stress and mechanical surface treatments, current trends and future prospects. *Proceedings of the 4th International Conference on Residual Stresses*, 1154–63.
54. See, D. W., Dulaney, J. L., Clauer, A. H., and Tenaglia, R. D. 2002. The Air Force manufacturing technology laser peening initiative, *Surface Engineering* 18 (1): 32–6.
55. Vaccari, J. A. 1992. Laser shocking extends fatigue life, *American Machinist*, 62–4.
56. Tran, K. N., Hill, M. R., and Hackel, L. A. 2006. Laser shock peening improves fatigue life of lightweight alloys, *Welding Journal* 85 (10): 28–31.
57. Kaysser, W. 2001. Surface modifications in aerospace applications, *Surface Engineering* 17 (4): 305–12.
58. Tenaglia, A. D., and Lahrman, D. F. 2001. Preventing fatigue failures with laser peening, *The AMPTIAC Quarterly* 7 (2): 3–5.
59. Nowotny, S., Scharek, S., and Schmidt, A. 2007. Advanced laser technology applied to cladding and buildup, *Welding Journal* 86 (5): 48–51.
60. Gapontsev, V. 2006. Industrial high power fiber laser systems. *Proceedings of the 2nd International Workshop on Fiber Lasers*, Dresden, Germany.

61. Andrews, J. 2007. Laser scanning improves dimensional accuracy of automotive gas tanks, *Photonics Tech Briefs*, IIa–4a, PTB, Tech Briefs Media Group, 1466 Broadway, Ste. 910, NY, NY 10036.
62. Mudge, R. P., and Wald, N. R. 2007. Laser engineered net shaping advances additive manufacturing and repair, *Welding Journal* 86 (1): 44–8.
63. Saida, K., Song, W., and Nishimoto, K. 2005. Tandem laser brazing of heat-resistant alloys using precious filler metals. Paper #2204. *Laser Materials Processing Conference, Proceedings of ICALEO-2005 Congress*, 1016–25.
64. http://link.abpi.net/l.php?20080108A6, Advanced Nanolasers, 01/08/2008, page 3 of 5.
65. http://www.morristech.com/services/rapid/dmls.asw, Direct Metal Laser Sintering (DMLS), 04/18/2008, page 1 of 2.
66. Sigillo, D. Laser soldering: A turning point. SEICA, Tech Solutions: Selective Soldering, SMT, p. 30, 06/28/2008, page 1 of 2.
67. Gillner, A. Noncontact laser soldering joins arrays of solar cells, Fraunhofer Institute for Laser Technology, Aachen, Germany, http://www.fraunhofer.de, *AM&P* (December 2007): 9.
68. http://link.abpi.net/l.php?20080701A2, Laser Microscalpel, 07/01/2008, page 1 of 5.
69. http://link.abpi.net/l.php?20090421A2

Bibliography

Blomquist, P. A. 2008. Lasers in U.S. Shipbuilding, http://www.industrial-lasers.com/display_article/148298/39/ARCHI/none; 1/31/2008, page 1 of 4.

Disk lasers weld doors and body reinforcements of Peugeot 3008, *Welding Journal* (July 2009): 12.

Farson, D., Cho, M. H., and Choi, H. W. 2006. Hybrid laser + GMAW process for fatigue-resistant welding, Ohio State University, Columbus, OH; *Fabtech International and American Welding Society*, October 2006, Session 6, Laser Welding and Processing 1: Dallas, TX.

GKN Aerospace and SLCR develop laser repair process for composites. March 30, 2009. http://www.reinforced plastics.com/view/977/gkn-aerospace-and-slcr-develop-laser-repair-process-for-composites, page 1 of 2.

Guo, C. 2009. Laser doubles efficiency of traditional lightbulbs. University of Rochester, http:// www.rochester.edu/news/show.php?id = 3385. 06/09/2009, page 1 of 2.

International Thermal Spray & Surface Engineering Newsletter, *AM&P* (February 2008): 53–74.

Kelly, S. M., Brown, S. W., Tressler, J. F., Martukanitz, R. P., and Ludwig, M. J. 2008. Using hybrid laser-arc welding to reduce distortion in ship panels, *Welding Journal* 89 (3): 32–6.

Kocheny, S. A., and Miller, B. 2008. Laser welding: It's not just for metals anymore, *Welding Journal* 89 (3): 28–30.

Laser drilling of ceramic composites to form smooth-walled holes, *yet2.com*, October 1, 2007, page 1 of 2, 01.21.2008, http://www.techbriefs.com/techsearch/tow/200710a.html

Laser peening technology used on power generation components, *AM&P* (February 2008): 58.

Laser shot peening of landing gear components, National Center for Manufacturing Sciences (NCMS), Ann Arbor, MI, February 2008.

Miniature thorax airbag wraps leverage proprietary laser welding technology, Schreiner ProTech, Southfield, MI, http://www.optoiq.com/articles/display/s-_printArticles/s-articles/s-industrial-laser-solutions/s-industry-up..., page 1 of 1, 9/23/2009.

Orozco, N. J., Blomquist, P. A., Rudy, R. B., and Webber, S. R. 2004. Real-time control of laser-hybrid welding using weld quality attributes. *Proceedings of the 23rd International Congress on Applications of Lasers & Electro-Optics*, 10 p. Sanford, ME: Applied Thermal Sciences, Inc.

Penn, W. 2008. Basic laser cladding, *Welding Journal* 87 (2): 47–49.

Rofin delivers CO_2 slab laser number 5000 to Daimler, *Welding Journal* (July 2009): 12.

Schaeffer, R. D., and Kardos, G. Post-laser processing cleaning techniques. http://www.optoiq.com/articles/display/s-_printArticle/s-articles/s-industrial-laser-solutions/s-volume-23/, page 1 of 6, 6/29/2009.

12

Magnetic Pulse Welding

Mel Schwartz

CONTENTS

Introduction

The magnetic pulse welding (MPW) process, a cold solid-state welding process, is fast advancing toward industrial maturity with series production currently operating for a number of automotive projects [1–6]. MPW is accomplished by the magnetically driven, high-velocity, low-angle impact of two metal surfaces. At impact, the surfaces (which always have some level of oxidation) are stripped off and ejected by the closing angle of impact. The surfaces, which are now metallurgically pure, are pressed into intimate contact by the magnetic pressure, allowing valence electron sharing and atomic-level bonding. This process has been demonstrated in the joining of tubular configurations of a variety of metals and alloys.

Basic Concepts

Figure 12.1 [7] shows the essentials of the magnetic pulse system. A suitably high current is discharged through a cylindrical coil and as a result induces an eddy current on the outside surface of a conductive workpiece (outer part). A typical MPW system includes a power supply, which contains a bank of capacitors; a fast switching system; and a coil. The parts to be joined are inserted into the coil, the capacitor bank is charged, and the low inductance switch is

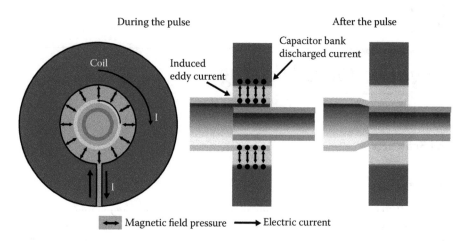

FIGURE 12.1
Basic concept of the magnetic pulse welding process.

triggered by a pulse trigger system and the current flows through the coil. When current is applied to the coil, a high-density magnetic flux is created around the coil, and as a result an eddy current is created in the workpieces. The eddy currents oppose the magnetic field in the coil and a repulsive force is created. This force can drive the workpieces together at an extremely high rate of speed and create an explosive or impact type of weld.

As in explosive welding, a jet is created between the two bonded surfaces by the impact force acting upon them. This jetting action removes all traces of oxides and surface contaminants, allowing the magnetic pressure caused impact to plastically deform the metals for a short instant and to drive the mating surfaces together. The quality of the bond at the interface is a product of many parameters, among them the magnetic force, the collision angle, the collision point velocity, and the initial standoff distance between the mating surfaces. Typically, the pressures at the collision point between the mating surfaces are of the order of 100,000 MPa (~15 million lb/in.^{2}), as measured by explosive welding researchers [8].

Joint Geometry

Typical part geometries have been produced for hermetically sealed capsules and tubes. The weld is a lap weld for all applications [7].

MPW Principles for Various Coils

Various types of coils have been designed and tried in the application of MPW for aluminum alloys and SPCC steel sheet joints [8]. Figure 12.2

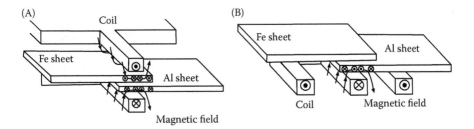

FIGURE 12.2
MPW coil structure. (A) Double layer, H-shaped flat coil; (B) one-layer E-shaped flat coil.

shows a one-turn flat coil instead of a solenoidal coil. This coil consists of the upper and lower H-shaped plates, which are called the double layer H-shaped coil. Lapped sheet workpieces were inserted between these two H-shaped plates. When the high current flows through the coil, it can create the magnetic field to both sides of the overlapped sheet workpieces, and as a result, the sheet metals were welded in the seam state. The magnetic flux produced by this type of coil is shown in Figure 12.2A. In this method, the eddy currents that flow in both sheets are considerably different when dissimilar sheet metals like Al/steel sheets are welded. The thickness of the workpieces was limited by the space between the two H-shaped plates. Therefore, for more applications, some contrivance or improvement was needed. Experimental results and welding characteristics for several samples such as Al–Al[4], Al–Cu[5], Al–Mg[6], Al–Ti[6], and Al–Fe[6] were also reported by Aizawa and Kashani [9].

Another coil, Figure 12.2B, was designed to improve the welding characteristics of aluminum alloy and SPCC steel sheet joints. This new coil is a one-layer E-shaped flat coil that the overlapped sheet workpieces were put on the one side of the coil, Figure 12.2B. This type of coil can be designed for applications ranging from short and small to large and long workpieces and also T-shaped joints with higher weld quality.

Typical Applications

MPW is particularly suitable for large series production and for automated feeding systems. Some of these applications and materials are:

a. Automotive air-conditioner receiver dryer, Al6061
b. Automotive fuel filter, Al 1060
c. High-capsule, Al7075-T6
d. Automotive earth connector, Al6060-T7 to SS304 or Al3003-H14 to SS304
e. Driveshaft, Al6061-T6
f. CO_2 accumulator, Al6082-T6

Weld Benefits

According to Shribman and Blakely [7], MPW provides the following:

- Simpler designs and designs that were previously impossible by conventional means (e.g., Al to steel and stainless steel or Al to Cu welds). It provides the advantage of higher strength-to-weight ratio.
- MPW provides a cold weld. A metallurgical bond is produced without any heat-affected zone (HAZ). Therefore, the original heat treatment properties are maintained by this cold weld and thus, better joint properties are achieved.
- Dissimilar material combinations are available with this method and thus complex joints may be replaced by simpler joints.
- Apart from higher process weld speed, there is a significant quality improvement with this system (e.g., there is no corrosion problem in the welded area), a fact proven by salt spray testing.
- Parameters are stable and reliable with less parameters to control than with conventional processes, minimal rework is required and minimal scrap is produced.

References

1. Birkhoff, G., D. P. McDougall, E. M. Pugh, and G. I. Taylor. Explosives with Lined Cavities, *Journal of Applied Physics* 18 (1948): 563.
2. Walsh, J. M., R. G. Shreffer, and F. J. Willig. Limiting Conditions for Jet Formation in High-velocity Collisions, *Journal of Applied Physics* 24 (1953): 249.
3. Allen, W. A., J. M. Mapes, and W. G. Wilson. An Effect Prodiced by Oblique Impact of a Cylinder on a Thin Target, *Journal of Applied Physics* 25 (1954): 675.
4. Shribman, V., J. D. Williams, and B. Crossland. Some Problems of Explosively Welding Tubes to Tubeplates. Presented to the Select Conference on Explosive Welding, Hove, Sussex, UK, September 1968.
5. Shribman, V., A. S. Bahrani, and B. Crossland. The Techniques and the Mechanism of Explosively Welding, *The Production Engineer*, February 1969.
6. Furth, H. P., and R. W. Wanick. Production and Use of High Transient Magnetic Fields, *Review of Scientific Instruments* 27 (1956): 195.
7. Shribman, V., and M. Blakely. Benefits of the Magnetic Pulse Process for Welding Dissimilar Metals, *Welding Journal* (September 2008): 56–9.
8. Aizawa, T., M. Kashani, and K. Okagawa. Application of Magnetic Pulse Welding for Aluminum alloys and SPCC Steel Sheets Joints, *Welding Journal* 86 (May 2007): 119s–24s.
9. Aizawa, T., and M. Kashani, *Proceedings of IIW International Conference on Technical Trends and Future Prospective of Welding Technology for Transportation, Land, Sea, Air and Space*, Osaka, Japan, 2004.

13

Casting

Mel Schwartz

CONTENTS

Squeeze Casting

Aluminum composites with excellent mechanical and physical properties have been produced at low cost by reinforcing the aluminum matrix with spherical ceramic particles derived from fly ash. These aluminum metal matrix composites (MMCs) produced by Cyco Systems Corp., Melbourne, Australia are a combination of aluminum–silicon (Al–Si) casting alloys and spherical ceramic reinforcing particles. The ceramic particles are derived from fly ash, a by-product of burning black coal in power stations and therefore are very low in cost. Fly ash is made up of tiny glass spheres, and consists primarily of silicon, aluminum, iron, and calcium oxides. The fly ash particles are collected by electrostatic precipitators from the exhausted gases in power stations. Fly ash particles can have two forms, either hollow spheres (called cenospheres) or solid particles, and typically the fly ash particles are in the range of 10–20 μm in size.

According to Graham Withers, "this new MMC called 'Ultalite' has significantly better properties than other aluminum/ceramic mixtures, including wear resistance, stiffness, hardness, and compatibility in friction couples. Ultalite can be made from secondary metal [and] is readily recyclable, and is easily machined with conventional machine tools." Additionally he claims that the above benefits can solve weight reduction challenges in parts such as brake components, for which they can successfully substitute for cast-iron brake drums and disk brake rotors. See Figure 13.1 [1,2].

Fabrication

Ultalite MMCs have been fabricated by adding fly ash particles to a molten aluminum alloy, and then casting that liquid metal/ceramic mixture into near net shape (NNS) components. In the past, Mr. Withers claims that when prototype quantities of the material were made, the fly ash particles were stirred into a crucible filled with the liquid aluminum alloy by a paddle-style stirring system. After the fly ash was well distributed, the composite alloy was poured into ingot molds and solidified. The ingots were then shipped to a casting plant. There they were remelted and squeeze-cast to NNS components such as the brake drums shown in Figure 13.1.

Squeeze casting is a modified die casting process in which a fully liquid aluminum alloy is injected into a reusable, hardened steel die. With

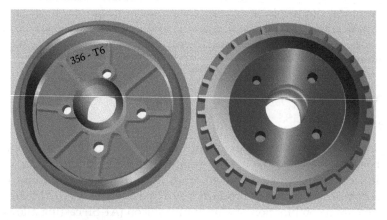

FIGURE 13.1
This brake drum was cast from the Ultalite composite.

conventional die casting, the liquid metal is injected into the die at a high speed (30–60 m/sec), producing parts having an excellent surface finish. However, the high-speed injection produces a significant amount of residual porosity in the casting, limiting mechanical properties.

In contrast, Mr. Withers says that "squeeze casting injects the liquid metal at slow speed (about 0.25 m/sec) through relatively massive runners and gates into the die cavity. The slow injection speed ensures that the liquid aluminum fills the cavity without turbulence, pushing the air in the cavity out through vents strategically placed at the last parts of the cavity to fill. Once the cavity is full, pressure is applied to the biscuit remaining in the shot sleeve to force additional liquid into the die to feed solidification shrinkage. In this manner, a high quality casting essentially free of porosity, can be produced." Figure 13.2 shows a photograph of the fly ash particles distributed in an A356 aluminum alloy [1,2].

Semisolid Casting

To facilitate high volume commercial production of components such as truck brake drums. Mr. Withers firm, Cyco Systems, has investigated semisolid casting. Rather than fully liquid aluminum, semisolid casting injects a partial liquid, partial solid mixture into the die. By ensuring that the solid phase is essentially globular in shape, and by controlling the solid fraction in the range of 25–50%, semisolid casting can also produce high-quality, NNS parts essentially free of porosity. In fact a semisolid casting process, designated as rheocasting, has been especially suitable for forming many

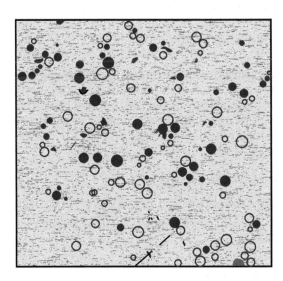

FIGURE 13.2
Microstructure of the ultalite composite material shows fly ash particles distributed in a matrix of A356 aluminum.

Ultalite parts. The reason is that the semisolid slurry is formed adjacent to the casting machine by stirring the liquid aluminum as it is cooled into the semisolid temperature range.

A semisolid aluminum casting process called SEED has been developed and has the advantage of being simpler and more flexible than similar processes, says the National Research Council of Canada. Developed by Alcan, SEED means Swirled Enthalpy Equilibration Device. The process consists of cooling molten metal in a swirling crucible. During cooling, the heat loss in the aluminum is mainly controlled by the crucible material and mass. Therefore, their proper selection is critical to ensuring that the aluminum has the correct combination of temperature and solid fraction. The SEED semisolid process minimizes molten metal casting problems, as the thicker consistency of the paste causes less agitation during mold filling. The reduced turbulence and lower molding temperatures also allow the parts to be strengthened by heat treatments, which generate even better properties. The process is also suitable for a greater variety of casting and wrought alloys [3].

Slurry on Demand

Another example of the semisolid casting technique and technology is "Slurry on Demand" [2]. This process, through bulk delivery tankers, provides molten metal that is transferred to holding furnaces. A slurry-producing unit that can contain enough liquid aluminum to produce one part is ladled full of molten metal from the furnace, where the fly ash particles are added. The charge is simultaneously stirred by a magnetic field and cooled in a controlled manner to form a semisolid billet consisting of a specific solid fraction, typically around 50%. The stirring not only breaks up the dendrites, producing the globular semisolid microstructure, but also distributes the fly ash particles in the aluminum. After the billet is cooled to the target solid fraction, the aluminum composite is too viscous for the ceramic particles to float out. At that point, the semisolid composite part can be ejected from the slurry-producing unit and transferred immediately into the cold chamber of a high-pressure die casting machine, where it is injected into the die cavity.

Advantages of Semisolid Processing (SSP)

- Enables high dimensional accuracy and metallurgical integrity
- Extends die life, and produces components to tighter tolerances than other types of castings
- Provides higher structural integrity, quality, and soundness than conventionally cast parts
- Parts may be heat-treated to develop properties similar to those of permanent mold or investment castings

Cost Advantages

Ultalite is lower in cost than other Al-MMCs because the reinforcement material (fly ash particles), as a by-product of electricity generation, is significantly less expensive than other ceramic reinforcement particles such as silicon carbide. Also, as 20% of the aluminum is replaced by fly ash particles, the Ultalite composite is 15–20% cheaper than monolithic aluminum alloys. Other advantages include:

- Die life is greatly extended because, with semisolid processing, the temperature is lower than needed for other production techniques.
- The spherical reinforcement particles aid the flow of metal through the die gate and they do not erode the steel surfaces.
- Less energy input is required because the process operates at lower temperatures.
- The manufacturing process can be arranged in a continuous production line from mixing to machining.
- Lower-cost secondary metal can be utilized for Ultalite castings.
- The casting runners, biscuits, and other scrap can be recycled directly into the secondary metal melt, reducing the requirement for new metal.
- Semisolid Metal (SSM) Processing

In another study covering SSM processing optimization of alloys was evaluated [4,5,6]. The evaluation considered SSM as a procedure in which SSM billets are cast to provide dense, heat-treatable castings with low porosity. The study evaluated four key factors that are essential for SSM alloy development/optimization:

1. Solidification range (ΔT): The temperature range between the solidus and the liquidus lines of the alloy.
2. Temperature sensitivity of fraction solid: For stable and repeatable processing conditions, the temperature sensitivity of the fraction solid should be as small as possible in the fraction solid range of commercial operations.
3. Temperature process window (ΔT): Depending on the application, for rheocasting ΔT is defined as the temperature difference between 0.3 and 0.5 fraction solid, whereas for thixoforming, ΔT is defined as the temperature difference between 0.5 and 0.7 fraction solid.
4. Potential for age hardening: To achieve high strength, the alloys designed for SSM processing need to have high potential for age hardening. Extensive thermodynamic calculations were conducted to evaluate the SSM processibility of commercial alloys, including 356/357, 390/383, 319, 206, and wrought alloys.

Salient Results

The A319 alloy has a similar SSM temperature process window for rheocasting as SSM A356 (24°C vs. 23°C), and a much longer temperature window thixocasting/thixoforging (12°C vs. 3°C). Moreover, the alloy has very small dfs/dt values in the fraction solid range of commercial forming. This makes it an excellent material for semisolid processing. Compared to SSM A356, the SSM temperature process window of 380 for rheocasting is somewhat small. The SSM processibility of the alloy A380 can be improved by optimizing/modifying the alloy composition. The 206 alloy has a fairly poor SSM processibility. The alloy has a quite small SSM temperature process window, and a high temperature sensitivity of fraction solid for rheocasting applications [5,6].

Applications

The commercialization of the MMCs has been applied to several automotive applications, such as truck brake drums, brake rotors, engine blocks, cylinder heads, pistons, pulleys, and transmission components. A considerable amount of testing has been carried out on brake drums squeeze cast from an aluminum A356/20% fly ash alloy. The results of testing can be summarized as follows:

- Weight: The Ultalite composite brake drums are nearly three times lighter than conventional drums produced from cast iron. (This can correspond to a total weight savings of 14 kg for four brake drums.)
- Heat: Under simulated braking conditions, the superior heat dissipation properties of the Ultalite composites mean that the peak temperature reached by the Ultalite brake drums was considerably lower at 200°C than cast-iron drums, which reached a peak temperature of 300°C.
- Linings: Unlike brake drums fabricated from other aluminum MMCs, Ultalite brake drums can be used with conventional linings.
- Wear: The Ultalite brake drums showed lower wear rates than cast-iron drums. In tests the Ultalite drums were observed to wear 0.02 mm, compared with 0.03 mm for cast-iron drums.
- Pollution: The Ultalite drums produce less black dust on wheels than iron drums.

Other Squeeze Casting MMCs

A viable technique for making magnesium matrix composites by a squeeze casting technology was reported by J. Lo and R. Santos, National Resources Canada [7]. The report suggested that the use and selection of squeeze casting in combining gravity with die casting and with closed die forging in a single operation resulted in composite products with properties superior to those produced by stir casting, pressureless infiltration, investment casting, or gravity casting. In fact, they found that pore-free casting was possible under an applied pressure of 100 MPa (14.5 ksi) for magnesium alloy AM50. Furthermore, transmission electron microscopy showed that the reinforcement particles were well bonded to both the reaction layer and the matrix. In the comparison below the magnesium MMC far exceeds the AZ91 cast material [7]:

- Property AZ91 as cast SiC/AZ91
- Density, g/cm^3 1.8 2.4
- Young's modulus, GPa 45 120
- Shear modulus, GPa 17 48
- Tensile strength, MPa 196 326
- CTE, ppm 27.3 13.8

Robocasting

The process provides moldless fabrication from slurry deposition and this free-form fabrication method produces dense ceramics in less than a day without requiring organic binders.

Free-form Fabrication

Free-form fabrication is the NNS processing of materials by sequentially stacking thin layers until complicated and 3-D shapes are produced. The operation is computer-controlled and requires no mold. This exciting new field of technology provides engineers with the ability to rapidly produce prototype parts directly from CAD drawings, and little or no machining is necessary after fabrication [8].

Techniques for free-form fabrication with several types of plastics and metals are already quite advanced [9,10]. Very complicated plastic models can be fabricated by stereolithography, selective laser sintering, fused deposition modeling, or 3-D ink-jet printing. Metals may be free-formed by the LENS technique, and porous ceramic bodies by 3-D printing into a porous powder bed. The techniques that are being developed for the free-form fabrication of dense structural ceramics primarily revolve around the sequential layering of ceramic loaded polymers or waxes. Laminated object manufacturing and computer-aided manufacturing of laminated engineering materials (CAM-LEM) [11] processing use controlled stacking and laser cutting of ceramic tapes [9,12]. Similar to fused deposition modeling, ceramic-loaded polymer/wax filaments are being used for the fused deposition of ceramics [9,10]. Extrusion free-form fabrication uses high-pressure extrusion to deposit layers of ceramic-loaded loaded polymer/wax systems.

Modified stereolithographic techniques use ceramic loaded ultraviolet curable resins [9]. Green parts made with any of these techniques typically have

40–55 vol.% polymeric binder. In this regard, these techniques are analogous to powder injection molding of ceramics; very long and complicated burnout heat treatments are necessary to produce a dense ceramic free of organics. Heating rates of 0.2°C per minute are common [13]. So, while a part may be rapidly prototyped within a few hours, it may still take several days to densify the part.

Robotics

A technique developed at the Sandia National Laboratories for the free-form fabrication of dense ceramics does not require organic binders. Therefore, a dense ceramic part may be free formed, dried, and sintered in less than 24 hours. This technique, called robocasting [8], uses robotics for computer controlled deposition of ceramic slurries through an orifice and is useful for the NNS fabrication of components with simple or complex shapes that are completely solid or thin walled. The process uses highly loaded ceramic slurries that are typically 50–65 vol.% ceramic powder, less than 1 vol.% organic dispersants, and 35–50 vol.% volatile solvent (usually water). The ceramic slurries are deposited in sequential layers. Any conceivable 2-D pattern may be "written" layer-by-layer into a 3-D shape (Figure 13.3) [8]. Orifice openings can range from several millimeters to tenths of millimeters [8].

In some regards, robocasting is analogous to the ceramic NNS processing techniques, slip casting and gel casting [14], however, robocasting is moldless and binderless, and fabrication times can be quicker. To maintain structural

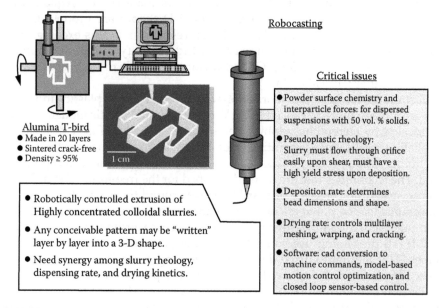

FIGURE 13.3
Schematic and description of the robocasting process with an inset photo of an aluminum oxide thunderbird made in 20 layers and sintered crack-free to 96% theoretical density.

integrity while building a component, robocasting relies on a rheological transition of the deposited slurry induced by partial drying of the individual layers. The slurry is deposited upon a heated plate and undergoes a pseudo-plastic to dilatant transition during the build process. In contrast to gel casting and other free-form fabrication techniques, robocasting does not require organic polymerization reactions or solidification of a polymeric melt to maintain the shape of components.

Optimization of Processing

Robocasting is no more complicated than caulking a bathtub—except that during robocasting, the substrate actually moves underneath the point where material is dispensed from a stationary orifice. While conceptually simple, transforming this concept into reality for the manufacture of ceramics requires a synergistic control of the viscosity and rheology of the slurry, percentage of solids in the ceramic powder slurry, dispensing rate of the slurry through the orifice, drying kinetics of the dispensed bead of slurry, and computer code for optimal machine instructions.

The slurry viscosity must be tailored during processing for optimal performance. Additionally, the rheological dependence of viscosity with shear rate must be controlled. During dispensing, the slurry will experience high shear conditions while flowing through the orifice and as the moving substrate interacts with the dispensing slurry. However, immediately following this process, the slurry experiences a shear rate near zero. Therefore, to control the shape of dispensed beads, the slurry rheology must be extremely pseudoplastic with a yield point (shear-thinning) so that the material can flow smoothly during dispensing but then solidify in place once shear stresses are removed (similar to latex paint). If the slurry is too fluid, beads will spread uncontrollably.

If the slurry is too viscous, beads lay down like rope and maintain rounded tops. When the proper rheology is obtained, beads yield nearly rectangular cross sections with relatively straight walls and flattened tops. Perfectly rectangular cross sections would be optimal for filling space when additional layers are sequentially dispensed. For optimal robocasting, it is also important to have a high content of solid ceramic powder in the slurry. The solids cement affects the viscosity and rheology of the slurry and the final bead dimensions. Additionally, high solids contents minimize the amount of drying shrinkage and sintering shrinkage so that final bead dimensions and final NNS tolerance can be more accurately controlled [8].

Slurry Characteristics

For slurry processing, a typical ceramic powder may be considered to have an average diameter on the order of several microns and possess a relatively monosized distribution. Ceramic powders with this character that are dried from a dispersed slurry typically pack into a consolidated structure that is approximately 65% of the theoretical density. However, the character of flow-able slurries with solids loadings just below the consolidated density is what is important for robocasting.

The behavior of a typical dispersed ceramic powder slurry is described as follows; at low solids loadings, dispersed slurries have very low viscosity and are rheologically Newtonian. Around 40 vol.% solids, the slurries begin to show pseudoplastic shear-thinning behavior, even though the viscosity is still relatively low. As the solids content approaches 60 vol.%, interpar-ticle interactions and interparticle collisions become dominant; viscosity begins to increase appreciably, and the rheological behavior becomes highly shear-thinning. At approximately 63 vol.% solids, particle mobility becomes restricted, and the slurry locks up into a distant mass. Therefore, it is desir-able to robocast with slurries that have solids loadings close to dilatant tran-sition, so that with minimal drying, a robocasted layer becomes structurally sound and a foundation upon which more layers may be deposited.

Controlled drying kinetics of the dispensed bead, as well as the time between subsequent layer deposition, are also important for the minimiza-tion of cracking, warping, and defects due to delamination. Under optimal drying conditions, the moisture content of a previously deposited bead is low enough to induce the permeation of fluid into its structure from a wet, freshly dispensed bead. The fluid transport naturally drags some powder particles with it and creates a finite interpenetrated region between the two beads that is free of defects or delaminations [8]. If the drying rate is too fast, warping, cracking and delamination may occur. When the drying rate is too slow, the pseudoplastic/dilatant transition is delayed and accumulated weight from several layers can induce slumping and the creation of nonuni-form walls. When the pseudoplastic/dilatant transition is maintained close to the top of the part being casted, thick components may be produced, as well as a relatively uniform thin-walled structure [8].

In addition to rheology and solids loading, the initial dimensions of the dispensed beads (prior to drying) are largely determined by the rate at which the slurry is dispensed. Since the slurry is dispensed through a stationary orifice that is suspended above a moving X–Y table, the initial size of the bead is determined by the volumetric rate of dispensing (ml/s) and the travel speed of the X–Y table (mm/s). For a given volumetric dispensing rate, the bead will be thick if the table speed is slow. If the table speed is fast, the bead will be stretched thin. These are very important parameters when consider-ing the fabrication of an actual part. Bead dimensions will determine the optimum Z-axis increments for subsequent layer deposition and the opti-mum X–Y axes increments for filling in solid regions within a component.

Casting Al_2O_3

When the material issues discussed previously are properly controlled, robo-casting can be used to make intricate ceramic bodies that sinter into relatively strong, dense, and defect-free parts. This has already been demonstrated for the robocasting of aluminum oxide (Al_2O_3). Aqueous slurries were made according to Cesarano et al. [15] with alumina contents of 58–61 vol.%. The slurries were gently mixed on a ball mill until the proper shear-thinning rheology was obtained. Examples of some cast parts are shown in Figure 13.4. The part with tailored porosity in Figure 13.4b is particularly interesting. It demonstrates the precision capability of robocasting and the possibilities for fabricating structures that are not currently possible by traditional manufacturing methods. Advantages of robocasting are seen in Table 13.1 [8].

Future for Process

Developmental techniques have been centered on multimaterial deposition for the moldless manufacture of intricately shaped composites and hybrid electronic packages. This task requires a significant effort in the development of computer software that can analyze a CAD file and convert the file into a set of machine instructions that optimize processing parameters

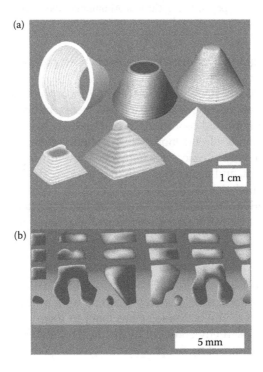

FIGURE 13.4
Examples of sintered aluminum oxide parts made by robocasting. (a) Shows thin walled and solid parts. Notice that with minimal machining, surfaces can be made smooth. (b) This cross section shows the symmetry and control that can be obtained in an intricate structure with tailored porosity. Due to regions with closed porosity and undercuts, this part could probably not be made by traditional molding or machining techniques.

TABLE 13.1

Cold-Spray Applications

Metal restoration and sealing	Castings, molds, dies, weld seams, autobody repair, HVAC equipment
Corrosion protection	Carbon steels, magnesium
Thermal barriers	Piston heads, manifolds, disc brakes, aircraft engine components
Thermal dissipation	Microelectronics
Soldering priming	Microelectronics
Wear-resistant coatings	
Electrically conductive coatings	Metal, ceramic surfaces
Dielectric coatings	Aerospace, automotive, and electronics packaging
Anti-stick properties	Deposits impregnated with PTFE or silicone
Friction coatings	Paper mills
Rapid prototyping/manufacturing	Additive manufacturing
Biomedical	Orthopedic implants, prostheses, dental implants
Decorative coatings	

Source: NADCA White paper, Using Custom Aluminum Alloys to Expand Die Casting Applications, http://www.diecasting.org/oem.htm

producing high-quality robocast components with precise control of feature resolution.

Additionally, efforts have gone forward to incorporate smart processing capabilities into robocasting with the aid of sensors and computer modeling. Sensors that monitor build conditions in real time are being incorporated into the system to provide closed loop, sensor-based feedback. Computer simulations of the relevant physical phenomena—flow during dispensing, bead shape development, and dimensional changes during drying—are also being developed for knowledge-based processing control. These simulations will be used in combination with real-time sensor feedback to interpret and calculate optimal adjustments of processing parameters during component fabrication.

Thixomolding

Thixomolding [16] is the semisolid injection molding of magnesium net-shape components, practiced in a one-step process in a unitized machine. The critical innovation of the technology has been to incorporate the semisolid process in an enclosed environment in the unitized machine. The following advantages are produced by the process:

- Environmentally friendly production—worker safe and friendly, no global warming SF_3 cover gas, no polluting scrap or waste
- Net-shape parts with little, if any, machining
- No heat treatment required

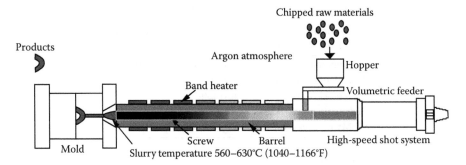

FIGURE 13.5
Schematic of thixomolding machine.

- Close dimensional tolerances
- Low porosity, hence improved ductility
- Longer die life
- Higher metal yield, hence lower costs
- New part design, consolidating several parts into one molding and integrating multiple functions

As seen in Figure 13.5 [16], the alloy granules are fed through a hopper into a heated barrel within which operates a rotating and reciprocating screw. The barrel, heated to about 600°C, partially melts the granules and the screw rotation homogenizes the semisolid mixture and may refine and round the solid phase. Periodically the screw injects the thixotropic alloy into a heated die, conventionally through a nozzle, sprue, and runners. The melt is protected by argon in the barrel and a vacuum can be used in the die, as both practices minimize oxygen pickup.

Types of Alloys

Many different magnesium alloys have been thixomolded. The dominant commercial alloy has been AZ91D. See Table 13.2 for other alloys that have been thixomolded [16].

New Developments

In the realm of new thixomolding [16] technology, hot sprues, hot runners, and thixoblending have been recently developed and used in operations. Keeping the sprue and runner and their encapsulated slurry hot saves making a scrap sprue and runner on each injection shot. This boosts the metal yield very significantly and lowers production costs. Thixoblending of granules of differing alloys and master alloys in the hopper enables the "paint

TABLE 13.2

Range of Magnesium Alloys and Composites that have been Thixomolded

Conventional Alloys	AM 50, AM 60, AZ 91D
Thixoblended Alloys	AZ 71, AZ 81, Mg-Zn-Al
Mg–Zn–Al Alloy	AM-Lite
High Temperature Alloys	AE 42, AJ 52, AJ 62, MRI 153
Mg Matrix Composites	Mg-SiC, Mg-Al$_2$O$_3$, Melram

TABLE 13.3

Thixomolded Applications in Various Markets

Electronic/Communications	Hand Held Tools
Laptop computers	Drills
Cell phones	Saws
Digital projectors	Chain saws
Digital cameras	Nailers
Camcorders	
TV surrounds	**Sporting Goods**
Walkman	Sun glasses
"Dog Tag" MP3 player	Gun scopes
Defense detectors	Fishing reels
Radar detectors	Snow board clamps
Check sorters	Motorcycle wheels and fairings "Go-Cycle" bicycle

store" concept of designing and agile production of new alloys on short notice.

Magnesium/plastic composites are another promising area of development, to meld the virtues of each material while minimizing the deficiencies. For example, overmolding magnesium with plastic provides the color and corrosion protection of plastic combined with the stiffness, strength, and electromagnetic shielding of magnesium.

Applications

Applications include the automotive industry: seat backs, steering column brackets, mirror parts and brackets, lazy susan bins, foldable car tops, windshield wiper gear boxes, lift gate mechanisms, cup holders, and brackets for trucks. The response in electronics/communication, hand tools, and sporting goods applications has been very vigorous (see Table 13.3) [16]. Cell phone components and laptop computer covers have been thixomolded at the rate of tens of millions per year. By replacing plastic, magnesium provides stiffening, electromagnetic shielding, and heat removal in electronic/communication devices.

References

1. Withers, G. Cyco Systems Corp., Melbourne, Australia; gwithers@optusnet. com.au; http://www.ultalite.com
2. Withers, G. Ultalite Aluminum Composites, *AM&P* (September 2005): 45–8.
3. Bouchard, D. Semisolid Coating Process Cuts Cost for Aluminum Parts, *AM&P* (April 2007): 15.
4. North American Die Casting Association, Advances in Die Casting Alloys, *AM&P* (January 2008): 47–9.
5. Twarog, D. I. NADCA, Wheeling, IL; twarog@diecasting.org; http://www. diecasting.org. Advanced Casting Research Center at WPI: http://www.wpi. edu/Academics/Research/ACRC/Research/
6. NADCA White paper, Using Custom Aluminum Alloys to Expand Die Casting Applications, http://www.diecasting.org/oem.htm
7. Lo, J., and R. Santos. Magnesium Matrix Composites for Elevated Temperature Applications, SAE Paper 2007-01-1028; 2007 SAE World Congress.
8. Cesarano III, J., R. Segalman, and P. Calvert. Robocasting Provides Moldless Fabrication from Slurry Deposition, Sandia National Laboratories, Albuquerque, NM and University of Arizona, Tucson, AZ, *Ceramic Industry* (April 1998): 94–102.
9. *Proceedings of the Solid Freeform Fabrication Symposium*, Vol. 1990, 1991, 1992, 1993, 1994, 1995, 1996, 1997. Edited by D. L. Bourell, J. J. Beaman, R. H. Crawford, H. L. Marcus, and J. W. Barlow, Austin, TX: The University of Texas.
10. Agarwala, M. K., et al. *The American Ceramic Society Bulletin* 75, no. 11 (November 1996): 60–5.
11. *Ceramic Industry* (March 1998): 42–8.
12. Griffin, E. A., D. R. Mumm, and D. B. Marshall. *The American Ceramic Society Bulletin* 75, no. 7 (July 1996): 65–8.
13. Remco Van Weeren, Allied Signal Research and Technology, Morristown, NJ, personal communication.
14. Young, A. C., O. O. Omatete, M. A. Janney, and P. A. Menchhofer. *Journal of the American Ceramics Society* 74, no. 3 (1991): 612–8.
15. Cesarano III, J., and I. A. Aksay. *Journal of the American Ceramics Society* 71, no. 12 (1988): 1062–7.
16. Decker, R. F., and S. E. LeBeau. Thixomolding, *Advanced Materials & Processes* (April 2008): 28–9.

Bibliography

Jorstad, J. L. Permanent Mold Casting Process, *Advanced Materials & Processes*, April 2008, 30–3.

NADCA, Die Casting Industry's Environmental Challenges, Responsibilities and Results, White paper, 2007.

Six parameters of cost-effective metal casting design, *American Foundry Society*, March 2009, 11 p.

14

Coatings (Thermal Spray Processes)

Mel Schwartz

CONTENTS

Introduction

In the thermal spray coating process, finely divided metallic or nonmetallic materials are deposited in a molten or semimolten condition to form a coating. The coating material may be in the form of powder, ceramic-rod, wire, or molten materials [1]. Based upon the two heat sources, a "family tree" of thermal spray methods is shown in Figure 14.1.

Flame Spraying

In the flame spraying process, oxygen and a fuel gas, such as acetylene or natural gas, are fed into a torch and ignited to create a flame. Either powder or wire is injected into the flame where it is melted and sprayed onto the workpiece. Flame spraying can be readily performed in the shop or on-site and is generally low cost. Some of the materials that are typically applied are stainless steels, nickel aluminides, Hastelloy alloys, tin, and babbit. With relatively low particle velocities, the flame spray process provides thicker build-ups for a given material than the other thermal spray processes. The low particle velocities result in coatings that are more porous and oxidized as compared to other thermal spray coatings. Porosity can be advantageous in areas where oil is used as a lubricant, because a certain amount of oil is always retained within the coating, thus increasing its life. Oxides in the coating increase hardness and enhance wear resistance. The flame spraying technique is also known as oxy-fuel. Other oxy-fuel methods include wire, powder (metallic and ceramic), molten metal, ceramic-rod, detonation, and high velocity oxy-fuel (HVOF).

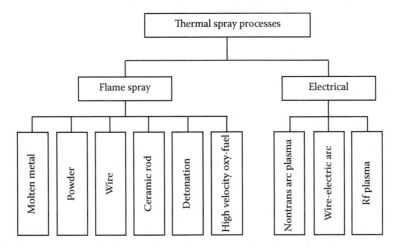

FIGURE 14.1
Thermal spray application methods by heat source schematic. (From Hermanek, F. J., *Thermal Spray Terminology and Company Origins*, ASM International, Materials Park, OH, 2001. With permission)

Arc Spraying

In the arc spray process, two wires are simultaneously brought into contact with each other at the nozzle. The electrical load placed on the wires causes the tips of the wires to melt when they touch. An atomizing gas such as air or nitrogen is used to strip the molten material off the wires and to transport it to the workpiece. Arc spraying is reasonably inexpensive and readily usable in the field. Low particle velocities enable high maximum coating thickness for a given material. Materials typically applied by arc spraying include stainless steels, Hastelloy, nickel aluminides, zinc, aluminum, and bronze. Recent advancements in nozzle and torch configurations provide greater control over coating quality and spray pattern. For example, the wires can be sprayed finely or coarsely. A "fine" spray leads to smooth, very dense coatings whereas a "coarse'" spray enables greater coating thickness.

Molten Metal Flame Spray

A thermal spraying process variation in which the metallic material to be sprayed is in the molten condition [1,2]. The molten metal process has advantages and disadvantages. Advantages include: cheap raw materials, use of inexpensive gases, and gun design is very basic. Noteworthy disadvantages are: gun is cumbersome to use in the manual mode, can only be held in a horizontal plane; high maintenance due to high temperature oxidation and molten metal corrosion; and useful only with low-melting temperature metals. Use for the molten metal thermal spray process include the fabrication of molds, masks, and forms for the plastics industry, using low-melting point bismuth-based alloys (the Cerro family of alloys); the deposition of solder alloys to joints that would be coalesced using torches or ovens; and the production of metal powders.

Powder Flame Spraying

A thermal spray process in which the material to be sprayed is in powder form [2] (see Figure 14.2). Powder flame spraying consists of feeding a powder through the center bore of a nozzle where it melts and is carried by the escaping oxy-fuel gases to the workpiece. Unfortunately, this approach yields coatings high in oxides and with void contents approaching 20 vol.% (v/o). However, coating quality can be improved by feeding air to the nozzle through a small jet, which reduces the pressure in a chamber behind the nozzle. This chamber is connected to the powder feed hopper. In this way a gentle stream of gas is sucked into the gun and carries powder with it [1,2]. If air is not used then the density of the supporting gas influences the feed rate and, for any particular powder, there is an optimum amount that can be carried in a gaseous stream. It depends upon the velocity and volume of the

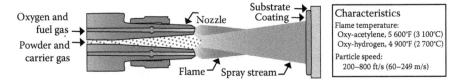

FIGURE 14.2
Powder flame spray process. (From Hermanek, F., What is Thermal Spray? Thermal Spray Engineering Consultant, ITSA, fhermanek@aol.com; www.thermal spray.org, 2001. With permission.)

gases used. The usefulness and criticality of flowmeters and pressure gauges are governing factors.

Wire Flame Spraying

A spray process in which the feed stock is in wire or rod form [1,2]. In about 1912, Dr. Max Ulrich Schoop developed the first device for spraying metal wires. The apparatus consisted of a nozzle in which a fuel, probably acetylene or hydrogen, was mixed with oxygen and burned at the nozzle's face [3]. A stream of compressed air surrounding the flame atomized and propelled the liquefied metal. Process continuation depended on feeding the wire at a controllable rate so it melted and was propelled in a continuous stream. Schoop approached this problem by using a turbine to actuate gears and drive rolls that pulled the wire into the nozzle. This apparatus appeared to him to be similar to a pistol or gun, so he (and now we do also) referred to the thermal spray devices as "guns" or "pistols" and never "torches." The wire flame spray gun has not radically changed since the days of Dr. Schoop. While there have been changes in nozzle and air cap design, replacement of the air turbine with an electrical motor and even the use of barrel valves, the basic principal, however, remains the same "push or pull a wire into a flame, melt and atomize it and deposit the molten droplets to form an adherent coating."

Ceramic Rod Flame Spraying

A spraying process in which material to be sprayed is in a ceramic rod form [1,2]. The spraying of ceramic rods dates back to the early 1950s when a demand rose for heat resistant refractory coatings. Plasma had not come into its own and flame sprayed powder coatings, due to their porous nature, lacked the integrity and protection required. The solution was rather simple—Coors Ceramic and Norton developed ceramic rods, referred to as Rokide, while the Metallizing Engineering Company (Mogul), modified a wire gun to spray rods. What differentiated the guns were the drive rolls. A wire gun had serrated steel rolls to grip and feed the wire while the Rokide gun employed "V" slotted fiber rolls that pinched the rods and fed them forward. The principle of operation in either gun is similar—the nozzle's flame is concentric to the wire or rod in order to maximize uniform heating. A coaxial sheath of compressed gas around the flame atomizes the molten material and accelerates it to the workpiece [1,2]. Particle velocities in both the wire and rod process are approximately the same—185 m/sec (600 ft/ sec) while coating densities have been measured at approximately 95 v/o. Ceramic rods are 61 cm (24 in.) in length and offered in three diameters 3.18, 4.75, and 5.72 mm (1/8, 3/16, and 1/4 in.). Compositions include several stabilized zirconias, white and gray alumina, and a spinel.

Detonation Flame Spraying

A thermal spray process variation in which the controlled explosion of a mixture of fuel gas, oxygen, and powdered coating material is utilized to melt and propel the material to the workpiece [2]. Union Carbide Corp., Linde Division, developed concepts in the early 1950s using explosions in a unique manner. Their concept was to introduce powdered materials into detonation or shock waves. The "waves" were produced by igniting a mixture of acetylene and oxygen into the detonation chamber that was opened to a 1 m (39.37 in.) long tube 2½ cm (1 in.) in diameter [2].

The system was complex. In operation, a mixture of spray material, acetylene and oxygen, was injected into the detonation chamber. Combustion gases were neutral, reducing, or oxidizing and had their temperature controlled by the addition of an inert gas, for cooling, or hydrogen to heat it. The procedure was initiated by a gas/powder metering system that measured and delivered the mixture to the chamber where it was ignited. The resulting shock wave accelerated the powder particles to over 731 m/sec (2400 ft/sec) and produced temperatures in excess of 4000°C (7232°F). Pressures from the detonation closed the controlling valves until the chamber pressure was equalized. When this occurred the cycle was repeated either four or eight times per second. There was a nitrogen purge between cycles. Each detonation deposited a dense and adherent layer several microns thick and about 2.54 cm (1 in.) in diameter. Repeating the cycle produced thicker coatings. Detonation coatings were designed for applying hard materials, especially carbides, on surfaces subject to aggressive wear. Praxair Surface Technologies uses the term "D-gun" for this process. Also, the equipment generates noise in excess of 150 dBA and must be acoustically housed confining noise emissions.

High Velocity Oxy-Fuel Spraying (HVOF)

A high velocity flame spray process [1] invented over 20 years ago, yet it has thrust the thermal spray application range into areas that were once unattainable. In HVOF spraying, a combination of process gases such as hydrogen, propane, propylene, and even kerosene, and oxygen and air are injected into the combustion chamber of the torch at high pressure and ignited. In the combustion chamber, burning by-products are expanded and expelled outward through an orifice at very high velocities. Often times they produce "shock diamonds" exiting the spray gun [1,2].

Powders to be sprayed via HVOF are injected axially into the expanding hot gases where they are propelled forward, heated, and accelerated onto a surface to form a coating. Gas velocities exceeding Mach 1 have been reported with temperatures approaching 2300°C (4172°F). The coupling of inertially driven, highly plasticized particles can achieve coatings approaching that of theoretical density. The results are the densest thermal spray coatings available.

Disadvantages include low deposition rates and in-flight oxidation of particles. Future efforts will focus on applying thick coatings and improvements in process control including in-flight transit time and exposure to atmospheric oxygen. The HVOF process is the preferred technique for spraying wear- and/or corrosion-resistant carbides as well as alloys of Hastelloy, Triballoy, and Inconel alloys. Due to the high kinetic energy and low thermal energy that the HVOF process imports on the spray materials, HVOF coatings are very dense, with less than 1% porosity, have very high bond strengths, fine as-sprayed surface finishes, and low oxide levels.

High-Velocity Plasma Thermal Spraying

With the industry trend toward faster [4], more heat-intensive processes, maintaining part temperature during thermal spray coating applications is becoming a greater challenge than ever before. Traditional cooling methods, such as forced air, are not only inadequate but also inefficient, resulting in wasted powder, process gas, and spray booth time. Recognizing the industry's need for a better solution, Air Products developed an automated thermal spray cooling system that enables users to maintain part temperature within a ± –7°C (±20°F) range using cryogenic, inert nitrogen vapor –196°C (–320°F) while eliminating the inefficiencies. This new technology allows high-quality thermal sprayed coatings to be applied faster and at a lower cost.

High-velocity plasma thermal spray is a heat-sensitive process used for coating parts, such as landing gear, bearing races, valves, and turbine components with wear, corrosion, and heat-shielding materials. As with many thermal spray coating operations, exposing parts to too much heat can negatively impact coating adhesion, part strength, fatigue life, corrosion resistance, and dimensional tolerances. During the plasma coating process molten metal, composite, or ceramic droplets are sprayed onto a part from a gun. The energy required to melt the feed powder, accelerate the molten materials in the gas jet, and deposit these particles onto the target surface results in significant heating of the part. The process is then repeated over and over again to build up the full coating thickness, generating additional heat into the part.

Although cryogenic cooling methods offer significant enhancements in the ability to remove heat quickly, they are rarely used in the thermal spray coating industry due to the risk of nonuniform cooling, which results in uncontrolled levels of residual stresses. Cryogenic nitrogen vapor significantly accelerates cooling over traditional air-cooled processes. This new cooling technology is compatible with existing thermal spray systems and offers a variety of system designs for application-specific use. The system uses a novel cryogenic spray nozzle that installs easily on robotically operated thermal spray guns. During spray application, the cryogenic vapor jet follows the thermal spray plume to maintain the part's temperature within a preset temperature range. The liquid nitrogen is atomized inside the nozzle by another stream of nitrogen gas to form rapidly boiling, microscopic droplets that turn into cryogenic nitrogen vapor within a short distance point. This prevents undesired "wetting" of the coated surface and steep temperature gradients in the coating [4]. The cryogenic cooling system efficiently and uniformly cools parts by monitoring temperatures over the entire part surface and varying the cooling intensity to match the heat generated in the spraying process.

Nontransferred Plasma Arc Spraying

A thermal spray process in which a nontransferred arc is a source of heat that ionizes a gas that melts the coating material and propels it to the workpiece [1,2].

Plasma

Plasma is an ionized gaseous cloud composed of free electrons, positive ions, neutral atoms, and molecules. Because of its unique properties, some have referred to it as the "fourth state of matter." Plasma is generated whenever sufficient energy is imparted to a gas to cause some of it to ionize. If a gas is heated above 5000°C (9032°F) chemical bonds are broken down and its atoms undergo violent random movements. This results in atomic collisions that cause some electrons to become detached from their nuclei. Electrons are the negatively charged constituents of atoms; so having lost an electron, the heavier nuclei, with any remaining electrons, become positively charged. When a gas undergoes this disruption it is said to be ionized and the cloud it has become is identified as plasma. Its behavior involves complex interactions between electromagnetic and mechanical forces. Plasma is present in any electrical discharge even one as in an ordinary arc or in a vacuum tube. It is cold plasma that excites the phosphors within a fluorescent tube.

Plasmas have been known for a considerable time. In commercial technology they are considered as hot streams of particles attaining temperatures greater than 10,000°C (18,032°F). Today's plasma guns are sufficiently robust to produce temperatures from 5000°C (9032°F) to 16,000°C (28,832°F) for long periods. These guns are referred to as "nontransferred arc plasma generators." The generator is essentially an electric arc working in a constricted space. Two electrodes, front (anode) and rear (cathode), are contained in a chamber, as is the arc through which the effluent (the operating gas) passes, a concept developed by H. Gerdien [5] of Germany in the 1920s. With the advent of the space age, scientists realized the need for the plasma process and workable systems were commercially introduced in the 1950s.

Plasma generators work on the concept that if sufficient voltage is applied to two electrodes, separated by a small gap, the difference in potential at that moment causes electrons to be extracted from the cathode. The electrons accelerate and speed toward the anode. If a gas is inserted in the gap between the two electrodes, its atoms will collide with the ensuing electrons and themselves, causing more electrons to detach and travel toward the anode. Meanwhile, the nuclei stripped of their electrons, and positively charged, move to the cathode. Thereby, the gas in the gap has been ionized, becoming electrically conductive—a plasma arc; it exits through an orifice in the anode as a plasma stream, containing only electrons and ionized gas is formed [1,2]. Meanwhile the issuing plasma stream, reaching temperatures exceeding 9000°C (16,232°F), begins to cool and the once ionized gas begins to recombine.

Plasma Spraying

Plasma spraying is generally regarded as the most versatile of all the thermal spray processes. During operation, gases such as argon and hydrogen are passed through a torch. An electric arc dissociates and ionizes the gases. Beyond the nozzle, the atomic components recombine, giving off a tremendous amount of heat. In fact, the plasma core temperatures are typically greater than 10,000°C (50,000°F)—well above the melting temperature of any material. Powder is injected into this flame, melted, and accelerated toward the workpiece.

Plasma spraying was initially developed and remains the preferred process for applying ceramic coatings such as chromia, zirconia, and alumina. However, metals can be readily sprayed with this method. The particle velocities for plasma are higher than for those of flame and arc spraying and result in coatings that are typically denser and have a finer as-sprayed surface roughness. The trade-off is that the maximum coating thickness for a given material is usually reduced.

Plasma Guns

Most commercial plasma guns are fundamentally simple in design, consisting of a chamber and front nozzle (anode) in which there is an orifice. The chamber and nozzle are water cooled. At the rear of the chamber is another electrode that is nonconsumable and is fashioned from thoriated tungsten [1,2]. A port, somewhere within the chamber, allows the high-pressure plasma forming gas, or gases, to enter. A high-frequency spark initiates operation and is discontinued upon ignition. It should be noted that the high-pressure gas cools the outer layer of the plasma arc so extreme heat is kept away from the nozzle bore.

Typical plasma forming gases include argon, nitrogen, hydrogen, and helium. They may be used either alone or in combination: viz, argon–hydrogen, argon–helium, nitrogen–hydrogen, and so on. Argon and nitrogen are generally utilized as primary plasma gases and hydrogen is favored as a secondary as it aids in producing a hotter plasma. Nitrogen is less expensive than argon so, based on economics, is more widely used than argon. Helium tends to expand the plasma and when used in combination with argon produces a "high velocity plasma" that exits the nozzle at about 488 m/sec (1600 ft/sec). Argon/hydrogen and nitrogen/hydrogen exit velocities have been measured at roughly 366 m/sec (1200 ft/sec).

As most plasma guns are designed to spray powders, the powder is introduced through an external port at the nozzle orifice. Hardware is also available for injecting powder internally upstream into the nozzle bore. The primary plasma forming gas is usually used as a carrier to transport the powder to the plasma stream.

Miniature Plasma Spray Guns

The plasma spray application of coating materials that include tungsten carbide has been investigated as an alternative to electroplating of hard chromium at the Naval Research Laboratory, Washington, D.C. Plasma spraying involves fewer process steps than does electroplating, and no hydrogen bakeout is necessary. Prime examples of wear surfaces are the inner walls of cylinders in aircraft hydraulic actuators and dampers. What makes it feasible to consider plasma spraying of carbides as an alternative to chrome plating according to J. Quets, Praxair Surface Technologies [1,2], is the recent commercial development of miniature plasma spray guns. The limitations of the process are primarily the minimum inner diameter and the maximum axial length that can be coated. For a given plasma gun, the minimum coatable inner diameter is defined by the size of the gun plus the standoff (the required distance between the gun and the surface to be coated). The coating materials investigated include a WC/Co mixture, two slightly different WC/Co mixtures incorporating a nickel-based self-fluxing alloy, a Co/Mo/Cr/Si alloy, and a WC/CrĊ/Ni mixture. In tests, the sliding and abrasive-wear performances of the plasma sprayed carbide-based materials were found to be equivalent or superior to those of electroplated hard chromium.

Electric Arc Spraying

A thermal spray process in which an arc is struck between two consumable electrodes of a coating material. Compressed gas is used to atomize and propel the material to the substrate [1,2]. The electric arc spray process utilizes metal in a wire form. This process differs from other thermal spray processes in that there are no external heat sources as in any the combustion gas/flame spray processes. Heating and melting occur when two electrically opposed charged wires, comprising of the spray material, are fed together in such a manner that a controlled arc occurs at their intersection. The molten metal is atomized and propelled onto the prepared workpiece by jets of compressed air or gas.

The gun is relatively simple. Two guides direct the wires to an arcing point. Behind this point a nozzle directs a stream of high-pressure gas or air onto the arcing point where it atomizes the molten metal and carries it to the workpiece [1,2]. Typically, power settings of about 450 A can spray over 59 kg/hr (110 lb/hr). Electric arc spray systems are offered that feed wire by either an air or electrical motor. Some units push the wire to the gun while others pull the wire into the arc. Controls include volt and ampere meters and air regulators.

Electric arc spraying has the advantage of not requiring the use of oxygen and/or a combustible gas; it has demonstrated the ability to process metals at high-spray rates; and is, in many cases, less expensive to operate than either plasma and/or wire flame spraying. Pseudo alloy coatings, or those constructed by simultaneously feeding two different materials, are readily fabricated. An example would be copper–tin coatings constructed by feeding pure copper and tin wires into the arc to produce a heterogeneous mixture of each in the coating. Also, the introduction of cored wires has enabled the deposition of complex alloys (such as MCrAlY) as well as carbide-containing metal alloys that were only attainable using powdered materials as feedstock. Some materials produce self-bonding coatings that are sprayed in a superheated condition. The overheated, hot particles tend to weld to many surfaces thereby increasing the coatings' adhesive strength.

RF Plasma Spraying

A system in which the torch is a water-cooled, high-frequency induction coil surrounding a gas stream. On ignition, a conductive load is produced within the induction coil, which couples to the gas, ionizing it to produce a plasma [1,2]. Inducing electricity to flow through a conductor causes heating to occur. This occurs primarily as a result of resistance to the flow of the induced current and is proportional to the square of the current (I) and directly proportional to resistance (R) and time (t) or (I^2Rt) [1,2].

Induction occurs when a conductor is placed in an alternating magnetic field. When the effect is sufficient, great eddy currents are set up in the conductor, which rapidly gets hot or even melts, the magnetic linkages necessary being increased with the frequency. To be used for thermal spraying, a water cooled helix of several turns is fashioned from oxygen-free high conductivity (OFHC) copper. It is wrapped around a quartz tube that is closed at its top end and fitted with two inlet ports to feed a spray material and a plasma forming gas. Releasing gas into the tube and energizing the copper helix by a high-frequency current that sets up an intense magnetic field inside the tube causing ionization of the gas. Continuous feeding of the gas causes it to escape through the open bottom of the tube. Powder fed into the plasma-filled tube is melted and relying on either gravity or the plasma flow is conveyed to the work surface. Coatings produced using radio frequency (RF) plasma has shown to be generally homogeneous and not porous. This method, using neutral atmospheres, can deposit reactive and toxic metals including calcium, uranium, columbium ,and titanium.

Cold Spray

Cold spray [6], or more precisely cold gas dynamic spray (CGDS), is a high-rate material deposition process in which powder particles (typically 1–50 µm) are accelerated to velocities in the range of 200–1000 m/s (40,000–200,000 ft/min) in a supersonic jet of compressed gas at a temperature far below the melting point of the feedstock powder (ambient temperature to 700°C or 1290°F). Upon impact with a target surface, the solid particles experience plastic deformation that disrupts thin surface films (such as metal oxides) and provides intimate conformal contact between the clean metal surfaces under high local pressure. This permits bonding to occur and rapid layer build-up of deposited material.

In addition, the cold spray being a new solid-state spraying process, is capable of providing protective deposits, surface modification, restoration, near-net shapes, and other applications without the undesirable effects of process temperature or metallurgical incompatibility among materials. Similar to conventional thermal-spray processes, the cold spray produces coatings or freestanding deposits for a large number of applications in a wide range of industries. However, unlike conventional thermal spray processes, cold spray technology can deposit metallic and nonmetallic materials onto diverse surfaces at much lower temperatures, virtually avoiding any thermal effects.

Cold spray is generally understood as part of the large family of thermal spray, which encompasses a number of coating processes used to apply metallic and nonmetallic coatings to a wide variety of substrates. The number of applications for thermal spray is countless and includes corrosion protection, thermal barriers, oxidation protection, wear resistance, surface lubricity, material restoration, and aesthetics coatings. Applications for thermal spray are found in almost any industry, including aerospace, automotive, power generation, biomedical, heavy equipment, nuclear, mining, chemical, and electronics. The aircraft industry is perhaps one of the largest beneficiaries of thermal spray technology, in applications such as plasma spray deposition of yttria-stabilized zirconia (YSZ) on engine components for thermal protection. Cold spray deposition is based on experimental results that revealed the possibility of producing dense coatings by accelerating micron-size feedstock powder against a substrate at velocities beyond a material-dependent critical density.

Process Principles

To achieve the required particle-impact velocity, feedstock powder particles are injected into a supersonic gas flow that ensures the proper momentum transfer to the particles. Due to their aerodynamic properties, inert gases such as helium or nitrogen are preferred as the propellant gases but dry air (79% N_2–21% O_2) can also be used as an alternative propellant. There are currently two commercially available variants of the cold spray technique: high pressure and low pressure.

In high-pressure systems, a new commercial system is used (CGT GmbH Kinetiks 4000, 47 kW) where the propellant gas (helium or nitrogen) is injected at a high pressure of 2.04–3.23 MPa (300–600 psi) and can achieve process gas temperatures up to 800°C (1470°F) and process gas pressures up to 40 bars. The feedstock powder can be heated to temperatures over 500°C (930°F), resulting in improved adhesion for some specific coating systems. The Active Jet Cold Spray Gun can be operated with or without preheating the gas. This enables the system to be configured for the application of tailored coatings. The powder feedstock is injected at high pressure, mixed with the propellant gas and axially fed into a convergent-diverging DeLaval nozzle.

In low-pressure systems, the propellant gas (dry air or nitrogen) is injected at a low pressure of 0.54–0.70 MPa (80–100 psi) and preheated up to 400°C (752°F), again to optimize aerodynamic flow conditions. The powder feedstock (which also includes mixtures of metal and ceramic particles) is introduced downstream into the diverging section of the DeLaval nozzle, thus eliminating the need for high-pressure delivery systems, which increases portability and dramatically reduces costs. At the nozzle, the particles are accelerated to supersonic velocities in the range of 300–500 m/s (60,000–120,000 ft/min).

Properties and Applications

The cold spray's lower thermal load results in lower porosity and oxygen content. The coated surface has deposition rates that can exceed 95%, and is durable with respect to surfaces coated using other technologies. Having fewer oxides implies that the surface will have electrical and thermal conductivity. Several materials have proven to be suitable for cold spray including:

- Metals (Al, Cu, Ni, Ti, Ag, Zn, Ta, Cb)
- Refractory metals (Zr, W, Ta)
- Alloys (steels, Ni alloys, MCrAlY)
- Composites (Cu–W, Al–SiC, Al–Al_2O_3)

Its ability to preserve the original microstructure of feedstock materials makes cold spray ideal for depositing temperature-sensitive materials such as nanostructured and amorphous materials, oxygen-sensitive materials, copper, titanium, and phase-sensitive materials. Since it works similarly to microshot peening, the cold spray also induces compressive residual stresses on the deposit; the production of thick 5–50 mm (0.19–1.9 in) coatings or self-standing shapes that display acceptable adhesion and precision are also possible. The high-energy/low-temperature deposition results in a wrought-like microstructure with near-theoretical density values. Cold spray is suitable for a wide number of applications, as illustrated in Table 14.1 [6].

Localized plastic deformation at the particle/substrate interface appears to be necessary for the kinetic energy transformation. For this reason, powders for cold spray are mostly metals with relatively high ductility (such as Al, Cu, Zn, Ag, Ni, and their alloys), as well as mixtures of these metals with ceramic particles. Ductile metals and metal/ceramic mixtures can be successfully applied to any metallic, glassy, or ceramic substrate. One huge potential for the cold spray process promises to address many of the shortcomings associated with classical methods for corrosion protection. Although corrosion rates of modern high-grade magnesium alloys are quite acceptable for interior applications, automotive exterior environments are extremely harsh for bare and even coated magnesium parts. The cold-spray application of an aluminum coating may provide enough protection to enable the exterior application of magnesium alloy parts.

TABLE 14.1

Cold-Spray Applications

Advantages of Robocasting
• No molds or binders are required
• Densified parts can be made in under 24 hours
• Complex shapes of variable thicknesses can be made
• Polymerization reactions are not required
• Parts are strong, dense, and defect-free
• Thick parts can be made unobtainable by slip casting
• Unique structures (tailored porosity) are possible
• Potential exists for multimaterial (composite) structures

Source: Villafuerte, J., *Canad. Weld. Asso. J.-IIW* Special Edition, pp. 42–3, 46, 2007. With permission.

Glasses

Glasses are amorphous materials with disordered atomic or molecular structures that result when any material (ceramic, metallic, or polymeric) is cooled from its molten state at a rate higher than what the material would require to transform into a more thermodynamically stable crystalline structure. The term glass is commonly associated with the glass derived from silica (SiO_2), limestone ($Ca.CO_3/Ca.Mg.CO_3$) [7].

Heated Glass

When the temperature differentials between the inside and outside surfaces of a glass window are extreme (e.g., during cold winters, hot summers, and in refrigeration systems) one side of the panel might be cold enough to promote water condensation (fogging) or even solidification (icing), if the exposed surface gets below the dew point of the surroundings. For many years, anti-fog and anti-icing glass have been made by creating, on the glass surface, a pattern of fine lines of an opaque, electrically conductive material. A screen printing of silver-based frit has been a common method to create patterns, as well as the interconnected bus bars. This method has been popular for anti-fog refrigerator doors and rear windows in automobiles. However, screen printing is cumbersome, often requiring surface preparation and postheat treatments at elevated temperatures. Additionally, soldering of metal lead-outs often requires stringent techniques to avoid damaging the printed pattern, and not to mention that screen-printed lines are visible and interfere with vision.

Thin-film deposition represents a new technique for making heated glass without compromising transparency and aesthetics. Thin-film heated glass is very reliable, as the generated heat is rather evenly diffused over a large surface rather than concentrated in a small area. The uniform radiation, especially at moderate temperatures, is one of this technology's strengths, often requiring less overall energy for equivalent performance.

Photovoltaic Glass

Photovoltaic glass windows incorporate a semitransparent photovoltaic film on the exterior glass of traditional double-panel glass units. The electric wires extending from the sides of each glass unit are connected to wires from other windows, building up the entire system. The technology is already available in Europe but it is still carving its way into the U.S. market through pilot programs by the U.S. Department of Energy.

Robotic low pressure cold spray has already been successfully used to create bus bars on heated glass using a proprietary blend of nonferrous materials. Cold sprayed bus bars proved to exceed the bond strength and electrical resistance requirements set forth by the industry. Since cold spray does not require any special pre- and/or posttreatment, cost savings associated with this process have sparked interest from glass manufacturers. Some of the advantages of cold spray over traditional screen printing processes are as follows:

- Minimum setup
- Low-cost materials
- No pre-or posttreatment
- No masking
- High flexibility and easy changeover
- Consistent bonding properties
- No residual stresses
- Low porosity
- Low-oxygen content
- Any glass surface profile
- No special procedure required for lead-free soldering of lead outs

Plasma Vapor Deposition (PVD)

Plasma vapor deposition (PVD) [8] is a coating process that first atomizes (vaporizes) a material from a solid source into a gas or vapor phase and then deposits it on to a substrate where it condenses. The coating doesn't penetrate the surface. Instead, it forms a strong metallic bond to the component surface. All PVD processes take place in a high vacuum and are plasma supported. The process begins during production with incoming inspection and ultrasonic cleaning of all components. Depending on the surface, several pretreatments can enhance the surface and make it tribologically fit. The coating process starts with evacuation of the coating chamber to high (4×10^{-5} Pa) vacuum necessary for cleanliness and process requirements.

Impurities and oxides are removed during and after the heating step and prior to the coating process. They are removed through intense argon-ion bombardment of the cleaned component surface. A strong bond between the substrate and the coating requires a metallically clean surface. Subsequent evaporation of coating material can be done by introducing thermal energy (electron beam, electric arc) or with atomic impact processes (sputtering). Coating by means of electron beam evaporation and sputtering results in coatings with fine surface topographies. Hence these techniques are candidates for coating more demanding (polished or structural) surfaces.

Arc coatings may require mechanical polishing before use. Tungsten carbide/carbon (WC/C) uses sputtering to vaporize tungsten. And diamond-like carbon coatings (DLC) are produced with plasma-activated chemical-vapor-deposition (PACVD) technology after applying a thin PVD adhesion layer. Reactive gases can be supplied to create TiN, for example.

Titanium (Ti) vaporizes and reacts with supplied nitrogen (N) to form titanium nitride (TiN). All described methods are line-of-sight processes; that is, the local coating thickness distribution depends on the position of the part in the coating chamber. Coating thicknesses of 3 μm (1/10 in.) are typical. PVD coatings have a fine structure and are under compressive stress, but are not necessarily brittle despite their high hardness of between 1500 and 3500 $HV_{0.05}$. The maximum coating temperature for carbon-based coatings is 249°C (480°F). All other coatings are applied at a maximum of 510°C (950°F). Low-temperature coatings are also available. And some newer coatings can withstand operation temperatures up to 1093°C (2000°F) and have a multi or nanolayered structure.

Coating Types

More than 15 PVD and PACVD coating types are currently available. The number of coatings specifically designed for component applications rose in recent years to meet the demands of different tribological situations. The PVD/PACVD coatings are generally separated into two groups: carbon based and noncarbon based. Hard coatings with carbon typically have lower coefficients of friction (CoF) and are applied at relatively low temperatures under 249°C (480°F). These coatings are effective in avoiding adhesive wear. Nitride coatings generally have a higher hardness, are applied at temperatures of 510°C (950°F) at most, and feature better wear resistance.

Tests comparing WC/C to a commercially available thin dense chrome (Cr) coating showed the different behavior of these two systems. WC/C has a stable 0.1 to 0.2 CoF, whereas thin, dense Cr coatings behave like a typical uncoated steel surface (0.7 CoF). Therefore, adhesive wear or galling was unavoidable. Coating thickness can be about the same for both technologies, but WC/C is generally harder, keeps counterbody wear to a minimum, and doesn't allow buildup on part surfaces.

A recent carbon-based coating system called Star is a combination of CrN (chromium nitride) and a carbon-based coating. It improves performance for high-impact applications and will work on softer substrates. With Star, hardness rises gradually from the base material through the CrN layer and on through the hard carbon-based coating at the surface. The CrN base works as an emergency layer. It supports the hardest top layer and better distributes the load into the substrate. As a result the coating depends less on the hardness of the substrate. And it maintains good load bearing capacity and avoids the "egg shell effect" or crazing.

PVD Coatings

Typical PVD hard coatings are 20 times thinner than a human hair, yet they can drastically improve performance, boost reliability, and extend service lives of tool and machine components. In addition to conventional PVD hard coatings such as TiN (titanium nitride), so called "tribological coatings" with optimized frictional properties have been developed in recent years. These coatings protect highly stressed components that see sliding or rolling contact. They handle a range of applications in motorsport, fluid power, medical, and aerospace including engine, transmission, pump, motor, and bearing components.

The bulk material provides different properties than the PVD-coated surface. The primary role of the tribological coating is to reduce friction so components perform better and last longer. It's the combination of an appropriate substrate material, surface topography, and PVD coating that spells the difference between design success and failure, see Figure 14.3 [8]. For example, various WC/C and DLC coated valve train and transmission components used in racing motorcycles reportedly improve power output and torque throughout the entire rpm range. Measurements using a dynamometer, or dyno for short, have shown that ceramic-like DLC and Balinit C (WC/C) coatings boost engine power and torque by 2 and 3%, respectively, while protecting components from wear. A racing motorcycle with Balinit-C-coated gears was in a racing competition and was able to finish the competition though there was a total oil loss during the race—an impossible feat without coatings.

Less wear improves component life and ensures more consistent performance throughout a race or season. Moreover, these thin hard coatings can be applied to off-the-shelf components with no need to redesign the parts with which they mate. Less friction, lower operating temperatures, and more

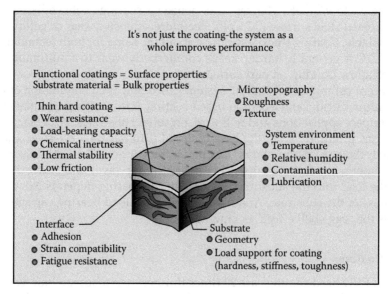

FIGURE 14.3
It's not just the coating—the system as a whole improves performance.

consistent power can be expected. Hard coatings also minimize adhesive wear between mating steel parts. Adhesive wear, also known as scoring, galling, or worse—case seizing, results when two solid surfaces slide over another under pressure. Surface projections, or asperities, plastically deform and eventually weld together under the high-localized pressure. As sliding continues, these bonds break. This creates cavities on one surface and projections on the other. Tiny abrasive particles can also form causing additional wear.

Nickel Vapor Deposition (NVD) Process

Weber Manufacturing Technologies Inc., of Midland, Ontario, Canada, is using an nickel vapor deposition (NVD) process to apply uniform coatings of nickel to powders. Nanoparticles as small as tens of microns can be coated, and the coating thickness can be controlled over a broad range (10–80% by weight, for example). Nickel-coated powders are used in electromagnetic interference shielding for electronics, in nickel-coated abrasives and cutting tools, and in thermal arc spray coatings. NVD turns nickel powder and carbon monoxide into nickel carbonyl gas. In the powder coater, the gas converts back to nickel on contact with the substrate material at temperatures ranging from 130 to 170°C (266–338°F). The carbon monoxide is returned to the system for reuse. Many different powders can be easily coated and in large quantities. Tests have been conducted by the company where aluminum and graphite powders have been test coated. An R&D coater can coat 30 kg of substrate with 6 kg of nickel in 3 hours. A full-scale production unit will be able to coat about 150 kg of substrate in the same time. In addition to achieving uniform coatings, the NVD process is about 20 times faster than electroplating. Also, there is no need for drying afterward, the powder is ready for use immediately after the coating.

Sputtering

Sputtering is a PVD process in which a material is transported from a source (target) to a substrate by means of the bombardment of the target by gas ions that have been accelerated by a high voltage. Atoms from the target are ejected by momentum transfer between the incident ions and the target. These ejected particles move across the vacuum chamber to be deposited on the substrate. In its simplest form, 0.13–13 Pa (1×10^{-3} to 100×10^{-3} torr). In most cases, the gas is argon because it has a higher mass than neon or helium; higher mass yields higher momentum, and more ejected particles.

The sputtering process begins when an electric discharge is produced and the argon becomes ionized. The low-pressure electric discharge is known as glow discharge, and the ionized gas is termed plasma. The argon ions hit the solid target, which may be in the form of a sheet or a tube. The target is the source of the coating material (and not to be confused with the substrate, which is the item to be coated). The target is negatively biased and therefore attracts the positively charged argon ions, which are accelerated in the glow discharge. This attraction of the ions to the target (also known as bombardment) causes the target to sputter, which means that material is dislodged from the target surface because of momentum energy exchange. The higher the energy of the bombarding ions, the higher the rate of material dislodgement (see Figure 14.4) [9–11]. Materials that can be coated by sputtering include pure metals, alloys, inorganic compounds, and some polymeric materials. A major restriction for the substrate material is the temperature of the process, which can range from 260 to 540°C (500–1000°F). Sputtering is often used for depositing compounds and materials that are difficult to coat by thermal evaporation techniques.

Chemical Vapor Deposition (CVD)

Chemical vapor deposition (CVD) is the method most often used to deposit DLCs. Adjusting deposition conditions allows the processor to change the coating from graphite to diamond-like. One process used can deposit the DLC in a gas atmosphere at reduced pressure without a fixed target. This plasma-assisted CVD allows large workpieces to be coated on all sides without

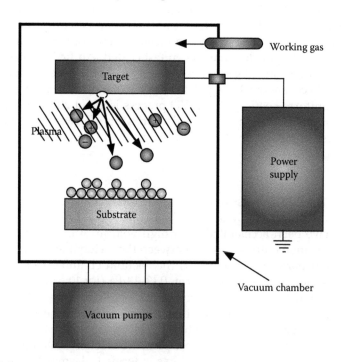

FIGURE 14.4
Physical vapor deposition. (From Doering, T., Tough Coatings, Oerlikon Balzers, Amherst, NY, *Mach. Des.*, 90–8, December 14, 2006. With permission.)

turning. However, substrates must be heated to roughly 800°C (1472°F) when using CVD. Reduced substrate temperatures are offered by dual ion-beam-enhanced deposition. Substrate temperature reaches only 66°C (151°F), and the dual ion-beam process does not rely on epitaxial growth for its formation as CVD does. Epitaxial growth requires a crystalline substrate; because dual ion-beam processing is free of this need, it enables amorphous materials to be coated as well.

Ion plating is one of the two methods used to deposit DLCs. A relative newcomer to the coating field, DLCs are commonly made from hydrocarbon (often methane) and H_2 gases heated to 2000°C (3632°F). The carbon coatings are prized for their wear resistance, as well as electrical and optical properties. DLCs represent a huge potential in their stage of commercialization. However, their wide range of properties, along with their relatively low cost, leads many to predict huge growth in DLCs. The coatings are being used to improve wear resistance in tool bits, as electronic heat sinks, and to boost wear and corrosion resistance in optical materials.

The use of a CVD diamond coating has proven to save millions in a process to fabricate dimension critical aerostructures more efficiently, more accurately, and faster. As part of its fabrication process, Lockheed-Martin Aerospace Company (LMAC) performs postmold machining of carbon fiber-reinforced wingskins to net edge shape and size. The cutting tool selected at the start of the project produced excessive delamination, thereby reducing overall quality of the wingskin.

After an exhaustive study, a resulting tool that relies heavily on DiaTiger, a CVD multilayered diamond coating produced by Diamond Tool Coating and designed specifically for machining composites, including fiberglass reinforced plastics, graphite, carbon fiber composites, and ceramics. The new design and new material increased tool life from 9 linear feet at 1/3 material thickness to 17.37 linear meters (57 linear feet) at full material thickness. Test coupons ultrasonically inspected by LMAC verified the integrity of the parts. LMAC now can machine a complete wingskin using only two cutting tools—one to rough and one to finish—instead of the 24 cutting tools used previously [11]. Cost savings per aircraft is ~$80,000. If LMAC were to manufacture 2783 F-35s as currently planned, total cost savings over the life of the project would be ~$222.6 million—not including conservation realized via scrap reduction and time savings.

A carbon nanotube fabrication tool has been designed for both CVD and plasma-enhanced CVD (PECVD) processing capability called NanoGrowth 1000n, the tool features an ultrahigh purity gas delivery system and flexible closed loop control systems that allow users to define target tolerances and achieve a high level of repeatability during all phases of the process. A very high degree of hardware modularity allows the tool to expand easily and configured to meet current and future processing techniques such as inductively coupled plasma (ICP) and dual sputter sources for catalyst deposition. The tool is controlled by unique, touch-screen SCADA-style software with supervisory control and data acquisition; this has been developed and refined over more than 7 years on high-end, thin-film deposition tools. This software provides an extremely user-friendly interface that sits between the user and the tool, making complex growth or deposition processes, easy to create and run.

Magnetron-Sputtering

In magnetron sputtering [10], the target is tubular, and the substrate to be coated is also tubular and located close to the target. Most of the plasma is confined to the near-target region. This plasma confinement is achieved by establishing strong magnetic fields above the target surface that reshape the trajectories of the secondary electrons ejected from the target surface into convoluted spiral-like patterns skipping across the surface of the cathode. In this magnetic arrangement, the secondary electrons are trapped and most of their energy is expended in the near-target region. This increases ionization and greatly improves the sputtering and deposition rates. This method is quite successful in producing high-quality, low-impurity films at reasonable deposition rates.

Magnetron sputtering is a vacuum coating process for depositing thin films on glass. Since their invention in the late 1960s, sputtering electrodes have undergone a developmental revolution. The most significant technological advances are rotating cylindrical magnetrons and advanced rotating cylindrical sputter targets. These two parallel developments have enabled manufacturers to boost coating throughput and reduce cost, while maintaining layer quality and thickness consistency. A metallic DLC coating deposited on cutting tools at ~200°C (392°F) by a magnetron sputtering process optimizes tribological systems, according to Metplas Ionon, Germany. As a multilayer structure, the coating, W-C:H_{mod}, efficiently counteracts tribological influences such as friction and adhesion. At the same time, it has a high degree of elasticity compared with other metallic DLC coatings.

Typical coatings are thin, with a thickness of only 2 to 4 µm; therefore, drawing changes will only be required in exceptional cases. Also, the roughness on smooth surfaces is hardly changed. Through the substantially inert surface chemistry of W-C:H_{mod} coatings, cold-welding effects can be avoided. This is especially well proven in the processing of zinc-plated panels with reforming tools, punching dies, and trimming knives. When cutting aluminum, smearing and formation of micron sized chips are significantly reduced when the punching tools are coated with W-C:H_{mod} coatings.

Through the inclusion of graphite structures within the coating, low friction is combined with excellent emergency running properties if lubrication is lost. The many shortcomings of planar magnetron sputtering techniques can be overcome by the adoption and implementation of rotating cylindrical technology. There are three significant advantages to adopting the rotating cylindrical magnetron sputtering method, they include: superior material inventory, a higher degree of utilization, and the possibility to triple the power density, resulting in much faster sputter rates or in more complex stacks.

Rotatable Sputter Targets

As the market interest in vacuum coating by magnetron sputtering grows, target manufacturing is consequently expanding. Thermal spray is the preferred technology to manufacture sputtering targets, because it offers a broad range of capabilities to meet these very complex manufacturing demands. Four parameters directly impact total cost of ownership:

- Material composition: Doped materials can be produced in both stoichiometric and nonstoichiometric compositions without the limits of phase diagrams, allowing operators to develop specific coatings that cannot be made via classic target casting technologies. Thermal spraying does not need to take possible restrictions of limited solubility into account with thermal spraying. Any mixture of two materials can be processed by simply mixing the appropriate fractions together before spraying.
- Expanded coverage: Nearly all materials can be sprayed, from low-melting point metals to high-melting point ceramics.
- Target flexibility: Long life (dog bone shaped) targets increase thickness of the material at both ends. As a result, high target material utilization is possible with most materials and for different target lengths [up to 386 cm (152 in.)], and are easily produced.
- Film composition: Typical thin films and coating stacks, such as SnO_2, TiO_2, SiO_2, and Si_3N_4, can be made via advanced cylindrical target tubes.

Silicon Aluminum Targets

Thin films of SiO_2 and Si_3N_4 are sputtered from Si(Al) targets. The successful production of Si(Al) targets by thermal spray takes advantage of key spray process features. Its inherent flexibility for target geometry allows a wide range of target diameter, length, and straight or dog bone target ends, while maximizing target sputter capacity by increasing the target layer thickness up to 9 mm (0.35 in.).

High Density Tin

Standard thermal sprayed tin targets have 90% of the required theoretical density, with an estimated oxygen content of 2000 ppm. However, advances in thermal spray technology have resulted in a new high density tin target, reaching more than 98% of the required theoretical density, combined with oxygen content below 250 ppm.

Titanium Oxide

A perfect illustration of how thermal spraying results in a value-added target product is the production of TiO_x targets. First, the high process temperatures allow the ceramic titanium oxide to melt. Simultaneously, the titanium oxide undergoes partial reduction with the process gases, transforming it into an electrically conductive phase. At high-cooling rates, it remains conductive at room temperature. This material greatly enhances stability during reactive processes, without requiring a feedback loop process control system, yet it still improves sputter deposition speed.

Indium Tin Oxide

Indium tin oxide is one of the top performing transparent conductive oxides available to the display market. Applications include flat panel displays, such as LCD, Panel Display Plasma (PDP), and Organic Light Emitting Diode (OLED), in which the indium tin oxide layer serves as a transparent electrode. Planar ceramic targets consist of one or more tiles bonded to a metallic backing plate. Reactive DC magnetron sputter deposition from a planar ceramic target is the most widely deployed technique for deposition of indium tin oxide (ITO) coatings on glass and plastic substrates. In spite of their popularity, planar targets have several intrinsic restrictions because of their planar structure. Rotating cylindrical ITO targets resolve many of the limitations of planar ceramic ITO targets. Some of its inherent advantages include:

- Larger useful target inventory and increased target material utilization, both of which lead to reduced machine downtime.
- Increased process stability for reactive deposition.
- Improved target cooling, which increase power density and raises the deposition rate.
- Preliminary field tests have shown that total cost of ownership can be reduced by more than 40% per square meter while doubling the utilization of targets.

Three Magnetic Effects

Control of the magnetic field is important because it effects three key characteristics of the sputtered coating; thickness uniformity, high-deposition rate, and maximized target utilization. The magnetic field at the target surface must be carefully controlled in order to combine the highest possible deposition rate, optimized thickness uniformity, and maximized target material consumption. To answer these three magnetic technical challenges, Bekaert [11] has introduced a new Adjustable Magnet Bar designed for large area rotating cylindrical magnetron sputtering. Incorporating recent magnetic and mechanical enhancements, the complete magnet bar is essentially a combination of magnets on a pole piece attached to a water conduction tube. The bar is robust, easy to handle, and qualified for an off-line measurement setup. The new design is suitable for horizontal as well as vertical applications, and offers enhanced and more flexible tuning of the magnetic field while optimizing target utilization. The magnet bar is critical for high-performance glass coaters operating at high-coating levels. Finally, magnetron sputtering is the most economical and results-driven process available today because of remarkable R&D advancements in technology, process, and engineering.

Vacuum Plasma Spraying (VPS)

Vacuum plasma spraying (VPS) [12] is a relatively new technique in which heat-softened or molten particles in sizes from 10 to 100 µm are accelerated toward a substrate, where they flatten and solidify. VPS coatings generally exhibit higher densities than possible with other thermal spray processes. Control of the chamber atmosphere during the process facilitates a reduction in the amount of impurities and contaminants, and results in higher-quality deposits. Because of the high-energy density of the thermal plasma, materials with very high-melting points, such as refractory metals and ceramics, can be deposited by this technique. Almost any material available in powder form with a stable liquid phase can be sprayed via this process (Figure 14.5) [12].

Recently, a variation of the technology known as vacuum spray forming (VPSF) has emerged as one of the leading technologies for manufacturing near net shape components. The molten particles hit the target surface and

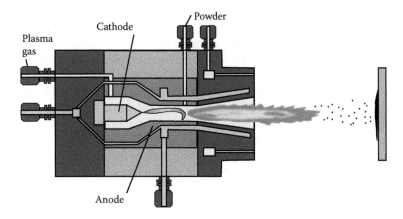

FIGURE 14.5
A schematic of the VPS process showing the plasma flame, powder injection, and splat formation. (From Azarmi, F., *AM&P*, 37–40, August 2005. With permission.)

rapidly cool to form a stable structure. After the deposit solidifies, the mandrel can be removed and the process is complete [12]. VPSF is also capable of producing multilayered, functionally graded structures, and composite materials by spraying different materials with separate nozzles.

Process Fundamentals

A conventional VPS coating system involves three different processes:

- Formation of plasma flame
- Powder injection
- Splat formation

The plasma consists of positive ions, free electrons, and neutral atoms. It can be produced by providing enough energy to a gas to cause localized ionization. Powders are injected into high temperature (~10,000 K) plasmas created by striking an electric arc between a nozzle-shaped copper anode and a tungsten cathode. In the VPS process, the entire operation is done in a vacuum of ~10 kPa (1.4 psi).

Argon is the most common gas for generating plasmas, and is the main component of many plasma gas mixtures. Thus, argon can be considered a primary plasma gas. Other gases such as nitrogen, hydrogen, and helium can be added to an argon plasma in smaller proportions to change the physical properties of the flame. In such a case, these gases are considered secondary gases. The plasma gas flows between the cathode and the anode. A very high-temperature plasma stream can be achieved by increasing the arc current, which results in a thicker arc and more ionization. A plasma stream created by argon at a very large arc current usually produces a temperature high enough to melt most materials. For spraying ceramics and refractory materials with very high-melting points—and where the arc current can not be increased further—a secondary gas (such as hydrogen) may be required to raise thermal energy of the plasma stream.

Powders are fed into the plasma flame where they are rapidly heated, accelerated, and directed toward a work surface. The distance between the plasma gun and substrate can be on the order of 100–400 mm (3.9–15.6 in.). The molten particles land, flatten, and solidify on the substrate or mandrel surface in the form of individual splats. In an ideal plasma spray process, all of the injected particles would reach the substrate while still in the molten state. Over time, the continuous deposition of the molten droplets forms a layer of coating consisting of millions of splats on the surface of the substrate.

Coating Microstructure

The microstructures resulting from the VPS process may differ from those resulting from conventional processing routes. VPS-fabricated coatings generally exhibit a layered type of structure, in which the lamellae of the deposit are almost parallel to the substrate surface. The coating is built up from splats resulting from the impact, flattening, and solidification of melted powder particles on the substrate. The splats form mutually interlocked lamellae that consist of columnar grains. The flattened droplets solidify in the direction of the temperature gradient after impact at a very high cooling rate ($\sim 10^4$ to $10^8 °C/sec$). Thus the thermal sprayed structure consists of billions of individual splats connected to each other by mechanical and chemical bonding. These deposits typically exhibit more uniform chemical composition with smaller grain size.

VPS coatings generally exhibit lower porosity than air plasma spraying and high velocity oxygen fuel (HVOF) coated samples. It could be the result of the presence of lower amounts of unmelted particles due to the high-temperature plasma flame. VPS-fabricated materials also can be considered to be oxide free, because of the vacuum condition during the spraying process. These two important factors are the main reasons for lower porosity within the microstructure. VPS is superior to air plasma spraying for materials sensitive to oxygen or nitrogen contamination, such as titanium alloys, and/or where improved adhesion and density are required.

Process Drawbacks

Every coating process has certain limitations, and VPS is not an exception. Some of the disadvantages are:

- The microstructure of deposit shows anisotropic behavior with respect to the substrate surface. It is caused by the presence of unmelted and partially melted particles, porocity, microcracks, and splat boundaries.
- A number of phase transformations may take place during plasma spraying, and the coating properties may be different from those of the initial powder because of the formation of metastable phases.
- The VPS process also produces coatings with larger residual stresses than those resulting from conventional methods.
- Substrate dimensions are dictated by the chamber size.
- It is still a relatively high cost and complex process.

Luckily, some of these disadvantages can be remedied by applying a successful postfabrication heat treatment under vacuum. Heat treatment may help to develop isotropic properties and reduce residual stresses. The preheating of the substrate inside the chamber is another technique to reduce residual stresses. In addition, thermomechanical treatments can modify final phase composition, grain size, and phase distribution.

VPS Applications

The applications for plasma spray coatings are virtually limitless due to the need for fabrication of high purity coatings with rapid deposition rates. At the present time, more than 300 different metal, ceramic, and alloy coatings are available. VPS coatings enhance the base material's resistance to corrosion, erosion, cavitation, friction, and abrasion. Lubrication can be provided by incorporating materials such as graphite into the coating. Oxide coatings produced by VPS in medical implants have shown outstanding properties. Recently, nanostructured coatings have been produced, resulting from the short dwell time at high temperature, which decelerates grain growth. In addition, the VPS process has successfully fabricated near-net shape and near-full density components for aerospace applications. Some components require excellent mechanical properties such as high-tensile strength, while at the same time they must resist wear and corrosion. It is almost impossible to find a single material that satisfies all these requirements; frequently, the best solution is to build up a component from multiple layers of different materials. As mentioned previously, one of the advantages of VSF is its capability to produce such a structure. This type of structure combines the strength and ductility of metals with the high resistance to corrosion and friction of ceramics.

Melt Infiltration (MI)

Composites of zirconium carbide matrix material reinforced with carbon fibers can be fabricated relatively rapidly in a process that includes a melt infiltration (MI) [13] step. Heretofore, these and other ceramic matrix composites (CMCs) have been made in a chemical vapor infiltration (CVI) process that takes months. The finished products of the CVI process are highly porous and cannot withstand temperatures above ~1600°C (3000°F). In contrast, the MI-based process takes only a few days, and the composite products are more nearly fully dense and have withstood temperatures as high as ~2400°C (4350°F) in a highly oxidizing thrust chamber environment. Moreover, because the MI-based process takes much less time, the finished products are expected to cost much less.

Fabrication begins with the preparation of a carbon fiber preform that, typically, is of the size and shape of a part to be fabricated. By the use of low-temperature, ultraviolet-enhanced CVD, the carbon fibers in the preform are coated with one or more interfacial materials, which could include oxides. The interfacial material helps to protect the fibers against chemical attack during the remainder of the fabrication process and against oxidation during subsequent use; it also enables slippage between the fibers and the matrix material, thereby helping to deflect cracks and distribute loads. Once the fibers have been coated with the interfacial material, the fiber preform is further infiltrated with a controlled amount of additional carbon, which serves as a reactant for the formation of the carbide matrix material.

The next step is MI. The preform is exposed to molten zirconium, which wicks into the preform, drawn by capillary action. The molten metal fills most of the interstices of the preform and reacts with the added carbon to form the zirconium carbide matrix material. The zirconium does not react with the underlying fibers because they are protected by the interfacial materials. The success of the MI step depends on the interface material selection and uniform coating of the fibers, infiltration with the correct amount of carbon, and careful control of temperature and rate of heating.

Applications of PVD Coatings

Piston Rings

Piston rings also perform better with engineered coatings [14–19]. Engine maker Scania AB, in Sweden, tested different piston-ring coatings against gray cast iron liners. The 6 hour test ran at a frequency of 10 Hz and put parts under an 8 MPA (1.2 ksi) load at 79°C (175°F). The chrome ceramic (CKS) and Balinit C coatings showed no significant wear on the piston ring OD surfaces. But the wear on the liner was significantly different. The hard and porous CKS surface doesn't run in as well as the Balinit C with its low CoF. This boosted the wear on both the CKS liner surface and the piston ring. The piston train is the biggest contributor of all engine systems to frictional loss with about 30%. Low-friction coatings can improve efficiency, longevity, and increase power output.

Another test done by a Japanese piston ring company compared galvanic-chrome coating, nitriding, PVD-CrN, and PVD-WC/C coated piston rings against a 12% silicon–aluminum alloy piston with oil as a lubricant. Of the coatings tested, the wear rate of WC/C was the lowest. This mainly comes from the coating's low CoF and ability to thwart adhesive wear. The choice of coating depends on the liner material and other aspects of the tribological system.

Bearings, Gears, and Transmissions

Unlike plain-journal bearings, wave-journal bearings have a wave profile circumscribed on the inner-bearing diameter. Appropriate materials ensure that bearing geometry doesn't change over time. And Balinit C and DLC coatings improve bearing performance under start/stop and oil out conditions. Design trends for transmissions include lightweight, more efficient constructions that can carry heavier loads, consume less lubricant, use fewer additives, and have longer maintenance intervals. These requirements tend to increase wear. Alternative materials alone won't always meet these challenges.

Surface fatigue or pitting takes place on tooth surfaces when the maximum number of load cycles or loading capacity is exceeded. The carbon-based PVD coating Balinit C (WC/C) is one candidate that has proven effective against all types of wear mechanisms over the years. The coating reduces local surface pressures (Hertzian pressure) and improves reliability of poorly lubricated gears by separating the two metallic gear surfaces with a hard ceramic-like layer.

A test of a helicopter transmission under emergency conditions (loss of transmission fluid) revealed the capabilities of WC/C. Two identical transmissions were tested in a rig setup, one with coated gears and the other without. The transmission with uncoated gears failed after about 1 hour. This was shown by its rise in temperature. The time before failure according to the test profile was not enough to ground the helicopter safely. The transmission with coated gears ran for more than 6 hours, three landing cycles could be simulated and the transmission was still in working condition. The WC/C coating reduced friction and avoided scuffing.

Hydraulic Pumps

Producers of hydraulic drives are confronted with lubrication and corrosion issues. The trend toward lighter weights and higher pressures and speeds means that hydraulic components must handle more severe tribological stresses. With coatings, it is possible to reduce abrasion, allow dry operation for pneumatic valves, and handle most acids and alkaloids safely. Uncoated vanes in an accelerated vane pump test (with additive-free hydraulic oil) reportedly scuff after a few minutes. Similar test results could be obtained with and CFC-free refrigerants as a medium in pumps, valves, and motors. CrN coatings are a frequent choice for these applications. Balinit CNI, for example, applies using a sputtering process with a maximum temperature of 249°C (480°F) so a broad range of materials can be coated. It provides a smooth, hard surface with a 50% lower CoF than other thin, dense galvanic Cr coatings.

Consideration of Processes Effected by Coatings

Surface Preparation

Bonds between surfaces and coatings depend on surface conditions. Surfaces must be metallically bright, ground, honed, polished, or lapped. Grinding cracks, burrs, oxide skins, and rehardening burns on surfaces must be avoided. Blind holes and inside contours must be contamination free. Plugs and screws should be removed before cleaning and coating. The white layer of EDM (electrodischarge machining) surfaces must be reduced by running several finish cuts and by microblasting before coating. For best results surface roughness should be less than the thickness of the coating. Typical surface roughness values for components are 0.4 μm (16 μin.) or better. Only oil-free polishing agents should be used before coating.

Heat Treatment

Heat treatment before coating must be such that the coating temperature does not cause loss of hardness or dimensional change. Typical maximum coating process temperatures are 249–510°C (480–950°F) depending on the coating system. Ball-bearing steels, case-hardening steels, and special-purpose tool steels with proper heat treatment can be coated with carbon-based coatings. Special low-temperature coatings are available. High-speed steels, higher tempering temperature tool steels, and stainless steels are coatable without restrictions. Nitrided surfaces can be coated after mechanical pretreatment, and plasma nitriding is the preferred method for subsequent coating.

General purpose construction steels can be PVD coated but lack strength to support a hard coating. So they should serve only in low-load applications. Nickel and titanium are readily coatable. Polishing of these softer materials before coating must not introduce stresses or trap any type of residue in the surface. Copper and magnesium can be coated with restrictions to applicable temperature, cleaning process, and the applied load they will see after coating. Chrome and nickel-plated metals can be coated but adhesion of the galvanic coating is typically not as strong as that of PVD adhesion. This double coating system has advantages at lower loads and for corrosion prevention. Cemented carbide can be coated with no problems. But if machined, coolants with cobalt inhibitors should be used to prevent cobalt leaching. Sintered metals with open pores cannot be coated because residues from the sintering process outgas in a vacuum, interfering with the plasma process. Metallized or conductive ceramics can be coated. Plastic is not an option for these functional PVD coatings.

Ceramic Coatings

Thermally sprayed dielectric ceramic coatings [14] are the primary means of attaching strain and temperature gauges to hot-section rotating parts of turbine engines. As hot-section temperatures increase, lifetimes of installed gauges decrease, and seldom exceed 1 hour above ~1100°C (2000°F), and the required high temperature lifetime is 10 hours minimum. Typically, to enable a ceramic coating to adhere to the smooth surface of an engine component, a thermally sprayed NiCrAlY or NiCoCrAlY bond, a coat is applied to the smooth surface, thereby providing a textured surface to which the ceramic coat can adhere. The main failure mechanism of this system is decohesion and/or delamination at the interface between the ceramic top coat and the bond coat and stresses from the mismatch between the coefficients of thermal expansion of the ceramic top coat and the metallic bond coat.

In order to increase the high-temperature lifetime of a gauge attached to an engine component by the method described above involves (1) selective oxidation of the bond coat by means of a heat treatment in reduced oxygen partial pressure followed by (2) the application of a noble metal diffusion barrier. In experiments to test this approach, heat treatments of NiCoCrAlY bond coats were carried out in a tube furnace in which, in each case, the temperature was alternatively (1) increased at a rate of 3°C (37°F) per minute and (2) held steady for 1 hour until the desired temperature was reached. The tube furnace was continuously purged with dry nitrogen gas. A final heat-treatment temperature range of 871–982°C (1600–1800°) proved most beneficial.

To provide a basis of comparison for evaluation of the relative merits of the various surface treatments and heat treatments, some of the NiCoCrAlY and NiCrAlY bond coats were incorporated into the coupons in the as-sprayed condition; that is, the affected coupons were not subjected to the heat treatment at reduced oxygen partial pressure. The heat treatment of the NiCoCrAlY bond coats at reduced oxygen partial pressure yielded a significant increase in lifetimes: Coupons heat treated to 954°C (1750°F) at reduced oxygen partial pressure exhibited more than double the cycle lives of those containing as-sprayed NiCoCrAlY. This considerable increase in life can be attributed to the fact that selective oxidation of the aluminum and chromium in the bond coat yielded a graded interface. The heat treatment of the NiCrAlY bond coats yielded little or no increase in lifetimes.

In an effort to reduce the extent of internal oxidation in the bond coats, platinum and rhodium coats were employed as diffusion barriers. Initially, as-sprayed NiCoCrAlY bond-coated coupons were coated with platinum to a thickness of 2 µm by PVD. The platinum-coated Inconel coupons were heat treated to 982°C (1800°F), then magnesium aluminate spinel top coats were thermally sprayed over the platinum coats. Rhodium diffusion barriers were applied to the surfaces of the NiCoCrAlY bond-coated coupons by pen electroplating. (Pen electroplating was investigated as a means of forming diffusion barriers because it is easy to perform and does not entail costly capital investment.)

The rhodium diffusion barriers yielded only a marginal increase in the lives of the NiCoCrAlY bond-coated coupons. However, platinum diffusion barriers applied by PVD in conjunction with reduced oxygen partial pressure heat treatment yielded substantial increase in lifetimes. The platinum films were thick enough to constitute oxygen diffusion barriers that slowed the growth of internal oxides by promoting the formation of alumina-rich scale at the interfaces between the top and bond coats. The best results achieved to date were realized by use of sputtered platinum diffusion barriers in conjunction with heat treatments to 982°C (1800°F) at reduced oxygen partial pressures. This combination yielded a fourfold increase in the fatigue lives of the NiCoCrAlY bond-coated coupons.

Other Ceramic Coatings

Doped pyrochlore oxides of a type described below are under consideration as alternative materials for high-temperature thermal-barrier coatings (TBCs). In comparison with partially yttria-stabilized zirconia (TSZ), which is the state-of-the-art TBC material now in commercial use, these doped pyrochlore oxides exhibit lower thermal conductivities, which could be exploited to obtain the following advantages:

- For a given difference in temperature between an outer coating surface and the coating/substrate interface, the coating can be thinner. Reductions in coating thickness could translate to reductions in the weight of hot-section components of turbine engines (e.g., combustor liners, blades, and vanes) to which TBCs are typically applied.
- For a given coating thickness, the difference in temperature between the outer coating surface and the coating/substrate interface could be greater. For turbine engines, this could translate to higher operating temperatures, with consequent increases in efficiency and reductions in polluting emissions.

TBCs are needed because the temperatures in some turbine engine hot sections exceed the maximum temperatures than the substrate materials (superalloys, Si-based ceramics, and others) can withstand. YSZ TBCs are applied to engine components as thin layers by plasma spraying or electron-beam PVD. During operation at higher temperatures, YSZ layers undergo sintering, which increase their thermal conductivities and thereby renders them less effective as TBCs. Moreover, the sintered YSZ TBCs are less tolerant of stress and strain and, hence, are less durable [19].

Patches for Ceramics and CMCs

A novel coating that is a patch for repairing ceramics and CMCs has been developed [15]. Patches consist of ceramic fabrics impregnated with partially cured polymers and ceramic particles and must withstand temperatures above the melting points of refractory metal alloys. These patches were conceived for use by space-suited, space-walking astronauts in repairing damaged space shuttle leading edges: as such, these patches could be applied in the field, in relatively simple procedures, and with minimal requirements for specialized tools. These design characteristics also make the patch useful for repairing ceramics and CMCs in terrestrial settings [15].

In a typical patch, as supplied to an astronaut or repair technician, the polymer would be in a tacky condition, denoted as an "A" stage, produced by partial polymerization of a monomeric liquid. The patch would be pressed against the ceramic or CMC object to be repaired, relying on the tackiness for temporary adhesion. The patch would then be bonded to the workpiece and cured by using a portable device to heat the polymer to a curing temperature above ambient temperature but well below the maximum operating temperature to which the workpiece is expected to be exposed. The patch would subsequently become pyrolized to a ceramic/glass condition upon initial exposure to the high-operating temperature. In the original space shuttle application, this exposure would be Earth atmosphere reentry heating to about 1600°C (3000°F).

Patch formulations for space shuttle applications include SiC and ZrO_2 fabrics, a commercial SiC-based preceramic polymer, and suitable proportions of both SiC and ZrO_2 particles having sizes of the order of 1 μm. These formulations have been tailored for the space shuttle leading edge material, atmospheric composition, and re-entry temperature profile so as to enable repairs to survive reentry heating with an expected margin.

Layered CBC/EBCs

Ceramic thermal and environmental barrier coatings (T/EBCs) [19] that contain multiple layers of alternating chemical composition have been developed as an improved means of protecting underlying components of gas turbine and other heat engines against both corrosive combustion gases and high temperatures. A coating of this type consists of the following:

- An outer, or top oxide layer that has a relatively high coefficient of thermal expansion (CTE) and serves primarily to thermally protect the underlying coating layers and the low CTE ceramic substrate structural material (the component that is ultimately meant to be protected) from damage due to exposure at the high temperatures to be experienced in the application.

- An inner, or bottom Si-containing/silicate layer, which is in contact with the substrate, has a low CTE, and serves primarily to keep environmental gases away from the substrate.

- Multiple intermediate layers of alternating chemical composition (and, hence, alternating CTE).

The intermediate alternating composition coating layers are chemically compatible with themselves as well as with the inner and outer coating layers. These intermediate coating layers act as an energy-dissipating interlayer;

they dissipate strain energy associated with the CTE mismatch between the inner and outer coating layers, thereby reducing stresses and helping to increase (relative to prior ceramic T/EBCs) coating resistance to cracking and delamination from the substrate surface.

Typically, there are between 4 and 10 alternating composition intermediate layers, comprising higher CTE oxide layers interspersed with lower CTE silicate layers, each layer having a thickness between 5 and 50 µm. The compositions of the oxide and silicate alternating layers can be the same as those of the outer and inner layers, respectively. Alternatively, different oxide and silicate compositions can be chosen to increase tolerance of strain, resistance to cracking, and/or protection against chemical attack by gases in any intended application.

During thermal cycling, the alternating layers become regions of alternating tension and compression. This stress and strain configuration facilitates microsegmentation in the oxide layers while maintaining effective compression in the silicate layers. As a consequence, the thermal expansion of the energy-dissipating interlayer is reduced, stresses are reduced, and tolerance of strain is greatly enhanced. Cracking that starts in the outer oxide layer of the coating is arrested within the alternating layers because of the compressive stress in the silicate alternating layers and the tendency toward deflection and/or bifurcation of cracks at the interface between the alternating layers. Moreover, during cooling, the compression in the silicate alternating layers helps to ensure the integrity of the overall coating system in its role as an environmental barrier by helping to prevent penetration of combustion gases to the surface of the substrate.

The thickness of the alternating oxide and silicate layers are quite dependent on the intended engine application. A thicker silicate layer/thinner oxide layer structure could increase the strain tolerance of the coating and protect the substrate (or engine component in application) from the hot gases in the engine environment; however, on the other hand, a slightly thinner silicate layer next to a slightly thicker oxide layer structure could increase the coating's resistance to stress and penetration of any damaging gas constituents through cracks, potentially reacting with the substrate.

Nano Coatings

Scientists and engineers working with university researchers [16,20–23,27] have significantly advanced the understanding of super tough nanocomposite coatings—coatings that could improve the performance and durability of aircraft engines. Through careful experimentation, the team identified a mechanism of macroscopic ductility unique to extremely hard nanocrystalline/amorphous composites. Laboratory researchers initially developed a new class of wear-resistant materials comprised of very hard, 3–5 nm grains of carbides or oxides embedded in an amorphous matrix of either DLC or a metal/ceramic mixture. During this preliminary characterization stage, the new materials exhibited an unusual combination of high hardness (exceeding that of ceramics) and fracture strength (similar to that of tough metal alloys).

To better understand these findings, TEM studies discovered a new mechanism of macroscopic ductility—one not based on dislocation mobility or diffusion-related boundary reconstructions. The new mechanism is a unique feature of nanocrystalline/amorphous composite design, resulting from a large number of 1–2 nm shifts of nanograins inside the amorphous matrix. The individual crystalline and amorphous phases, although very hard, do not deform plastically. This work [16] has led to the exploration of a wide range of possible nanocomposite coating applications for military and commercial aerospace. In one specific application, engineers are investigating nanocomposite coatings for short takeoff and vertical landing propulsion system components. The components are heavily preloaded friction pairs and could substantially benefit from surface strengthening.

Miscellaneous Applications of Coatings

The University of Michigan College of Engineering researchers [24] have developed a coating that could be painted or sprayed on structures to sense their stability over time [22–34]. The coating is an opaque, black material made of layers of polymers and networks of carbon nanotubes that run through the polymers. One layer tests the pH level of the structure, which changes as steel corrodes. Another layer registers cracks by actually cracking under the same conditions that the structure would. The perimeter of the carbon nanotube skin is lined with electrodes that are connected to a microprocessor. To read what is going on underneath the skins, inspectors send an electric current through the embedded carbon nanotubes.

An experimental study [26] has been performed to learn about the physical and chemical mechanisms of self-lubrication of the coatings that comprise nanostructured composites of yttria-stabilized zirconia (YSZ), silver, and molybdenum. These and other YSZ-based nanocomposite coatings have received increasing attention in recent years because they offer a combination of hardness, toughness, resistance to wear, and low-friction carbon properties that make them attractive for reducing wear and friction and increasing the lifetimes of hot, sliding components of mechanical systems. In addition to the excellent mechanical and thermal stability of the basic YSZ ceramic material, the nanocomposite structures of these coatings, consisting of combinations of amorphous and crystalline phases, provide a "chameleon" surface adaptation, in which different phases turn into lubricants in response to different test environments, contact loads, sliding speeds, and temperatures. Moreover, proper sizing of nanocrystalline grains can restrict crack sizes and create large volumes of grain boundaries, thereby increasing the toughness and contact load bearing capabilities of these coatings.

The YSZ–Ag–Mo composite coatings for the experimental study [26] were deposited on steel and nickel alloy substrates in a hybrid process that included deposition of 100 nm thick titanium adhesion layers by use of a filtered titanium arc plasma, followed by the pulsed layer deposition of YSZ from a YSZ target. Ag and Mo were added to the coatings by magnetron sputtering from Ag and Mo targets. All of the coatings were grown to a thickness of about 2 μm. In some cases, the coating was formed as two 1 μm thick YSZ–Ag–Mo layers, and a 100 nm thick TiN diffusion barrier layer was deposited between the YSZ–Ag–Mo layers by introducing a flow of nitrogen into the deposition chamber during operation of the filtered titanium arc plasma source. In other cases, TiN barrier layers, containing pinholes were deposited on the surfaces of YSZ–Ag–Mo coatings to limit through-the-thickness diffusion of silver [26].

The study revealed different chameleon-like, high-temperature, adaptive lubrication mechanisms in the nanocomposite coatings. Coefficients of friction of about 0.4 or less were found to be maintained at all temperatures from 25 to 700°C (77–1292°F). The as-deposited coatings were found to include silver nanograins embedded in amorphous/nanocrystalline YSZ–Mo matrices. At high temperatures, heating-induced diffusion and coalescence of silver were found to result in microstructural and chemical changes that included formation of silver films on surfaces with silver-depleted YSZ–Mo layers left underneath. Crystallization of zirconia matrices was found to occur simultaneously with the diffusion of silver to surfaces when the coatings were heated. It was confirmed that the diffusion of silver on the surfaces of YSZ–Ag–Mo nanocomposite coatings plays an important part in high-temperature lubrication.

Silver was determined to be an effective lubricant at temperatures below 500°C (932°F), and the coalescence of silver on surfaces was found to isolate Mo inside the composites from ambient oxygen. At temperatures above 500°C (932°F), the silver surface layers were found to be rapidly removed from wear tracks and, hence, the reactive Mo inside the silver-depleted YSZ–Mo layers was exposed to ambient air. Contact tribochemistry was found to result in the formation, in wear tracks, of Mo oxides, which provided lubrication at 700°C (1292°F).

In the case of specimens containing the internal TiN barrier layers, these layers were found to preserve lubricants underneath, thereby providing for continuous replenishment of lubricants. The TiN layers were also found to force subsurface silver to diffuse laterally toward wear scars, once the TiN layers were broached by wear. This behavior affords an adaptive response, which includes an on-demand supply of lubricant from storage volumes inside YSZ–Ag–Mo composites to surface contact areas. The YSZ–Ag–Mo coatings that contained the internal TiN barrier layers were found to maintain CoF of approximately 0.4 during more than 25,000 cycles, while the monolithic YSZ–Ag–Mo coatings lasted fewer than 5000 cycles. The specimens having

TiN surface layers with pinholes were found to have wear lifetimes greater than 50,000 cycles.

Scientists at the U.S. Department of Energy's Brookhaven National Laboratory have developed a method for coating metal surfaces with an ultrathin film containing nanoparticles—particles measuring *billionths* of a meter—which renders the metal resistant to corrosion and eliminates the use of toxic chromium for this purpose. "Our coating is produced right on the metal using a simple two- or three-step process to produce a thin film structure by crosslinking among the component compounds," said chemist Toshifumi Sugama, guest researcher at Brookhaven Lab. "The result is a layer less than 10 nanometers thick that protects the metal from corrosion, even in brine conditions" [27].

Corrosion resistance is essential for metals used in a wide range of applications, from electronics to aviation to power plants. Traditionally, compounds containing a toxic form of chromium have provided the best corrosion resistance. Scientists looking to develop chromium-free alternatives have been able to achieve the thin layers desirable for many applications. "Ultrathin coatings reduce the amount of material needed to provide corrosion resistance, thereby reducing the cost," Sugama explained. Sugama's approach achieves several goals—low toxicity and excellent corrosion resistance in a film measuring less than 10 nanometers that can be applied to a wide array of metals, including aluminum, steel, nickel, zinc, copper, bronze, and brass. According to Sugama, the coating should be of specific interest to industries that produce coated valves, pumps, and other components, as well as the manufacturers of aluminum fins used in air-cooled condensers at geothermal power plants, where preventing brine-induced corrosion is a high priority.

The coating can be made in a variety of ways suited to a particular application. In one embodiment, it starts as a liquid solution that can be sprayed onto the metal, or the metal can be dipped into it. The metal is then subjected to one or more treatment steps, sometimes including heating for a period of time, to trigger cross-linking reactions between the compounds, and simultaneously, to form corrosion-inhibiting metal oxide nanoparticles, such as environmentally benign cerium-based oxides. "Among the key factors that ensure the maximum corrosion-mitigating performance of these ultrathin coating films are the great water-repellency, the deposition of metal oxide nanoparticles over the metal's surface, and their excellent adhesion to metal. The combination of these factors considerably decreased the corrosion of metals," said Sugama.

"The corrosion resistance of these coatings can be comparable, and even superior, to chromium-based coatings," Sugama said, "In fact, these new coatings provide even better coverage of metal surfaces than chromium coatings." Sugama added, "This is particularly advantageous when the metal to be coated possesses fine structural detail." Because the method deposits such a thin coating of material, it is highly economical and efficient.

Work by Shawn-Yu Lin, Professor of Physics at Rensselaer Polytechnic Institute, Troy, New York, has been developing the darkest manmade material with the potential of the carbon nanotube array that absorbs light could boost solar energy conversion [34]. The material, a thin coating comprised of low-density arrays of loosely vertically aligned nanotubes, absorbs more than 99.9% of light and one day could be used to boost the effectiveness and efficiency of solar energy conversion, infrared sensors, and other devices. "It is a fascinating technology, and this discovery will allow us to increase the absorption efficiency of light as well as the overall radiation-to-electricity efficiency of solar energy conservation," says Professor Lin. "The key to this discovery was finding how to create a long, extremely porous vertically aligned carbon nanotube array with certain surface randomness, therefore minimizing reflection and maximizing absorption simultaneously."

All materials, from paper to water, air, or plastic, reflect some amount of light. Scientists have long envisioned an ideal black material that absorbs all the colors of light while reflecting no light. So far they have been unsuccessful in engineering a material with a total reflectance of zero. The total reflectance of conventional black paint, for example, is between 5 and 10%. The darkest manmade material, prior to the discovery by Lin's group, boasted a total reflectance of 0.16–0.18%. Lin's team created a coating of low-density, vertically aligned carbon nanotube arrays that are engineered to have an extremely low index of refraction and the appropriate surface randomness, further reducing its reflectivity. The end result was a material with a total reflective index of 0.045%—more than three times darker than the previous record, which used a film deposition of nickel-phosphorous alloy.

"The loosely packed forest of carbon nanotubes, which is full of nanoscale gaps and holes to collect and trap light, is what gives this material its unique properties," Lin says. "Such a nanotube array not only reflects light weakly, but also absorbs light strongly. These combined features make it an ideal candidate for 1 day realizing a super black object."

"The low-density aligned carbon nanotube sample makes an ideal candidate for creating such a super-dark material because it allows one to engineer the optical properties by controlling the dimensions and periodicities of the nanotubes," says Pulickel Ajayan, the Anderson Professor of Engineering at Rice University in Houston, Texas, who also worked on the project. The array has been tested over a broad range of visible wavelengths of light, and showed that the nanotube array's total reflectance remains constant.

"It's also interesting to note that the reflectance of the nanotube array is two orders of magnitude lower than that of the glassy carbon, which is remarkable because both samples are made up of the same element—carbon," says Lin. This discovery could lead to applications in areas such as solar-energy conversion, thermal photovoltaic electricity generation, infrared detection, and astronomical observation.

The Spire Corporation, Bedford, MA [20], is developing nanophase calcium phosphate coatings loaded with bone morphogenic proteins (BMPs).

The coating will be applied to improve bone integration into dental implants, leading to more rapid and reliable device fixation. A unique nanocrystalline film deposition technology is the key to produce coatings that control the release rate of BMPs. The coatings will maintain critical threshold concentrations of BMPs only at the implant site, and will provide a scaffold over which bone can grow.

The incorporation of BMPs into medical device coatings will provide significantly enhanced integration compared to that which can be achieved with ceramic coatings alone. The platform is flexible enough to permit incorporation of BMPs as well as other proteins and biologics. The coatings also have applications in a wide range of orthopedic, dental, and other types of medical implants. Using improved, nanostructured coatings could be used to measure rapid pressure fluctuations. This has led to the development of nanorod-based, fast response pressure-sensitive paints (PSPs) [28]. This development has been devoted to the exploitation of nanomaterials in PSPs, which are used on wind tunnel models for mapping surface pressures associated with flow fields.

A PSP contains a dye that luminesces in a suitable wavelength range in response to photoexcitation in a shorter wavelength range. Most PSPs include polymer-based binders, which limit the penetration of oxygen to dye molecules, thereby reducing responses to pressure fluctuations. The incorporation of nanomaterials (nanorods) would result in paints having nanostructured surfaces that, relative to conventional PSP surfaces, would afford easier and more nearly complete access of oxygen molecules to dye molecules. One measure of greater access is an effective surface area. For a typical PSP applied to a given solid surface, the nanometer scale structural features would result in an exposed surface area more than 100 times that of a conventional PSP, and the mass of proposed PSP needed to cover the surface would be less than a tenth of the mass of the conventional PSP.

One aspect of the development would be to synthesize nanorods of Si/SiO_2, in both tangle-mat and regular-array forms, by the use of CVD and wet chemical processes, respectively. The rods would be coated with a PSP dye, and the resulting PSP signals would be compared with those obtained from PSP dye coats on conventional support materials. Another aspect of the development would be to seek to exploit the quantum properties of nanorods of a suitable semiconductor (possibly GaN), which would be synthesized by CVD. These properties include narrow wavelength band optical absorption and emission characteristics that vary with temperature. The temperature sensitivity might enable simultaneous measurement of fluctuating temperature and pressure and to provide a temperature correction for the PSP response.

Commercial Coatings

Metallic surfaces, special effect colors, and reflective finishes have increasingly found their way into product design. The metallic trend shows no sign of abating. Designers of appliances have taken advantage of economical materials such as porcelain-enamel coatings that have emerged. These engineered coatings give trendy metallic looks with a substantially lower sticker price. Cost savings per square foot for a typical kitchen range serves as an example. With porcelain-enamel coatings called Evolution, designers can fabricate the range from 24 gauge plain steel instead of 300 Series stainless [31]. The coating would go on as a 2 mm (0.002 in.) ground coat and a 4 mm (0.004 in.) cover coat. The porcelain-enamel coat would have cut costs by nearly 60%. The porcelain coating also withstands scratches that would relegate a significant percentage of stainless steel parts to the scrap heap.

Unlike their strictly metal counterparts, porcelain-enamel coatings don't discolor from heat and resist stains, scratches, and chemical cleaners. Fingerprints also easily come off. In addition, the metallic-look coatings for refrigerators, ranges, cooktops, sinks, and plumbing can also give cabinetry matching or contrasting hues. And the coating resists heat well enough for use on exterior surfaces of cookware, barbecue grills, and fire bowls that must take not only heat from flames, but weather and corrosion.

The coatings can help differentiate products because they have a wider range of custom metallic color possibilities. Metallic pigments and metal-flake sparkle effects can combine with most enamel colors. Brushed, satin, or highly reflective finishes are also possible. Porcelain enamels are glassy coating materials that protect substrates while sprucing up their looks. They bond to metals (typically carbon steel, stainless steel, cast iron, or aluminum) at temperatures from 538 to 871°C (1000–1600°F).

There are two general types: ground coats and cover coats. Ground coats contain adherence-promoting oxides and are used for oven cavities, stove grates, hot water tanks, and dual purpose finish coats. A pyrolytic ground coat is an extremely heat-resistant enamel for self-cleaning oven cavities exposed to operating temperatures of about 538°C (1000°F). Cover coats provide additional chemical, physical, or cosmetic properties. They need a ground coat as a primer layer. Aluminum enamels can be considered cover coats. They contain low-temperature glasses that fuse directly onto aluminum and commonly serve on cookware. Performance benefits of porcelain enamels include:

Sanitary qualities: The hard, dense surface is an excellent barrier to odor and bacteria.

Easy cleaning: The hardness and abrasion resistance makes for easy cleaning with mild cleaning solutions. High gloss and surface lubricity also lets graffiti wipe off.

Scratch and abrasion resistance: Hardness also gives substantially more resistance to scrapes than the hardest organic coating.

Chemical and corrosion resistance: Surfaces stand up to acids, alkalis, water, solvents, oils, and UV light, as well as to corrosive industrial atmospheres, salt, air, gas, smoke, and soil.

Flameproof: Heat doesn't change physical or chemical properties or appearance.

Color stability: Coating colors are used as physical color standards for the ink, plastics, and textile industries. Glossy, acid-resistant porcelain enamel has shown no change in color or gloss after 15 years of exposure to weather. And it will not peel, blister, or delaminate from the metal surface if applied correctly.

Environmentally friendly: The application process requires no solvents.

Coating with Porcelain Enamel

Evolution cover coats come as powders that get reconstituted with water (100:47) in a high-shear blender. The resulting enamel slip is put through 60–100 mesh screen and applied with a wet spray gun onto a suitable metal substrate. It goes on top of a fired-enamel ground coat that can be applied either electrostatically or by wet spray. The cover-coated part is fired at temperatures from 804 to 854°C (1480–1570°) for 2–3 minutes. The furnace must be at the proper firing temperature from the beginning, not ramping up. It's also important that parts be correctly spaced on hangers. Parts should be designed to hang with the longest dimension down to reduce warping. All forming and assembly of the product takes place before finishing. There are no postfinishing steps, although panel controls or other decorations are sometimes added by screen printing. The fired thickness of the cover and ground coats is 125–150 µm (5–6 mils).

To prevent warping during firing, steel should be at least 14–24 gauge, thicker for large parts. Any attachments should be made of the same material and two gauges lower than the base part. If the ground coat is to be applied electrostatically, it's important to avoid creating Faraday cage areas or corners that can keep the powder enamel from sticking. Observing minimum radius requirements and making tabs less than 12.7 mm (0.050 in.) long will prevent Faraday cage defects. More detailed information, including guidelines, is available in the technical manual PEI-101: *Design & Fabrication of Metal of Porcelain Enamel.* This and other technical manuals, coaters, and other suppliers can be found on the Porcelain Enamel Institute Web site (www.porcelainenamel.com) [32].

Sophisticated Coatings

The process that allows the formation of sea shell interiors and pearls has been examined for use to deposit nacre-like coatings onto metal surfaces.

Scientists at the University of Dayton Research Institute [33], working under a U.S. Air Force contract were looking for biological ceramic coatings that are naturally derived and that don't involve high-temperature, high-pressure procedures now required for building ceramic coatings. The goal is to produce lightweight, durable coatings able to protect aircraft from impact and corrosion damage, but to do so at room temperature and pressure. Researchers recently discovered that oysters use blood cells to deposit crystals rather than precipitating calcium carbonate from seawater. So they have used the blood cells to deposit crystals in an ordered manner on a variety of metals to build a multilayered ceramic coating.

Coatings on the next generation military fighter aircraft are critical to maximizing airflow and minimizing thermal buildup at high speeds. On a sophisticated aircraft such as the F-35, also known as the Joint Strike Fighter (JSF), those coatings must cover a complex 3-D shape at different thicknesses and still maintain the aircraft's stealth invisibility to radar. Because of the demanding requirements, the aircraft has been one of the most heavily modeled and simulated programs in U.S. military history, with many of those simulations focusing on the facilities, systems, and methods that will apply coatings.

To achieve the precision of these coatings, robots are used extensively. Seven bays are employed to prepare aircraft surfaces and apply coatings. The five largest bays are the size of barns and two of those, for the fully assembled wing, are even bigger. Robotic systems that apply the coatings use two customized material-handling robots and two customized positioners. The coating methods, off-line programming (OLP) and verification techniques are still in development. Simulations are performed with the aid of different software packages. The one for coatings is ULTRAPAINT.

The sophisticated coatings serve a variety of purposes. They play an unspecified role in providing invisibility to enemy radar (stealth); the F-35 is the first supersonic, multirole, radar evading aircraft. The coatings also provide abrasion resistance on the nose and the leading edges of the wings, elevators, and twin tails. One abrasion-resistance test is a sea-level ride over the Gulf of Mexico at Mach 1.4. The supersonic impacts of millions of water droplets can leave leading-edge coatings with a sandblasted appearance (which can also alter the radar signature). The coatings also deliver thermal protection at the fuselage rear and F-35 turbine exhaust. To add to the complexity of applying these coatings, three-dimensionally different versions of the F-35 are being developed, with coatings varying with intended missions.

Among the many factors that have to be optimized and balanced with simulations are constantly varying application rates—speeds, feeds, and flows. Engineers summarize the challenges as a combination of verifying precise coating thicknesses and meeting stringent time spans. They must deal with different coatings on specific areas of the aircraft; the application of multiple coatings, some requiring many passes; variations in coating thicknesses over aircraft surfaces; some with precise gradients; the tendency of liquids on concave surfaces to rise slightly around the edges; leaving a coating surface that is very slightly concave (OLP calculations must take the meniscus into account). Since air is constantly moving through the aircraft bays at 60 m/s (200 feet per second) during coating, the "plume" of each coating must be calculated as it is applied. Along with these challenges are the geometric complications, since almost no surface on the F-35 is flat. Many components to be coated have compound contours, eliminating the use of constant-rate programming for the coating application.

Because the coating process requires a short standoff from the working surface, the operation of robotic systems must be controlled to avoid damage to the aircraft surface, which can affect its aerodynamics. One must not affect the aircraft's OML or outer mold line, which determines aeronautical capabilities. The need for precision is critical and very tight control of the hardware by the OLP software. Some of the most demanding simulations, were studies of the robots' reach and the interactions of the coating tools and part orientations while optimizing the placement of parts within the cell. These simulations accommodate arc-like travel in any two planes and sometimes all three plus roll, pitch, and yaw of the R-2000s' robot's wrists. Programmers have been able to do an entire simulation, including multiple robots, fixtures, and material handling devices by using a simple graphical programming interface. Therefore, the key advantage of simulation is that it helps make the decisions on how soon after a part enters a bay that coating can start and on the order in which surfaces are coated. Production coating has been underway since 2005 and the prospects of success have been emerging continually.

Thermal Barrier Coatings (TBC)

Thermal barrier coatings (TBCs) [25,29] are widely used surface treatments typically consisting of a top coat of Y_2O_3 air plasma sprayed partially stabilized ZrO_2, and a MCrAlY bond coat (where M is Ni, Co, or NiCo), which can be deposited using different thermal spray processes such as air plasma spray (APS), vacuum plasma spray (VPS), and HVOF [35–37]. Another example is a top coat of Y_2O_3 partially stabilized ZrO_2 deposited using electron beam physical vapor deposition (EBPVD), and an aluminum alloy bond coat deposited using CVD. For many metallic components, adequate creep properties (creep resistance, thermal efficiency, thermal shock resistance, etc.) can only be achieved with the use of nickel and cobalt superalloy base material plus adequate coatings and surface treatments to obtain resistance to oxidation and corrosion at high temperature. Repair of components means restoring their original properties. Therefore, these surface treatment processes are not only used during manufacture, but also during repair.

Mechanics of Process

Thermal spray processes are widely used to coat components because they can deposit a range of materials and obtain thicknesses from 80 to 2000 μm. The main stages of a thermal spray coating deposition are:

- Feeding the coating material into the spray gun
- Heat transfer from plasma or combustion gas to melt coating material
- Transport of melted particles on to the substrate surface
- Heat transfer from the melted particles to the substrate and solidification
- Coating formation by means of overlapping of several coating material layers

The particles assume a splat configuration and solidify and transfer the heat to the base material. The final coating consists of several layers of coating materials. The bonding is generally mechanical, created by local melting and diffusion.

Surface preparation is critical so that the spray particles can adhere and establish an integral coating. For thermal spray, the most common surface preparation method is grit blasting the surface using a suitable abrasive material such as corundum or silicon carbide. The main parameters to be considered are the type of blasting machine, blasting media (type, grain size distribution, shape, and hardness), working pressure, and hardness of substrate materials. Tuning of blasting parameters is crucial to obtain good surface preparation without causing contamination from the blasting particles.

In the case of plasma spraying under a vacuum, the coating process is carried out at reduced pressure (about 10–50 mbar). The jet velocity is higher because of the absence of air resistance, and the lack of oxygen results in oxide-free coatings [37,38]. The higher jet velocity and less cooling result in a higher energy deposition, which creates coatings that are denser and have higher chemical purity.

In turbine applications, air plasma spray is used to spray yttria partially stabilized zirconia (YPSZ) (functioning as the TBC) and also aluminum on the "fir-tree" sections of blades to serve as a seal with the rotor when the part is assembled. (The fir-tree section refers to the bottom part of a blade that slots into the rotating assembly.) VPS is used to spray MCrAlY alloys as a bond coat and a high-temperature corrosion- and oxidation-resistant coating. The HVOF process is used to deposit several antiwear and anticorrosion coatings. The coatings are dense and the splat on the surface is ideal. The powder is injected into the flame by a suitable carrier gas (usually nitrogen), melted, and projected on to the substrate surface. The most common materials sprayed using this technology are metals, cermets, carbides, and some polymers [35–42].

Applying MCrAlY Coatings

Studies have been carried out to compare HVOF and low pressure plasma Spray (LPPS) VPS technologies by evaluating their MCrAlY bond coats [35,37]. Commercial powder (Amdry 995, a widely used coating) has been used in tests and deposited film quality was assessed. Coatings from the two processes were compared in terms of the microstructural (porosity, oxide concentration, presence of unmelted particles) and mechanical characteristics (hardness). The surface composition and morphology of the coatings were also determined, and specific efficiency tests were performed for the three technologies.

Vacuum plasma sprayed technologies are currently the state-of-the-art for the production of MCrAlY alloy coatings, which serve as oxidation-resistant material and the bond coat in TBC systems. The quality of the coatings obtained by VPS is better than that obtained by means of HVOF. Nevertheless, HVOF coatings demonstrate a good porosity level due to the high-flame velocity, but the oxide content is slightly higher than in VPS coatings. Hardness is the same for both processes. Prior to coating, serviced gas turbine components usually go through repair processes. Factors that should be considered before thermal spraying include:

- The extent and quality of potential damage to the base material.
- The presence of various kinds of materials that are carried over from any repair processes. For example, the bulk material, and welding and brazing materials having different thermal expansion coefficients will influence the properties of the coating.
- The risk of cracks or damage from preheating by plasma spray.

Possible solutions are HVOF technology, or using an improved preheating procedure for LPPS coatings. The HVOF process is a relatively cold process; that is, preheating the components to high temperature is not needed, and the coatings have less oxides (similar to LPPS). This process is also more economical. Preheating to a high temperature is needed for VPS processes. The plasma jet generates a local overheating and the inhomogeneous temperature sometimes reopens cracks or damage in repaired areas. Therefore, an improved preheating procedure is needed to minimize the contact between the plasma components whereby an inert gas furnace is used for preheating.

Applying Top Coats

After deposition of metallic MCrAlY bond coats, a ceramic layer could be deposited to reduce the life service temperature of components. Usually yttria partially stabilized zirconia (YPSZ) is used as a coating material and deposited by APS. Typically, Original Equipment Manufacturer (OEM) specifications classify the ceramic top coat according to thickness. A thin TBC, usually applied to blades and vanes, is defined as a 200–800 μm thick coating having from 5 to 20% porosity. A thick TBC, usually applied to stationary gas turbine components such as the combustion chamber or segments and liner that cover turbine internal walls, is generally a 1.5–2.0 mm (0.059–0.078 in.) thick zirconia coating having up to 28% porosity.

Zirconium oxide (ZrO_2) is selected due to its properties low heat conduction coefficient (making it useful as a thermal barrier) and a higher coefficient of thermal expansion than that of most oxides (such as alumina and chromia), which makes the deposition of the ceramic coating on a metal alloy substrate easier. Furthermore, the relatively poor mechanical stability of ZrO_2 caused by a 4% volume increase during transformation from tetragonal to monoclinic phase, has been substantially improved in recent years. This is accomplished by partially stabilizing the tetragonal phase at room temperature by adding suitable stabilizers such as calcium oxide (CaO), magnesium oxide (MgO), and yttrium oxide (Y_2O_3). Yttria is normally preferred since CaO and MgO tend to vaporize at high temperatures [43,44].

YPSZ is a multiphase material. At room temperature, the stable cubic phase coexists with the tetragonal phase and a small percentage of the monoclinic phase. This structure produces better mechanical properties in terms of mechanical strength and fracture toughness [45]. Powders having low-monoclinic content are preferred, and phase content in the coating after deposition has to be checked carefully. While thermal spray processes are versatile and widely used, they cannot be used in some situations such as the coating of internal cooling passages. In such cases, an aluminum coating is deposited using the CVD process.

Thermal spray processes are widely used in gas turbine applications thanks to their versatility. In particular, VPS technologies currently remain the state of the art for the production of MCrAlY alloy coatings to be used as oxidation-resistant material and bond coats in TBC systems. Nevertheless, HVOF coatings demonstrate an acceptably low porosity level due to the high impact conditions during impingement and, additionally, the oxide content is slightly higher with respect to VPS. On the other hand, the HVOF process is less expensive than a VPS process. Zircotec of Didcot, Oxfordshire, U.K., says it has developed a ceramic coating that allows composites to be used in high-temperature environments such as in motor sport, aerospace, and military applications, with minimal weight gain.

The ceramic thermal barrier technology enables the use of composites in environments with temperatures above their melting point. In tests for a typical application, the composite surface temperature was reduced by more than 125°C (257°F). Zircotec believes it is the only product of its type to be commercially available. "The thermal environment in motorsport and aerospace applications is not always conducive to the use of composites," says Peter Whyman of Zircotec. "This results in the use of heavier materials or bulky heat shields, which can reduce overall performance. The solution allows engineers to access a range of materials previously unavailable."

The coating can be engineered to suit specific customer requirements, adjusting the coating properties both through thickness and across the surface so the component can cope with hot spots. The coating can also be customized for different forms of heat transfers such as radiant, conductive, or convective heating. One can apply a reflective surface layer to help protect against radiant heat, or one can increase the thickness of the ceramic in areas where a hot spot can occur. Also one must have the ability to build in a conductive sublayer that will help to dissipate heat away from any hot spots, and can also help deal with transient heating solutions. This means applying just the right amount of coating to deliver the necessary protection while minimizing the weight impact of the coating (as low as $0.03 \ g/cm^2$ for some applications). Now it's possible to vary and control parameters within coatings in all three dimensions. Zircotec has already supplied examples of the coatings to several Formula I teams for race applications. The technology is based on Zircotec's Thermohold technology, and has been designed to be robust and have high resistance to vibration and mechanical damage. It also adheres well to the underlying composite structure.

Thermal Protection Systems (TPSs)

NASA Ames Research Center [46] scientists and technologists are endeavoring to develop durable, oxidation-resistant, foam TPSs that would be suitable for covering large exterior spacecraft surfaces, would have low to moderate densities, and would have temperature capabilities comparable to those of carbon-based TPSs reusable at ~1659°C (3000°F) with an application of suitable coatings. These foams may also be useful for repairing TPSs while in orbit. Moreover, on Earth as well as in outer space, these foams might be useful as catalyst supports and filters. Preceramic polymers are obvious candidates for use in making foams. The use of these polymers offers advantages over processing routes followed in making conventional ceramics. Among the advantages are the ability to plastically form parts, the ability to form pyrolyzed ceramic materials at lower temperatures, and the ability to form high-purity microstructures having properties that can be tailored to satisfy requirements.

Heretofore, preceramic polymers have been used mostly in the production of such low-dimensional products as fibers because the loss of volatiles during pyrolysis of the polymers leads to porosity and large shrinkage (in excess of 30%). In addition, efforts to form bulk structures from preceramic polymers have resulted in severe cracking during pyrolysis. However, because the foams in question would consist of networks of thin struts (in contradiction to nonporous dense solids), these foams are ideal candidates for processing along a preceramic-polymer route.

The present research has explored the feasibility of forming ceramic foams using sacrificial blowing agents and/or sacrificial fillers in combination with preceramic polymers. The possibility of using reactive fillers in combination with the aforementioned ingredients was also investigated. The use of such reactive fillers as Ti or Si reduced the large shrinkage observed in the pyrolysis of polymers. The fillers also react with excess carbon that, in the absence of such reaction, would be present in the foam pyrolysis products. A reactive filler becomes converted to a ceramic material with an expansion that reduces overall shrinkage in the pyrolyzed part. The expansion of the reactive filler thus compensates for the shrinkage of the polymer if the appropriate volume fraction of filler is present in a reactive atmosphere (e.g., N_2 or NH_3).

Previously, this reactive filler approach yielded limited success in efforts to make fully dense structural composite materials (in contradistinction to foams). However, in the present research, this reactive filler approach has been modified to enable processing of foams with minimal shrinkage. The representative foam microstructures that have been produced shows in all cases that the foams are isotropic, open-celled structures. Foams processed by use of a polyurethanes blowing agent have large cell sizes (50–500 μm), whereas foams processed by incorporating sacrificial fillers (e.g., polymer microspheres) generally have much smaller cell sizes (as low as 3 μm, depending on the diameter of the sacrificial filler particles). It is evident from the examination of the foam that the original unpyrolized structure is retained after pyrolysis, without the loss of spherical cell shape.

Sol-Gel Coatings

Paint systems [47] often provide the first layer of defense against the effects of environmental action to underlying substrates. Modern day systems are the products of decades of research and development requiring a variety of properties to fulfill a range of applications and environmental conditions. Environmental legislation is now beginning to impact upon the nature of the chemistry that can be adopted to produce paint systems. One recent major change in environmental legislation is that of the reduction and eventual ban on the use of chromate-based coatings and pretreatments. Due to the carcinogenic nature of Cr^{IV+} ions much effort is now being devoted to developing alternative replacement systems (e.g., anodized coatings, sol-gel (SG) coatings, plasma coatings etc.).

Organic–inorganic hybrid coatings are of increasing interest to the industry due to their potential widespread applications. Such interest has led to a variety of SG derived systems being developed at Sheffield Hallam University (SHU) [48,49] to improve the corrosion resistance of substrates such as; steel, zinc, aluminum and magnesium. These systems include inorganic forms [50] and organically modified hybrid formulations, for example, Ormosil, an organically modified silicate material [51]. The organic component in the matrix offers the advantages of mechanical toughness and flexibility while the organic component provides the coating with its hardness, wear resistance, and thermal stability properties. However, excessive inorganic content of the coating has the disadvantage of long cure times and a limitation in the coating thickness, less than 1 µm for a single-coat system. These limitations are in part due to the buildup of intrinsic coating stresses and the nature of the processing of these coatings; namely, the requirements of a high-temperature heat treatment. In comparison, organically modified hybrid coatings exhibit increased flexibility and thickness and can be cured at temperatures down to and including room temperature. However, low-temperature–cure systems have a lower hardness than the high-temperature cured inorganic counterparts.

Many studies on SG preparations of Al_2O_3 thin films or coatings have been undertaken [52–54]. However, it is generally very difficult using conventional SG techniques to produce a thin film with a thickness greater than 1 µm using a single coat; more specifically the thickness of single-coat Al_2O_3 films is of the order of 100–200 nm. An alternative coating system allowing a low-temperature cure route is that based upon a silica/alumina inorganic/polymer organic hybrid. This approach has the advantage that thicker coatings can be applied and fast-cured by varying the inorganic/organic ratio. Furthermore functionality can be incorporated into the formulation, for example, antibacterial properties, hydrophobic surfaces and varying electrical conductivity. Akid and Wang [47] reflects on a selection of results showing the anticorrosion performance of SG coatings applied to a variety of substrates including hot dip galvanized steel and Galfan coated steel, aluminum, magnesium, and mild steel.

Sol-Gel Preparation and Test Methods

The SG coating systems currently being applied to substrates are based upon one of two metal alkoxide pre-cursors, namely SiO_2 or Al_2O_3.

Preparation

The generic method of preparation of the SG system involves making the organic precursors in ethanol in the ratio 10:6, respectively. Deionized water is added dropwise to the solution at 50–80°C (122–176°F). Glacial acetic acid (CH_3COOH) is then added as the catalyst to promote hydrolysis and condensation reactions. Additional components can be added to vary the functionality or properties of the coating, for example the addition of alumina or corrosion inhibitors to provide improved coating hardness and adhesion and corrosion inhibition (self-healing properties).

The composite sol may then be applied to the substrate by a variety of application methods, notably, dipping, roller or spin coating, and spraying. Coating thickness values vary depending upon sol gel formulation and coating application method. Generally coating thickness values of 300 nm–5 µm are produced. The coating is then cured at temperatures ranging from room temperature up to 250°C (482°F), depending upon substrate/formulation, for a period typically around 30 seconds–15 minutes [26].

Corrosion Resistance

The corrosion resistance of the SG coatings was evaluated using three different electrochemical methods under simple immersion coatings. Several substrates have been coated and the results given below are a selection from the tests conducted to date.

Mg substrates: Results of simple immersion tests conducted on a Mg substrate coated either with a SG coating or a proprietary commercial Cr-free pretreatment. The results show the corrosion resistance of the commercial pretreatment is compromised after only 3 hours immersion and multiple site damage occurs. This damage increases dramatically as time increases to 24 hours, with the development of corrosion filaments spreading across the surface of the sample.

Zn substrates: Zinc and zinc alloy (Galfan) hot dip coatings, on steel substrates, are also suitable candidates for SG coating prior to top coating with paint. The results of EIS tests on selected SG formulations are given in by Akid and Wang [47]. EIS is a recognized method of assessing the film properties, notably resistance, of coated substrates immersed within a conducting electrolyte. As immersion time increases the chromate treated surface degrades more quickly than that of the SG and after around 100 hours the chromate and SG coated substrates show similar resistance values, around 5×10^5–1×10^6 Ω cm^2. One reason currently being considered for this continued corrosion protection via the SG coating is that the coating effectively seals and penetrates into the defects/pores in the Zn coating.

Al alloy substrates: Sol-gel hybrid coatings have also been applied to Al alloy substrates, for example AA2024-T3 and AA5005. Evaluation of the Al alloy corrosion performance has also been conducted using EIS when immersed in 3.5% NaCl. As with the results obtained from the Zn substrates, the impedance of the coating is around 10^6 Ω cm^2 after 24 hours immersion in the solution.

Mild steel: Initial attempts to coat mild steel with SG coatings were unsuccessful due to the reactivity of the steel surface and the formulation chemistry of the SG. Recent modifications to the formulation, combined with a suitable cure procedure now allows steel substrates to be coated.

Functional coatings: Hybrid SG coatings are considered to be an ideal carrier matrix allowing encapsulation of a variety of components thus enabling coatings to be made having a wide range of functionality. To date bioactive molecules/organisms have been added to the coating for the application of either biosensors or as an antimicrobial induced corrosion (MIC) coating. This latter type of coating is being developed for applications where microbial activity within the external environment can lead to accelerated localized corrosion. The capability to produce biologically active systems is further being explored for applications such as antibacterial coatings. Two different bacteria types have been reported for the Al substrate, notably vegetative and endospore types. To date only the vegetative type bacteria has been used with the stainless steel substrate.

In conclusion, a nontoxic, environmentally friendly silane-based hybrid SG system has been developed for use as a pretreatment or stand-alone coating for a variety of metal substrates. The SG system has the advantage of fast cure times at low temperatures, providing surface coatings of thickness up to 5 microns. The SG-coating system has also proved an acceptable surface for subsequent organic top coat treatments.

Comparison of the SG coating on Mg substrates has been made with a commercial Cr-free pretreatment. Immersion and EIS test results show the SG system to have significantly improved performance in terms of corrosion resistance. Hot dip Zn-based coatings were also found to be suitable substrates for the application of the hybrid SG system showing comparable performance with conventional chromate treated surfaces. Formulation variations of the SG system have opened up the possibility to produce novel bacteria-containing coatings for use where MIC causes material failure. Two different types of bacterial strain have, individually, been encapsulated within the SG each providing improved corrosion performance when tested in nutrient-laden artificial seawater.

Laser Cladding

Typically, drilling rigs have four to six tensioners, and oil production platforms have 24 to 60, depending upon the specific density. Historically tensioners have been fabricated from various carbon steel alloys, which are then coated with chrome, 316 stainless steel or other corrosion resistant alloys. Those coatings are generally applied via conventional means in thicknesses up to 0.5 mm (0.02 in.). The service life expectancy is only 2–3 years due to cracks that tend to occur with these relatively rigid coatings applied on these components that tend to flex, as well as other degradation of the coatings, which results in the cylinders being subject to corrosive attack, which, in turn, reduces the service life of the tensioners. Rig and platform manufacturers have requested an improved material/fabrication process to significantly extend tensioner service life. One such solution is laser cladding the tensioner with a cobalt based alloy.

Laser cladding enables rapid heat transfer to form a metallurgical bond between an alloy, and its specified properties, and the component substrate, which is something that does not occur with the mechanical bond of a coating. It is a high density process that uses a 4–20 kilowatt laser system—depending upon the part size. It results in minimal intermetallic dilution, a narrow heat affected zone and low thermal distortion. Micro-Melt CCw is a powder metal cobalt base material with high wear and corrosion resistance that has been produced via gas atomization. It has outstanding wear resistance in salt water environments thanks to the addition of controlled amounts of carbon and nitrogen, selected carbide forming elements and chromium to its cobalt matrix.

Laser cladding produces finished thicknesses up to 1.3 mm (0.05 in.) of this alloy on the tensioner and up to 25 mm (1 in.) on the seal groove area. These capabilities are critical in achieving the desired service life of the equipment. For the alloy to produce the high levels of corrosion and wear resistance, it must have a refined, homogeneous microstructure. The powder metallurgy (P/M) material has virtually no segregation and contains small, uniformly distributed carbide particles and a fine grain size compared with its cast-wrought counterpart. This enables the material to exhibit excellent corrosion and wear resistance properties once it is properly applied, as with the laser cladding process. For additional information on laser cladding and Micro-Melt material see Carter and Hunter [55] and Chapter 11 on Lasers.

PCB Coatings

The ideal PCB surface finish does not exist today. Every surface finish has an Achilles heel that effects fabrication, solderability, testability, reliability, or shelf life. The five common surface finishes: hot air solder leveling (HASL); organic solderability preservative (OSP); electroless nickel immersion gold (ENIG); immersion silver (ImAg); and immersion tin (ImSn), have advantages and disadvantages. Surface finish selection should be based on the specific application and a comparison of the advantages and disadvantages.

Historically, HASL and OSP are the most widely used surface finishes. OSP coatings have been around for a long time. They are gaining in popularity as an alternative to HASL in lead-free applications. OSP is an antitarnish organic compound that is applied over exposed copper surfaces to prevent the copper from oxidizing and/or tarnishing. OSP coatings are usually water-based organic compounds that form an organic/metallic layer over the copper surface. Common OSP coatings include benzotriazol, imidazol, and benzimidazol. Applied by spraying or dipping, the finish provides consistent coverage. Coating thickness will range from approximately 0.01 μm for a thin coating to approximately 0.5 μm for a thick coating. OSP is a cost-effective coating.

PCBs must be compatible with higher soldering temperatures and longer dwell times. The OSP coating process does not expose the PCB to high application temperatures as is the case with HASL. The high application temperatures associated with HASL may warp the pCB and/or crack plated thru-holes (PTHs). One disadvantage of OSPs is a concern caused by moisture absorption. Work continually goes on in this area of development. See Rowland and Kleinfeldt [56].

Coating Properties and Materials

The important properties of a coating include the lamellar or layered splat structure, entrapped unmelted or resolidified particles, pores, oxide inclusions, grains, phases, cracks, and bond interfaces [57]. Splat is the term given to a single impacted droplet or particle. Many overlapping splats solidify and adhere to one another to form a continuous layer. Thus, the splat is the basic structural building block in thermal spray coatings. Splats are created when the accelerated, molten particles impact a prepared surface. The arriving molten droplets are generally spherical, and on impact with the substrate they spread over and fill the underlying interstices (spaces). The droplets become flattened, disk-like structures.

Thermal spray processes are also characterized by rapid solidification. As the relatively small individual particles impact the more massive substrate, their heat is liberated quickly. Solidification rates (for metals) are in the range of 10^5–10^8°C/s. Such rapid cooling rates produce a wide range of material states, from amorphous to metastable. Two structures are generally present within a coating: splat structures and intersplat structures [57].

Thermal Spray Coating Properties

Thermal spray coatings are used to address an ever increasing variety of surfacing needs. These properties are usually expressed in terms of:

- Bond strength
- Hardness
- Corrosion/oxidation resistance
- Thermal properties
- Electrical properties, such as conductivity, resistivity, and dielectric strength
- Magneto-optical properties, such as absorptivity and reflectivity
- Machinability for finish

The relative importance of these properties is based on the intended coating function. Coating characteristics such as porosity, splat cohesion, and oxide content all have significant bearing on these properties. Bonding of the coating to the substrate and cohesion between consecutive splats is affected, in rough order, by;

- Residual stresses within the coating
- Melting and localized alloying at the contact surface between particles and between the substrate and adjoining particles
- Diffusion of elemental species across splat boundaries
- Atomic level attractive forces (van der Waals forces)
- Mechanical interlocking

For other aspects of coating properties such as measurable bond strength, mechanical interlocking, and cohesion see Davis [57], Berndt and Berndt [58], Granger and Blunt [59], and Tucker [60].

Thermal Spray Materials

Generally, materials that are suitable for thermal spray processing are stable at elevated temperatures. Materials that dissociate, decompose, or sublime tend to be poor candidates for coatings. Thus, most metals, intermetallics, alloys, all forms of ceramics (including oxides, borides, silicides, etc.), cermets, and some polymers are sprayable by one or more of the thermal spray processes. Again, these materials are usually applied in powder, wire, or rod form.

Coating Functions

The flexibility of thermal spray processing offers the possibility of applying coatings of almost any material onto almost any substrate to address a wide range of applications. The following is a brief list of common applications. The list is in no way complete. Once the attributes of the various thermal spray methods are known and their effects on coating or materials structure are understood, then only one's imagination limits the use of thermal spray.

- Wear coatings [57–60]
- Thermal insulation [57–60]
- Corrosion resistance [57]
- Abradables and abrasives [57–60]
- Electrically conductive coatings [57–60]
- Electrically resistive/insulating coatings [57–60]
- Dimensional restoration coatings [57–60]
- Medical coatings [57–60]
- Polymer coatings [57–60]

In summary, thermal spray coatings have become critical to enhancing the performance of many base materials used in spacecraft, aircraft engines, gas turbines, chemical reactors, metalworking mills, textile guides, bridges, pumps, compressors, medical prostheses, and even household items such as frying pans and toilet seats. An important surfacing method, this technology is heavily relied upon by manufacturers to provide wear, corrosion, and/or thermal protection; to improve electrical properties; and/or to refurbish and maintain many products. Historically, thermal spray has been used as a coating technology. However, recent process developments now extend thermal spray to the production of freestanding structures, composite metal structures, metal/ceramic matrix materials, superconducting oxides, and near-net shape parts. Thus, thermal spray has become more widely used as a materials processing technology for a broad range of manufacturing applications.

References

1. Hermanek, F. J. *Thermal Spray Terminology and Company Origins*, 1st Printing, Materials Park, OH: ASM International, 2001.
2. Hermanek, F. What Is Thermal Spray? Thermal Spray Engineering Consultant, ITSA, fhermanek@aol.com; www.thermal spray.org, 2001.
3. *The Metals Handbook*, 8th ed., Vol 1., Materials Park, OH: American Society for Metals, 1961.
4. Air Products, Allentown, PA, *Welding Journal* (August 2008): 27–9.
5. Ballard, W. E. *Metal Spraying and the Flame Deposition of Ceramics and Plastics*, 4th rev. ed. London: Charles Griffin and Company Limited, 1963.
6. Villafuerte, J. Cold Spray—A New Solid-State Material—Processing Technology, *Canadian Welding Association Journal*-IIW Special Edition (2007): 42–3, 46.
7. Villafuerte, J. Cold Spray: A Solution for Architectural Glass, *Welding Journal* (August 2008): 40–2.
8. Doering, T. Tough Coatings, Oerlikon Balzers, Amherst, NY, *Machine Design* (December 14, 2006): 90–8.
9. Physical Vapor Deposition, Tech Notes, Metals Handbook Desk Edition, Surface Engineering, Physical Vapor Deposition, Sputtering, *Advanced Materials & Processes (AM&P)* (March 2006): 59.
10. Luys, S. Advances in Sputter Coating, *AM&P* (March 2006): 35–40.
11. Bekaert Corp., President Kennedypark 18, B-8500 Kortrijk, Belgium. http://www.bekaert.com/bac
12. Azarmi, F. Vacuum Plasma Spraying, *AM&P* (August 2005): 37–40.
13. Williams, B. E., and R. E. Benander. Rapid Fabrication of Carbide Matrix/Carbon Fiber Composites, Ultramet, NASA-MSFC, *NASA Tech Briefs* (September 2007): 21.
14. Gregory, O. J., and M. Downey. University of Rhode Island; S. Wnuk and V. Wnuk, HPI Inc., Increasing Durability of Flame-Sprayed Strain Gauges, Glenn Research Center, *NASA Tech Briefs* (May 2007): 44–6.
15. Hogenson, P. A., G. R. Toombs, S. Adam, et al. Patches for Repairing Ceramics and Ceramic-Matrix Composites, Boeing Co., *NASA Tech Briefs* (September 2006): 28–30.
16. Voevodin, A. A., and P. Meltzer, Jr. Super-Tough Nanocomposite Coatings, Anteon Corp. & AFRL Matls. & Mfgturing Directorate, *AFRL Technology Horizons* (April 2005): 43.
17. Laha, T., K. Rea, T. McKechnie, S. Seal, and A. Agarwal. Synthesis of Bulk Nanostructured Aluminum Alloy Component through Vacuum Plasma Spray Technique, *Acta Materialia* 53, no. 20 (2005): 5429–38.

18. Seal, S., S. C. Kuiry, P. Georgiva, and A. Agarwal. Making Nanocomposite Components: Present Status and Future Challenges, *MRS Bulletin* 29, no. 1 (2004): 16–22.
19. Bansal, N. P., and D. Zhu. Glenn Research Center and U.S. ARL, Lower-Conductivity Ceramic Materials for Thermal-Barrier Coatings, *NASA Tech Briefs* (September 2006): 30–2.
20. Spire Receives NIH Grant to Develop Nanophase Coatings, Spire Corp., Bedford, MA 01730-2396, *AM&P* (January 2006): 83.
21. Koch, C. *Nanostructured Materials: Processing, Properties and Potential Applications.* Boca Raton, FL: Taylor & Francis, 2002.
22. Vishwanathan, V., A. Agarwal, V. Ocelik, N. Sobczak, J. De Hosson, and S. Seal. The Art of High Energy Density Processing of a Free Form Ni-Alumina Free Form Bulk Nanocomposite, *Journal of Nanoscience and Nanotechnology* 6 (2006): 651–60.
23. Tian, Z. R. Nanowire Coating for Bone Implants, Stents, *AM&P* (November 2007): 104.
24. Temperature Correction for Pressure-Sensitive Paint, LEW-16915, *NASA Tech Briefs* 24, no. 1 (January 2000): 50.
25. Scrivani, A., and G. Rizzi. Thermal Barrier Coatings: The Basics, *AM&P* (November 2007): 111–3.
26. Hu, J. J., C. Muratore, and A. A. Voevodin. Self-Lubrication of Hot YSZ-Ag-Mo Nanocomposite Coatings, AFRL, *Defense Tech Briefs* (June 2007): 27–8.
27. Sugama, T. Scientists Patent Corrosion-Resistant Nano-Coating for Metals, Brookhaven National Laboratory News, March 25, 2009; http://www.bnl.gov/bnlweb/pubaf/pr/PR_display.asp?prID = 929; p. 1 of 2, 4/8/2009
28. Bencic, T., and R. L. VanderWal. Nanorod-Based Fast-Response Pressure-Sensitive Paints, *NASA Tech Briefs* (September 2007): 28.
29. Miller, S. Carbon Nanotubes Fabrication Tool Combines DVD and PECVD, *AM&P* (August 2008): 8.
30. Casting Pearls, *Av. Wk. & Sp. Technol.* (April 21, 2008): 19.
31. Baldwin, C., and T. Poplar. Metallic Looks that Last, *Machine Design* (September 13, 2007): 106–14.
32. Porcelain Enamel Institute, Inc., PO Box 920220, Norcross, GA 30010. Available at online http://www.porcelainenamel.com
33. Thornton, J. Simulations Speed & Smooth Automated Coatings on F-35 Aircraft, *Robotics World* (January/February 2006): 12–6.
34. Lin, S-Y. Researchers Develop Darkest Manmade Material, http://www.manufacturingcenter.com/enews/2008, March 28, 2008; p. 1 of 3.
35. Scrivani, A., et al. A Comparative Study on HVOF, Vacuum Plasma Spray and Axial Plasma Spray for CoNiCrAlY Alloy Deposition, *Proceedings of ITSC2001,* Singapore, May 28–30, 2001.
36. Scrivani, A., et al. On the Experimental Correlation Between the Properties of Yttria Partially Stabilized Zirconia Coatings and the Process Characteristics, *Proceedings of ICCE/8,* Tenerife, Spagna, August 5–11, 2001.
37. Scrivani, A., et al. A Comparative Study of High Velocity Oxygen Fuel, Vacuum Plasma Spray (VPS) and Axial Plasma Spray (AxPS) for the Deposition of CoNiCrAlY Bond Coat Alloy, *Journal of Thermal Spray Technology* 12, no. 3 (September 2003): 1.

38. Scrivani, A. Thermal Spray Coatings: Introduction and Application to the Power Generation Field, *Proceedings of II Course, International Summer School on Advanced Materials Science & Technology, Advanced Coating Technologies*, 124–33. Jesi-Ancona, Italy, August 28–September 1, 2000.

39. Scrivani, A., et al. Resistance of Thermal Spray Coatings in Sour Environments: A Comparison of Tungsten Carbide, Chromium Carbide and Inconel 625, *Proceedings of EUROPM 99, Advances in Hard Materials Production*, Torino, Italy, November 8–10,1999.

40. Scrivani, A., et al. A Contribution to the Surface Analysis and Characterization of HVOF Coatings for Petrochemical Application, *Wear* 250, no. 1–12 (2001): 107–13.

41. Scrivani, A., et al. A Comparative Study of the Deposition of Spray and High Velocity Oxygen Fuel, *Proceedings of ITSC2001*, Singapore, May 28–30, 2001.

42. Scrivani, A., M. Rosso, and L. Salvarani. Performances and Reliability of Thermal Spray Coatings WC Based Materials, *Proceedings of the 15th Plansee Seminar*, Reutte, Austria, May 28–June 1, 2001.

43. Scrivani, A., et al. On the Experimental Correlation Between Plasma Spray Process Conditions and Yttria Partially Stabilized Zirconia Coating Properties, Thermal Spray, Surface Engineering via Applied Research, *Proceedings of ITSC2001*. Edited by C. Berndt, 1207–10. Materials Park, OH: ASM International, 2001.

44. Suhr, D. S., T. E. Mitchell, and R. J. Keller. Microstructure and Durability of Zirconia Thermal Barrier Coatings, *Advances in Ceramics*, Vol. 3. Edited by A. H. Hueur and L. W. Hobbs, 503–17. Westerville, OH: American Ceramic Society, 1982.

45. Heintze, G. N., and R. McPherson. Structures of Plasma-Sprayed Zirconia Coatings, *Advances in Ceramics*, Vol. 24. Edited by S. Somiya, H. Yanagida, and N. Yamamoto, 431–37. Westerville, OH: American Ceramic Society, 1988.

46. TPS Coatings, NASA Ames Research Center, Rept. ARC-15260-1, 6-13-2008.

47. Akid, R., and H. Wang. Sol Gel Coatings: An Alternate Corrosion Protection System, Materials & Engineering Research Institute, Sheffield Hallam University, *Corrosion Management* (May/June 2006): 9–12.

48. Wang, H., and R. Akid. Patent Application GB2006/001620.

49. Wang, H., and R. Akid. Patent Application GB2006/001527.

50. Metroke, T. L., R. L. Parkhill, and E. T. Knobbe. *Progress in Organic Coatings* 41 (2001): 233.

51. Mackenzie, J. D. Structures and Properties of Ormosils, *Journal of Sol-Gel Science and Technology* 2 (1884): 8.

52. Masalski, J., J. Gluszek, et al. *Thin Solid Films* 349 (1999): 186.

53. Parola, S., M. Verdenelli, et al. *Journal of Sol-Gel Science and Technology* 26 (2003): 803.

54. Velez, K., and J. F. Quinson. *Journal of Sol-Gel Science and Technology* 19 (2000): 469.

55. Carter, M., and J. D. Hunter. Laser Cladding a Cobalt Based Powder Metal Alloy Provides Long Life in Offshore Drilling and Production Applications, http://www.cartech.com/news.aspx?id = 3498; pp. 1–3, 5/6/09.

56. Rowland, R., and H. Kleinfeldt. The Lead-free PCB Evolution, Surface Mount Technology, March 2009; http://smt.pennnet.com/display_article/356147/35, 5/13/09, pp. 1–4.

57. Davis, J. R., ed. Coating Structures, Properties, and Materials as published in *Handbook of Thermal Spray Technology*, 2004. Revised by D. E. Crawmer, 47–53. Materials Park, OH: ASM International and the Thermal Spray Society, Thermal Spray Technologies.
58. Berndt, M. L., and C. C. Berndt. Thermal Spray Coatings, Corrosion: Fundamentals, Testing, and Protection, *ASM Handbook*, Vol. 13A, 803–13. Materials Park, OH: ASM International, 2003.
59. Granger, S., and J. Blunt, eds. Thermal Spray Processes, *Engineering Coatings: Design and Application*, 2nd ed., 119–66. Norwich, NY: William Andrew Publishing, 1998.
60. Tucker, R. C. Thermal Spray Coatings, Surface Engineering, *ASM Handbook*, Vol 5, 497–509. Materials Park, OH: ASM International, 1994.

Bibliography

Assadi, H., F. Gartner, T. Stoltenhoff, and H. Kreye. Bonding Mechanism in Cold Gas Spraying, *Acta Materialia* 51, no. 15 (2003): 4379–97.
Balani, K., G. Gonzalez, R. Hickman, A. Agarwal, S. O'Dell, and S. Seal. Synthesis, Microstructural and Mechanical Property Evaluation of Vacuum Plasma Sprayed Tantalum Carbide, *Journal of the American Ceramic Society* 89, no. 4 (2006): 1419–25.
Branagan, D. Enabling Factors Toward Production of Nanostructured Steel on an Industrial Scale, *Journal of Materials Engineering and Performance* 14, no. 1 (2005).
Brückner, F., D. Lepski, and E. Beyer. Modeling the Influence of Process Parameters and Additional Heat Sources on Residual Stresses in Laser Cladding, *AM&P* (August 2007): 88.
Ceramic Coating Allows Composite Use at High Temperatures, Zircotec; www.zircotec.org.uk
Champagne, V. K. *The Cold Spray Materials Deposition Process: Fundamentals and Applications*, U.S. Army Research Laboratory, 376 p. Cambridge.: Woodhead Publishing Ltd., September 2007.
Chandra, S., and P. Fauchais. Formation of Solid Splats During Thermal Spray Deposition, *Journal of Thermal Spray Technology* JTTEE5 (February 4, 2009): 1–33.
Chen, H. Metallic Glasses, *Chinese Journal of Physics* 28, no. 5 (1990): 10.
CVI vs MI for Carbide Matrix/Carbon Fiber Composites, *NASA Tech Briefs* (September 2007): 21.
DiCarlo, J. A. Glenn Research Center and R. T. Bhatt—ARL, Improving Thermomechanical Properties of SiC/SiC Composites, *NASA Tech Briefs* (September 2006): 24–5.
Froning, M., P. F. Ruggiero, and R. Bajan. Thermal Spray Techniques for Deep Bore Applications, *Welding Journal* (August 2008): 34–5.
Heinrich, P. 7th HVOF Colloquium Generates High Interest, *AM&P* (August 2007): 83–6.
Jeandin, M., S. Barradas, V. Guipont, R. Molins, M. Arrigoni, M. Boustie, C. Bolis, L. Berthe, and M. Ducos. Laser Shock Flier Impact Simulation of Particle-Substrate Interactions in Cold Spray, *AM&P* (November 2007): 116.

Koivuluoto, H., J. Lagerbom, and P. Vuoristo. Microstructural Studies of Cold Sprayed Copper, Nickel, and Nickel-30% Copper Coatings, *AM&P* (November 2007): 116.

Mattern, N. Structure Formation in Metallic Glasses, *Microstructure Analysis in Materials Science*, Freiberg, June 15–17, 2005.

Moody, D. R. Off-Line Development of Robot Motion Programs, *Welding Journal* (August 2008): 44–7.

Phani, P. S., D. S. Rao, S. V. Joshi, and G. Sundararajan. Effect of Process Parameters and Heat Treatments on Properties of Cold Sprayed Copper Coatings, *AM&P* (August 2007): 89.

Price, T. S., P. H. Shipway, D. G. McCartney, E. Calla, and D. Zhang. A Method for Characterizing the Degree of Inter-Particle Bond Formation in Cold Sprayed Coatings, *AM&P* (November 2007): 117.

Safe Practices for Thermal Spraying, *Welding Journal* (August 2008): 66; Datasheet 298, excerpted from the *Welding Handbook*, 9th ed., Vol. 3, Part 2.

Salimijazi, H. R., L. Pershin, T. W. Coyle, J. Mostaghimi, S. Chandra, Y. C. Lau, L. Rosenzweig, and E. Moran. *AM&P* (November 2007): 117.

Scrivani, A., H. K. Pugsley, and G. Rizzi. Development of Thick Thermal Barrier Coating with Varying Porosities and Continually High Functional Properties, *Proceedings of ITSC2004*, Osaka, Japan, May 10–12, 2004.

Thermal Spray Coatings Protect Petro-Chemical Components from Corrosion, *AM&P* (August 2008): 71–3.

Thermal Spray Coatings Protect Steel Structures from Corrosion, *AM&P* (November 2007): 114–5.

Thermal Spraying Provides Corrosion Protection to Wind Turbines, *AM&P* (August 2007): 80.

Unger, R. H., R. D. Cook, and W. C. Mosier. Comparison of Deposits of Wires Applied by Welding, Thermal Spraying, and Spray and Fuse, *Welding Journal* (August 2008): 50–3.

Villafuerte, J., and W. Zheng. Corrosion Protection of Magnesium Alloys by Cold Spray, *AM&P* (September 2007): 53–4.

15

Ballistic Protection for Ground Vehicles, Human Personnel, and Habitats

Mel Schwartz

CONTENTS

Introduction

Why Armor Development

One thing the war in Iraq has taught the military and its suppliers is that the threat—and the enemy that poses it—is no longer clearly defined. Armor manufacturers have been called upon to design protective systems for vehicles and personnel that can meet not only increasing levels of threat, but also withstand damage delivered by a variety of weapons wielded by unlikely combatants. Pressure is always on, as well, to reduce armor system weight for security forces (military and civil) that require high mobility and maneuverability in environments as varied as the open desert to the congested urbanscapes. As a result, fiber-reinforced composites are earning a larger role in protective systems, supplanting or supplementing legacy systems that rely on metals and ceramics.

Karl Chang, a research associate at Dupont Advanced Fiber Systems (Wilmington, Delaware), reports today's armor design must meet multiple functional requirements in addition to ballistic performance, the system must meet weight limits and, when incorporated into vehicles, also fulfill structural requirements. Further, armor systems must account for what the industry calls "over-matching threat." "Over-matching threat means if I design an armor system, I have to design for a specific threat, but can't guarantee that the enemy will only shoot at me at the specification I am designing to, so I must be prepared to deal with that," Chang says. "That" can be armor-piercing bullets (called *rounds* by insiders) or terrorist bombs and incendiary devices. The latter, which armor designers categorize as improvised explosive devices (IEDs), not only deliver a blast load and projectile fragments but often include a fireball as well, says Chang, who notes that flammability has become a significany issue in armor design.

Dr. Leo Christodoulou, program manager for the Defense Advanced Research Projects Agency (DARPA), adds that armor also must withstand repeated hits without catastrophic failure and remain environmentally stable; that is, maintain its performance properties when exposed to the elements. Here, says Christodoulou, it is important to remember that armor is still as much art as science.

A third requirement, and perhaps the most difficult, is that armor design must reflect the changing nature of war. By way of illustration Chang notes that high mobility multipurpose wheeled vehicles (HMMWVs), commonly referred to as *Humvees*, are lightweight, lightly armored vehicles that were originally designed to perform behind-the-frontline duties. In Iraq, however, they are highly vulnerable and today, he says, "all vehicles need protection." Due to the nature of current and future threats to humans and vehicles, Chang believes extensive attention should be given to military armor protection elaborating on new armor technologies, challenges, and developments in protective equipment and recent research on armor components.

Not only should the above be highlighted and discussed but also how military armor protection will help to identify and reduce armor shortcomings and increase performance and ensure that warfighters and vehicles have the best possible armor protection against the persistent threats. How capabilities and technology can be optimized to offset current and future strategies and solutions. These future strategies explore solutions for mobility, survivability, and adaptability in hostile terrain, and gain insights into the future of military armor in supporting the warfighter through vehicle survivability measures and personal protective armor.

Finally, the following areas of protection require attention:

- Unidirectional body armor systems to defeat explosively formed projectiles (EFPs) and IEDs
- Innovations in survivability of ground vehicles
- Design and performance improvements for individual ballistic protection
- Developing flexible body armor with hard armor ballistic characteristics
- Development of next generation "smart armor"
- Commercialization of next generation structural composite armor

---

Armor Protection for Vehicles

There has been a need to provide rocket-propelled grenades (RPG) protection for HMMWVs. In live testing, to validate the performance and optimize the engineering design, a modular design proved effective at preserving the integrity of the vehicle and safety of the crew. An aluminum-alloy, modular bar armor system was selected and provided protection against rocket-propelled grenades without compromising the operational capabilities of the vehicle. This was installed on a U.S. Army RG31 vehicle and reported by K. Spiller [1], BAE Systems. At less than half the weight of comparable steel designs, the modular unit bolts onto the vehicle without welding or cutting and can be repaired in the field.

The U.S. Army has introduced the next generation of combat vehicles, which are currently being fielded for the first time in Iraq. The new Mine Resistant Ambush Protected (MRAP) vehicle is the latest development in troop protection. What makes the vehicle impressive is its height, and the fact that it can drive about 20–30° on its side without rolling over. Additionally, the fabricators of the MRAP gave considerable attention to comfort for the soldiers in their design. The MRAP design was based on Textron's Armored Security Vehicle (ASV), and is being produced for the U.S. Army. The ASV is a 4×4 wheeled armor vehicle that offers exceptional crew protection through multiple layers of armor that provides defense against small arms fire, artillery projectile fragments, and land mines. This advanced armor is exceedingly lightweight and allows the vehicle to be able to "roll-on/roll-off" C-130 military transport aircraft. The ASV possesses superior mobility, agility, handling, and ride quality through the utilization of a four-wheel independent suspension system.

There are two categories of MRAPs: Category one holds up to six soldiers and will replace the Humvee: Category two is a longer version, which can hold a crew of 10 soldiers. Both vehicles feature an angled bottom shell to deflect mine blasts, as well as pneumatic doors and suspended seats. The MRAPs were designed to be both safe and effective for soldiers conducting patrols, convoy security, and missions in all types of terrain. They are currently being fielded by units in areas with the highest threat and will become the new standard U.S. Army vehicle (Figure 15.1).

Two new glass fiber composite materials, Featherlight and Quicksilver, have become a key component in the battle to keep military vehicles such as the MRAP and HUMVEE from becoming hugely overweight. The new Featherlight glass fibers are engineered to deliver ultrahigh performance against severe threat levels. The advanced Featherlight fibers provide an increase in protection of 5–10% over standard S-2 Glass fiber composite armor. The new Quicksilver glass fibers reportedly enable significantly stronger, stiffer, and higher composite parts than traditional E-Glass reinforcements and are designed to be a cost-effective solution where weight is deemed to be less of a concern. The Featherlight and Quicksilver fibers have been used in a variety of ways in composite armor systems including:

FIGURE 15.1
The MRAP is the next-generation army combat vehicle.

- Compression molded laminates made from phenolic resin and woven roving. An armor system using S-2 Glass fibers and used as a fire retardant armor solution deployed on several military vehicles.
- Thermoplastic molded laminates made from unidirectional fibers and thermoplastic resins. These laminates, designed to be thermoformable for more complex armor shapes, have a potential to be combined with other thermoplastix fibers to provide hybridized system solutions with synergy performance in terms of structural ballistic performance.
- Hybrid laminates made from a combination of reinforcement materials and designed to exploit the full range of ballistic performance.
- Very thick compression molded laminates designed to stop the most current and advanced threats such as EFPs.
- Three-dimensional or noncrimp fabrics, which have been designed to facilitate complex shapes via easy processing using infusion technology while maintaining ballistic performance.

A new concept has been formulated and put into action on the M915 A2/A3 tactical vehicle and the HMMWV. The lightweight armored doors have been used to reinforce high-hard (HH) steel with a laminated, woven, high-tensile strength glassfiber/polyester-matrix composite that has performed well as armor material in previous military applications. A fabrication procedure for implementing the concept can be summarized as follows:

- Prepregs (sheets of the woven glass fibers preimpregnated with the uncured polyester (resin) are cut to sizes and shapes to fill their assigned place in the door.
- Cut prepregs are stacked as multiple plies to obtain the required thickness.
- Using the HH steel door panel as a mandrel or mold, the prepreg stack is vacuum-bagged and cured into the final composite form by heating according to a prescribed temperature versus time schedule.
- The HH steel door panel and the composite panel are prepared for adhesive bonding, then assembled with a polysulfide adhesive, then vacuum-bagged for room temperature or optional elevated temperature curing of the adhesive.
- Other steps (e.g., installation of weather stripping) in the ordinary fabrication of the door are then completed [2].

"This armor exceeds IED specifications," said Robert Codney. "The U.S. Army took the HMMWV onto the streets of several Iraqi cities where actual IED attacks resulted in minor abrasions to the troops using this armor. It has exceeded expectations." Additionally other engineered composite technologies for armor applications have been installed on the M915 A2 and A3 cab as well as underbody armament and the fully armored HMMWV, which also uses ceramic systems containing aramid fibers.

An all-composite version of the HMMWV military vehicle has now been unveiled. The composite HMMWV weighs 400 kg (900 lbs) less than a current steel and aluminum model required to carry the same heavy armor. This weight savings means the composite vehicle can carry more equipment or soldiers. The patented Seeman composites resin infusion molding process (SCRIMP) technology, which is a vacuum-assisted resin transfer molding (RTM) process has been used to fabricate the vehicle. The frames and bodies are made of a combination of fiberglass, balsa wood, foam, and carbon reinforcements held together with resin. The chassis has extra carbon for added strength and stiffness, while other composite parts of the vehicle are lighter and more pliable. This prototype vehicle (Figure 15.2) is part of the U.S. Army's All Composite Military Vehicle program. This vehicle serves as a demonstration of how we can apply composites technology to other military vehicles or vehicle components. The Army continues to identify technological advancements and attain technological capabilities with lighter, stronger, and/or more producible vehicles and survivability systems.

Initially, vehicle retrofit kits were made chiefly of metal; however, the trend now is to use composites with metal or ceramic strike faces to augment protection and reduce weight. Dupont's aramid fiber (Kevlar) has been used to produce a composite backing, known as a spall liner, specifically designed to deal with over-matching threats and to reduce damage caused by fragments. The Kevlar fiber line possesses a tensile strength ranging from 525 ksi to more than 560 ksi. The highly resilient fibers "catch" projectiles that are slowed down but ultimately pierce the "hard" armor.

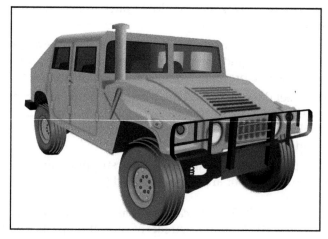

FIGURE 15.2
The all-composite HMMWV military vehicle.

Kevlar spall liners are used in a number of the U.S. military's vehicle armor systems, including the M113, the Low-Signature Armored Cab (LSAC) for the Army's Family of Medium Tactical Vehicles (FMTV), and the Armor Security Vehicle (ASV). Kevlar is an ideal material for vehicle armor systems facing IED threat because it is inherently nonflammable, offering high temperature resistance at continuous temperatures as high as 150°C (302°F). Kevlar does not burn but instead chars or decomposes when the temperature rises above 500°C (932°F).

Another similarly employed material is Honeywell's Spectra fiber. The high-tenacity, high-medium polyethylene (HDPE), made into sheet/rolls with unidirectional tapes, can stop, without any additional materials, up through a NIJ Level III threat, as set by the U.S. National Institute of Justice (NIJ). The NIJ Standard identifies threat thresholds, which range from Level I to Level IV. Although written specifically for personal body armor, the standard is frequently used to compare other armor products. A key point to note is that, once a threat reaches above NIJ Level III, composites must be combined with another material (i.e., ceramics or metals) to adequately diffuse the threat. The excellent performance of the Spectra fiber is due to the gel spinning process. Unlike other fiber spinning processes, gel spinning involves not only melting the polyethylene polymer, but also dissolving it with solvent to facilitate realignment of the molecular chains and then extracting the solvent, according to Lori Wagner, Honeywell International, Morris Township, New Jersey. The result is a very high degree of orientation with minimum folding of molecule chains, which optimizes fiber strength.

While retro kits for vehicles already in the field have accounted for a significant portion of the armor market in the past 5 years, the trend today is to integrate armor into new vehicle structure during the design stage, both to save weight and reduce overall cost. A notable example is the *David* armored vehicle from the Armor Division of Arotech Corp. (Auburn, Alabama). Shorter, lighter and narrower than an armored Humvee, the 3298 kg (7400 lb) *David* holds a crew of seven and is designed to maneuver easily in urban combat situations. It offers integrated 360° protection against point-blank assault rifle fire and short-range, armor-piercing rounds as well as limited IED protection. A mix of low-cost, fiber-reinforced polymer (FRP) and advanced composite armor components—including Dupont Kevlar and Dyneema high-modulus, high-strength polyethylene fibers from DSM (Heerlen, The Netherlands)—have been designed into the main cabin, including front and rear doors of the *David* vehicle.

Meanwhile, other companies are outfitting vehicles with a trademarked S-2 Glass armor system in Europe. First used for U.N. peacekeepers in need of protection against sniper fire and land mines in Bosnia, the CAV 100 armored vehicles, for instance, seat six in an all S-2 Glass-reinforced compartment. Made by high-pressure compression molding, the cabin components provide substantial protection against threats from fragmentation, high-velocity small arms and blast threats.

S-2 Glass—the only S-Glass fiber produced in the United States—entered the ballistic armor market in the 1980s. The defense market looked at the S-2 Glass fibers as a replacement for spall liners made with aramid.

Some scientists claim that S-2 Glass fibers offer better fire/smoke/toxicity (FST) characteristics and no water absorption at a lesser cost than aramid. The S-2 Glass is suitable for integrated vehicle armor because it can, in a composite, support structural loads. The magnesium aluminosilicate chemistry of S-2 Glass yields more desirable mechanical properties than E-Glass. The S-2 is about 2% lighter, 40% stronger, and 20% stiffer than E-Glass, with strain-to-failure of 5.5%.

Other alternative materials being used include more conventional and less expensive fibers in easily processed thermoplastic matrices. There is a ThermoBallistic line of structural laminates that consists of three offerings: ThermoBallistic-E combines continuous E-Glass with polypropylene (PP) thermoplastic polymers, Thermoplastic-S is reinforced with S-Glass, and an aramid ballistic-grade laminate. Some engineers and chemists believe thermoplastics are an attractive material for the ballistic armor market because they emit no volatile organic compounds (VOCs) and are recyclable. Additionally, thermoplastics can answer environmental concerns.

Creating innovative armor systems for military vehicles and identifying promising new armor systems is an active program under DARPA (Defense Advanced Research Projects Agency). One of these programs utilizes a hybrid metallic/composite solution that reportedly offers the same or better protection performance as existing metal zrmor systems but at a lighter weight. The system uses a patented unidirectional high-tensile reinforcement fabric that offers mechanical properties in line with that of carbon fiber. Initial data indicates that using a composite, which incorporates a fine metal wire reinforcement, can obtain ballistic performance that can give the same protection as conventional metal armor at a reduced weight and may not only be suitable for vehicle armor and other various ballistic applications. The hybrid system can be processed by a variety of methods, from hand layup to vacuum infusion to pultrusion, depending on the application.

Another new material and process receiving interest is the manufacture of 2- and 3-D fabrics for armor applications. To achieve multiple hit performance, especially in the area of 3-D fabrics, 3-D woven fabrics can withstand multiple hits because their z-directional fibers tie all the plies together, which makes the material a candidate for next-generation vehicle systems. A high-speed rapier-type weaving machine has been capable of weaving fabrics as thick as 38.1 mm (1.5 in.). Typical standard woven roving used in ballistic applications is a 5×5 construction, however, in development is the 300×300 construction. In tests, 3-D multilayer glass fiber fabrics have been used to reinforce Humvee hoods using the SCRIMP process, a patented variation on the vacuum-assisted resin transfer molding (VARTM) process.

Another development is a newer fiber, known as M5, and is said to offer greater compression strength than aramid fibers, at a lighter weight.

Developers claim that, based on U.S. Army tests, the M5 fiber could offer a weight savings of up to 40% in body armor systems, yet provide greater protection than other fibers.

Current commonly deployed armor materials are either inexpensive but heavy, such as steel-based rolled homogeneous armor (RHA); inexpensive but volumetrically inefficient, such as aluminum-based armor alloys (5083, 7039); or very lightweight and very expensive, as in the case of advanced ceramic plates with composite backing. Each material approach has historically served as effective protection. However, either the weight penalties, high cost, limited performance, difficulties forming/shaping, or negative operational limitations of these traditional approaches in the new threat environment (ballistic, IED fragmentation, and blast) drive the need for research into new armor approaches.

Lockheed Martin Missiles and Fire Control has been developing physics-based, armor-as-a-system solutions the address the above tactical needs [3]. The new armor, called Tekshield 3, will provide armor-piercing, bullet, fragment, shrapnel, and blast protection with tactical theater durability and maintainability at a very low cost. It is being developed in consideration of coupled threat effects of armor in tactical environments (i.e., blast followed by a swarm of projectiles) and not a single point threat protection. Although developed with the intent of direct incorporation as vehicle structure (not as an add-on), the Tekshield system is also suitable for the traditional A-kit/B-kit uparmor approach common today. Ballistic testing, which has been underway for months, has a high degree of correlation with modeling and simulation results.

The system is based on a metallic/polymer/ceramic system of low-cost elements, with the intent of combining the best characteristics of each class of material into a highly mass-efficient and cost-effective armor solution. Metallic and/or composite face and backing sheets provide handling durability combined with strength and toughness. Polymers provide repairable blast overpressure protection and internal shock wave attenuation. Discrete ceramic elements in the body of the armor provide the hard strike surface needed to shatter armor-piercing cores and to absorb the energy of ballistic threats, while maintaining multi-hit performance and repairability.

Testing and simulation have focused on the 0.30 caliber armor piercing bullet at high velocity, and on the 0.50 caliber fragment simulating projectile (FSP), which represents a class of fragments encountered with fragmenting bombs. Configurations tested to date include systems with metallic facesheets of titanium (Ti-6-4), aluminum (5083, 7039, scandium/aluminum), steel (HHS, 17-4 PH), bimetallics (titanium/aluminum), and metal matrix composites (reinforced titanium). Back plates evaluated include materials the same as the facesheets listed previously, plus composites: aramid fiber-based composites, ultrahigh molecular weight polyethylene (UHMWPE) fiber-based composites, toughened graphite epoxy, liquid crystal polymers (LCPs), and carbon nanotube reinforced aramid-based fiber composites. Figure 15.3

FIGURE 15.3
The mule and its armor protection, tekshield.

shows the Mule, a 2.5 ton class unmanned ground vehicle for the Future Combat System (FCS).

The military has studied the tactics and actions of Alexander the Great, King of Macedonia, who spent 8 years moving his massive military forces to victories across 11,250 miles. His victories were won through superior logistics, achieved by transporting troops and supplies on ships able to carry 400 tons, rather than horses carrying 90.9 kg (200 lbs) and requiring 9.1 kg (20 lbs) of food daily. By addressing similar weight and transportation issues, the U.S. military is developing a "logistical footprint" designed to reduce the vehicle weight of its ground forces. Because 70% of combat vehicle weight is composed of structure and armor, designers have been working with metals manufacturers to develop lighter-weight solutions.

The changing of armoring concepts has produced FCS. The new systems will transform military logistics through highly responsive land, air, and sea operations into a coordinated offensive. The United Kingdom will achieve similar types of "network-centric" logistics through its program called Rapid Effects System. Finding the logistical equivalent to Alexander's ships in its C-130 Hercules aircraft, the DOD is restricting FCS combat vehicle weight to 19 tons, a weight that permits transport in C-130 cargo planes. This smaller logistical footprint reverses a trend toward heavier armored vehicles, and effectively imposes limitations that will "compel a fundamental shift—in the armor community," writes Retired Lt. Col. Andrew F. Krepinevich Jr. The U.S. Revolution in Military Logistics initiatives require a 75% reduction in support and supply vehicle fuel consumption. Increased fuel efficiency is logistically critical, because fuel comprises 70% of the U.S. Army's total weight shipped to battle zones. Under these conditions, lightweight armor solutions yield an immediate return on investment.

To reach weight reduction goals, armor designers are turning to metals with higher mass efficiencies, represented by *Em*. The value of this number serves as a way to compare the mass efficiency of armoring materials, with RHA defined as 1. For example, dual hardness steel has an *Em* value of 1.78, enabling it to reduce armor weight by 56% over RHA. The high *Em* value makes dual hardness steel ideal for armored light tactical vehicles and a good candidate for armor on support and supply vehicles. Titanium armor also has proven highly successful in several current applications. The Bradley A3/FIST variants take advantage of titanium armor. The M1A2 tank commander's hatch, GPS cover, turret blowoff panels, and armor skirts are made of titanium to reduce vehicle weight and prevent corrosion. Forged titanium is in the M2 Bradley commanders hatch cover, and cast titanium is in the Expeditionary Fighting Vehicle sprocket and idler wheel.

The U.S. Army's ARD&EC identifies titanium as an "excellent alternative to steel," because it's 30% lighter. Selected for its weight reduction characteristics and structural integrity, titanium gunner protection shields are in the Stryker Fire Support Vehicle and Reconnaissance Vehicle. To help reduce the overall weight of the Mobile Gun System and meet C-130 weight limits, titanium has replaced steel armor in several applications on this vehicle. Additionally, CP titanium was selected for the construction of explosive reactive armor boxes. These boxes protect vehicles from RPG threats by exploding outward, effectively redirecting shaped charges away from the vehicle.

The U.S. Marines will likely adopt titanium alloys for their future combat vehicles. These vehicles are expected to weigh between 10 and 20 tons, and be capable of surviving attacks from ballistic and blast threats. Besides evaluating titanium (6Al-4V) as a candidate material the Marines are considering aluminum and polymer-based composites as structural materials for FCS vehicles.

Whether for stopping cars or bullets, titanium is the material of choice, but it has always been too expensive for all but the most specialized applications. That could change, however, with a nonmelt consolidation process being developed by Oak Ridge National Laboratory (ORNL) and industry partners. The new processing technique could reduce the amount of energy required and the cost to make titanium parts from powders by up to 50%, making it feasible to use titanium alloys for brake rotors, artificial joint replacements, and of significant interest now, armor for military vehicles.

A titanium alloy door made by ORNL for the Joint Light Tactical Vehicle combat vehicle reduced the weight of the vehicle while decreasing the threat of armor-piercing rounds. The lightweight titanium alloy also improves the operation of the door and increases mobility of the vehicle, making it even more useful to the military. The nonmelt approach includes roll compaction for directly fabricating sheets from powder, and press and sinter techniques to produce net shape components and extrusion. Instead of using conventional melt processing to produce products from titanium powder, the powders remain in their solid form during the entire procedure, saving energy required for processing and reducing the amount of scrap.

One other threat that will not be discussed here in any detail is that of corrosion. Combat vehicles capable of surviving minefields and direct fire from enemy RPGs are still vulnerable to the same corrosion threats as poorly armored support vehicles. According to a study by NACE International, "corrosion is potentially the number one cost driver in lifecycle costs" for the DoD. "Corrosion impacts vital aspects of the military by reducing combat readiness, increasing maintainance costs, and reducing the morale and training time of soldiers strapped with the responsibility of 'rust busting,' or tending to rust prevention," says Rich Hays, U.S. Marine Corrosion Branch Head. Vehicles are highly susceptible to corrosion during transport, often being exposed to marine environments while aboard ships. To develop better materials, the Office of Naval Research (ONR) is conducting a 6-year study (2013), which will assist in material selections that can help achieve the goal of "vehicles capable of more than 20 years of service with minimal corrosion impact."

Clearly the collaboration between materials manufacturers, armor engineers, and combat vehicle designers will continue to produce lightweight, corrosion-resistant solutions to protect our vehicles.

It should be noted here that around $9 billion of material—including fiber reinforced composites, steel, aluminum, titanium, and ceramics—will be required to meet the U.S. Department of Defense armor procurement rates for military ground vehicles between 2008 and 2013 according to recent studies. "Material requirements in the 2008 surge in Iraq alone created the demand especially due to MRAP (mine resistant ambush protected) vehicle armor procurement and EPP (explosively formed projectile) protection for HMMWVs and MRAPs," claims Marcia Price, Vector Strategy Inc. "In addition the implementation of the Long Term Armor Strategy within the Family of Medium Tactical Vehicles and within HEMTT A4 production mid year also increased demand for armor materials."

Looking at the future, the material market for armor for military ground vehicles will decline by 2011–2012, but then grow in 2013 and beyond due to Joint Light Tactical Vehicle (JLTV) and FCS Manned Ground Vehicle armor, combined with a move from lower cost metallic armor to higher cost nonmetallic solutions. The JLTV family of vehicles being developed under contract by the DoD with Northrop Grumman has successfully undergone armor ballistic and mine-blast testing. Oshkosh (Oshkosh, Wisconsin) and Plasan USA (Bennington, Vermont), have both been selected and involved in the design of the vehicle's armor. The vehicle's prototype armor in the first round of testing passed all threshold capability and achieved objective-level, force-protection requirements. The armor is an advanced composite-technology armor system that maximizes protection while minimizing weight impact.

The designers have a vehicle of unique performance and protection that can provide value to the warfighter today yet is flexible enough to meet the combat requirements of tomorrow. The designer's particularly took on the challenge of armor volume, applying some of the most innovative thinking to the balance of performance, protection, and payload. The results of the armor testing validate the design especially with the incorporation of a diesel-electric drive system, which eliminates the need for a transmission and conventional drivetrain. This allows for the creation of improved blast protection for the crew. "The innovative use of a diesel-electric system reduces the number of vehicle components and frees up space to allow for increased survivability for the soldiers in these vehicles," said John Stoddart, Oshkosh President of Defense.

A new lightweight, all-composite truck cab for the U.S. Army's tactical wheeled vehicle fleet was recently unveiled. The all-composite cab is light, durable, and strong enough to carry the heaviest armor and mine-blast protection. It will let soldiers carry more armor, ammunition, and equipment because it weighs hundreds of pounds less than cabs constructed with conventional materials. For example, it can accomodate 182 more kg (400 lbs) than its aluminum-based counterpart. The cab is also said to be able to carry the latest armor, something the aluminum design can't do. This would allow for a weight savings far exceeding 182 kg (400 lb).

The current technology, along with advanced armor has allowed for an increase in reliability, performance, and protection of platforms on which it is installed. It is a leap forward in vehicle manufacturing and now the prototype vehicle undergoes road testing prior to full acceptance and use for the HEMTT A-3 (Heavy Expanded Mobility Tactical Truck).

With the goal of protecting military people who face the threat of armor-piercing weapons on the battlefield, the Air Force Research Laboratory (AFRL) engineers have been treating a new type of transparent armor that is stronger and lighter than traditional glass materials. The investigation has led to aluminum oxynitride (ALON) as a replacement for the conventional, multilayered glass transparencies in existing ground and air armored vehicles. The Army is exploring ALON transparent armor as a replacement for ground vehicle windows, while the Air Force is interested in the material for slow, low-flying aircraft such as the C-130, the A-10, and helicopters.

Aluminum oxynitride is a ceramic that has high compressive strength and durability. When polished, it is the premier transparent armor for use in armored vehicles. "The substance itself is light years ahead of glass and offers higher performance and lighter weight," asserts Lt. Joseph La Monica, transparent armor subdirection lead for AFRL's Electronic and Optical Materials Branch. Whereas manufacturers create traditional transparent armor by bonding thick layers of glass together, they are creating ALON transparent armor by combining the ALON piece (as a strike plate), a middle section of glass, and a polymer backing. The ALON armor is notably thinner than traditional armor. In addition, ALON is virtually scratch- and impact-resistant providing better durability and protection against armor-piercing threats at roughly one-half the weight and thickness of traditional glass transparent armor.

According to Ron Hoffman, UDRI (University of Dayton Institute) researcher, the ability to establish the necessary level of protection with only a small amount of material is very advantageous. "When looking at higher-level threats, you want the protection, not the weight," he explains. "Achieving protection at lighter weights will allow the armor to be more easily integrated into vehicles." Mr. Hoffman also emphasizes the benefit of ALON's durability; "eventually, with a conventional glass surface, degradation takes place and results in a loss of transparancy. Things such as sand have little or no impact on ALON, and it probably has a life expectancy many times that of glass."

Though the possibilities of this material seem limitless, researchers must address issues such as manufacturability, size, and cost before they can transition ALON armor to the field. "Traditional transparent armor costs a little over $3 per square inch; when you look at ALON transparent armor, the cost is $10–$15 per square inch," Lt. La Monica ezplains. "The difficulties arise with heating and polishing processes, which, in turn, lead to higher costs, but we are looking at more cost-effective alternatives." Experimenting with the polishing process has proven beneficial. By polishing in a certain way, the strength of the material was increased two-fold.

Currently, size is also a limitation, because the equipment needed for heating larger pieces is expensive. To help lower the costs, researchers are studying design variations that tile smaller pieces of the armor together to form larger windows. Once manufacturers can produce the material in sufficiently large quantities to meet the military's needs and the cost comes down, ALON's durability and strength will prove beneficial to the warfighter. As summed up by Lt. La Monica, "It might cost more in the beginning, but it's going to cost less in the long run because you're going to have to replace it less."

The GKN Aerospace company (Redditch, UK) is developing instantly dimmable bullet-resistant windows. The development program will combine GKN's bullet-resistant-glazing expertise with suspended-particle-device (SPD) technology—also called SmartGlass—from Research Frontiers, Inc. (Woodbury, New York). The SPD technology will also offer vehicle occupants instant "on-demand" light control and attenuation, privacy, and protection from heat, glare, and UV-light penetration.

Coated on or between layers of glass or plastic, SPD technology allows light transmission to be electrically controlled with a quick response time and high uniformity across the clear aperture. A film contains randomly oriented microscopic particles that absorb light; when a voltage is applied to the film, the particles all align, allowing light to pass through. By varying the voltage, the amount of light passing through the film can be varied. The initial application for this new development will be in the global counterterrorism market for government very important person (VIP) armored personnel vehicles but also has real value in the civilian VIP market. These "SmartShade" windows will give a level of control, protection, and privacy for vehicle occupants not available until now.

The development program will develop and refine a manufacturing process that will enable SPD technology to be effectively incorporated into armored windows. Secondly, the program will ensure that these new windows perform effectively in a range of extreme environmental conditions, and will satisfy stringent ballistic-testing performance requirements against a variety of urban threats.

Armor Protection for Personnel

Introduction

In past wars, armies were generally made up of people who would just as soon be somewhere else. They were blacksmiths and bakers and storekeepers and farmers, and were mostly drafted into service. Today's armies of terrorists want to destroy civilization and are made up of quite different people. They are fanatics whose goal in life is to cause death. Because any person, building, or vehicle that represents civilization could be a target, the effort to develop effective, low-cost, easily manufactured armor becomes especially urgent. To provide its soldiers with a higher level of protection, the U.K. Ministry of Defence is incorporating a superstrong, lightweight Dyneema unidirectional polyethylene sheet into the inserts for its new Osprey body armor. The UHMWPE plastic combines exceptional strength with very low weight, and serves as the backing material for the ceramic insert, increasing resistance to ballistic impact.

The Dyneema material was selected for its ballistic performance at low weight, has consistent quality, and security of supply. The material combines exceptional strength with very low weight and adds significant protection with minimal additional weight. The Dyneema UD backing not only increases the vest's resistance to ballistic impact, but also helps protect the wearer against wounds in the event the ceramic shatters. The material already is in use in the United States as the backing material in SAPI and E-SAPI (Enhanced- Small Arms Protective Inserts).

Hard plate hybrid composite armor tiles that are lower in cost and lighter in weight than existing alternatives have passed key ballistic tests, reports CPS Technologies (Chartley, Massachusetts). The tiles are suitable for personal protective armor, light vehicle armor, and heavy vehicle armor. In the CPS structures, the ceramic inserts are as thin as 0.47 cm (0.185 in.) in SiC and 0.52 cm (0.205 in.) in 96% alumina. In existing armor alternatives, the typical ceramic insert thickness needed to pass the NIJ Level IV is 0.83 cm (0.325

in.) or greater. With CPS armor plates, the thickness of the ceramic can be adjusted to withstand higher threat levels than NIJ Level IV.

CPS structures tested to date place dense, hard ceramic at the interior of the hybrid composite structures with aluminum at the exterior of the structure. Other components can also be included in the hybrid. An infinite number of combinations can be created with this technology and are expected to meet various applications in military and private applications, at various price and performance points.

Ceradyne makes hot-pressed boron carbide and silicon carbide for weight-critical body armor. In addition, the integration of hot-pressed boron carbide designs with advanced helmet technology provides higher protection levels without weight gain. The development and widespread deployment of the above types of armor as well as other advanced armors are key elements to frustrating terrorists to kill and maim. Not only do military vehicles require armor protection but civilian automobiles are finding that armor protection is a must for the police in many cities. Researchers have been working on the development of new lightweight multifunctional hybrid armor systems capable of providing superior ballistic protection without compromising performance. Such armor systems combine a ceramic-tile strike surface: epoxy-matrix, glass-fabric-reinforced laminates with multiwalled carbon nanotube interlacing mats.

Material Systems and Materials for Personnel

Spectra Shield composite utilizing Spectra fibers discussed previously has helped fill a growing need for expanded small arms protective liners (E-SAPI)—strike plates that fit into pockets sewn into soldiers' vests. The plates, which are standard issue for every soldier in combat, consist of a ceramic strike face with a Spectra Shield spall liner. The ceramic breaks up a bullet or turns it on its side to diffuse its impact force and rob it of its velocity, while the liner serves as a "fragmentation energy catcher." Spectra Shield helps absorb and dissipate energy and catches fragments. The Spectra Shield composite is a flexible, cross-plied (0°/90°) nonwoven, impregnated with thermoplastic resin to optimize its impact absorption characteristics. Thermoset resins have been used to make helmets and other applications requiring more rigidity. One area of growth is in side body panels made from Spectra Shield, which extends the coverage beyond a breast and back plate.

Gold Shields, according to Lori Wagner, is another armoring material that is used in protective body applications. The product is created by laying parallel strands of aramid fiber side by side and then bonding them in place with an advanced resin system. Since May 2008, more than 100,000 vests featuring the Gold Shield product have been introduced into the field.

Dupont's Kevlar also is used in protective vests for the military. Kevlar, combined with a thermoset resin, is a material of choice for helmets because, according to the company, helmets made of Kevlar are 25–40% more resistant to fragments than steel helmets of equal weight. Helmets must hold their shape and withstand impact, but also support a soldier's weight if he sits on it. Structural performance is not always associated with ballistic performance, but it is still important for function.

Traditional glass-only transparent armor systems can be quite thick and heavy due to the amount of glass needed to stop high-powered projectiles. These systems consist of laminated glass layers with a polymer backing to stop ballistic projectiles, which often consist of armor-piercing (AP) threats with a soft outer jacket and hard AP core. In glass systems, where thickness of individual glass layers ranges from 4 to 12 mm (0.15–0.47 in.), the glass needs to absorb the energy from the projectile, and slow down the AP core enough for subsequent glass or backing layers to catch the core. As threats and specifications become tougher to defeat, ceramic composites are becoming a leading solution to reduce thickness and weight, while improving protection and transparency.

Saint-Gobain Crystals and Saint-Gobain Sully (glass-based armor lamination) initiated a codevelopment program to evaluate the effectiveness of sapphire sheets as a component of its Transparent Armor laminated system. As shown in Table 15.1, Sapphire Transparent Armor systems offer a thickness and weight savings of greater than 50%, and for certain systems the weight savings exceeds 60%. An important feature of the Sapphire Transparent Armor systems is the ability of the sapphire to break apart the core of the armor-piercing projectile. This means that subsequent interlayers, glass, and backing materials need only defeat smaller, lower-energy projectiles. As shown in Figure 15.4, when compared with that of glass, the Knoop hardness and fracture toughness of sapphire are far superior. These systems have been extensively tested and partial results are provided in Table 15.2.

Sapphire Transparent Armor systems also utilize a novel interlayer material with industry-proven weathering durability under a variety of environmental conditions, and backing material that offers excellent scratch resistance, chemical resistance, and durability. The combination of an extremely durable sapphire strike face, nonyellowing interlayers, significantly reduces the need

TABLE 15.1

Comparison of Saint-Gobain Sapphire Transparent Armor to Glass Armor

Threat	Number of Shots	Glass Thickness mm	Glass Areal Density, kg/m²	Sapphire Armor Thickness, mm	Sapphire Armor Areal Density, kg/m²	Thickness Savings with Sapphire Armor	Weight Savings with Sapphire Armor
7.62 × 39 API-BZ	3	58	133	20.78	55.6	64%	58%
7.62 × 54R B32 API	3	104	248	33.54	86.48	67%	65%

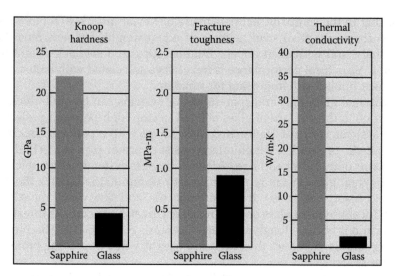

FIGURE 15.4

These graphs compare selected material properties of glass and sapphire.

TABLE 15.2

Sampling of Ballistic Tests of Saint-Gobain Sapphire Transparent Armor

ID Number	Sample Size, mm	Thickness, mm	Areal Density, kg/m2	Projectile	Projectile Velocity, m/s	Penetration
1	150 × 150	21.1	52.12	7.62 × 51 M-80 Ball	835	Partial
2	150 × 150	21.1	52.12	7.62 × 39 API-BZ	776	Partial
3	150 × 150	21.1	52.12	7.62 × 51 AP (M61)	768	Partial
4	150 × 150	21.1	52.12	7.62 × 51 AP (M61)	846	Complete
5	150 × 150	29.4	72.78	7.62 × 54R B32 API	857	Partial
6	150 × 150	24.87	67.46	7.62 × 54R B32 API	864	Partial
7	150 × 150	43.78	159.21	7.62 × 51 AP-WC	917	Partial
8	150 × 150	46.3	169.11	7.62 × 51 AP-WC	921	Partial

TABLE 15.3

Optical Measurements on Saint-Gobain Sapphire Transparent Armor System

Thickness, mm	Areal Density, kg/m²	Luminous Transmission, %	Haze, %
29.4	72.78	85.9	.99
29.4	72.78	85.3	1.25
41.1	101.11	84.4	1.00
41.1	101.11	84.3	1.12

for maintenance/replacement, and allows for consistent high performance during service.

Sapphire comes in many sizes and shapes: tube, rod, and sheet. Most recently, a very large sapphire sheet has been introduced with the development and volume production of CLASS Sapphire (Clear Large Aperture Sapphire Sheet). The excellent properties of CLASS (strength, hardness [resistance to erosion], chemical inertness, and high optical transmission) are very beneficial for transparent armor, making sapphire an enabling solution for lightweight, high-performance transparent armor solutions for both military and commercial applications (see Table 15.3).

The ultimate goal of ceramic armor is to provide better ballistic protection at a lighter weight. the U.S. military, in its drive to provide the ultimate protection for its soldiers, is borrowing from the Olympic ideals: Lighter, Stronger, Mobile. Therefore, the U.S. military stated goals are:

- To be able to transport an armored brigade anywhere in the world within 96 hours (FCS).
- To lighten the load of the combat soldier (Land Warrior).

Ceramics are a critical material to enable the U.S. Army to achieve its goals for a lighter force with the same lethality and survivability of today's heavy forces. This "lighter survivability" can realistically be achieved only with ceramic armor.

Ceramic Armor Materials

Ceramic materials for composite armor protection systems include SiC, B_4C, sapphire, and alumina. The various properties of these materials, such as high hardness, compressive strength, and elastic modulus, provide superior ballistic protection capabilities to help defeat high-velocity projectiles. The products made from these materials range from high-volume net shape tiles (square, rectangular, and hexagonal) to complex engineered components. Examples of complex products include torso plates (SAPI, ESAPI, XSAPI), side plates (ESBI, MSAP, and SSAPI), aircraft (AC-130U gunship panels), and vehicular (HMMWV doors and seats). Materials and their characteristics include:

- Hexoloy sintered SiC: high hardness, compressive strength, and lightweight
- Norbide hot-pressed B_4C: high hardness, low density, and ultralow weight
- Saphikon sapphire: transparent armor with high flexural strength and modulus of elasticity
- Silit SKDH reaction bonded SiC: high modulus and high sonic velocity

The toughest job in the armor system is reserved for ceramic materials. They must directly absorb the frontal force of the armor piercing threat, blunting the penetrating nose or breaking up the projectile altogether. Ceramics carry out this same function whether for a soldier or on the side of a vehicle.

Hexoloy SA SiC is a pressureless-sintered form of alpha SiC. It is sintered at a high temperature with the aid of boron and carbon in a vacuum or an inert atmosphere such as argon, helium, or nitrogen. Boron reduces grain boundary energy, while carbon increases the surface energy of the SiC particles and promotes sintering. Hexoloy has very low porosity, with density greater than 98% of theoretical, and has very fine grains, with grain size of 4–10 microns. It is extremely hard, with a high flexural strength of 380 MPa. In addition, it has no free silicon, and is highly resistant to corrosion. It also has high thermal conductivity and a low coefficient of thermal expansion.

Some ceramic armor is composed of a two-part ballistic system: ceramic and backing. For example, a Hexoloy plate is adhered to a polymer backing and when a bullet strikes the Hexoloy plates, the bullet shatters and the plates absorb and disperse the force, protecting the wearer:

- Ceramic (SiC) absorbs the force of a projectile and breaks the core.
- Backing (aramid or PP) keeps the ceramic in compression and catches any fragments from the broken projectile.

A process that increases the hardness and improves the ballistic performance of the material used by the U.S. military for body armor has reportedly been developed by researchers at the Georgia Instutute of Technology. Professor Robert Speyer and his research team have developed a B_4C formation process that yields higher relative densities than currently available B_4C armor. This pressureless sintering method yields a 98.4% relative density, and hardness greater than possible by hot pressing. The process is also faster and cheaper than hot pressing. For the most demanding applicatiions, postsintering hot isostatic pressing (HIP) increases the relative density of the part to 100% through the hydrostatic squeezing action of a high-temperature, high-pressure gas. The technology also enables fabrication of complex, curved shapes for helmets and form-fitting body armor [6].

Reinforced Polymeric Composite Fiber Materials

Today's scenario has moved from the open battlefield to house-to-house urban warfare, demanding armor protection from more lethal ammunition and resistance to multiple bullet strikes. Threats ranging from sniper fire to the infamous IEDs have spurred a rapid evolution in armor made from traditional reinforced polymer composites. Materials must be the moving target of optimizing weight, ballistic resistance, and cost. Innovations in the fiber chemistry affect both soft and hard armor, though composite material requirements are different for each. Soft armor is typically flexible, strong and lightweight, with high modulus of elasticity to dissipate impact energy over as large an area as possible in the armor. Hard armor is typically rigid and thicker, and may involve combining ploymer composites in laminate or fabric form with steel and ceramics. A good example of composite soft and hard armor in a single product: a strong and flexible protective vest that incorporates one or more small arms protective inserts (SAPI) in rigid panel form. Larger composite panels are used as combat shields, in vehicle armor, and in blast protection for infrastructure.

Aramids

Today's Kevlar para-armid fiber is at least 50% stronger than the fiber tested 31 years ago in protective vests. The highly oriented, long molecular chains of poly-paraphenylene terephthalamide provide secure interchain bonding, giving Kevlar its five-times-greater-than steel strength properties. Among Dupont's recent Kevlar-based ballistic products, Kevlar Comfort XLT has reduced weight by 25% in protective vests as compared to all-aramid vests, and simultaneously reduced backface laminate layers and deformation. In fact, Kevlar Comfort XLT approaches the performance of a unidirectional structure by combining fiber and fabric so as to fully utilize the fiber strength in both the warp and fill direction. This maximizes load sharing across all yarn crossover points, while maintaining the structure's stability and improving ballistic performance at less weight.

Twaron

Teijin Twaron, of Arhem, the Netherlands, supplies Twaron para-aramid fiber and yarn globally and fulfills requests for ballistic protection material. When reinforcing a rubber-modified phenolic resin, Twaron fibers are successful for armor in both strike face laminates and in spall barriers behind or in complete composite encasement of steel and ceramic strile faces. Teijin reports that the weight savings from a combination of Twaron composite and steel to commercial vehicle armor, as compared to an all steel armor, have affected top vehicle speed by a decease of only 5 km/h. A new fiber that has been developed is Twaron CT and it contains 50% more microfilaments. This brings higher modulus of elasticity in protective vests, and 23% lower weight than previous Twaron fibers. In German army helmets, Twaron CT improves ballistic performance by 100% over steel.

Tensylon

BAE Systems in Cincinnati, Ohio in conjunction with the U.S. Army's Heavy Tactical Vehicle Armor Program have developed Tensylon. It is planned to use this material by supplying Tensylon composite armor panels for Heavy

Equipment Transporter (HET) vehicles. The HET vehicles transport, deploy, recover, and evacuate combat-loaded main battle tanks and other heavy tracked and wheeled vehicles to and from the battlefield. Tensylon is an ultrahigh modular weight polyethylene tape and composite material used to improve the readiness and sustainment of survivability products, including ground vehicles such as the Mine Resistant Ambush Protected (MRAP) vehicles, and individual soldier equipment. The material development, testing, and evaluation was initiated in 2007 and was aimed at composite armor solutions to defeat current and emerging threats in Iraq and Afghanistan. These tests and evaluations resulted in the development of a lightweight applique armor kit using Tensylon.

"Tensylon is a truly unique material that ultimately increases the protection provided to our men and women in uniform through a cost-effective, lightweight survivability solution," said Tony Russell, President of Security and Survivability Division, BAE Systems. Tensylon reported significant advantages when compared to other high-performance materials. It offers distinct benefits in both survivability and commercial applications and has a "green" environmental profile, requiring no environmental permits to produce.

Teijin's Twaron, Dupont's Kevlar, and DSM's Dyneema fibers have been selected for use in the U.S. armed forces' Interceptor Body Armor as the most up-to-date modular, multithreat personnel armor available. They are being utilized in the Enhanced Side Ballistics Insert (ESBI) carrier as well as in the SAPI (ESAPI) components.

Spectra

Honeywell International's Specialty Materials uses a gel spinning and drawing process to manufacture Spectra ultrahigh molecular weight polyethylene (UHMWPE) fiber and Spectra Shield technology in ballistic composites. Ten times stronger than steel and lighter than most aramid fibers, Spectra conforms well in soft and hard armor. The Spectra Shield line encompasses noncrimped 0/90° cross-plied tapes; reinforced with Spectra fiber are the standard Spectra Shield, Spectra Shield Plus, and Spectra Shield II; reinforced with aramid fibers are the Gold Flex and Gold Shield materials.

Honeywell (Morris Township, New Jersey) has introduced, in 2009, a new Gold Shield material with improved ability to stop bullets and fragments in military and police armor applications. The new ballistic material, Honeywell Gold Shield GN-2117, has demonstrated up to a 10% weight reduction when compared to Honeywell's traditional Gold Flex material, which is also used in soft armor applications. The new product also is said to provide increased surface durability and chemical resistance, allowing it to meet the toughest global body armor standards for military and law enforcement applications.

The Gold Shield GN-2117 builds on Honeywell's proven Gold Flex ballistic material, a soft armor material that combines Honeywell's patented Shields technology with aramid fiber. For more than 10 years, Gold Flex has been one of the most widely used ballistic materials in police and military vests. The new Gold Shield GN-2117 incorporates a proprietary resin and coating system, which provides increased environmental and chemical resistance, as well as improved fragnent protection. The technology helps Gold Shield GN-2117 meet stringent global body armor standards, including those of the NIJ, which is the research, development, and evaluation agency of the U.S. Department of Justice. Multiple vest models containing Gold Shield GN-2117 have successfully completed certification testing under the new NIJ 0101.06 standard for body armor.

Spectra Shield products are also applicable in armor for the Humvee, U.S. Marine Corps' Sea Knight helicopters, protective vests, ESAPI and ESBI plates, and Blastshield fabric.

Dyneema

The other leading high modulus, high performance polyethylene (HHPE) is a Dyneema fiber from DSM, manufactured in the Netherlands and in the United States. Recent ballistic materials in the Dyneema line include HB25 and HB26 unidirectional (UD) reinforced composites for hard armor that are notable for faster, more efficient transfer of impact energy along the fibers than other conventionally woven fabrics. Manufacture of armor with HB26 cuts handling time in half and speeds up production. The Danish armed forces have tested HB26 composite in vehicle armor and found it effective. The HB25 composite works well in cab armor for medium tactical vehicles.

The DSM has also hybridized Dyneema HPPE fibers with steel cord material for slab protection in armored vests that meet the U.K. standard HGIA/KRI and are worn by Ireland's police force. The U.K. Ministry of Defence will soon incorporate Dyneema UD in the backface of ceramic plates for the new Osprey body armor. The DSM has also created a hybrid laminate sheet material of Dyneema HPPE fibers with Dyneema UD tape that provides resistance to ammunition, knives, and needles.

According to Jan Grinberg, "The carbon/carbon bonds in Dyneema HPPE create one of the world's strongest lightweight fibers—up to 75% less real density compared to steel and we'er only using 20% now of Dyneema's fiber strength, so there's greater potential for it in ballistic products." The ultrastrong Dyneema HB26 unidirectional composite fiber has reportedly been developed by DSM Dyneema, Evansville, Indiana. The Denver Police Department in Colorado is the first U.S. organization to adopt the new Protech Delta LT helmets. This model meets the National Institute of Justice IIIA specifications and is 15% lighter in weight than comparable 100% aramid helmets.

The Dyneema HB26 composite is made from HHPE fiber. It provides protection against handgun fragments and rifle threats at extremely low weights in both personal and vehicle hard ballistic armor applications. The Dyneema

UD is made of several layers of Dyneema fibers, with the direction of fibers in each layer placed perpendicular to those in the adjacent layers. This unidirectional configuration allows the energy transfered from the impact of a bullet or other threats to be distributed along the fibers much more rapidly and efficiently than in conventional woven fibers. The composite provides excellent mechanical rigidity and toughness, as well as resistance to temperature extremes and moisture.

Hybrids

Recent success in hybrid-fiber composite SAPI plates comes from armor makers Homeland Security Group Intl., Del Mar, California and Protective Enterprises LLC, Dulles, Virginia. Hybridizing Dyneema and Spectra fibers in SAPI plates have resulted in a composite component that weighs an average 1.5 kg (3.2 lbs) compared to other composite plates at 2.6 kg (5.8 lbs). Other companies report that they have produced 5 mm thick composite SAPI panels for protective vests that weigh only 4.2 kg/m^2.

Glass

Considered the heaviest of the ballistic reinforcing fibers, glass nonetheless holds its own niche in composite armor. AGY in Aiken, South Carolina, the only global producer of trademarked S-2 Glass fibers, maintains that its fibers provide a balance among tensile, compressive, stiffness, and fatigue properties that offer both ballistic and structural performance in a composite laminate. The company's HJ1 hard armor panels, made from woven fabric based on S-2 Glass reinforced phenolic, are used in armor for the Humvee M114 and the CAV-100 European peacekeeping vehicle. The company believes that engineered fiber sizing to enhance the S-2 Glass fiber with a matrix resin bond, or in bonding with other reinforcing fibers, offers advantages such as lower finished part cost, thinner parts, easier fabrication, improved structural capability, and a higher degree of consistent ballistic performance.

Montrose, a Colorado-based company who manufactures Polystrand's ThermoBallistic-H armor panels that utilize the company's trademarked X-Ply hybridized E- and S-Glass/PP sheet material. The composite panels are created from a 0/90° laminate with the E-Glass in the strike face and the S-Glass on the backface. "Glass fiber may be heavier than aramid and carbon fiber, but it is also less expensive and widely available," says Polystrand's president Ed Pipel. He points out that X-Ply requires no special storage, shelf life is indefinite, and costs less in laminate form than other ballistic composites. The panels can be tailored to thickness and in the ratio of S- to E-Glass. "Compared to traditional thermoset composite ballistic panels, Polystrand materials can reduce pressing time from 40 to 60%, and eliminate degassing and release of volatile organic compounds," says Pipel.

3WEAVE

In ballistic applications, 3WEAVE fabric reduces blunt trauma transmitted to the wearer, and increases the armor's ability to withstand multiple strikes. With virtually no crimp, delamination is eliminated, even in edge hits. On shipboard armor, 3WEAVE fabric panels made with phenolic matrix resin can stop high velocity, high caliber ammunition. In pultruded blast resistant panels integrated 3WEAVE fabrics have been tested to withstand 2.3 kg (5 lbs) of C4 explosives at 0.9 m (36 in.). The 3WEAVE is not a traditional 3-D fabric, such as angle interlock, so it can deliver better mechanical properties, a high degree of conformabilty, and rapid composite fabrication speed in resin infusion processes.

New Fibers and Nano Materials

New reinforcing fibers for composite armor has been under development at a rapid pace. Dupont Advanced Fiber Systems and Magellan Systems International, Richmond, Virginia, see a step-fiber for ballistics in M5, a complex polyphenylene fiber chemistry that Magellan describes as "a rigid rod polymer featuring bi-directional hydrogen bonding that creates a three-dimensional honeycomb network."

Another thermoplastic composite new to the ballistic market is high tensile PP tape yarn reinforcing a PP matrix as part of the moldable fabric technology (MFT) from Milliken Fabric, Spartanburg, South Carolina. Commercial in vehicle armor, the supplier has explored MFT for helmets and personnel armor as well.

Evolution Armor, Carlsbad, California, is developing Evo-Flex 2000 and 3000 ballistic laminate armor that Allan Bain calls "the first water-resistant aramid laminate." He adds that Spectra and Gold Flex reinforcements have been sampled in the laminates, and carbon-fiber nanotubes may also be added in a hybrid with the aramid fibers in further research.

Researchers at the University of Delaware's Center for Composite Materials and the Weapons and Materials Research Directorate of the U.S. Army Research Laboratory have made headway with shear thickening fluid (STF), a unique "liquid armor." The SFT technology exists in a flexible, fluid-like state under normal conditions but becomes rigid upon impact to resist penetration within milliseconds.

Nanotechnology R&D at the Hong Kong University of Science and Technology has investigated reinforcing UHMWPE films with multiwalled carbon nanotubes in composite form and has revealed significant improvement in strength and modulus by the addition of only 1% by weight of the nanotubes. This hybridization also appears to reduce UHMWPE fiber slippage by knots formed by the nanotubes in the spun fiber.

Lifetek Armor, New Bern, North Carolina, is currently using polymer reinforced composites in their products, for electronic weapons that may be the next combat threat. According to Terry Nelson of Lifetek, "Nanomaterials could provide the armor solution for such weapons."

New Metal–Ceramic Hybrid

The AFRL material scientists worked with scientists from Excera Materials Group, of Columbus, Ohio, to develop a novel metal–ceramic hybrid material for use in higher-performance, lighter-weight, lower-cost SAPI for body armor vests. These strike plates exceed the Army's performance standard for fielded applications. The material scientists created a lightweight, layered composite panel with angled external tiles that would cuase bullets to tumble and stop instead of piercing armor plating on tanks and aircraft. Thus a lighter-weight, metal-infused ceramic laminate panel for warfighter flak vests was developed.

To stop an assault rifle bullet, armor must have properties that will flatten, crack, or blunt the point of the bullet. The armor must also have a fiber backing that will catch bullet fragments and absorb the pressure wave generated when a bullet strikes the armor. Developing ceramic materials that will retain high hardness in the shapes required for body armor applications is a challenging task. All current SAPI plates consist of press-sintered ceramic materials, which are very hard but produce strike plates that are heavier and more fragile than desired.

The AFRL scientists found Excera, a manufacturer of a unique ceramic material called ONNEX. Excera creates ONNEX by infusing a ceramic material with liquid aluminum, and the material was originally intended for use in high-temperature applications such as brake rotors and rocket motor impellers. The material exhibits the high hardness of B_4C, but it also demonstrates fracture toughness 10 times that of other pressed ceramic materials. The hardness of an ONNEX armor plate will shatter and stop a striking bullet, and because the material's fracture toughness confines damage to a small area, the armor can tolerate multiple strikes to the same region. To complete the panel design, the AFRL/Excera tean reinforced the ONNEX material with a ballistic backing made from Dyneema polyethylene fiber and an elastomeric polyurethane coating. The ONNEX-based SAPI is lighter and less expensive than currently deployed systems.

After passing all U.S. Army ballistic tests, the material has been manufactured in production as body armor for both front and back strike plates. Continued development efforts have investigated applications for the new armor material that support the Defensive Fighting Position concept (an individual, reusable, and transportable ballistic barricade), the Deployable Defensive Panel System, and appendage armor (inserts that slip into uniform pockets to protect the arms or legs of a soldier).

Engineering researchers at the University of California–San Diego (UCSD) [4], are using the shell of the red abalone, a seaweed-eating snail, as a guide for developing bullet-stopping armor. The colorful oval shell of red abalone is highly prized as a source of nacre, or mother-of-pearl, used in jewelry. But the UCSD researchers are most impressed by the shell's ability to absorb heavy blows without breaking. Red abalone creates its helmetlike home with 95% calcium carbonate "tiles" and 5% protein adhesive. Calcium carbonate, or chalk, is ordinarily weak and brittle, but research has shown that the mollusk creates a highly ordered bricklike tiled structure that is the toughest arrangement theoretically possible. Of course, the abalone shell can's stop an AK47 bullet. But it looks more promising than laminates and other materials, which have been disappointing as armor. Researchers figure they have exhausted conventional possibilities, so they are turning to biology-inspired or biomimetic structures. Biomimetic researchers interested in tough materials have discovered that mollusk shells, bird bills, deer antlers, animal tendons, and other biocomposite materials have recurring building plans that yield a hierarchy of structures from the molecular level to the macroscale.

Specifically, abalone shell at the nanoscale is made of thousands of layers of calcium carbonate "tiles," about 10 micrometers across and 0.5 micrometer thick, or about one–one hundredth the thickness of a strand of human hair. The irregular stacks of thin tiles refract light to yield the characteristic luster of mother-of-pearl. The mother-of-pearl growth surface of the abalone shell is colored because of how light refracts as it strikes tiny terraces of calcium carbonate. Engineering researchers have shown that the terraced, Christmas treelike surface of abalone shell has evenly spaced nucleation sites from which stacks of hexagonal "tiles" of calcium carbonate begin to blow. The top and bottom surfaces of each layer of tiles are separated by a protein adhesive, but the adhesive does not bind the edges of tiles to adjoining tiles. Under stress, tiles of calcium carbonate can slide, absorbing energy. Because of this microstructure, the abalone shell can absorb a great deal of energy without failing.

A key to the strength of the shell is a positively charged protein adhesive that binds to the negatively charged top and bottom surfaces of the calcium carbonate tiles. The glue is strong enough to hold layers of tiles firmly together, but weak enough to let the layers slip apart, absorbing the energy of a heavy blow in the process. Abalones quickly fill in fissures within their shells that form due to impacts, and they also deposit "growth bands" of organic material during seasonal lulls in shell growth. The growth bands further strengthen the shells. The adhesive properties of the protein glue, together with the size and shape of the calcium carbonate tiles, explain how the shell interior gives a little without breaking. In contrast, the whole structure is weakened when a conventional laminate material breaks.

Armor used to conjure images of knights in heavy metal suits, barely able to move. Today, technology has allowed armor manufacturers to increase protection while decreasing weight and bulk, making armor safer while allowing more mobility for users. One example is Max Pro-Police's newly released Level IV body armor that enhances performance while weighing approximately 30 kg (7.5 lbs) per plate. The new armor can sustain multiple hits without shattering and has a cover material that traps shattered bullets, preventing collateral casualties from ricocheted pieces. The armor is made of preshaped, triple-curve, alumina-ceramic plate, which is bonded inside an advanced thermoplastic composite cover and backing.

Diaphorm Technologies manufactures the armor parts. The proprietary molding technology produces an interface bonding between the two material layers, which provides the armor with excellent ballistic performance. The interface bond, in conjunction with the molded cover and backing, allow the plate to remain intact beyond first impact. As a final source of protection, the plates are encapsulated with a durable black outside layer to prevent the surface from scratching or chipping. The plates are molded with a triple curve shape for optimum fit and are designed to fit standard tactical vests. The front plate weighs 30 kg (7.5 lbs) and the back plate weighs 28.4 kg (7.9 lbs). The plates have been tested and inspected by the National Law Enforcement and Corrections Technology Center and have been certified to Threat Level IV based on the National Institute of Justice 2005 Interim Requirements for bullet resistant body armor [5].

A new patented technology, DuPont Kevlar XP, provides ballistic and trauma protection in a more comfotable body armor solution that has been successfully tested. Both DuPont and independent tests show that Kevlar XP consistently stops bullets within the first three layers of a vest designed with a total of 11 layers. The remaining layers of Kevlar XP absorb the energy of the bullet, resulting in less trauma, or backface deformation, to the vest wearer. Based on DuPont experience, significantly more layers are typically required to stop a bullet in other commercially available lightweight technologies.

Kevlar XP is currently available, however, DuPont is developing additional ballistic applications for the future. "In our experience, we have never seen a technology that works as effectively as Kevlar XP at stopping bullets and reducing body trauma in a lightweight solution. This is an exciting new ballistic technology platform and the latest example of continued innovation in personal protection products from DuPont," said Thomas G. Powell, vice president and general manager, DuPont Advanced Fiber Systems. "Kevlar XP technology has been designed to provide end-users with higher performance body armor that is more comfortable, while meeting the most demanding requirements of current and pending global standards."

Kevlar XP typically provides a 15% reduction in backface deformation and at least a 10% lighter-weight vest design against the most challenging NIJ Level IIIA threat, a 0.44 magnum bullet. Additionally, Kevlar XP provides excellent layer-to-layer abrasion resistance and Kevlar XP is targeted to the growing global market for ballistic protection.

The U.S. Army in conjunction with M Cubed Technologies, Trumbull, Connecticut are developing and commercializing an advanced ceramic armor for troops in Iraq and Afghanistan. The ceramic armor plates have improved durability and multihit performance through the use of a multiphase microstructure that results in reduced cracking. The new armor of ceramic plates go into pockets in the vests of the Interceptor Body Armor System worn by troops with two plates per vest, one in front and one in back. The System is used by the U.S. Army and Marines in Iraq and Afghanistan. The addition of the new plates to the traditional armor vest significantly increases its protection level. The vest alone can only defeat fragment and handgun munitions. With the plates, armor-piercing rifle rounds are defeated.

Also in advanced development at M Cubed Technologies is a unique ceramic/metal matrix composite (MMC) armor to protect vehicles from roadside bombs and direct fire. The ceramic constituent provides the hardness to defeat armor-piercing threats while the MMC provides durability and multihit resistance. With this new composite, M Cubed can produce complex-shaped structures up to 2.4 × 2.4 m (8 × 8 ft) in a single piece. Large panels make assembly more efficient and cost effective for armor integrators compared to the use of many small tiles that require complex assembly fixtures and excessive handling.

Engineers at the University of Leeds, U.K. are working on a new type of body armor made from cement. The new vests will combine super-strong cement with recycled carbon fiber materials to make a material tough enough to withstand most types of bullets. By using cement instead of alumina the engineers feel that they can deliver a cost-effective level of protection for many people at risk. It should be good enough for people like security guards, reporters and aid workers who are worried about the odd pot shot being taken at them.

Many of the armored vests sold currently are overengineered for the threats they face. Cement-based body armor would not only create a whole new market but it would also take some of the pressure off the demand for high-specification alumina models so that people like soldiers, who really need this kit, can get it.

Currently, available high-specification body armor is constructed with alumina plates—the raw material used to make aluminum—which is heated to 1600°C (2912°F) for up to 2 weeks in a process called "sintering" in order to make them ultrahard. Enhanced combat body armor (ECBA) as supplied to U.K. troops uses sintered alumina plates. In the past U.K. and U.S. soldiers serving in Iraq and Afghanistan have faced shortages of EBCA as production has struggled to keep up with soaring global demand. Cement vests are just of a range of novel uses for the 2000 year old material that the Leeds engineers are investigating. Other ideas include cement-based, pumpless refrigerators, a new type of catalytic converter, and improved bone replacements.

In a sad, but sobering, commentary on contemporary life, BackPack-Shield Mfg., Austin, Texas (backpackshield.com) has launched a series of bullet-resistant portable shields for student backpacks, computer bags, briefcases, and other portable cases and bags. Impermeable to most handgun bullets, the shields can be moved between backpacks and bags.

The patented BackPackShield is a lightweight armor panel insert consisting of more than 10 layers of Dupont Kevlar aramid fiber fabric. The layers are bound and epoxy-impregnated into a sheet less than 13 mm (0.5 in.) thick and weighing less than a typical textbook. A standard size insert, which fits most student backpacks is available in more than 10 colors. Unlike weaker NIJ (National Institute of Justice) Level II ballistic curtains that are permanently sewn into bags, BackPackShields are manufactured to the much higher Level IIIA specification. Armor that meets Level IIIA specifications, can stop even the hardened, full-metal jacketed, high-velocity handgun bullets including 9 mm and 0.44-caliber Magnum that pass right through a Level II bulletproof liner. The BackPackShield Mfg. Co., realized that a light, thin, semirigid armor panel that stops virtually every handgun round could be discretely added to student and collegiate backpacks, providing both protection and peace of mind for students and families.

Future Developments

Increased ballistic threats mandate armor system upgrades:

- 7.62 mm (Tungsten armor piercing)
- 14.5 mm (armor piercing)
- 20 mm FSPs
- IED (improvised explosive device)
- EFP (explosively formed projectile)

New and improved alternative materials include hot-pressed B_4C, the material of choice for body armor because of its high hardness and low weight. However, hot-pressed B_4C exhibits distinct performance deterioration against next-generation body armor ballistic threats ("shocked-induced amorphization"). Sintered SiC, of which Hexoloy is the benchmark material, has demonstrated good performance against these next-generation ballistic threats.

The U.S. Army Natick Soldier Center (Natick, Massachusetts) is sponsoring research to develop proprietary carbon nanotube technology for the purpose of improving body armor. Protective body armor is one of the major contributors to the weight a soldier must carry in combat. The Army is continually seeking and striving to reduce the weight of body armor while improving its ability for protection from ballistic threats including bullets and fragments from IEDs. Today's soldiers carry heavy combat loads. For the past several years, researchers have been working with the Natick Soldier Center to develop nanomaterials with the objective being to create a new generation of lightweight, ballistic protection systems.

The object for using carbon nanotubes is to utilize their very long lengths—up to 1 mm. As a result the product is said to be significantly stronger, more conductive, and safer to use in end applications when compared to short, powder-like nanotubes that have previously appeared in the market. Nanocomp Technologies, Concord, New Hampshire, working with the Army has been producing large-area carbon nanotubes to demonstrate their value to the Army as well as aerospace and electronics applications. Carbon nanotubes are "cousins of diamonds and graphite." They can be made into long carbon nanotubes that bond to each other and can be used to create yarn or sheet material. The material's potential has a breaking strength 2.5 times better than what is available today. The elasticity of carbon nanotubes can enable development of protective body armor that not only stops bullets, but also can prevent blunt force trauma, reported engineers from the Centre for Advanced Materials Technology at the University of Sydney, Australia.

The elasticity of carbon nanotubes means that blunt force trauma may be avoided, and is the reason engineers in Sydney have undertaken experiments to find the optimum point of elasticity for the most effective "bullet-bouncing" gear. By investigating the force-repelling properties of carbon nanotubes and deciding on an optimum design it may be possible to produce far more effective bulletproof materials. The dynamic properties of the materials found to date means that a bullet can be repelled with minimum or no damage to the wearer of a bulletproof vest [6].

Finally, The Air Force Office of Scientific Research (AFOSR), is supporting a team from the University of Wyoming that is investigating spider silk proteins to create biomaterials for military purposes. Producing useful quantities of natural spider silk has proven unrealistic because of challenges inherent in managing large numbers of small spiders, which are typically cannibalistic. As a result, researchers have been creating artificial spider silk that is stronger than the polymer Kevlar and more flexible than nylon. To produce new kinds of spider silks, the team has made its own spider silk genes and put them into bacteria to produce chemically identical spider silk proteins for use in experiments (see Figure 15.5). "These researchers spin the proteins into fibers and test them for better properties," said Dr. Randy Lewis, team leader. "The team has also produced genetically-modified goats that produce milk containing the spider silk proteins to aid in the research."

The proteins derived from the goat's milk can be spun into strong, lightweight, and extremely elastic silk to be used in the construction of light, bulletproof vests for the military. The fibers can also be used for much stronger parachutes enabling larger payloads to be delivered. They can also be used to create artificial ligaments. To make a 45 kg (5 lb) bulletproof vest, a producer would use 600 gallons of goat milk containing the silk protein. The milk production from 200 goats in 1 day would be used for just one vest. According to Dr. Lewis, "spider silk body armor will be expensive with a price that is two times as much as Kevlar body armor." However, he's confident the military will be willing to make the investment because silk body armor is light and elastic and therefore adaptable to different needs.

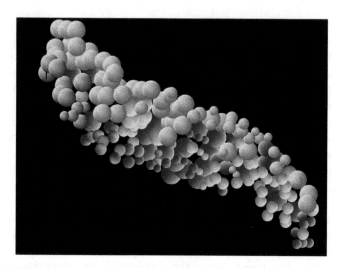

FIGURE 15.5
This image shows a small part of two spider silk protein molecules interacting like two sides of a zipper. The "teeth" of the zipper can be seen in the slots of the other molecule. These zippers on hundreds of thousands of protein help form the spider silk fiber and give it its extraordinary strength.

Armor Personnel

For the past two decades, high-tenacity fibers such as S-Glass, Aramid, and UHMWPE have been the materials of choice for making lightweight armor. But today's soldiers, police officers, and other security forces now have a 10% lighter helmet that far exceeds Advanced Combat Helmet (ACH) specifications for impacts. The helmets also meet all ballistic protection specifications and greatly reduce shock transference to the person they protect. In standardized ACH tests, the helmet transferred one-third less impact to the head than all other helmets tested. The composite combat helmet is molded from continuous-fiber-reinforced thermoplastic composites (CFRTPCs). The helmet shell construction resembles that of tactical helmets meeting NIJ Level IIIA. More than 10,000 Level IIIA helmets have been worn in Iraq and Afghanistan, and by U.S. police units.

A proprietary quick-cycle molding process combines the inherently tough, lower-density TPs with ballistically rated continuous-fiber reinforcement. This creates monolithic, hybrid, and hard-faced rigid CFRTPC helmet shells that are 10% lighter than other ACH -grade helmets. Thermoplastic advanced composites are a rapidly growing field of advanced materials. Comprised of reinforcing fibers embedded in a matrix of thermoplastic resin, these materials offer high specific strength and stiffness and low density. In addition, thermoplastics' high toughness makes their use appealing in applications that require energy absorption and strength after impact. Thermoplastic advanced composites also offer potential benefits of reduced cycle time and in-plant air quality for manufacture.

A recent investigation into the fabrication of a thermoplastic antiballistic infantry helmet was undertaken by Fiberforge Corp. (Glenwood Springs, Colorado) [7]. The objectives of this investigation were to assess the potential for manufacturing a high-quality helmet via thermoforming and to measure the cycle time for each processing step. The helmet construction included an inner aramid composite antiballistic liner and an outer carbon-fiber-reinforced thermoplastic shell. Using innovative layup and forming techniques, Fiberforge was able to show the potential of coforming the structural and ballistic sections of a hybrid helmet design. The initial processing steps took an excessive number of minutes, however, automation and selection of more appropriate processes than those used in an initial study showed that the times were competitive for the selected processes. In the future the evaluation of the selected processes would be required and this would demonstrate conclusively the benefits of thermoplastic composites in this helmet application.

Armor Protection for Habitats

In Iraq and Afghanistan, fighting insurgents often require frequent, shorter-duration deployments in remote regions where it is logistically difficult to provide traditional ballistic protection for tent camps such as sandbags and concrete barriers. The modular ballistic protection system (MBPS) provides lightweight, affordable, modular, rapidly erectable, reusable ballistic protection for soldiers where it never existed before—in their tents where they work, eat, and sleep. The dual-use blast/ballistic composite materials provide a low-cost, lightweight solution that meets the transportation and mobility requirements for Forward Operating Base construction.

The MBPS consists of composite ballistic panels that are mounted to the inside of a military tent frame using an energy-absorbing connection system. Requiring no tools, the MBPS can be used by four soldiers to up-armor a 600×960 cm (20×32 ft) tent in less than 30 minutes. Developed at the University of Maine, the up-armored tent meets specified Army requirements for mitigation of fragmentation threats and blast overpressure. The advanced thermoplastic composite panels are lightweight, reusable, and inexpensive. The MBPS provides immediate protection to troops in tents at the beginning of a deployment before sandbags and concrete barriers are erected; it also provides mobile protection to units on the move.

The MBPS panels feature a sandwich design, with a wood composite core and thermoballistic composite skins. Unique energy-absorbing connections attach the panels to the tent frame. The connections allow the panels to move under the blast forces, reducing the blast loads on the tent frame. Even though the tent frame can only withstand 50 mph winds, the MBPS allows the tent to survive blast pressures of significantly higher energy. The panels are installed under the fabric of the tent, eliminating any visual signature, while accommodating electrical HVAC. The panels are ruggedized with a polymer coating protecting them against abrasion, UV light, and moisture.

A proprietary compression molding process is used to produce the thermoplastic sandwich panel in one step, complete with rugged abrasion resistant edging material, protective surface coating, handholds, and hardware inserts. This process results in a product with no out-gassing. it has low tooling costs. it allows the use of many different materials combinations. and it provides a finished product complete with surface and perimeter protective material.

Several product variations have been developed using different glass and aramid fibers in a thermoballistic matrix. The ballistic and core layer materials and the material layering technology were engineered to optimize both ballistic performance and blast response. The result is a dual structural/ballistic panel capable of resisting mortar fragment penetration, yet with the ductility and strength to withstand significant blast forces.

A Rapid Armor Shelter System from FY Composites (Nokia, Finland) is a modular ballistic protection system. The military-certified, lightweight fortification wall system can be used to protect fast-moving troops. The walls can be built and dismantled in a matter of minutes, allowing troops to be protected even during short operations. The panels protect troops from fragments; for bullet protection, a multipanel system with a sand filling can be used. Two basic shapes, squares and triangles, are equipped with quick-coupling devices and support legs that enable the construction of a large number of combinations. All components can withstand UV radiation, fuel and oil spatters, and ambient temperatures from −40 to 60°C. The panels also do not absorb moisture.

Advanced Armor Protection for Bridges and Other Infrastructures

A bridge armor is under development that is a multilayered, multifunctional armor designed to provide protection for critical elements of bridges against various threats including IEDs, ballistic, blast, fire, and other destructive forces [8].

Architectural Applications and Bullet-Resistance Armor

The ballistic protection industry requires high performance thermoset sheets, prepregs, and molded shapes that can withstand the most destructive military and civilian security applications. Major considerations include (1) resistance to a projectile or blast, fire, and forced entry; (2) ease of installation; (3) weight; and (4) cost and availability. A new advanced fiberglass-based material offers a solution to meet these demanding specifications.

Norplex-Micarta manufactures the only commercially available Class 1-A fire- and smoke-rated building material (ASTM E84) that is resistant to projectiles, ricochet, heat, and fire. Norplex–Micarta's ShotBlocker product is a self-extinguishing thermoset composite that will not catch on fire or give off toxic smoke when exposed to intense heat, making it ideal for all types of civilian and military defense applications. ShotBlocker can be manufactured on five levels of bullet-resistance depending on the weapon threat and security function. The material can be cut and drilled in the field with commercial quality circular, table, or panel saws and readily covered with decorative plastic laminates using mechanical fasteners or adhesives. For aesthetic conformity, ShotBlocker can be surfaced on site with veneer, drywall, wallpaper, or other wall covering.

Glass fabric-based, antiballistic materials, such as ShotBlocker, are one-third the weight of ballistic steel armor plate, but can still provide the same level of protection. The lower weight makes the material less cumbersome to fabricate, handle, and install. The combination of lightweight and robust threat protection makes ShotBlocker well-suited for some architectural applications that would not be possible with heavier ballistic steel. Antiballistic fiberglass composites are also much less expensive than many conventional armor alternatives. ShotBlocker material is specified for courtroom, check-cashing, gas station, and convenient market armor, providing superior security and production in walls, doors, and counters. Additionally, ShotBlocker can be used for military/defense architectural applications in the field to fortify guard stations and other military structures. Overall, the performance of the new composites is similar to that of conventional S-2 Glass laminates however, they can do considerably more.

Armor Protection Using Textiles and Gels

Textiles

The Dow Corning Active Protection System is a unique, "smart," technical textile. It is an alternative to hard, rigid armor systems for protection against high-energy impacts, and is the only system available today that offers the same combined level of impact protection and comfort in the versatile form of a textile. The creative combination of dilitant silicone material and spacer fabric is soft and flexible under normal conditions but instantly undergoes a temporary transformation to a rigid solid when stressed by a high impact force. Independent testing shows that the Active Protection System meets or exceeds EU protection standards for a variety of sports apparel and equipment. The successful development of this product required a number of technical innovations, including optimization of the spacer textile to ensure maximum force absorption while remaining breathable, and modification of the dilitant silicone chemistry to allow for repeated cleaning in domestic washing machines without a significant loss of performance. The Active Protection System has been successfully taken from laboratory-scale production to full-scale, wide-width manufacture.

The Dow Corning Active Protection System presents a significant advance in the development of impact protection fabric—one that can substantially improve the quality of life for people all over the world in many different ways. Any activity where body impacts could or do occur would benefit from the superior protection, breathability, and flexibility provided by Active Protection System technology. Such properties can mean less exertion, greater performance, and increased safety for the wearer. In addition to its benefits for those engaged in motorcycling, skiing, equestrian sports, mountain biking; and contact sports such as cricket, rugby, baseball, lacrosse, hockey, and soccer, the Active Protection System offers improved quality of life for people with injuries or disabilities who need to wear protection 24 hours a day. It also has the potential to increase the safety of industrial workers, police, and emergency rescuers.

Gels

A gel that contains nanoporous energy-absorbing particles has been designed to help mitigate traumatic brain injury by Agilenano, San Diego, California. Called Agilezorb, sudden impact pressure rapidly forces the liquid into the normally empty nanopores, absorbing a tremendous amount of energy. The first use of this technology, will likely be in military helmets, to help mitigate traumatic brain injury (TBI). Agilezorb came about in an investigation of energy absorbing nanocomposites for helmets, bumpers, and body armor. The experiments demonstrated that energy absorption could be amplified hundreds or thousands of times by suspending nanoporous particles in a liquid or a gel. "The research suggests that Agilezorb is the only known material that reacts fast enough to smooth and lower the explosive shock front generated by IEDs, converting the shock wave into a slow rising, non-shock wave," says Doug Giese, AgileNano CEO. Recent tests at a U.S. government facility showed that Agilezorb reduced blast wave pressure by more than 85% [9].

References

1. Spiller, K. Alexandria, VA: BAE Systems; karen.spiller@baesystems.com
2. Wolbert, J. P., and D. M. Spagnuolo. Fabrication of Lightweight Armored Doors for HMMWVs, ARL-0016; http://www.defensetechbriefs.com/tsp under Manufacturing & Prototyping Category.
3. Hunn, D. Lockheed Martin Armor Development, *AM&P* (October 2006): 23.
4. http://link.abpi.net/l.php?20050118A2. *NASA Tech Briefs* (January 19, 2005).
5. http://www. maxpropolice.com; http://www.diaphorm.com
6. Mylvaganam, K. Carbon Nanotubes Build Better Protective Body Armor, Sydney, Australia: University of Sydney; kausala@aeromech.usyd.edu.au; http://www.usyd.edu.au

7. Campbell, D. T., and D. R. Cramer. Thermoplastic Composite Ballistic Helmet: Automated Fabrication & Materials Study, Glenwood Springs, CO: Fiberforge Corp.; *SAMPE Journal* 44, no. 6 (November/December 2008): 18–23.
8. Hardwire LLC Opens New Plant, Hardwire LLC, Pocomoke City, MD, Composites World, July 2, 2008.
9. Qiao, Y. Nanoporous Particles in Gel Absorb Energy From Explosives, La Jolla, CA: University of California at San Diego; yqiao@ucsd.edu; http://www.ucsd.edu; *AM&P* (December 2008): 17.

Bibliography

Best, D. W., and K. E. Gaskill. MAGTF Expeditionary Family Fighting Vehicles (MEFFV), Power Point Presentation, 2005; http://www.onr.navy.mil

Brown, A. Eyes on the Defense Market, *Composites Manufacturing* (September 2008): 13–5, American Composites Manufacturers Association (ACMA), Arlington, VA, USA.

Clements, E. L. Sapphire Transparent Armor Laminates Protect Vehicles, Valley Forge, PA: Saint-Gobain Corp.; http://www.saint-gobain.com/us

Composite Armor Portfolio Expanded to Meet Urgent Need for Enhanced Lightweight Ballistic Protection, e mail—AGY (VACS), 3/18/2009.

Cooper, M. Engineers Develop Cost-Saving Repair for Damaged Helmets, http://www.af.mil/news/story.asp?id = 123128555, 2 p.

Dixon, C. Blast-Proof Wheels for the Mean Streets of War Zones, *The New York Times* (February 24, 2008): 11.

Dorr, J. Carbon Nanotube Technology to Build Improved Body Armor, Concord, NH: Nanocomp Technologies; *AM&P* (December 2008): 22; http://www.nanocomptech.com

Honeywell Introduces New Composite Armor Applications, *Composites World*, 6/30/2009, p 1 of 1.

Hunn, D. Lockheed Martin Armor Developments, *AM&P* (October 2006): 23.

Kevlar Conveyor Takes the Heat, *Machine Design* (October 11, 2007): 41.

Marz, S. J. Marines Get a New Assault Vehicle, *Machine Design* (December 11, 2008): 70–9.

McConnell, V. P. A Stitch in Its Time: Fabrics Sew Up Diverse Composites Markets, *Reinforced Plastics* (December 2007): 20–2.

New Body Armor Made from Cement and Recycled Fibers, http://www.leeds.ac.uk/media/press_releases/current09/bullet.htm; 7/8/2009, p 1 of 3.

Rioux, J., C. Jones, M. Mandelartz, and V. Pluen. Transparent Armor, *AM&P* (October 2007): 31–33.

Speyer. R. Atlanta, GA: Georgia Institute of Technology; Robert.speyer@mse.gatech.edu; http://www.gatech.edu

Wakeman, M. D., D. R. Cramer, and J. E. Manson. Technical Cost Modeling of an Automated Net-Shape Preforming Process for Thermoplastic Composites: A Case Study, *Proceedings of the 26th SAMPE Europe International Conference*, Paris, France, April 2005.

Walsh, S. M., B. R. Scott, and D. M. Spagnuolo. The Development of Hybrid Thermoplastic Ballistic Material with Application to Helmets, Aberdeen, MD: Army Research Labs, December 2005.

Walsh, S. M., B. R. Scott, D. M. Spagnuolo, and J. P.Wolbert. Composite Hekmet Fabrication Using Semi-Deformable Tooling, *2006 SAMPE Proceedings*, Long Beach, CA, 2006.

16

New Innovative Heat Treating Processes

Mel Schwartz

CONTENTS

Methods

To effectively describe diffusion surface treatment see Tables 16.1 and 16.2 [1] that describe the temperatures, diffusion surface treatments, and elements of diffusion into steels.

Microwave Carburizing

The AtmoPlas microwave atmospheric plasma technology was officially launched by Dana Corp. in 2004 [2]. It is unique in its ability to generate and sustain plasma at atmospheric pressure. Plasma, a partially ionized gas, is crucial because it is an efficient absorber for microwaves (using up to 95% of the energy) and can rapidly reach very high temperatures. Once the plasma is ignited and sustained at atmospheric pressure, plasma temperatures can exceed 1200°C (2190°F) in seconds. In heat-treating applications, this offers an advantage over traditional processes that require longer heating times to reach high temperatures. Also, the heat flux can be custom-shaped by

TABLE 16.1

Surface Modification

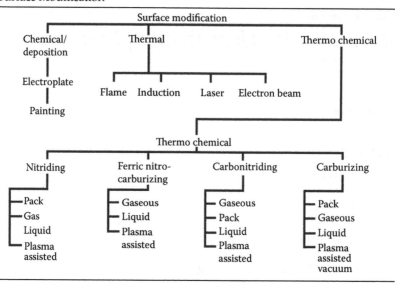

TABLE 16.2

Surface Treatments

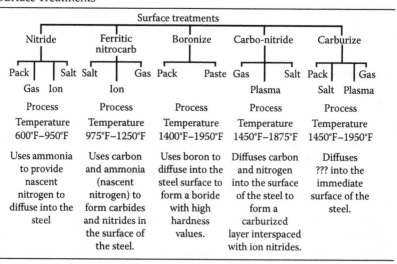

controlling the plasma thickness, which is accomplished through the proper placement of the workpiece.

The AtmoPlas technology improves heat-treating processes by delivering more precise heating control and higher temperatures, while reducing cycle times and energy costs. It also offers the capability to be used as an alternative method of carburizing. In carburizing applications for steel, the process produces properties that are equal or superior to those of conventional processes in terms of surface hardness, surface carbon concentration, case depth uniformity, retained austenite, and consistency of microstructures. AtmoPlas carburizing produces properties equal to or superior to those of conventional processes in terms of surface hardness, surface carbon concentration, case depth uniformity, austenite retention, and consistency of microstructure.

A significant amount of research has been done involving the use of the AtmoPlas technology for the carburizing process. Carburization using the technology also improves retained austenite and yields finer grain size. Carburizing tests performed on AISI 8620 alloy steel gears show significantly improved cycle time and case depth through AtmoPlas heating compared with conventional gas and vacuum carburized samples as shown in Table 16.3.

Using microwave atmospheric plasma technology, carburization is surprisingly simple. After a workpiece is placed inside a processing cavity inside the processing chamber, a gas such as argon is allowed to flow into the cavity. Plasma is ignited using a proprietary method, and the temperature rapidly increases. Acetylene gas is introduced into the processing cavity when the temperature of the workpiece reaches 930°C (~1700°F). The temperature is maintained at a fixed level by adjusting the microwave power. After treating the part for the required time, the part can be quenched and tempered if necessary.

TABLE 16.3

Characteristics of Carburizing Methods for Processing AISI 8620 Alloy Steel

Characteristic/ Property	Conventional Gas Carburized	Vacuum Carburized	AtmoPlas Carburized
Total carburization time	142 min. boost + 110 min. diffuse + 20 min. temp. drop = 272 min.	Carburizing zone time = 205 min.	112 min. boost + 80 min. diffuse + 20 min. temp. drop = 212 min.
Effective case depth	~0.035 in. (~0.9 mm)	~0.035 in. (~0.9 mm)	~0.045 in. (1.14 mm)
Microstructure (% and depth of retained austenite)			
Corner microstructure	~15%–30% to ~0.0125 in. (~0.319 mm)	~10%–15% to ~0.0067 in. (~0.172)	~5%–20% to ~0.0067 in. (~0.148 mm)
Surface microstructure	~10%–20% to ~0.0046 in. (~0.119 mm)	~5%–15% to ~0.0095 in. (~0.243 mm)	~5%–20% to ~0.0058 in. (~0.148 mm)
ASTM E112-96 Grain size No. (comparison method)			
Case	8–10 (22.5–11.2 μm)	8–9 (22.5–15.9 μm)	10–12 (11.2–5.6 μm)
Core	8–9 (22.5–15.9 μm)	9–10 (15.9–11.2 μm)	10–12 (11.2–5.6 μm)

Acetylene is used as the carbon source gas because it breaks down efficiently when exposed to the microwave plasma. The deposition of carbon is accurately controlled by the acetylene flow, the level of microwave energy, and the size of the cavity used to contain the plasma. With its ability to operate at higher temperatures, AtmoPlas can obtain a higher diffusion of carbon atoms into steel. An additional benefit of the technology is that the heat is confined around the part being processed, and that the system does not contain high-maintenance items like power feed-throughs or heating elements. The result is a carburizing process with a high level of flexibility and control.

Steam Treating

Steam treating [3] is a controlled oxidation treatment of metals to produce a thin layer of oxide on the surface of the component. The process is used to impart increased corrosion and wear resistance, to increase surface hardness, and to provide an aesthetically pleasing surface finish. In the case of porous materials, such as powder metallurgy parts, the process seals porosity and increases the density. Similar to other processes, the time, temperature, and atmosphere relationship is critical to the success of the process. In processing a ferrous component, the first step is to heat the component in air to a temperature above 315°C (600°F); a rule of thumb is to heat the part above 370°C (700°F) before exposing it to steam to ensure that the entire furnace charge is above 315°C (600°F). After the load is at temperature, it is brought into contact with dry steam, which means that any condensate, as a result of the steam cooling while going between the boiler and steam treating unit, is allowed to flow to a drain and not into the steam treating unit.

The water vapor in the steam begins to react with the iron in the part to form magnetite (Fe_3O_4) in the reaction:

$$4\ H_2O\ (gas) + 3\ Fe < -- > Fe_3O_4 + 4\ H_2\ (gas),$$

The Fe_3O_4 is an oxide of iron that is blue to black in color and has a microhardness of −50 hardness Rockwell C (HRC). Heating continues to a temperature of about 540°C (1000°F), at which temperature the reaction and the reaction rate for the formation of Fe_3O_4 by water vapor are optimized. The time the components are held at 540°C in the steam is a function of application. Applications, such as sealing, require that the part be held in dry steam for about 60 minutes; while other applications, such as increasing corrosion resistance or hardness, may only require a retention time of about 30 minutes to achieve the desired result.

The final step in the process is to allow the component to exit the steam treating unit. Even though the component is still very hot, exposure to open air is generally not a problem. Many producers collect the part in a container and allow them to cool naturally. Technology has continually evolved for the steam treating equipment, which initially was carried out in a batch type device. Batch steam treating is best suited for lower volume production requirements. The need to heat and cool the entire system for each cycle results in a loss of energy and time. The loading style of this type of system also adds to the inefficiencies due to the need to soak to achieve temperature uniformity throughout the load.

As cost-competitive production requirements increase, steam treating has moved toward continuous processing. These systems allow a component to be loaded onto a belt and passed through heating zones of a furnace that contain either air or steam. This technology produces an excellent steam-treated product. The first continuous belt steam treating units were humpback designs. More recent technology for steam treating uses a straight-through belt furnace that eliminates the need to pull the belt up and down inclines. The quality of steam-treated components is equal to that of parts treated in humpback and batch style units. However, straight-through belt furnace technology optimizes production capacity and reduces the need for maintenance.

Fluidized Bed Treatment

The fluidized bed is not only a more efficient way to heat treat aluminum parts than conventional methods such as salt baths, air chamber furnaces, induction heaters, and infrared heating, but also it is environmentally friendly and significantly reduces the heat treat cycle time [4]. Processes used to heat treat aluminum components include salt baths, air chamber furnaces, induction heaters, infrared heating, and fluidized bed furnaces, each with its advantages and limitations. Commonly used air circulation furnaces offer greater flexibility in operating temperatures and have no hazardous effects, but their relatively slow heating rates result in long heat treat cycles. Molten salt baths provide relatively fast, uniform heating and minimize distortion during heating, but they are potentially hazardous and require special precautions. Induction heating is an efficient method for in-line heating of flat rolled products, but their efficiency and power factors are significantly low for many nonferrous materials including aluminum alloys and especially for castings with varying thickness. Now, what about fluidized bed processing?

A typical fluidized bed furnace consists of a bed of dry, finely divided solid particles, which is made to behave like a fluid by feeding gas/air through orifices beneath the bed. The gas flow rate is sufficiently high to exceed the terminal velocity of the solid particles, and the bed goes into motion. Fluidized bed furnaces (batch and in-line continuous types) can be used to solutionize, quench, and age a part. The heat transfer coefficient in a fluidized bed is high, typically between 120 and 1200 $W/m^2{}^\circ C$ [5], and the bed offers excellent temperature uniformity (± 2–$3^\circ C$). The uniform heating of parts reduces distortion during heating.

A fluidized bed furnace offers significant advantages over salt baths, air chambers, and induction furnaces. However, the conventional fluidized bed is not optimized for heat treatment of metallic components. Two major drawbacks of conventional fluidized beds used for heat treating operations are: (1) inefficient heat transfer from the heat source to the fluidized bed due to the use of a single hot air system to provide both heating and fluidization, and (2) inefficient recovery of the temperature drop in the fluidized bed upon part loading. More gains in process cycle time and fuel efficiency are possible by optimizing the design of fluidized beds for heat treatment. To address inefficient heat transfer, new designs continue to evolve like the one that decouples bed fluidization and bed heating by using radiant heaters at the bottom of the bed just above separate fluidizing air tubes [6]. The bed is heated with direct radiant energy from the heaters independent of the fluidizing air. To address inefficient temperature recovery, the fluidized bed is designed with separate zones for part loading/heating and holding. Holding zones do not exchange heat energy with the loading/heating zone, thus minimizing overall temperature drop during part loading.

The process is continually being explored for future potential and evaluated like the use of fluid bed (FB) as a quenching medium, solution heat treat castings at a temperature close to the solidus temperature, and reduction of aging times for aluminum alloys.

Controlled Nitriding Using the Zeroflow Process

Traditional nitriding, which has limited control of the growth of the nitrided layer, is still very much in use worldwide [7]. Controlled gas nitriding, performed for more than 60 years, represents an advancement (but still used today) over the traditional method using 100% ammonia atmospheres. Two-component atmospheres consisting of ammonia and disassociated ammonia (NH_3 and dissociated NH_3) [8,9], as well as ammonia and molecular nitrogen (NH_3 and N_2) came into use about 50–60 years ago [10–12].

The concept of controlled nitriding using a Zeroflow process assumes carrying out the process of controlled nitriding using NH_3 alone. This is a simpler process than that using two-component mixes of ammonia diluted with N_2 or dissociated NH_3. The concept is based on experimental and theoretical investigations in terms of the thermodynamics and kinetics of gas nitriding. Investigations were carried out in a laboratory furnace using a quartz tube and also in an industrial furnace using a steel retort and a circulating fan. Nitriding was performed using NH_3, $NH_3 + H_2$, $NH_3 +$ dissociated NH_3 mixtures. In the industrial furnace, the supply of NH_3 into the retort was periodically stopped though with an operating circulation fan.

The studies demonstrated that the growth rate of the nitrided layer depends only on the composition of the atmosphere or K_N in the retort. It does not depend on either the type of atmosphere introduced into the furnace (NH_3,

NH_3 + dissociated NH_3 or NH_3 + H_2) or the type of furnace (e.g., laboratory with a quartz tube or industrial with steel retort and fan). The results also prove that stopping the supply of NH_3 to the retort does not affect the growth rate of the nitrided layer. A detailed analysis of the results is given by Anil [13] and Solar Atmospheres [14].

By performing the Zeroflow process using NH_3 alone, it is possible to produce nitrided layers identical to those obtained using standard processes in two-component atmospheres. In such a process, the control of composition of the atmosphere in the retort is carried out by occasionally stopping NH_3 flow into the furnace, thereby obtaining precise control of the kinetic growth of the nitrided layer similar to that available using NH_3 + dissociated NH_3 atmospheres. The Zeroflow process offers practical, economical, and environmental benefits over processes using two-component atmospheres including:

- Low consumption of gas (up to eight times less than in processes using NH_3 + dissociated NH_3 and NH_3 + N_2 atmospheres).
- Easier, less expensive nitriding installation for the Zeroflow process. Only one simple gas inlet valve and a gas analyzer is required to precisely regulate and control the chamical composition of the atmosphere obtained from NH_3 compared with two inlet valves, a high quality gas flow meter, and a gas analyzed required for processes using NH_3 + dissociated NH_3 and NH_3 + N_2 atmospheres.
- The Zeroflow process can be carried out in a furnace fitted with a steel retort, which is significantly cheaper than that made of a Ni-base heat resistant alloy required for processes using a two-component atmosphere.

Vacuum Carburize

It should make no difference if a carburized case is developed by the atmosphere or vacuum carburizing, since the end result is the same, right? Wrong! Almost all products today are being engineered to achieve increased performance at lower cost in smaller and smaller footprints. For example, transmission specifications call for higher horsepower ratings produced from smaller packages. Extending life and improving performance are necessities for today's manufacturer.

Vacuum carburizing is a modified gas carburizing process in which the carburizing is done at pressures far below atmospheric pressure (760 torr). The typical range for low pressure vacuum carburizing is 3–20 torr. The advantage of this method is that the steel surface is cleaned during heatup and the vacuum environment makes the transfer carbon to the steel surface faster (higher carbon transfer values) since atmosphere interactions such as those found in the water gas reaction do not take place. In addition, intergranular oxidation cannot occur. The hydrocarbon currently being used for vacuum carburizing are acetylene (C_2H_2), acetylene/hydrogen (C_2H_2/H_2), acetylene/ethylene/hydrogen $C_2H_2/C_2H_4/H_2$, propane (C_3H_8), propane/methane (C_3H_8/CH_4), and cyclohexane (C_6H_{12}), a liquid methane (CH_4) alone is still used in certain applications usually at temperatures above 940°C (1750°F).

Advancements in computer process control, the development of new steels, continuous improvement in high-pressure gas quenching technology, and vacuum equipment design innovations are the driving forces behind the resurgence of vacuum carburizing as a process capable of carburizing production workloads. Using a unique combination of ethylene (C_2H_4), acetylene (C_2H_2), and H_2 at pressures of 1.3 to 40 mbar allows this technology at low cost to the heat treat industry.

Solar Atmospheres has carburized an AISI 4329 sun pinion that connects to a generator as a planetary speed increaser for a hydroelectric plant. The 102.2 kg (225 lb) pinion needs selective carburizing, therefore stop-off paint was applied to the shaft to prevent carburizing, which enables further machining. The gear required 2.3–2.8 mm (0.090–0.110 in.) effective case depth and a surface hardness of 56–62 HRC. Vacuum carburizing reduces furnace time for the deep case compared with that using traditional atmosphere carburizing [15].

Low-Pressure Vacuum Carburizing (LPVC)

Low-pressure vacuum carburizing, a technology developed in the mid-1990s, has helped revolutionize the heat treating industry, primarily because of its capability for precise process control. Tight control enables absolute repeatability, optimized part microstructure, enhanced mechanical properties, and reduced manufacturing costs. Demand for vacuum carburizing systems equiped with high-pressure gas quenching, oil quenching, or both has quadrupled since 2000, as the technology has replaced atmosphere carburizing and traditional oil quenching in many cases. Low-pressure vacuum carburizing has been applied to aircraft parts such as braking systems, actuator systems, flight controls, and guidance systems.

New materials (Table 16.4) [15] designed specifically for high-temperature service retain their hardness and mechanical properties well into a service range of 315–510°C (600–960°F) and higher. Many are similar in chemistry to stainless and tool steels to take advantage of higher resistance to corrosion and wear, but they have better core microstructures. Temperature uniformity is critical for vacuum carburizing, especially in the low-temperature ranges, and a maximum spread of ±5.5°C (±10°F) is required to maintain tight case depth control throughout the load. This is equally important for carbonitriding cycles, which are run as low as 775°C (1425°F). Highly loaded gears and shafts, such a those in off-road vehicles under extreme race conditions,

TABLE 16.4

Advanced Carburizing Alloys

Material	Carbon, %	Significant Alloying Additions, %	Material	Carbon, %	Significant Alloying Additions, %
XD15NW	0.37	Cr (15.5), Mo, V, Ni	CSS-42L	0.12	Cr (14.0), Co, Mo, Ni, V, Cb
X13VDW	0.12	Cr (11.5), Ni, No, V	CSB-50NIL	0.13	Mo (4.25), Cr, Ni, V
AF1410	0.15	Co (14.0), Ni, Cr, Mo	Pyrowear 675	0.07	Cr (13.0), Ni, Mo, V
CBS 223	0.15	Cr (4.95), Co, Mo	Pyrowear 53	0.10	Mo (3.25), Ni, Cr, Si
CBS-600	0.19	Cr (1.45) Mo, Si	AerMet 100	0.23	Co (13.4), Ni, Cr, Mo
BG42	1.15	Cr (14.5) Mo, V	HP 9-4-30	0.30	Ni (9.0), Co, Cr
N360 lso Extra	0.33	Cr (15.0), Mo, Ni	HY-180	0.13	Ni (10.0), Cr, Mo
N695	1.05	Cr (17.0), Mo, Si	R250	0.83	Mo (4.30), Cr, V
EN30B	0.32	Cr (1.30), Mo, Si	R350	0.14	Mo (4.30), Cr, Ni
Cronidur 30	0.31	Cr (17.0), Mo, Si, Ni	C61	0.16	Co (18.0), Ni, Cr, Mo
CS62	0.08	Co (15.0), Ni, Cr, V	VascoMax C-250	0.03 max	Ni (18.5), Co, Ni, Cb
C69	0.10	Co (28.0), Cr Ni, Mo, V	VascoMax C-350	0.03 max	Ni (18.5), Co, Ni, Cb

Source: Herring, D. H., and Otto, F. J., *AM&P*, 165, no. 3, 31–33, March 2007.

require fracture toughness values two or three times those of conventional vehicles. Tightly controlled, low-pressure vacuum carburizing processes deliver these high-performance values.

Materials such as those in Table 16.4 require better control of hardness and carbon distribution (case/core hardness, surface/near surface), optimized microstructures, and control of such factors as retained austenite, carbide size (type and distribution), and nonmartensitic phases. Property control is also critical, especially for characteristics such as residual stress patterns and surface finish, and mechanical properties such as toughness, impact strength, and wear resistance. To improve part performance throughout the industry, commercial heat treaters are beginning to share their knowledge with respect to successes and lessons learned:

- Part cleanliness has improved because of the development of aqueous, semiaqueous, and solvent-based cleaning technology.
- Elimination of manufacturing steps such as preoxidation is helping to improve quality.

- Lead times are being reduced and process control improvements are better documented.
- Preventive maintenance practices result in up-time reliability greater than 95%.

Typical commercial and military aircraft applications of low-pressure vacuum carburizing include components such as bearings, ball screws and nuts, planetary gears, pinions, and shafts. The debate about whether low-pressure vacuum carburizing technology is superior to atmosphere carburizing and whether parts traditionally oil quenched can be replaced, in the majority of cases, by high-pressure gas quenching techniques has been answered in the affirmative. Demand for vacuum carburizing systems equipped with high-pressure gas quenching, oil quenching, or both, has quadrupled since 2000 [16]. Commercial heat treating has seen the second largest increase in the number of installed low-pressure vacuum carburizing units. The breakdown of the various industrial segments is automotive (73.6%), commercial heat treating (15.6%), industrial products (6.9%), and aerospace (3.9%) [17].

A new development being evaluated is the vacuum nitriding process that is applicable for high-temperature solution nitriding of martensitic stainless steel at vacuum partial pressures of nitrogen. Surface hardening of stainless steel is currently one of the hottest topics in the heat treating industry including nitriding, carburizing, and nitrocarburizing processes. Companies have successfully performed high temperature solution nitriding of grade 422 martensitic stainless steel at subatmospheric pressures, obtaining surface hardnesses of 59–60 HRC. Other programs include solution nitriding 410, 420, 422, and 440C evaluating variables of nitriding partial pressure, temperature, and time. Tests were run in a 10 bar high pressure gas quenching vacuum furnace, allowing for direct quenching parts from the nitriding temperature 1050–1103°C (2000–2100°F) to prevent nitride precipitation on cooling. It is expected that the study will provide information for the development of new surface treatments for improved part performance [18].

Gas Carburizing

Regulatory standards are driving process control and quality higher up the agenda for the users of heat treated products, and advances in furnace control are enabling heat treaters to meet these demands in a cost efficient manner. As Henry Ford, the grandfather of today's automotive industry once observed, "Quality means doing it right when no one is looking" [19].

The ability to measure and control all process parameters enables a company to ensure the quality of its products. The gas carburizing process is a cornerstone in the modern production of automobiles and aircraft, as well as many other industries. However, the critical process of dissolving carbon atoms into the surface of the steel component occurs in a furnace when no one is looking. Contemporary control systems allow the user to control many key process variables, but they work on the principle that the levels of certain unmeasured furnace gases, such as methane (CH_4), remain constant throughout the process. The nature of the carburizing process means that the furnace atmospheres varies during the process, and the level of uncracked CH_4 in the furnace has an effect on the end result. This can be overcome by using a process or agitation factor; a constant calculated when the furnace is commissioned and applied to furnace recipes, which enables the user to produce the rquired case depth.

Errors can creep into the process whereby an oxygen probe calculates the carbon potential (C_p) of a furnace atmosphere based on the assumption that the level of CO in the furnace remains constant throughout the process. However, the CO level is lower at the start of the process, giving a C_p reading higher than reality.

In a vision for the gas carburizing process, a new atmosphere controller takes the quality assurance of gas carburizing furnaces to a new level. Designed to eliminate scrap by giving the furnace control system the most comprehensive vision of the furnace atmosphere. The atmosphere controller measures the levels of CH_4, CO_2, and CO in the furnace atmosphere and uses these to apply a compensation to the oxygen probe reading to calculate the true instantaneous carbon potential in the furnace. The analysis of CO, CO_2, and CH_4 is carried out by infrared gas analyzers. Trials have shown that a furnace using an atmosphere controller is three times more accurate than a furnace controlled using an oxygen probe alone, with a tolerance of ±6% of C_p. Integrating an atmosphere controller within the furnace control system enables highly repeatable treatments, with quality built into the process. In an industry where quality is paramount, an atmosphere controller adds a new dimension of accuracy and repeatability to process control [20].

To meet the increasing demand for improved parts quality and economical operations, new and modern gas carburizing technology has been introduced into the automotive and aerospace industries, for example. Due to significant efforts made during the past 20 years to improve process technology and furnace design of conventional gas carburizing plants, carburizing with endothermic gas, nitrogen/methanol, or natural gas/air is a proven technology, which is securely entrenched in the field of heat treatment. Continuous pusher furnaces have become indispensable in many sectors of the automobile and truck industries, and are considered as state-of-the-art technology. Increased demands in cost effectiveness have led to the development of a new generation of plants capable of producing improved product quality combined with highly reproducible heat treatment results.

Three different modern furnace designs shown in Figures 16.1, 16.2, and 16.3 are a three track pusher furnace equipped with a bottom inlet and outlet sluice, an automatic gas carburizing and fixture rotation hardening plant, and an improved ring hearth furnace with separated heating, carburizing, and

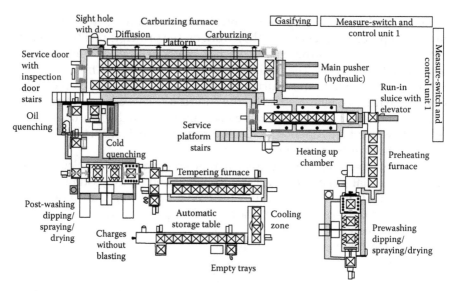

FIGURE 16.1
Schematic of three-track pusher furnace with bottom inlet and outlet sluices.

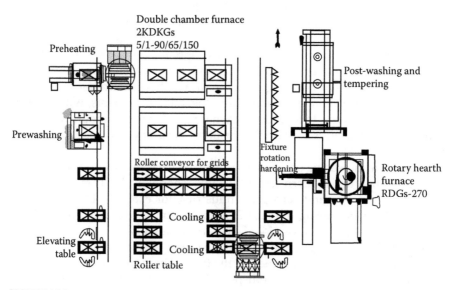

FIGURE 16.2
Schematic of automatic gas carburizing and fixture rotation hardening plant for TDI camshafts.

diffusion zones. Several different plants were developed recently, all having the characteristic design features of pusher furnaces, differing according to specific customer requirements. For example, high temperature furnaces were designed based on the required throughput capacity, the material grade to be heat treated, and the desired case hardening depth (CHD). In applications having high-throughput capacity and high case depth, two- and three-track pusher furnaces are usually used. Furthermore, different CHD values can be processed simultaneously (see Figure 16.1).

Figure 16.2 shows the plant layout for a heat treatment line for a four-cyclinder TDI engine designed to allow cost-effective case hardening of camshafts for a medium throughput capacity production line. Of paramount importance in the design was the need to reduce distortion and rejection of the camshafts.

The ring hearth furnace design combines the advantages of a rotary hearth furnace and a pusher furnace, allowing a separation of heating, carburizing, and diffusion zones. Therefore, different carbon levels and temperature profiles can be processed nearly independently in the different chambers.

Figure 16.3 illustrates equipment for carburizing plus press quenching. To minimize dimensional changes and distortion, very sensitive gear parts, such as synchronizing gears, crown wheels, and camshafts should be press quenched. Depending on throughput capacity, these parts are usually carburized in chamber furnaces or small continuous furnaces and gas cooled, followed by reheating in a rotary furnace and press quenched. The new ring hearth concept allows carburizing and direct hardening using press quenching in a single step because of the separated zones in the furnace, which can be used at different temperatures and C-levels without interaction. This can save up to 30% of the process time compared with conventional process (single hardening).

Figure 16.4 shows a typical furnace layout for carburizing plus oil quenching. As an alternative to pusher furnaces, full loads can be carburized and subsequently quenched in big ring hearth furnaces. The load is pushed directly onto the ring hearth, which rotates all loads without any further push or pull through the entire furnace. Loading and unloading is done at the same position (sluice). The advantages of this concept are the reduced strain and longer life of trays due to the rotation of the loads in the furnace, savings of basic trays and furnace motor drives, and a significant shortening (up to 25%) of the process time. The reason is a quicker heating up and an improved gas transport possibility between the different loads, because the loads are not pushed together, leaving some space in between.

Nitrocarburize and Nitriding

There are three basic nitriding processes [20]:

- Liquid nitriding (salt baths)
- Gas nitriding (dissociated ammonia)
- Plasma nitriding

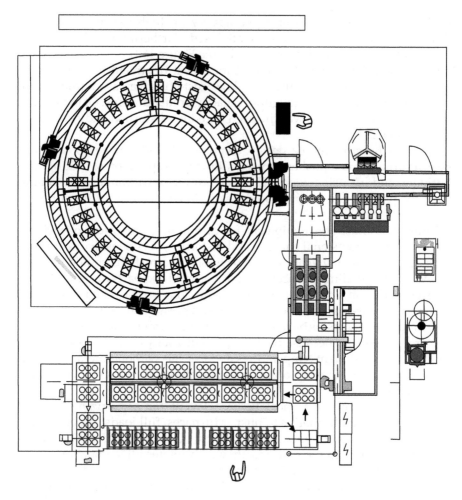

FIGURE 16.3
Schematic of ring hearth furnace system with press quenching.

Of these, plasma nitriding typically exhibits a number of process advantages including:

- Precise phase control of the nitrided surface layers
- Relatively low energy and gas consumption
- No environmental pollution
- Lower surface treatment temperatures [from 350°C (666°F)]

Gas nitriding is a process where nitrogen is introduced into the surface of the steel at a suitable temperature in contact with a nitrogenous gas, usually ammonia. The nitriding temperature for all steels is between 495 and 565°C

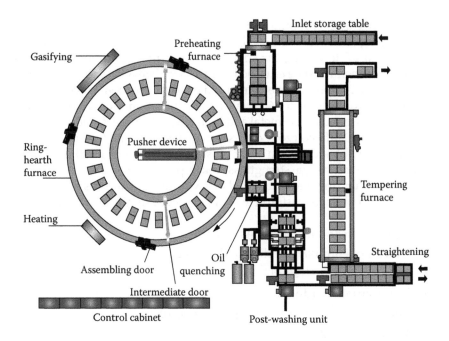

FIGURE 16.4
Schematic of ring hearth system with oil quenching. (Herring, D.H. and Otta, F.J., *AMFP*, 165, no. 3, 31–3, March 2007. With Permission.)

(943 and 1076°F). Quenching, if necessary, is performed before the nitriding treatment. Liquid nitriding is carried out in a molten salt bath at temperatures in the range of 510–580°C (971–1104°F). Salt baths for nitriding are typically composed of molten cyanides or cyanates [20].

Salt bath nitrocarburizing is a thermochemical process for improving the properties of ferrous metals. However, some tool and other high-alloy steels are susceptable to reductions in core hardness after standard nitrocarburizing. To prevent such losses, a low-temperature salt bath nitrocarburizing process has been developed. With treatment temperatures as low as 480°C (914°F), this process not only maintains core hardness, but also can sometimes increase core hardness. In addition, recent advances have made salt bath nitrocarburizing environmentally friendly. Modern salt bath nitrocarburizing involves noncyanide salts, a completely regenerable bath (no waste salts produced), and typical process times of 60–90 minutes. Fully automated equipment, designed with zero water discharge, can control and document the process as well as eliminate wastewater. By comparison, the old process operated with high cyanide baths, required very long treatment times, and generated large amounts of waste salts.

During salt bath nitrocarburizing, the part is immersed in a vessel of molten salt. Nitrogen and carbon in the salt react with the iron in the surface, forming a compound layer with an underlying diffusion zone. The compound layer consists of iron oxides, chromium nitrides, or other such compounds, depending on the alloying elements in the steel, and small amounts of carbides. Ranging in depth from 2.5 to 20 µm, the compound layer provides improvements in wear and corrosion resistance, as well as in service behavior and hot strength. Hardness of the compound layer, measured on a cross section, ranges from 700 HV on unalloyed steels, up to 1600 HV on high-chromium steels. Note that this layer is formed from the base metal and is an integral part of it, and is therefore not a coating. The diffusion zone can extend as deep as 1 mm (0.04 in.), depending on the steel. This diffusion zone causes an increase in rotating-bending strength and rolling fatigue strength as well as pressure loadability.

Salt bath carburizing may be applied to a wide range of ferrous metals, from low-carbon to tool steels, cast iron to stainless steels, specifically, the process:

- Improves wear and corrosion resistance
- Reduces or eliminates galling and seizing
- Increases fatigue strength
- Raises surface hardness
- Provides highly predictable, repeatable results
- Performs consistently, even with varying contours and thicknesses within the same part or load
- Maintains dimensional integrity
- Shortens cycle times
- Offers flexibility and ease of operation

Conventional treatment temperatures are in the range of 580°C (1104°F), but for highly alloyed steels as well as stainless and tool steels, this temperature can cause a reduction in core hardness. The above benefits, derived both from the nitrogen and carbon diffused into the metal surface, as well as the processing in a liquid bath, are often necessary for applications in which a reduction in core hardness is not acceptable.

It should be noted that nitrocarburizing is a modification of nitriding, not a form of carburizing. In the process, nitrogen and carbon are simultaneously introduced into the steel while it is in a ferritic condition; that is, at a temperature below that at which austenite begins to form during heating. A very thin white layer is formed during the process, as well as an underlying "diffusion" zone. Like nitriding, rapid quenching is not required. Examples of gear steels that are commonly nitrocarburized include SAE 1018, 1141, 12L14, 4140, 4150, 5160, 8620, and certain tool steels.

Nitrocarburizing is normally performed at 560–600°C (1025–1110°F) and can be used to produce an equivalent 58 HRC minimum hardness, with this value increasing dependent on the base material. White layer depths range from 0.0013 to 0.056 mm (0.00005 to 0.0022 in.) with diffusion zones from 0.03 to 0.80 mm (0.0013 to 0.032 im.) being typical.

Plasma Nitrocarburizing

In plasma nitrocarburizing, carbon is introduced into the substrate together with nitrogen using a treatment atmosphere containing 78% H_2, 20% N_2, and 2% CH_4. Although carbon occupies the same interstitial lattice positions as nitrogen, carbon supersaturation in solid solution is much lower than that of nitrogen (5.6 at.% versus 25 at.%). Nitrogen and carbon are randomly distributed in the middle of the edges of the cubic crystal structure resulting in an expanded face-centered cubic (fcc) unit cell. The degree of expansion depends on the amount of nitrogen or carbon in solid solution, which is dependent on the treatment parameters and depth within the layer.

Plasma or Ion Nitriding

Plasma, or ion nitriding [20], is a method of surface hardening using glow discharge technology to introduce nascent (elemental) nitrogen to the surface with subsequent diffusion into the material. The process is conducted in a vacuum under high voltage and the ions in the plasma that is formed are accelerated for impingement on the workpiece. This ion bombardment process heats the workpiece and cleans the surface providing active nitrogen under the influence of the glow discharge to form the nitrided case. The treatment can be performed at temperatures as low as 350°C (366°F), due to plasma activation (which does not exist in gas nitriding).

An advantage of plasma nitriding is the potential for the formation of monophase layers that are less fragile than the multiphase layers typically formed during liquid and gas nitriding. In addition, it is also possible to prevent compound layer formation and, therefore, only produce a diffusion zone that has greater substrate adhesion. Such a layer also can subsequently be used in conjunction with physical vapor deposition (PVD) to produce multiple surface coatings. Also, thin-film hard coatings including chromium nitride and titanium carbonitride can be applied over a plasma nitrided coating. Depending on the number of coatings, these are called duplex and multiplex coatings.

Fernandes and colleagues [20] show a schematic illustration of equipment used for plasma nitriding or nitrocarburizing treatments. The process involves evacuation of the chamber and then a mixture of 80% H_2 + 20% N_2 gases are injected for nitriding. A dc voltage of approximately 800–1000 V is then applied between the chamber and the material to be nitrided using a suitable power supply. This results in the formation of a plasma around the parts to be nitrided. A plasma is neither a solid nor a liquid or a gas, but is a fourth state of matter containing an ionized gas, which is formed when sufficient energy is applied to free electrons from atoms or molecules. This results in a high-energy atmosphere where ions and electrons coexist. This is important because the diffusivity of the excited nitrogen atoms may be up to 10 times faster than conventional gas or liquid nitriding. Therefore, plasma nitriding results in faster nitriding times since the plasma process produces faster surface saturation resulting in faster diffusion. Furthermore, the plasma nitriding sputtering process results in cleaner surfaces.

Two constituents of nitrided surface layers are: (1) a compound layer composed of iron nitrides of the Fe_3N (ε) and Fe_4N (γ) types or nitrides of the alloying elements present in the steel and (2) a diffusion zone located beneath the compound layer, which is a nitrogen-enriched region containing chromium nitride precipitates. The sequence of nitride layer formation is the deposition of FeN on the parts and diffusion of N_2 into the part, resulting in Fe_3N and Fe_4N formation in the most external region of the layer [21].

Austenitic, ferritic, and duplex stainless steels have applications that are used in many industries and austenitic stainless steel is among the most commonly encountered. However, these materials typically have relatively low hardness and, consequently, poor wear, fatigue, and antigalling resistance. Therefore, stainless steel is one of the most promising applications for the plasma nitriding treatment to impart substantially improved surface properties, and thus significantly expand potential applications.

Plasma nitriding has been particularly promising for the production of hard corrosion and wear-resistant layers in stainless steels because of the relatively low treatment temperatures; typically lower than 450°C (862°F) relative to gas nitriding, which requires a minimum treatment temperature of approximately 500°C (931°F). The higher treatment temperatures lead to the formation of chromium nitrides, which despite producing high hardness, reduce corrosion resistance. Plasma nitriding carried out in a temperature range of 350–450°C (666–862°F) produces nitride layers having high hardness without loss of corrosion resistance, thereby extending the potential applicability of these steels.

References

1. Pye, D. The Heat Treat Guy, *Heat Treating Progress* (July 2005): 33.
2. Cherian, K. Microwave Carburizing Shows Promise, *Heat Treating Progress* 6, no 1 (January/February 2006): 46–7.
3. Feldbauer, S. L. Consider Steam Treating to Enhance Surface Properties of Metal Components, *Heat Treating Progress* 7, no. 6 (September/October 2007): 51–6.
4. Chaudhury, S. K., and D. Apelian. Fluidized-Bed: High Efficiency Heat Treatment of Aluminum Castings, *Heat Treating Progress* 7, no. 6 (September/October 2007): 29–33.
5. Reynoldson, R. W. Heat Treatment in Fluidized Bed Furnace, Materials Park, OH: ASM International, 1993.
6. Maldziński, L. Controlled Nitriding Using a Zeroflow Process, *Heat Treating Progress* 7, no. 5 (August 2007): 53–5.

7. Floe, C. F. A Study of the Nitriding Process, *Transactions, ASM* 32 (1944): 134–49.
8. Bever, M., and C. F. Floe. Case Hardening of Steel by Nitriding, *Surfaces Protection Against Wear and Corrosion*, 123–43, Materials Park, OH: American Society for Metals, 1953.
9. Minkevich, A. N. Thermochemical Treatment of Metals and Alloys, *Mashinostroenie*, 331, Moscow: Publishing House Moscow, 1965.
10. Yu Sorokin, V., and A. N. Minkevich. Nitriding Steel in a Mixtire of Nitrogen and Ammonia, *MiTOM* no. 5 (1966): 49–52; Polish Patent No. 85924, Method of Gas Nitriding, 1977.
11. Maldziński, L., and J. Tacikowski. Concept of an Economical and Ecological Process of Gas Nitriding of Steel, HTM Z. Werkst. Wärmebeh. Fertigung, 61 (2006): 296–302.
12. Heat Treating, *ASM Handbook*, Vol. 4, 387–425, Materials Park, OH: ASM International, August 1991.
13. Anil, K. S. *Physical Metallurgy Handbook*, 16–89, New York: McGraw-Hill, 2002.
14. Solar Atmospheres, Inc., Souderton, PA, 2008. William Jones, Chairman of the Board.
15. Herring, D. H., and F. J. Otto. Low-Pressure Vacuum Carburizing, *AM&P* 165, no. 3 (March 2007): 31–33.
16. Otto, F. J., and D. H. Herring. Advancements in Precision Carburizing of New Aerospace and Motorsport Materials, *Heat Treating Progress* 7, no. 3 (May/June 2007): 35–40.
17. Otto, F. J., and D. H. Herring. Vacuum Carburizing of Aerospace and Automotive Materials, *Heat Treating Progress* 5, no. 1 (January/February 2005): 33–37.
18. www.solaratm.com
19. www.eurotherm.com
20. Fernandes, F. A. P., A. Lombardi Neto, L. C. Casteletti, A. M. de Oliveira, and G. E. Totten. Ion Nitriding and Nitrocarburizing, *Heat Treating Progress* 8, no. 4 (July/August 2008): 41–43.
21. Heat Treating, *ASM Metals Handbook*, Vol. 4, 944, 1001, Materials Park, OH: ASM International, 1991.

Bibliography

Altena, H., and F. Schrank. Modern Gas-Carburizing Technology for the Automotive Industry, *Heat Treating Progress* 7, no. 2 (March/April 2007): 17–22.
Chaudhury, S., and D. Apelian. *Heat Treating Progress* 7, no. 2 (March/April 2007): 12–13.
Chaudhury, S., and D. Apelian. *International Journal of Cast Metals Research* 19, no. 6 (2006): 361–69.
Chaudhury, S., and D. Apelian. *Metallurgical and Materials Transactions A* 37A (July 2006): 2295–2311.
Chaudhury, S., and D. Apelian. *Metallurgical and Materials Transactions A* 37A (March 2006): 763–78.

Gallo, A., S. A. Gallo, and A. Vitiello. Steam Oxidation of Ferrous Sintered Parts, *Powder Metallurgy* 46, no. 3 (2003).

Herring, D., and G. Lindell. Heat Treating Heavy-Duty Gears, *Heat Treating Progress* 8, no. 4 (July/August 2008): 49–50.

Pritchard, J., and S. Rush. Oil Versus Gas Quenching, *Heat Treating Progress* 7, no. 3 (May/June 2007): 19–23.

Rolinski, E., G. Sharp, and A. Konieczny. Plasma Nitriding Automotive Stamping Dies, *Heat Treating Progress* 6, no. 6 (September/October 2006): 19–23.

17

Resin Transfer Molding and Associated Closed Molding and Infusion Processes

Mel Schwartz

CONTENTS

Introduction

"Many firms utilized open molding processes 100% of the time. Currently, these firms have dedicated and converted their production lines to closed molding processes. We think that as the future goes along and the pressure builds regarding environmental issues and emissions, regulations are only going to get tighter and drive the industry to closed molding anyway," says Peter Jeffrey, president of FormaShape, a British Columbia, Canada-based firm. "Our idea was to get ahead of the curve" [1].

Regulation wasn't the only factor that drove FormaShape to adopt closed molding. More importantly, company executives felt strongly that closed molding could offer improvements in other areas of the business; namely productivity, quality, and employee retention. FormaShape like many companies in the industry took a step-by-step approach to its transition from open to closed molding. The company decided to begin by incorporating the cost of closed molding into its bids on projects that required tolerances, uniformity, and volume. That way any new orders that they won were automatically cost-effective for manufacturing with closed molding.

However, the company had to phase certain specified clients into the process more gradually. "The vast majority of the molds were owned by the customer, so it's not as if you could just transition everything from open to closed mold overnight," Jeffrey says. For example, there might be one part that's repeated a half a dozen or dozen times, and another part that's only repeated once. It wasn't as if the company could just turn around to their customers and say they were going to closed molding, so there is an increase in capital needed for the change in processes. Instead, when a customer retired a family of molds or developed a new product, FormaShape would submit a proposal based on closed molding, so that as each year passed, the total amount of closed molding in the plant would increase.

The effect on productivity is very important in switching processes. One reason more companies don't make the transition to closed molding is that they don't know if it's worth the investment. When FormaShape started investing in closed molding processes, it was on faith that the move would be successful in the long term. "The company couldn't find the stats to show how cost effective a move this would be," Jeffrey says. "We couldn't find other people doing it in a volume shop. We only saw some people doing one-offs, where they can take days to perfect the part." Because there were no statistics, the company collected its own data and measured it against expected results. This example is true for many companies in the industry.

Achieving that level of productivity came after an enormous amount of experimentation. "We couldn't find another volume producer that was doing closed molding, so there was a huge amount of trial and error with the resins, gel coats, and especially with the glass mat and finding a flow media that would work fast," Jeffrey says. "We're always looking for resins and mats that improve the process, increase the throughput and reduce the production times, so we can drill down both on labor cost and material cost." The size of parts that a company manufacturers also poses a production challenge. If parts are big then you need lifting devices, which in many cases don't exist, therefore, these devices require new designs. Finally, there is a difference and higher skill set for closed molding versus open molding if you plan to use closed molding and you intend on using employees that have always worked with open molding. Ensure that you put in a good training program and get those individuals to see the closed molding techniques working successfully in an existing plant.

Basics of the Various Processes

Engineers have been looking for alternative processing methods that can reduce costs while maintaining the high performance of autoclave-cured components. Within the last few years, liquid composite molding (LCM) technologies have advanced to the point where they can provide that alternative. LCM processes are characterized by the injection of a liquid resin into a dry fiber preform and include resin transfer molding (RTM) and vacuum-assisted RTM (VARTM).

In conventional RTM, the preform is placed into a closed, matched tool and resin is injected under pressure on the order of 0.69–1.38 Mpa. This low-pressure molding process contains the mixed resin and catalyst, which are injected into a closed mold containing a fiber pack or preform. When the resin has cured, the mold can be opened and the finished component removed. A wide range of resin systems can be used including polyester, vinylester, epoxy phenolic, and methyl methacylates, combined with pigments and fillers including Al trihydrates and Ca carbonates if required. The fiber pack can be C-glass, aramid, or a combination of these.

RTM

Early RTM processes lacked die consistency needed for aerospace components, in both dimensional tolerances and mechanical properties. Fiber volume fractions were significantly lower than the 60–65% typical of prepregs. Problems with predicting flow fronts as well as flaws that were introduced into the preform when closing the matched metal molds often led to high-void contents and dry spots. Additionally, there has been a constant battle between open and closed mold processes for leading the composites market and with the new EPA and MACT standards. There is also a growing trend in switching to closed molding processes. Marine, wind, and some other industries, once dominated by open mold processes, are now moving to closed molding processes such as VARTM and RTM. A recent market study "Global Composites Market 2003–2008" breaks down the global composites industry as well as many of the market segments such as marine, automotive, and construction and provides a clear picture on the size of open and closed molding processes in various market segments [2].

Improvements in both materials and processes have recently made RTM a viable option for aerospace manufacturing, where it normally takes 10–15 years for a new technology to become accepted. With RTM, the breakthrough began when Lockheed Martin selected RTM for many of the F/A-22 Raptor's structural components. Composites comprise approximately 27% of the F/A-22's structural weight (24% thermoset and 3% thermoplastic). RTM accounts for more than 400 parts, made with both BMI (bismaleimide) and epoxy resins. The wing's sine-wave spars were probably the first structural application of RTM composites in an aircraft. For a vertical tail on another Lockheed Martin aircraft, the RTM process reduced the part count from 13 to 1, eliminated almost 1000 fasteners, and reduced manufacturing costs by more than 60% [3].

Because the RTM process is more complex than autoclave curing, it is more difficult to develop a general qualification methodology. With prepregs, the material manufacturer mixes the resin and impregnates the tape or fabric under highly controlled conditions. Once a material is qualified, for example, an end user just has to demonstrate site equivalency of its manufacturing process. With RTM, however, both the resin mix and the resin content are more variable. In particular, the final resin content depends on maintaining a good flow front.

The high cost of tooling also limits the adoption of RTM. A set of production tools can cost on the order of $500,000. Although the price is competitive with autoclave tools, autoclave programs can usually get by with a single set of tooling for both development and production. With RTM, the resin flow front (and hence the part quality) is highly dependent on the tooling geometry. Often it is necessary to build one or more sets of prototype tools, to develop and test the process, before the production tooling can be built. Although prototype tooling is less expensive than production tooling, it is not so inexpensive that it can be considered expendable. Simulation of the molding process can predict flow fronts, allowing engineers to virtually test different mold designs without building expensive hardware. For example, you can model the resin flow and be able to identify preferred injection sites and sequences to achieve complete mold filling with no dry spots.

Generally, dry preforms for RTM are less expensive than prepreg material and can be stored at room temperature. The process can produce thick, near-net shape parts, eliminating most post-fabrication work. It also yields dimensionally accurate complex parts with good surface detail and delivers a smooth finish on all exposed surfaces. It is possible to place inserts inside the preform before the mold is closed, allowing the RTM process to accommodate core materials and integrate molded infittings and other hardware into the part structure during the molding process. Moreover, void content on RTM parts is low, measuring in the 0–2% range. Finally, RTM significantly cuts cycle times and can be adapted for use as one stage in an automated, repeatable manufacturing·process for even greater efficiency, reducing cycle time from what can be several days, typical of hand lay-up, to just hours or even minutes.

If one would ask 100 people to define RTM, you will get 100 different answers. Each manufacturer has its own version of RTM, and in many cases the differences are great enough to consider the process unique. The variations are even greater for VARTM, which can include a boat hull manufacturer using Seemann's Composite Resin Infusion Molding Process (SCRIMP) and an aerospace manufacturer using all steel molds with a vacuum only. For example, V System Composites (VSC) has three versions: classical RTM, HyPerVARTM, and HyPerRTM each tailored to a different need.

HyPerVARTM

In addition to classical RTM, VSC also has developed the HyPerVARTM process, a resin infusion system targeted to the aerospace industry. Most VARTM systems use a flow medium to get good resin coverage. This works well for relatively simple structures like boat hulls, but cannot handle the features amid complexities common in aerospace components such as multiple build-ups and ply drop-offs. HyPerVARTM is a single point-of-inspection system. Resin delivery is designed into the tool, eliminating the infusion medium and significantly reducing other consumables and touch labor, which translates into much lower costs. Selective control of permeabilities throughout the part provides excellent control over fiber volume fractions. The process is ready for production and has been proven on the CH-47 helicopter's forward pylon deck and on a Ka band deep space reflector built for NASA's Jet Propulsion Laboratory. Airbus is currently evaluating HyPerVARTM for production use, as well.

HyPerRTM

The third system in VSC's process development is the HyPerRTM process, which the company says combines the best of HyPerVARTM with classical RTM [4]. HyPerRTM molds are built using the HyPerVARTM process and can incorporate some metallic details where necessary. Resin distribution details also are built into the HyPerRTM mold. HyPerRTM enables one to produce parts with the high quality of classical RTM but at the price of HyPerVARTM.

CoRTM

A new innovative process has been developed through the efforts of government and industry that reduces fabrication and assembly costs. The process is called CoRTM (co-curing of an uncured skin to a resin transfer molded substructure). It was developed by Northrup Grumman and produces large, integrated, weight-efficient, precise, and repeatable structures [5]. A vertical stabilizer from the F-35 was used to demonstrate the technology. These results, using CoRTM in the manufacturing of the part, revealed that nearly $14,000 in savings could be derived through reduced tooling, part count, fastener count, and the associated fit-up, liquid shimming, and surface mold line treatments for air vehicles. Traditional aircraft structures consist of multiple piece assemblies that are prefit together; gaps between mating surfaces are filled with shim materials to create a snug fit and then mechanically fastened in place. This results in very lengthy manufacturing flow times and high-acquisition costs.

Now, the CoRTM process has been proven to be a viable and promising alternative for affordable composite structures. CoRTM combines two cost-effective processes: fiber placement (the automated placement of bands of high-strength fibers combined with resin onto a tool) for skin structures, currently used on the F-35, F-18, V-22, F/A-22, and so on; and RTM (the injection of high-strength resin into a mold containing high-strength fibers formed to a specified shape) for substructures [6]. Instead of fastening the skin to the substructure, the CoRTM process enables the skin and the substructure to be designed and fabricated as a single component, eliminating the need to fasten them together. This creates structures with fewer parts and minimal fasteners, resulting in reduced assembly costs. The savings versus the baseline construction costs for the F-35 tail represent a 52% reduction in part count, a 38% reduction in tool count, a 7% reduction in weight, and a 17% overall cost reduction when compared to the typical F-35 construction process.

VARTM

Vacuum-assisted resin transfer molding (VARTM) refers to a variety of related processes that represent the fastest-growing new molding technology. The salient difference between VARTM-type processes and standard RTM is that in VARTM, resin is drawn into a preform through the use of a vacuum, rather than pumped in under pressure. VARTM does not require high heat or pressure. For that reason, VARTM operates with low-cost tooling, making it possible to inexpensively produce large, complex parts in one shot. VARTM is best described as a complementary process to RTM. VARTM mold costs are basically half the price of equivalent RTM molds but they produce, at best, at half the rate of RTM; however, the process provides molders an attractive introductory route into closed mold production.

In VARTM, resin flow rates cannot be speeded up above an optimum level in order to fill the mold more quickly, as the recommended VARTM mold construction and the atmospheric mold clamping pressures limit overall in-mold pressures to less than 0.5 bars. As with any composite closed-mold production technique, VARTM is no exception to the rule in demanding high-quality, accurate composite molds in order to provide good mold life and consistent production of good parts (see Figure 17.1). In the VARTM

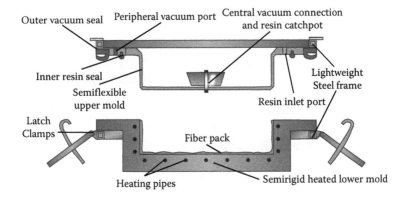

Outer vacuum seal   Peripheral vacuum port   Central vacuum connection and resin catchpot

Inner resin seal

Semiflexible upper mold

Lightweight Steel frame

Resin inlet port

Latch Clamps

Fiber pack

Heating pipes   Semirigid heated lower mold

FIGURE 17.1
VARTM process. (From Heider, D. and Gillespie, Jr., J. W., *J. Adv. Mater.,* October 1994. With permission.)

process, fiber reinforcements are placed in a one-sided mold, and a cover (rigid or flexible) is placed over the top to form a vacuum-tight seal. The resin typically enters the structure through strategically placed ports. It is drawn by a vacuum through the reinforcements by means of a series of designed-in channels that facilitate wet-out of the fibers. Fiber content in the finished part can run as high as 70%. Current applications include marine, ground transportation, and infrastructure parts.

Recently, an advanced VARTM system called Sequential Multi-Port Automated Resin Transfer Molding (SMART-Molding) was developed to fully automate the infusion process. The system incorporates sequential injection VARTM processing, actuators to control the flow, sensors to detect the flow behavior during the resin injection, and online control to optimize the opening and closing of the actuators. The system has been successfully demonstrated during laboratory-scale process trials [7] (see Figure 17.2). For the first time, the SMART-Molding approach completely automates this process. The injection gates are opened in sequence when mold-mounted flow sensors detect the tool-surface flow front at a perpendicular gate. The process is completely autonomous and does not depend on resin permeability variations, changes in preform characteristics, or other processing factors. In addition, the SMART-Molding system does not require costly trial-and-error development of the conventional sequential scheme and therefore does not demand the engineering knowledge normally required during resin impregnation. However, the number of inlets and sensor locations must be specified [7].

Liquid Injection Molding Simulation

The progression of resin through a mold during infusion is a complex and often inscrutable phenomenon. The seat-of-the-pants mold design based on

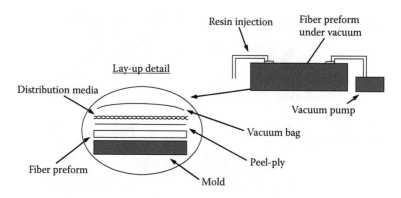

FIGURE 17.2
Schematic of VARTM process with lay-up detail. (From Heider, D., and Gillespie, Jr., J. W., *J. Adv. Mater.*, October 1994. With permission.)

past experience might work for simple part geometries, but as shape complexity increases it becomes more difficult to position mold vents and gates for the best results. Getting it wrong can mean incomplete wet-out of the reinforcements, nonuniform cure, poor resin/fiber bond, and residual stress.

"There is a real need for analytical tools to supplement experience in manufacturing design," says Dr. Jeffrey M. Lawrence of VSC, Chester, Pennsylvania and Anaheim, California) [8]. The UD-CCM's (University of Delaware Center for Composite Materials) Drs. Suresh and Pavel Simacek as well as Dr. Lawrence have developed the liquid injection molding simulation (LIMS) tool, which enables the user to digitally model and predict a resin's in-mold flow behavior in RTM and VARTM processes. The LIMS software has become a key tool in VSC's efforts to optimize tool, part, and manufacturing process designs. In addition to the simulation model, the VSC system team also uses a resin flux process monitoring method, which allows the infusion process to be monitored in real time. Together, the modeling and monitoring ensure high-quality repeatability for better parts.

VSC employed LIMS to help validate VARTM as an alternative process for fabricating a complex aerospace part. VSC replaced a multipart metallic design with a single, infused carbon/epoxy part that included monolithic laminate blade stiffeners arranged in a grid pattern, supporting an integral monolithic skin, co-infused and co-cured as a single part. LIMS played a key role in helping the company demonstrate the manufacturability, performance, and affordability of infusion for high-performance aerospace parts. The company believes that VARTM is sufficiently mature to be implemented in aerospace manufacturing programs. The key benefit to being able to simulate the VARTM process is that an entire part, despite the complexity, can be successfully tooled and infused in one shot, instead of having to secondarily bond features such as stiffening ribs, saving time, and labor, concludes Dr. Lawrence.

RARTM

The RARTM process uses trapped rubber and a pressure relief valve with a closed mold. This provides:

- Easy tool closing
- Improved resin flow and fiber wetting
- High fiber volume, low void content laminate

Figure 17.3 reflects the RARTM process that is compared to conventional RTM (Figure 17.4), which is a closed-mold, low-pressure process. A progressive injection method is used in the RARTM process with multiple injection/vent ports. The part is positioned vertically during resin injection, and the resin is injected at the lowest geometric portion of the part and vented from the highest position. When the resin reaches the port, the vacuum is removed and the resin is injected, and the procedure is repeated as the resin progresses up the part.

Light RTM

Light RTM is a variant of RTM that's growing in popularity. Low injection pressure, coupled with vacuum, allows the use of less-expensive, lightweight two-part molds. See below sections.

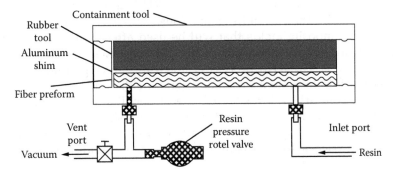

FIGURE 17.3
RARTM process.

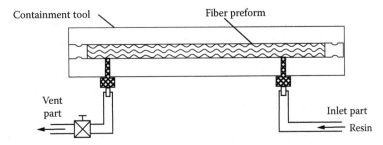

FIGURE 17.4
Conventional RTM.

Vacuum-Assisted Resin Infusion Process

The liquid composite molding process variant known as vacuum-assisted resin infusion (VARI) was first introduced by Marco [9]. This VARI approach presents an interesting alternative to classical closed mold RTM injection molding. It enables successful manufacture of large parts at a relatively low cost. In this process, a stack of dry fibrous reinforcement is placed between a stiff mold half and a plastic bag. The resin is injected by gravity after partial or total vacuum has been achieved in the cavity containing the reinforcement. If vacuum-driven techniques are now commonly used in many industrial applications, usually their development is based on trial and error testing. Williams et al. [10] present an interesting review on the main developments in this field. From the VARI process developed by Group Lotus Cars Ltd [11] to SCRIMP [12], are many variants of resin infusion that are now used in production. The main advantage of these techniques is their low tooling cost compared to closed mold RTM.

The VARI process is suitable for low production series. The vacuum-assisted, LCM process variant known as RTM Light consists of injecting a resin at low pressure in a composite shell mold. The resin is injected after a partial vacuum has been achieved in the cavity in order to increase the pressure gradient that drives the resin flow. RTM Light is suitable for larger production series than resin infusion. Thus, the VARI process can manufacture the composite molds that will be used afterward for RTM Light manufacturing. The above-mentioned LCM processes complement each other quite well. The RTM process is suited for larger production series. The resin is injected into a closed and rigid mold. The VARI process performs a vacuum driven infusion under a plastic film; it is appropriate for low production series or to manufacture a composite shell mold for RTM Light. Finally, RTM Light represents a low cost alternative that combines the two approaches: low pressure injection together with partial vacuum in a deformable mold.

TERTM Process

The Thermal Expansion RTM (TERTM) is a hybrid process within the resin infusion processes as there are in any other manufacturing category [13]. TERTM may utilize either a resin infused into the primary preform structure by the RTM or a Vacuum infusion processing (VIP) method. The next step after resin infusion is created by the internal structure within the part or tooling. TERTM relies on an expanding core material (caused by the increase in the cure temperature) to provide preform/resin compaction within the tooling. The expanding core material then squeezes out excess resin and drives the preform to a net higher fiber volume while conforming nicely to the inner tooling surface.

Infusion Processes

Resin Film Infusion (RFI)/DIAB Method

Although there are many variations of the infusion process, some propri-etary and some generic, the basic principle is the same. Dry materials (rein-forcements and core) are placed into a mold and then covered with a vacuum bag. A vacuum is then applied, resin is fed into the mold and the vacuum draws it through the pan until saturation occurs. The vacuum is then main-tained until the part is cured. Normally the process yields high-performance parts; that is, low weight due to a high-fiber fraction. It is a repeatable process that is easy to control in a quality management system. Thus, the process is not reliant on the individual skill of the workers. It produces parts with a consistent and even quality. The working environment can be drastically improved and if done utilizing a mixing and dispensing machine, emissions can be decreased by more than 95% compared with open mold methods.

Resin film infusion (RFI) is a process whereby a resin film is impregnated in a preform by melting and infusing it into the preform. The infusion pro-cess utilizing grooved core materials is an innovative, faster, and cheaper closed manufacturing method for producing sandwich constructions with fiber composite skins [14]. The grooved core method, however, utilizes an optimized fine groove pattern in the core surface to facilitate resin distribu-tion. One RFI bagging configuration used is shown in Figure 17.5. A resin film was placed under the five-axis 3-D fabric to effectively remove the air inside the five-axis 3-D fabric. The resin film was prepared beforehand by

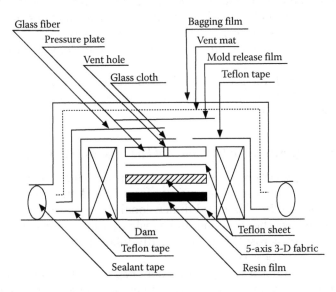

FIGURE 17.5
Bagging configuration for RFI processing. (From Williams, C., Summerscales, J., and Grove, S., *Composites*, 27A, 517–24, 1996. With permission.)

melting bismaleimide at approximately 100°C (212°F) and molding into the shape of the flat panel. A bismaleimide in the melt state has very low viscosity of 100 centipoise or less, and it is very easy to impregnate. Because it may leak from a clearance around dams, the clearance is scaled by the Teflon tape. The amount of resin was adjusted so that the fiber volume content of the fabricated composite might become 50%. The cure process was carried out at a temperature of 191°C (375°F) and at a pressure of 0.62 MPa for 6 hours [15].

DIAB Infusion Method

The DIAB Infusion Method [16] is an innovative, faster, and cheaper close manufacturing process for producing sandwich construction with composite skins. The DIAB method has the following implications:

- Grooved core = > faster flow and no extra distribution media
- Faster flow = > larger panels are possible
- Larger panels = > less extra materials and minimum waste
- Less extra materials = > faster lay-up and a less costly product

The DIAB method has the advantage of being extremely fast, making large structures feasible in one shot. It also does not need additional consumables like resin distribution materials, release films, and peel-plies, which makes it more environmentally friendly and less costly. Significant cost savings can also be made compared with traditional methods if molds and production equipment are adapted accordingly. Examples of successful implementation in applications are boat hulls, wind turbine nacelles, shelters, containers, and storage tanks [17].

C-fiber composite panels are an established technology in motor racing and luxury sports cars. This type of material allows lighter and stiffer components for low-volume production. New materials have been developed to offer low-cost alternatives to conventional prepreg processing and a route to higher production volumes.

Seemann's Composite Resin Infusion Molding Process (SCRIMP)

Bill Seemann's process represents a major step forward in fabricating very large composite structures using a version of RTM that involves "hard" tooling on one side for shape and dimensional control, a specially designed vacuum bag as the other tool side, and a resin distribution system that accelerates the infusion across the surface rapidly, followed by saturation through the composite thickness [18].

The process requires a very good vacuum bag integrity in order to achieve the resin infusion driving force required. A specially designed bagging system provides a "resin distribution media" that pulls the resin rapidly across the composite preform surfaces at several locations. The process has been successfully used with polyester, vinyl ester, and epoxy resin systems in a variety of composite structures. Large marine structures and yachts have been of interest. Marine deck structures, masts, and above deck structures have generated considerable interest within the industry. Refrigerated rail cars have been SCRIMP'ed using this process to achieve the above-mentioned part count reduction, insulation integrity, and improved manufacturing (e.g., lower cost products).

The SCRIMP process [18,19], and the ultraviolet (cure) RTM process, termed UVRTM, exhibit similar characteristics as far as the resin distribution approach is concerned. SCRIMP infusion allows for all production FRP materials to be used; that is, reinforcements such as E-Glass, S-Glass, Kevlar, and carbon, and the resins mentioned above, as well as structural cores. The success of large-scale infusion and the SCRIMP process depends upon a combination of practical commercial practices with just the right amount of aerospace practices and quality control procedures [20].

The first part built by Seemann Composites for the U.S. Navy was a test module composite deckhouse, which became an unqualified success and resulted in the Navy realizing that high-quality composites could be built at a price that was affordable for large scale structures on Navy ships. The key was to show that shipyards could build the structures themselves, so the technology was transferred to the shipyards from Seemann Composites. Today Northrup Grumman SS Avondale Division is building the AEMS type masts in full production for the latest U.S. Navy ship, the LPS17. The next generation ship design, the DDX destroyer is being designed with an all composites deckhouse. This structure has currently been designed as an all carbon fiber structure. To give the reader an idea of the scale of this structure, each deckhouse will use almost 450,000 kilos of carbon fiber. Additionally, the U.S. Coast Guard Fast Response Cutter, or FRC has been dramatically redesigned and the original steel-hulled FRC with an expected service life of 15–20 years has been replaced with an all composite-hulled, SCRIMP infused, craft with a 40-year hull life. Production, sail and power vessel hulls, decks, and structural grids based on SCRIMP infusion are being constructed across the world. SCRIMP has spawned a whole range of infusion variations that are marketed throughout the world [21].

Infusion Categories and Advantages

Estimates in the past year claim that about 700 U.S. composites manufacturers initiated a changeover to some form of resin infusion. Yet the results are often disappointing because the move can be made with a minimum of process modification. Boatbuilders, for example, can transition to vacuum infusion simply by widening the flange on existing open molds and adding vacuum bagging, resin transport lines, and a vacuum pump. While this minimalist approach cuts worker exposure to and plant emission of hazardous air pollutants (HAPs) and other volatile organic compounds (VOCs), it may do nothing to increase process efficiency.

How does one make the process faster? We hear that question about every day, reports Keith Bumgartner, technical service representative. For resin manufacturer Ashland Specialty Chemical-Composite Polymers Division (Columbus, OH), answers to that question abound, in part, because the number of composites manufacturers who employ infusion have reached the critical mass necessary to make supplier development of infusion-friendly products profitable. The main suppliers work very closely with the various manufacturers to the extent that customers work with experts in dedicated labs, developing an infusion process for their operations without taking time away from the production at their facilities. The suppliers learn the chemistries needed: laminate schedules, dimensions of the application, and then return to their labs and start infusing small parts and panels to hone in on optimal physical properties and process parameters. The suppliers also look at the entire operation including secondary lamination operations tabbing, bonding, reinforcing ribs, hardware attachment points, and incorporating these into a single lamination process. These are the steps that make it truly "closed molding" and not just "laminating under a bag." To optimize infusion performance, the fabricator must first select the most practical method. Resin infusion describes a fairly wide range of closed molding methods. To be categorized as infusion, a molding process must enclose dry reinforcement via bagging or a matched mold, followed by injection of liquid resin into the reinforcement. Today, that description fits a number of processes—23 in all including developed-enhanced variations denoted by acronyms like SCRIMP and SQRTM (Radius Injection System) [21].

For discussion's sake, however, infusion methods may be classified in three major categories. The first, under the general heading of vacuum infusion processing (VIP), includes methods that use negative pressure (i.e., vacuum) to draw the resin through the reinforcement. VIP molds can be closed relatively inexpensively, using disposable vacuum bagging film on the previously open or side of the part [22]. The second category includes all varieties of RTM, where positive pressure is used to push resin into the reinforcement. Methods that employ only positive pressure require a more expensive matched mold, with hard B-side tooling, to resist what can be very high injection pressures. The third category covers any method that employs both vacuum and positive pressure, including VARTM and light RTM. These methods may require hard B-side tooling as well, but depending on the balance of positive and negative pressure, the B-side may be formed from lighter, thinner, and therefore, less expensive materials. In some cases, reusable silicone bagging materials may be sufficient to support very light positive pressure.

It is estimated that, overall, RTM is three to four times faster than VIP, using the same resin and reinforcement. Each type of the B-side equipment also is characterized by different ranges of infusion rates. Frequently comparisons have been made of light RTM, VIP (with nylon bagging) and closed-cavity bag molding (CCBM, with silicone bagging). Working with one resin system and the same reinforcements for all three it has been shown that depending on how one manipulates the pressures, infusion speed varies significantly. CCBM, with some positive resin injection pressure, reportedly enables a pre-measured quantity of resin to be injected in a short period of time, typically less than a minute. As the injected resin flows through the laminate, fully impregnating the reinforcements, the operator can continue on to the next part. This results in significant time savings, since one operator can handle multiple parts, but it also requires a structural change in shop layout. CCBM is most suitable for medium-volume production with minimal cosmetic requirements for the B-side of the part [18]. See Table 17.1.

Engineered Textile Preforms for RTM

Net shape textile preforms have considered attention for RTM composite consolidation [18]. Among the processes that are used for producing preforms are: 2-D weaving, 3-D weaving, 2-D braiding, 3-D braiding, knitting, and stitching. There are currently three main limitations in preforming technology for advanced composite RTM markets [23]:

- Meeting performance requirements for engineered structures
- Meeting shape requirements for complex parts
- Manufacturing economics

Recognizing these, a number of technology-oriented companies have pursued preform development with a variety of technologies. The different textile preforming technologies suitable for RTM are described in Benjamin and Beckwith [18] and Joubaud, Trochu, and Le Corvec [19] as well as the factors that must be taken into account in order to transition these textile processes from R&D rarities to the actual production of primary load bearing structures.

The manufacture of all textile product forms start with raw fiber. Discrete length fibers, known as staple fiber, can be processed into random or semi-oriented mats, known as nonwovens. The raw fibers can also be twisted together to form a spun yarn. Continuous filament yarns, however, are the most common in the aerospace industry. These multifilament continuous yarns are typically converted into fabric structures using weaving, knitting, or braiding technology. Benjamin and Beckwith [18] have an overview of the textile preforming techniques and the advantages and limitations of each process. Finally, several case studies are provided that compare the applicability of these techniques to make specific parts.

Tooling

A new heat transfer technology has been announced that provides "in mold" solutions to heating and cooling of complex cores and cavities in composite, steel, nickel, and cast resin molds [24]. These are heat pipes that are

TABLE 17.1

Deciphering the "Alphabet Soup" of RTM-Type Processes

RTM: Type Process Terminology	Description or Attributes of RTM-Type Process
Resin Transfer Molding (RTM)	• Developed from urethane technology • Resin injected into preform under pressure into closed mold • Low-to-moderate fiber volumes (20–45% by vol typical)
VARI, VARTM, VRTM, and VIP: • Vacuum-assisted resin transfer injection • Vacuum-assisted RTM • Vacuum RTM • Vacuum infusion process	• Vacuum used to pull resin into preform • Process *may* use pressure to push resin in at same time • Vacuum levels typically 10–28 inch Hg • Void content typically lower • Higher fiber volumes possible
VIMP: • Variable infusion molding process	• Resin transfer occurs from *interior* within mold • Preformed with vacuum or gravity resin feed transfer
TERTM: • Thermal-expansion RTM	• Internal core material with preform • Resin is infused and mold/structure is heated • Heating causes core to expand and compact lamination • High compaction forces material to mold surfaces
RARTM: • Rubber-assisted RTM	• Like TERTM except rubber materials used in place of core • Rubber tooling materials create extremely high compaction • Very low void contents but heavy tooling required • Very high fiber volumes possible (60–75% by vol)
RIRM • Resin injection recirculation molding	• Combination of vacuum and pressure infusion used • Resin injection into desired multiple ports sequentially • Resin recirculation until preform satisfactorily wet-out
CRTM (Hexcel Corporation Process): • Continuous RTM	• Preform pulled into pultrusion die and mold system • Resin injected into stationary mold system • Upon cure completion, process repeated after part removal
CIRTM: • Coinjection resin transfer molding	• Multiple resin injection (different resins) process • Multiple preforms possible, co-curing achieved • Utilizes vacuum bag and soft-sided tooling
RLI: • Resin liquid infusion	• Liquid resin injected or placed into lower mold cavity • Preform, shaping tooling, and upper mold cavity assembled • Heat and external mold pressure forces resin to fill preform
RFI: • Resin film infusion	• High viscosity resin *film* or *sheet* placed in mold cavity • Preform, forming tooling, and upper mold cavity assembly • Heat and autoclave mold pressure sequentially applied • Required heat lowers resin viscosity to assure flow/wet-out
SCRIMP (SCRIMP systems process): • Seemann's composite resin infusion molding process	• Patented process involving vacuum bag resin distribution • Resin quickly distributed *across* large part surface area • Resin then saturates through the preform thickness • Vacuum bag, soft-sided tooling required
UVRTM • Ultra-violet (cure) RTM	• Like SCRIMP process in terms of rapid resin saturation • UV-transparent vacuum bag system required to effect cure • UV cure energy, vacuum bag, soft-sided tooling required

Source: From Benjamin, W. P., and Beckwith, S. W., *SAMPE Monograph No.3*, May 1999. With permission.

superthermal conductors that have the capacity to transfer large amounts of heat at high speeds in both heating and cooling applications. They can transfer thermal energy at speeds thousands of times faster than a copper rod of the same dimensions. They can be used with RTM, resin infusion molding (RIM), and vacuum-assisted infusion processes. Mounted within the mold structure, they provide heat transfer to the mold face from either heater banks and/or cooling channels located in accessible regions of the molds. The technology can be used alone or in engineered matrices to normalize the thermal energy of the cure cycle in a specific location or throughout the complete mold. It can be designed or retrofitted into many mold applications. The heat pipes achieve isothermal conditions along their entire length within seconds thereby avoiding random hot or cold spots within the mold.

A new solution for composite tooling being introduced by Nova-Tech [23] is known as Integrated Composites Structure or ICS. This composite structure is a multilayer material, sandwiching carbon foam between layers of carbon fiber. It has many advantages over other, more conventional materials such as Invar, bismaleimide (BMI) laminate, and the VARTM process for use in fabricating large composite tools. It is lightweight, and has high rigidity. It meets the requirements for thermal and structural stability, and has shown exceptional reworkability. This ICS tooling system is ideally suited for larger tooling structures such as aircraft wing skins, fuselage skin panels, and nacelle cure and bond fixtures for fiber placement applications and also hand lay-up mandrels for smaller composite parts. The tools are lighter weight, more energy efficient, and more cost effective to build than tools made of metallic materials. One use that is currently being examined for this material is a larger fuselage barrel mandrel for aircraft. The mandrel is approximately 1524 cm (50 ft) in length, 579 cm (19 ft) in diameter, and is built in six segments. The weight of each segment is calculated at 3517 kg (7755 lbs) with a total tool weight of 21,060 kg (46,530 lbs). This is significantly less than the Invar tool that had been previously built and weighed 124,173 kg (273,750 lbs).

RTM & VARTM Process Materials

A benzoxazine resin system for RTM and VARTM processing of aerospace components is being evaluated as an effective and efficient alternative technology to epoxy chemistry. Epsilon 99100 is the first of the family of aerospace composite resin systems to be introduced in 2008. It has a cure temperature of 180°C (356°F) and is said to be stable at ambient temperatures for over a year as a one-part resin, which reduces the cost of storage and shipment. Other characteristics include: a broad processing window, which is good for large parts and complex shapes; low heat release during cure; low cure shrinkage; and good thermal resistance. The resin is also fire retardant and suitable for use in aircraft fuselage and interiors.

Another low viscosity resin is Crestapol 1210 designed for use in RTM processing. This low viscosity resin is said to be suitable for the production of railway furniture, domestic household doors and garage doors, general building and construction projects, as well as modular buildings. Fire resistance or other specific properties can be incorporated with the use of selective fillers. Production can be optimized to achieve a cycle time of 6–7 minutes for RTM. According to specialists at the U.K. company, Scott Bader, rapid mold fill and de-mold can also be attained with large composite parts. The resin requires the addition of a catalyst plus two accelerators to start the curing reaction. The levels of catalyst and accelerators used are dependent upon the gel times. The resin can be used with a variety of fillers, including aluminum trihydrate (ATH) for fire retardant applications.

Applications

SCRIMP

Seemann Composites Inc. (SCI) of Gulfport, Mississippi [25] has fabricated two 1250 cm (41 ft) research catamarans. These catamarans have been designed and built for the National Oceanic and Atmospheric Administration's (NOAA) Office of Marine Sanctuaries. The boat was built with E-Glass fiber and Seemann's custom-toughened vinyl ester resin system STVE-5. The hull below the water line is solid laminate, above the water line the hull is a sandwich structure using Core-Cell foam. The decks and deckhouse were also made using Core-Cell sandwich laminates. The whole composite structure was fabricated using the SCRIMP process. The boat measures 1250 cm (41 ft) and 457 cm (15 ft) wide and is said to be an ideal platform for missions such as dive coordination, whale observation/research, and enforcement.

VARTM

Composite structures for space payload vehicles generally are fabricated from carbon/epoxy prepreg systems that are typically cured under high pressure and temperatures in an autoclave. As these structures increase in size, the availability of suitable autoclaves to process these large structures has become limited. The aerospace community has actively been searching for lower cost out-of-autoclave processing techniques suitable for large carbon fiber based structures. As part of a technology demonstration program for the AFRL [26], a VARTM process boat tail section was successfully fabricated by ATK Aerospace Structures, Clearfield, Utah to demonstrate the feasibility of out-of-autoclave processing for large space payload structures. This VARTM boat tail part was integrated in a complete ATK fairing design being qualification tested as a full scale test article.

Light RTM (RTM Light)

Light RTM is now firmly established as a major closed mold manufacturing process. Light RTM, or RTM Light [27] is a process employing two lightweight matched composite mold faces between which dry reinforcement is positioned before the two halves are clamped together with a vacuum, thereby locking the mold set closed prior to the resin being introduced. Resin is then injected into the peripheral cavity to fill the dry reinforcement within the mold. Two sets of mold seals are employed; one to exclude atmospheric air and the other to ensure that no resin leaks from the mold cavity during mold fill. This operating principle is simple and is the main reason why the process has attracted so many production molders to adopt it as their preferred closed mold, molding method. However, aborted or scrap moldings occasionally occur, often due to the operator's general misconception that the process is totally robust and needs little attention paid to mold maintenance regimes or correct manufacturing procedures. Without doubt, air entrapment in the guise of incomplete fill, dry patches, or voids is the most common fault.

Air Voids

The use of a vacuum—in other words subatmospheric pressure—inherently creates a major cause for potential air voids, or bubbles in the finished molded part. During the molding process, air at the atmospheric pressure is eagerly attempting to enter the mold and as it is some 17,000 times less viscous than resin, it can do so with remarkable ease through even the smallest holes. Even when the mold is completely filled with resin it still remains under a vacuum condition before gelling and curing. It may be surprising to learn that leaky mold vacuum seals are not, as would be expected, the reason for air voids. The mold fittings, inserts and pipe connections, or cracked molds are, however, the usual source of such leaks. As described earlier, Light RTM mold technology uses two seals, one seal preventing resin from leaking out of the mold and the other, set further away on the edge of the mold flange, sealing against the ingress of atmospheric pressure.

Dry Patches

The discovery in a filled mold of patches of dry glass that consistently appear in the same area is in the main caused by mold cavity inaccuracy. Another reason for dry patches, especially near, but not at the vent/catchpot is basically due to the incorrect positioning of the catchpot point. Either this point can be relocated, or alternatively another can be retrofitted in the area of dry glass to "mop up" the last vestiges of trapped air.

Corner Cracks

Cracks in the gel coat of a molded part usually observed in external radii demonstrates a classic problem associated with resin rich areas, and can again be traced back to inaccurate mold build, whereby the second mold half does not faithfully follow its counterpart and therefore has a local corner thickness greater than the desired main cavity thickness. As a result it may sometimes be appropriate to compensate for the unintentional over thickness by applying extra glass strips in these thicker corner areas to prevent the product cracking.

Three examples of Light RTM fabricators include Jeanneau, one of Europe's leading boat builders who manufacture in excess of 1000 yachts and motor cruisers per year ranging in size from 1067 cm (35 ft) to 1372 cm (45 ft) [28]. Their boat designs and the developments that they have made in their closed mold production and processing have promoted them to the position of world leaders in the supply of large marine craft and to the forefront of twenty-first century composite technology.

One of their most impressive Light RTM closed mold parts—the Sun Odyssey 42 DS—is a 1280 cm (42 ft) long yacht deck of 55m^2 surface area and complex geometry. Over 800 of these decks have been molded using Jeanneau's own variant of the Light RTM closed mold process. This large complex part has precisely positioned layers of glass fabric and mat combined with a host of various shaped cores loaded into the mold cavities prior to vacuum closing and injection. Once the automatic injection machine is signaled to start, all details of resin flow, programmed automated catalyst ratio adjustment, temperature, resin volume, and precise cavity injection pressure are all monitored and written to data logging files. This helps if there is a need to transfer this information to a PC for detailed analysis.

The second item is manufactured by Mandiola Composites S.L. in Bilbao, Spain [28]. They successfully met a challenge to design and fabricate a large 72,000 liter fish farm tank using the Light RTM method and installed the tank on site. The key was to design the tank to be free standing and strong enough to withstand 72 tons of water when installed. The tanks were designed to be made in sections to allow them to be efficiently installed on site. Mandiola has subsequently and successfully completed the production of 27 fully assembled tanks using the Light RTM process, with an overall composite weight of 58 tons. A subsequent order for two more similar fish farm tank systems was placed and was delivered in 2008 [28].

The third application is a composite bridge and the arguments for the selection of materials and process are very compelling: rapid construction affords minimal traffic disruption; there is no metal to corrode and, thus, no long-term maintenance; and light weight allows an increase in a span's load classification. With more than half a million brides in the United States alone, and a fifth of those now rated as substandard, a vast market for composite bridge designs has seemed tantalizingly near for years but, to date, has not materialized.

The primary barrier is up-front cost: Composite materials cost more than steel and concrete and constitute a greater proportion of total project costs. "That paradigm might be shifting," says Scott Reeve, president of Composite Advantage, Dayton, Ohio. The company has completed the fabrication of 1300 m^2 (14,000 ft^2) of composite decking for the Anacostia River Walk pedestrian bridge. The pedestrian bridge will span an active railroad corridor near Washington, D.C., connecting bicycle and pedestrian trails that run roughly parallel to the tracks on the opposite sides.

In contrast to some recent bridge designs, the Anacostia pedestrian bridge deck is a cost-competitive hybrid design that incorporates an integrally molded, composite sandwich deck supported by steel girders. "Molded bridge deck products offer maximum design flexibility," says Reeve. See Figure 17.6 [29]. "With the stiffness and deflection targets dictated," says Reeve, "the deck panels were designed with stitched laminate skins and thick TYCOR G6 fiber-reinforced, closed-cell foam core." The trademarked

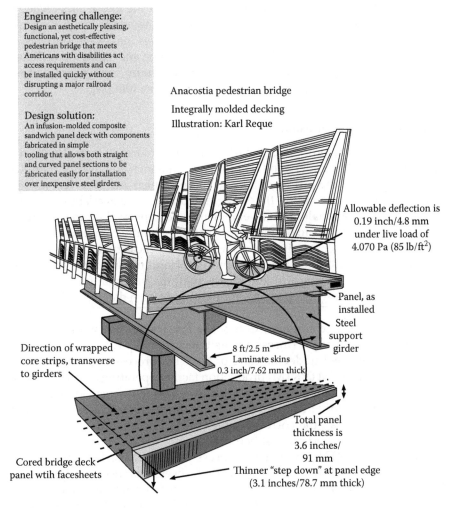

Engineering challenge:
Design an aesthetically pleasing, functional, yet cost-effective pedestrian bridge that meets Americans with disabilities act access requirements and can be installed quickly without disrupting a major railroad corridor.

Design solution:
An infusion-molded composite sandwich panel deck with components fabricated in simple tooling that allows both straight and curved panel sections to be fabricated easily for installation over inexpensive steel girders.

Anacostia pedestrian bridge
Integrally molded decking
Illustration: Karl Reque

Allowable deflection is 0.19 inch/4.8 mm under live load of 4.070 Pa (85 lb/ft²)

Panel, as installed
Steel support girder

Direction of wrapped core strips, transverse to girders

8 ft/2.5 m
Laminate skins
0.3 inch/7.62 mm thick

Total panel thickness is 3.6 inches/ 91 mm

Cored bridge deck panel wtih facesheets

Thinner "step down" at panel edge (3.1 inches/78.7 mm thick)

FIGURE 17.6
Anacostia pedestrian bridge integrally molded decking. (From Black, J., *Composites Technol.,* 46–48, December 2007. With permission.)

TYCOR materials are a combination of multiaxial and unidirectional stitched fabrics. "Further," notes Reeve, "TYCOR is intended to work in concert with the infusion process." A key factor in the selection of a molded sandwich construction for the deck was the manufacturing and structural advantages of the WebCore TYCOR product and infusion process.

To produce the Anacostia panels, simple tooling was created, using dimensional lumber and concrete. The removable steel edge strips from the panel step-downs. Panels are infused with vinyl ester resin and the WebCore Infusion Process (WIP) that combines the WebCore WIP process and RTM light, takes about 40 minutes. The panels are allowed to cure at ambient temperature for 4 hours. "With this type of design, you use only the amount of material you need," Reeve notes, pointing out that the lower panel weight, when compared to concrete, brings benefits that cascade through the project, allowing the use of lighter steel girders and less massive abutments as well as reducing the cost of both installation labor and equipment. Taken together, all of these factors overcome the FRP premium, making this hybrid bridge concept competitive with steel-reinforced concrete [29].

Crossover to Resin Infusion

The resin infusion process is rapidly becoming more mainstream as the technology is better understood and the environmental and technical limitations of open molding processes become more problematic. New technology and techniques surrounding the "bagging" process, the availability of specialist infusion materials, and developments in computer flow analysis have enabled manufacturers to optimize production of highly complex structures on a very large scale. Despite these developments, the actual supply of resin into the mold is still often a manual process. Hand mixing large quantities of resin can be both wasteful and messy, with the potential for very costly mistakes. Bulk mixing of epoxy resin can bring its own dangers of bulk exothermic reaction.

To pioneer these techniques, Composite Integration has worked closely with the U.K. company, Princess Yachts International plc. [30]. Princess has developed a detailed understanding of the RTM process with an in-house vacuum-assisted RTM (VRTM) production facility operating over 200 molds. The injection equipment is fitted with an radio-frequency identification (RFID) mold recognition system that allows the operator to scan each mold to automatically load the correct settings, including resin quantity, output speed and pressure and catalyst percentages.

Alongside its VRTN production, Princess yachts has also been at the forefront of large-scale infusion development. Princess has successfully infused hulls up to 2438 cm (80 ft) using equipment specifically developed for its requirements, with resin quantities up to 1500 kg. The advantages of this optimized process have been a dramatic reduction in waste, rapid, and controlled fill times, with a far more controllable production sequence. During the infusion process, detailed data were recorded to enable the entire process to be accurately controlled and to further develop the control system for long-term production use [30].

Finally, a new Club Swan 42, by Nautor [31] is being manufactured using infusion technology. Near the end of 2004, the New York Yacht Club began considering the development of a new "Corinthian" one-design class for its members. Nautor together with the Frers design office was successful in its bid to build the yacht and the official launching of the first NYYC42 took place in 2006. Core infusion was chosen for the production of both the deck and hull moldings in order to achieve high-quality laminates with good fiber fractions and a high order of weight consistency from boat to boat. To speed the lay-up process and to ensure an excellent core fit core materials maker DIAB was selected as the core material. This new yacht is designed to take advantage of the latest thinking in design and construction. It is light, fast, and features a retractable bowsprit. Although intended to be sailed as a one-design boat, it has been designed to be competitive when sailing under the IRC rule and to also offer good cruising capabilities.

References

1. Brown, A. Close in on new processes, *Composites Manufacturing* (March 2009): 26–7.
2. Patrick@e-composites.com; p. 2 of 5 and *Composites Week* 5, no. 48 (December 2, 2003).
3. Berenberg, B. Liquid composite molding achieves aerospace quality, *High-Performance Composites* (November 2003): 18–22.
4. Schwartz, M. M. *New materials processes, and methods technology*, 686 p., Boca Raton, FL: Taylor & Francis, CRC Press, 2006.
5. CoRTM process reduces fabrication & assembly costs, *AF Man Tech Highlights* (Summer 2003): 10.
6. Black, S. New approaches to cost-effective tooling, *High-Performance Composites* (July 2003): 30–5.
7. Heider, D., and Gillespie, Jr., J. W. Automated VARTM processing of large-scale composite structures, *Journal of Advanced Materials* (October 1994).
8. Black, S. Inside analysis: Simulating VARTM for better infusion, *High-Performance Composites* (January 1, 2008): 5 p; and http://www.composites world.com/articles/inside-analysis, p 1 of 5, February 13, 2009.
9. Marco method, U.S. Patent No 2495640 (1950).
10. Williams, C., Summerscales, J., and Grove, S. Resin infusion under flexible tooling: A review, *Composites* 27A (1996): 517–24.
11. Group Lotus Car Ltd., Vacuum molding patent, GB Patent No. 1432333, March 30, 1972.
12. Seaman, W. H. Plastic transfer molding techniques for the production of fiber reinforced plastic structures, U.S. Patent No. 4902215, filed March 30, 1989.

13. Beckwith, S. W. Resin infusion technology: Part 2 process definitions and industry variations, *SAMPE Journal* 43, no. 3 (May/June 2007): 46.
14. Reuterloev, S. Grooved core materials aid resin infusion: Influence on mechanical properties, *SAMPE Journal* 39, no. 6 (November/December 2003): 57–64.
15. Uchida, H., Yamamoto, T., and Takashima, H. Development of low-cost, damage-resistant composite using RFI processing, *SAMPE Journal* 37, no. 6 (November/December 2001): 16–20.
16. Reuterloev, S. Infusion: The DIAB method, http://www.rpasia.com/rpasia/conf_sesle.html; p. 1 of 1, 07/28/02.
17. Black, S. An elegant solution for a big component part, *High-Performance Composites* (May 2003): 45–8.
18. Benjamin, W. P., and Beckwith, S. W. Resin transfer molding, *SAMPE Monograph No.3* (May 1999).
19. Joubaud, L., Trochu, F., and Le Corvec, J. Analysis of resin flow under flexible cover in vacuum assisted resin infusion (VARI), *JAM* 37, no. 3 (July 2005): 3–10.
20. Raybould, K. The success of SCRIMP? *Reinforced Plastics* (April 2006): 55–6.
21. Haberkern, H. Tailor-made reinforcements, *Reinforced Plastics* (April 2006): 28–33.
22. Mason, K. Cutting infusion time and cost, http://www.compositesworld.com/articles/cutting-infusion...2/13/2009, p. 1 of 9.
23. Scrivo, J. V. Composites laboratory to commercial reality, *Proceedings of the 3rd Advanced Composites Conference and Exhibition*, 16–23, California, September 15–7, 1987.
24. Carver, L. New solutions for composite tooling, *SAMPE Journal* 45, no. 1 (January/February 2009): 48–54.
25. Seemann composites expands and launches catamaran, *Reinforced Plastics* 52, no. 9 (October 2008): 17; http://www.seemanncomposites.com
26. Berg, J. S., and Higgins, J. VARTM infusion: Processing large carbon/epoxy space structures out of the autoclave, *SAMPE Journal* 44, no. 6 (November/December 2006): 40–7.
27. Light RTM-A review of process problems and their remedies, *Plastech Thermoset Tectonics Ltd.* (Winter 2008): 4–8.
28. Jeanneau's closed mould production puts wind in their sails, *RTM Today* (Winter 2005): 1–2.
29. Black, J. Pedestrian bridge deck makes business case for composites, *Composites Technology* (December 2007): 46–8; http://www.compositesworld.com
30. Leonard-Williams, S. The crossover from RTM to resin infusion, *Reinforced Plastics* (November 2008): 28–29.
31. Infusion technology chosen for new Swan, *Reinforced Plastics* (February 2007): 5; and www.nautorgroup.com and DIAB AB; http://www.diabgroup.com

Bibliography

186 ft composite yacht under construction, *Reinforced Plastics* (February 2008): 6.

A new process for fabricating random silicon nanotips, *NASA Tech Briefs* (November 2004): 62–3.

Automated RTM process optimizes cost reduction in VW automotive plant, http://www.compositesworld.com/tech_zone/show/13, 2/7/2008, p. 1 of 3.

Bayer receives grant for resin infusion research, *Composites World*, http://www.compositesworld.com/news/bayer-receives -grant-for-resin-infusion-research/aspx, August 7, 2009, p 1 of 1.

Beckwith, S. W. Resin infusion technology: Part 1 Industry highlights, *SAMPE Journal* 43, no. 1 (January/February 2007): 61.

Beckwith, S. W. Resin infusion technology: Part 3 A detailed overview of RTM and VIP infusion processing technologies, *SAMPE Journal* 43, no. 4 (July/August 2007) 6, 66–70.

Berenberg, B. RTM process turns out high-quality jet engine blades, *High-Performance Composites* (July 2004): 38–41.

Black, S. All-composite hovercraft rises to performance challenge, *Composite Technology* (February 2007): 48–50.

Black, S. Innovative composite design may replace aluminum chassis, *Composites Technology* (February 2006): 44–8.

Boat builder develops infusion processes, Marlow Yachts, www.marlowyachts.com, *Reinforced Plastics* (September 2007): 15.

Bogdanovich, A. E., Wigent, III, D. E., and Whitney, T. J. Fabrication of 3-D woven preforms and composites with integrated fiber optic sensors, *SAMPE Journal* 39, no. 4 (July/August 2003): 6–15.

Castro, F. Effective mould release for RTM processes, *Reinforced Plastics* (November 2006): 30–1.

Christou, P. Advanced materials for turbine blade manufacture, *Reonforced Plastics* (April 2007): 22–4.

Chivers, P. Paris air show review, *High-Performance Composites* (September 2007): 34–7.

Closed Mold Alliance launches website, www.reinforcedplastics.com/view/1850/closed-mold-alliance-launches-website, May 27, 2009, p. 1 of 2.

Composite bridge beats alternative designs, *Reinforced Plastics* (February 2008): 5.

Composite caravan, *General News* (February 26, 2008): 1.

Composite car gears up for Panasonic World Solar Challenge, *Reinforced Plastics* (September 2007): 5.

Composite railroad bridge, *Reinforced Plastics* (March 2008): 48–9.

Composites Europe, RTM—4th European Trade Fair & Forum for Composites, Technology and Applications, Neue Messe Stuttgart, Germany, October 27–9, 2009.

Coppens, P. Infusion helps laser performance achieve one-design advantages, *Reinforced Plastics* (January 2008): 30–1.

Cutting infusion time and cost, http://compositesworld.com/tech_zone/show/13/12, 2/12/2008, p 1 of 3.

Dawson, D. K. Composite spoilers brake Airbus for landing, *High-Performance Composites* (July 2006): 52–5.

EDO to produce composite ducts for General Electric's GEnx engine, *High-Performance Composites* (September 2006): 15.

FormaShape makes world's largest RTM'd waterslide part, http://www.compositesworld.com/news/cwweekly/2007/May/111525, May 16, 2007, p. 1 of 2.

Gaetzi, R. Why vacuum infusion benefits your quality, budget and environment, *Reinforced Plastics* (January 2008): 28–9.

Goodell, B., Lopez-Anido, R., Herzog, B., Qian, Y., and Souza, B. Tooling for the composites pressure resin infusion system: ComPRIS, *SAMPE Journal* 41, no. 4 (July/August 2005): 20–5.

Griffiths, B. Rudder new twist with composites, *Composites Technology* (August 2006): 60–2.

Harper, A. RTM past, present and future, *Reinforced Plastics* (November/December 2009): 30–3.

Heider, D. Resin infusion process to build carbon composite ship hulls, *AM&P.* (October 2007): 18.

Inside analysis: Simulating VARTM for better infusion, http://composites world. com/tech_zone/show/12, 2/12/2008, p. 1 of 3.

Jacob, A. RTM suits Czech manufacturer, *Reinforced Plastics* (November 2007): 22–6.

Japanese project achieves 10-minute RTM cycle time, *Composites World*, 9/23/2008, p. 1 of 1.

Kruckenberg, T., Qi, B., Falzon, P., Liu, X. L., and Paton, R. Experimental and predicted in-plane flow height measurements for stiffened structures made using the resin film infusion process, *SAMPE Journal* 37, no. 3 (May/June 2001): 28–34.

Large aircraft tail structure by RTM, *Reinforced Plastics* (January 2006): 18.

LeGault, M. New dimensions in tooling, *High-Performance Composites*, http://www. compositesworld.com, January 2008.

Lesko, J. J., Peairs, D. M., Zhou, A., Mutnuri B, Zhang W. Rapid prototyping and tooling techniques for pultrusion development, *SAMPE Journal* 44, no. 1 (January/ February 2008): 65–8.

Ma, L., Mitschang, P., Ogale, A., and Schlarb, A. K. Water disposable core technology for the manufacturing of hollow-structure parts, *SAMPE Journal* 43, no. 5 (September/October 2007): 24–33.

Mack, P. Volitization and the vacuum infusion process, *Composites Manufacturing* (March 2008): 30–3; http://www.cmmagazine.org

Mason, K. F. Composite ships: Building a new paradigm, *Composites Technology* (August 2005): 60–3; http://www.compositesworld.com

McConnell, V. P. Composites in North America, *Reinforced Plastics* (December 2005): 26–33.

Michaeli, W., and Tucker, J. RTM for railway applications: Development of a tram front end bumper, *SAMPE Journal* 37, no. 2 (May/June 2001): 69–74.

Mills, A., Patel, Z., Dell'Anno, G., and Frost, M. Resin transfer moulding: Novel fabrics and tow placement techniques in highly loaded carbon fibre composite aircraft spars, *SAMPE Journal* 43, no. 3 (May/June 2007): 67–72.

Mohamed, M. H., Bogdanovich, A. E., Habil, I., Dickinson, L. C., Singletary, J. N., and Lienhart, R. B. A new generation of 3D woven fabric preforms and composites, *SAMPE Journal* 37, no. 3 (May/June 2001): 8–17.

Musselman, M. Europe's infusion pioneer simplifies process with bottom up approach, *Composites Technology* (October 2003): 34–8; http://www.compositesworld.com

Nida-core Structiso in infusion and lite-RTM applications, Nida-Core Advertisement, http://www.NIDA-CORE.com, Port St. Lucie, FL, 2004.

Pederson, C., La Faro, C., Aldridge, M., and Maskell, R. Epoxy-soluble thermoplastic fibers: Enabling technology for manufacturing high toughness structures by liquid resin infusion, *SAMPE Journal* 39, no. 4 (July/August 2003): 22–9.

Pegorari, S. A customized technique to produce high performance boats, *Reinforced Plastics* (January 2007): 32–7.

Reichl, M. Composites meet aviation requirements, *Reinforced Plastics* (June 2007): 38–40.

Resin-transfer-molding of a tool face, *NASA Tech Briefs* (November 2004): 63.

Reuterloev, S. Improved foam core materials for advanced composites processing, *SAMPE Journal* 42, no. 1 (January/February 2006): 6–9.

RocTool adapts cage system for RTM, *Reinforced Plastics* (June 2007): 15.

Russell, J. D. Composites affordability initiatives, *AM&P* (June 2007): 29–32.

Russell, J. D. Composites affordability initiatives: Successes, failures—Where do we go from here? *SAMPE Journal* 43, no.2 (March/April 2007): 26–36.

Serrano-Perez, J. C., and Vaidya, U. K. Modeling and implementation of VARTM for civil engineering applications, *SAMPE Journal* 41, no. 1 (January/February 2005): 20–31.

Sloan, J. Fabric forms get sophisticated, *Composites Technology* (October 2007): 44–9; http://www.compositesworld.com

Snowdon, I., and Rigby, M. New developments in release agent technologies, *Reinforced Plastics* (November 2006): 24–28, Elsevier Ltd., The Boulevard, Langford Lane, Kidlington, Oxford OX5 1GB, UK.

Sourcebook 2008, Expanded Version, Volume 16-A, January 2008, 348 p.

Steenkamer, D. A., Karbhari, V. M., Wilkins, D. J., and Kukich, D. S. An overview of the resin transfer molding process, CCM Report 94-02, 40 p., Newark, DE: University of Delaware Center for Composite Materials, March 1994.

Structural composite batteries, Army Research Laboratory, http://www.defensetech-briefs.com/index2.php?option = com_content7task = view&id = 1220&   ..., 4/2/2008, p 1 of 2.

Terzi, A., and Ockels, W. Superbus concept relies on lightweight composite construction, *Reinforced Plastics* (February 2008): 28–33.

Vacuum infusion guide, http://compositesworld.com/ct/issues/2004/April/433, 2/12/2008, p 2 of 2.

Vaidyanathan, R., Campbell, J., Lopez, R., Halloran, J., Yarlagadda, S., and Gillespie, J. W. Water soluble tooling materials for filament winding and VARTM, *SAMPE Journal* 41, no. 4 (July/August 2005): 49–55.

18

Vacuum-Assisted Resin Transfer Molding

James L. Glancey

CONTENTS

Introduction Liquid Composite Molding and Vacuum Infusion Processes

Liquid composite molding (LCM) is a class of manufacturing processes used to produce polymer composite components and structures (materials consisting of high-strength fibers embedded in a polymer matrix). In any LCM process, a fibrous preform material is placed into a mold, and then the mold is closed and sealed to prevent leakage before a liquid resin is injected. The resin is introduced with either positive pressure or by drawing a vacuum (Advani and Sozer 2003). The most common LCM process that uses two-sided rigid tooling is called resin transfer molding (RTM). As seen in Figure 18.1a, the rigid tooling in RTM encompasses the fibers and compresses them to the desired fiber volume fraction. Then resin is injected through ports located in the rigid tooling to fill the empty spaces between the stationary fibers. Once the preform is fully wetted and the resin cures, the mold is opened and the finished part removed. The vacuum-assisted resin transfer molding (VARTM) process, shown in Figure 18.1b, uses only a single-sided tool, and the mold is sealed by enveloping the preform fabric with a nonrigid polymer film adhered to the tooling surface with a sealant tape. A vacuum is used to compact the preform and draw the resin into the mold from a reservoir at atmospheric pressure. Many variations of this process have been introduced to overcome the disadvantage of lengthy filling times, which is especially protracted when manufacturing large structures. The most favored

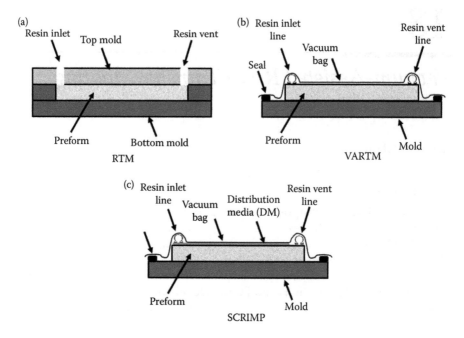

FIGURE 18.1

Cross sectional views of the RTM, VARTM, and SCRIMP mold configurations. (a) An RTM mold consisting of rigid mold top and bottom; (b) a VARTM mold composed of only a single-sided tooling surface at the bottom with bag serving as the top surface to ensure mold closure; (c) a SCRIMP lay-up that includes the addition of distribution media on the preform.

variation of VARTM that reduces the filling time by a significant amount is called the Seemann's composite resin infusion molding process (SCRIMP; Seemann 1990, 1991a, 1991b). SCRIMP involves placement of a layer of highly permeable fabric, commonly referred to as the distribution media (DM) on top of the preform (under the vacuum bag) to increase the flow rate of resin into the part. This allows the resin to quickly spread along the top surface of the part and then penetrate through the thickness to impregnate the tool side of the preform, dramatically reducing the manufacturing time of large structures (Marsh 1997). Figure 18.1c shows a typical SCRIMP lay-up that includes DM.

Manufacturers of composite materials have been working toward low-cost processes like LCM. For processes like VARTM and SCRIMP that require only a single-sided mold, costs are inherently lower compared to RTM. In addition, the development of resin systems that allow out-of-autoclave manufacturing (i.e., resins that cure at room temperature) has facilitated the use of these processes for making large scale structures at costs that were not possible with high temperature processing. As a result, VARTM has grown in popularity for a variety of applications including aerospace, army, naval, civil infrastructure, and automotive.

The purpose of this chapter is to provide an overview of the VARTM and SCRIMP processes. The basic elements and subsystems of these processes are presented, and the theories that describe resin infusion into a VARTM preform are reviewed. The state of the art in process automation and control for vacuum infusion is presented, and finally, current and future trends in VARTM and SCRIMP manufacturing are discussed.

The VARTM Process

The Vacuum-assisted resin transfer molding (VARTM) is a manufacturing process that offers advantages over other processes due mainly to low tooling costs. One of the disadvantages of this process is the labor associated with laying-up a part to be manufactured. Lay-up refers to the manufacturing steps that occur prior to infusion when a mold surface is created and the mold and preform are prepared for infusion. In order to inject resin into the preform, the preform is cut into the desired shape and placed on a one-sided mold surface and enveloped with an air tight bagging film, which is then sealed to the tooling surface with a double-sided adhesive tape. As seen in Figure 18.2, a vent line and resin injection line are placed under the bagging film; the vent line is connected to a vacuum source and the resin injection line is connected to a reservoir of resin. A pressure difference of one atmosphere (at most) is seen between the vacuum source and the resin reservoir, which drives resin into the preform, impregnating the empty volume between the fibers of the preform. The atmospheric pressure acts on the entire lay-up, thus compacting the preform during infusion (Figure 18.3). The injection is allowed to continue until the entire preform is completely infused at which time the resin injection line is closed, thus allowing the preform to bleed

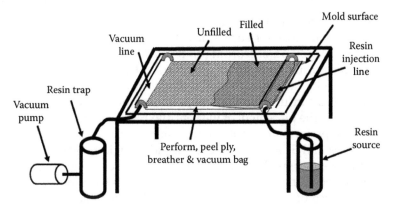

FIGURE 18.2
Schematic of a typical VARTM manufacturing setup.

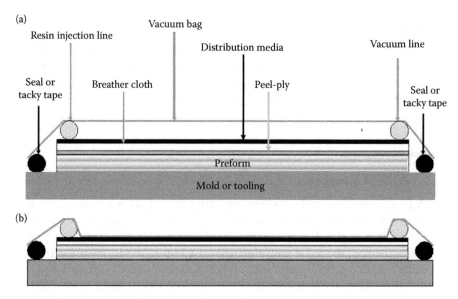

FIGURE 18.3
Cross section of a typical VARTM lay-up (a) before and (b) after a vacuum is applied.

additional resin through the vent line. This bleeding practice is typically performed to insure that any remaining air in the preform has been expelled. For larger and more complex structures, multiple resin and vent lines are used to insure complete infusion and reduce total infusion time.

LCM processes including VARTM still do not share the same level of reliability and repeatability seen previously with traditional manufacturing techniques. Due to the variations of the materials and lay-up, it is common upon examining a cured composite structure fabricated by LCM, to find portions of the fabric left devoid of resin. Such parts, as seen in Figure 18.4, must be rejected due to their poor mechanical properties. A lack of understanding of the flow behavior through complex geometries during the processing is the most common cause for such a defect. In industry this has often led to a trial and error method of learning to properly fill each unique structure. Such a manufacturing approach is expensive in time and money, especially when dealing with large structures. To improve the ability to consistently manufacture fully infused composite parts, numerical simulations can be used to optimize the placement of the vent and resin lines to insure complete infusion of the preform, thus enhancing process yield.

Both VARTM and SCRIMP utilize single-sided, rigid molds. The mold (or tooling) must be designed and manufactured to meet several performance requirements including part tolerances, surface finish requirements, temperature, and service life. For higher volume production, steel and aluminum are the most common metals used to fabricate molds. Nonmetallic materials like wood and composites can be used for low production volumes. One

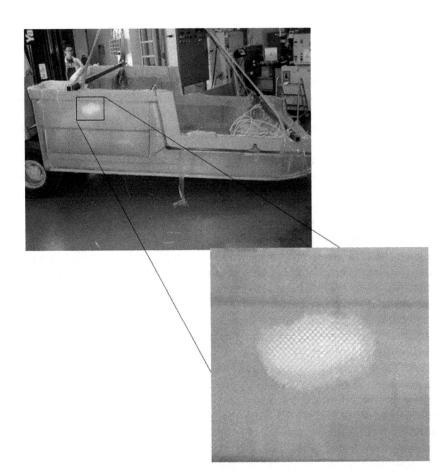

FIGURE 18.4
A cured composite part that contains a void that was not saturated with resin during infusion.

concern with any mold surface is wear due to temperature induced fatigue, solvents, and scratching; coatings are used to prevent (or repair) damaged mold surfaces. Historically, differences in thermal expansion characteristics of metal molds and the fibrous preform have been an important consideration is designing molds for composite manufacturing (Schwartz 1992). This issue is less of a concern with VARTM since many parts now cure at room temperature without the use of a high-temperature autoclave.

The most common fiber reinforcement materials used in VARTM manufacturing are fiberglass, carbon, and aramid (e.g., Kevlar). These materials can be used in a variety of configurations including woven, stitched, knitted, braided, and as a random mat. The type and configuration of the reinforcement will determine the strength and stiffness of the finished structure. It should be noted that these attributes will also dictate how resin flows through the preform during infusion, thus illustrating the importance of the coupled nature of composite material design in which the structural and processing characteristics must be considered together. The permeability of the preform is a measure of its resistance to resin flow; lower permeability indicates increased resistance to flow thus increasing time to fully infuse the preform. Permeability of the preform is a function of both the configuration of the fibers and the fiber volume fraction. The fiber volume fraction is the ratio of the volume of reinforcing fibers to the total volume of the part. Prior to infusion, the preform can be debulked by applying, then relaxing, full or partial vacuum pressure to the mold; depending on the number of debulking cycles, the type and configuration of reinforcement material, and the thickness of the preform, debulking can improved the thickness uniformity as well as increase the fiber volume fraction of the finished part.

Core materials can be integrated into VARTM lay-ups as a means to improve the structural properties, reduce weight, and lower cost. Various foams and wood have both been used as core materials for larger composite structures. Metal inserts have also been used, especially to provide a means of connectivity to other parts or structures. When using both metallic and nonmetallic inserts, the surfaces that will be in contact with the polymer matrix must be processed to insure good bonding.

Today, the polymer matrix is usually formed using thermosetting resins; for VARTM these include epoxy, polyimide, phenolic, vinyl ester, urethane, and polyester. This group of resins has relatively high stiffness and strength, dimensional stability, and low viscosities. They are also chemically and thermally stable. Prior to infusion into the preform, the resin is mixed with a catalyst to induce the exothermic chemical cure reaction that induces polymerization, thus forming the polymer matrix around the reinforcing fibers. The cure kinetics, and in particular the curing time, can be modified with the use of additional additives. Often inhibitors or retarders are added to the resin mix to extend the cure time in order to insure the part is fully infused before the matrix forms. An inhibitor causes polymerization to cease, while a retarder reduces the reaction rate. Commonly used inhibitors and retarders for VARTM include 1,4-benzoquinone, hydroquinone, and catechol. Other materials such as oxygen, water, and certain metal salts can also act as inhibitors or retarders (Li and Lee 2001).

Although relatively low viscosity resin is used, the VARTM infusion process is very slow due to the relatively low pressure gradient (one atmosphere at most) and lack of DM. The SCRIMP process is much faster and the DM can be seen as a form of off-line resin flow control. As a result, most research and commercialization have been in the area of the material lay-up and DM in order to assure complete impregnation of the fabric without excessive waste or time. This off-line approach to flow control can be designed well with extensive experience and with numerical modeling and simulations (Hsiao, Devillard, and Advani 2004). However, even with proper manufacturing process design, any off-line strategy lacks the ability to account for unexpected and unpredictable behavior that is often exhibited during the vacuum infusion processes due to variations in materials and the human error that is introduced during the lay-up and the bagging step (Devillard et al. 2003). As discussed in more detail later, an on-line control with feedback is necessary in the process in order to design a truly robust manufacturing process that can adapt to these variations.

Theory of Resin Infusion in VARTM

The resin flow in all LCM processes including VARTM and SCRIMP is based on flow through porous media theory in which one uses Darcy's Law (Equation 18.1) to describe the relationship between the flow rate and the pressure drop within the media.

$$\mathbf{v} = -\frac{\mathbf{K}}{\mu} \cdot \nabla P. \tag{18.1}$$

Here v is the average velocity, μ is the viscosity of the resin. $\mathbf{K}$ is the permeability of the preform, and P is the pressure gradient. In RTM and VARTM most of the flow is in-plane as seen from Figure 18.5a so one needs to characterize the preform permeability only in the in-plane directions and flow is usually modeled in two dimensions. For SCRIMP, the flow is three dimensional as illustrated in Figure 18.5b, hence, one needs to characterize the permeability of the fabric in three dimensions, along with the permeability of the DM. The average flow velocities are used in the mass conservation as shown in Equation 18.2 to obtain the governing equation for the pressure of the resin in the mold.

$$\nabla \cdot \left(\frac{\mathbf{K}}{\mu} \cdot \nabla P \right) = 0. \tag{18.2}$$

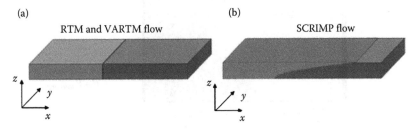

(a) RTM and VARTM flow (b) SCRIMP flow

FIGURE 18.5
(a) Flow pattern typically seen in RTM and VARTM processes where resin flows only in the *x–y* plane; (b) flow pattern typically seen in the SCRIMP process where resin also flows through the thickness of the fabric.

In RTM, there is no change in thickness so mass conservation is inherently satisfied. However, in VARTM and SCRIMP, the preform may change in thickness, and hence, one must account for this change by performing a mass balance on a control volume that allows for the change in the preform thickness as seen in Equation 18.3.

$$\nabla \cdot \left(\frac{\mathbf{K}}{\mu} \cdot \nabla P \right) = \dot{\varepsilon}. \qquad (18.3)$$

Here, ε is the rate of change of control volume with time divided by the original control volume. To solve for flow, one must solve for the pressure using boundary conditions that are illustrated in Figure 18.6. As the resin flow advances through the mold, one must solve this moving boundary problem at each time step. The methodology used to solve Equation 18.3 is a finite element/control volume approach in which the pressure field is solved for using finite elements and flow front advancement is tracked by the status of the fill factors associated with each node. A fill factor of a node is the region surrounding the node that is obtained by joining all the centroids of the elements that are connected to that node. If the region of the node is filled with resin, the fill factor is one and if the node has no resin it is zero. The node represents the resin front if its fill factor is between zero and one. The pressure equation is solved for filled nodes. The fill factors are updated using Equation 18.1 and the time step is selected to bring into the solution domain one more node. The pressure equation needs to be resolved as the domain has changed. This process is repeated until the mold is full. This

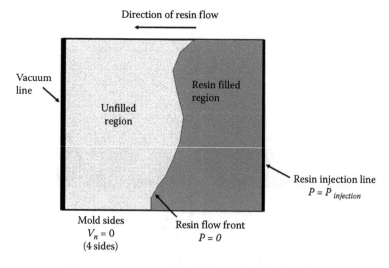

FIGURE 18.6
The boundary conditions used to simulate the VARTM and SCRIMP processes.

quasi-steady state assumption of solving for the pressure equation at each time step has been proven to represent the actual flow that occurs in such processes. Many validations of this solution method have been reported (Simacek and Advani 1996, 2004) and software is currently available to perform VARTM and SCRIMP mold filling simulations (Simacek, Sozer, and Advani 1998; Simacek, Advani, and Binetruy 2004).

Process Automation and Control

Current lay-up methods used for VARTM that introduce multiple distribution lines for resin injection and point or line sources of vacuum, provide very limited controllability of the resin flow within the mold. As a result, the potential for voids continues to limit the complexity of the mold designs as well as production cost and process efficiency. Previous research has demonstrated several different methods capable of modifying and controlling resin flow in VARTM; these techniques include using feedback of resin position (Heider et al. 1999; Walsh and Mohan 1999), regulation of resin flow rate (Heider, Epple, and Gillespie 2001; Mogavero, Sun, and Advani 2001), intelligent process control (Heider et al. 2003), localized induction heating (Johnson and Pitchumani 2003) and differential vacuum (Hsiao et al. 2001). In addition, real-time simulations using a finite-difference approach to simulate resin flow and control injection pressure in RTM has been shown to be very effective in eliminating defects (Nielsen and Pitchumani 2002). A similar real-time control strategy was demonstrated using artificial neural networks trained via simulations (Nielsen and Pitchumani 2001). For both the finite-difference and neural network control strategies, resin flow behaviors were modified to achieve the desired flow pattern, thus reducing or eliminating part defects. Also for RTM, off-line flow simulations have been utilized to design the flow control system (Hsiao and Advani 2004). With this approach, an evaluation function is formulated to optimize the flow sensing network design, and a multitier genetic algorithm is used to optimize the locations of vents and gates. In this case, a numerical evaluation for testing the computer-generated flow control solutions demonstrated reductions in injected part defects. This methodology was automated and the results were validated using a lab-scale mold (Devillard, Hsiao, and Advani 2005). In addition, a new resin delivery system has been developed that allows for the independent delivery of resin through a multisegmented single injection line (Nalla et al. 2007).

A great deal of progress has been made in controlling resin flow and injection behavior during traditional RTM injections, however similar progress has been limited for VARTM. As these methods evolve and begin to be adapted by manufacturers, additional control methods that complement and expand these techniques to VARTM will be critical for improving process yield and reducing manufacturing costs.

Current VARTM Applications and Future Developments

VARTM has been gaining popularity because of the low-initial tooling cost, less than one-half that of RTM, due to having only one surface that can be less rigid than in RTM. Also, the vacuum tends to reduce void formation (as compared to RTM), since it draws out the air that could potentially cause voids, whereas RTM processes sometimes push air around in the mold. VARTM is also capable of making much larger parts, since the process is only constrained by one tool surface. The Navy, for example, is manufacturing 12 m tall ship hulls with the VARTM process. Other typical applications for VARTM include large panels as replacements for sheet metal components in military vehicles, armor plating for military vehicles, blades for wind turbine power plants, residential and commercial roofing, floor, and wall panels. Being a closed mold process, VARTM is also virtually emission free, and thus brings with it this benefit over open molding.

VARTM does have its drawbacks, although there have been marked improvements in the last 5 years. Since VARTM utilizes only one mold side, only one surface of the part produced is smooth, the other has small variation caused by the fiber weave. Although this is superficial and does not affect part quality or structural integrity, it is nevertheless an additional aesthetic, and in the case of air flow, possibly dynamic, concern. VARTM is also limited in its manufacturing application by the length of time the lay-up and injection takes. Currently, VARTM can only be applied in larger-scale, smaller-number productions because of the intensive time required for setup prior to injection. The largest current limitation with VARTM however, is the controllability of the process. If the injection and vent lines are not placed in a manner to optimize the distribution of resin, the time the process takes, the resin amount used, and the possibility of voids is greatly enhanced. Thus, large parts can often take many hours to inject and require extensive preparation of the injection lay-up, as well as post-injection processing time to ensure part quality requirements have been met.

One of the primary areas of VARTM-related research continues to focus on higher part yield and quality through automation and control during infusion. In the last few years, port-based resin delivery systems have been developed and tested. Unlike conventional VARTM setups, in which injection lines are placed manually on top of fiber preforms before bagging, the port injection process, illustrated in Figure 18.7, utilizes ports that are built into the tooling surface, as well as sensors imbedded into the mold to monitor flow behavior. The ports are controlled through a combination of

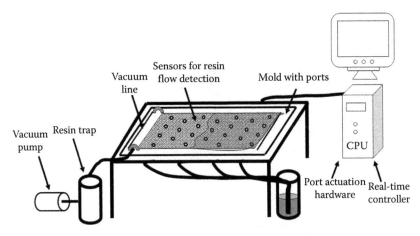

FIGURE 18.7
Illustration of a computer controlled, port-based resin delivery system for VARTM.

sensors, control valves, and pneumatic actuators. Along the bottom of the mold are grooves that connect each port directly to a single resin source. Each port is capable of being independently opened or closed, enabling a wide variety of resin delivery and control during an injection both along the length and width of a part. With this ability, the injection process is no longer limited to just a straight injection line path, but instead can be applied at points in any region, accounting for possible permeability change and void formation.

Research into this new injection method has an end goal of eliminating the issues of both controllability and extensive lay-up time and skill. Thus far, the port injection method removes much of the guess work and trouble of placing injection lines to obtain a quality part that is fully infused. An injection can be started with little thought put into injection line placement, and as flow front issues arise, the control system is in place to address the situation and turn on and off ports as necessary. This allows for both better part quality due to the reduction in dry spot formation, but also an increase in repeatability of an infusion. The same type of part can be injected many times, and if one part has some unexpected variation in permeability, the injection system should recognize and adjust for it, ensuring that the injection is still successful. Furthermore, a decrease in injection time of over 300% has been seen through the application of this new injection process in respect to the currently practiced injection line lay-up (Fuqua and Glancey 2007).

One limitation with port-based resin delivery in VARTM is the relatively low permeability of the preform and its effect on the flow from a resin port. To enhance the flow from a port, the Vacuum Induced Preform Relaxation (VIPR) process has been developed. The VIPR process is a variant of the VARTM process that uses a secondary vacuum chamber to create a seal on the bag surface of a vacuum infusion process mold (Alms et al. 2009;

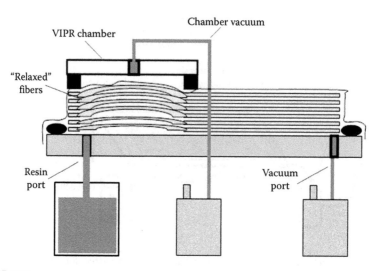

FIGURE 18.8
Vacuum induced preform relaxation (VIPR) method used to increase preform permeability and enhance flow from a resin port.

Kasperzak et al. 2006). This extra vacuum chamber is relatively small compared to the total size of the mold and can be placed directly over an injection port drilled into the tooling surface. During a vacuum infusion process the VIPR chamber, shown in Figure 18.8, is placed onto the mold to form an airtight seal against the bag surface. Once the secondary vacuum source (~0.2 bars) is applied, the vacuum causes the fabric under the chamber to become less compact and thus more permeable. When the chamber is specifically placed over an injection port the VIPR chamber can have a profound effect on where and how resin flows into the mold. Results clearly demonstrate that the sequential deployment of the VIPR chamber accelerates the flow introduced through a port by increasing the preform permeability. This approach provides the potential for flow manipulation and control for successful filling despite the variations in the materials or the process, thus improving reliability and repeatability of vacuum-based processes.

References

Advani, S. G., and M. Sozer. 2003. *Process Modeling in Composite Manufacturing*. New York: Marcel Dekker.
Alms, J. B., A. Catry, J. L. Glancey, and S. G. Advani. 2009. Flow modification process for vacuum infusion using port-based resin flow control. *SAMPE Journal* 45 (2): 54–63.

Devillard, M., K. T. Hsiao, and S. G. Advani. 2005. Flow sensing and control strategies to address race-tracking disturbances in resin transfer molding—Part II: Automation and validation. *Composites Part A: Applied Science and Manufacturing* 36 (11): 1581–9.

Devillard, M., K. T. Hsiao, A. Gokce, and S. G. Advani. 2003. On-line characterization of bulk permeability and race-tracking during the filling stage in resin transfer molding process. *Journal of Composite Materials* 37:1525–41.

Fuqua, M., and J. L. Glancey. 2007. Design and performance of a closed loop control, port-based resin delivery system for vacuum-assisted resin transfer molding. *Proceedings of the International SAMPE Symposium and Exhibition, SAMPE '07,* Vol. 52, Baltimore, MD, June 3–7, 2007.

Heider D., A. Graf, B. K. Fink, and J. W. Gillespie, Jr. 1999. Feedback control of the vacuum assisted resin transfer molding (VARTM) process. *Proceedings of SPIE—The International Society for Optical Engineering,* Bellingham, WA, 3589:133–41.

Heider D., J. W. Gillespie, Jr., T. L. Pike, G. E. Thomas, T. Steele, and J. Florence. 2003. Intelligent process control for affordable VARTM processing of DOD structures. *Proceedings of the International SAMPE Symposium and Exhibition, SAMPE '03,* Vol. 48, 657–69, Long Beach Convention Center, CA, May 11–15, 2003.

Heider D., S. Epple, and J. W. Gillespie, Jr. 2001. Flow rate control during Vacuum-Assisted Resin Transfer Molding (VARTM) processing. *Proceedings of the International SAMPE Symposium and Exhibition, SAMPE '01,* Vol. 46, 1061–71, Long Beach Convention Center, CA, May 6–10, 2001.

Hsiao K. T., and S. G. Advani. 2004. Flow sensing and control strategies to address race-tracking disturbances in resin transfer molding—Part I: Design and algorithm development. *Composites Part A: Applied Science and Manufacturing* 35 (10): 1149–59.

Hsiao K. T., J. W. Gillespie, Jr., S. G. Advani, and B. K. Fink. 2001. Role of vacuum pressure and port locations on flow front control for liquid composite molding processes. *Polymer Composites* 22:660–7.

Hsiao K. T., M. Devillard, and S. G. Advani. 2004. Simulation based flow distribution network optimization for vacuum assisted resin transfer molding process. *Modeling and Simulation in Materials, Science and Engineering* 12:175–90.

Johnson R. J., and R. Pitchumani. 2003. Enhancement of flow in VARTM using localized induction heating. *Composites Science and Technology* 63 (15): 2201–15.

Kasprzak, S., M. Fuqua, J. Nasr, and J. L. Glancey. 2006. A robotic system for real-time resin flow modification during vacuum-assisted resin transfer molding of composite materials. *Proceedings of the 2006 ASME International Annual Meeting,* Chicago, IL. IMECE2006-14416.

Li, L., and L. J. Lee. 2001. Effects of inhibitors and retarders on low temperature free radical crosslinking polymerization between styrene and vinyl ester resin. *Polymer Engineering and Science* 41 (1): 53–65.

Marsh, G. 1997. Scrimp in context. *Reinforced Plastics* 41 (1): 22–6.

Mogavero, J., J. Q. Sun, and S. G. Advani. 2001. A nonlinear control method for resin transfer molding. *Polymer Composites* 18 (3): 412–7.

Nalla, A., M. Fuqua, J. L. Glancey, and B. Leleiver. 2007. A multi-segment injection line and real-time adaptive, model-based controller for vacuum assisted resin transfer molding. *Composites Part A: Applied Science and Manufacturing* 38 (3): 1058–69.

Nielsen, D., and R. Pitchumani. 2001. Intelligent model-based control of preform permeation in liquid composite molding processes, with online optimization. *Composites Part A: Applied Science and Manufacturing* 32 (12): 1789–803.

Nielsen, D., and R. Pitchumani. 2002. Closed-loop flow control in resin transfer molding using real-time numerical process simulation. *Composites Science and Technology* 62 (2): 283–98.

Schwartz, M. M. 1992. *Composite Materials Handbook,* 2nd ed. New York: McGraw-Hill.

Seemann, W. H. 1990. Plastic transfer molding techniques for the production of fiber reinforced plastic structures. U.S. Patent No. 4, 902, 215.

Seemann, W. H. 1991a. Plastic transfer molding apparatus for the production of fiber reinforced plastic structures. U.S. Patent No. 5, 052, 906.

Seemann, W. H. 1991b. Unitary vacuum bag for forming fiber reinforced composite articles. U.S. Patent No. 5, 316, 462.

Simacek, P., and S. G. Advani. 2004. Desirable features in a mold filling simulation for liquid molding process. *Polymer Composites* 25 (4): 355–67.

Simacek, P., and S. G. Advani. 1996. Permeability model for a woven fabric. *Polymer Composites* 17:887–99.

Simacek, P., E. M. Sozer, and S. G. Advani. 1998. Numerical simulations of mold filling for design and control of RTM process. *Annual Technical Conference—ANTEC Conference Proceedings*. Atlanta, GA.

Simacek, P., S. G. Advani, and C. Binetruy. 2004. Liquid Injection Molding Simulation (LIMS) A comprehensive tool to design, optimize and control the filling process in liquid composite molding. *JEC-Composites*, no. 8. Paris, France, 58–61.

Walsh, S. M., and R. V. Mohan. 1999. Sensor-based control of flow fronts in vacuum-assisted RTM. *Plastics Engineering* 55 (10): 29–32.

19

VARTM Processing

Myung-Keun Yoon

CONTENTS

Introduction

The vacuum-assisted resin transfer molding (VARTM) is a composite fabrication process where a vacuum draws resin into a fiber preform placed on a one-sided hard tool covered by a flexible bag on the other side. This process is also called by several different names such as the vacuum infusion process (VIP), vacuum infusion molding (VIM), vacuum-assisted resin injection

(VARI), vacuum-assisted resin infusion molding (VARIM), and Seemann composites resin infusion molding process (SCRIMP). This technique is a type of liquid composite molding (LCM) process and differs slightly from a conventional resin transfer molding (RTM).

A conventional RTM process consists of a few manufacturing steps: (1) placing a dry preform inside a closed and matched mold, (2) injecting resin into the preform by a typical pressure of about 0.7 MPa, (3) curing the resin, and (4) demolding the final part. Meanwhile a typical VARTM process involves the following steps: (1) laying up dry fabrics or a preform with a highly permeable layer on a single-sided tool, (2) bagging the layup with a vacuum bag and sealing the bag, (3) applying vacuum to compact the layup and to draw resin into the layup, (4) curing the resin, and (5) taking out and trimming the final part. Both processes use similar reinforcements and matrix systems. RTM needs to trim edges at the injection and vent locations, while VARTM needs to trim the whole edges.

Table 19.1 compares the characteristics between RTM and VARTM processes. A typical RTM uses a CNC-machined double-sided mold made from metals and ceramics [1]. Preforms are placed directly in the double-sided mold and small- to medium-scale composite parts ranging from a few inches to a few feet can be manufactured. VARTM uses a one-side tool typically made from fiber reinforced composites laid up on a CNC-machined master mold. A large-scale VARTM tool can be easily made by connecting several small-scale tools in series and composites can be cured under atmospheric pressure, which allows VARTM to manufacture large-scale composite structures. RTM uses compressed air to push resin from a resin bucket to the mold. A typical injection port pressure is about 0.7 MPa. VARTM pulls a vacuum at vents to draw resin into the mold from a resin bucket at atmospheric pressure. Since RTM applies a higher compaction pressure onto the preform, the fiber volume fraction

TABLE 19.1

Characteristics Between RTM and VARTM

Parameter	RTM	VARTM
Mold	A closed and matched tool	A single-side hard tool with a flexible vacuum bag
Part size	Small- to medium-scale	Large-scale
Resin infusion by	Air compressor	Vacuum pump
Injection/vent pressure	0.6–1.5 MPa/0.1 MPa	0 Pa/0.1 MPa
Compaction pressure	1–100 MPa	0.1 MPa
Fiber volume fraction	0.60–0.70	0.50–0.60
Part production	Low- to medium-volume	Low-volume
Finishing	Trim only a few edges	Trim the whole edges

can reach greater than 0.65. Meanwhile, a typical fiber volume fraction of VARTM is about 0.55.

VARTM processes have advantages including: (a) lower tooling costs and short start-up time compared to the traditional autoclave based manufacturing process, (b) short manufacturing time for large-scale structures with a relatively low pressure difference due to a use of highly permeable distribution medium, (c) reduced volatile emissions compared to a hand-lay-up process due to a use of closed process, and (d) reduced assembly costs by manufacturing a larger and integrated net-shape part.

Disadvantages may involve: (a) potential risk of creating voids in the final parts resulting from introduction of air by failure of vacuum sealing during the resin infusion process, (b) variation of thickness in the preform due to the pressure gradient between the injection ports and vents during the curing and relaxing cycles, (c) creation of rough or wavy surfaces on one-side of the preform contacted to the flexible sealing bag due to the nature of fibrous preforms, and (d) requirement of an additional trimming process to finish cured parts.

A whole VARTM manufacturing process includes designing a part and process, manufacturing the part, and assessing quality of the finished part. The first step involves designing the part (i.e., determining geometry and configuration and performing structural analysis). The next step is predicting the resin progression in the preform and designing the tool. Many composite manufacturers have adopted finite element analysis (FEA) tools [2–11] to predict the resin flow fronts and design the process, which reduces both the time and cost to obtain optimum process conditions. This study addresses the fundamentals, design processes, and processing issues in VARTM manufacturing processes. The next two sections describe the materials (fibers and resins), fabrication setup, and procedure in a typical VARTM process. That will be followed by sections covering the fundamentals and theory, design and processing issues of parts and tools, properties of parts, and their applications.

Materials

Fibers

Composites have two major ingredients: reinforcement and resin. Reinforcement is made of fibers since a fibrous form is much stiffer and stronger, and has fewer internal defects than bulk. It gives all the necessary stiffness and strength to the composites. The most common reinforcement includes glass, carbon, aramid, and high molecular weight polyethylene (HMWPE) fibers. Glass fibers are the most widely used and account for almost 90% of the reinforcement in thermosetting resins. E-Glass is commonly used and least expensive of all fibers. S-Glass has higher strength and stiffness and is more expensive than E-Glass. Carbon fibers have low density, high strength and stiffness, and are more expensive than glass fibers. PAN fibers are the most commonly used fibers since they have higher strength and failure strain than pitch fibers. Carbon fibers are most compatible with epoxy and extensively used in aircraft and aerospace applications. The coefficient of thermal expansions (CTE) in the fiber and transverse directions are typically $-1.4 \times 10^{-6}/°C$ and $11 \times 10^{-6}/°C$, respectively, for high-modulus fibers [12]. This makes possible the design of composite products with zero to very low CTE by using the carbon fibers as the main reinforcement or just an additive. Aramid fibers are tough fibers since they can undergo some plastic deformation before fracture [13]. They are resistant to fire, impact, and abrasive damage. Kevlar is the brand name of an aramid fiber developed by DuPont. The modulus and strength of Kevlar is roughly comparable to that of glass, yet its density is almost half that of glass. Kevlar can therefore be substituted for glass where lighter weight is desired. Kevlar is sensitive to moisture and salts and thus requires waterproofing. Kevlar has a negative CTE (i.e., $-3 \times 10^{-6}/°C$) in the fiber direction but positive CTE (i.e., $70 \times 10^{-6}/°C$) in the radial direction. The HMWPE fibers have the lowest density of all fibers above. The commercial products include Spectra and Dyneema. Aramid and HMWPE fibers are common in the use of impact and ballistic applications such as helmets and body/vehicle armors [14]. Table 19.2 shows the characteristics of various fibers [15].

VARTM utilizes continuous fibers that easily conform to various shapes. Typical diameters range from 5 to 20 μm. Fiber reinforcements are woven and knitted fabrics, and braided fabrics with three-dimensional construction. Translaminar reinforcement can be added where delamination is concerned [16]. Note that VARTM uses dry preforms that can be formed more easily on a complicated mold than prepregs. In addition, the dry preforms cost less than prepregs and do not need freezer for storage.

TABLE 19.2

Properties of Commonly Used Fibers

Fiber	Density (g/cm³)	Tensile Modulus (GPa)	Tensile Strength (MPa)	Break Elongation (%)	CTE ($10^{-6}/°C$)	Estimated Cost ($/kg)
E–Glass	2.6	74	2500	3.5	5	2
S–Glass	2.5	86	3200	4	3	14
Kevlar	1.45	130	2900	2.3	–3 (70)	70
C–HT	1.75	230	3200	1.3	–0.4	70
C–HM	1.8	390	2500	0.6	–1.4 (11)	140
PE	0.96	100	3000			

HT: high tension, HM: high modulus, and PE: polyethylene.

Matrix Systems

Matrix, as a binder, maintains the fiber orientation, protects the fiber network from environmental degradation, and transfers applied loads to the fibers. VARTM uses wet thermoset resin systems that have low viscosity (about 100 cps) and high wettability. Thermoset resins are thermally and chemically stable but brittle. Resins can be toughened by adding modifiers such as nano-particles and rubber additives, which however, increases viscosity and restricts resin flow. In VARTM processes, toughening tactifiers or thermoplastic fibers woven with the reinforcing fibers can be used [17] to increase impact resistance. Resin systems for VARTM include polyester, vinyl ester, phenolic, and epoxy resins. They are typically cured at an elevated temperature but some of them are cured at room temperature. The most commonly used thermoset resin is unsaturated polyester typically dissolved in a liquid reactive monomer (usually styrene). Polyester resins dominate the commercial markets mainly because of their low cost [18]. Vinyl ester can replace the polyesters when a higher temperature capability is required [13]. Epoxy provides the best structural characteristics of all the resins and thus is used in most military applications as the matrix material. Phenolic resins are used in nonstructural applications that require fire-resistant properties [13]. Some phenolic resins produce volatile substances as by-products during the curing process, which may lead to voids or porosity in the cured part. Polyimides and cyanate esters can be used for applications that require high-operating temperatures up to 600°F. Table 19.3 shows the properties of various thermoset resins [15].

Most resins, except epoxies, are mixed with a catalyst to initiate polymerization. The rate of polymerization can be controlled by the mix-ratio of an inhibitor and accelerator to the catalyst so the period of a process cycle can be controlled. For example, polyester resin cures by an exothermic reaction created by promoters (cobalt and amine) and an activator (MEKP). The resin and any additives must be carefully stirred to disperse all the components evenly before the catalyst is added. Careful stirring is required so as not to introduce air into the resin mix, which affects the quality of the

TABLE 19.3

Properties of Commonly Used Thermoset Resins

Polymer	Density (g/cm³)	Tensile Modulus (GPa)	Tensile Strength (MPa)	Break Elongation (%)	CTE (10⁻⁶/°C)	Processing Temperature (°C)	Estimated Cost ($/ kg)
Polyester	1.2	4	80	2.5	80	60–200	2.4
Epoxy	1.2	4.5	130	2	110	90–200	6–20
Phenolic	1.3	3	70	2.5	10	120–200	
Polyimide	1.4	4–19	70	1	80	250–300	expensive
Cyanate ester	1.2	3	60	2–3.8	70	200–300	expensive

final part. Air voids can be removed further by placing the resin mix under a vacuum pressure for a few minutes. Epoxy is a two-part resin system and differs from other resins in that a hardener, rather than a catalyst, is used for occurrence of polymerization when the two components are mixed. If the hardener and epoxy are not mixed in the correct ratio, unreacted resin or hardener will remain within the matrix, which will negatively affect the final properties after cure. Hardener should be carefully selected by taking into account the minimum temperature expected during the processing along with the required pot life. Curing time will vary with temperature of the part.

During the resin filling cycle, little reaction occurs. As resin cures, the viscosity increases and gel starts. Beyond the gel point, the resin flow becomes impossible and cross-linking continues. Resin evolves from a rubbery (i.e., gel) state to a glassy state at the glass transition temperature that limits the maximum service temperature of composites. During the curing and relaxing cycles, resin experiences shrinkage involving thermal shrinkage (i.e., CTE) and chemical shrinkage. Low shrinkage resins are preferred for less thermal stresses and distortion. The shrinkage data are sometimes available from resin manufacturers. If a low-temperature resin is infused into a high-temperature tool, differential curing rates across the part can also distort the part. In this case, the temperature of input resin should be controlled.

Description of Fabrication Procedure

This section will describe a typical procedure of fabricating a simple composite panel.

Laying-Up Preform

Fabric plies can be cut manually by a knife or scissors typically on a self healing cutting mat. The plies need to be cut oversize to the finished dimensions because of the inaccuracy in cutting and the fraying of fabric ends. An automatic cutting machine controlled by a computer can be used to reduce the waste of fabrics by trimming. Preform cutting is made generally in a separate room from the processing room; otherwise the trimmed fibers may fly over the processing room and land over the sealing lines, resulting in leak of vacuum. Note it is required to use appropriate eye and breathing protections throughout the fabrication process.

Figure 19.1 shows a schematic diagram of a typical VARTM processing setup. Before placing the prepared plies, the tool surface should be cleaned. Cured resin films left over on the tool surface from the previous fabrication

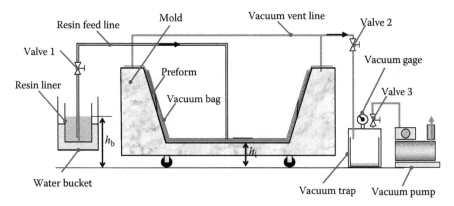

FIGURE 19.1
A schematic diagram of a VARTM processing setup.

step should be removed with a scraper and any residual dirt or grease should be wiped away by a cloth with acetone on it. After cleaning the tool surface, the entire surface of the tool is coated with a mold release agent, which prevents adhesion between the part and tool surface.

Figure 19.2 shows a schematic diagram of a VARTM lay-up. After the release agent is dry, the whole plies are placed on the tool surface as per the designed sequence. Next, a sheet of peel ply is placed between the preform and distribution layer to separate them easily after the resin is cured. Any wrinkles in the peel ply will imprint to the surface of the part and should be avoided. The distribution medium that is placed on top of the peel ply is a mesh-like shaped, highly porous, flow medium made from a knitted nylon material or a thermoplastic biplanar net. The length of distribution medium should be cut shorter than that of the preform by about two to three times the preform thickness as shown in Figure 19.3. Spray adhesive can be used to keep the preform and distribution layer in place. Smooth weights can also be used to temporarily hold the distribution layer in place.

Bagging Preform

After the preform is laid out on the tool surface, run an acetone wetted cloth along the line where the sealing tape is to be placed. This will remove the release agent from the tool surface and permit the sealing tape to adhere well to the tool surface. Place a strip of sealing tape to form a perimeter around the entire preform, allowing spacing of a few inches to the edge of the preform. Place short pieces of resin feeding and vent lines (e.g., aluminum wound coils, PE spiral wrap, or Omega flow lines from Airtech Advanced Material Group) on top of the distribution layer and breather layers, respectively, as shown in Figure 19.2. The omega tubes can be used only on a flat surface of preform but the wound coils or spiral wraps are better adapted to for any

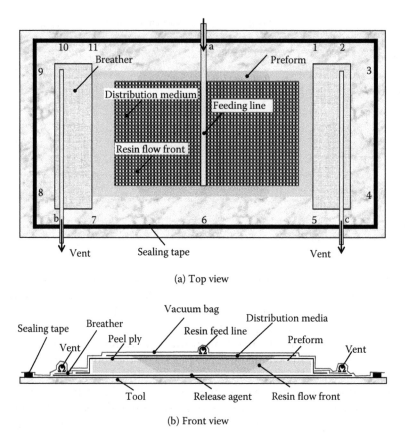

FIGURE 19.2
Schematic diagrams of a VARTM lay-up. (a) Top view and (b) front view.

surfaces including a curved surface of preform. The breather layer makes an air channel under a vacuum bag when compaction pressure is applied. There should be no sharp edges or points remaining on the end of the feed/ vacuum lines to protect the vacuum bag. Once the feed and vacuum lines are placed, the next step is to seal the whole preform with the vacuum bag. The vacuum bag should conform to the edges of the preform without bridging. Figure 19.3 shows two vacuum bagging cases: (a) with, and (b) without, forming a bridge at an edge of the preform. Bridging can have a gap formed between the vacuum bag, and the preform and tool surface such as the dashed circle as shown in Figure 19.3a. The gap has less flow resistance than the preform, which causes resin to flow around the preform through the gap rather than through the porous preform during the resin infusion. This is called race tracking: resin may arrive at the vent line through the highly permeable race tracking channel before fully filling out the preform, which may result in an unexpected dry spot in the cured preform. To avoid the

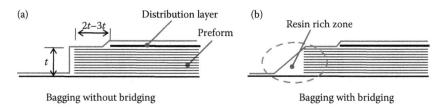

FIGURE 19.3
Vacuum bagging (a) without bridging and (b) with bridging.

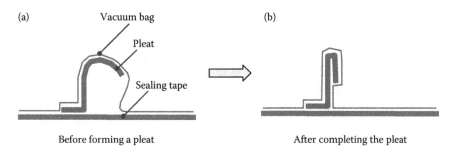

FIGURE 19.4
Forming a pleat with a sealing tape in the vacuum bag. (a) Before forming a pleat and (b) After completing the pleat.

bridging, the size of the vacuum bag should be greater than the surface area inside the sealing tape. Note the bridging results in a resin-rich zone that has potential to generate thermal stresses resulting in structural distortion and micro-cracks. Pleats are accordingly needed to provide extra spacing along the sealing tape at the locations of preform steps and the feed and vent lines (at the locations numbered by 1–11 in Figure 19.2a). Figure 19.4 shows how to make a pleat at a location on the sealing tape. A pleat is formed by adding a piece of sealing tape to the surrounding sealing tape. A piece of sealing tape is wrapped around the feed and vent lines at the locations where the lines meet with the sealing tape (at the location marked by a, b, and c in Figure 19.2a). Finally, press the bag down around the whole perimeter of the sealing tape a few times and remove any wrinkles on the bag, especially on the tool surface to secure sealing.

Applying Vacuum

The short feed line is connected to a long tube that reaches to the resin liner. The tube should not be too long since a large flow resistance in the tube creates a small pressure drop between the gate and vent in the preform. The end of the tube is cut with an angle to avoid blocking the resin flow when it contacts the bottom of the resin liner. Close valve 1 in Figure 19.1 by folding the feed line or clamping it off with a pinch valve. The vacuum line is also connected to a long vacuum line that is inserted into the vacuum trap. Close valve 2 and open valve 3 in Figure 19.1 and turn the vacuum pump on. The sealing of the vacuum trap can be checked by monitoring the vacuum gauge when valve 3 is closed. Open valve 2 fully, and valve 3 partially, and draw air from the preform. When the bag starts to conform to the shapes of the part and tool, position the vacuum bag manually. Ensure the feed and vent lines are positioned appropriately. After the bag is positioned, apply full vacuum by fully opening valve 3 to seal the vacuum bag. Press the bag down to the entire perimeter of the sealing tape to avoid leaking, especially from the wraps and pleats. Listen for individual leaks. An acoustic leak detector can be used to easily locate the leaking spots. Seal any pin holes or leaking spots on the bag by using small pieces of sealing tape. After an adequate vacuum is established, a leak test should be conducted. Close valve 3 and read the vacuum gauge to check air leaking in the part. The leak test is successful if the vacuum level decreases less than 1 inch Hg during a period of 5 minutes [19]. Use a second bag on top of the first bag (i.e., double-bagging) to secure vacuum integrity when all efforts do not work to seal the leaking. Note that establishing and maintaining an appropriate vacuum level during the process is one of the most critical factors to ensure a quality part.

Infusing Resin

After the vacuum integrity on the part is established, resin should be properly prepared. The resin is typically put in a disposable liner placed in a water bucket (see Figure 19.1) to prevent melting of the liner since an exothermic reaction typically occurs during the polymerization of the resin. Open valve 1 to allow the resin to impregnate into the preform. The resin will flow through the distribution layer first and down to the fibrous preform until it fills the entire preform. At this stage, a typical pressure at the injection gate is about the atmospheric pressure and that at the vent is about the vacuum pressure. Note this pressure gradient causes thickness variation of the preform along the line from the injection gate to the vent. If the resin liner is located above the injection gate, the pressure at the injection gate will be greater than the atmospheric pressure. In this condition, the sealing may fail near the gate location. To avoid this situation, the resin liner should always be placed beneath, but close to, the injection gate. Close valve 1 when the resin fills the vacuum lines. Now it is time for the resin to start to gel. Note that the resin gel time should be predetermined at the resin preparation step. Generally, it is a good idea to defer the resin gel time until the pressure gradient is reduced enough to obtain a uniform thickness in the preform [20]. When the resin starts to gel and exothermic heat is generated, the part will start to cure. For some resins, toxic gases such as styrene might be produced during the curing time, which requires a ventilation hood placed over the vacuum pump. The part is sometimes left overnight for complete (room temperature) cure. Once the resin is cured in the entire part, the pump can be turned off.

Finishing Part

Once the part is cured, remove the bagging material and take the part off from the tool. Since the resin is not fully cured, post curing is generally needed in an oven. Information about the curing conditions can be obtained from the resin manufacturers. Since a finished part is fabricated to near net-shape, additional machining operations such as trimming edges and drilling holes are needed to obtain the final dimensions. If the finished part is flat and has straight edges, a circular saw with a diamond blade is preferred for most composite materials. If the finished part has curved surfaces and edges, a portable diamond-plated router can be used to trim the edges. Extreme caution should be exercised to avoid overheating the cured resin since overheating destroys the part locally. Using a coolant extends the tool life without affecting the cut edge quality. A wet surface grinder can be used for cutting ASTM test coupons. Water jet and laser cutting have become an industrially accepted production method that ensures high-quality cuts. Water jets do not introduce heat in contrast to more traditional carbide or diamond tools [21]. Several critical points are listed in the References [21–23] about machining, drilling, and cutting of composites. After machining, opaque paint or urethane clear coat can be sprayed on the part for UV protection or just cosmetic purposes.

Theory and Fundamentals

Governing Equations

This subsection will review the fundamentals of the resin flow in a fibrous porous preform having simple geometry. It is assumed that the resin is a Newtonian fluid, the resin flow is incompressible, and the resin has constant density and temperature during the infusing cycle until the resin starts to gel. Resin flow through the porous media is governed by the continuity equation and Darcy's Law:

$$\nabla . \vec{u}_D = 0, \tag{19.1}$$

$$\vec{u}_D = -\frac{\tilde{K}}{\mu}(\nabla P - \rho g \vec{e}_v), \tag{19.2}$$

where $\vec{u}_D$ is the Darcy velocity vector; $\tilde{K}$ is the permeability tensor; P is the resin pressure; g is the acceleration due to gravity; $\vec{e}_v$ is the unit vector in the gravity direction; and ρ and μ are the resin density and viscosity, respectively. Assuming the gravity term and lengths of injection and vent lines/tubes are negligibly small, Equation 19.2 can be transformed to give one-dimensional flow front location (s) in a simple homogenous preform:

$$\phi \dot{s} = -\frac{\bar{K}}{\mu}\left[\frac{P_o - P_i}{s}\right],$$

(19.3)

where $\dot{s}$ is the physical velocity of the resin at the flow front location (i.e., the average velocity of the resin fluid particles, which can be measured by a flow sensor); ϕ is the porosity of the preform; $\bar{K}$ is the spatially average permeability [20]; P_o is the pressure at the resin flow front and typically considered to be the vacuum pressure (i.e., zero). P_i is the pressure at the injection gate.

The amount of resin to wet out the preform is calculated by the equation below:

$$W_m = \rho_m (1 - v_f) V_t,$$

(19.4)

where W_m is the weight of the resin; ρ_m is the density of the resin; v_f is the fiber volume fraction; and V_t is the total volume of the part. Note an extra amount of resin will be needed to fill the distribution media, breather, and the feed and vent lines in the real process.

The resin fill time is the time required for a resin to completely fill the preform. If a part is a rectangular flat plate, the fill time can be calculated analytically. Integrating Equation 19.3 gives the relationship between the flow front location and infusion time:

$$t = -\frac{\phi\mu}{2\bar{K}P_i}s^2,$$

(19.5)

where t defines the resin infusion time. Equation 19.5 can be used to estimate the resin fill time by replacing the flow front location (s) by the entire length of the preform. The fill time is proportional to the preform length squared, the porosity, and viscosity, but inversely proportional to the average permeability and injection pressure. Note the resin fill time is not dependent upon the cross-section area (i.e., width and height) of the preform. The resin should completely fill the preform before it starts to cure, so that the fill time should be less than the gel time (i.e., time required to start the polymerization process).

Combining Equations 19.1 and 19.2 gives the Laplace equation for the fluid pressure field in a preform:

$$\nabla.\left(\frac{\bar{K}}{\mu}.\nabla P\right) = 0.$$

(19.6)

Note that applying the divergence operator (∇) on a constant term such as $\vec{g}$ gives zero in linear coordinate systems (i.e., not in a curved coordinate system). Once the preform geometry, resin, and preform properties are given, Equation 19.6 can be used to solve the pressure field that is substituted into Equation 19.2 to provide the Darcy velocity. Finally the flow front location can be obtained by integrating the Darcy velocity multiplied by the porosity over time. Equation 19.6 can be solved by FEA tools when the part has a complex shape.

Effective Parameters for Hybrid Layers

Many VARTM parts have very large in-plane dimensions compared to the thickness. The incorporation of a highly permeable medium (HPM) on the surface of a low permeable medium (LPM), i.e., a fibrous preform, results in significant flow in the thickness direction. A conventional flow simulation models HPM as a discrete layer, which requires a large number of elements resulting in a significant increase in computational time and memory size due to the aspect ratio constraint. Those issues were partly solved by using 3-D expanded thickness elements [6] or 2-D plane elements [5] for HPM in FEA. In order to further reduce the computational time and complexity, a mass-average approach has been introduced that uses a homogenous model using the equivalent porosity and permeability. Figure 19.5 shows a hybrid model consisting of HPM and LPM and the corresponding equivalent homogenous model. Using the equivalent model reduces the number of dimensions that

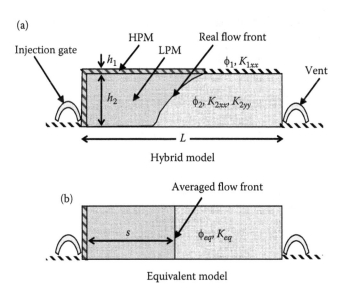

FIGURE 19.5
Equivalent homogenous model for a preform with hybrid layers. (a) Hybrid model and (b) equivalent model.

needs to be modeled (e.g., 3-D is reduced to 2-D in-plane and the 2-D cross section is reduced to a 1-D flow problem). Balancing the masses of infused resins between the hybrid and equivalent model produces an explicit expression of the equivalent porosity and permeability [24]:

$$\phi_{eq} = \frac{\phi_1 h_1 + \phi_2 h_2}{h_1 + h_2},$$ (19.7)

$$K_{eq} = \frac{K_{1xx} h_1 + K_{2xx} h_2}{h_1 + h_2},$$ (19.8)

where ϕ_1 and ϕ_2 are the porosities of HPM and LPM, respectively; K_{1xx} is the in-plane permeability in HPM; K_{2xx} and K_{2yy} are the permeabilities in the in-plane and thickness directions, respectively, in LPM; both HPM and LPM have constant porosity and permeability in each layer; the thickness of HPM (h_1) is assumed to be much thinner than that of LPM (h_2). Figure 19.6a and b shows the contour plots of resin flow front profiles between the two models and Figure 19.6c compares the average resin flow front locations with time obtained from the two models and analytic solution. The average resin flow front location was obtained by dividing the total volume of resin infused in the preform by the cross sectional area of the preform at each

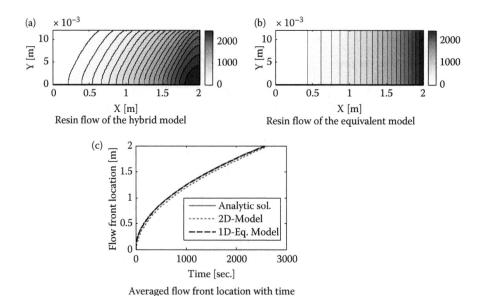

FIGURE 19.6
Resin flow profiles and the averaged flow front location with time. (a) Resin flow of the hybrid model, (b) resin flow of the equivalent model, and (c) averaged flow front location with time.

time step. The average flow front location shows a good agreement with each other throughout the resin infusion process. Note the input data about the simulation conditions are adopted from the Reference [24]. The mass-average approach estimated the resin fill time within a limited level of errors and significantly reduced costs in computational time and complexity, especially in modeling large-scale composite shell structures [24].

Sequential Injection Scheme

If the preform has a long distance between the injection and vent lines due to the limitation in its geometry, the resin fill time may increase too much (sometimes greater than the resin gel-start time) to make it impossible to completely wet out the preform. A sequential injection scheme allows one to solve this problem. Sequential injection schemes using multiple injection ports have been studied [25,26] in RTM processes in order to reduce the resin fill time and process cycle time. Figure 19.7a and b shows two injection schemes using single and multiple injection lines, respectively. F_i and V_o show the injection and vent lines, respectively, in the single injection scheme. F_{i1}, F_{i2}, and F_{i3} show the multiple injection lines that open in sequence when the resin reaches each injection line in the sequential injection scheme. In VARTM, not like in RTM, each injection line should open when the resin reaches the injection line, not at HPM but at the bottom of LPM [27]. This scheme allows an elevation in resin pressure and hence, resin flow rate. Figure 19.7c shows the resin flow front location with time. The input data

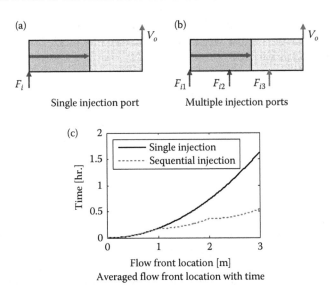

FIGURE 19.7
Flow front locations by the single and sequential injection approaches. (a) Single injection port, (b) multiple injection ports, and (c) averaged flow front location with time.

about the simulation conditions are adopted from Reference [24]. The resin fill times in the single and sequential injection schemes are 1.8 hours and 0.6 hours, respectively. By adding two more injection lines, the sequential injection approach reduced the fill time by 67% compared to the single injection approach. A sequential injection scheme with N injection lines decreases the resin fill time to $1/N$ of that by a single injection scheme. Note another injection method called a simultaneous injection scheme (e.g., F_{i1} and F_{i3} are simultaneously injected and F_{i2} and V_o are vented at the same time) may form numerous air voids where the flow fronts merge.

Thickness Variation

VARTM uses a flexible vacuum bag and elastic preform so that resin pressure in the preform changes the thickness of the preform. Recently, the thickness variation and compaction in the preform have been an active issue in the research area of VARTM processes [28–32] since they are related to fiber volume fraction and thus mechanical properties and dimensional tolerances of final parts. If the resin flow is quasi-steady and one-dimensional, the governing equation can be obtained by combining the continuity and Darcy's Law averaged over the thickness direction in the preform as follows:

$$\frac{\partial}{\partial x}\left(\frac{\tilde{K}.h}{\mu}\frac{\partial P}{\partial x}\right) = 0. \tag{19.9}$$

The preform thickness (h) and the fiber volume fraction (v_f) are correlated by the number of layers (n), the weight per unit area of a fabric layer (a), and the fiber density (ρ_f) as follows:

$$h = \frac{na}{\rho_f v_f}. \tag{19.10}$$

The fiber volume fraction is a function of the compaction pressure as follows:

$$v_f = aP_{comp}^b, \tag{19.11}$$

where the constants a and b can be obtained from a curve fit to experimental data. Note the compaction pressure is the difference between the atmospheric pressure and the resin pressure in the preform (i.e., $P_{comp} = 1 - P$). Combining Equations 19.10 and 19.11 gives the relationship between the preform-thickness and the compaction-pressure. Figure 19.8 shows experimental data between the preform thickness and compaction pressure for a homogenous preform (i.e., random mat). The data show path-dependent

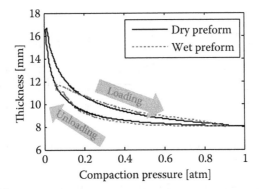

FIGURE 19.8
Thickness variation with compaction pressure.

behaviors between the loading and unloading of the compaction pressure. The thickness variation was smaller when the preform was wet by resin than when it was dry. The permeability can be modeled as a function of the fiber volume fraction using the Kozeny-Carman equation as follows:

$$\widetilde{K} = k\frac{\left(1-v_f\right)^2}{v_f^2},$$ (19.12)

where k is the Kozeny constant. Equations 19.9 through 19.12 with the mass balance equations (i.e., a total amount of resin infused in the preform is equal to the resin infused through the injection gate over the filling period, and a total removed resin in the preform is equal to the resin bled through the vent over the bleeding period) will give the thickness variation of the preform during the resin filling and bleeding steps.

Figure 19.9a through c depict the typical three process steps that include: (a) compacting the dry preform before the resin infusion, (b) releasing the wet preform during the resin infusion, and (c) recompacting the preform after the resin infusion. Before the resin infusion step, the injection gate is closed and the vent is open to apply vacuum to the entire dry preform. In this case, the preform is compacted with a uniform thickness by the pressure difference between the inside and outside of the vacuum bag. During the resin infusion step, both the injection gate and vent are open. The injection gate has the atmospheric pressure and the vent has the vacuum pressure. The pressure gradient from the gate and vent is the driving force for the resin to fill the preform. The pressure gradient, however, results in different thicknesses in the preform. After the resin infusion step, the injection gate is closed and the vent is open. In this case, the pressure gradient in the preform decreases as the resin keeps bleeding at the vent into the resin trap until it completely cures, which reduces the thickness gradient of the preform. At this step, the resin starts to gel and cure. If the resin curing starts too early,

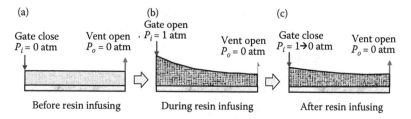

FIGURE 19.9
Thickness variation during a VARTM process. (a) Before resin infusing, (b) during resin infusing, and (c) after resin infusing.

the thickness gradient will exist in the final part. It was reported [31] that a panel made from woven glass fabrics with epoxy by a VARTM process had thicknesses of 2.8 mm and 1.4 mm at the gate and vent, respectively, at the beginning of curing and 2.2 mm and 2.0 mm at both ends of the finished part at the end of curing. Another study reported [32] that the thicknesses of a flat VARTM-processed panel were 25.1 mm at the feed line and 22.5 mm at the vent line, respectively. One way to avoid this situation is to control the initiation time of resin curing to be long enough to have an expected uniformity in the preform thickness. However, a longer initiation time makes a longer process cycle time. A way to reduce the cycle time as well as the gradient in the preform thickness is to make another vent at the injection gate. In this case, vacuum pressure will be applied at both ends of the preform after the resin infusion step, which will reduce the resin bleeding cycle period.

Structural Deformation

Structural distortion in a final part can occur due to the residual stresses caused by the effects of CTE and chemical shrinkage during the resin curing/relaxing period. The distortions and dimensional tolerances in the final part have been great concerns especially to the aircraft and aerospace community and users [32]. Deformation greater than a permissible level in parts results in difficulties in assembly, leading to increased costs. Figure 19.10 shows examples of structural deformations caused by different mechanisms during the resin curing/relaxing cycles. The dashed and solid lines show the original and deformed shapes, respectively, of the parts. CTE of polymer resins is usually much higher than that of the fibers. Fibers and a matrix resin have high temperature due to exothermic reaction during the curing cycle. However, the resin shrinks more when it is cooled down during the relaxing cycle. Thus, residual stresses form around the fibers. In addition, fiber orientation makes CTE in a unidirectional fabric ply as a function of the orientation angle. Figure 19.10a shows the distortion of a flat plate due to the different CTEs between two plies when the upper ply has a higher CTE than the lower one. This phenomenon normally happens when lay-ups are not balanced or not symmetric. Figure 19.10b shows a part distorted due to the shear interaction

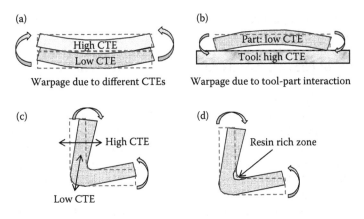

FIGURE 19.10
Deformations in the final part in the curing/relaxing cycle. (a) Warpage due to different CTEs, (b) warpage due to tool-part interaction, (c) spring-in due to different CTEs, and (d) spring-in due to resin rich zone.

between the tool and part when the tool has a higher CTE than the part. This phenomenon happens, for example, when the tool material is made of metallic material (i.e., high CTE) and the part is carbon composites (i.e., low CTE). Figure 19.10c shows the spring-in phenomenon caused by the difference in CTEs in the fiber and transverse directions in a laminated part. The spring-in phenomenon happens also when the corner of an L-shaped plate has a resin rich zone as shown in Figure 19.10d. Different temperature distributions in a part during the exothermic reaction also contribute to part distortion since more shrinkage occurs at a higher temperature region when the part cools down. The dimensional variation can be predicted and controlled based on the thermal stress analysis, which enables one to effectively assemble parts and achieve tight dimensional tolerances in high end applications. Note the finished or assembled parts can be deformed not only by the processing conditions of parts but also by the operating conditions of an assembled structure such as moisture absorption and temperature to affect CTE.

Design of Part, Process, and Tool

Part Design

The configuration of a structure should be determined such that it best meets the needs of customers. The configuration means the shape and geometry of the structure such as composite laminated plates, sandwich plates, and beams and tubes. A part configuration can be designed using CAD software that includes SolidWorks (Dassault System), Pro-E (PTC), I-DEAS (EDS), and CATIA (Dassault System). Once the fiber and resin systems are determined, the number of plies are calculated from the thickness of each ply. The next step is to create the lay-up of a composite laminate, for example, stacking sequence and ply orientation of laminates. The lay-up of laminates needs to be verified by stress and strain analyses [33]. A few commercial simulation codes are available to provide information about micromechanics and macromechanics for relatively simple structures based on the laminated plate theory: CompositePro (Peak Composites, Inc.), CompositeStar (Material), CMAP [34], and CADEC [35]. If a part has complex shapes, FEA software can be used: ABAQUS (Simulia), ANSYS (Ansys, Inc), MSC.NASTRAN (MSC Software), NEiNastran (Noran Engineering, Inc.), CATIA V5 Composites Design (Dassault Systems), GENOA (Alpha STAR Corp.), and LUSAS Composites (LUSAS). Sometimes, the ply-by-ply analysis helps engineers to create a layering and orientation strategy that provides appropriate strength in appropriate places with a minimum of fiber reorientation or interference. Examples of the ply-by-ply analysis software include FiberSIM (VISTAGY), Laminate Tools (Anaglyph Ltd.), ESAComp (Componeering, Inc.), and PAM-QUIKFORM (ESI) [33].

General rules of thumb in part design include:

1. Building laminates to orient fibers relative to applied loads that have balanced properties.
2. Laying up laminates to be symmetric and balanced.
3. Minimizing distortion of a part by selecting or using low-warpage polymers and fibers; appropriate part materials should be selected especially for assembly features such as male and female parts to provide a clearance fit.
4. Designing wall thicknesses as thin as possible and only as thick as necessary with an aim for uniform wall thickness; a thick wall increases the rigidity but also the weight of the part; a thicker portion cools slowly and thus shrinks more, which leads to distortion of a final part or creation of shrinkage cavities in the part due to the temperature difference between the cold surface and hot core; changes in wall thickness should be gradual and well filleted to avoid stress and shrinkage distortions; optimized wall thickness distribution influences material costs and can reduce the production time; use of ribs and sandwich constructions is an effective means of increasing the stiffness while reducing the weight to overcome the problems caused by thick walls.
5. Avoiding sharp corners and angles in a part; resin flow channels are highly likely to form between the sharp corner and the vacuum bag. This distorts the resin flow to create dry spots or resin rich zones

that result in thermal distortion in the part; in addition, the sharp corners themselves are sources of stress concentrations that may cause cracking.

6. Designing for a draft or taper in a part; a draft facilitates removal of a part from the tool.

7. Placing the good quality surface of a part facing the tool side since the other surface is rough.

8. Designing not to have greater precision than required; note excessively tight tolerance requirements increase the reject rate and quality control costs.

9. Designing multifunctional components that consider economic assembly techniques; designs with multiple integrated functions reduce the number of individual parts through integration of several functions in one part.

Flow Simulation Software

Understanding progression of the resin flow front is required to obtain optimum process conditions in order to assure complete impregnation of the resin in the preform. A poorly designed resin infusion setup may cause dry spots formed in a final part, which results in significant degradation of the mechanical properties or rejection of the parts after inspection. An analytic method can be used to predict resin flow and fill time for a simple VARTM process. A VARTM process for complex parts can be simulated and designed by resin flow analysis software just as structural analysis helps to avoid overdesigned parts and excessive safety factors. Examples of the flow simulation FEA/CV software include LIMS developed by the University of Delaware [7], RTMsim by UCLA [8], RTM-Worx by Polyworx [9], and PAM-RTM by ESI Group [10]. LIMS can run as an independent or slave module controlled by other software such as MATLAB® and Labview. The simulation software takes input data such as preform geometry, material data (e.g., resin viscosity and preform permeability), and processing conditions (e.g., temperature and pressure), and produces a resin flow front and cure profile. Reliability in predicting the resin progression is significantly dependent upon accurate input process parameters such as porosity and permeability [36–39]. The simulation software can import fiber orientations from a draping analysis software (e.g., FiberSIM and MSC.Patran Laminate Modeler) to provide more accurate results [40]. The simulation software can estimate the resin fill time and potential dry spots, which will provide preferred feed and vent line sites or sequences to achieve complete resin filling. The software enables engineers to virtually test different preforms and tools without building expensive hardware [40], which improves predictability of a VARTM process and reduces the risk of failure during actual production [11].

Tool Design

Geometry and tolerances of a part are affected not only by a part-manufacturing process but also by a tool-manufacturing process and tooling material. Thus the part, process, and tool designs must proceed in parallel. Comments made for the part design also generally apply to the tool design. A tool can be designed and fabricated typically by using CAD/CAM systems that include the same CAD systems used for the part design together with MasterCAM (CNC Software). A tool can be made directly by a CNC machine from any machinable materials such as polyurethane foam, medium density fiber (MDF) board, plastic, ceramic, aluminum, and steel. Figure 19.11a shows a picture of an upper half-mold milled out from a block of a few bonded MDF boards by a CNC machine. The machined surface was coated with Duratec primer to fill, level, and protect the surface, and sanded and polished later. The tool was used to manufacture small-scale wind turbine blades. This type of tool can fabricate only a few parts. The number of fabricated parts increases if stronger material is used for the tool. A tool can also be fabricated indirectly by a composite manufacturing process such as the spray lay-up, hand lay-up, and VARTM process over a CNC-machined master-mold. Figure 19.11b shows a picture of a tool made from random-mat glass/epoxy composites fabricated by a VARTM process over a CNC-machined master-mold made from polyurethane foam coated with the gel coat above. The tool was used to fabricate a Formula SAE car chassis. This type of tool can produce over 20 parts as long as the surface of the tool is well maintained. If the size of a master model is greater than the stroke limit of a CNC-machine, a few sections are subsequently fabricated and later bonded together to make the large master model as intended.

Table 19.4 compares various tooling materials [1]. Since only the atmospheric pressure is applied on the tool, the VARTM tool is not required to use high

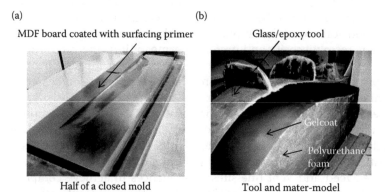

(a)   (b)

MDF board coated with surfacing primer   Glass/epoxy tool

Half of a closed mold   Tool and mater-model

FIGURE 19.11
Examples of various CNC machined mold and master-pattern. (a) Half of a closed mold and (b) tool and master-model.

TABLE 19.4

Characteristics of Tooling Materials

Material	Specific Gravity	Modulus (GPa)	Strength (MPa)	CTE (10⁻⁶/°C)	Thermal Conductivity (% of Al)	Estimated Cost ($/kg)
Polymer	0.6–2	0.5–5	100	30–100	0.1–0.5	2–3
GRP	1.5–2	7–20	150–400	15–20	0.5	2–3
CFRP	1.5	35–50	–500	4	0.5	3–10
Aluminum	2.7	71	50	23	100	1.5
Steel	7.9	210	300	15	30	0.2
Ceramic	2.3–3.3	14–200	150	7–12	35	20–40
Foam	0.4–0.6	1	20	30–100	Perfect insulator	10
MDF	0.6	3	1 (IB*)			1

IB*: Internal bonding.

strength tooling materials. Polymer has relatively high CTE and low-thermal conductivity, which may lead to a loss of tolerances if the tool temperature is not well controlled. Glass reinforced plastics (GRP) are normally made by hand lamination onto a gel-coated master-model. In this case, the tool is quite robust but the gel coat is easily damaged by scratches. Carbon fiber reinforced plastics (CFRP) can be made by hand lay-up or VARTM processes with a gel coat. CFRP tools have especially poor toughness and thus impacts are likely to cause delaminations or loss of vacuum integrity. Aluminum is a soft metal and it is easily scratched, dented, and otherwise marked, which restricts the use of metal scrapers. Aluminum tools have high-thermal conductivity and ensure uniform temperatures across the tool. Problems can occur when the tool can cool down around a CFRP part leading to high loads to the part. Steel can be considered for small and very complex tools, where tolerances are absolutely critical, or if production volumes are very high. Steel tools are heavy and have a problem of corrosion with acid catalysts. Ceramics are expensive tooling materials and machinable and castable ceramics are available. These tools are not tough at all, but scratch resistant. Foam tools are easily cut but large compressive forces are likely to deform the foam. Some foam is porous and have to be sealed prior to use. Foam and MDF materials are typically used for fabricating a master-model or prototype. Note there are many other innovative tooling materials available for customized applications in commercial markets.

Figure 19.12a shows a case of a part with high CTE and a tool with low or medium CTE. If the part cools down after curing, it shrinks away from the tool and thus can be easily removed from the tool. Figure 19.12b shows a case of a part with low CTE and a tool with high or medium CTE. If the part cools down after curing, it shrinks less than the tool and compressive stresses apply between the part and tool. This makes it difficult to take the part off from the mold. In the worst case, the tool has to be destroyed to save

(a)   (b)

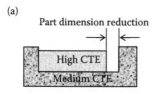

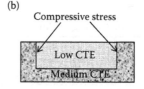

Higher CTE/shrinkage in the part   Lower CTE/shrinkage in the part

FIGURE 19.12
Different shrinkage/CTE between part and tool. (a) Higher CTE or shrinkage in the part and (b) lower CTE or shrinkage in the part.

the part when the part is more valuable than the tool. Note that carbon fiber based composites generally have low CTE and thus, either the corresponding tool should have similar CTE or the tool side surfaces should be slightly tapered in order to easily take the part off.

The design of gate and vent lines is also important since their locations determine the resin filling behaviors that could create the weld line and dry spots, leading to warpage and poor mechanical properties in the final part. The gate and vent should be placed away from highly stressed areas. Distortion of a tool should be minimized by selecting low-warpage tool materials or by appropriately controlling the heating and cooling cycles with embedded heating and cooling systems. Since a nonuniform thick wall of a tool may cause warping due to shrinkage, it is preferred for a tool to have a uniform wall thickness or use reinforcing ribs where strength is required. A sharp corner should be avoided so that a resin flow channel between the tool surface and preform is not made. Basically, the tool surface should be smooth enough for the preform to follow the shape of the tool surface so as not to produce resin rich zones, or wrinkles and folds of the preform itself. Note the longevity and costs of the tooling make a major contribution to the production economics [1].

Properties

A huge number of studies have been published on the mechanical properties of composites made by conventional composite processes such as autoclave with prepregs and RTM but not many on those by VARTM [41–44]. Nevertheless, the mechanical properties of composites made by VARTM can be successfully predicted by the micromechanics since the density, modulus, and strength are a function of the fiber volume fraction. The modulus of elasticity of a unidirectional composite ply along the direction of the fiber is given by the rule of mixtures:

$$E_1 = v_f E_{1f} + (1 - v_f)E_m, \tag{19.13}$$

where E_{1f} and E_m are the longitudinal fiber and matrix moduli, respectively. Note the modulus of the ply depends essentially on the longitudinal modulus of fiber because $E_{1f} \gg E_m$. The transverse modulus is a matrix-dominated property and sensitive to the local state of stress. Since the approaches based on assumptions of simplified stress states do not yield accurate results, the Halpin-Tsai semiempirical relation can be used to predict the transverse modulus of a unidirectional lamina [45]:

$$E_2 = E_m \frac{(1+v_f)E_{2f} + (1-v_f)E_m}{(1-v_f)E_{2f} + (1+v_f)E_m}, \tag{19.14}$$

where E_{2f} is the transverse modulus of the fiber. Assuming the fibers break before the matrix during loading along the fiber direction, the axial strength of a unidirectional lamina is given by:

$$F_{1f} = \sigma_{fu}\left[v_f + (1-v_f)\frac{E_m}{E_{1f}}\right], \tag{19.15}$$

where σ_{fu} is the tensile strength of the fiber. CTE in the longitudinal direction (α_1) of a unidirectional lamina is given by:

$$\alpha_1 = \frac{\alpha_{1f}v_f E_{1f} + \alpha_m(1-v_f)E_m}{v_f E_{1f} + (1-v_f)E_m}, \tag{19.16}$$

where α_{1f} is the axial CTE of the fiber. Note the fiber properties dominate over the matrix properties in the estimated longitudinal strength, stiffness, and CTE. CTE in the transverse direction (α_2) can also be obtained by the equation for orthotropic fibers and isotropic matrix [45]. Table 19.5 shows the

TABLE 19.5

The Predicted Mechanical Properties of a Unidirectional Lamina

Process	v_f	E_1 (GPa)	E_2 (GPa)	F_{1t} (MPa)	α_1 (10^{-6}/°C)	α_2 (10^{-6}/°C)
RTM	0.66	255	4.5	1635	−0.71	59
VARTM	0.55	217	4.5	1388	−0.36	72

predicted mechanical properties of a unidirectional carbon/epoxy composite lamina using Equations 19.14–19.16 where the fiber volume fraction of composites made by RTM and VARTM are assumed to be 0.65 and 0.55, respectively. Note the properties of fiber (C-HM) and resin (epoxy) are adapted from Tables 19.1 and 19.2. As the fiber volume fraction decreases by 15%, the axial modulus and strength also decreases by about 15% but the axial and transverse CTEs increased by 50 and –22%, respectively.

Applications and Products

VARTM processes are best fit to fabricate large-scale composite structures, which include various applications such as marine, transportation, military, civil infrastructure, aircraft, and aerospace areas. Examples of the marine applications include boats [11], racing yachts [46,47], and docking fenders for seaports, an ambulance boat and 123-foot mega-yacht built by sandwich composites using grooved cores [48,49], a half-scale ship section with a mass of 10,433 kg with UV curing resins, and a rudder for a Navy ship [50]. Examples of the transportation applications include rail cars [51], buses [52], and customized car bodies. Examples of the military applications include fenders for military trucks [52], a deckhouse module for the Navy with a stiffened skin construction, and a prototype Apache Helicopter fuselage for the Boeing Corporation [46,47]. A large complex structure integral to a fleet ballistic missile has been demonstrated [53] and reduced the part count and eliminated many fasteners.

VARTM has also been applied to the civil infrastructure areas. The examples include bridge deck prototypes [51], pedestrian decks [54], all-composite bridges [55,56], repair of cracked girders in a bridge [57], and a composite railroad bridge over which a full-size locomotive traversed while pulling 26 heavy axle load coal cars [58].

The European Wind Energy Association (EWEA) launched an initiative to achieve 12% of the world's electricity from wind power by 2020. The blades account for 15–20% of the total cost of a wind turbine [59] and are typically fabricated by VARTM. As the blade lengths grow greater than 70 m, carbon fibers are expected to replace the glass fibers as the primary reinforcements due to their high strength and stiffness [60].

Low dimensional tolerances (due to the inherent thickness gradient) and low fiber volume fraction (compared to autoclave processing) have limited the VARTM process in aircraft and aerospace applications. A few applications have been successful such as the CH-47 working platform, A380 components [61], the composite aft pressure bulkhead for the Boeing 787 Dreamliner, A400M transport cargo door for Airbus Industries [62], and satellite dishes [63].

Acknowledgment

This work was supported by the National Aeronautics and Space Administration (NASA) through a grant provided by the Experimental Program to Stimulate Competitive Research (EPSCoR) program in 2008 (No. NNX07AT52A).

References

1. Potter, K. 1997. *Resin Transfer Moulding*, London: Chapman & Hall.
2. Hieber, C. A., and Shen, S. F. 1980. A finite element/finite difference simulation of the injection mold filling process, *Journal of Non-Newtonian Fluid Mechanics* 7: 1–31.
3. Bruschke, M. V., and Advani, S. G. 1990. A finite element/control volume approach to mold filling in anisotropic porous media, *Polymer Composites* 11 (6): 398–405.
4. Correia, N. C., Robitaille, R., Long, A. C., Rudd, C. D., Simacek, P., and Advani, S. G. 2004. Use of resin transfer molding simulation to predict flow, saturation, and compaction in the VARTM process, *Transactions of ASME, Journal of Fluid Engineering* 126: 210–5.
5. Simacek, P., and Advani, S. G. 2004. Desirable features in mold filling simulations for liquid composite molding processes, *Polymer Composites* 25 (4): 355–67.
6. Chen, R., Dong, C., Liang, Z., Zhang, C., and Wang, B. 2004. Flow modeling and simulation for vacuum assisted resin transfer molding process with the equivalent permeability method, *Polymer Composites*, 25 (2): 146–64.
7. Advani, S. G., and Simacek, P. 2002. *LIMS-5.0 User manual*, Center for Composite Materials, Newark, DE: University of Delaware.
8. Lin, M., Hahn, H. T., and Huh, H. 1998. A finite element simulation of resin transfer molding based on partial nodal saturation and implicit time integration, *Composites Part A: Applied Science and Manufacturing* 29 (5–6): 541–50.
9. http://www.polyworx.com/pwx/cvi
10. http://www.esi-group.com/products/composites-plastics/pam-rtm
11. Brouwer, W. D., Herpt, E. C. F. C., and Labordus, M. 2003. Vacuum injection moulding for large structural applications, *Composites, Part A* 34: 551–8.
12. Kulkarni, R. S. 2004. Characterization of carbon fibers: Coefficient of thermal expansion and microstructure, M.S. Thesis, College Station, TX: Texas A&M University.
13. Schwartz, M. 1996. *Post Processing Treatment of Composites*, Covina, CA: SAMPE.
14. Lane, R. 2005. High performance fibers, *The AMPTIAC Quarterly* 9 (2): 3–9.
15. Gay, D., Hoa, S. V., and Tsai, S. W. 2002. *Composite Materials: Design and Applications*, Boca Raton, FL: CRC Press.
16. Dickinson, L. 2005. Using fiber to fight delamination, *Composites Manufacturing* November/December: 16–24.

17. McCarvill, W., and Strong, B. 2007. Toughening composites, *Composites Manufacturing* September: 78–80.
18. Military Handbook-MIL-HDBK-754(AR). 1991. Plastic matrix composites with continuous fiber reinforcement, U.S. Department of Defence.
19. Lewit, S. M., and Reichard, R. P. 2008. Pass the leak test, *Composites Manufacturing* September: 21–61.
20. Yoon, M. K., Baidoo, J., Heider, D., and Gillespie, Jr., J. W. 2005. A closed form solution for the vacuum assisted resin transfer molding (VARTM) process incorporating gravitational effects, *Journal of Composite Materials* 39 (24): 2227–42.
21. Black, S. 2004. Abrasive machining methods for composites, *High-Performance Composites* November: 32–37.
22. Billups, E. 2006. VARTM processing applied to M939 fender fabrication, *Composites 2006 Convention and Trade Show*, American Composites Manufacturers Association, 1–6, October 18–20, St. Louis, MO.
23. Gilpin, E. 2008. Machining composites, *Composite Manufacturing* December: 24–28.
24. Yoon, M. K., and Dolan, D. F. 2008. Homogenous modeling of VARTM processes with hybrid layered media, *Journal of Composite Materials* 42 (8): 805–24.
25. Kang, M. K., Jung, J. J., and Lee, W. I. 2000. Analysis of resin transfer moulding process with controlled multiple gates resin injection, *Composites Part A: Applied Science and Manufacturing* 31 (5): 407–22.
26. Kim, B. Y., Nam, G. J., and Lee, J. W. 2004. Optimization of filling process in RTM using a genetic algorithm and experimental design method, *Polymer Composites* 23 (1): 72–86.
27. Heider, D., and Gillespie, Jr., J. W. 2004. Automated VARTM processing of large-scale composite structures, *Journal of Advanced Materials* 36 (4): 11–7.
28. Acheson, J. A., Simacek, P., and Advani, S. G. 2004. The implications of fiber compaction and saturation on fully coupled VARTM simulation, *Composites: Part A* 35: 159–69.
29. Lopatnikov, S., Simacek, P., Gillespie, Jr., J. W., and Advani, S. G. 2004. A closed form solution to describe infusion of resin under vacuum in deformable fibrous porous media, *Modeling Simulation and Materials Science and Engineering* 12: 191–204.
30. Song, Y., and Youn, J. 2008. Modeling of resin infusion in vacuum assisted resin transfer molding. *Polymer Composites* 29 (4): 390–5.
31. Li, J., Zhang, C., Liang, R., Wang, B., and Walsh, S. 2008. Modeling and analysis of thickness gradient and variations in vacuum-assisted resin transfer molding process, *Polymer Composites* 29 (5): 473–82.
32. Tackitt, D., and Walsh, S. M. 2005. Experimental study of thickness gradient formation in the VARTM process, *Materials and Manufacturing Processes* 20: 607–27.
33. Sloan, J. 2007. FEA roundup: Design, simulation & analysis converge, *Composites Technology* 13 (2): 32–5.
34. http://www.ccm.udel.edu/Tech/CDSindex.html
35. Barbero, E. J. 1999. *Introduction to Composite Materials Design*, New York: Taylor & Francis.
36. Pavel, B. N., and Advani, S. G. 2002. A method to determine 3D permeability of fibrous reinforcements, *Journal of Composite Materials* 36 (2): 241–54.
37. Tari, M. J., Imbert, J. P., Lin, M. Y., Lavine, A. S., and Hahn, H. T. 1998. Analysis of resin transfer molding with high permeability layers, *Journal of Manufacturing Science and Engineering* 120: 609–16.

38. Gokce, A., Chohra, M., Advani, S. G., and Walsh, S. M. 2005. Permeability estimation algorithm to simultaneously characterize the distribution media and the fabric preform in vacuum assisted resin transfer molding process, *Composites Science and Technology* 65: 2129–39.

39. Yoon, M. K., Barooah, P., Berker, B., and Sun, J. Q. 1999. Permeability and porosity estimation in resin transfer molding process, *Journal of Materials Processing and Manufacturing Science* 7 (2): 173–84.

40. Berenberg, B. 2003. Liquid composite molding achieves aerospace quality, *High-Performance Composites* November: 18–22.

41. Wu, T. J., and Hahn, H. T. 1998. The bearing strength of E-glass/vinyl-ester composites fabricated by VARTM, *Composite Science and Technology* 58: 1519–29.

42. Khattab, A. 2005. Exploratory development of VARIM process for manufacturing high temperature polymer matrix composites, Ph.D. Thesis, Columbia, MO: University of Missouri-Columbia.

43. Tekalur, S. A., Shivakumar, K., and Shukla, A. 2008. Mechanical behavior and damage evolution in E-glass vinyl ester and carbon composites subjected to static and blast loads, *Composites Part B: Engineering* 39: 57–65.

44. Song, Y. S. 2007. Multiscale fiber-reinforced composites prepared by vacuum-assisted resin transfer molding, *Polymer Composites* 28 (4): 458–61.

45. Daniel, I. M., and Ishai, O. 2006. *Engineering Mechanics of Composite Materials*, New York: Oxford University Press.

46. Greene, E. 2006. Bill Seemann lifetime innovator, *Composites Manufacturing* August: 24–57.

47. Livesay, M. A. 1999. UV-VARTM fabrication of low cost composite structures, *44th International SAMPE Symposium*, 57–66, May 23–27, Long Beach, CA.

48. Reuterlöv, S. 2002. Cost effective infusion of sandwich composites for marine applications, *Reinforced Plastics* 46 (12): 30–4.

49. Greene, E. 2006. Infusion infusion: How boatbuilders can approach change, *Composites Manufacturing* April: 20–6.

50. Lweit, S. L., and Jakubowski, J. C. 1997. Low cost VARTM process for commercial and military applications, *42nd International SAMPE Symposium*, 1173–87, May 4–8, Anaheim, CA.

51. Juska, T. D., Dexter, H. B., and Seemann, III, W. H. 1998. Pushing the limits of VARTM, *43rd International SAMPE Symposium*, 32–43, May 31–June 4, Anaheim, CA.

52. SCRIMP offers a cleaner alternative. 2002. *Reinforced Plastics* 46 (5): 26–9.

53. VARTM process improves missile fabrication. 2001. *Process Engineering*, March 16.

54. Black, S. 2007. Engineering insights: Pedestrian bridge deck makes business case for composites, *High-Performance Composites*, December.

55. Chajes, M. J., Gillespie, Jr., J. W., Mertz, D. R., Shenton, III, H. W., and Eckel, II, D. A. 2000. Delaware's first all-composite bridge, *ASCE Conference*, Philadelphia, PA.

56. Crane, R. M., Ratcliffe, C. P., Gillespie, Jr., J. W., Heider, D., Eckel, II, D. A., and Yoon, M. K. 2001. Monitoring an all-composite road bridge for cumulative damage, *2001 ASCE Structures Congress*, Washington, DC.

57. Uddin, N., Vaidya, U., Shohel, M., and Serrano-Perez, J. C. 2004. Cost-effective bridge girder strengthening using vacuum-assisted resin transfer molding (VARTM), *Advanced Composite Materials* 13 (3–4): 255–81.

58. Composite railroad bridge. 2008. *Reinforced Plastics* 52 (3): 48–9.
59. Dawson, D. K. 2003. Big blades cut wind energy cost, *Composites Technology* February: 28–32.
60. DeLong, D. J. 2007. Global market outlook for carbon fiber: 2006 and beyond, *SAMPE Journal* 43 (1): 39–46.
61. http://www.airframer.com/journal_story.html?story = 8036
62. Brosius, D. 2007. Boeing 767 update, *High-Performance Composites*, May: 56–9.
63. Mazumdar, S. K. 2002. *Composites Manufacturing: Materials, Product, and Processing Engineering*, Boca Raton, FL: CRC Press.

20

Modeling of Infusion-Based Processes for Polymer Composites

Sylvain Drapier and Jean-Michel Bergheau

CONTENTS

Composite Materials and Associated Manufacturing Processes

Composite materials result from the assembly of components that are very different in nature and properties. Structures based on composite materials can be finely designed regarding the loadings expected in service. However, such optimal designs cannot be obtained from mechanical studies only, since it is the manufacturing stage that will lead, at the end, to the targeted structural piece and will consequently also control the final properties obtained. Indeed, it is one of the main features of using composite materials that the composite itself is elaborated during the structure manufacturing, in a single stage, enhancing the "final properties-manufacturing conditions" issue.

Here, composite materials are presented and associated manufacturing processes are then introduced, focusing on infusion-based processes for structural high performance composites.

Structural Composite Materials

Definitions

A composite material results from the assembly of at least two components of different but complementary nature, leading to a material whose global properties are superior to those of the components considered separately. The aim is to tailor materials whose specific properties; that is, with respect to their density,* are usually better than other materials.

Usually, composite materials are made up from the association of two components: reinforcements and matrix. The reinforcements bring up the main mechanical properties of the assembly and their distribution, and more importantly their orientation will be chosen regarding the mechanical loading the structure will have to withstand. The matrix main function consists of ensuring the load transfer between reinforcements and to prevent the composite from external attacks. The matrix also gives the final form of the composite piece. To these two main components, additives can be appended that permit modifying the aspect or the final material characteristics: improvement of the fiber-matrix adhesion, coloring pigments, surface

---

* Terminology density will be used along with specific mass, both representing the mass to volume ratio expressed in adequate unit.

finishing gel-coat, UV-insulating agents, fire-retardant particles, thermal or acoustic insulation, and so on.

As indicated, composite material properties result from the combination of the component properties, from their spatial distribution, and from their volume fraction as well. Unlike classical materials, the mechanical properties of composite materials are reached only after the final piece is manufactured since the product (structure) and the material are elaborated in the same stage. This is obviously of great interest, but the counterpart designing of composite structures may be quite complex since not only does the material and geometry have to be selected, but also the internal organization has to be defined. The anisotropic nature of composite materials is the real key issue in using composite; it offers the possibility to control the targeted properties by an adequate design.

Composite Properties

Organic matrix composites represent 99% of the whole composite used in the industry. They are made up of either thermoplastic or thermosetting resins and structural reinforcements that are generally glass, carbon, aramide, or natural materials. Other composites based on nonorganic matrices exist, such as metal or ceramic matrix composites, the diffusion of which remains minor (Dopler, Michaud, and Modaressi 1999).

Composites Types

Composite materials are generally classified within two main categories: Large diffusion (LD) composites that offer both reduced costs and midtechnical properties and are used for large series applications (RP 2002b, 2002c, 2003) and high performance (HP) composites whose high specific structural properties allow using them for structural (primary) parts (RP 2002a).

Large diffusion composites have some mechanical properties very often lower than those of reference materials such as steel. They are mainly used for economic reasons and for their low density. Manufacturing processes employed for these composites (injection, pultrusion, etc.) allow obtaining complex shaped parts in a single step. Therefore, they are well suited for large series and are usually based on glass fibers. They can be found mainly in nonworking applications for transportation (see Figure 20.1a) or capital goods.

High performance composites are characterized by mechanical properties that are higher than those of LD composites in exchange for a higher global cost due to material and handling, as well as associated manufacturing processes (autoclave, RTM, etc.). These materials are mainly based on long fiber (orientated) reinforcements, and will be considered for the rest of this chapter. These materials are employed in leading-edge applications such as in Formula 1, aeronautics (see Figure 20.1b), competition boats, leisure,

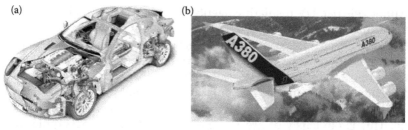

Carbon fiber/Composite materials

FIGURE 20.1
Example of LD composites: (a) Aston Martins V12 (From RP. *Reinforced Plastics* 46(5), 30–2, 2002.) and HP composites, (b) Airbus A380.

and so on. Historically, HP composites have brought the best benefits to the aeronautics sector considering the economy induced (paying loads) as well as the dimensions of structural parts that have been redesigned. This sector has originated a large number of technological breakthroughs that have been since transposed in other domains. Today, the use of HP composites is continuously expanding owing to the improvement and control of design techniques for new composite structures. The last years have seen the emergence of a so-called integral design that consists in minimizing the number of subassemblies. This is achieved through a redesign of products, relying on the use of the mandatory composite curing stage as an assembly stage, for example, the French High Speed Train power transmission developed at ENSM-SE in collaboration with Gec Alsthom-Cerisier (1998).

Advantages

As indicated previously, the main characteristic of composite materials with respect to more classical materials lies in their anisotropic nature. Engineers must build on this particular feature during the design of new structures against the loadings already undergone. Composite materials are in addition light, with densities between 1 and 3.5 g/cm³. This property is the very key in using composites for aeronautics and space transportation and more generally for lighter primary structures. Let us compare the composite material properties with properties of other materials. This implies both to defining their specific properties; that is, properties with respect to density, and the precise orientation and fiber volume fraction of the considered composites. Table 20.1 illustrates that specific moduli and specific strengths of unidirectional composites (long fibers orientated in the same direction) are from 1 to 3 times greater than those of other structural materials. This means that a unidirectional composite material loaded in the fiber direction would be up to 3 times lighter, as equivalent mechanical properties than the same structure made up of aluminum alloy.

Eventually, let us notice that manufacturing costs of a composite structure are interesting. First, the composite structure permits a simplification

TABLE 20.1

Comparison of Materials (Unidirectional Plies with 60% of fiber Volume Fraction) and Metals Properties

Mechanical Properties Metals and Composite Materials	Steel 35 NCD 16	Aluminum AU 4 SG	Carbon HR Epoxy Resin	Glass R Epoxy Resin
Specific mass (kg/m³)	7900	2800	1500	2000
Young's modulus (GPa)	200	72	130	53
Strength (MPa)	1850	500	1000–1300	2000
Specific modulus (MN m kg⁻¹)	25	26	87	26.5
Specific strength (MN m kg⁻¹)	0.24	0.18	0.65–0.85	0.9–1
Coefficient of thermal expansion				
—longitudinal (°C⁻¹)	12.10^{-6}	23.10^{-6}	$-0, 2.10^{-6}$	6.10^{-6}
—transversal (°C⁻¹)	12.10^{-6}	23.10^{-6}	35.10^{-6}	31.10^{-6}

Source: From Negrier, A., and Rigal, J. C., *Techniques de L'ingénieur* A7790, 1991. With permission.

by integrating functions. Second, manufacturing processes associated with composites are very often less expensive than those employed for metal forming.

Limitations

Due to forming limitations (drapability; *see* Boisse, Gasser, and Hivet 2001, for instance), connections with other pieces (Cerisier 1998), and stability in compression (Drapier, Grandidier, and Potier-Ferry 1999; Drapier and Wisnom 1999a; Drapier, Grandidier, and Potier-Ferry 2001) and shear (Drapier and Wisnom 1999b; Dufort, Drapier, and Grédiac 2001), the weight savings with respect to metal alloys can be reduced to some tens of percentage. However, this weight saving is a real advantage for all the applications where it is the driving issue, particularly for aeronautics, aerospace, energy, transportation, and so on. Moreover, when several directions have to be reinforced to face multidirectional loading, the specific characteristics drop. For instance, if two perpendicular directions are to be reinforced, the specific properties in both directions are roughly divided by two.

Since the basic components are different in nature, inhomogeneities along with dry resin regions can remain after manufacturing. These dry regions are responsible for a decrease in the mechanical properties, macroscopically as well as locally under the form of a weakness against ply cohesion (i.e., delamination). Other effects such as water sorption (Jedidi 2004) or chemical degradations are facilitated by the presence of such dry sites. These dry zones are induced during manufacturing, and one of the main objectives today is setting up processes limiting their presence.

Eventually, in-service temperatures of structural composites cannot, today, exceed 200°C, or sometimes 300°C regarding the organic nature of the matrix. This may be a limitation that should be overcome using thermoplastic specific matrices. Let us focus on this matrix component.

Matrix Component

Properties

As stated earlier, the matrix component maintains the reinforcements together, ensuring the role of a binder while maintaining them in an appropriate position and orientation. Consequently, the resin used can be the dimensioning element in the composite, controlling both strengths and partial stiffnesses in a different direction from the reinforcement direction. It also serves as a shield against external nonmechanical loadings and sets the highest in-service temperature that can be withstood by the composite. Besides these intrinsic characteristics, the quality of the impregnation of the reinforcement network by the resin will play a central role in the mechanical properties of the composite. Indeed, in order to obtain optimal composite properties, in its ultimate configuration the matrix must fill in all the interfiber spaces and be free of air (gas) bubbles. To achieve this impregnation, resin must reach a liquid state and then be polymerized by thermal and/ or chemical activation. This chemical reaction and both activation and inservice temperatures hence depend on the type of resin employed. It is then naturally that polymeric matrix-based composites are classified regarding the resin type considered: thermoplastics or thermosettings.

Resin Types

Two types of polymeric resin exist, both reaching their solid state by creation of bonds at the level of the monomer chains during the reaction stage. Both types of resin are distinct by the energetic level of the bonds created: Van der Waals low energetic bonds in thermoplastic resins allows for reversible melting-solidification cycles, and hence recycling, while thermosetting resins present an irreversible cross-linked state based on covalent bonds.

   Thermoplastic resins, usually high molecular weight polymers, are economically interesting. However, their mechanical and thermomechanical properties are low. Activation temperatures are of the order of a few hundred degrees; this requires heavy heating manufacturing facilities. But this crosslinking temperature yields interesting in-life temperatures, higher than for thermosetting resins. At the moment, thermoplastic resins are mainly used for LD composites, even though some applications can be found for heat-resistant pieces, and research is being conducted for the use of thermoplastic resin in structural applications regarding recycling issues. As for thermosetting resins, the more widely used are the polyester resins. But at the moment, epoxy resins are natural candidates as a matrix in structural HP composites due to their fairly good properties.

Reinforcements

Reinforcements bring to composite materials most of their mechanical performances of interest, both strength and stiffness in the fiber direction and some of the thermal, electrical, or chemical properties while yielding weight savings with respect to metals. Families of fibers are numerous; they can be classified according to their nature, their properties, or their architecture (arrangement).

Mechanical Properties

The main mechanical properties of the three most common fiber types used in structural composites are gathered in Table 20.2. These characteristics may vary depending on the supplier. These common fiber types are carbon fibers, glass fibers, and aramide fibers:

- Carbon fibers remain the reference for the continuous reinforcement event if their cost is still relatively high with respect to other fibers. These fibers possess excellent properties in tension and compression while offering a low density.
- Glass fibers are widely used in LD composites for their excellent mechanical properties to price ratio. However, their mechanical properties remain lower than those of carbon fibers and they are sensitive to humidity.
- Aramid fibers have some mechanical properties similar to carbon fibers for a slightly lower density, but above all for a price from 3 to 5 times lower. The most famous of these fibers is Kevlar from DuPont de Nemours. They possess very good properties of shock absorption and fatigue resistance, but conversely yield problems of adhesion to the matrix. This leads to very poor properties under compression.

TABLE 20.2

Mechanical Properties of High Resistance Carbon, E-Glass, and Aramid Fibers

Mechanical Properties of the Fibers	Carbon (HR)	Glass (E)	Aramid (Kevlar 49)
Specific mass ρ (kg/m^3)	1750	2600	1450
Diameter (μm)	5–7	10–20	12
Young's modulus E_f (GPa)	230	73	130
Ultimate stress σ_{fu} (MPa)	3000–4000	3400	3600
Specific Young's modulus E_f/ρ (MN m kg^{-1})	130	28	90
Specific ultimate stress σ_{fu}/ρ (MN m kg^{-1})	1710–2290	1300	2480

Source: From Berthelot, J. M. *Matériaux Composites: Comportement Mécanique et Analyse des Structures.* Tec & Doc Lavoisier, Paris, 2005. With permission.

Thermal Properties

One of the main characteristics of fibers is their negative, or even null, coefficient of thermal expansion in the axial direction. Depending on the thermomechanical behavior of the matrix, this can induce detrimental interfacial stresses (Gigliotti 2004). Conversely, using judicious fiber arrangements it is possible to conceive materials with a null thermomechanical expansion (Grosset 2004).

Different Types of Architecture

Fibers used as reinforcements are a few μm in diameter. In order to simplify handling operations, semiproducts are used that contain thousands of fibers placed side by side along controlled directions. This orientation can be achieved from different architectures (see Figure 20.2): Unidirectionals (UDs), bidirectionals (woven, knitted, braided, etc.) or multiaxial reinforcements. Reinforcements are then marketed under the form of blankets (or fabrics) that can be preimpregnated with resin (so-called prepregs) or not (dry blankets) regarding the process considered.

Unidirectional reinforcements (see Figure 20.2a) present some mechanical properties that depend on the type of reinforcement (carbon, glass, etc.) in the fiber direction. Fibers can be presented under the form of fiber rovings, or flat fabrics where a few fiber bundles are placed transversally, about 1% in volume fraction, maintaining all the fibers in the privileged direction to form blankets. From these blankets, stacking sequences are elaborated depending on the mechanical properties targeted. Let us notice that UD prepregs still permit achieving the best manufacturing quality and mechanical performance. But

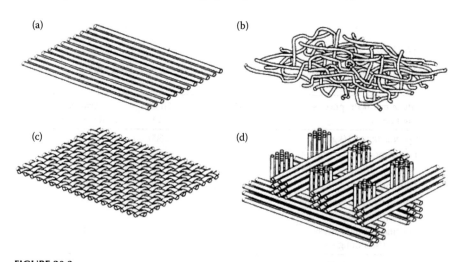

FIGURE 20.2
Different types of reinforcements: (a) unidirectional blanket, (b) mat, (c) plain weave, and (d) tridimensional orthogonal braids.

the handling costs required to form stacking sequences from UDs lead the composite industry to introduce multidirectional reinforcements.

Multidirectional reinforcements (see Figure 20.2b, c, and d)   present at least two fiber directions (woven fabric: Figure 20.2c). They can also be reinforced in a third direction (3-D fabrics: Figure 20.2d). The most current format of multi-directional reinforcement is the fiber mats (see Figure 20.2b) that are formed from fibers placed randomly in a plane. This structure does not present any preferential direction of reinforcement and leads to quasi isotropic or rather low mechanical properties. These reinforcements are used for nonworking structures. Woven fabrics offer a response to both handling reductions and reinforcement in the transverse direction to overcome delamination weak-ness, while allowing shear deformation required for material forming. Many types of weaves can be considered and yield corresponding fiber bundle wav-iness that translates in a property drop, especially under destabilizing load-ings. To overcome this decrease of properties, new multiaxial reinforcements have been developed, these are the so-called noncrimped fabrics (NCF) from Airbus and NC2 (noncrimped new concept) from Hexcel Reinforcements.

Concept from Reinforcements

Multiaxial Reinforcements   Multiaxial reinforcements denote materials elab-orated from UD blankets stacked and stitched together across the thickness assembly. These reinforcements are particularly suited for use with dry route processes. Contrary to woven fabrics, they don't exhibit any waviness as their name suggests, at least theoretically as far as NCF are concerned (Drapier and Wisnom 1999a).

NCF   elaborated on Liba machines are made up from a continuous lay-up of fiber bundles (or tows) containing only a few thousand fibers (3000–24,000) that may be orientated along four directions. The 2–4 blankets stacking are then stitched across the fabric thickness. The tow spreading remains compli-cated regarding the technology itself; this leads to properties reductions in compression (Drapier and Wisnom 1999a) and interlaminar shear (Drapier and Wisnom 1999b), due to both waviness and the presence of resin pockets in the intertow spaces. These intertow spaces permit, however, facilitating the resin infusion in plane and across the fabric thickness. Eventually, the reduced number of fibers in the tows makes this reinforcement rather expensive.

The NC2   is elaborated from fiber tows, containing from 24,000 to 80,000 fibers, which are spread and then placed side by side in order to form UD blankets. Up to four blankets are then laid up continuously to build the mul-tiaxial reinforcement. The blankets obtained from the tow spreading are very homogeneous, this leads to high mechanical characteristics but may cause problems of resin infusion during manufacturing. Fortunately, the stitching yarn permits the resin infiltration into the fabrics (Drapier et al.

(a)   (b)

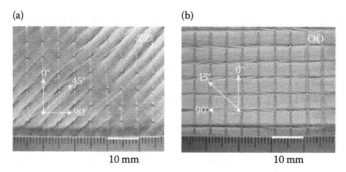

10 mm   10 mm

FIGURE 20.3
Picture of glass [+45, –45, 90] NC2 for wind energy applications (a) straight stitching face ZZ, and (b) loop stitching face OO. (From Drapier, S., Monatte, J., Elbouazzaoui, O., and Henrat, P., *Composites Part A: Applied Science and Manufacturing* 36, no. 7, 877–92, 2005. With permission.)

2002), although it may induce fiber misalignments that tend to reduce the mechanical properties (see Figure 20.3). Finally, though a further step of tow spreading must be achieved in NC2 elaboration, the total expense reduction is from 20 to 30% with respect to NCF.

Pressure–Temperature Cycle

Manufacturing composite parts consists in achieving final dimensions and fiber volume fraction while reaching a full infusion of resin in inter-reinforcement spaces before it solidifies. This is realized through a pressure–temperature cycle that both components—in fact their mixing—will undergo. As indicated previously, this cycle is specific to the resin-hardener system considered and is provided by the resin supplier. The pressure cycle allows bubbles to be removed that can be present in the liquid resin, either air entrapped or gas resulting from the thermochemical curing reaction. An example of the pressure–temperature cycle is given in Figure 20.4 for an epoxy resin. While this cycle depends on the resin system used, the way the resin-reinforcements mix is obtained depends on the process considered.

Processes

Historically, the highest mechanical properties have been obtained by using unidirectional preimpregnated reinforcements despite high costs for labor, stocking, material, and facilities. Since the late 1980s, dry route manufacturing processes have permitted considerably reducing the manufacturing costs provided dry reinforcements such as multiaxial fabrics are used. Today, new solutions permit reaching the properties obtained from preimpregnated materials. The main limitations lie now in material forming processes. Then, new processes are constantly being developed in

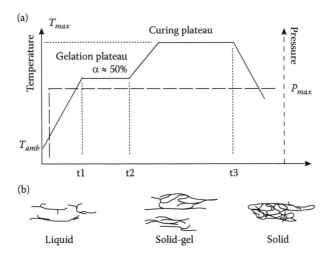

FIGURE 20.4
Curing cycle for an epoxy resin: (a) pressure and temperature versus time and (b) corresponding schematic changes in the thermosetting resin.

industry to improve the impregnation quality and master the characteristics of the final piece.

Wet Route Processes and Preimpregnated Materials

Blankets and preimpregnated fabrics result from the association of fibers and resin in a state of polymerization suited for forming but also for its conservation. The advantage they exhibit is one of convenience, this permits easy control of the final product properties, particularly the fiber volume fraction. In principle, the stacking sequence is realized and the piece is put under a vacuum before placing in an autoclave that will prescribe a pressure-temperature cycle specific to the preimpregnated system used. The piece obtained, present some excellent mechanical properties resulting from low porosity, high fiber volume fraction, and a homogeneous resin distribution. But this high quality has a price. Stocking and raw material costs are high, and they could be 40% lower if the resin and fibers could be stocked separately. Added on top of these costs are manufacturing and process facility expenditures.

Dry Route Processes

The apparition of these dry route processes respond to industrial demands for cost reduction while preserving identical final properties. Such processes are employed mainly for HP composites in the leading edge applications such as race cars, aeronautics, naval construction, wind energy, and so on.

There mainly exist two process families from which derive many variants. Injection processes consist in injecting liquid resin in reinforcement preforms placed in a closed rigid mold. These processes are known as resin transfer molding (RTM), vacuum-assisted resin transfer molding (VARTM), or injection compression resin transfer molding (ICRTM). As a response to filling problems met in large structures and in the strategy of continuous cost reduction, infusion-based processes have been developed during the past 20 years. These are better known as vacuum-assisted resin infusion (VARI), resin infusion under flexible tooling (RIFT), Seemann composites resin infusion molding process (SCRIMP), resin film infusion (RFI), and liquid resin infusion (LRI). It should be noted here that new processes are constantly being developed in composites industry, even if the basic principles remain unchanged. We propose describing briefly, the major principles underlying the more widely used processes, namely the RTM injection-based processes and the infusion-based processes LRI and RFI, the modeling of which will be detailed further.

RTM Process

RTM injection process was introduced in the 1980s. It consists of forming dry fabrics and placing them in a closed mold equipped with an event and an inlet (Barbero 1998). Resin is then injected under a controlled pressure or velocity before a temperature cycle is prescribed (see Figure 20.5). It is also possible to create an air vacuum thanks to the event, which corresponds to the vacuum-assisted RTM (VARTM). Resin injection takes place mainly in the fiber (longitudinal) direction. Manufactured pieces can have complex shapes with smooth faces associated with a controlled thickness. The use of rigid molds in two parts can yield over-cost but induces short time cycles and low-handling costs. Moreover, fiber and resin volume fractions are well controlled. Conversely, it is complex to reach a perfect mold filling for large size structures. These filling difficulties have generated many studies about the placement of injection points, taking into account the resin viscosity during filling (Kang, Jung, and Lee 2000).

LRI and RFI Infusion Processes

Infusion-based processes have been developed during the last years in order to solve the filling problems met in the RTM processes for long distances and also to reduce the production costs. These processes consist, in a general

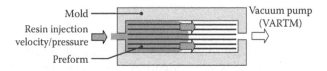

FIGURE 20.5
Schematic of the RTM process: resin transfer molding.

manner, to infuse resin across the reinforcement thickness rather than in the plane directions, under a driving pressure induced by the vacuum previously realized. The basic idea of these processes was that alternate stacking of dry reinforcements and resin could permit the air bubbles to be evacuated during the stage of curing and infusion in an autoclave (Blest et al. 1999). This technique, improved since then, permits producing thick structures of good quality, while saving costs with respect to techniques relying on preimpregnated materials or RTM processes. Among these processes, we propose modeling two process types that gather the whole particularities of the infusion-based processes: RFI and LRI processes. The models developed in this way will be easily applicable in the framework of new variants relying on the same physical mechanisms.

In order to avoid any confusion in the text hereafter, the following vocabulary will be employed. Wet preforms will be qualified for impregnated fabrics or laminates. The dry preforms will be referred to as fiber network or dry fabrics. Last, the zone where resin stands alone will be designated by purely fluid zone.

Principle The RFI process consists in laying fiber preforms on top of the solid resin layer (Qi et al. [1999], Han et al. [2003], Antonucci et al. [2003]). In order to ensure a proper finishing of the top surface, a punched plate can be placed on top of the preform/resin stacking. A bleeder, made up generally of glass fibers, can also be used to absorb excess resin. The stacking is then isolated thanks to an antiadhesive plastic film, and introduced in a system that will permit prescribing a pressure cycle (vacuum pump) and temperature cycle (autoclave or heating table). After temperature is applied, the resin film viscosity drops, this leads to the resin infusion across the preform thickness under the action of the pressure cycle. In a few words, infiltration and consolidation (i.e., resin cross-linking) takes place in a single step (see Figure 20.6).

More recently, techniques using liquid resin beds have been developed. In that case, the resin layer is achieved thanks to a distribution medium, highly permeable and placed on top of the preform stacking. The distribution medium is supposed to be infinitely stiff regarding the deformations

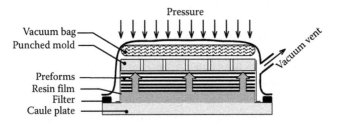

FIGURE 20.6
Schematic of the RFI process: resin film infusion.

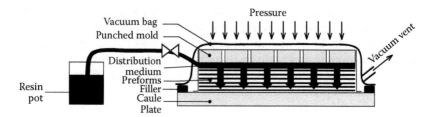

FIGURE 20.7
Schematic of the LRI process: liquid resin infusion.

inside the dry and wet preforms. The pressure differential induced between the resin inlet located at the level of the distribution medium and the event located at the bottom of the preforms will induce the resin infusion in the distribution medium first, and then across the dry preform thickness (see Figure 20.7). This two-stage flow scenario has been recently validated experimentally in-situ by thermocouple and fringe projection techniques (Wang et al. 2010).

NOTE: Highly permeable peel plies can be placed on the top and bottom faces in order to get better surface finishing. The use of a countermold is facilitative, it is often employed to obtain a homogeneous thickness distribution and leads to fiber volume fractions relatively constant.

Performances Infusion in the fabric transverse direction exhibits a double advantage. First in terms of cycle times, composite pieces usually have very low thickness to planar dimensions ratios. Distances that resin has to travel are reduced with respect to distances into play in classical injection processes. Second, in terms of the quality of infusion these small distances resin has to travel permit dissociating infusion and cross-linking stages; allowing these processes to produce composites characterized by high-mechanical properties resulting from both a proper impregnation of the reinforcements by the resin and high fiber volume fractions, up to 60% under certain pressure–temperature conditions. It is relatively cost-efficient with respect to RTM due to the use of semirigid molds and the reduced handling operations. Eventually, problems of dry zones and air bubbles present in the injection processes can be almost eliminated due to the short distances the resin has to travel.

Limits Globally, despite the basic idea that underlies the infusion-based processes, the physical phenomenas that come into play are quite complex. Controlling both thickness and fiber volume fractions is extremely tricky. The development of new composite solutions manufactured by infusion can induce large R&D costs. It lies essentially on experimental trial and errors analyses of the process since at the moment no numerical tool exists that could permit predicting the thickness variations observed. This lack of numerical tool comes from a deficit in the knowledge of the mechanisms that control the interactions between resin and fiber components, as well as between the different zones that are the resin alone and the wet and dry preforms. On top of that, one must point out that certain parameters of the process are not easily assessed, in particular the viscosities, capillary phenomenas, and transverse permeabilities.

Conclusion

The advantages provided by infusion-based processes such as LRI or RFI are numerous and promising, in terms of cost reductions as well as in terms of the final mechanical properties of the composite structures manufactured. However, mastering these processes implies controlling some complex physical phenomenas that make it quite hard to predict simultaneous cycle times, thicknesses, and fiber volume fractions from the processing parameters. At the moment, using this type of process is restricted to leading edge engineering fields regarding the expenses related to the development phases. Migrating these processes toward other industrial domains such as transportation or energy requires some improvements in the understanding of the effect of both the process conditions and the physical phenomenas ruling the various mechanisms implied, as well as their interaction. At this stage, one can evaluate the major interest of developing a multiphysics model for the study of composite material processing through infusion-based processes.

Modeling Infusion: State of the Art

Introduction

Infusion-based processes are very poorly controlled, although they can appear as very promising in terms of cost reduction while they permit maintaining or even surpassing the mechanical properties achieved with pre-impregnated materials or injection-based processes. Mainly, predicting the preform and resin response during the infusion stage—thanks to a numerical model—has not yet been achieved properly despite the numerous studies that have tried to yield some responses by decoupling resin flow and preform deformation. It is clear that only a general model, which would couple all the mechanical phenomenas (fluid and solid) along with the thermal and chemical phenomenas associated with cross-linking, would permit predicting efficiently the in-life physical and mechanical properties of the manufactured structure. A literature review of the recent studies regarding infusion modeling is presented in this section.

Difficulties and Interests of the Modeling Approach

Analyzing the infusion-based processes permit drawing certain physical phenomenas that a model will have to account for. At the moment, the main obstacle met in modeling infusion consists in representing the fluid flow (resin) in a medium undergoing large transformations, especially during the vacuum pulling that corresponds to the stacking compaction. In literature, experimental studies propose characterizing the interactions between the resin and the fiber network deformation inside the saturated porous medium. In these analyses, empirical or semiempirical models have been elaborated, but they remain limited to specific resins and multiaxial reinforcements and can be applied only under certain conditions. Coupling both resin flow and medium deformation induces further complexity since these mechanisms appear while the preform undergoes large strains and even large rotations for complex-shaped pieces (Drapier et al. 2005; Kelly, Umer, and Bickerton 2006). In addition, mechanical tests achieved in the direction transverse to the plane of dry or wet preform highlight a strong nonlinear behavior (Breard et al. 2004b) of the dry stacking, and even show the presence of irreversible phenomena only mentioned in Lopatnikov et al. (2004) and Comas-Cardona et al. (2007). So far, reinforcement deformations during infusion have not been accounted for, mainly due to a lack of knowledge. However, this is a key in understanding and predicting thickness variations and fiber volume fractions. Moreover, it appears that these deformations must also be considered in the representation of the fluid flow since the preform domain will deform during the preforms filling.

Modeling approaches as well as numerical simulations represent, in this framework, an efficient way to control finely infusion-based processes regarding the complexity of the physical phenomena into play. Such approaches will also help in designing proper experimental studies, necessary and complementary in terms of accessible variables. Beyond the thicknesses and fiber volume fraction control, such approaches will also bring efficient ways of controlling and optimizing infusion times by adjusting process parameters. Moreover, numerical analyses permit prediction of residual stress concentrations and hence achieve more reliable mechanical analyses on in-service structures. For instance, Ruiz et al. (2005) studied the influence of the temperature, fiber volume fraction, and resin cure advancement onto the mechanical properties of some composites. Eventually, and it is undoubtedly the main interest of such a modeling approach, it permits drastically reducing the effort of prototyping and testing to set up new composite solutions thanks to a numerical preselection, which is industrially with very little expense.

Infusion appears as a very complex phenomenon to be represented with models. Indeed, one must couple mechanical problems formulated classically in displacements for the preform response and in the velocity for the resin behavior. In the following, a state of the art is presented regarding the three main difficulties to be overcome by setting up a multiphysic model for studying infusion. First, the interaction between the resin flow and the deformation of the porous preforms must be taken into account. This interaction regards the mechanical action exerted by the fluid onto the fiber stacking as well as accounting for the displacement and deformation of the fabrics during resin flow. Boundary conditions to be prescribed take into account the coupling conditions between flows inside and outside the preforms representing the second difficulty. Last, and this is general, studying injection-based processes such as the RTM exhibit mass conservation issues at the fluid front. These numerical problems are related to the methods employed, the filling algorithms, and will have to be solved in these infusion-based processes approach, which a priori will rely on the same algorithms.

Literature Review

This literature review presents the main physical problems faced during composite manufacturing processes. Difficulties related to the characterization of the physical properties of the material during infusion are also presented and then related to setting up a multiphysical model describing infusion. For this purpose, some numerical difficulties will be detailed.

Main Physical Problems

Setting up and analyzing the dry route composite manufacturing processes reveal some difficulties, related mainly to the presence of the resin flow during the preform compaction. These difficulties translate into the formation of air bubbles and dry zones during resin infusion or injection, corresponding to the weakened zones. Controlling the fiber volume fraction is also essential in order to predict the structural mechanical characteristics. At the moment, the response to these difficulties in controlling and assessing the preform fill-in, from an experimental point of view, relies mainly on permeability measurements (Elbouazzaoui 2004) and stress/strain relationships (Morel, Binetruy, and Krawczak 2002). Measurements, thanks to optical fibers, have also been achieved in order to study deformations during the cross-linking cycle and subsequently during mechanical testings (Vacher 2004). These physical simulations have also permitted proposing some methods to follow the fluid front (Antonucci et al. 2003).

Filling and Air Bubble Issues

Problems of filling are located on two different scales: the micropores scale inside tows and the macropores scale made up mainly of the intertow spaces (see Figure 20.8). In order to suppress these undesired effects two techniques exist. First, Afendi, Banks, and Kirkwood (2005) propose an experimental protocol based on the resin filtration due to a fibrous membrane, and degassing in order to eliminate micro/macrovoids formation. Second, the fluid

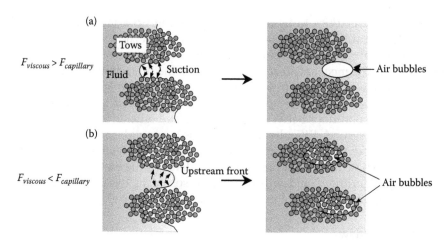

FIGURE 20.8
Mechanisms of air bubbles formation in transient regime: (a) $F_{viscous} > F_{capillary}$ (b) $F_{viscous} < F_{capillary}$. (From Lee, D. H., Lee, W. I., and Kang, M. K., *Composites Science and Technology*, 66, 3281–9, 2006. With permission)

velocity can be modulated. Depending on the fluid velocity, viscous forces induced will be lower or larger than capillary forces generated by surface tensions (Lee, Lee, and Kang 2006). For high fluid velocity, fiber tows are not saturated and air bubbles form inside tows (see Figure 20.8a). On the contrary, if the fluid velocity is reduced, capillary-induced forces locally increase the fluid velocity, and consequently air bubbles form in the intertow spaces (see Figure 20.8b). Capillary effects are closely related to tension surface effects (see Figure 20.8). Studies by Ruiz et al. (2005) and Lee, Lee, and Kang (2006) propose to optimize the injection rate in order to avoid relative flows between the resin inside and the outside tows, so that eliminated micro and macro void formation. Lim, Kang, and Lee (1999) propose a mathematical model in order to predict the void dimensions from a capillary number introduced in this same study.

Permeability Determination

As described previously, the fluid flow in fibrous preforms is quite complex, mainly due to the multiscale character of the preform architecture where fluid/structure interactions are not ruled by the same fiber/fluid characteristics. Very classically, the capability of the reinforcements to be infiltrated by fluids is considered at a macroscopical or mesoscopical scale. Then, permeability measurements in the case of fibrous reinforcements are, quite rightly, very much studied in literature. Indeed, knowing permeability is essential for predicting, in a realistic manner, the process of physics, and hence the main characteristics of the composite structure (thicknesses, fiber volume fraction, etc.). A literature review splits the permeability analysis into two categories. Analyses in saturated and transient regimes are achieved, respectively, on preforms fully impregnated and being impregnated. Permeabilities are then obtained by feeding pressure measurements and flow rates directly into the theoretical relationships describing flows in the porous media (for instance the Darcy's Law; Breard, Henzel, and Trochu 1999; Drapier et al. 2002). Some other authors utilize semiempirical relationships determined from theoretical models such as the Carman-Kozeny's model, in order to propose a permeability measurement (Charpentier 2004). Capillary phenomena highlighted in the previous paragraph play a predominant role onto the permeability. Reinforcement's saturation will have to be taken into account in determining permeability (Breard, Henzel, and Trochu 1999). Michaud and Mortensen (2001) propose separating permeability into two parts $K = K_{effective}\ K_{relative}$. The effective permeability $K_{effective}$ is the permeability measured in saturated regime while $K_{relative}$ is a value comprised between 0 and 1, depending on the saturation level denoted by s.

As a conclusion, the behavior in the vicinity of the flow front cannot be described by associating a Darcy's Law with an effective permeability. Capillary forces acting on the fluid flow front must be taken into account due to a relative permeability depending mainly on the saturation level.

Numerical Difficulties

Flow in a Compressible Porous Medium

One of the main difficulties met in the modeling approach is related to coupling conditions between resin flow fabric compressibility. Preform compressibility depends on the transient equilibrium between the pressure applied by the vacuum bag and the action of the fluid onto the fiber network or equivalently on the skeleton if preforms are seen as porous media. The pressure distribution and the fluid flow must moreover take into account the domain geometrical change with time. In this type of analysis, this pressure distribution is mainly controlled by the permeability that depends on the same time of the preforms porosity (Park and Kang 2003) and their degree of filling or saturation (Acheson, Simacek, and Advani 2004). Porosity, and consequently the fiber volume fraction, is mainly a function of the preform deformation and hence directly associated with the preform mechanical response (Gutowski et al. 1987; Morel, Binetruy, and Krawczak 2002). From an experimental point of view, these interactions have been clearly established in literature where significant deformations can be measured (Drapier et al. 2002) and depend at the same time on the flow regime studied and boundary conditions applied (Morel, Binetruy, and Krawczak 2002). However, the lack of knowledge regarding the integration of the preform compressibility leads to simulating the infusion of viscous resins in nondeformable media to solve this type of problem. In that case, the model employed represents the fluid flow through an infinitely stiff porous medium, using for instance the Darcy's Law. In the aim of improving these simple models, during the last decade some approaches have emerged, more or less empirical, to deal with the problem of preform compression during infusion.

In the framework of modeling infusion processes the first models developed by Loos and MacRae (1996) take into account the porosity changes, and hence the permeability variation, with the compaction prescribed by the vacuum bag, but do not account for the action of the resin onto the reinforcements deformation. Studies by Preziosi, Joseph, and Beavers (1996) and Ambrosi, Farina, and Preziosi (2002) propose the use of linear models for studying the injection of a viscous resin into an elastic porous preforms for 1-D problems. This simplified model is based on modified mass and momentum balance equations for both fluid and solid phases. More recently, some authors recommend using a constitutive law of Terzaghi's type; that is, a Kelvin-Voigt model in which the action of the liquid part (resin) onto the solid part (preforms) is introduced through its hydrostatic pressure (Hubert, Vaziri, and Poursartip 1999; Park and Kang 2003). In that case, an experimental law is used for the dry preform response. In their work Deleglise, Binetruy, and Krawczak (2006) focus on the influence of the induced or forced deformations on the pressure distribution for unidimensional problems. Blest et al. (1999) propose a 1-D model based on writing conservation equations for resin, preforms, and impregnated preforms, in the case of alternate resin film—preform stackings. However, these 1-D analyses, like the previous ones, are not suited for integration in industrial solvers based on the finite element method. Indeed, in these models the conservation equations do not appear clearly under the form of partial differential equations for mass and momentum. They are presented under an integral form and take into account further thermo–physico–chemical phenomena. Works by Correia et al. (2005) deal with the development of an analytical model for infusion in a compressible domain, but the compressibility analysis is limited only to transverse directions and relies on the Gutowski's model (Gutowski et al. 1987). The same remarks can be formulated for the works by Kessels, Jonker, and R. Akkerman (2007) regarding the modeling approach of RIFT processes.

On the contrary, studies by Joubaud, Trochu, and Le Corvec (2002) deal with a numerical analysis of the effect of wet preform compression on the stress distributions (VARI process) and introduce new finite elements for that (Darcy elements).* These new elements permit minimizing the void formation during injection by an optimization of the injection rates in the case of compressible preforms (Trochu et al. 2006). More generally, current studies are mainly based and formulated from experimental considerations. They mainly use semiempirical laws to calculate the stress distribution inside wet preforms. Although some references synthesize—in an exhaustive manner—the conservation equations for multiphase flows (Schrefler and Scotta 2001),

---

* So called Darcy finite elements integrated by ESI-GROUP in the PAM-RTM solver for the numerical simulations of VARI type processes.

the associated numerical tools to solve the strongly coupled multiphysics problem are not presented.

As a summary, the modeling approach and numerical simulation of a flow inside the preforms undergoing finite strains is considered in literature. However, the proposed models are dedicated to unidimensional cases, specific materials and particular boundary conditions. It seems well-justified then, on the basis of these observations, to formulate in another way the interactions between these two mechanisms; namely, the resin flow in a compressible medium on one hand and the preform deformation under the action of the fluid inside the pores of these preforms on the other hand. On top of this first difficulty, comes other types of problems when the flow also takes place outside the preforms. A complementary condition must connect the response of a purely fluid zone to the response of a preform zone partially or fully saturated.

Coupling between a Purely Fluid Region and a Partially Saturated Porous Medium

Coupling conditions between the resin flow inside the preforms and the resin flow alone are very seldom studied in literature, restrained mainly to experimental analyses leading to empirical models. At the moment, models and numerical simulations of infusion do not take into account the resin zone alone. This zone is usually replaced by a simplified boundary condition, corresponding to constant pressure (Ouahbi et al. 2007) or flow rate (Park and Kang 2003) on the boundary. Some studies focus on taking into account the resistance to the flow during the filling stage, through an update of injection pressures or flow rate (Deleglise, Binetruy, and Krawczak 2005). Although this coupling condition is not currently studied in the framework of composite manufacturing process modeling, some references relative to water flow in soils and streaming in surface propose a synthesis of boundary conditions to be applied so that continuity of conservation equations be satisfied (Layton, Schieweck, and Yotov 2003). These conditions, commonly referred to as Beaver-Joseph-Saffman conditions, have been exposed by many authors as nondeformable porous media (Jäger and Mikelic 2001; Rivière and Yotov 2005; among others). In these approaches, the stress vector continuity at the interface, as well as the normal velocity continuity, must be satisfied. This coupling condition also introduces a sliding factor; that is, a condition on tangential velocities that has to be verified by an experimental analysis (Porta 2005). Particularly, this sliding factor will play a major role when curvatures are introduced in the model. It is worth noting that the current coupling condition does not take into account the porous medium deformation. Consequently, it is necessary to thoroughly study a new coupling condition in the case of composite materials infusion processes. In parallel, an experimental characterization of the sliding factor, introduced for the condition on the tangential velocity, must be carried out to describe more precisely the edge effects.

In dry route composites process simulations, more and more studies are carried out in order to couple media of very different permeabilities. Two types of applications are in scope in that case. First, numerical analyses must be properly carried out for structured media with low-permeabilities running channels, so-called *runners*, which are used in practice to ensure a better distribution of injection points (Trochu et al. 2006; Hammami, Gauvin, and Trochu 1998). Second, infusion-based processes of the LRI type require using distribution media with low permeabilities that have to be represented also. Works by Diallo, Gauvin, and Trochu (1998) indicate a good correlation between experimental results and numerical ones for preform layers of different permeabilities. However, when permeabilities are very different, as in the case of the association of a distribution medium (large permeability $\approx 10^{-6}\ m^2$) with some reinforcements for LCM processes (low permeability $\approx 10^{-13}\ m^2$), the current numerical solvers do not allow satisfactory results regarding the strong scale effects. Srinivasagupta et al. (2003) propose in their simulation of "sandwich" structures, few dimensionless parameters in order to eliminate this effect.

Mass Balance and Fluid Front Tracking

Beyond the aspects intrinsic to resin infusion in preforms, problems on mass balance verification remain even in the case of injection into nondeformable porous media. These difficulties are mainly related to the filling algorithms currently used, and hence to the interface connecting the saturated porous medium with the dry preforms (i.e., the fluid front). The current software for simulating flows rely on two-steps for filling algorithms based on the finite element method (commercial software: PAM-RTM "ESI-GROUP"[*] and RTMWorx "PolyWorx";[†] academics LIMS Brouwer et al. [2003] "Delaware"[‡] and FE/CV code "SNU"[§]). For a given geometry, the first step consists in determining a group of elements and a filling time for this group. The pressure field in this group is computed using specific boundary conditions. The second step consists in computing, from this pressure field, the resin flux at the fluid front. Knowing these fluxes and the filling time at the previous filling iteration, new elements are filled and the algorithm is continued until complete filling of the discretized geometry.

Currently, two types of filling algorithms are used. The so-called finite element method/control volume (FEM/CV) used in most of the commercial or academic (LIMS, RTMWorx, etc.) codes relying on the construction of Voronoï cells from a finite element mesh (Frey and George 1999; Okabe, Boots, and Sugihara 1992). Solving the flow problem on this mesh yields the pressure field. Then, every Voronoï cell is associated with a degree of

---

[*] http://www.esi-group.com/
[†] http://www.polyworx.com/
[‡] http://www.ccm.udel.edu/
[§] http://www.snu.ac.kr/engsnu/

freedom in pressure that permits computation of the fluxes across the volume boundaries (Shojaei, Ghaffarian, and Karimian 2003). Associated with the filling time, this flux permits determining new elements filled during the following iteration (Advani and Bruschke 1994; Lim and Lee 2000). Some approaches in transient and quasi-static regime have been proposed, taking into account (or not) the saturation. Although it is very popular, this first proposed method does not satisfy precisely the mass balance. Those errors met in 2-D as well as in 3-D are mainly related to the association of a constant pressure to a control volume (Voronoï cell). Consequently, when fluxes are computed across the volume boundaries, they lead to a poor estimation of the filling of new elements (Joshi, Lam, and Liu 2000). Particularly, an error decrease is observed when refined meshes are used.

In order to overcome the problems met and improve the mass balance verification, some methods based on pretreatments or different solving methods have been proposed. Lin, Hahn, and Huh (1998) suggest an implicit integration of the equations by introducing saturation in the finite element problem. Although promising, this approach faces the lack of both knowledge and experimental data to characterize the relationship between saturation and pressure. More recently a method was developed based on element sorting followed by a mass balance inside every saturated element such that fluid in excess is reallocated (Joshi, Lam, and Liu 2000). These methods remain, however, very expensive in terms of numerical operations since they require some pre- or posttreatments. The nonstructured finite element method introduced by Trochu, Ferland, and Gauvin (1997) in LCM FLOT and later in PAM-RTM via PRO-FLOT libraries, propose an alternative to the FEM/CV method. In that case, construction of the control volume through Voronoï cells is no longer necessary since the control volume is associated with a finite element. In counterpart, the pressure field obtained due to a nonconform approximation is no longer continuous. Nevertheless, this method leads to a better mass balance verification since fluxes are computed from the pressure gradient known at every face center.

Whatever the method used to deal with the filling, it is not possible to apply the boundary conditions directly onto the fluid front since it is located inside elements. Kang and Lee (1999) propose a fluid front refinement by moving the nodes of the elements located on the fluid front, without changing the problem size (i.e., the number of degrees of freedom remains unchanged). Bechet, Ruiz, and Trochu (2003) propose, similarly, an adaptative method relying on a remeshing of the fluid front during filling.

The X-FEM method (extended finite element method) or the level set approach could also be adapted to the flow front tracking (Simone 2004). This method offers a way of managing discontinuities inside finite element meshes (Belytschko et al. 2003). In that case, from a transport equation (advection problem solved, for instance, thanks to a Lesain Raviart approximation; Fortin, Béliveau, and Demay 1995; Abbès, Ayad, and Rigolot 1999; and Bird, Stewart, and Lightfoot 2002) or the current filling algorithm based on the control volume FEM/CV, the fluid front is located and the elements on the fluid front are enriched. The boundary conditions relative to the fluid front are applied very precisely inside the finite elements of the mesh themselves through the X-FEM method.

On the basis of these observations, a new method that will permit dealing more efficiently with the mass balance will be proposed. This new method will be necessary, as in the mean time the porous medium deformation will have to be accounted for.

Conclusion

This short overview has shown that simulating infusion-based processes faces major numerical difficulties related to fluid flow in a compressible medium with variable permeability. At the moment, very few models have permitted addressing these problems, even individually. However, only a global approach solving at the same time (1) the resin flow in preforms undergoing compaction, (2) the coupling of a purely fluid region and a partially saturated porous medium, and (3) mass balance and fluid front tracking issues can lead to a meaningful predictive tool. In the following section, a complete numerical model able to represent all the main phenomena previously described that govern infusion-based processes is presented.

Numerical Simulation of Infusion-Based Processes LRI/RFI

In the following, a fully coupled model is set in a general framework for the simulation of nonisothermal infusion processes. To achieve this properly, two major types of problems must be tackled. The first problem is a thermo-physico-chemical one and involves the coupling between heat transfer and resin curing, it can be solved in a rather straight manner (*see* next section). The second problem, the very heart of our approach, is mechanical. It requires characterizing simultaneously both resin flow, inside and outside the preform, and fibers network compression during the resin infusion.

The proposed solution has been constructed by taking into account the interactions between all the solid and fluid components directly in mass balance and momentum conservation equations. This is similar to the method proposed by Schrefler and Scotta (2001) for nonmoving media but with two-gas flows. Here, to develop these new equations, an ALE formulation is introduced to handle the action of the fibrous medium response onto the fluid flow. To take into account the contribution of the fluid pressure onto the preform mechanical response, a Terzaghi's Law is used. This study also deals with the coupling condition between the pure resin region and wet preforms. In addition, mass transfers between these two regions must be carefully controlled, and this is of prime importance for a global model of infusion processes. Therefore, a mechanical mixed formulation between velocity and pressure fields leading to accurate mass balance is used to study the fluid flow both inside the purely fluid region (resin) and the porous medium (wet preforms).

From this model, a specific implementation has been achieved and validated in an industrial code PAM-RTM and the associated libraries PRO-FLOT. An updated Lagrangian formulation for solid mechanics has been implemented in this Eulerian-based software from ESI Group, along with a Darcy ALE formulation for the fluid part that permits studying flows in a deformable medium.

Model Overview

To model the various phenomena involved, one has first to select the relevant scale of observation. If a local approach is chosen, a very fine description can be proposed but local data are quite tricky to assess and a geometrical description of the (fiber) network must be realized, inducing very costly computations (Bechtold and Ye 2003). In turn, one can represent, for instance, the void formation during infusion (Lim, Kang, and Lee 1999). However, from an industrial point of view, macroscopical approaches are more easily considered, yielding reasonable computation times. The drawback of such approaches is to rely on macroscopical properties that may be hard to assess, since they depend on local phenomena, and quite often are represented through semiempirical laws.

A macroscopical approach is then chosen, even if some material parameters, such as the preform permeabilities (Breard, Henzel, and Trochu 1999; Drapier et al. 2002), are on their own complex to acquire. Then, simulation of the infusion processes will rely on a representation of the resin/preform stacking through three homogeneous regions (Figure 20.9):

- The dry preforms
- The wet preforms
- The resin

These three domains are connected with moving boundaries through specific conditions described subsequently. In the following, the modeling approach is presented in the most general manner. For this reason, RFI processes will be considered since the LRI process can be seen as a derivation of the latter.

Compaction and Infusion Stage: Mechanical Modeling

The fibers network compressibility during the resin flow is controlled by both resin pore pressure and mechanical external pressures. The result is

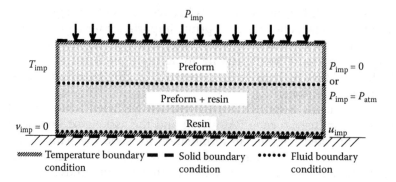

FIGURE 20.9

Splitting of the studied domain into three different domains connected with moving boundaries for the RFI process.

that the liquid and porous media will be described by different models depending on whether resin and preform stand alone, respectively, in the fiber-free domain and dry preform domain, or are mixed in the wet preform area (see Figure 20.9). The second difficulty of the present approach concerns the proper selection of boundary conditions on moving interfaces. A systematic model can be built by considering mass (continuity equation) and momentum balance equations of both fluid and preform media.

Resin Flow

ALE Formulation for Fluid Flow in a Mobile Domain   The Arbitrary Lagrangian Eulerian (ALE) approach is well suited to study flows in deformable moving domains (Rabier and Medale 2003). This formulation requires a virtual intermediary mobile domain, called subsequently, reference domain $\hat{\Omega}$, where computations are performed and that must undergo topological variations to account for the real material domain changes such as mechanical and physical properties changes (Belytschko, Liu, and Moran 2000). In our case an ALE formulation allows a precise flow front tracking while fluid sources can be represented.

Constitutive laws for the fluid behavior (Newtonian incompressible fluid) are classically expressed in terms of Cauchy stress tensor (σ):

$$\sigma(x,t) = 2\eta D(x,t) - p(x,t)I, \tag{20.1}$$

with η, the fluid dynamic viscosity, $D(x, t)$, the strain rate tensor, $p(x, t)$, the hydrostatic pressure, x, the current position in the material frame Ω, and I, the second-order identity tensor.

Momentum and mass balance equations for an incompressible flow therefore is written in an Eulerian formulation:

$$\text{div}_x \sigma + f_v = p(x,t)\frac{D_v(x,t)}{Dt} \quad \text{and} \tag{20.2a}$$

$$div_x v(x,t) = 0, \tag{20.2b}$$

where f_v are the volumetric external forces, $\rho(x, t)$ is the medium density, and $v(x, t)$ is the current velocity.

However, this Eulerian formulation must be modified to yield an appropriate ALE formulation (i.e., include a reference domain displacement). The relationship between the Eulerian formulation and the ALE one appears through the time material derivative that can be rewritten in terms of an Eulerian gradient ∇_x (i.e., calculated with respect to the material frame). This is achieved by introducing the convective particles velocity $c(x,t) = v(x,t) - \hat{v}(X,t)$, equal to the difference between the particles velocity $v(x, t)$ and the reference domain velocity itself $\hat{v}(X,t)$. An ALE material derivative for the function f writes:

$$\frac{Df(X,t)}{Dt} = \frac{\partial f(X,t)}{Dt}\bigg|_X + c(x,t)\cdot\nabla_x f(x,t), \tag{20.3}$$

where χ is the current position in the reference domain $\hat{\Omega}$ and t is time.

Using this new expression of the time derivative (Folch Duran 2000), the so-called quasi-Eulerian momentum equation writes as:

$$\text{div}_x \sigma + f_v = \rho(X,t)\frac{\partial v(X,t)}{Dt}\bigg|_X + \rho(x,t)c(x,t)\cdot\nabla_x v(x,t), \tag{20.4}$$

while the mass balance equation (Equation 20.2b) remains unchanged.

Finally, the reference domain motion has to be defined. This is achieved, for instance, by using an elastic constitutive law (Belytschko, Liu, and Moran 2000). In our case, this domain is totally arbitrary for resin regions and coincide with preforms in the case of Darcy's zone.

Resin Model in the Resin Domain   Due to both the infusion processes duration and the behavior of the resin (Newtonian incompressible fluid), low-resin Reynolds numbers can be considered at a macroscopic level. Hence, in the particular framework presented here, inertial forces can be neglected compared with viscous forces (Kaviany 1995; Loos and MacRae 1996). Nevertheless convective terms coming from the quasi-Eulerian formulation must be kept in equations. Neglecting volumetric forces allows the Stokes equations to be written as follows (momentum and continuity equations):

$$\rho_r^r c_r^r \cdot \nabla_x v_r^r = -\nabla_x p_r^r + \eta\nabla_x v_r^r, \tag{20.5a}$$

$$\text{div}_x v_r^r = 0, \tag{20.5b}$$

where variables are now defined for a zone (superscript r = pure resin, f = wet or dry fiber preform), and medium (subscript r = resin, f = fiber, d = reference domain), $c_r^r = v_r^r - v_d^r$ is the resin convective velocity in the purely fluid region; that is, the difference between the material velocity v_r^r and the reference domain velocity, v_d^r which results from the domain deformation prescribed to map its boundary displacements (Celle, Drapier, and Bergheau 2008a). When the RFI process is considered, the resin film fading is followed with a remeshing of the resin area (Figure 20.7).

Resin Model in the Preform   Since the applied external pressures override surface tension effects (Ahn, Seferis, and Letterman 1990), capillarity effects between fibers and resin can be neglected in the wet preform region. Then, as classically achieved, the fluid flow across the preform is represented through a Darcy's approach. In this macroscopic model, the Darcy's Law (Darcy 1856) expresses the superficial velocity $\bar{v}$ proportionally to the pressure gradient ∇_p. However, Darcy's Law fails in representing properly sticking conditions for a macroscopical viscous flow. To overcome this problem, in 1947, Brinkman proposed a generalization of the Stokes equation for viscous flows inside porous media by adding a correction term accounting for transitional flow between boundaries. Brinkman's equation writes:

$$f_v - \frac{\eta}{K} \cdot \bar{c}_f^f - \nabla_x p_f^f + \eta \nabla_x \bar{v}_f^f = 0, \qquad (20.6)$$

where K is the porous medium permeability tensor, $\bar{v}_f^f = \phi s v_f^f$ is a superficial or average velocity, ϕ is the porosity, s is the saturation, and $\bar{c}_f^f = \phi s c_f^f$ is a superficial average convective velocity. For low permeability media, advection and diffusion terms become predominant with respect to the $\eta \nabla_x \bar{v}_f^f$ term (Farina and Preziosi 2000a; Preziosi and Farina 2002), and then Brinkman's equation (Equation 20.6) equates the classical Darcy's equation. The mass continuity equation remains unchanged with respect to the Stokes and Darcy mass continuity equations (Equation 20.5b).

In the case of permeability lower than 10^{-3} m^2, such as for preforms in infusion-based processes, it has been shown that Darcy's Law is a good assumption of Brinkman's relation (Equation 20.6; Celle, Drapier, and Bergheau 2008a). On the contrary, for components with larger permeability, for instance in the distribution medium (resin layer) for the liquid resin infusion processes (Figure 20.7), Brinkman's equation should be used.

Eventually, it must be pointed out that the quality of the experimental determination of the permeability tensor will strongly influence the reliability of the results obtained. Recently, several studies were conducted on techniques to accurately measure the permeability tensor (Buntain and Bickerton 2003; Drapier et al. 2005). It was shown that permeabilities strongly depend on both porosity and saturation level (i.e., the filling level of the volume part considered. It is well-known that permeability characterization is a major problem since saturated and unsaturated permeabilities may differ by a factor 6–10 for multiaxial stitched fabrics for instance and may also depend on the face receiving the fluid (Drapier et al. 2005). Regarding the state of the art upon permeability measurements, in the present study, the Carman Kozeny's equation has been employed to relate permeability and porosity (Park and Kang 2003) as a first approximation:

$$K_{ij} = \frac{d_f^2}{16 h k_{ij}} \frac{\phi^3}{(1-\phi)^2},$$

(20.7)

where d_f is the fiber diameter and h_{kij} is the Kozeny's constant. However, the Carman Kozeny's model remains a coarse assumption of the permeability since the saturation is not taken into account. When reliable expressions for permeability variation according to porosity and saturation are made available from experimental studies, the proposed model can be updated in a straight forward manner to integrate this.

Preform Response

Lagrangian-Based Formulation The preform stacking stage in our manufacturing processes induces solid nonlinear deformations. A Lagrangian formulation is well suited to follow the large strains associated with compaction. Numerically, this type of nonlinear analysis is usually achieved through an iterative procedure that aims at solving the nonlinear equilibrium at each loading increment (the total loading is divided into subloadings) as the ultimate solution of a sequence of linearized states. However, in such an incremental procedure, the configuration chosen to compute the solution is not unique. An iterative updated Lagrangian formulation (Belytschko, Liu, and Moran 2000) is chosen here that consists in updating the geometry for every linearized state of the iterative scheme. This choice of the formulation has been driven by the representation of the action of the fluid on the solid media. Indeed, fluid stresses will be used to modify the solid medium mechanical response through a Terzaghi's model (see Equation 20.10). Since working with fluid is much straighter when Cauchy stresses are considered (Equation 20.1) an iterative updated Lagrangian formulation is more suitable to assimilate the current Cauchy stress tensor to the stress tensor known in the last updated configuration (Zienkiewicz and Taylor 2000; Bathe and Zhang 2002).

Conservation Equations In a material Lagrangian formulation, mass balance is implicitly verified since both computational and material domains coincide. Then, in our case the mass balance equation relates masses at times t and $t + \Delta t$ via the macroscopic preform density $\bar{\rho}_f^f$ and Jacobian of transformation J at these times. In our case, the preforms are assumed to be deformable but composed of incompressible fibers. This leads to an explicit relationship between the deformation states and associated porosities φ:

$$J(x,t+\Delta t)(1-\varphi(x,t+\Delta t)) = J(x,t)(1-\varphi(x,t)). \tag{20.8}$$

This relationship is an original feature of the proposed model compared to those exposed in literature where empirical approaches are used to relate the fiber fraction and the applied pressure (Gutowski et al. 1987; Farina and Preziosi 2000b). Through this global approach, when the new configuration has been obtained with the iterative updated Lagrangian formulation for the current load increment, the preform volume change gives the porosity variation assuming the incompressibility of the fibers. As for the momentum balance equation, without volume forces it is classically written in terms of Cauchy stresses in the preforms $\sigma_f^f(x,t)$:

$$\mathrm{div}_x\sigma_f^f(x,t)=0. \tag{20.9}$$

Constitutive Equations   The constitutive law of the wet preform will depend on both fiber network behavior and resin flow. For a general use, unlike more specific responses for given fiber preforms under very particular boundary conditions (Gutowski et al. 1987; Kessels, Jonker, and Akkerman 2007), the constitutive law can be formulated following the Terzaghi's hypothesis. In this model, the influence of the resin on the preform response is taken into account through the hydrostatic resin pressure (Terzaghi, Peck, and Mesri 1967; Gutowski et al. 1987; Kempner and Hahn 1998):

$$\begin{vmatrix} \sigma_f^f = \sigma_{ef}^f - sp_f^f\ \mathbf{I} & \text{in the wet preform} \\ \sigma_f^f = \sigma_{ef}^f & \text{in the wet preform} \end{vmatrix}, \tag{20.10}$$

where σ_{ef}^f is the effective stress in the preform skeleton (reinforcement), s is the saturation level, p_f^f is the resin pressure in the wet preform, and $\mathbf{I}$ is the identity tensor. A Biot's model could also be considered, this does not change fundamentally the formulation presented here, but in that case a further coefficient introduced must be identified from proper experimental models.

The response of the preform under compression can be considered as transversely isotropic, the isotropy plane being the preform plane. The preform stiffness in the plane direction is well known and assumed constant. For the normal direction, in order to express a consistent constitutive law from experimental data, and because Cauchy stresses are necessary, the conjugated logarithmic strains are naturally chosen. Finally, a co-rotational formulation must be considered in order to deal properly with large rotations and large displacements (Celle, Drapier, and Bergheau 2008a).

Preform Filling

The inability to account for the change in saturation level is one of the main limitations encountered at the moment with traditional infusion or injection simulations. This lack may be due to the difficulty in measuring the saturation itself, defined as the ratio of pore volume occupied by the resin over the total pore volume. From a physical point of view, this saturation is likely to be progressive. According to Spaid, Phelan, and Frederick (1998), the relation between pressure and saturation is governed by tension-surface related effects. Therefore, in a transient approach a further relationship between the pressure field and the saturation degree must be characterized from saturated and unsaturated flow behaviors in porous media (Breard, Henzel, and Trochu 1999). Currently, the lack of information concerning the relation between pressure and saturation leads to use the so-called slug-flow assumption. This hypothesis yields a direct binary relationship between the saturation level $s(x, t)$ and the hydrostatic pressure $p_r (x, t)$:

$$\begin{vmatrix} s(x,t) = 1 \text{ for } p_r(x,t) \neq 0 \\ s(x,t) = 0 \text{ for } p_r(x,t) = 0 \end{vmatrix}. \tag{20.11}$$

This hypothesis eliminates one degree of freedom in the finite element formulation (Michaud and Mortensen 2001). The corresponding numerical approach relies on a control volume associated with a degree of freedom in saturation (or pressure) that permits determining the flux between the boundaries of this volume (Loos, Rattazzi, and Batra 2002). As stated in the literature review section, mainly two types of numerical methods are associated with the slug-flow approach. Finite element/control volume is widely used since continuous pressure fields can be represented but require a further "mesh" with Voronoï cells, and more importantly yields mass balance problems. On the contrary, the nonstructured elements method, used in PAM-RTM (Trochu, Ferland, and Gauvin 1997), relies on the existing finite element mesh but leads to discontinuous pressure fields. Some other methods have been reported in the literature to solve these mass balance problems issues. At the node/element numbering level, Joshi, Lam, and Liu (2000) proposed an element sorting, Kang and Lee (1999) suggested a refinement of the flow front by node replacement without any increase in the global system size. A mesh refinement can also be operated, such as suggested in the literature for other engineering fields: X-FEM (Chessa and Belytschko 2003) and ALE (Belytschko, Liu, and Moran 2000).

However, in all these methods the fluid velocity field cannot be computed directly. In the present model, the quantity of resin must be assessed precisely, especially the amount of resin transferred from the resin zone to the preform zone. This is the key in coupling fluid and porous mechanics. For that reason, a mixed formulation for the Darcy equations has been implemented that consists in computing simultaneously both pressure and velocity

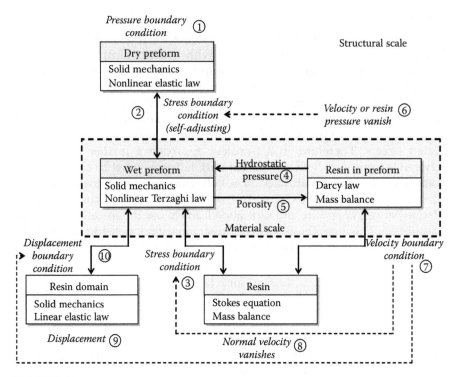

FIGURE 20.10
Boundary conditions and interactions at material and structural scales. (Courtesy of Elsevier.)

fields using P1-bubble/P1 or P1+/P1 finite elements (see Implementation and Numerical Aspects section and Celle, Drapier, and Bergheau 2008b).

Boundary Conditions

Assessing boundary conditions is a major issue, especially on the moving interface between resin and wet preforms. These boundary conditions at the structural scale along with coupling conditions between material regions are reported in Figure 20.10.

The structural boundary conditions can be split into two different categories. The first type of conditions is specific to the stress vector continuity (② and ③). The interface conditions between the resin and the wet preform area are mentioned in the literature as Beaver–Joseph–Saffman conditions for porous elastic soils. Classically, these conditions include the continuity of the normal velocity for the resin mass balance (⑦), and the stress vector continuity for the momentum conservation between the wet preform and the purely fluid area (③). A further condition is usually appended that concerns the sliding condition on the interface; that is, the tangential velocity (Jäger and Mikelic 2001; Layton, Schieweck, and Yotov 2003; Rivière and Yotov 2005). Here, the sliding effect between both areas is not constrained since normal infusion dominates the flow. These Beaver–Joseph–Saffman conditions are completed with a continuity of the resin hydrostatic pressure.

The second type of boundary conditions comes directly from the physics of the LRI/RFI process. Concerning the resin in the pure resin area, the mass balance implies continuity of the velocity field at the interface (⑦) with the wet preform domain, and a zero normal velocity on the bottom mold surface (⑧). The vacuum bag creates a mechanical boundary pressure on the surface of the dry preform (①), and this pressure must tend toward zero on the flow front (⑥). Finally, the resin domain in the pure resin area, introduced to take into account the flow of resin in a moving frame, is bounded with Dirichlet conditions in displacement (⑨ and ⑩).

Coupling Conditions at the Material Scale

Interactions between preforms and resin are directly integrated in mass and momentum balance equations. At this material scale, these coupling conditions are first related to the pressure obtained from the ALE formulation of the mass balance for Darcy/Brinkman flows (Equation 20.5b; ④) and integrated in the Terzaghi's or Biot's model. Second, it is the influence of the pressure on the porosity change and therefore on the corresponding permeability (⑤).

Curing Stage: Thermochemical Modeling

In the curing stage, unlike for the infusion stage, three homogeneous domains are considered. Currently, the wet preform is modeled as a single homogeneous equivalent material since the resin flow is very slow. The properties of this theoretical material are obtained using appropriate mixture rules (Hassanizadeh 1983a; Hassanizadeh 1983b; Loos, Rattazzi, and Batra 2002). From there, classically the thermochemical phenomenas are governed by two macroscopic equations: a heat transfer equation and a curing equation.

Thermochemical Equations

The heat equation for the LRI/RFI modeling rests upon the first law of thermodynamics, by considering a Fourier's law for conduction and convective terms. It includes a source term representing the energy released by the curing reaction (Bergheau and Fortunier 2008):

$$\rho c \frac{DT}{Dt} = \sigma : D_{irrev} + \mathrm{div}(\lambda.\nabla T) + \Delta H \frac{D\alpha}{Dt}, \tag{20.12}$$

where c is the specific capacity, T is the temperature, λ is the thermal conductivity tensor, ΔH is the heat of reaction, α is the resin degree of cure, and D_{irrev} is the irreversible part of the strain-rate tensor. The mechanical dissipative term ($\sigma : D_{irrev}$) is usually neglected for that kind of flow.

In the literature, several models have been used to express the rate of the resin degree of cure. They are constructed by coupling Arrhenius' Laws with power laws. One of the most popular model is of Kamal-Sourour's type (Farina and Preziosi 2000b):

$$\frac{D\alpha}{Dt} = \frac{\partial\alpha}{Dt} + \mathbf{v}\cdot\nabla\alpha = \left(A_1 e^{\frac{E_1}{RT}} + A_2 e^{\frac{E_2}{RT}}\alpha^p \right)(1-\alpha)^q, \tag{20.13}$$

where A_i are preexponential constants, E_i are activation energies, and p and q are model constants. For the wet preform area, where both resin and preforms are present, equivalent thermal parameters can be evaluated from mixture rules (Hassanizadeh 1983a; Loos and MacRae 1996).

Relation With Mechanical Problem

Apart from the transport phenomena, the viscosity controls the coupling condition between mechanical and thermochemical phenomena. Park and Kang (2003) proposed an empirical equation giving viscosity as a function of both temperature and degree of cure along with resin specific constants.

Thermochemical Boundary Conditions

The boundary conditions for this thermochemical model are of two types: conditions imposed by the autoclave and by the process itself; that is, the temperature of the autoclave and conditions on the moving interfaces to guarantee the continuity of the temperature field. The degree of cure associated with the resin must be prescribed as an initial condition.

Implementation and Numerical Aspects

Accounting for the Pure Resin Region

Resin Flow

The incompressible fluid flow in the pure resin zone must be treated due to a particular formulation in order to satisfy the incompressibility (i.e., the mass balance). Here a mixed method is employed, based on the simultaneous resolution of the velocity and pressure fields in the resin using P1 + /P1 or P1-buble/P1 triangle or tetrahedron finite elements. In order to satisfy the Brezzi-Babuska stability condition (Arnold, Brezzi, and Fortin 1984; Pierre 1995), this specific family of finite elements is considered. These elements are based on a further velocity degree of freedom, introduced at the center of every element in the mesh that permits satisfying the Brezzi-Babuka condition and to overcome numerical problems due to null diagonal terms (Perchat 2000). The shape function associated with this new degree of freedom referred to as bubble velocity v_b equates 1 at the element centroid and vanishes on its edges. Then, the pressure (Equation 20.14) and displacement or velocity fields (Equation 20.15) approximations write for an element:

$$P(x) = \sum_{i=1}^{n^e} N_i(x) p_i, \tag{20.14}$$

$$\{v(x)\} = \{v_l(x)\} + \{v_b(x)\} = \sum_{i=1}^{n^e} [N_i(x)]\{v_{l_i}\} + [N_b(x)]\{v_b\}, \tag{20.15}$$

where the N_i's are the standard shape functions at the vertices of the triangle (tetrahedron), N_b is the shape function associated with the bubble velocity and p_i's and v_{li}'s are, respectively, the nodal pressures and velocities.

Figure 20.11 presents two examples of such bubble functions. The polynomial bubble function that corresponds to the shape functions product as the linear triangle is the most basic (Figure 20.11c). Arnold, Brezzi, and Fortin (1984) show that the hierarchical function (Figure 20.11b), equal to the minimum of the shape functions for the linear triangular element, yield excellent results even for coarse meshes. For some bubble functions that verify the orthogonality property, the elementary system to be solved for a linear element writes at the element level, after static condensation of the further nodal velocity:

$$\begin{bmatrix} [K_{ll}^e] & [K_{lp}^e] \\ [K_{pl}^e]^T & [C_{pp}^e] \end{bmatrix} \begin{Bmatrix} \{v_l^e\} \\ \{p^e\} \end{Bmatrix} = \begin{Bmatrix} \{F^e\} \\ \{0\} \end{Bmatrix}.$$

With $[C_{pp}^e] = [K_{pb}^e][K_{bb}^e]^{-1}[K_{bp}^e]^T$ and where $[K_{ll}^e]$ is the stiffness matrix associated with the linear velocity field, $[K_{lp}^e]$ is the incompressibility matrix associated with the linear velocity field, $\{v_l^e\}$ is a vector containing the

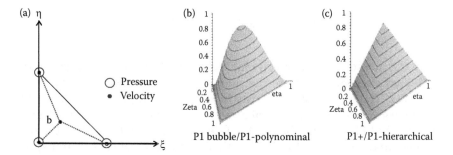

FIGURE 20.11
P1+/P1 and P1/bubble P1 finite element: (a) discretization, (b) P1–bubble/P1 element using a polynomial function, and (c) P1 + /P1 element using a hierarchical function.

element nodal velocities, $\{p^e\}$ contains the nodal pressures, $\{F^e\}$ contains the nodal loadings, and $\left[K^e_{bp}\right]$ and $\left[K^e_{bb}\right]$ contain, respectively, incompressibility and stiffness matrices associated with the bubble velocity field.

The bubble formulation employed in the treatment of the Stokes equations in stationary regime has also been used to deal with Darcy's equations in the saturated porous medium. Using the same approach simplifies the coupling between the Stokes and Darcy zones, and to propose the same method to solve for the flow in those two regions. Numerical results computed through these methods have been successfully validated by comparison with analytical results for test cases (Celle, Drapier, and Bergheau 2008a).

Accounting for Curvature Effects

The 2-D sections and 3-D models associated with composite pieces usually present curvature effects. These curvatures lead to zero normal velocity and displacements boundary conditions corresponding to impervious walls; that is to say, absence of normal flow. In order to take into account this type of condition, a penalty method has been first implemented in the Eulerian FE code. This method maintains the system size constant. Later, this method will improve by proposing a resolution of the kinematic condition directly in local frames associated with the nodes on boundaries. In this case, further operations are required for local/global referential switching. The resulting matrix system, however, is better conditioned and well suited for the use of iterative methods due to the disappearance of cross penalty terms.

Remeshing

Vanishing of the resin zone in the case of the RFI process must be represented. This leads to a major modification of the resin flow in the resin zone that must absolutely be integrated in order to obtain boundary conditions suitable to describe the flow in preforms. To deal with this disappearance, a remeshing technique is used. It is made up of three steps. The first step consists in defining a remeshing criterion Q_{mesh}. Here, the following angular criterion was chosen for triangular linear elements:

$$Q_{mesh} = min(Q_{elt}) = min\left(\frac{\alpha_{min}}{\beta_{max}}\right), \tag{20.16}$$

where α_{min} and α_{max} are, respectively, the smallest and largest angles of the considered element. The more the shape factor Q_{elt} is close to 1, the better its quality. The second step consists in the remeshing itself. To do so, the contour associated with the initial mesh is exported from the finite element libraries PRO-FLOT, remeshed in the free mesher GMSH and then the new mesh is imported back into PRO-FLOT. The last step consists in transporting the nodal values from the old mesh into the new mesh. Every degree of freedom of the new mesh is projected in the old mesh and the element shape functions are used to recalculate the new nodal values.

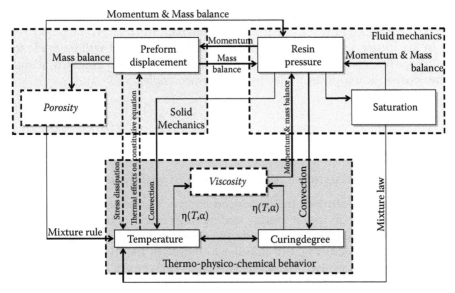

FIGURE 20.12
Coupling between thermo–physical–chemical, fluid and solid mechanical behaviors.

Overview: Numerical Implementation

Figure 20.12 schematically depicts a synthesis of the proposed model, for a better understanding of the relationships between the various phenomenas described in the previous sections.

The bold boxes contain independent variables, either scalar or vectorial. Each variable is connected with conservation equations and physical laws. The viscosity is not an independent variable since there is an explicit dependence upon temperature and degree of cure. Similarly, there is a direct dependence of the porosity on the displacement (Equation 20.8. Summarizing, this model consists of seven independent variables (three displacements, resin pressure, saturation, temperature, and degree of cure) for a 3-D problem. If velocities are to be taken into account, for a pressure–velocity formulation, an adequate numerical scheme must be implemented. It must be pointed out that solving simultaneously for nonlinear solid as well as for fluid mechanics leads to severe numerical problems, mainly due to variable scaling difficulties.

Concerning the thermochemical problem, the strong coupling between temperature and degree of cure with transport phenomena requires a suitable formulation to treat the convection–diffusion equation (Brooks and Hughes 1982).

One of the innovative features of the presented model is the coupling between nonlinear poroelasticity and fluid mechanics in a strong form. The iterative updated Lagrangian and Darcy ALE formulations implemented in PAM-RTM have been strongly coupled using an iterative method to study the

LRI/RFI processes, and a mixed ALE Stokes formulation has been employed to deal with fluid flow outside the porous media.

In the future, in order to improve the model an attempt will be made to reinforce this coupling. As for the strong coupling between mechanical and thermochemical aspects, it may be even more complex.

Examples of Infusion-Based Processes Simulation

Since the main feature of the developed approach is its ability to simulate the infusion of resin into preforms undergoing compaction, infusion-based processes are considered now. In these processes, many variants are used for actual manufacturing, but in order to highlight the main characters of industrial processes, two representative types are considered: they will be referred to as LRI and RFI; (Figures 20.6 and 20.7).

The main difference between both process types comes from the location of the resin zone with respect to the preforms and consequently with respect to the external mechanical pressure applied onto the vacuum bag surrounding the stacking. In LRI-like processes (Figure 20.7), resin is placed on top of the preforms lay-up and then undergoes a mechanical pressure from the vacuum bag. In RFI, the neat resin is placed at the bottom of the stacking and then is submitted to a pressure resulting from the transient equilibrium of forces induced by the external pressure on top of the stacking (Figure 20.6). This difference will change the way the process can be simulated and the corresponding numerical operations.

Using the numerical approach presented briefly above and detailed in Celle, Drapier, and Bergheau (2008a, 2008b), dry-route processes can be simulated. It has to be noticed that injection-based as well as infusion-based processes, or any process combining infusion and injection, can be simulated by the presented model. To illustrate this, some results of resin injection in T-shaped preforms under the vacuum bag are presented below. Then the LRI and RFI-like manufacturing simulation of a curved piece are detailed.

Vacuum Bag RTM of a T-Shape

Simulating such light processes using one-sided molds and vacuum bagging is a first step toward more complex process such as infusion. The processing conditions considered here may not be exactly the ones used in reality in processes such as the vacuum bag resin transfer molding (Kang, Lee, and Hahn 2001) or vacuum-assisted resin transfer molding processes (Brouwer, van Herpt, and Labordus 2003), but are very close. The light process described

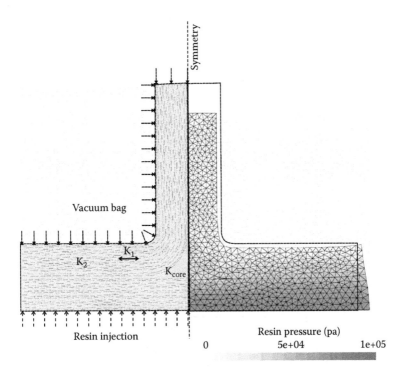

FIGURE 20.13
T-shape obtained with vacuum bag resin injection: (a) initial configuration, (b) final piece and resin pressure distribution.

here consists in injecting resin, under constant pressure or velocity, in preforms placed on a mold and simply vacuum-bagged (Figure 20.13). From a mechanical point of view, only solid and porous medium mechanics are required, this permits evaluating the robustness of this central piece of the simulation tool. Indeed, resin is simply represented through a pressure or velocity prescribed as a boundary condition, and the filling representation during compaction can be highlighted.

The resolution scheme for this simple case relies on both elasticity and poroelasticity problems in finite strains, solved implicitly. Convergence is evaluated regarding the total residual that is the sum of the residuals associated with the degrees of freedoms considered (i.e., displacements, velocities, and pressure). The iterative resolution scheme is presented in Figure 20.14, it consists in solving first the filling stage of the preform, knowing the geometry and permeability fields. This yields resin velocities and resin pressure inside the preforms. Then, the updated mechanical properties of the wet preforms can be considered in the global solid mechanics resolution that will give the updated geometry and permeability field.

In the presented 2-D case, a T-shape is made up of orthotropic fabrics, and a solid core is present in the inner part of the preform. Consequently,

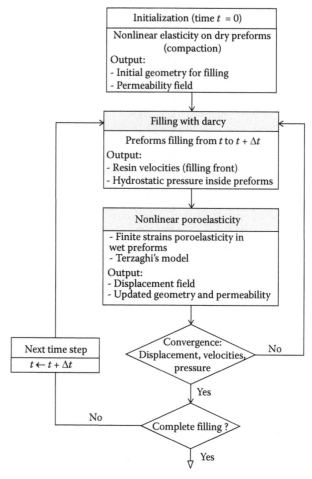

FIGURE 20.14
Resolution scheme for vacuum bag RTM-like process.

permeabilities are orientated and they are represented thanks to a Carman-Kozeny model. Kozeny constants are $h_{k1} = 100$, respectively, $h_{k2} = 10$, leading to initial permeabilities of $K_1 = 2.41E^{-13}\,\mathrm{m}^2$ in fabrics plane ($= K_{core}$), respectively, $K_2 = 2.41E^{-14}\mathrm{m}^2$ in the fabric transverse direction. These data are extrapolated from measurements achieved in previous studies on Hexcel's NC2 materials (Drapier et al. 2002; Elbouazzaoui 2004; Drapier et al. 2005). The dry preform orthotropic mechanical response is assessed using a classical loading machine and the corresponding nonlinear Caushy stress/logarithmic strains response is extracted. The initial porosity is 61.3%. Resin viscosity is considered constant, equal to 0.027 *Pa.s*, it corresponds to RTM6 resin viscosity at 120°C. In the considered case, the injection pressure equilibrates the atmospheric external pressure applied on the vacuum bag. Regarding the filling stage of interest here, isothermal conditions are considered.

One can verify that the compaction of the preforms leads to dimension reduction while the resin injected tends to make the piece swell on its lateral edge, which is left free. This dimension increase is in agreement with literature observations (Ahn, Seferis, and Letterman 1990; Kang, Lee, and Hahn 2001). There is clearly a competition between external mechanical pressure and the resin pressure induced by the forced flow into the preforms. Even in that simple case, as stated previously, finite strains and filling mechanisms have to be solved simultaneously.

LRI-Like Process Simulation

Liquid resin infusion is rather straight to simulate, in terms of boundary conditions and resin mass balance. Indeed, for any time step, the external pressure can be applied directly onto the pure resin region, representing the saturated distribution medium considered infinitely stiff. Solving the Stokes problem in this region yields resin pressures on the resin/preform interface that can then be used to evaluate the resin infusion into the preform through solving a Darcy's problem. Once infusion is finished for the considered time step, the local resin pressure is deduced. Solid mechanics is solved with updated mechanical equivalent properties (Terzaghi's model) and yields the new configuration verifying the equilibrium. The corresponding updated permeability is then easily deduced and can be used for the following time step. With respect to the algorithm used for the vacuum injection (Figure 20.14), this corresponds with solving Stokes equation for the distribution medium, prior to the filling stage with Darcy and solid mechanics equilibrium (nonlinear poroelasticity).

Half Curved Piece Infusion

Results of such simulation are illustrated for a half curved piece modeled in 2-D. As previously, orthotropic fabrics are considered. Figure 20.15a presents the initial configuration. It must be noticed that resin is considered here to be placed in sufficient quantity prior to the whole infusion process (i.e., no resin is fed in the system during filling). This corresponds to the schematic of Figure 20.7 with the resin pot that is left disconnected from the rest of the circuit. One can notice that after compression, the preforms thickness is reduced (Figure 20.15b). Infusion then takes place, and resin tends to move toward the concave place while saturated preforms undergo maximum loading in the convex lower location (Figure 20.15c).

Considering this resin pressure distribution, it can be inferred that filling will be rather long due to the resin heterogeneous distribution. Furthermore, industrial manufacturing conditions are far more complex than this first approach of the LRI-like processes and this can turn into large variations in terms of filling times.

To illustrate this, let us consider the same piece, but elaborated using a resin feeding system; that is, resin can be drawn from the pot in schematic of

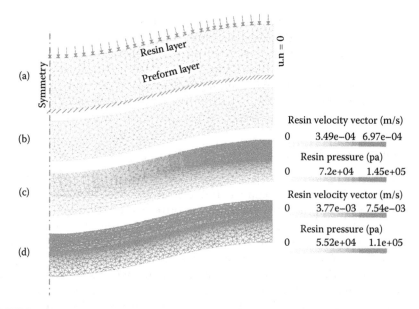

FIGURE 20.15
Modeling of a half-curved piece infusion with the LRI process: (a) preform and resin at initial time, (b) compression time, (c) final stage of infusion with resin initially placed, and (d) resin drawn from a resin pot.

Figure 20.7. Then boundary conditions will represent an impervious upper face of the distribution medium placed on top of the preforms and pressure of the inlet is controlled. This yields different pressure distribution (Figure 20.15d), and corresponding filling times are reduced by a factor of 2 due to the homogeneity of the filling (Figure 20.16b). On the contrary, porosities, or more generally the infused configuration, depend mostly on both fabrics and resin properties (Figure 20.16) and consequently will not change significantly.

Experiments Versus Simulation for a Flat Plate

In order to correlate the predictions of the presented model with experimental measurements, a simple geometry has been selected. A flat plate [90₆0₆]ₛ* made up of 24 "UD fabric" reference G1157 E01 produced by Hexcel Corp. was employed. These carbon fabrics are plain weave with 96% of weight in the warp direction and 4% of weight in the weft direction. For the resin, the experimental LRI tests have been performed using an epoxy resin (HexFlow© RTM-6). Before injection, the resin is preheated at 80°C in a heating chamber. The preform is heated by a heating plate located below the semirigid mold. As for the external pressure prescribed over the stacking, it is uniform and equal to the local atmospheric pressure, induced by the vacuum ensured in the sealed system. Figure 20.17 shows

---

* Following the standard notations, this corresponds to 6 plies at 90° wrt for the main direction, then 6 plies at 0°, and the symmetrical stacking wrt for the midplane.

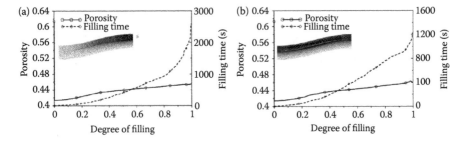

FIGURE 20.16
Filling times and porosities for the LRI process for a curved piece (Figure 20.15), representing a process with (a) resin placed initially in and (b) resin pot feeding the system.

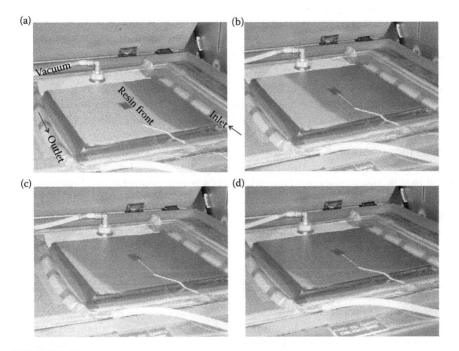

FIGURE 20.17
Infusion of a plate carried out by the LRI process at Hexcel Reinforcementsí facilities (Les Avenières, France): flow front advancement from (a) to (d). (From Wang, P., Drapier, S., Molimard, J., Vautrin, A., and Minni, J.-C., *Composites Part A: Appl. Sci. Manuf.*, 41(1), 36–44, 2010. With permission.)

an example of the infusion of a plate carried out by the LRI process, the resin inlet and outlet are indicated in this figure. The filling temperature is 120°C and the curing temperature is 180°C maintained for a period of 2 hours. The resin front advancement can be observed in the flow enhancement fabric at the macroscopic scale. This front is not homogeneous, this can have various origins such as an heterogeneous temperature field, resin entrance, or preform properties.

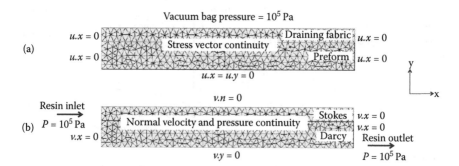

FIGURE 20.18

Mesh and associated boundary conditions for the simulation of the case realized experimentally (Figure 20.17): (a) solid and (b) fluid.

A comparison can be made between experiments and simulations of the filling stage for this case, as detailed in Wang et al. (2010). Simulations were achieved with 1180 triangle mixed velocity–pressure elements, considering a constant resin viscosity, and representing as precisely as possible the process boundary conditions (see Figure 20.18). Starting from the initial thickness measured, 9.8 mm, computations yield a thickness after compaction of 6.17 mm while the average value obtained from the thickness variation measured at the four edges of the plate with a Vernier caliper is 6.18 mm. After the filling stage, computations show a 0.63 mm expansion while a mean value of 0.59 mm is assessed experimentally with a fringe projection technique with a coefficient of variation of 21.2% for the measurements. Eventually, the mass of resin received by the system (plate + distribution layer + pipes) is 470 g experimentally and 410 g numerically for the plate. The final thickness of the cured plate, measured with a contact method at 25 locations over the surface, is 6.24 mm with a coefficient of variations of 8.9%.

Regarding the filling times, there is an order of magnitude between measurements and simulations. This is a critical issue of simulations, since it is related at the same time to simulations (mesh dependency, filling algorithm, macroscopical approach, boundary conditions representation, etc. Celle, Drapier, and Bergheau (2008a, 2008b) and experiments [permeability measurements; Breard et al. (2004a), Drapier et al. (2005), temperature control, boundary conditions control, etc.].

RFI-Like Process Simulation

Resin film infusion (RFI) processes are still more complex to simulate. The major difficulty consists in the proper determination of the resin pressure that controls the preform infusion. Indeed, this pressure results from the transient mechanical equilibrium of the preforms undergoing compaction, submitted to the external pressure and to the resin pressure induced by the preform filling. This implies solving three implicit problems; namely, filling, nonlinear poroelasticity, and Stokes before pressure is determined. This may lead to some convergence issues. Moreover, a further operation of remeshing of the resin zone is mandatory to follow the resin zone shrinkage, corresponding to the resin volume decrease that compensates the volume of resin that has entered the saturated preforms.

As one can verify in Figure 20.19c, the resin distribution is very close to the one for LRI (Figure 20.15c). However, resin pressures are not the same in both processes. Indeed, in RFI-like processes resin pressure is not controlled, it is a result, and it varies during filling as illustrated in Figure 20.20 for a flat plate. Even in this simple case, resin pressure drops after compaction, and keeps on decreasing as filling is going on. Final resin pressure value is about half of the resin pressure that one would consider in a classical simulation where resin would be replaced by a constant pressure boundary condition.

To illustrate this, consider this resin pressure as constant at the resin/ preform interface leading to the resin pressure distribution presented in Figure 20.19d. In that case, accordingly to the LRI-like process changes

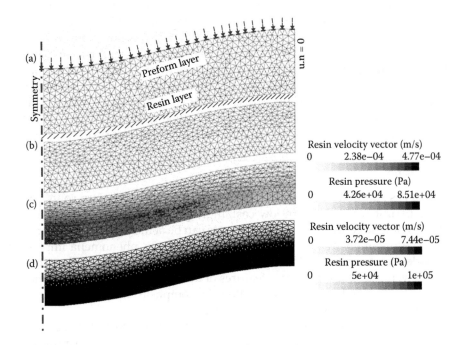

FIGURE 20.19
Modeling of a half-curved piece infusion with the RFI process: (a) preform and resin at initial time, (b) compression time, (c) final stage of infusion, and (d) resin replaced by a constant pressure.

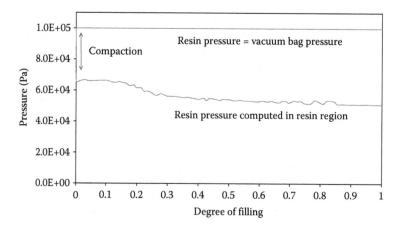

FIGURE 20.20
Pressure at the resin/preform interface in the RFI process during infusion of a flat plate—given pressure and pressure resulting from the complete simulation.

(Figure 20.15c and d), filling times will obviously be changed for 1 atm constant pressure, dropping from 2900 s to 1200 s, but here porosities will also be changed, from 0.41 to 0.45, since preforms will tend to swell under this excessive pressure. Indeed, in RFI-like processes, the final thickness of the preforms, and hence porosities, result from the competition of both the external pressure that tends to compact them, and the internal pressure induced by the resin filling them in. This is illustrated in Figure 20.21 where one can verify that after the compaction stage resin will infuse in preforms. Consequently, the resin zone will become thinner while preforms will swell, leading to larger porosities for higher pressures. In that case, resin is in excess at the end of the filling stage, accordingly to manufacturing processes where bleeders and vents permit evacuating bubbles by further circulation of the resin.

Conclusion

Industry is continually in need of HP composite structures. Dry route processes permit realizing, with low cost, high-quality complex shaped parts. At the moment, controlling these processes can be achieved only by simulating experimentally and numerically the multiphysics phenomena implied. A model has been presented here for simulating injection as well as infusion-driven processes. The main features of this model is to represent simultaneously dry preforms, the resin flow in composite fabrics undergoing compaction and fluid in pure resin regions.

Running these dry-route process simulations give access to important data that one would evaluate in order to optimize the process parameters regarding both process requirements and the final properties targeted. For instance, in the LRI and RFI-like processes, the external pressure required to properly fill in the preforms and final dimensions of the parts elaborated can

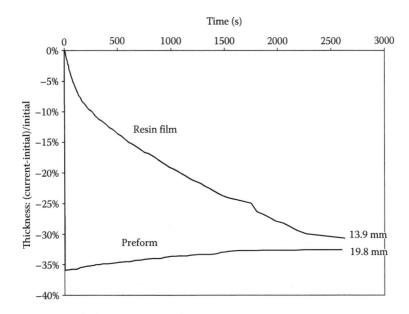

FIGURE 20.21
Preform and resin film thickness variation during the infusion process corresponding to
Figure 20.19c.

be computed since the global mechanical equilibrium is sought, resulting
from the competition between preform compaction under mechanical loading and preform swelling due to resin infusion. This numerical approach
must be completed by experimental characterization of the same physical
phenomena, measuring the most critical physical quantities. At the same
time, experimental approaches under progress will allow improving the
numerical tool, especially by refining the boundary conditions representation that were shown to drastically change filling times.

References

Abbès, B., R. Ayad, and A. Rigolot. Une méthode de pseudo-concentration pour la simulation 3D volumique du remplissage de moules d'injection. *Revue Européenne
des Eléments Finis* 8, no. 7 (1999): 695–724.
Acheson, J. A., P. Simacek, and S. G. Advani. The implications of fiber compaction
and saturation on fully coupled VARTM simulation. *Composites Part A: Applied
Science and Manufacturing* 35, no. 2 (2004): 159–69.
Advani, S. G., and M. V. Bruschke. A numerical approach to model non-isothermal
viscous flow through fibrous media with free surfaces. *International Journal for
Numerical Methods in Fluids* 19 (1994): 575–603.

Afendi, M., W. M. Banks, and D. Kirkwood. Bubble free resin for infusion process. *Composites Part A: Applied Science and Manufacturing* 36, no. 6 (2005): 739–46.

Ahn, K. J., J. C. Seferis, and L. Letterman. Autoclave resin infusion process: Analysis and prediction of resin content. *S.A.M.P.E. Quarterly* 21, no. 2 (1990): 3–10.

Ambrosi, D., A. Farina, and L. Preziosi. Recent developments and open problems in composites materials manufacturing. In *Progress in Industrial Mathematics*, 475–87. New York: Springer, 2002.

Antonucci, V., M. Giordano, L. Nicolais, A. Calabro, A. Cusano, A. Cutolo, and S. Inserra. Resin flow monitoring in resin film infusion process. *Journal of Materials Processing Technology* 143–144 (2003): 687–92.

Arnold, D. N., F. Brezzi, and M. Fortin. A stable finite element for the stokes equations. *Estratto da Calcolo* 21, no. 4 (1984): 337–44.

Barbero, E. J. *Introduction to Composite Materials Design.* Boca Raton, FL: Taylor & Francis Group, 1998.

Bathe, K. J., and H. Zhang. A flow-condition-based interpolation finite element procedure for incompressible fluid flows. *Computers & Structures* 80, nos. 14–15 (2002): 1267–77.

Bechet, E., E. Ruiz, and F. Trochu. Adaptive mesh generation for mould filling problems in resin transfer moulding. *Composites Part A: Applied Science and Manufacturing* 34, no. 9 SU (2003): 813–34.

Bechtold, G., and L. Ye. Influence of fibre distribution on the transverse flow permeability in fibre bundles. *Composites Science and Technology* 63, no. 14 (2003): 2069–79.

Belytschko, T., W. K. Liu, and B. Moran. *Nonlinear Finite Elements for Continua and Structures.* New York: John Wiley, Ltd., 2000.

Belytschko, T., C. Parimi, N. Moes, N. Sukumar, and S. Usui. Structured extended finite element methods for solids defined by implicit surfaces. *International Journal for Numerical Methods in Engineering* 56, no. 4 (2003): 609–35.

Bergheau, J. M., and R. Fortunier. *Finite Element Simulations of Heat Transfers.* New York: ISTE-J. Wiley, 2008. ISBN 9781848210530.

Berthelot, J. M. *Matériaux Composites: Comportement Mécanique et Analyse des Structures.* Paris: Tec & Doc Lavoisier, 2005.

Bird, R. B., W. E. Stewart, and E. N. Lightfoot. *Transport Phenomena.* New York: John Wiley, Inc., 2002.

Blest, D. C., S. McKee, A. K. Zulkifle, and P. Marshall. Curing simulation by autoclave resin infusion. *Composites Science and Technology* 59, no. 16 (1999): 2297–313.

Boisse, P., A. Gasser, and G. Hivet. Analyses of fabric tensile behaviour: Determination of the biaxial tension-strain surfaces and their use in forming simulations. *Composites Part A: Applied Science and Manufacturing* 32, no. 10 (2001): 1395–414.

Breard, J., Y. Henzel, and F. Trochu. A standard characterization of saturated and unsaturated flow behaviors in porous media. In *12th International Conference on Composite Materials ICCM12*, Paris, 1999.

Breard, J., Y. Henzel, F. Trochu, and R. Gauvin. Analysis of dynamic flows through porous media. Part I: Comparison between saturated and unsaturated flows in fibrous reinforcements. *Polymer Composites* 24, no. 3 (2004a): 409–21.

Breard, J., Y. Henzel, F. Trochu, and R. Gauvin. Analysis of dynamic flows through porous media. Part II: Deformation of a double-scale fibrous reinforcement. *Polymer Composites* 24, no. 3 (2004b): 409–21.

Brooks, A. N., and T. J. R. Hughes. Sreamline upwind/Petrov-Galerkin formulation for convection dominated flow with particular emphasis on the incompressible Navier-Stokes equations. *Computer Methods in Applied Mechanics and Engineering* 32 (1982): 199–259.

Brouwer, W. D., E. C. F. C. van Herpt, and M. Labordus. Vacuum injection moulding for large structural applications. *Composites Part A: Applied Science and Manufacturing* 34, no. 6 (2003): 551–8.

Buntain, M. J., and S. Bickerton. Compression flow permeability measurement: A continuous technique. *Composites Part A: Applied Science and Manufacturing* 34, no. 5 (2003): 445–57.

Celle, P., S. Drapier, and J. M. Bergheau. Numerical aspects of fluid infusion inside a compressible porous medium undergoing large strains. *European Journal of Computational Mechanics* 18, nos. 5–7 (2008a): 819–27.

Celle, P., S. Drapier, and J. M. Bergheau. Numerical modelling of liquid resin infusion into fiber preforms undergoing compaction. *European Journal of Mechanics/A* 27 (2008b): 647–61.

Cerisier, F. *Conception d'une structure travaillante en matériaux composites et étude de ses liaisons*. PhD Thesis, Ecole Nationale Supérieure des Mines de Saint-Etienne, 1998.

Charpentier, J.C. Elements de mécanique des fluides: Application aux milieux poreux. *Les Techniques de L'ingénieur* J1-065 (2004): 12.

Chessa, J., and T. Belytschko. An extended finite element method for two-phase fluids. *ASME Journal of Applied Mechanics* 70, no. 1 (2003): 10–7.

Comas-Cardona, S., P. Le Grognec, C. Binetruy, and P. Krawczak. Unidirectional compression of fibre reinforcements. Part 1: A non-linear elastic-plastic behaviour. *Composites Science and Technology* 67 (2007): 507–14.

Correia, N. C., F. Robitaille, A. C. Long, C. D. Rudd, P. Simacek, and S. G. Advani. Analysis of the vacuum infusion moulding process: I. Analytical formulation. *Composites Part A: Applied Science and Manufacturing* 36, no. 12 (2005): 1645–56.

Darcy, H. *Les Fontaines Publiques de la Ville de Dijon*. Paris: Dalmont, 1856.

Deleglise, M., C. Binetruy, and P. Krawczak. Simulation of LCM processes involving induced or forced deformations. *Composites Part A: Applied Science and Manufacturing* 37, no. 6 (2006): 874–80.

Deleglise, M., C. Binetruy, and P. Krawczak. Solution to filling time prediction issues for constant pressure driven injection in RTM. *Composites Part A: Applied Science and Manufacturing* 36, no. 3 (2005): 339–44.

Diallo, M. L., R. Gauvin, and F. Trochu. Experimental analysis and simulation of flow through multi-layer fiber reinforcements in LCM. *Polymer Composites* 19 (1998): 246–56.

Dopler, T., V. Michaud, and A. Modaressi. Simulation of non isothermal low pressure infiltration processing. In *ICCM12 Conference*, Paris, 1999.

Drapier, S., J.-C. Grandidier, and M. Potier-Ferry. A structural approach of plastic microbuckling in long fibre composites: Comparison with theoretical and experimental results. *International Journal of Solids and Structures* 38 (2001): 3877–3904.

Drapier, S., J.-C. Grandidier, and M. Potier-Ferry. Towards a numerical model of the compressive strength for long fibre composites. *European Journal of Mechanics* 18 (1999): 69–92.

Drapier, S., J. Monatte, O. Elbouazzaoui, and P. Henrat. Characterization of transient through-thickness permeabilities of non crimp new concept (NC2) multiaxial fabrics. *Composites Part A: Applied Science and Manufacturing* 36, no. 7 (2005): 877–92.

Drapier, S., A. Pagot, A. Vautrin, and P. Henrat. Influence of the stitching density on the transverse permeability of non-crimped new concept (NC2) multiaxial reinforcements: Measurements and predictions. *Composites Science and Technology* 62, no. 15 (2002): 1979–91.

Drapier, S., and M. R. Wisnom. Finite-element investigation of the compressive strength of non-crimp-fabric-based composites. *Composites Science and Technology* 59, no. 12 (1999a): 1287–97.

Drapier, S., and M. R. Wisnom. Finite element investigation of the interlaminar shear strength of non-crimp fabric based composites. *Composites Science and Technology* 59, no. 16 (1999b): 2351–62.

Dufort, L., S. Drapier, and M. Grédiac. The cross section warping in short beams under three point bending: An analytical study. *Composites Structures* 52, no. 2 (2001): 1233–46.

Elbouazzaoui, O. *Caractérisation de la perméabilité transverse de nouveaux renforts multiaxiaux cousus pour composites structuraux.* PhD Thesis. Saint Etienne, France: Ecole Nationale Supérieure des Mines de Saint-Etienne, 2004.

Farina, A., and L. Preziosi. Infiltration of a polymerizing resin in a deformable preform for fiber reinforced composites. In *Applied and Industrial Mathematics "Venice 2,"* 259–71. The Netherlands: Kluwer, 2000a.

Farina, A., and L. Preziosi. Non-isothermal injection molding with resin cure and preform deformability. *Composites Part A: Applied Science and Manufacturing* 31, no. 12 (2000b): 1355–72.

Folch Duran, A. *A numerical formulation to solve the ALE Navier-Stokes equations applied to the withdrawal of magma chambers.* PhD Thesis. Barcelona, Spain: Universitat Politècnica de Catalunya, 2000.

Fortin, A., A. Béliveau, and Y. Demay. Numerical solution of transport equations with applications to non-Newtonian fluids. In *Trends in Applications of Mathematics to Mechanics,* 311–22. Edited by M. D. P. M. Marques and J. F. Rodrigues. Pitman Monographs and Surveys in Pure and Applied Mathematics. London: Longman Group, 1995.

Frey, P. J., and P. L. George. *Maillages: Applications aux éléments finis.* London: Hermes Sciences Publication, 1999.

Gigliotti, M. *Modelling, simulation and experimental assessment of hygrothermoelastic behaviour of composite laminated plates.* PhD Thesis. Saint Etienne, France: Ecole Nationale Supérieure des Mines de Saint Etienne, 2004.

Grosset, L. *Optimization of composite structures by estimation of distribution algorithms.* PhD Thesis. Saint Etienne, France: Ecole Nationale Supérieure des Mines de Saint Etienne, 2004.

Gutowski, T. G., Z. Cai, S. Bauer, D. Boucher, J. Kingery, and S. Wineman. Consolidation experiments for laminate composites. *Composites Materials* 21 (1987): 650–69.

Hammami, A., R. Gauvin, and F. Trochu. Modeling the edge effect in liquid composites molding. *Composites Part A: Applied Science and Manufacturing* 29, no. A (1998): 603–9.

Han, N. L., S. S. Suh, J. M. Yang, and H. T. Hahn. Resin film infusion of stitched stiffened composite panels. *Composites Part A: Applied Science and Manufacturing* 34, no. 3 (2003): 227–36.

Hassanizadeh, S. M. General conservation equations for multi-phase systems, 1. Averaging procedure. In *Flow Through Porous Media,* 1–14. Edited by G. F. Pinder. Southampton: CML Publications, 1983a.

Hassanizadeh, S. M. General conservation equations for multi-phase systems, 2. Momenta, energy, and entropy equations. *Flow Through Porous Media*, 17–29. Edited by G. F. Pinder. Southampton: CML Publications, 1983b.

Hubert, P., R. Vaziri, and A. Poursartip. A two-dimensional flow model for the process simulation of complex shape composite laminates. *International Journal for Numerical Methods in Engineering* 44 (1999): 1–26.

Jäger, W., and A. Mikelic. On the roughness-induced effective boundary conditions for an incompressible viscous flow. *Journal of Differential Equations* 170, no. 1 (2001): 96–122.

Jedidi, J. *Etude du comportement en cyclage hygrothermique d'un matériau composite épais.* PhD Thesis. Saint Etienne, France: Ecole Nationale Supérieure des Mines de Saint-Etienne, 2004.

Joshi, S. C., Y. C. Lam, and X. L. Liu. Mass conservation in numerical simulation of resin flow. *Composites Part A: Applied Science and Manufacturing* 31, no. 10 (2000): 1061–8.

Joubaud, L., F. Trochu, and J. Le Corvec. Simulation of the manufacturing of an ambulance roof by vacuum assisted resin infusion (VARI). *Composites 2002 Convention and Trade Show, Composites Fabricators Association*, Atlanta, GA, 2002.

Kang, M. K., J. J. Jung, and W. I. Lee. Analysis of resin transfer moulding process with controlled multiple gates resin injection. *Composites Part A: Applied Science and Manufacturing* 31, no. 5 (2000): 407–22.

Kang, M. K., and W. I. Lee. A flow-front refinement technique for the numerical simulation of the resin-transfer molding process. *Composites Science and Technology* 59, no. 11 (1999): 1663–74.

Kang, M. K., W. I. Lee, and H. T. Hahn. Analysis of vacuum bag resin transfer molding process. *Composites Part A: Applied Science and Manufacturing* 32, no. 11 (2001): 1553–60.

Kaviany, M. *Principles of Heat Transfer in Porous Media.* New York: Springer-Verlag, 1995.

Kelly, P. A., R. Umer, and S. Bickerton. Viscoelastic response of dry and wet fibrous materials during infusion processes. *Composites Part A: Applied Science and Manufacturing Selected Contributions from the 7th International Conference on Flow Processes in Composite Materials held at University of Delaware* 37, no. 6 (2006): 868–73.

Kempner, E. A., and H. T. Hahn. A unified approach to manufacturing simulation for composites. In *Proceedings of the First Korea-U.S. Workshop on Composite Materials*, Seoul, Korea, 1998.

Kessels, J. F. A., A. S. Jonker, and R. Akkerman. Fully 21/2d flow modeling of resin infusion under flexible tooling using unstructured meshes and wet and dry compaction properties. *Composites Part A: Applied Science and Manufacturing* 38, no. 1 (2007): 51–60.

Layton, W. J., F. Schieweck, and I. Yotov. Coupling fluid flow with porous media flow. *SIAM Journal on Numerical Analysis* 40, no. 6 (2003): 2195–218.

Lee, D. H., W. I. Lee, and M. K. Kang. Analysis and minimization of void formation during resin transfer molding process. *Composites Science and Technology* 66 (2006): 3281–89.

Lim, S. T., M. K. Kang, and W. I. Lee. Modeling of void formation during resin transfer moulding. In *12th International Conference on Composite Materials ICCM12*, Paris, 1999.

Lim, S. T., and W. I. Lee. An analysis of the three-dimensional resin-transfer mold filling process. *Composites Science and Technology* 60, no. 7 (2000): 961–75.

Lin, M., H. T. Hahn, and H. Huh. A finite element simulation of resin transfer molding based on partial nodal saturation and implicit time integration. *Composites Part A: Applied Science and Manufacturing* 29, nos. 5–6 (1998): 541–50.

Loos, C., and J. D. MacRae. A process simulation model for the manufacture of a blade-stiffened panel by the resin film infusion process. *Composites Science and Technology* 56, no. 3 (1996): 273–89.

Loos, A. C., D. Rattazzi, and R. C. Batra. A three-dimensional model of the resin infusion process. *Journal of Composite Materials* 36, no. 10 (2002): 1255–73.

Lopatnikov, S., P. Simacek, J. Gillespie, Jr., and S. G. Advani. A closed form solution to describe infusion of resin under vacuum in deformable fibrous porous media. *Modelling and Simulation in Material Science and Engineering* 12, no. 3 (2004): 191–204.

Marsh, G. RP. Bigger blades—The carbon option. *Reinforced Plastics* 46, no. 3 (2002a): 20–2.

Marsh, G. RP. Composites on the road to the big time? *Reinforced Plastics* 47, no. 2 (2003): 33–6.

Michaud, V., and A. Mortensen. Infiltration processing of fibre reinforced composites: Governing phenomena. *Composites Part A: Applied Science and Manufacturing* 32, no. 8 (2001): 981–96.

Morel, S., C. Binetruy, and P. Krawczak. Compression des renforts dans les procédés lcm 1. analyse mécanique et microstructurale. *Revue des Composites et des Matériaux Avancés* 12, no. 2 (2002): 243–63.

Negrier, A., and J. C. Rigal. Présentation des matériaux composites. *Techniques de L'ingénieur* A7790 (1991).

Okabe, A., B. Boots, and K. Sugihara. *Spatial Tessellations: Concepts and Applications of Voronoi Diagrams.* New York: John Wiley, 1992.

Ouahbi, T., A. Saouab, J. Bréard, P. Ouagne, and S. Chatel. Modelling of hydro-mechanical coupling in infusion processes. *Composites Part A: Applied Science and Manufacturing* 38, no. 7 (2007): 1646–54.

Park, J., and M. K. Kang. A numerical simulation of the resin film infusion process. *Composites Structures* 60, no. 4 (2003): 431–7.

Perchat, E. *MINI-Element et factorisations incomplètes pour la parallélisation d'un solveur de Stokes 2D. Application au forgeage.* PhD Thesis. Paris: Ecole Nationale Supérieure des Mines de Paris, 2000.

Pierre, R. Optimal selection of the bubble function in the stabilization of the p1-p1 element for the stokes problem. *SIAM Journal on Numerical Analysis* 32, no. 4 (1995): 1201–24.

Porta, P. F. *Heterogeneous domain decomposition methods for coupled flow problems.* PhD Thesis. Augsburg, Germany: Universitat Augsburg, Mathematisch-Naturwissenschaftliche Fakultät, 2005.

Preziosi, L., and A. Farina. On Darcy's law for growing porous media. *International Journal of Non-Linear Mechanics* 37 (2002): 485–91.

Preziosi, L., D. D. Joseph, and G. S. Beavers. Infiltration of initially dry, deformable porous media. *International Journal of Multiphase Flow* 22, no. 6 (1996): 1205–22.

Qi, B., J. Raju, T. Kruckenberg, and R. Stanning. A resin film infusion process for manufacture of advanced composite structures. *Composites Structures, Special Issue Tenth International Conference on Composite Structures* 47, nos. 1–4 (1999): 471–6.

Rabier, S., and M. Medale. Computation of free surface flows with a projection FEM in a moving mesh framework. *Computer Methods in Applied Mechanics and Engineering* 192, nos. 41–2 (2003): 4703–21.

Rivière, B., and I. Yotov. Locally conservative coupling of Stokes and Darcy flows. *SIAM Journal on Numerical Analysis* 42, no. 5 (2005): 1959–77.

Ruiz, E., V. Achim, S. Soukane, F. Trochu, and J. Breard. Optimization of injection flow rate to minimize micro/macro-voids formation in resin transfer molded composites. *Composites Science and Technology* 66 (2005): 475–86.

Schrefler, A., and R. Scotta. A fully coupled dynamic model for two-phase fluid flow in deformable porous media. *Computer Methods in Applied Mechanics and Engineering* 190, no. 24–25 (2001): 3223–46.

Shojaei, A., S. R. Ghaffarian, and S. M. H. Karimian. Simulation of the three-dimensional non-isothermal mold filling process in resin transfer molding. *Composites Science and Technology* 63, no. 13 (2003): 1931–48.

Simone, A. Partition of unity-based discontinuous elements for interface phenomena: Computational issues. *Numerical Methods in Engineering* 20 (2004): 465–78.

Spaid, M. A. A., J. Phelan, and R. Frederick. Modeling void formation dynamics in fibrous porous media with the lattice Boltzmann method. *Composites Part A: Applied Science and Manufacturing* 29, no. 7 (1998): 749–55.

Srinivasagupta, D., B. Joseph, P. Majumdar, and H. Mahfuz. Effect of processing conditions and material properties on the debond fracture toughness of foam-core sandwich composite: Process model development. *Composite Part A: Applied Science and Manufacturing* 34 (2003): 1085–95.

Stewart, R. RP. SCRIMP offers a cleaner alternative. *Reinforced Plastics* 46, no. 5 (2002b): 26–9.

Swiatecki, S. RP. Vacuum moulding route chosen for rescue vessels. *Reinforced Plastics* 46, no. 5 (2002c): 30–2.

Terzaghi, K., R. B. Peck, and G. Mesri. *Soil Mechanics in Engineering Practice*. New York: John Wiley & Sons, 1967.

Trochu, F., P. Ferland, and R. Gauvin. Functional requirements of a simulation software for liquid molding processes. *Science & Engineering of Composite Materials* 6, no. 4 (1997): 209–18.

Trochu, F., E. Ruiz, V. Achim, and S. Soukane. Advanced numerical simulation of liquid composite molding for process analysis and optimization. *Composites Part A: Applied Science and Manufacturing Selected Contributions from the 7th International Conference on Flow Processes in Composite Materials held at University of Delaware* 37, no. 6 (2006): 890–902.

Vacher, S. *Capteurs à fibres optiques pour le contôle de l'élaboration et la caractérisation mécanique des matériaux composites*. PhD Thesis. Saint Etienne, France: Ecole Nationale Supérieure des Mines de Saint-Etienne, 2004.

Wang, P., S. Drapier, J. Molimard, A. Vautrin, and J.-C. Minni. Characterization of liquid resin infusion (LRI) filling by fringe pattern projection and in-situ thermocouples. *Composites Part A: Applied Science and Manufacturing* 41, no. 1 (2010): 36–44.

Zienkiewicz, O. C., and R. L. Taylor. *The Finite Element Method*, Vols 1–3. Oxford: Butterworth-Heinemann, 2000.

21

Low Melt Viscosity Imide Resins for Resin Transfer Molding

Kathy C. Chuang

CONTENTS

Introduction

The use of high temperature polymer matrix composites in aerospace applications has expanded steadily over the past 30 years, due to the increasing demand of replacing metal parts with lightweight composite materials for fuel efficiency and bigger payloads in aircraft and space transportation vehicles. Polyimide/carbon fiber composites, especially, have been regarded as major high temperature matrix materials, based on their outstanding performance in terms of heat resistance, high strength-to-weight ratio, and property retention, relative to epoxies (177°C/350°F) and bismaleimides (232°C/450°F) [1]. Traditionally, thermoplastic polyimides were prepared from dianhydrides and diamines in N-methyl-2-pyrrolidinone (NMP) at room temperature to form polyamic acids, which were then imidized at 150°C to yield high molecular weight polyimides. However, the high-boiling solvent (NMP, BP = 202°C) is very difficult to remove, often leading to the formation of voids during composite fabrication.

PMR-15 Polyimide for Prepreg Application

In the early 1970s, PMR addition curing polyimides with reactive endcaps were developed at the Lewis Research Center (renamed NASA Glenn) to ensure the easy processing of imide oligomers in methanol during composite fabrication. Using the **PMR** approach (in-situ polymerization of monomer reactants), PMR-15 [2] was formulated from 3,3′,4,4′-benzophenonetetracarboxylic dimethyl ester (BTDE), methylene dianiline (MDA) with endo-cis-bicyclo[2.2.1]-5-heptene-2,3-dicarboxylic acid, methyl ester (nadic ester, NE) as the reactive endcap (Figure 21.1). The ratio of BTDE:MDA:NE corresponded to n:n + 1:2, where n is the repeat unit of the oligomer. For PMR-15, n = 2.087, which essentially yields a formulated molecular weight (FMW) of 1500 g/mole. The FMW of a PMR polyimide can be calculated as follows:

$$\text{FMW} = 2(\text{MW of endcap}) + n(\text{MW of dianhydride derivative})$$
$$+ (n + 1)(\text{MW of diamine})$$
$$- 2(n + 1)(\text{MW of water} + \text{MW of alcohol}).$$

As shown in Figure 21.1, the monomers were first dissolved in low-boiling methanol to form a solution, which was then painted onto the surface of various carbon fibers or fabric reinforcements to form a prepreg. The monomers were polymerized in-situ within the stacks of prepregs upon heating to form

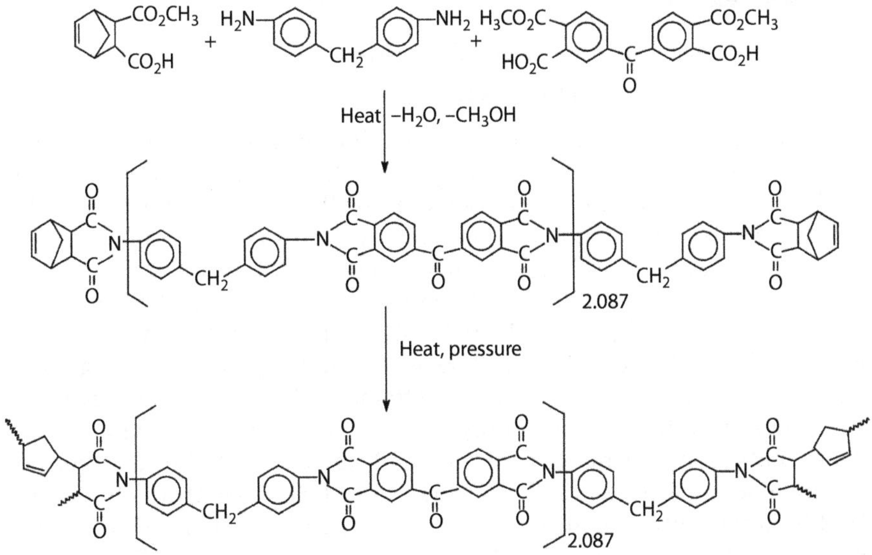

FIGURE 21.1
Composition and processing of PMR-15. (*Source*: Chuang, K. C., Waters, J. E., Hardy-Green, D., *Soc. Adv. Mat. Proc. Eng. Ser.*, **42(2)**, 1283, 1997.)

low molecular weight oligomers [3], which facilitate easier processing of laminates. At the final stage of curing, the reactive nadic endcaps of the imide oligomers were crosslinked under pressure (200 psi) and heat (316°C/600°F) to form polyimide composites. The ease of alcohol removal during processing is in the order of methanol > ethanol > isopropanol, following the vapor pressure of these alcohols. Solution stability of the monomer solutions and the shelf life of prepregs follow in the order of isopropyl ester > ethyl ester > methyl ester, because of the slower reaction between isopropyl esters of dianhydrides and the nadic endcap with diamines that prevents the aging and precipitation of the resin solution [4]. Nevertheless, isopropanol is more difficult to remove than either ethanol or methanol. The curing of nadic endcaps is very complicated and is believed to involve several possible pathways including the retro-Diels-Alders reaction, the addition of cyclopentadiene to either bismaleimides or the unreacted nadic unit [5–9], and a simple curing of double bonds of the nadic endcap [10].

PMR-15 offers easy processing and good property retention at a reasonable cost; thus, it is widely used in aircraft engine components and has been recognized as the state-of-the-art composite material for long-term use (thousands of hours) at 288°C (550°F). However, MDA in PMR-15 is a known toxin to the liver; therefore, it requires stringent safety regulations, which increases the manufacturing cost of PMR-15 composites.

Second Generation of PMR-Polyimides
Based on 6F-Dianhydride (6FDA)

To increase the high temperature stability of PMR-15 beyond 288°C (550°F), a second generation of PMR polyimide, PMR-II-50 [11], was developed for 315°C (600°F) applications. PMR-II-50 was formulated with 4,4′-(hexafluoroisopropylidene)diphthalic acid, dimethyl ester (HFDE), *p*-phenylenediamine (*p*-PDA) with n = 9, and the NE as the endcap. Upon curing, PMR-II-50 yielded a backbone similar to DuPont's thermoplastic polyimide Avimid N (i.e., NR-150B2) that was based on HFDA and a 95/5 mixture of *p*-PDA and *m*-PDA [12]. However, the approach of using nadic endcapped oligomers produced in methanol offered improved processability over the thermoplastic polyimide in NMP. Since it is known that the aliphatic components of the nadic endcap contributed to the thermooxidative degradation of the PMR polyimides [13], other efforts to modify PMR-II-50 concentrated on changing the endcap (Table 21.1) from the NE to other endcaps containing aromatic moieties; such as 4-amino-[2.2]-*p*-cyclophane (CYCAP) [14], *p*-aminostyrene (V-CAP) and 4-phenylethynylphthalic acid, methyl ester (PEPE) [15], and 3-phenylethynylaniline (PEA) [16]. Another approach was to reduce the amount of nadic endcaps used to half of that in PMR-II-50 as demonstrated in AFR700-B polyimide [17]. A systematic comparison of several 6F-polyimide composites terminated with NE, p-aminostyrene (V-CAP), 4-phenylethynylphthalic ester (PEPE), and 3-phenylethynylaniline (PEA) endcaps, and processed under the same conditions were presented in Tables 21.2 and 21.3. The T_gs of these polyimides ranged from 330 to 388°C after postcured at 371°C for 16 hours. (Table 21.2) Since these polyimides were all composed of 6F-dianhydride and *p*-PDA as the backbone except variations with ~10% of different endcaps, they all exhibited comparable weight loss [18] and similar mechanical properties. As shown in Table 21.3, higher T_g and optimal mechanical strength could be achieved in polyimide composites by air postcure at 371°C for 20 hours followed by nitrogen postcure at 399°C (750°F) for an additional 20 hours [19–22]. Nitrogen postcure was believed to involve reactions of free radicals trapped within the polyimide composites [23]. Nevertheless, prolonged nitrogen postcure at 399°C for 40 hours eventually resulted in lower mechanical strength, due to the degradation of polyimides at elevated temperatures.

The ease of processing for the polyimide composites with various endcaps followed the order of phenylethynyl > *p*-aminostyrene > nadic endcap. The imide oligomers usually melt around 220–240°C and the nadic endcap can be fully cured at 315°C (600°F) in 2 hours. However, the curing process of the phenylethynyl group (315–371°C) is slower than the nadic endcap and often requires 2 hours of curing at 371°C (700°F). In addition, the oligomers terminated with

TABLE 21.1

Second Generation PMR Polyimides Based on HFDE/*p*-PDA/Endcap

Resin Name	Formulated Oligomer Structure	$T_g{}^a$ (°C)
PMR-II-50 FMW[b] = 5044		345 (Ref. 15)
N-CYCAP-60 FMW[c] = 4982		341 (Ref. 14)
V-CAP-78 FMW[d] = 7874		341 (Ref. 15)
PEPE-II-78 FMW[e] = 7800		345 (Ref. 15)
HFPE-II-52 FMW[b] = 5216		362 (Ref. 19)

(Continued)

TABLE 21.1 (CONTINUED)

Second Generation PMR Polyimides Based on HFDE/*p*-PDA/Endcap

Resin Name	Formulated Oligomer Structure	$T_g{}^a$ (°C)
AFR-PE-4 FMW[f] = 3134		430[i] (Ref. 17)
PEA-II-75 FMW[g] = 7502		331
AFR-700B FMW[h] = 4382		380[a] 405j

[a] T_g determined by dynamic mechanical analysis (DMA) based on the onset decline of the storage modulus (G′), using a Rheometric 800 at a heating rate or 5°C/min in a torsional rectangular geometry at 1 Hz and 0.05% tension after specimens were postcured at 371°C for 16 hours in air.

[b] Monomer stoichometry: 2 endcap/9 HFDE/10 diamine.

[c] Monomer stoichometry: 2 endcap/8 HFDE/9 diamine.

[d] Monomer stoichometry: 2 endcap/15 HFDE/14 diamine.

[e] Monomer stoichometry: 2 endcap/14 HFDE/15 diamine.

[f] Monomer stoichiometry: 2 endcap/4 HFDE/5 diamine.

[g] Monomer stoichometry : 2 endcap/10 HFDE/9 diamine.

[h] Monomer stoichometry: 1 endcap/8 HFDE/9 diamine.

[i] Postcured at 343°C/2 hours, 357°C/2 hours, 371°C/2 hours, 385°C/2 hours, 400°C/2 hours.

[j] T_g after postcure for 24 hours at 427°C in air.

TABLE 21.2

T_gs of PMR Polyimide/T650-35 Composites Based on HFDE/p-PDA/Endcaps

Condition	No Postcure T_g (°C) G′[c] Tan δ	Air PC[a]/371°C/16 hours T_g (°C) G′[c] Tan δ	N₂ PC[b]/399°C/20 hours T_g (°C) G′[c] Tan δ
Resin	G′[c] Tan δ	G′[c] Tan δ	G′[c] Tan δ
PMR-II-50	332 365	346 421	408 451
V-CAP-75	322 348	341 366	397 433
HFPE-II-52	345 369	362 390	390 434
PEPE-II-78	319 352	345 379	371 410
PEA-II-75	317 342	331 388	371 432

[a] Air postcure = the polyimide composites were postcured in air at 371°C for 16 hours·

[b] Nitrogen postcure = the polyimide composites were postcured in air at 371°C (700°F) for 16 hours. followed by nitrogen postcure at 399°C (750°F) for 20 hours.

[c] G′ = the onset decline of storage modulus.

TABLE 21.3

Mechanical Properties of Polyimide/Unidirectional T650-35 Carbon Fiber Composites after Air and Nitrogen Postcure

Property Resin	Flexural Strength (MPa) 3-Point Bending			
Treatment	N₂ PC[a]/20 hours	N₂ PC/20 hours	N₂ PC[b]/40 hours	N₂ PC/20 hours
Test Temp	RT	316°C/600°F	316°C/600°F	371°C/700°F
PMR-II-50	1455 ± 41	862 ± 28	744 ± 50	676 ± 50
V-CAP-75	1296 ± 70	738 ± 50	703 ± 62	469 ± 4
HFPE-II-52	1317 ± 76	834 ± 80	689 ± 14	—
PEPE-II-78	1400 ± 62	910 ± 89	800 ± 28	324 ± 20
PEA-II-75	1413 ± 96	765 ± 48	717 ± 62	—
Flexural Modulus (GPa)				
PMR-II-50	119 ± 3	117 ± 7	97 ± 7	110 ± 14
V-CAP-75	114 ± 4	112 ± 3	103 ± 7	56 ± 3
HFPE-II-52	103 ± 6	102 ± 5	83 ± 7	—
PEPE-II-78	119 ± 3	117 ± 7	110 ± 7	32.4 ± 1
PEA-II-75	124 ± 14	97 ± 7	103 ± 14	—
Short Beam Shear Strength (MPa)				
PMR-II-50	79 ± 1	45 ± 3	51 ± 2	32 ± 2
V-CAP-75	70 ± 5	38 ± 3	43 ± 3	28 ± 1
HFPE-II-52	103 ± 6	43 ± 4	41 ± 7	—
PEPE-II-78	95 ± 6	48 ± 4	41 ± 7	—
PEA-II-75	96 ± 14	46 ± 3	46 ± 2	—

[a] The polyimide composites were postcured in air at 371°C (700°F) for 16 hours followed by nitrogen postcure at 399°C (750°F) for 20 hours to get optimal mechanical properties.

[b] The polyimide composites were postcured in air at 371°C (700°F) for 16 hours followed by nitrogen postcure at 399°C (750°F) for 40 hours.

either phenylethynyl or vinyl groups and exhibited more plasticity than the nadic endcap during processing. Moreover, the *p*-aminostyrene endcap cures around 150°C and usually produce polyimides with lower T_gs than the corresponding nadic endcap [15]. Due to the wider processing window and higher thermo-oxidative stability, phenylethynyl endcaps have gained a distinctive advantage for composite applications. In recent years, several formulations based on 6F-dianhydride, *p*-phenylenediamine, and 4-phenylethynyl endcap have been investigated, including HFPE-II-52 ($n = 9$) [19] and AFR-PE-4 ($n = 4$) [20,24,25], which is marketed as Cytec's 5700-4.

Solvent-Assisted Resin Transfer Molding (SARTM) and Resin Infusion (RI)

Although PMR-type polyimides have been successfully used in prepregs for composite applications, the aerospace industry often prefers low-cost manufacturing processes, such as resin transfer molding (RTM), vacuum-assisted RTM (VARTM), or resin film infusion (RFI). These methods have been successfully used for epoxy and bismaleimide composite fabrication. While the viscosity of PMR polyimides are too high to be amenable to RTM, PMR polyimide composites, prepared from the esters of benzophenone dianhydride (BTDA) [26] and 6F-dianhydride (6FDA) in alcohols, have been adapted for use with solvent assisted RTM (SARTM) and resin infusion (RI) to produce composites with 288–315°C performance capability. For example, a 70–75% monomer reactant solution in methanol was used to fabricate a GE center vent tube by SARTM [27]. Concentrated monomer solutions (>90%) have also been used to fabricate lightweight composite components with complicated geometry by RI [19]. However, SARTM processed PMR polyimide composites often contain a higher void content (≥5–10%) than what is acceptable for primary structures (<2%).

PMR Imide Powders for Resin Transfer Molding (RTM)

To avoid the use of solvent in PMR-type polyimides, another approach used the diester diacid of benzophenonetetracarboxylic dianhydride (BTDE) along with the NE endcap, similar to the making of the PMR-15 imidized powder, but substituted 4,4'-methylene dianiline (MDA) with a combination of a flexible 4,4'-[1,3-phenylene-bis(1-methylethylidene)] (i.e., Bisaniline

M) and rigid *m*-phenylenediamine (*m*-PDA) at a lower molecular weight formulation. These RTM processable PMR-type polyimides exhibited melt viscosities as low as 3–10 poise as measured by a Brookfield viscometer at 260°C. After curing at 315°C, these resins displayed glass transition temperatures (T_gs) in the range of 260–300°C, making them suitable for 260°C (500°F) applications [28].

Phenylethynyl Terminated Polyimides

Besides the nadic endcap used in PMR polyimides, other approaches to improve the processability of polyimides for composite applications include using imide oligomers terminated with acetylene (-C $\equiv$ CH) [29] and benzocyclobutane [30] as reactive crosslinking groups. However, the reactive acetylene terminal groups started to crosslink around 160–180°C [31], very close to the melting region of imide oligomers (195–200°C). This resulted in rapid molecular weight build-up and a very narrow processing window in systems such as the commercial Thermid series [32], making composite fabrication with these resins difficult.

To prevent premature curing, phenylethynyl terminated imide oligomers with 3-aminophenylethynylaniline (PEA) [33–36] and 4-phenylethynylphthalic anhydride (PEPA) endcaps [37–39] were developed to raise the curing temperature of the oligomers and thus widen the processing window. Since phenylethynyl terminated oligomers usually exhibited an exothermic maximum around 350–400°C [15] as indicated by differential scanning calorimetry (DSC), they provided a processing window about 100°C wider than acetylene endcaps. During the 1980s, NASA Langley had successfully developed a series of phenylethynyl terminated imide oligomers (most notably, PETI-5) and other phenylethynyl pendant oligomers [40–41] for long-term (60,000 hours) airframe application at 177°C (350°F). This work was supported under NASA's High Speed Civil Transport (HSCT) program with Boeing to build a Mach 2.4 supersonic commercial aircraft. PETI-5 polyimide resin is composed of 91 mole% of 3,3′,4,4′-biphenyltetracarboxylic dianhydride (*s*-BPDA), 85 mole% of 3,4′-oxydianiline (3,4′-ODA), 15 mole% of 1,3-bis(3-aminophenoxy)benzene (1,3,3-APB) and 18 mole% of PEPA at a formulated oligomer molecular weight of about 5000 g/mole (Figure 21.2). PETI-5 (T_g = 260°C) exhibited excellent toughness and adhesive properties (Table 21.4) as well as long-term mechanical property retention at 177°C (Table 21.5) [42,43]. However, PETI-5 prepregs contain about 22 weight% NMP, which often raised concerns about the volatiles and voids generated in building large airframe structures.

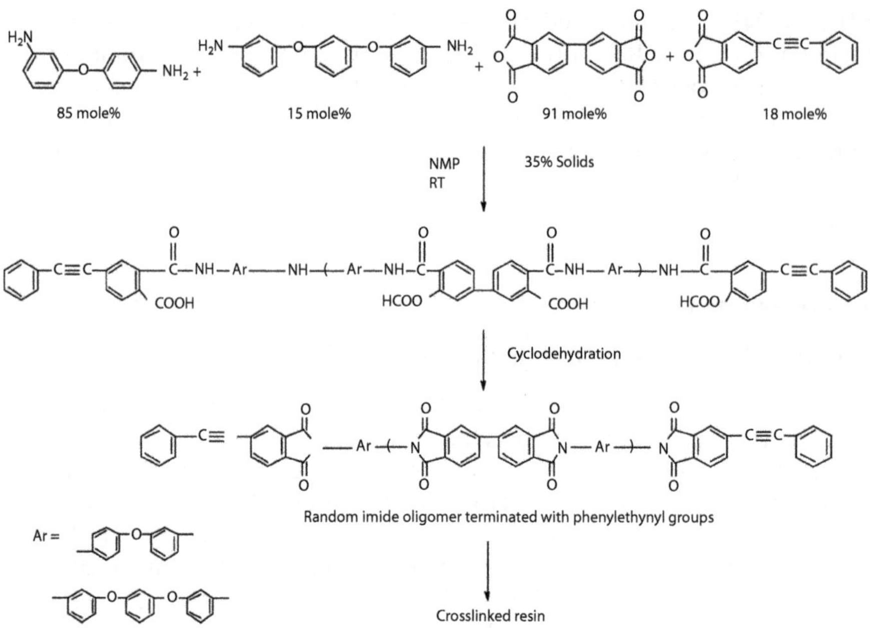

FIGURE 21.2
Synthetic route to PETI-5 polyimide resin. (*Source:* Hergenrother, P. M., *SAMPE J.*, **36(1)**, 30, (2000)).

TABLE 21.4

Lap Shear Adhesive Strength of PETI-5

Formulated MW	2500 g/mole	5000 g/mole	1000 g/mole
T_g (cured 1 hour @ 375°C)	275°C	270°C	271°C
Lap Shear Strength MPa (% of Cohesive Failure)			
Cured 1 hour @ 350°C			
RT	38 (70%)	53 (70%)	29 (20%)
177°C	31 (30%)	34 (30%)	20 (5%)
Cured 1 hour @ 375°C			
RT	40 (30%)	36 (80%)	14 (5%)
177°C	30 (30%)	26 (80%)	22 (5%)
Cured 1/2 hour @ 325°C			
and 1/2 hour @ 375°C	45 (70%)	44 (70%)	29 (20%)
RT	33 (30%)	26 (50%)	21 (70%)
177°C			
Cured 2 hours @ 316°C			
RT	45 (90%)	35 (10%)	29 (0%)
177°C	35 (20%)	34 (50%)	26 (30%)

Source: Cano, R. J. and Jensen, B. J., *J. Adhesion*, **60,** 113, (1997).

TABLE 21.5

IM-7/PETI-5 Laminate Properties

Mechanical Property (Normalized to 62% fiber volume)	2500 g/ mole	5000 g/mole	Lay-up
Open hole tension strength (KSI)			
RT (dry)	64.3	66.9	$(+45, 0, -45, 90)_{4S}$
177°C (dry)	63.3	65.5	(25/50/25)
Open hole tension strength (KSI)			
RT	83.8	80.8	$(+45, -45, 90, 0, 0, +45,$
177°C (dry)	82.1	81.9	$-45)_S$
			(38/50/12)
Open hole compression strength (KSI)			
RT	49.6	48.6	$(+45, 0, -45, 90)_{4S}$
177°C (wet)	31.8	34.5	(25/50/25)
Open hole compression strength (KSI)			
RT	54.6	53.5	$(+45, -45, 90, 0, 0, +45,$
177°C (wet)	38.2	42.9	$-45)_S$
			(38/50/12)
Open hole compression strength (KSI)			
RT	66.5	65.3	$(\mu45, 0,0, \mu45, 0, 0, \mu45,$
177°C (dry)	57.3	49.7	$0)_{2S}$ (58/34/8)
177°C (wet)	49.9	50.0	
Compression after impact strength (KSI)			
177°C (dry)	47.6	45.9	$(+45, 0, -45, 90)_{4S}$
177°C (wet)			
Compression after impact, modulus (MSI)			
RT (dry)	8.4	8.1	$(+45, 0, -45, 90)_{4S}$
Compression after impact, microstrain (µin./ in.)	5908	5986	$(+45, 0, -45, 90)_{4S}$ (25/50/25)
0° compression strength (KSI)			
RT (dry)	256	241	$(0)_{8t}$
0° Compression, modulus (MSI)			
RT (dry)	19.0	19.3	$(0)_{8t}$
0° Tension strength (KSI)			
RT (dry)	342.7	332.7	$(0)_{8t}$
0° Tension modulus (MSI)			
RT (dry)	21.6	22.8	$(0)_{8t}$
0° Tension strain, microstrain (µin./in.)	152.59	13860	$(0)_{8t}$
In-plane shear modulus (MSI)			
RT (dry)	0.77	0.61	(0/100/0)
177°C (dry)	0.62	0.50	
Interlaminar shear strength (KSI)			
RT (dry)	18.8	20.6	$(0)_{16t}$
Compressive interlaminar shear (KSI)			
RT (dry)	13.9	12.5	$(0)_{30t}$
177°C (wet)	8.6	6.8	
Thermal Cycling Microcracks/in.²	0	0	$(\mu45, 90, 0, 0, \mu45, 0, 0,$ $\mu45, 0)_{2S}$ (58/34/8)

Source: Hergenrother, P. M., *SAMPE J.*, **36(1)**, 30, (2000).

Phenylethynyl Terminated Imide Oligomers for Resin Transfer Molding (RTM)

Imide oligomers based on 3,3′,4,4′-biphenyl dianhydride (s-BPDA): PETI-298

After NASA's HSCT program ended, efforts were carried on to reduce the melt viscosity of PETI-5 imidized powder to meet the requirements of RTM by lowering the molecular weight of the oligomers [44,45], adding plasticizer [46], and changing the ratio of the monomers in PETI-5. Eventually, a solvent-free PETI-RTM resin powder (75% 1,3,3-APB and 25% 3,4′-ODA, $M_n = 750$ g/mole, $T_g = 258°C$) was demonstrated to be adaptable to low-cost processing [47]. To further increase the T_g and the use temperature, modified resins formulated with s-BPDA, PEPA along with a mixed ratio of the diamines, selected from 1,3-bis(3-aminophenoxy)benzene (1,3,3-APB), 1,3-bis(4-aminophenoxy)benzene (1,3,4-APB), 3,4′-oxydianiline (3,4′-ODA), and 3,5-diamino-4′-phenylethynylbenzophenone (DPEB) as a crosslinkable pendant group (Figure 21.3), were shown to exhibit low melt viscosity (<10 poise at 280°C for 1–2 hrs) that are amenable to low-cost RTM and resin infusion (RI) processes [48]. The use of a second crosslinkable pendent

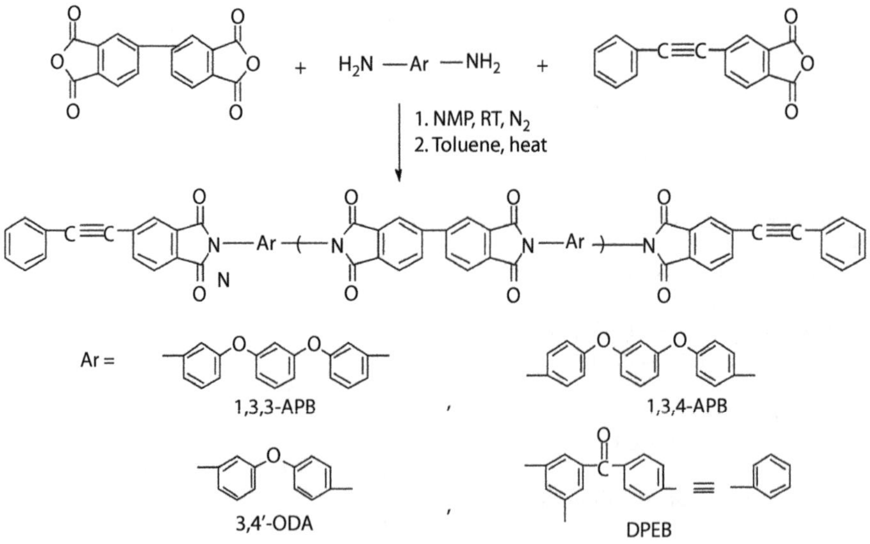

FIGURE 21.3
Synthesis of phenylethynyl containing imide oligomers based on 3,3′,4,4′-biphenyl dianhydride (s-BPDA). (*Source:* Smith, J. G., Jr., Connell, J. W., and Hergenrother, P. M., *Soc. Adv. Mat. Proc. Eng. Ser.*, **46**, 510, (2001)).

phenylethynyl monomer (DPEB) further raised the T_g of polyimide resins up to ~320°C, however, the increased crosslink density often led to microcracks (Table 21.6). Among these s-BPDA based resins, PETI-298 ($T_g = 298°C$ by DSC), formulated from s-BPDA (0.48 mole), 4-PEPA (1.04 mole), 1,3,4-APB (0.75 mole), and 3,4'-ODA (0.25 mole), displayed a low melt viscosity (6–10 poise) that is suitable for RTM processing and yielded the best overall mechanical performance without microcracking. PETI-298 possessed outstanding toughness at room temperature as evidenced by a very high open hole compression strength (421 MPa), because of multiple ether linkages (-O-) present in two diamines (Table 21.7). However, due to its T_g of 298°C, PETI-298 is limited to 450°F (232°C) applications, even though it has better properties than BMI and is less prone to microcracks.

In another attempt, low molecular weight resins ($n = 1$) were made from 6FDA, *m*-phenylenediamine, and PEPA in a ratio of 1:2:2 for RTM application. However, the viscosity of the 6F-RTM is too high unless the resin powder is made in NMP. While the residual NMP in the resin helped to lower its viscosity for RTM, the NMP is difficult to remove, and the void content of the resulting 6F-RTM composite is about 10% [49].

TABLE 21.6

Composition and Properties of Phenylethynyl Imide Oligomers Based on S-BPDA

Oligomer	Diamine Composition (%)	η^* @ 280°C[a] Pa-sec	Initial T_g[b] °C	Cure T_g[c] °C	Microcrack Cracks/cm
PETI-RTM	1,3,3-APB (75), 3,4-ODA (25)	0.6	132	258	—
P1	1,3,3-APB (65), 3,4-ODA (15), DPEB(20)	0.4	129	295	33
P2	1,3,4-APB (100)	16	123 (246)	302	—
P3	1,3,4-APB (75), 13,3-APB (25)	14	134	283	—
P4(PETI-298)	1,3,4-APB (75), 3,4-ODA (25)	0.5	139	298	0
P5	1,3,4-APB (75), 3,4- ODA (15), DPEB (10)	0.6	143	313	43
P6	1,3,4-APB (85), DPEB (15)	1.0	136 (236)	320	67

Source: Smith, J. G., Jr., Connell, J. W., Hergenrother, P. M., and Criss, J. M., *Soc. Adv. Mat. Proc. Eng. Ser.*, **46**, 510, (2001).

[a] Complex melt viscosity (η^*) of oligomers were measured by parallel plates at angular frequency of 100 rad/sec by rheometrics at a heating rate of 4°C/min.

[b] Initial T_g determined on oligomer powders by DSC at a heating rate of 20°C/min.

[c] Cured T_g determined on samples held in the DSC pan at 371°C for 1 hour.

TABLE 21.7

Properties of PETI-298/AS4-5HS Carbon Fabrics[a]

Properties	Test Temp. (°C)	PETI-298
Compression strength (MPa)	23	421
Compression modulus (GPa)	23	76
Open hole compression strength (MPa)	23	264
Open hole compression modulus (GPa)	23	45
Open hole compression strength (MPa)	288	178
Open hole compression modulus (GPa)	288	43
Short beam shear (MPa)	23	46.5
Short beam shear (MPa)	232	38.2
Short beam shear (MPa)	288	29.7

Source: Smith, J. G., Jr., Connell, J. W., Hergenrother, P. M., and Criss, J. M., *Soc. Adv. Mat. Proc. Eng. Ser.,* **46**, 510, (2001).

[a] Fabricated from unsized, AS4 carbon fabrics, 5 harness satin weave.

Imide Oligomer based on 2,3,3′,4′-biphenyl dianhydride (a-BPDA): (PETI-330, PETI-375)

Recent work indicated that replacing 3,3′,4,4′-biphenyl dianhydride (s-BPDA) with 2,3,3′,4′-biphenyl dianhydride (a-BPDA) in polyimides generally increased their solubility, reduced the viscosity, and raised T_gs of the polyimides, without affecting their performance and thermo-oxidative stability [50–53]. Therefore, NASA Langley reformulated PETI-298 by substituting a-BPDA with s-BPDA and confirmed the general observation that the a-BPDA formulation yielded a lower melt-viscosity in the oligomers and a higher T_g (312°C) in the cured resin (Figure 21.4 and Table 21.8). To increase the use temperature to > 300°C while preserving the toughness without sacrificing processsability, rigid 1,3-phenylenediamine (1,3-PDA) was used to replace the more flexible 3,4′-ODA in PETI-298 to achieve a higher T_g of 330°C. PETI-330, composed of a-BPDA (0.53 mole), PEPA (0.94 mole), 1,3,4-APB (0.5 mole), and 1,3-phenylenediamine (0.5 mole), displayed a low complex melt viscosity (η^*) of 1.8–3.0 Pa.s (18–30 poise) held at 280°C for 1 hour (Figure 21.5), and retained better mechanical properties [54] than PETI-298 at 288°C (550°F). However, after hot-wet cycles that simulate an actual engine environment, the composite properties of PETI-330 dropped significantly, because the wet T_g of PETI-330 is about 100–125°C lower than the corresponding dry T_g [55]. As a result, the properties of PETI-330 at 288°C after hot-wet cycles could not meet the performance standard of PMR-15 composites for engine applications. To further increase the T_g to meet 288–315°C use temperature demand, 2,2′-bis(trifluoromethyl)benzidine (TFMBZ) was introduced into the PETI-330 formulation in place of *m*-phenylenediamine (*m*-PDA) to stiffen the backbone. The twisted noncoplanar biphenyldiamine moiety of TFMBZ [56] not only increased the rigidity of the imide backbones, but the two trifluoromethyl groups on the biphenyl ring also raised the rotational barrier

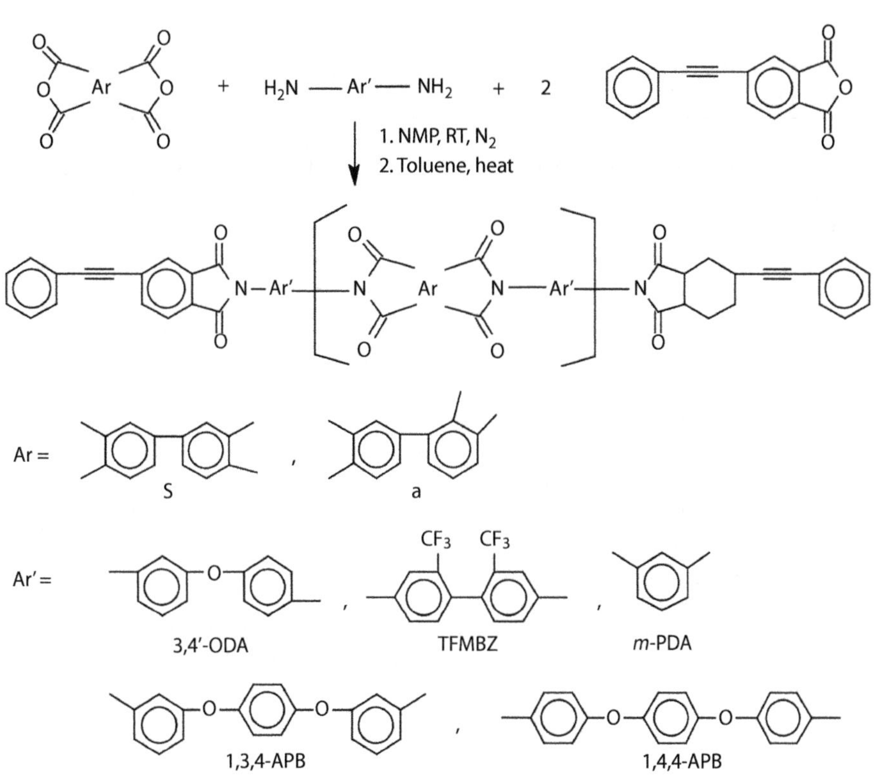

FIGURE 21.4
Synthesis of phenylethynyl terminated PETI-imide oligomers based on a-BPDA and s-BPDA. (From Smith, J. G., Jr., Connell, J. W., and Hergenrother, P. M., Yokota, R., Criss, J. M., *Soc. Adv. Mat. Proc. Eng. Ser.*, 47, 316, 2002. With permission.)

of the cured resins, which resulted in higher T_g. PETI-375 ($T_g = 375°C$ by DSC), consisting of a-BPDA (0.906 mole), PEPA (2.19 mole), 1,3,4-APB (1.0 mole), and TFMBZ (1.0 mole), yielded composites with outstanding property retention at 316°C (600°F) relative to PETI-330 [57]. However, the fluorine-containing TFMBZ diamine is very expensive and high resins costs have impeded the commercial viability of PETI-375.

Composite Fabrication and Mechanical Properties of PETI-298, PETI-330, and PETI-375 by RTM and Vacuum-Assisted RTM (VARTM)

PETI-298, PETI-330, and PETI-375 composite panels were fabricated by RTM, according to Lockheed Martin specifications with a flow rate of 500 cc/min, pressure of 2.75 MPa, and 288°C temperature capability [58], using a high

TABLE 21.8

Comparison of Complex Melt Viscosity [η*] and T_g of PETI-Imide Oligomers Based on **a**-BPDA vs. s-BPDA

Oligomer	Diamine Composition (%)	BPDA	η* @ 280°C/2 hours Pa–s[a]	Initial[b] T_g (T_m)/°C	Cured T_g[c] °C
PETI-298	1,3,4-APB (75), 3,4'-ODA (25)	s	0.6–1.4	139	298
P1	1,3,4-APB (75), 3,4'-ODA (25)	a	0.4–3.0	147	312
P2	1,3,4-APB	s	13.5–26	123 (246)	298
P3	1,3,4-APB	75% s, 25% a	0.4–0.7	149 (239)	301
P4	1,3,4-APB	50% s, 50% a	0.8–1.0	168 (222)	307
P5	1,4,4-APB (75), 3,4'-ODA (25)	a	41–480	ND[d]	339
P6	1,3,4-APB (75), m-PDA (25)	a	1.2–18	151	318
P7	1,3,4-APB (50), m-PDA (50)	a	0.8–3.0	ND (182)	330
P8	1,3,4-APB (75), TFMBZ (25)	a	0.3–1.4	ND (179)	320
P9	1,3,4-APB (50), TFMBZ (50)	a	0.6–2.0	ND (164)	345

Source: Smith, J. G., Cornell, J. W., Hergenrother, P. M., Yokota, R., and Criss, J. M., *Soc. Adv. Mat. Proc. Eng. Ser.*, **47**, 316, (2002).

[a] 1 Pa–s equal to 10 poise.
[b] Initial T_g determined on powdered samples by DSC at a heating rate of 20°C/min.
[c] Cured T_g determined on samples held in the DSC pan at 371°C for 1 hour.
[d] ND = not detected.

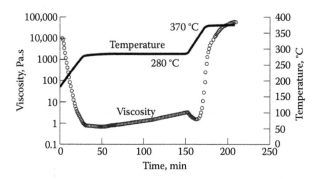

FIGURE 21.5
Complex melt viscosity of PETI-330 held at 280°C. (From UBE Web site for PETI-330. Courtesy of Dr. John W. Cornell at NASA LaRC.)

temperature injector made by Radius Engineering. Preforms were made from T650-35, 8 harness satin weave (8HS) or AS4 carbon fabrics with an 8 ply quasi-isotropic lay-up [+45/0/90/–45]$_s$. The sizing on the fabrics was removed by heating at 400°C under vacuum for 2 hours. The tool and the injector were preheated to approximately 315°C for 1 hour and held at temperature for 1.5 hours prior to injection. Then 600 g of the resin was charged into the chamber, heated to 288°C and held for 1 hour while degassing.

Subsequently, the molten resin was injected at 288°C at a rate of 200 cc/min and a hydrostatic pressure of 1.34–1.38 MPa into an Invar tool containing the preform. The PETI composite panels were cured at 371°C for 1 hour.

PETI-298, PETI-330, and PETI-375 all displayed outstanding toughness as evidenced by a very high initial open hole compression strength of 264, 270, and 297 MPa (Table 21.9), respectively. Owing to its lower T_g, the mechanical performance of PETI-298 (Table 21.7) based on s-BPDA could not match either PETI-330 or PETI-375, derived from a-BPDA [59]. PETI-330 displayed good mechanical properties up to 288°C whereas PETI-375 retained 53–55% of its initial room temperature properties at 315°C, due to its high T_g (Table 21.8). Furthermore, PETI-298, PETI-330, and PETI-375 all exhibited no microcracks and possessed better mechanical properties than BMI-5270-1.

The VARTM process has been used successfully in the processing of epoxy, cyanate esters, and bismaleimide resins with 150–200°C performance capability. However, it has not been able to process polyimide resins until the development of PETI-298 and PETI-330 imide resins with low complex viscosity (η^*) of 6–10 poise. PETI-298 has been fabricated by VARTM according to the set-up shown in Figure 21.6. A comparison of porosity of PETI-298 laminates produced by RTM and VARTM revealed that the void content is 1–2% by RTM and 4–7% by VARTM. Due to the higher porosity, the open hole compression strength of composites made by VARTM is slightly lower than those fabricated by RTM [60].

Recently, PETI-330 was also incorporated with carbon nanotubes [61] and nanofibers [62] to improve the thermal conductivity and electrical properties of the nanocomposites. However, the addition of nanofillers generally

TABLE 21.9

PETI Resins/Unsized T65-35 Laminate Properties

Property	Test Temp. (°C)	PETI-375	PETI-330
	23	297	270
	288	235	200
OHC strength (MPa)	316	160	—
	23	46	47
	288	45	44
OHC modulus (GPa)	316	42	—
	23	52	56
	232	40	43
	288	32	35
SBS strength (MPa)	316	29	—

Source: Connell, J. W., Smith, J. G., Jr., Hergenrother, P. M., and Criss, J. M., 49th International SAMPE Symposium, May 16–20, Long Beach, CA, 2004.

Note: OHC = Open-hole compression strength
SBS = Short-beam shear strength

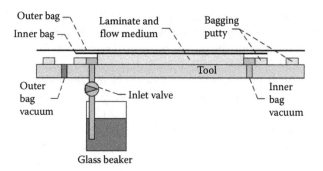

FIGURE 21.6
Diagram of VARTM set up. (Copy from Criss, J. M., Koon, R. W., Hergenrother, P. M., Connell, J. W., and Smith, J. G., Jr., *Soc. Adv. Mat. Proc. Eng. Ser.*, **33**, 1009, (2001). With permission.)

increases the viscosity, and the uniform dispersion and alignment of these nanofillers presents challenges in achieving optimal physical and mechanical properties.

RTM Imide Resins Based on Asymmetric Dianhydrides by a Solvent Free, Melt Process

PETI-330 and PETI-375 have been processed by RTM with T650-35 carbon fibers to yield composites with outstanding mechanical properties and low void contents (<2%) for applications between 260–315°C. However, the process to synthesize these imidized oligomers requires refluxing the monomers in NMP at ~185°C and removing water formed during the imidization by distillation of a toluene azeotrope. The oligomer solution was then poured into water to precipitate out the imide oligomer as a solid, which was subsequently dried in a forced air oven at 135°C for 24 hours to insure complete removal of any residual NMP (Figure 21.7). Such multistep preparation of PETI-resins is time consuming and costly.

RTM Imide Resins Based on 2,3,3′,4′-Biphenyl Dianhydride (a-BPDA)

To simplify the resin synthesis for cost saving, a different approach was taken at NASA Glenn by selectively using a series of kinked aromatic diamines and asymmetric biphenyl dianhydride (a-BPDA), along with PEPA to produce low melt viscosity imide resins by a solvent-free, melt process (Figure 21.8) [63]. The kinked diamines, 3,4′-oxydianiline (3,4′-ODA), 3,4′-methylenedianiline (3,4′-MDA) and 3,3′-methylenedianinline (3,3′-MDA), were mixed with a-BPDA and PEPA homogeneously as powders, and then melted between

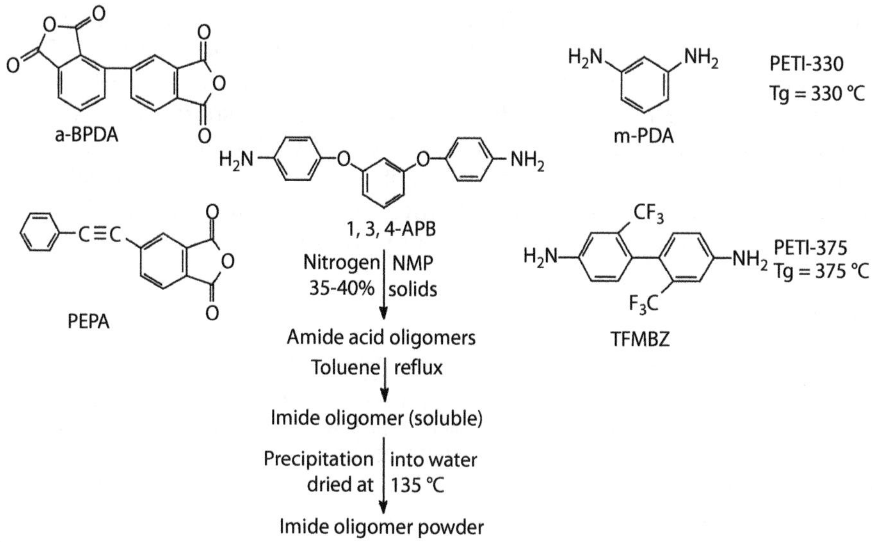

FIGURE 21.7
Synthesis of PETI-330 and PETI-375 imide resins.

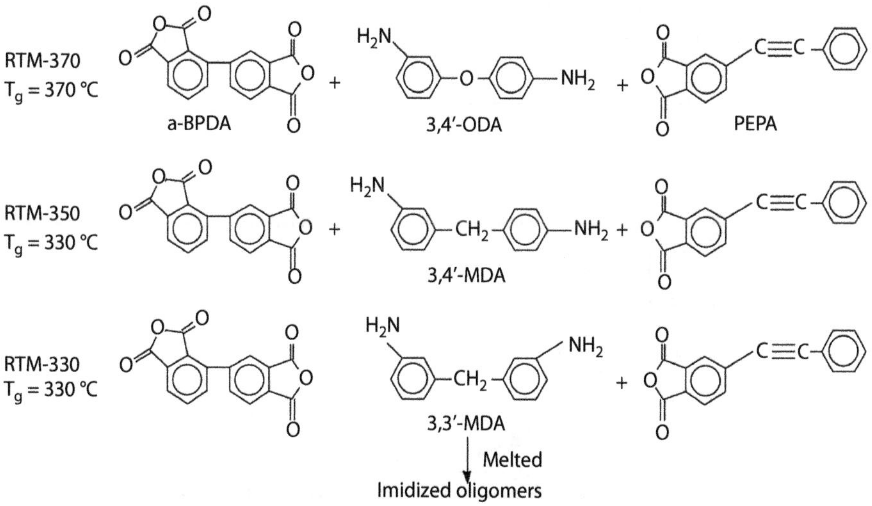

FIGURE 21.8
Synthesis of RTM imide resins based on a-BPDA. (From Chuang, K. C., Criss, J. M., Mintz, E. A., Scheiman, D. A., Nguyen, B. N., and McCorkle, L. S., Proceedings of Int'l SAMPE Symposium, Baltimore, MD, June 3–7, 2007.)

210°C and 270°C to afford imide resins with low melt viscosity ($\eta = 10$–30 poise by Brookfield viscometer at 280°C). Upon curing for 2 hours at 371°C and a postcure, a-BPDA based resins displayed T_gs in the range of 330–370°C (Table 21.10). The RTM resin containing 3,4'-MDA had a short pot-life due to rising viscosity compared to 3,4'-ODA and 3,3'-MDA (Figure 21.9). The methylene unit (-CH$_2$-), which can generate a stable benzylic radical, is believed to induce crosslinking, thus, contributing to the increase in viscosity [64]. The additional kink in 3,3'-MDA apparently contributes to its lower viscosity

TABLE 21.10

Physical Properties of Imide Oligomers/Resins Based on a-BPDA and PEPA

Resin	Diamine	Oligomer Min. η @ 280°C by Brookfield[a] (Poise)	Oligomer Min. Complex [η]* @ 280°C[b] (Poise)	Cured Resin T_g (°C) NPC[c] by TMA	Cured Resin T_g (°C) PC[d] @ 650°F by TMA[e]
RTM370	3,4'-ODA	14	11	342	370
RTM350	3,4'-MDA	7.4	20	338	350
RTM330	3,3'-MDA	1.5	6	288	330

Source: Chuang, K. C., Criss, J. M., Mintz, E. A., Scheiman, D. A., Nguyen, B. N., and McCorkle, L. S., 52th Int'l SAMPE Symposium, June 3–7, Baltimore, MD, 2007.
[a] Absolute viscosity measured by Brookfield Viscometer at 280°C.
[b] Complex viscosity measured by Aries Rheometer, using parallel plates.
[c] NPC = No post cure.
[d] PC = Post cured at 343°C (650°F) for 16 hours.
[e] TMA = Thermal mechanical analysis heated at 10°C/min, using expansion mode.

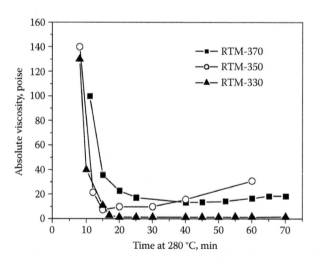

FIGURE 21.9

Absolute melt viscosities of RTM imide resins by Brookfield viscometer at 280°C. (From Chuang, K. C., Criss, J. M., Mintz, E. A., Scheiman, D. A., Nguyen, B. N., and McCorkle, L. S., Proceedings of Int'l SAMPE Symposium, Baltimore, MD, June 3–7, 2007.)

over 3,4'-MDA and 3,4'-ODA. On the other hand, RTM370 resin based on 3,4'-ODA exhibited higher T_g (370°C) than that of 3,4'-MDA ($T_g = 350$°C), and 3,3'-MDA ($T_g = 330$°C) containing resins. According to the molecular modeling of diamine monomers in polyimides, the bond angle of -O- linkage in oxyd-ianiline is 123° while the -CH$_2$- linkage in MDA is 111° [65]. When the reactive phenylethynyl endcaps in the oligomers are crosslinked, a higher rotational barrier is created for RTM 370 polymer chains within the network, due to the strain generated by the larger bond bending of 3,4'-ODA, relative to 3,4'-MDA, or 3,3'-MDA. RTM370 consisting of a-BPDA and 3,4'-ODA, especially, shows great promise for 315°C use temperature in aerospace applications.

RTM Imide Resins Based on 2,3,3',4'-Oxydiphthalic Anhydride (a-ODPA)

Similarly, asymmetric 2,3,3',4'-oxydiphthalic anhydride (a-ODPA) [66] could be formulated with 3,4'-ODA, 3,4'-MDA, 3,3'-MDA, and 3,3'-diaminobenzo-phenone (3,3'-DABP) to afford low melt viscosity ($\eta = 2$–4 poise at 280°C) imide resins (Figure 21.10) with T_gs in the range of 266–329°C (Table 21.11).

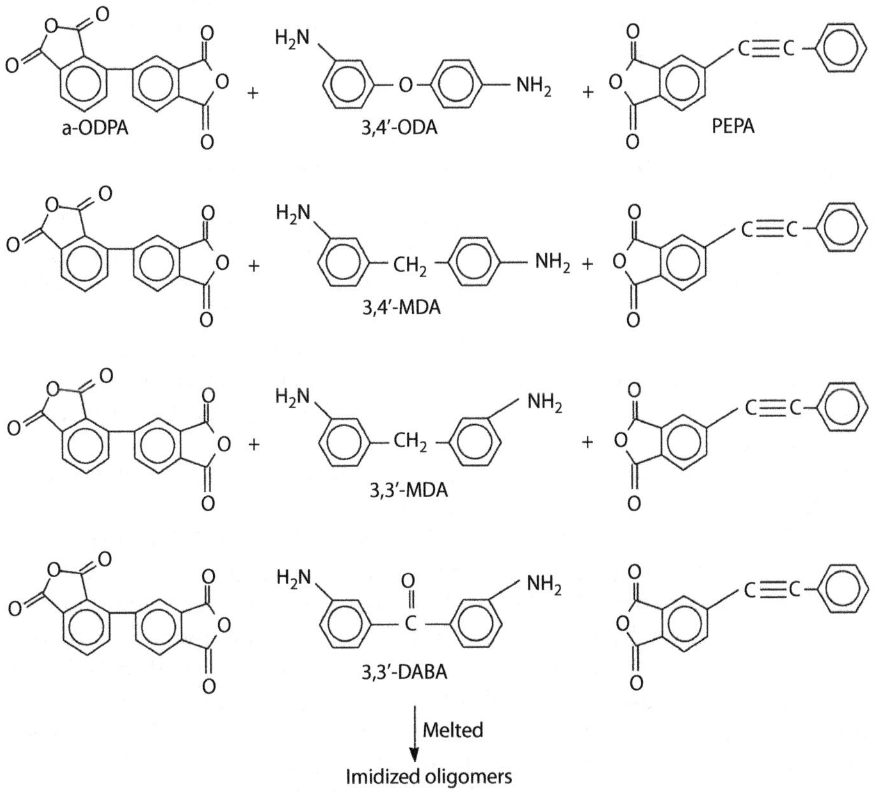

FIGURE 21.10
Synthesis of RTM imide resins based on a-ODPA.

TABLE 21.11

Physical Properties of Imide Oligomers/Resins Based on a-OPDA and PEPA

Dianhydride	Diamine	Oligomer Min. η @ 280°C by Brookfield[a] (Poise)	Oligomer Min. Complex [η]* @ 260°C[b] (Poise)	Cured Resin T_g (°C) NPC[c] by TMA	Cured Resin T_g (°C) PC[d] @ 650°F by TMA[e]
a-ODPA	3,4'-ODA	3.5	15.0	296	329
a-ODPA	3,4'-MDA	4.0	14.0	270	294[6]
a-ODPA	3,3'-MDA	2.5	3.0	273	266[6]
a-ODPA	3,3'-DABP	3.0	4.0	270	297

Source: Chuang, K. C., Criss, J. M., and Mintz, E. A., *Proceedings of 54th International SAMPE Symposium, May 18–21*, Baltimore, MD (2009).

[a] Absolute viscosity measured by Brookfield Viscometer at 280°C.
[b] Complex viscosity measured by Aries Rheometer, using parallel plates.
[c] NPC = No post cure.
[d] PC = Post cured at 343°C (650°F) for 16 hours.
[e] TMA = Thermal mechanical analysis heated at 10°C/min, using expansion mode.

The additional ether linkage (-O-) between the two phenyl rings in a-OPDA provides more flexibility in the imide backbones as compared to a-BPDA. This results in lower viscosity in the oligomers and lower T_gs in the cured a-ODPA resins.

Composite Properties of a-BPDA and a-ODPA Imide Resins

RTM processable imides based on a-BPDA and a-ODPA were infiltrated into T650-35/8 HS carbon fabrics composites with an 8 ply quasi-isotropic lay-up [+45/0/90/−45]$_s$ at 288°C, using a similar RTM technique used in processing PETI-330. After the tool and the injector were preheated to approximately 288°C, 600 g of the resin was injected at 1.38 MPa and then cured at 371°C for 2 hours. The resulting composite panels were post-cured in an oven at 343°C (650°F) for 8 hours to achieve the optimal mechanical properties at elevated temperatures.

RTM370, RTM350, and RTM 330 made from a-BPDA all exhibited outstanding open-hole compression strength ranging from 306, 285 to 252 MPa, relative to the state-of the-art RTM processable BMI-5270-1 (245 MPa) in the aerospace industry. Most noteworthy is the fact that RTM350 and RTM330, derived from 3,4'-MDA and 3,3'-MDA, respectively, retained better properties than RTM370 containing 3,4'-ODA. This evidence further confirms that methylene units (-CH$_2$-) enhanced the thermal stability of these polymers. However, the shorter pot-life of RTM350 and RTM330 induced by the methylene linkage made them more difficult to process and often resulted in poor composite quality. In contrast, RTM370 based on 3,4'-ODA was very processable and exhibited the best overall performance up to 315°C (Table 21.12).

TABLE 21.12

Composite Properties of T650-35 carbon fiber and Imide Resins Based on a-BPDA/PEPA compared to BMI-5270-1

Property	Test Temp. (°C)	RTM370 3,4'-ODA	RTM350 3,4'-MDA	RTM330 3,3'-MDA	BMI-5270-1 Cytec Resin
OHC strength	23	306	285	252	245
(MPa)	288	223	216	220	148
	315	166	199	185	—
OHC modulus	23	50	43	43	51
(GPa)	288	47	44	45	38
	315	42	45	50	—
SBS strength	23	62	58	57	37
(MPa)	288	43	32	38	14
	315	32	31	33	—

Source: Chuang, K. C., Criss, J. M., Mintz, E. A., Scheiman, D. A., Nguyen, B. N., and McCorkle, L. S., 52th Int'l SAMPE Symposium, June 3–7, Baltimore, MD, 2007.
Note: OHC = Open-hole compression
SBS = Short-beam shear

TABLE 21.13

Composite Properties of T650-35 Carbon Fiber/a-ODPA Imide Resins

Property	Test Temp. (°C)	Diamine 3,4'-ODA $T_g = 329$°C	Diamine 3,3'-DABP $T_g = 297$°C	BMI-5270-1 Cytec Resin
OHC strength	23	306	327	245
(MPa)	232	—	—	—
	288	281	221	148
OHC modulus	23	51	53	51
(GPa)	232	—	—	—
	288	49	53	38
SBS strength	23	70	58	37
(MPa)	232	56	46	—
	288	33	36	14

Source: Chuang, K. C., Criss, J. M., and Mintz, E. A., *Proceedings of 54th International SAMPE Symposium, May 18–21*, Baltimore, MD (2009).
OHC = Open-hole compression
Note: SBS = Short-beam shear

Composites prepared from a-ODPA/3.4'-ODA resins displayed better open hole compression strength (281 MPa) at 288°C [67] than those made with a-BPDA/3,4'-ODA resin (223 MPa), due to the additional ether (-O-) linkage that provided the flexibility and toughness (Table 21.13). However, a-ODPA composites softened when tested at 315°C, because of their lower T_gs of 300–320°C whereas the a-BPDA/3,4'-ODA (RTM370) composite still retained substantial mechanical properties at 315°C, due to higher T_gs of 370°C (Figures 21.11 and 21.12).

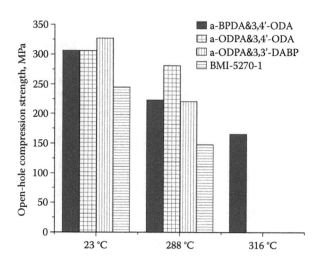

FIGURE 21.11
Open-hole compression strength of T650-35/a-ODPA composites versus a-BPDA composite (RTM370). (From Chuang, K. C., Criss, J. M., and Mintz, E. A., *Proceedings of 54th International SAMPE Symposium, May 18–21*, Baltimore, MD (2009)).

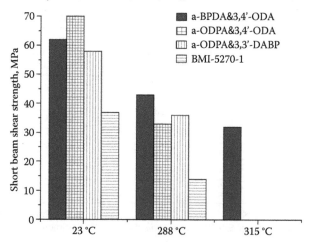

FIGURE 21.12
Short-Beam Shear Strength of T650-35/a-ODPA composites versus a-BPDA composite. (From Chuang, K. C., Criss, J. M., and Mintz, E. A., *Proceedings of 54th International SAMPE Symposium, May 18–21*, Baltimore, MD (2009)).

RTM Imide Resins Based on 2,3,3′,4′-Benzophenone Dianhydride (a-BTDA)

Furthermore, asymmetric 2,3,3′,4′-benzophenone dianhydride (a-BTDA) could also be formulated with 3,4′-ODA, 3,3′-MDA, and 3,3′-diaminobenzophenone (3,3′-DABP) to afford high T_g (330–400°C), lowmelt viscosity (η = 10–20 poise at 280°C) imide resins that are amenable to RTM [68].

Other Low Melt Viscosity Imide Resins

Additionally, the symmetric 2,2′,3,3′-biphenyl dianhydride (i-BPDA) can also yield low-viscosity resins with kinked diamines, but the T_gs of the resulting resins tend to be lower because of the geometric configuration of i-BPDA. Surprisingly, high molecular thermoplastic polyimides derived from i-BPDA are reported to have higher T_gs than a-BPDA. The melt viscosity of low molecular weight oligomers and the solubility of thermoplastic polyimides in NMP increases in the order of i-BPDA < a-BPDA < s-BPDA [69–71].

Comparison of Composite Properties of a-BPDA Based RTM Imide Resins: RTM370, PETI-330, and PETI-375

RTM370, PETI-330, and PETI-375 are all made from a-BPDA with the 4-phenylethynyl endcap. PETI-330 and PETI-375 consist of a flexible diamine and a rigid diamine whereas RTM370 uses kinked 3,4′-ODA. The mechanical properties, namely open hole compression strength, modulus, and short beam shear strength of these three RTM processable imide resins are shown in Figures 21.13 through 21.15 for comparison. PETI-330 displays good mechanical properties only up to 288°C, because of the limitation of its 330°C T_g. In contrast, RTM370 and PETI-375 exhibited comparable mechanical properties up to 315°C, due to their respective high T_g of 370°C and 375°C. However, RTM370 is easier to make from a melt process and costs less than PETI-375 that contains an expensive diamine, 2,2′-bis(trifluoromethyl)benzidine.

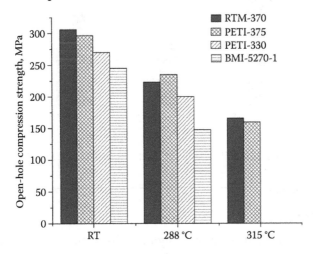

FIGURE 21.13
Open-hole compression strength of T650-35/a-BPDA imide resins: RTM370, PETI-375, and PRTI-330.

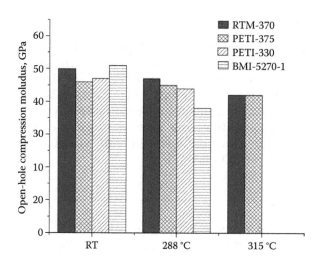

FIGURE 21.14
Open-hole compression modulus of T650-35/a-BPDA imide resins: RTM370, PETI-375, and PRTI-330.

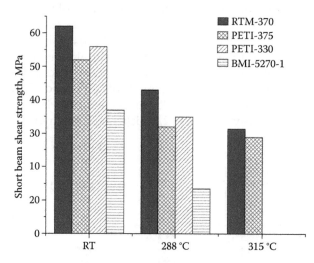

FIGURE 21.15
Short-beam shear strength of T650-35/a-BPDA imide resins: RTM370, PETI-375, and PETI-330.

Conclusions

In the polyimide field, reactive oligomers have been applied more successfully in composite applications than their thermoplastic counterparts. Using a monomer reactant solution, PMR-15 allows the use of oligomers to yield crosslinked polyimides with outstanding thermo-oxidative stability for long-term (thousands of hours) applications at 288–315°C (550–600°F). The introduction of a phenylethynyl endcap into polyimide resins provides a wider processing window and better thermo-oxidative stability than the nadic endcap. Attempts to make PMR polyimide composites by SARTM yielded matrices with a high void content, although lower void content (4–5%) composites can be made by RI, using very concentrated PMR monomer solutions (>95% solid content).

The real breakthrough in the development of low melt viscosity imide oligomers hinges on the discovery of a-BPDA with unique properties that not only yield low melt viscosity but also higher T_g in the cured polyimide. Coupled with the phenylethynyl endcap, these a-BPDA containing imide oligomers with low melt viscosities (~10 poise at 280°C) are amenable to low-cost processes of RTM and RI, commonly used for processing epoxy and bismaleimide (BMI) composites. As PETI-330 displays 288°C (550°F) use temperature and PETI-375 and RTM370 exhibit outstanding property retention up to 315°C (600°F) without microcracks, the innovation in polyimide chemistry has advanced polyimide composite fabrication and performance capability beyond the 232°C (450°F) limitation of BMI and its microcracking problem.

However, the costs of asymmetric dianhydrides are relatively high compared to the commercially available symmetric dianhydrides. After over 30 years of polyimide development, the approach of using commercially available symmetric dianhydrides with different diamines and reactive endcaps have exhausted all the variations and reached its ultimate limitations. The future of polyimide innovation relies on these new asymmetric dianhydrides to produce thermoplastic and thermoset polyimides with novel properties, such as colorless polyimide films for optical applications and easily processable imide resins for composite fabrications.

The future of polymer composites for lightweight aerospace components depends on the continued material and fabrication development of high T_g (≥350°C) and low melt viscosity imide resins (<10 poise) that can be readily fabricated into aerospace components, using low-cost processes such as RTM or VARTM. In addition, the preferred polyimide resins also require a combination of performance, including high fracture toughness, no microcracks, and good processability at reasonable material and manufacturing costs. Furthermore, the desirable polyimide composites also need to possess good thermo-oxidative stability to sustain continued service at 288–315°C (550°–600°F) for hundreds or even thousands of hours, similar to or exceeding the performance of PMR-15 and AFR-PE-4 composites while improving processability. Recent advances in low melt imide resins based on asymmetric dianhydrides not only eliminates the health and environmental hazards of dealing with the toxic diamine monomer in PMR-15, but also offers alternative low-cost RTM process for making high temperature polyimide composites. The future challenge is to make asymmetric dianhydrides affordable so that imide resins with even lower melt-viscosity would enable the manufacture of lightweight composites into complex shapes and large structures by VARTM for aerospace applications, similar to the current commercial fabrication of epoxy, bismaleic composites.

References

1. Lin, S.-C., and E. Pearce. *High Performance Thermosets: Chemistry, Properties, Applications*, 13–63, 247–66. New York: Hanser Publishers, 1993.
2. Serafini, T. T., P. Delvigs, and G. R. Lightsey. *J. Appl. Polym. Sci.* 16, no. 4 (1972): 905.
3. Wilson, D. Polyimides as Resin Matrices for Advanced Composites. In *Polyimides*, 187–218. Edited by D. Wilson, H. D. Stenzenberger, and P. M. Hergenrother. New York: Chapman and Hall, 1990.
4. Alston, W. B., D. A. Scheiman, and G. Sivko. High Temple Workshop XX, J1–J11, Jan. 24–27, 2000, San Diego, CA.
5. Lauver, R. W. J. Polym. Sci., Polym. Chem., 17 (1979): 2529.
6. Wong, A. C., and W. M. Ritchey. *Macromolecules* 14 (1981): 825.
7. Wilson, D. *Brit. Polym. J.* 20 (1988): 405.
8. Alston, W. B. Cyclopentadiene Evolution during Pyrolysis-GAS Chromatography of PMR Polyimides. In *Advances in Polyimide Science and Technology, Proceedings of the 4th International Conference on Polyimides*, 290–310. Edited by C. Feger, M. M. Khojasteh, and M. S. Htoo. Lancaster, PA: Technomic Publishing Co., 1993.
9. Iratcabal, P., and H. Cardy. *J. Org. Chem.* 60 (1995): 6717.
10. Meador, M. A. B., J. C. Johnston, and P. J. Cavano. *Macromolecules* 30, no. 3 (1997).
11. Serafini, T. T., R. D. Vannucci, and W. B. Alston. NASA Report. TM-X-71894, E-8680 (1976).
12. Gibbs, H. H. *J. Appl. Polym. Sci.* 35 (1979): 297.
13. Vannucci, R. D., D. C. Malarik, D. S. Papadopoulos, and J. F. Waters. *International SAMPE Technology Conference* 22 (1990): 175.
14. Waters, J. F., J. K. Sutter, M. A. B. Meador, L. J. Baldwin, and M. A. Meador., *J. Polym. Sci.: Part A: Polym. Chem.* 29 (1991): 1917.
15. Chuang, K. C., and J. E. Waters. *Soc. Adv. Mat. Pro. Eng. Ser.* 40, no. 1 (1995): 1113.
16. Myer, G. W., T. E. Glass, H. J. Grubbs, and J. E. McGrath. *J. Polym. Sci.: Part A: Polym. Chem.* 33 (1995): 3141.
17. Rice, B. P. HITEMP Review 1997, Paper 8, NASA CP-10192 (1997).
18. Sutter, J. K., Jobe, J. M., Crane, E. A., and Tanikella, M. S., High Temple Workshop XIV, Cocoa Beach, FL, January 31–February 3 (1994).
19. Chuang, K. C., Bowman, C. L., Tsotsis, T. K., Arendt, C. P. *High Perform. Polym.* 15, no. 4 (2003): 459.
20. Rice, B. P. HITEMP Review 1997, Paper 8, NASA CP-10192 (1997).
21. Vannucci, R. D., Malarik, D. C., Papadapouslos, D. S., and Water, J. F., *22nd International SAMPE Technical Conference*, 22, (1990): 175.
22. Bowles, K. J. *International SAMPE Technology Conference* 20 (1988): 552.
23. Ahn, M. K., T. C. Stringfellow, J. Lei, K. J. Bowles, and M. A. Meador. *Mater. Res. Soc. Symp. Pro. (High Perf. Polym. & Polym. Matrix Comp.)* 305 (1993): 217.
24. Arnold, F. E., T. Gibson, B. Price, R. Trejo, A. Drain, A. Jacques. High Temple Workshop XXVII, Sedona, AZ, Paper L (2007).

25. Whitley, K., and T. Collins. AIAA/ASME/AHS Adaptive Structure Conference, Schaumburg, IL, April 7–10 (2008)

26. Chuang, K. C., D. S. Papadopoulos, and C. P. Arendt. *Soc. Adv. Mat. Pro. Eng. Ser.* 47 (2002): 1175.

27. Vannucci, R. D., R. Gray, and D. A. Scheiman. *Proceedings of HITEMP* Review 1999, NASA/CP-1999-208915, Vol. 1, Paper no. 4, pp. 1–9

28. Gray, A., and L. R. McGrath. *SAMPE J.* 40, no. 6 (2004): 23.

29. Lin, S.-C., and E. Pearce. High Performance Thermosets: Chemistry, Properties, Applications. Chapter 5 in *Acetylene-Terminated Resins*, 137–85. New York: Hanser Publishers, 1993.

30. Lin, S.-C., and E. Pearce. High Performance Thermosets: Chemistry, Properties, Applications. Chapter 4 in *Benzocyclobutene Resins*, 109–135. New York: Hanser Publishers, 1993.

31. Ratto, J. J., O'Conner, S. R., and Distler, A. R. *J. Polym. Sci. Chem. Ed.*, 18 (1980): 1035.

32. Stenzenberger, H. Chemistry and Properties of Addition Polyimides. In *Polyimides*, 79–128. Edited by D. Wilson, H. D. Stenzenberger, and P. M. Hergenrother, New York: Chapman and Hall, 1990.

33. Harris, F. W., A. Pamidimukkala, R. Gupta, S. Das, T. Wu, and G. Mock. *J. Macromol. Sci.-Chem.*, A21, no. 869 (1984): 1117.

34. Unroe, M. R., and B. A. Reinhardt. *J. Polym. Sci., Polym. Chem.* 28 (1990): 2208.

35. Paul, C. W., R. A. Schultz, and S. P. Fenelli. High Temperature Curing End Caps for Polyimide Oligomers. In *Advances in Polyimide Science and Technology*, 220–44. Edited by C. Feger, M. M. Khojasteh, and M. S. Htoo. Lancaster, PA: Technomic Publishing Co. Inc., 1993.

36. Lin, S.-C., and E. Pearce. High Performance Thermosets: Chemistry, Properties, Applications. Chapter 7 in *Arylethynyl Resins*, 221–46. New York: Hanser Publishers, 1993.

37. Hergenrother, P. M., and J. G. Smith, Jr. *Polymer* 35, no. 22 (1994): 4857.

38. Johnston, J. A., F. M. Li, F. W. Harris, and T. Takekoshi. *Polymer* 35, no. 22 (1994): 4865.

39. Bryant, R. G., B. J. Jensen, and P. M. Hergenrother. *J. App. Polym. Sci.* 59, no. 8 (1994): 1249.

40. Connell, J. W., J. G. Smith, Jr., and P. M. Hergenrother. *High Perform. Polym.* 18, no. 3 (1998): 273.

41. Smith, J. G., J. W. Connell, and P. M. Hergenrother. *Polymer* 389, no. 18 (1997): 4657.

42. Cano, R. J., and B. J. Jensen. *J. Adhesion* 60 (1997): 113.

43. Hergenrother, P. M. *SAMPE J.* 36, no. 1 (2000): 30.

44. Smith, Jr., J. G., J. W. Connell, and P. M. Hergenrother. *Soc. Adv. Mat. Pro. Eng. Ser.* 43 (1998): 93.

45. Smith, Jr., J. G., J. W. Connell, and P. M. Hergenrother. *J. Comp. Matls.* 34, no. 7 (2000): 614.

46. Connell, J. W., J. G. Smith, Jr., P. M. Hergenrother, and M. L. Rommel. *Intl. SAMPE Tech. Conf.* 30 (1998): 545.

47. Criss, J. M., C. P. Arendt, J. W. Connell, J. G. Smith, Jr., and P. M. Hergenrother. *SAMPE J.* 36, no. 3 (2000): 32.

48. Smith, Jr., J. G., J. W. Connell, and P. M. Hergenrother. *Soc. Adv. Mat. Pro. Eng. Ser.* 46 (2001): 510.

49. Criss, J., R. Koon, and E. Mintz. 29th High Temple Workshop, February 9–12, Napa, CA (2009)
50. Yokota, R., S. Yamamoto, S. Yano, T. Sawaguchi, M. Hasegawa, H. Yamagucchi, H. Ozawa, and R. Sato. *High Perform. Polym.* 13 (2001): 861.
51. Hasegawa, M., Z. Shi, R. Yokota, F. He, and H. Ozawa. *High Perform. Polym.* 13 (2001): 355.
52. Hergenrother, P. M., K. A. Waton, J. G. Smith, Jr., J. W. Connell, and R. Yokota, *Polymer* 43 (2002): 5077.
53. Hasegawa, M., N. Sensui, Y. Shindo, and R. Yokota, *Macromolecules* 32 (1999): 387.
54. Connell, J. W., J. G. Smith, Jr., P. M. Hergenrother, and J. M. Criss. *Soc. Adv. Mat. Pro. Eng. Ser.* 48 (2003): 1076.
55. Bain, S., H. Ozawa, and J. M. Criss. *High Perform. Polym.* 18 (2006): 991.
56. Chuang, K. C., J. D. Kinder, D. L. Hull, D. B. McConville, and W. J. Youngs. *Macromolecules* 30 (1997): 7183.
57. Connell, J. W., J. G. Smith, Jr., P. M. Hergenrother, and J. M. Criss. *Proceedings of 49th International SAMPE Symposium*, May 16–20, Long Beach CA (2004), CD Version.
58. Smith Jr., J. G., J. W. Connell, P. M. Hergenrother, and J. M. Criss. *Soc. Adv. Mat. Pro. Eng. Ser.* 45 (2000): 1584.
59. Connell, J. W., J. G. Smith, Jr., and P. M. Hergenrother. *High Perform. Polym.* 15 (2003): 375.
60. Criss, J. M., R. W. Koon, P. M. Hergenrother, J. W. Connell, and J. G. Smith, Jr. *Soc. Adv. Mat. Pro. Eng. Ser.* 33 (2001): 1009.
61. Ghose, S., K. A. Watson, K. J. Sun, J. M. Criss, E. J. Siochi, and J. W. Connell, *Comp. Sci. & Tech.* 66, no. 13 (2006): 1995.
62. Ghose, S., K. A. Watson, D. C. Working, E. J. Siochi, J. W. Connell, and J. M. Criss, *High Perform. Polym.* 18, no. 4 (2006): 527.
63. Chuang, K. C., J. M. Criss, Jr., E. A. Mintz, B. Shonkwiler, D. A. Scheiman, B. N. Nguyen, L. S. McCorkle, and D. Hardy-Green. *Proceedings of 50th International SAMPE Symposium*, May 1–5, Long Beach, CA (2005), CD Version.
64. Alston, W. B. *Polymer Preprints* 27, no. 2 (1986): 410.
65. Kang, J. W., K. Choi, W. H. Jo, and S. L. Hsu. *Polymer* 39, no. 26 (1998): 7079.
66. Li, Q., X. Fang, Z. Wang, L. Gao, and M. Ding. *J. Polymer Science, Part A: Polymer Chemistry* 41 (2003): 3249.
67. Chuang, K. C., Criss, J. M., and Mintz, E. A. *Proceedings of 54th International SAMPE Symposium*, May 18–21, Baltimore, MD (2009), CD Version.
68. Chuang, K. C., Criss, J. M., and Mintz, E. A. *Proceedings of 55th International SAMPE Symposium*, May 17–20, Seattle, WA (2010), CD Version.
69. Fang, X., Z. Wang, Z. Yang, L. Gao, Q. Li, and M. Ding. *Polymer* 44 (2003): 264.
70. Chen, C., M. Hasegawa, M. Kochi, K. Horie, and P. Hergenrother. *High Perform. Polym.* 17 (2005): 317.
71. Kochi, M., C. Chen, R. Yokota, M. Hasegawa, and P. Hergenrother. *High Perform. Polym.* 17 (2005): 335.

22

Environment and Safety

Mel Schwartz

CONTENTS

Introduction

When we talk and discuss the environment, every variety of industry must examine its various operations to determine the health and people issues that face the company as well as the employees and the outside world.The environment is and can be effected by a multitude of industries. This chapter will cover the numerous industries and their fabrication, processing, and manufacturing technologies and in many cases what has occurred in the past, present, and what is being done, or will be done in the future to reduce the detrimental effects on our precious environment. It would be very politically incorrect if manufacturing or any other sector of the economy did not espouse the "greening" of the United States or others in the world. It's a good thing, and the manufacturing community has long been involved in taking steps to being nature's friend.

The environmentalists would have you believe that we can never be too green. People in the heat treating community have to say there has got to be a point where greener pastures hit the pocketbook (greenbacks) so hard that thousands of companies stand to be out of business and their millions of employees without jobs. That point may be now! Hidden in the Security Act (S.2191) is injurious language that could spell financial disaster for some types of industries. Two U.S. senators are attempting to substantially reduce U.S. greenhouse gas (GHC) emissions over the period of 2012–2050. Their approach would establish a cap and trade system, limiting greenhouse gas emissions while allowing companies to buy and sell the right to emit specified amounts of pollution. Forgetting for the moment the trading aspects of the legislation, let's turn to the economic impact the bill would have on the industry. After all, Congress requires that an economic impact study be conducted on such sweeping measures as this. Whether they take to heart the results of such studies is a topic for another day. The government's own study forecasts that most energy prices are projected to increase under S.2191, particularly, coal, oil, and natural gas, directly reflecting the impact of increasing CO_2 allowance prices.

Green Initiatives and Strategy

The green movement sweeping across companies worldwide may be politically and environmentally correct, but it also is creating some concerns among electronic manufacturers who are at the front line of these initiatives. At issue are increasing regulations that limit or restrict the use of materials in the assembly of electronic circuit boards, reinventing the wheel for most manufacturers. Materials like lead, mercury, and cadmium—all core components to the most traditional development processes—now are deemed hazardous and unacceptable. Alternatives to these materials include aluminum ion deposition, zinc/nickel, and tin/nickel, although using them can severely impact how manufacturers reduce the risk on their product development cycles. New materials can disrupt testing and manufacturing procedures, and strain quality control (QC) efforts [1].

Executives from a range of electronics manufacturers have discussed the issue and have addressed recommendations to expand the existing Restriction of Hazardous Substances (RoHS) directive from the current list of six materials nongrata to an extensive menu of restricted substances. This updated list includes a flame retardant that protects up to 80% of all PCBs and organic compounds containing chlorine and bromine, according to reports. The hazardous materials recommendations come from a study sponsored by the European Union Commission, which is charged with reviewing requests for materials exemptions, the use of alternative materials, and the impact these restrictions might have on the performance and QC of the final products. Efforts to eliminate the use of specific materials in the European electronics industry date back to the year 2000, although RoHS rules were enacted in mid-2006 with a detailed directive that banned the use of certain hazardous substances in all electronics sold in the EU.

In the United States, California has been at the forefront in the battle to limit or eliminate hazardous waste both at the production and manufacturing end and disposal. Targeted were photovoltaic (PV) modules, components of which include thin-film silicon wafers that many consider a hazardous waste during disposal. Further east, in August 2008, China reportedly approved a draft regulation covering the management of electronic waste, setting up a formal system to encourage recycling and centralized treatment. The new regulations stem from China's early involvement (2004) in WEEE (Waste Electrical and Electronic Equipment) regulations and its own rules that govern the import and use of hazardous materials. While beneficial for the environment, all of these efforts are putting manufacturers in a tricky spot, between the need to adopt green policies and meet the key factors driving competitive business today, such as maintaining QC, meeting shrinking time-to-market windows, and cutting costs.

As seen in many companies, improved and intuitive lifecycle testing techniques are an essential ingredient to the development of an effective green manufacturing strategy, especially in light of the increasing restrictions on the use of hazardous materials. A holistic cradle-to-consumer approach to manufacturing quality and proactive testing can result directly in:

- Improvements in product design and performance that not only reduce work cycles and manufacturing costs, but result in a more competitively positioned final product.

- A significant reduction in the use of material waste and toxins, which complies with current and pending regulations and eliminates hazardous working conditions for employees.

Greener aircraft engines will rely upon high-performance composites to achieve significantly reduced carbon emissions. Engine production is expected to transform civil aerospace and triple production by 2026. The U.S. Army has embarked on a service-wide effort to measure its carbon footprint with the aim of reducing the effect it has on the environment while at the same time optimizing its use of fossil fuels. The Army recently completed a proof-of-concept study [2] at 12 installations to measure the amount of greenhouse gases it puts into the environment as a result of its activities. The total amount of gases put into the environment by an organization constitutes its carbon footprint.

The study looked at emissions that included water vapor, carbon dioxide, methane, nitrous oxide, ozone, and chlorofluorocarbons. The most predominate of those emissions are carbon dioxide and then methane, said Tad Davis, the deputy undersecretary of the U.S. Army for environment, safety, and occupational health. From that proof of concept, the Army has kicked off a series of similar studies at all Army installations. These studies categorize the greenhouse gases produced by the activities of an installation into three scopes, Davis said. Emissions of buildings, on-post generators, tactical vehicles including tanks and helicopters, and nontactical vehicles, including privately owned vehicles and government vehicles are included in Scope 1, Davis said.

Scope 2 includes greenhouse gas emissions that are the result of energy used on an installation but produced off the installation. Finally, Scope 3 measures emissions from contractor-related activities on an installation and also emissions related to things like employee travel. For a soldier traveling on temporary duty, for instance, the Army could calculate the greenhouse gas emissions generated by the travel. The White House put into place a directive to reduce greenhouse gas emissions by some 30% by 2015, Davis said. And the Army is working to meet that goal, though he said that the service remains mindful that the current administration could change the goal or the target date. Reducing the amount of carbon the Army puts into the atmosphere through the burning of fossil fuels will require a reduction in the use of fossil fuels, Davis said. That can be accomplished by finding nonfossil fuel sources of energy to power the Army mission and also by making more efficient those parts of the mission that will continue to rely on fossil fuels.

A reduction in fuel use also results in decreased mission costs to the Army and in increased safety for soldiers, Davis said. "If we are able to reduce the amount of energy consumed, then that is going to probably reduce the amount of fuel that is going to be used, in the case of the forward deployed forces, we are able to reduce the convoys and the resupply which is one of the primary targets of a lot of the IEDs and ambushes taking place in Iraq and Afghanistan," Davis said. Reduction in energy use, and subsequent reduction in greenhouse gases can come from finding new sources of energy and also by reducing the energy the Army uses through efficiency.

At one U.S. Army installation, Fort Carson, Colorado, for instance, the Army partnered with a local energy provider to do an enhanced-use lease there. The energy provider built a PV solar array on top of a closed landfill. That site now provides energy to some 450 homes, Davis said. "That's a visible way we can address looking at more renewable sources of energy versus the nonrenewable sources which produce greenhouse gasses," Davis said. According to the International Panel on Climate Change there are five technologies recommended for meeting global greenhouse gas reduction targets. These include gas-fired plants, nuclear energy, fuel-efficient cars, ethanol production, and carbon capture and sequestration.

For more than 30 years, green engineering has empowered engineers and scientists to measure, diagnose, and solve some of the world's most complex challenges. Now, through a graphical system, engineers and scientists can design a platform and use modular hardware and flexible software to not only test and measure but also fix inefficient products and processes by rapidly designing, prototyping, and deploying new machines, technologies, and methods. Today, a number of the world's most pressing issues are being addressed through green engineering applications (ni.com/greenengineering).

- Acquire environmental data from thousands of sensors
- Analyze power quality and consumption
- Present measured data to adhere to regulations
- Design and model more energy efficient machines
- Prototype next-generation energy technologies
- Deploy advanced controllers to optimize existing equipment

Another green initiative is halogen-free (HF) products. The move to HF product in the near term, 2009–2010, is being driven by the marketing arm of the major computer manufacturers as part of their green initiative. Dell, HP, Apple, and Lenovo see all new products being HF as inevitable and irreversible. They are focused on every aspect of the product including PCBs, soldermask, underfill, solder flux/paste, components, cables, connectors, plastics, and so on. Now that the environmentalists have taken the lead out of solder, they want to remove halides from electronics [3].

To begin with, the term HF is a misnomer. Fluorine, chlorine, bromine, iodine, and astatine are all halogens, but the only elements considered in this discussion are chlorine and bromine. Teflon is safe for now. Are we to remove halogens from our everyday lives as well? That would mean no more salt on your French fries, and no more chlorine used to kill microorganisms in your backyard pools. And what about the chloride ions that are needed as part of your nerve impulse mechanism? We can still keep the fluoride in our toothpaste because this is not one of the targeted halogens.

Halogen-free also is inaccurate because low levels of halogen are allowed. The amount is defined differently by different groups: Japan Electronics Packaging and Circuits (JPCA) define it as having bromine (Br) levels lower than 900 ppm, and chlorine (Cl) below 900 ppm; International Electrotechnical Commission (IEC) and Association Connecting Electronics Industries (IPC) prefer Br 900 ppm max, Cl 900 ppm max, and Br + Cl 1500 ppm max. Perhaps we should call this "halogen-light". At present, the only halogen compounds restricted by RoHS are polybrominated biphenyls (PBB) and polybrominated diphenyl ethers (PBDE). There is no legislation announced or in place for the broad range of applications promoted by the computer manufacturers.

Intel is committed to the change and has implemented HF materials in most of their packages and will complete implementation for it in the core laminate by the end of 2009. Other companies doing the same are ON Semiconductor and Amkor. There are serious implications related to this proposed change. To begin with, the implementation of lead-free solder was preceded by more than 10 years of research on the properties of these new alloys and numerous reliability studies that were undertaken. HF laminate development began in the late 1990s and still is under development with new and improved materials from vendors each year. It is impossible to do reliability studies when the characteristics of the material supply keep changing.

Another area of concern is HF-sheathed cables. There is only one HF material available for cables and this only became available in 2008. Material characteristics will need to be examined by all of those who use cables. Cost is another factor. HF materials cost more than traditional laminates. That cost could be 1.5–3 times the original, and more, depending upon the application, the number of layers in the board, the complexity of the design, and so on. Manufacturing processes for HF materials must be adjusted to their properties. For example, HF laminates are more rigid and the drill shape needs to be modified. Also, the number of hits before the drill bit is replaced will drop. This increases the overall cost of polychlorinated biphenyl (PCBs). Most component suppliers do not want to have an HF line as well as a traditional technology line. Thus, we could all be forced to use HF parts or pay a 10 time premium to get traditional parts.

In conclusion, it is interesting to note that the original drive for HF materials came from Japan cell phone makers who claimed that the appearance of green technology gave them a marketing and sales advantage. When it comes to the environment, Washington's attention is fixed these days on the Congressional battle over legislation to control greenhouse gas emissions. But there are other pollutants, so-called ground level pollutants, as opposed to those that rise into the atmosphere, that also need urgent attention, starting with toxic mercury emissions from coal-fired power plants.

For various reasons, mainly heavy industry lobbying, these emissions have escaped federal regulation, whereas mercury emissions from other sources like incinerators and cement kilns have not. But the prospects for regulating power plant emissions have greatly improved since the 2008 election and the new administration. Under the EPA, there has begun a rule-making process that could require some power plants to reduce mercury emissions by as much as 90%. The GAO has just produced a report showing that such reductions are not only technologically possible but affordable, refuting the industry's longstanding claim that mercury controls would be too expensive. This is good news for the environment and for consumers. Mercury is a toxin that has been found in increasingly high concentrations in fish and poses human health risks, including neurological disorders in children. The nation's coal-fired power plants produce 48 tons of it per year, a little more than 40% of the total mercury emitted in the United States. Previous administrations have tried to pass weak legislation that could not be supported in court. For example, power companies were allowed to escape controls by purchasing emissions credits from power plants in other parts of the country. A trading system can make very good sense for greenhouse gas emissions, which disperse widely into the atmosphere. But mercury tends to deposit locally, and the legislation did nothing to reduce the pollution in local lakes and streams.

Fortunately, 18 states have laws or regulations requiring mercury reductions at coal-fired power plants. However, the GAO studied 25 boilers at 14 plants with advanced technologies, found that, in some cases, mercury emissions had been reduced by as much as 90% at a cost of pennies a month on consumer bills. That is a mere fraction of the cost of the equipment necessary to control other ground-level pollutants like SO_2, the acid rain gas. It is very important that the EPA issue a tough rule to control mercury, knowing that it is essential to protect Americans and that the power companies can certainly afford to do what is needed. It is significant to discuss the environmental issues in the material science and technology industries, focussing on the relationship between the materials science, the various multitude of industries, the processes used, and the environment. Solutions for the environmental impact of advanced materials and processes and products include: alternative technologies with lower environmental impact, waste minimization technologies, green materials, green processing, nanotechnology for environmental remediation, and protection materials recycling.

Processing, Product Manufacturing, and the Environment

One of the emerging issues in manufacturing performance and reliability is lead-free solders. Issues that effect performance and reliability of lead-free solders in diverse environments include:

- Pick-and-place machines
- Lead-free solder development
- Cleaners
- Tin whisker prevalence

Material Challenges in Alternative and Renewable Energy

Leaders in materials science and energy are continually sharing information on the latest developments involving materials for alternative and renewable energy sources and systems. The materials challenges in alternative and renewable energy addresses solar, wind, hydropower, geothermal, biomass, nuclear, hydrogen, and battery technology. Since the Industrial Revolution, fossil fuels including natural gas, coal, and oil, have provided the major sources of energy for powering new technologies and improving standards of living for countries around the world. The United States continues to be the largest consumer of these fossil fuels and currently must import more than half of its needed inventory, of which about two-thirds goes into transportation. That demand is anticipated to more than double in the next 50 years. In addition to the increasing demand of this diminishing resource and its insecure sources, there are also growing environmental concerns involving pollution and global warming threats. It is imperative that all nations develop alternative and renewable energy resources for the future. Most likely, not one, but an arsenal of these sources will be used, depending on the natural resources and capabilities available in the various regions.

The overall efficiency, effectiveness, and practicality of potential future energy sources and systems are directly related to many materials related factors. Some of these key features include materials costs, availablity and improvements in chemical, mechanical, electrical, and/or thermal properties of materials now being considered, as well as the ability to produce and fabricate materials in forms and shapes that work effectively in areas of energy generation, storage, and distribution.

Solar: Solar power is energy derived from sunlight and can be converted into various forms of energy such as heat and electricity. The conversion to electricity can take place by PV or solar cells, as well as by the use of solar power plants. There are currently more than a dozen major solar plants in the United States, with most of these facilities located in California.

Wind: Wind power plants or wind farms often consist of many individual units. The largest wind farm located in Texas, consists of over 400 wind turbines that generate enough electricity to power about a quarter of a million homes each year. The United States is ranked second in the world in wind power capacity, only following Germany. In countries such as Denmark, about 20% of its electricity is generated from the wind.

Hydropower: Hydropower is the most often used form of renewable energy in the United States. Mechanical energy is produced and used by harnessing moving water. Over half of the U.S. hydroelectric capacity to generate electricity is located in three states: Washington, California, and Oregon, with the largest facility being the Grand Coulee Dam in Washington. Hydropower currently accounts for about 6% of the total electricity generated in the United States.

Geothermal: The United States. produces more geothermal electricity than any other country, but this still amounts to less than 1/2 of 1% of all energy generated. Most geothermal reservoirs are deep underground but can find their way to the surface as volcanoes, hot springs, and geysers. California has almost three dozen geothermal power plants that produce the largest fraction of U.S. energy from this source.

Biomass: Biomass is energy derived from organic plant and animal matter including wood, crops, manure, and municipal solid wastes. When burned, the energy in biomass is released as heat but it can also be converted to other forms of energy like methane gas, ethanol, and biodiesel. Biomass fuels currently account for about 3% of the energy used in the United States.

Nuclear: Nuclear power extracts usable energy from atomic nuclei by controlled nuclear reactions and most often, through nuclear fission. On a global scale, there are more than 300 operating nuclear power plants in more than 30 countries, which generate about 30% of the energy produced in the European Union and almost 20% of the energy produced in the United States. Among the advantages of nuclear energy are no greenhouse emissions.

Hydrogen: Hydrogen is the simplest element known to man and like electricity, is primarily an energy carrier compared to an energy source. Hydrogen can be produced from a variety of domestic sources, including fossil fuels as well as from renewable resources and can be stored in gas, liquid, or solid forms. There is considerable work in progress on the development of materials and systems for effective hydrogen storage. This alternative is considered a promising energy concept of the future, but like many alternatives, there currently is no infrastructure in place to produce, store, transport, or distribute hydrogen effectively.

Battery technology: Batteries are devices that convert chemical energy into electrical energy. There are many types of batteries available, representing a multibillion dollar industry. Among the battery types of much interest are standard lead acid batteries and Li-ion batteries. Materials improvements are critical in making these energy systems more effective in the future.

Electronics

In July 2006 the European Union's RoHS Directive took effect. Suppliers of electronic equipment have to comply with regulations guaranteeing that the entire unit is free of lead, cadmium, mercury, hexavalent chromium, and PBB/PBDE fire retardants. Failure to do so results in stiff fines and a possible loss of business. Although suppliers to electronic original equipment manufacturers (OEMs) are not directly responsible for compliance, their customers have required that they meet the RoHS guidelines [4].

What is RoHS?

The RoHS Directive says: Member states shall ensure that, from July 7, 2006, new electrical and electronic equipment put on the market does not contain lead, mercury, cadmium, hexavalent chromium, PBB, or PBDE. Kluk and de Krom [4] and the Directive RoHS have a more complete and detailed account of the document. See Table 22.1 [4].

What RoHS Means to You

RoHS is a product-based initiative; therefore, the responsibility for compliance falls on the producer of the electronic product. However, if a company rebrands products and imports them into Europe, it is considered to be the producer.

Types of Testing

Chemical testing of materials may be required for a number of reasons. First, it provides a more reliable alternative to supplier material declarations. It also provides a way to confirm supplier compliance and a basis to assess compliance. Some test methods to determine the content of restricted materials already exist, but most are not appropriate for testing electronic products. In addition, most methods are not recognized internationally, and not all are accepted by every country in the EC. Table 22.2 [4] outlines some of the test methods available to assess the presence of restricted materials.

Effects of Europe Going Green

An example of a U.S. company and how it coped with RoHS is described:

> In October 1999, well before the ink dried on the EU's recent environmental legislation, engineers at GD Land Systems, Sterling Heights, Michigan [5], faced a daunting environmental mission of their own. Eliminating or

TABLE 22.1

Materials Affected by RoHS

Material	Pb, Cd, Hg	Cr(VI)	PBB/PBDE
Metals	**Testing needed**	Testing needed	Not relevant
Ceramics	**Testing needed**	Less relevant	Not relevant
Polymers	**Testing needed**	Less relevant	Testing needed

TABLE 22.2

Verification Test Procedures

		Test Methods		
Test techniques	Substances	Polymers	Metals	Homogeneous Electronic Components
Mechanical preparation	All	Direct measurement Grinding	Direct measurement Machining	Grinding
Chemical preparation	All	Microwave digestion Acid digestion Dry ashing Solvent extraction	Acid digestion	Microwave digestion Acid Digestion Solvent Extraction
Analytical methods	PBB/PBDE	Gas Chromatography/ mass spectroscopy (GCMS) liquid chromatography (LC)		Gas chromatography/ mass spectroscopy (GCMS) Liquid chromatography (LC)
	Cr (VI)	Colorimetry	Colorimetry	
	Hg	ICP/atomic emission spectroscopy; ICP/mass spectroscopy; cold vapor atomic absorption spectroscopy; atomic absorption spectroscopy		
	Pb/Cd	X ray fluorescence; ICP/atomic emission spectroscopy; ICP/ mass spectroscopy; atomic absorption spectroscopy X ray fluorescence; ICP/atomic emission spectroscopy		

Note: To bring some order to the process, ASTM Subcommittee F-40 is examining the regulations with the goal of developing a test specification, including limits and test methods.

reducing hazardous materials from the family of fast-paced combat vehicles for the Army's Stryker Brigade Combat Team (BCT). Their mandate was as follows: "Greening the government through leadership in environmental management that stipulates that the heads of each Federal agency establish strict environmental management, and so on."

But unlike engineers faced with transitioning existing products to meet EU environmental requirements, GD engineers could work from a clean slate and setup a Green Procurement Plan (GPP) for the Stryker. When the first vehicles hit the field in 2002, the team had eliminated about 0.9072 kg (2 lb) of pigment containing the restricted substance hexavalent chromium (Cr^{+6}) for each vehicle. They also found eco-friendly processes to replace Cr^{+6}-containing pretreatments on nonelectrical aluminum components and swapped out cadmium-plated fasteners and hardware with zinc-plated alternatives. Zinc plating also let designers specify a less-caustic trivalent chromium (Cr^{+3}) postrinse in place of conventional Cr^{+6} treatments.

All in all, the use of alternative green components and processes affected about 850 parts in every Stryker. And thanks to the lessons learned, there are other DoD programs including the F-35 (Joint Strike Fighter), T-45, and Advance Amphibious Assault Vehicle that have eliminated or minimized components containing heavy metals. Unfortunately, many OEMs planning to ship products to the EU won't have access to a well-staffed DoD logistics agency to help navigate the complicated process of meeting strict environmental restrictions. Environmental restrictions, including those imposed by RoHS and Waste Electric Electronic Equipment (WEEE) along with impending U.S., Japan, South Korean, and China green legislation, will soon burden OEM engineering departments.

Because RoHS and WEEE are directives and not regulations, they only outline objectives to be met. Most details on reaching compliance have been left for the market to iron out. OEMs (i.e., producers) were given no accepted global test procedures to evaluate EE products against the imposed restrictions or standards for documenting compliance. The EU also provides no enforcement body, which in effect leaves competitors to police each other's products. Pamela Gordon, Technology Forecasters, Inc. (Alameda, California), feels that companies working in RoHS/WEEE compliance generally have software tools with extensive databases that can often more easily pinpoint noncompliant parts and find possible replacements. They can also help facilitate lab analysis if needed, locate vendors with compliant parts (on a global scale), and lend assistance with meeting WEEE obligations in Europe.

According to Gordon, "RoHS compliance should take a one-time investment of between 2% and 3% of the cost of hardware goods sold. This includes the investment of product-tracking software, scrubbing the bill of materials (BOM), supplier and part-replacement activities, testing, planning, legal advice, and so forth." The cost could be less, however, if OEMs use contract manufacturers employing design-for-RoHS or design for environment (DFE) strategies that are ahead of the curve. Gordon further states that to increase profits in the face of escalating environmental restrictions, electronic OEMs need a visionary champion (at the highest executive level).

In a recent Surface Mount Technology Association (SMTA) International meeting [6], RoHS, which was adopted in February 2003 by the EU and took effect on July 1, 2006, was addressed. In a series of talks the impact of RoHS on substrates, alloys, and finishes was covered. The discussion covered the effect of RoHS across a variety of materials used in electronics manufacturing. Examples included removal of lead from solder alloys and surface finishes, eliminating bromine as a flame retardant in PCBs, and so on. In turn, the elimination of the dreaded six spurred new alloy and process developments in a very short span also exposed the industry to problems with which they had little understanding, such as tin pest, Sn whiskers, higher reflow temperatures, and an assortment of reliability concerns.

The experience base continues to expand with respect to Pb-free solutions within the electronics community. Successful assembly processes are being used by a growing number of OEMs and contract manufacturing services (CMSs) in the United States. as well as overseas. Moreover, Pb-free finishes and assembly processes have been investigated by many companies in the high end segment of the industry, which includes medical, military, and space electronics. In the latter case, solder joint reliability continues to be the major hurdle for Pb-free implementation and has led to the development of new solder alloys.

Back in the late 1980s and early 1990s, the elimination of CFC-based defluxing solvents dominated the covers of industry trade magazines [7]. Emissions from Freon and Trichlorethene-based defluxing solvents threatened the earth's ozone layer and would soon become banned from use. The electronic assembly industry responded with two alternative strategies: no-clean fluxes/pastes and aqueous-based defluxing. CFC emissions, as they relate to defluxing, are a distant memory. While in-line defluxing technologies have been environmentally attractive compared to CFC-based solvent cleaning systems, the standards by which we define "green" have changed. Today, water is a precious commodity in many parts of the world, particularly in southern California.

Over the past 2 years, assembliers who require clean (flux-free) assemblies in medium to high quantities have begun to embrace new high-yield batch defluxing technology. Unlike traditional low-volume batch defluxing processes, high-yield batch processes are capable of high-volume defluxing. High-yield batch processes gained popularity in Europe where environmental regulations carry considerable weight in process and equipment selection. Users of high-yield batch defluxing processes are able to process equal quantities of electronic assemblies while consuming only a fraction of the water required by in-line processes. Less water in, translates to less water out, thus reducing the volume of effluent discharge and associated liability. Because less water is required, zero-discharge configurations utilizing evaporative technology are implemented easily. Also, because most defluxing applications require a chemical additive as a percentage of wash water, less chemical input is required, reducing consumable expenses.

All of North America has begun to embrace high-yield batch defluxing technology for a combination of reasons. Like Europeans, North Americans are increasingly cognizant of the environmental impact of modern manufacturing techniques as well as the liability associated with environmental negligence. Any process that reduces product liability and environmental liability, while reducing a company's carbon footprint and reliance on natural resources, is valuable indeed.

The world price of solder is at an all-time high, so reducing solder consumption, therefore reduced incurred costs, should be at the top of every electronics manufacturer's priorities list. On average, electronic manufacturing system (EMS) companies throw away 75% of their solder as dross. By automating the solder dross recovery from their wave soldering process, a company can reduce dedrossing time by up to 80% and solder purchases by up to 50%. Recycle, reduce, reuse is set to become as much an industry mantra as smaller, faster, cheaper, as issues such as energy and resource consumption and waste management are pushed to the fore [8].

Safety in the workplace is paramount, not only for the operating procedures but also for routine maintenance and cleaning functions. Because of the nature of the process, extensive safety features must be built into solder dross recovery systems to protect against misuse. Ergonomics, simple instructions, and machine construction all contribute to a recovery safety process. The high cost of solder frequently drives EMS providers to invest in capital equipment that can reduce solder consumption. Lead-free solder is at a 200–300% premium over lead solder, and lead-free adoption is spreading. The electronics industries have more and more companies actively pursuing efficiency, effectiveness, and cost reduction in their manufacturing strategies.

Cleaning Processes for Solder Paste

It is very important to be able to determine printed circuit board assembly's (PCBA) cleanliness in a cleaning process. Secondly, cleaning materials change with environmental restrictions. Therefore, what is the best type of cleaning to choose for a chosen solder paste? To determine the answers, it is suggested that experiments be run that include contaminants, detergents, and process set-up parameters. An additional parameter, cleaning solutions that are environmental friendly, should also be considered. Cleaned electronic circuits represent only a small part of the electronic assembly market, as most companies use a no-clean process. In several sectors, aeronautics, medical, and automotive, cleanliness and reliability are crucial. Post-assembly cleaning of no-clean solder paste residue is necessary. These sectors have demanding requirements for cleanliness and chemical reliability after PCB assembly.

Requirements vary with the particular product assembly. For example, the aerospace industry needs long-term reliability, while the medical industry prioritizes highly reliable circuits for implants and other devices. Some products are covered with a protective varnish or conformal coating that requires the surface to be thoroughly cleaned before the sealant is applied. From a chemical point of view, flux and solder paste residues, common to high-reliability/long-lifecycle products, must be compatible with solvents or detergents used to clean them. Codevelopment of materials in this area leads to global solutions for assembly cleaning [9].

Contaminants

Contaminants on PCBs are mainly organic, brought about during PCB manufacturing and during the assembly process. Fluxes have a deoxidizing function; they deoxidize the pads and components. In the case of solder paste, they deoxidize the solder powder, ensuring good wettability of the alloy. Fluxes contain organic acids, resins, and other materials to perform this function. The term no-clean implies that, according to current norms, residues are chemically safe and can be left on the PCB. For high-reliability applications, no residues or other contaminants can remain on the circuit board.

Be certain that cleaners used on these PCBAs are compatible with the residues left by the soldering processes. In fact, if halides are hidden or trapped under components, partial cleaning can cause disastrous consequences by releasing these ions.

How can one determine the best cleaner to use for a chosen solder paste? The only variable here is the cleaner itself, a downstream parameter. A common test can determine if product A cleans better than product B for a chosen solder paste, with a given reflow profile, and with similar cleaning conditions for both products. For test results and procedures, see Abidh [9]. As a result of all the various parameters, experimental plans, cleaning products, and assembly products together, each production line is different and therefore, a combination of knowledge gained from experimentation and chemistry with knowledge from experience yields successful cleaning processes.

Biocomposites

A great deal of significant interest has been shown in the development of natural fibers (NFs) and bio-derived plastics to replace the oil-derived polymers and mineral reinforcement fibers, such as glass, curently used in the reinforced plastics industry. This was the theme of a recent conference, BIOCOMP in the UK [10], where for example, Ekotex, a Polish company that grows flax, from the stems of which it extracts NFs that can be used as a reinforcement in biocomposites—reinforced plastics wholly or partly derived from plants.

Natural Fibers

The agricultural nature of the reinforcements impresses itself still further as more samples illustrate the various stages of planting, harvesting, retting, scutching, and so on, that extract from the raw straw, cellulosic fibers that have significant mechanical and other properties. Samples of unidirectional tows and nonwoven mats made with the fibers provide the persuasive clincher to the idea that one can actually grow useful fiber reinforcements in fields, see Tables 22.3 and 22.4 [10]. Fibers can be cut to lengths to suit particular client uses. For instance, cleaning a flax tow, carding it to produce 18–30 ktex slivers, then cutting the slivers to lengths of 1 mm, 2 mm, 4 mm, and so on (0.039 in., 0.078 in., 0.156 in.) serves to prepare the fibers for injection molding. Fiber properties can be improved by further processes such as boiling, bleaching, and plasma treatment.

Expounding the virtues of NFs as compared with E-glass, Marek Radwanski cites that the density is about half, lower cost, lower embodied energy, and the fibers' nonabrasive nature. Natural fibers (NFs) can, he declares, be recycled, are a source that is renewable, and are carbon neutral since carbon dioxide

TABLE 22.3

Types of Natural Fiber

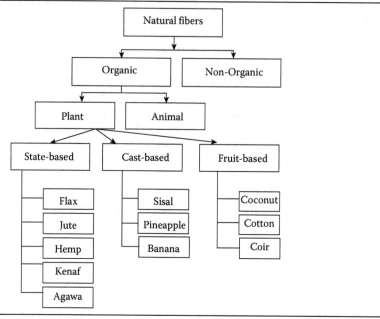

Source: Ekotex.

TABLE 22.4

Comparison Between Natural Fibers and Glass Fibers

	Natural Fibers	Glass Fibers
Density	Low	Twice that of natural fibers
Cost	Low	Low, but higher than natural fibers
Renewable	Yes	No
Recyclable	Yes	No
Energy consumption	Low	High
Distribution	Wide	Wide
CO_2 neutral	Yes	No
Abrasion to machines	No	Yes
Health risk when inhaled	No	Yes
Disposal	Biodegradable	Not biodegradable

Source: Ekotex.

(CO_2) emitted during production is reabsorbed by new plant growth. They pose no risk to human health when fiber particles are inhaled, and at the end-of-life (EOL) they are biodegradable. A downside is that, unlike glass, NFs are hydrophilic and can, if left unprotected, absorb moisture, swelling in the process. Although concentrating on flax, Radwanski also mentions the role of hemp, since this too can be grown in Europe, though it is somewhat more difficult to cultivate. On the other hand, hemp requires less soil than flax and has greater resistance to drought. Fibers from flax and hemp, also other plants such as jatropha, ramie, coir, sisal and kenaf, can be used in thermosets and thermoplastics.

Resins

Resins, too, can be grown in fields, says Hans Hydockx, business development manager with TransFurans Chemicals bvba (TFC), Belgium. TFC starts one of its core activities with bagasse, a biomass waste product from sugar cultivation, and uses it to produce furans-based resins for exploitation in composites, modified wood, polymer concrete, and other products. Furans is a basic ring-configured organic chemical that gives rise to a number of process streams.

TFC produces thousands of tons per year of furfural, a hemicellulose, which is then converted using a high-pressure steam process into furfuryl alcohol, a key platform chemical from which resins can be obtained. Although TFC exploits bagasse, other agro wastes containing the pentosan (C5) form of sugar can be used: these include corn cobs, oat hills, flax slivers, cotton seed hulls, almond husks, beech and birch wood, hazelnut shells, and others. From furfuryl alcohol chemists can, by the use of various coreactants and additives, derive natural thermoset resin that can be used with various organic and inorganic fibers to form composites. The biomatrix, which is notably fire and chemical resistant, can also be used with rockwood fibers in horticultural growing substrates, in wood to make a durable wood that can resist fungal and microbial attack, and in other products. One fire and chemical-resistant plastic made from it is Furolite, a resin that, when NF reinforced, results in an entirely plant-derived composite.

Furolite thermoset resin can be reinforced with natural, glass, and carbon fibers plus natural and mineral fillers to produce engineering materials. It can, says Hoydockx, be used with hand lay-up, spray up, prepreg, vacuum infusion, pultrusion, and filament winding techniques. Composites can be made by spraying the resin into NF mat that is then compression molded into a finished part, or by passing a NF textile through a resin bath, see Figure 22.1. Hoydockx explained that the resin adheres well to a range of fibers and has been used in prototype biocomposite floor modules, trunks, head liners, and similar parts for passenger cars. Advantages include solvent-free part production with no harmful volatile organic compounds (VOCs), easy machining with negligible abrasion of cutting tools, reduced reliance on oil feedstock, and of course, the fact that the composite is made entirely from renewable resources. Cycle times of below 60 seconds at temperatures of around 180°C (356°F) make the system suitable for automatic high-volume production. Physical properties are said to broadly be equal to those of NF-reinforced polyester composites, while high resistance to corrosion means that the resin can usefully be a substitute for epoxy in corrosive organic acid environments.

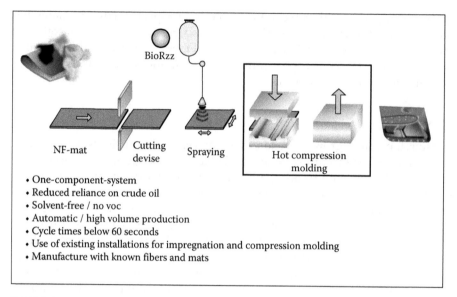

FIGURE 22.1
Processing of natural fiber reinforced composites. (From Transfuran Chemicals bvba, TFC, Industriepark, Leukaard 2, B- 2440 Geel, Belgium.)

Bo Madsen from the Materials Research Department of the Risø National Laboratory for Sustainable Energy at the Technical University of Denmark conceptually brought the two composite phases, fibers and matrix, together in his consideration of biocomposite properties [10]. Madsen claimed that plant-derived, cellulosic fibers, such as the bast fibers from flax and hemp, can provide good properties, but that there is a natural variability within a plant population, and even within individual plants according to where in the plant the fibers are taken from. The most useful fiber bundles are found in the outer stem layers just beneath the epidermis. Other influential factors include the degree of crystallinity, fiber density, and volume fraction, plus the cross sectional area and shape of the fiber bundles. Overall, Madsen claimed that new-generation biocomposites can be viable alternatives to traditional composites in selected applications.

Embedding the fibers into resins such as polypropylene (PP), polyethylene teraohthlate (PET), and polylactic acid (PLA) provides complete dual-phase biocomposites that are suitable for efficient manufacturing processes. Production articles can, for instance, be made by interleaving films of PLA (made from fermented corn starch) and nonwoven jute mat layers in a stack and then compression molding the result, all in a single continuous manufacturing step. As with conventional composites, properties are highly dependant on fiber alignment, fiber volume, and weight fractions, but also on porosity. Madsen and his Risø researchers have devised a mathematical tool for modeling the properties of biocomposites [10]. The result differs from standard models (i.e., for standard composite materials) in allowing for the porosity (and hence moisture absorption) of NFs that can have a deleterious effect on tensile stiffness and other mechanicals.

Applications

In replacing oil-based composites with plant-derived substitutes, most engineers agree that the mechanical properties of biocomposites can compare well with those of glass reinforced plastics (GRP), in some instances exceeding them. There remain, however, issues around aspects of availability, quality consistency, and degradation through moisture absorption. Even so, there is great interest in the potential for biocomposites, especially given the growing socioeconomic imperatives, exemplified by legislation such as Europe's EOL directives and the way these will affect automotive and other industries (see Figure 22.2). As a result, there are emerging engineering applications for biocomposites. Examples include vehicle door and head liners and other interior semistructural parts. The car maker Lotus, for instance, uses a NF/polyethylene matrix composite in plastic profiles that resemble wood in appearance.

There is a foreseen potential for thermoplastic polymers such as polyhydroxyl-butyrate (PHB), PLA and starch, as well as thermosets based on oils from crops such as linseed, castor, and sugar (furans).Other flax/PLA and wood/PLA decking, flooring, railings, and other products can be extruded using short fibers, while longer fibers lend themselves to the production of semistructural items such as foot rests in Smart cars. Producing a composite by spraying a bioresin onto an NF mat and then hot compression molding the result, can yield useful structural items such as door panels.

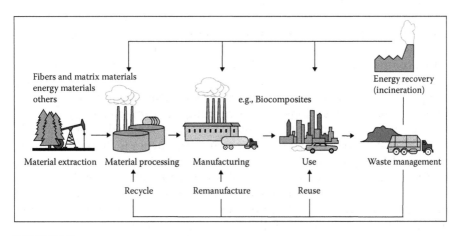

FIGURE 22.2
Life Cycle assessment. (From Technical University of Denmark, Dept of Management Engineering DTU, Anker Engelunds Vej 1, Bldg. 101A, DK - 2800 Kgs. Lyngby.)

Future

These are early days for biocomposites but, from these and other examples, one can draw positive and encouraging conclusions. In summary, it has been shown that NF/PLA compounds can be injection molded and extruded successfully into viable products, and that NF/furans composites compare well with conventional glass-based equivalents, while additionally offering superior fire and chemical resistance. The fact that crop oil-based resins can replace formaldehyde in MDF products raises other market prospects. All in all, the results bode well for future commercial exploitation of biocomposites, particularly in commodity and automotive sectors, where there is pressure to reduce environmental impact, weight, and cost. Further research and program investments running into the year 2012, are aimed at delivering complex structural and multifunctional parts from enhanced wood-based composites and to further the state-of-the-art in natural aligned fibers and textiles for use in structural composite applications, on-going from 2008 to 2012.

The BIOCOMP Conference showed that biocomposites are more than just a straw in the wind and that twenty-first century drivers are hastening their development and exploitation. A material study is underway to understand biomaterials. The project led by Net Composites of Chesterfield, and BRE of Watford, UK has been given the task to assess the environmental credentials of naturally derived construction materials, including fiber reinforced plastics (FRP). The BioCompass [11] project addresses the limited understanding of the environmnetal performance of these relatively new materials, which at the moment are hard to compare like for like with traditional materials. The project team, consisting of experts from the construction sector and natural materials industry, will provide information, guidance, and tools for the new materials. The project has focused on issues including: raw material supply, including crop production and land-use; energy requirements for primary and secondary processing; durability of these naturally derived materials compared to conventional alternatives; and EOL issues, including the possibility for recycling, composting, recovery, and resue [11].

In a companion program, COMBINE [12,13], a UK-based consortium, has been examining and developing composite materials for use in creating completely biodegradable NF reinforced plastics. Plant fibers have been used for millennia in applications such as clothing, building materials, and basket weaving; however, synthetic equivalents have replaced these materials in many cases and are used for more structural purposes. To date, NF have been used with polymers to form lightweight injection molded components for the automotive industry, but the fibers are short and randomly oriented so mechanical properties are relatively poor. Bio-polymers are a relatively new material and little is known about their mechanical behavior when mixed with bio-fibers. The COMBINE project, therefore, covers new ground in terms of developing materials that are fabricated from entirely renewable sources.

The COMBINE project plans to convert NFs into long, aligned reinforcements to exploit the inherent and mechanical properties of plants in structural applications, with the added advantage of having a lower weight than conventional reinforcements such as glass fiber. The project has focused primarily on using PLA, a biodegradable polymer derived from corn as the matrix material for a biocomposite. As a thermoplastic material, wetting and impregnation of fibers may prove to be difficult, as is often the case using petroleum based materials. Testing on different processing temperatures and chemical compatibility with NFs are being performed also. PLA has been considered a suitable matrix material. Polypropylene is also being considered as a partial nearer-to-market solution and candidate for evaluation. These two materials are to be combined with NF reinforcements such as flax and hemp. Spinning and weaving techniques have been developed to optimize material properties, while further work will include process optimization, painting, bonding, and molding [12].

Polymers derived from natural sources such as starch and corn are a new and exciting development, especially with growing concern over waste management and dwindling oil reserves. These materials can be found as a thermoplastic or thermoset, offering properties similar to petroleum-based polymers such as PP. Products such as disposable drinking cups and food trays made from biodegradable plastics are now on the market and are very successful in this industry sector. COMBINE's objectives are to develop high performance bio-derived composites for structural applications. The plan is now to manufacture three industrial demonstrator parts. COMBINE partners, Fairline Boats and Lightweight Medical [12], have begun to develop a marine component, marine bartop lid, marine wheelhouse roof, and a mobile baby incubator, respectively [12].

Composites: Health, Safety, and Environmental Concerns

Fabricators and OEMs must address health, safety, and environmental concerns when producing and handling composite materials. Their methods for maintaining a safe workplace include periodic training, adherence to detailed handling procedures, maintenance of current toxicity information, use of protective equipment (gloves, aprons, dust-control systems, and respirators), and development of company monitoring policies. Both suppliers and OEMs are working to reduce emissions of highly VOCs by reformulating resins and prepregs and switching to water-dispersible cleaning agents.

The U.S. Environmental Protection Agency (EPA) has continued to strengthen its requirements to meet the mandates of the Clear Air Act Amendments, passed by Congress in 1990. Specifically, the agency's goal is to reduce the emissions of hazardous air pollutants or HAPs, a list of approximately 180 volatile chemicals considered to pose a health risk. Some of the compounds used in resins and released during the curing process contain

HAPs. In early 2003, the EPA enacted regulatory requirements specifically for the reinforced plastic composites industry, requiring emission controls using maximum achievable control technology, or MACT. The regulations took effect in early 2006 [14].

Repair Considerations

As more composite materials find a place on aircraft, boats, bridges, and hundreds of other applications where part replacement is difficult and expensive, OEM engineers are considering the repairability of structural and secondary composite components during the initial design phase. According to recent reports by various consulting firms, spending by air transport maintenance and repair organizations (MROs) on air transport maintenance materials was $16 billion in 2008 and likely to increase to $20 billion by 2011. Today, composite repair materials account for a fraction of that: $15–$25 million, annually. MROs soon will have to repair planes that are 50% composites by weight. One answer may be automated repairs. Publicly unveiled in 2005, the Inspection and Repair Preparation Cell (IRPC) concept is the result of an initiative begun in 1999 to improve aircraft repair practices and reduce repair time through automation. The IRPC is designed to integrate into a single, automated work cell, a variety of tasks regularly performed manually by MROs. An IRPC cell, as currently conceived, will include a 3-D digitizing station, nondestructive testing (NDT) capability, radio frequency identification (RFID) tagging, automated machining equipment (to remove damage and cut out repair plies and core plugs), and advanced in-cell collision avoidance technologies.

Other composite components that require repair include heavy truck front ends and other automotive and marine products. Whether damage is minor or extensive, training and specific repair materials are available. A number of materials suppliers provide kits containing low- or room-temperature curing adhesives and potting compounds formulated especially for on-site repair, along with low-quantity dry fiber or prepreg and vacuum-bagging materials. At least a dozen training companies nationally offer from 1 to 10 days courses in composite repair. Composites are increasingly being used to repair structures with other materials, such as bridge beams, bridge decks, parking garages, and pipelines, including underground systems. Composites can also be used to repair product pipelines, such as natural gas and petroleum pipes at oil refineries and offshore.

Changes in Processes

One change is a trend away from open molding to closed molding. Due to emissions regulations and the high VOCs that result from the open molding process, heavy truck customers are primarily looking to the closed molding process to mitigate this environmental concern. In fact, some customers are even looking at converting existing components from the open mold process to the closed mold process before product model changes. Semipermanents, which are preferred for better control over VOC emissions, have been formulated specifically to meet the needs of resin transfer molding (RTM) and other closed molding processes. Internal release agents, added to the resin or gel coat and used instead of, or in addition to, external agents on the mold surface, further reduce emissions, and have negligent effect on a part's physical properties and surface finish. Internal release formulations are required for pultrusion processing, because the part is pulled continuously through the die, allowing no opportunity for intermittent application of external releases to the die surface. The composites industry also is developing advanced compounds using carbon fibers, nano-fillers, and bio-based resins to increase mechanical properties, reduce mass, and improve exvironmental friendliness.

Looking to the future, it is critical that suppliers reinforce to customers the importance of upfront design involvement to fully utilize the benefits of composites. Looking at the proposed customer application from just a component standpoint is not nearly as valuable as looking at the overall system to provide the most financially advantageous solutions through system integration. Composites have the ability to offset heavy trucks' weight increases resulting from the addition of components that are necessary to meet EPA emission requirements. With the increase in diesel prices and in aluminum material costs, this is an opportunity to reinforce the importance of composites as the most viable solution. Our materials directly impact the bottom line of owner-operators and fleets by reducing diesel consumption, and we must continue to strive to develop unique and innovative materials that reduce weight.

The Next Generation Composite Wing (NGCW) initiative is being heralded as one of the most significant aircraft research and technology programs launched in the UK for several decades [15]. Its aim is to develop technologies that will lead to the design and manufacture of aircraft that weigh less and are more environmentally friendly. Optimizing the weight and design of aircraft will improve their efficiency and performance, and burning less fuel will result in lower operating costs and lower gaseous emissions to the atmosphere. "Composite materials are aerospace's bright young things, offering huge potential for reducing CO_2 emissions and meeting EU emission targets," according to Neil Kirk, Managing Director of Atkins' Aerospace business. "Atkins engineers will be woking to develop design principles that maximize strength to weight, and directly impact on fuel efficiency and overall emissions levels from aircraft."

This NGCW program also takes into account the manufacturing facilities for the aircraft components. Morgan Professional Services (MPS) has been tasked with developing an efficient, sustainable and environmentally friendly factory of the future for the Airbus aircraft, which meets the manufacturing process requirements for aircraft structures production. Technology is required to reduce VOCs in binder/resins while maintaining or enhancing existing product properties. Potential solutions could include the addition of a compound or additive that reduces VOC demand, a new crosslinking system for polymers to enhance molecular network formation, or another technical solution that eliminates the existing issues with the low VOC resins currently available, such as poor appearance and high viscosity during drying (as typically encountered in a water borne approach), or slow drying and high application viscosity (as typically encountered in high-solids paint).

As part of an initiative to reduce and remove VOCs from fuel tank manufacturing processes, engineers at GKN Aerospace, Portsmouth, England, have developed the first single-skin flexible fuel bladder material that is able to offer crash resistance and puncture resistance. It is estimated that use of this material could lead to a 60% reduction in the use of VOCs and a 30% reduction in manufacturing time, while the weight of each bladder could be reduced by approximately 5% through the removal of adhesive coatings. This material is a composite of a woven aramid core with thermoplastic polyurethane coatings and an integrated proprietary fuel barrier layer. The fuel barrier provides the same level of fuel tolerance exhibited by traditional tanks. This means the crash-resistant layer is not applied with adhesives, and no final lacquer coating is required, significantly reducing VOC solvents. It is made by a hot-spread coating technology.

Architectural Composites

Manufacturers of architectural composites will be increasingly pressured in the coming years to report the sustainable attributes of their products to building teams. Green building projects have only represented approximately 10% of the commercial construction market up to now, but several factors ensure growth in this market segment. Many regulations adopt outright the once-voluntary green building rating systems developed by one of several nongovernmetal organizations (NGOs), such as the U.S. Green Building Council (USGBC), or The Green Building Initiative (GBI). In 2000, the USGBC introduced the leading national benchmark for sustainable buildings called Leadership for Energy and Environmental Design (LEED) Green Building Rating System. Several years later, the GBI followed with the Green Globes program. Here are just some examples of states, countries, and other entities taking a stand for green:

- Ohio requires LEED for all of its schools.
- Canada requires all building to be LEED-certified.
- The U.S. General Services Administration, the largest landlord in the country, requires project teams to produce a LEED Silver Certified federal office building, courthouse, or any government facility under the agency's responsibility.

- The U.S. Department of the Interior has formally recognized the Green Globes and other comparable systems as a sustainable construction tool for all its new construction projects.
- Minnesota adopted side-by-side legislation requiring its governent buildings to be certified by one of the two leading systems.

LEED has five categories: Sustainable Sites, Energy & Atmosphere, Water Efficiency, Materials and Resources, Indoor Environmental Quality and Innovation in Design. Green Globes has seven areas of assessment. Five are similar to USGBC's program, but two additional areas are covered: Project Management and Emissions. Some areas relevant to the composite industry would be the product's contributions to the overall recycled content of the building, the ways in which the product assists with energy efficiency, or test results showing the product has lower or eliminated off-gassing of harmful indoor air pollutants within an enclosed space. Green building is here to stay and is gaining momentum. The green writing is on the wall. Soon, sustainable building will be standard practice.

Polyvinyl-Chloride (PVC)

The Center for Health, Environment, and Justice (CHEJ), has been on a campaign against PVC use in consumer goods and packaging sold by various stores including Target Corp. Target has systematically reduced PVC in its own brand of products. It is collaborating with vendors whose PVC products it sells, to get the material out of its stores. PVC has been in the crosshairs of the CHEJ and other environmental organizations for years. "We target PVC because it is the worst plastic for our health and environment, releasing toxic chemicals that cause cancer and birth defects," says CHEJ PVC campaign coordinator Mike Schade [15]. "PVCs often contain toxic additives such as lead and phthalates. These chemicals are not bound to the plastic and can leach out over time. When lead isn't used, lesser-known metals such a sorgaotins are substituted. Studies have suggested links between organotins and suppression of the immune system, birth defects, and impacts on the liver, bile ducts, and the pancreas. For example, tributyltin (TBT), has been banned in the EU for use as an antifouling agent in boat paint, but has been found in PVC products."

Some factions say critics are using too broad a brush to paint all PVC as harmful. Lead does not have to be used to make vinyl products, and it should be deliberately avoided in packaging and products for children. Many polymers, including vinyl, require additives that serve as heat and light stabilizers and the principal metals used to make stabilizers include tin, barium, zinc, and calcium, which are accepted by government agencies and, to a much less degree, cadmium and lead, for which there are regulatory limits. Lead-based stabilizers principally serve in vinyl wire, cable jacketing, and insulation and are contained within the product. Even in these applications, companies are moving to new or different stabilizers.

TBTs are not used as vinyl stabilizers since these compounds have served as antifoulants in marine paints because they help prevent growth of micro-organisms, barnacles, and seaweed on the hulls of ocean vessels. Commercial tin stabilizers used by the vinyl industry exhibit no biocidal properties, and it is important that they not be confused with the antifouling TBTs. However, CHEJs Schade questions why lead is turning up in vinyl toys, lunch boxes, baby bibs, and other consumer goods if it isn't needed to make vinyl prod-ucts? "In the US," Schade continues, "lead in PVCs first came to light in 1996 when lead was found in vinyl mini-blinds. Soon after, many other vinyl products were found to contain lead. About 156,000 tons of lead are used each year in the worldwide production of PVC. And the world stock of PVC in use contains a staggering 3.2 million tons of lead" [15]. Importers need to set and enforce QC standards on the materials in their products no matter where they come from.

Reuse of Airliners and E-Waste

Airbus (the aircraft European Aerospace Co.) has been evaluating how best to build an industrial setup to handle its out-of-service aircraft with a mini-mum amount of waste. The company, along with several industrial partners, spent almost 3 years trying to figure out how to dispose of end-of-service air-craft more efficiently than the quick-and-dirty approach traditionally used. Several viable ideas have emerged and now the focus has shifted to devising a plan to establish a formal process that will be acceptable to airlines. The EU supported the project, which is aimed at demonstrating that recycling of aircraft can be done more effectively through processing of more than 60% of the material in terms of aircraft weight—the current standard. Of the 60%, half is actually recycled, the rest becomes waste.

Airbus conducted a study to evaluate a Model A300B4 on which various dismantling procedures were conducted. The bottom-line result was that the level of valorization (the amount of items that could be resold) reached 85% of the aircraft in terms of weight, and 70% of that could be reused or provided as a secondary raw material. One area where recycling is particu-larly attractive is aluminum, an element that requires a massive amount of energy in the production stage. The recycling process calls for about 90% less energy, and the end product is a high-grade quality that can be used again for aerospace applications. Although Airbus has identified a higher percent-age of the aircraft that can be recycled in a way that allows for materials to be sold on, the processes are more cumbersome and the costs are higher. Some issues, such as composite recycling, still need to be addressed. However, it will be years before this step becomes a major industry concern, given that aircraft with a high percentage of composites are only now starting to enter fleets in larger numbers.

Airbus itself is somewhat ahead of the game because its aircraft won't begin reaching end-of-service lives in significant numbers for several years. Right now, of approximately 1000 aircraft parked, Airbus types account for only 73 [16]. But over the next 20 years, the number will swell to 1000–1500 aircraft, or 20–25% of the total. The first peak is expected around 2017, with another by 2025. But by 2030, about half the retiring aircraft will be Airbus-built. Therefore a critical need for an EOL center becomes apparent. Airbus has not only been looking at the back-end but is considering how future aircraft design may be influenced and therefore lead to airplanes that are inherently more disposable in the end [16].

Another consideration concerning the airlines is greenhouse gas emissions. The EU plans to require airlines flying into and out of its member countries to partake in a carbon emissions cap-and-trade agreement starting in 2012. Under this program, airlines would be forced to buy carbon credits to offset their carbon dioxide emissions, or face possible restrictions on flying into Europe. While the EU claims the plan is needed to reduce greenhouse gas emissions from airlines, the United States opposes it, claiming the scheme would drive up costs and force already financially strapped carriers to hike airfares. Much discussion will likely continue on this subject.

Greenhouse Gas

Scientists at the National Oceanography Centre, Southampton, UK, have found that the warming of an Arctic current over the last 30 years has triggered the release of a potent methane gas, from methane hydrate, stored in the sediment beneath the seabed. The scientists, working in collaboration with researchers from the University of Birmingham, Royal Holloway London, and IFM-Geomar in Germany, have found that more than 250 plumes of bubbles of methane gas are rising from the seabed of the West Spitsbergen continental margin in the Arctic, in a depth range of 150–400 meters. Methane released from gas hydrate in submarine sediments has been identified in the past as an agent of climate change, and the likelihood of methane being released in this way was not expected this soon. Surveys have been designed to work out how much methane might be released by future ocean warming; however, it was not expected to discover the strong evidence that this process has already started.

A filigree wire is standardly used to slice blocks (ingots) into paper-thin wafers for solar cells. The wire can cut through the ingot at a speed of up to 60 km/h. Several hundred kilometers long, the wire is arranged in such a way that the ingot is sliced into hundreds of wafers simultaneously. The process takes around 6 hours and the resultant slices are approximately 180 μm thick. Researchers are setting out to reduce this thickness, and therefore reduce waste. Researchers at the Fraunhofer Institute for Mechanics of Materials (IWM) are aiming to reduce the saw gap width when slicing the wafers. The space between two wafers is governed by the thickness of the wire. The steel wire is wetted with a type of paste (slurry), a mixture of silicon carbide, and polyethylene glycol. This is harder than silicon and cuts through the ingot. The gap arises where the silicon is reduced to powder during cutting. Currently, gap widths are around 180 µm, which means that given a wafer thickness of 180 µm, the same amount of waste is generated for each silicon slice and that is inefficient. The researchers goal is to achieve smaller saw gap widths of around 100 µm, which are also suitable for industrial applications. In a project that has been funded by the German ministry for the environment, they are currently studying the abrasion process and contact regimes using a single-wire saw, and are principally interested in the interactions between the wire, slurry, and silicon. The researchers are striving to achieve gap widths of 90 µm, which would represent a huge increase in efficiency, as waste would be halved.

E-Waste and Composites

The European Composites Industry Association (EuCIA) has called for the disposal of composites waste by the cement kiln route to be accepted as recycling in new European legislation. Without this change, EuCIA says composites will not be able to meet the recycling targets of European waste directives like the end-of-life vehicle (ELV) and WEEE directives, and the composites industries in other regions of the world will leave Europe far behind. In its statement, EuCIA states its agreement on many of the principals of the directive but asks the European Parliament to take into consideration two facts. Initially, those composite materials combine the advantages of several materials into one material that results in very high strength, low weight, corrosion resistance, and other advantageous properties, which from the environmental point of view, lead to very significant energy savings and reductions of emissions over their whole long useful lifetime. And secondly, because of their specific nature as a combination of materials, composites need recycling solutions that allow material recycling in combination with some energy recovery.

The composites waste management process EuCIA proposes, starts with a recycling process carried out by European Composites Recycling Services Company (ECRC) in order to make the waste a suitable resource for cement kilns. The manufacturing process for making cement requires a lot of energy (which is overtly provided by the resin fraction of the composite material) and raw materials (provided by all other ingredients of the composite material). This energy turns the raw materials (present in the composite waste) into anhydrides, which give the cement its binding power. This means the energy is incorporated into the cement and becomes part of it (which is different from pure incineration). When mixed with water, and while recombining with water, the cement uses this energy to bind together the mortar and stones. Some energy recovery will therefore always be an inseparable integral part of the recycling process when composites are recycled via the cement kiln route. EuCIA believes that without the cement kiln route, complete recycling of all

FRP/composite waste will be impossible. Next is the relationship between FRP and cement (Source: EuCIA).

Typical FRP Composition Use in Cement

- 25–35% resin energy for making cement
- 25–45% glass fiber raw material for cement
- 20–50% inert filler $Al(OH)_3$/$CaCO_3$ raw material for cement
- Typical FRP waste = 70% raw material for cement + 30% energy for making cement

Therefore, the use of GRP/composites waste in cement kilns means 70% material recycling and 30% energy recovery = 100% useful application.

What do We do with E-Waste?

More than 1.5 million tons of e-waste (TVs, monitors, computers, cell phones, batteries, etc.) are thrown into American landfills and incinerators every year. As a result, the toxins that they contain, such as lead and mercury, end up being released into the air and water. While both government and industry agree that e-waste is a huge problem, there's no consensus on how to solve it. There are no federal regulations that address household electronic waste. Meanwhile, the EU has policies that make manufacturers responsible for recycling their products and decreasing the levels of harmful metals used. For now, nine states have passed e-waste laws in which communities collect and recycle worn-out devices and charge companies to cover the costs. For example, an old TV set can contain up to 4.54 kg (10 lbs) of lead. One must remember that e-waste is a public-health issue. Industry groups object to state laws, saying they're confusing since they differ from place to place and unfairly assign expenses to the manufacturers. So what should consumers do with unwanted stuff? First, donate any working equipment to a school or nonprofit. If it's broken, find out if your state has an e-waste disposal program. Your last option, send the machinery back to its maker. Many companies, including Apple, Dell, and Sony, will take back their products, although you may have to pay for shipping.

What if I showed you a garden pot, what would it have to do with an electronics assembly operation? In fact, it was created of a composite of 100% electronics waste (e-waste) like circuit boards and computer parts. Directives like RoHS and WEEE have been implemented with the idea that less-toxic electronics would be more easily recycled into the consumer and industrial cultures. Recyclable electronics have been the focus of many new product introductions (NPIs), primarily in the consumer electronics segment. Next are some do's and don'ts for electronics recycling.

EOL (End-of-Life) Do's

1. Before you even think about recycling, design solutions for your customers with minimal hardware—the first step in EOL management is designing for it.

2. Postpone recycling by designing products and your business model for reuse. Your products can live useful lives again and again with upgrades and efficient refurbishing.

3. Collect products that your customers are no longer using, and mine them for hard-to-find and/or valuable parts for refurbished units.

4. Design your products also for high-value recycling. Train engineers in DFE principles, including easy-to-disassemble modules for reuse and materials that are worth something.

5. Minimize the cost and environmental burden of product collection; design reverse logistics, according to minimal distance traveled and lowest carbon emissions.

EOL Don'ts

1. Don't assume that the photos you've seen of unsafe casual recycling in under-regulated regions are exaggerated; this practice really is as bad for human and environmental health as it looks.

2. Don't use a recycler that does not offer proof of where and how your products were recycled; the product has your name on it and publicity is given to brands whose companies irresponsibly recycle products.

3. Don't wait until the end of your product's design/manufacturing cycle before creating a reverse-logistics and recycling plan; design products for high-value recycling.

4. Don't think that no one wants products at the end of their first use; second- and third-hand sales are multimillion-dollar businesses for someone; it may as well be for your company than a broker.

5. Don't choose a recycler that outsources the recycling to some unknown-to-you entity.

Hazardous Waste

New powers that exist within the hazardous waste regulations have significantly increased the regulatory compliance burden for many companies and yet there are even more compliance issues to consider. Companies reviewing or implementing a legislative compliance strategy should take a broader look at the impacts of closely aligned legislation or best practice, such as storage, transport, and packaging requirements, for their organizations.

The FRP sector, like many others, has been significantly affected by environmental and waste legislation in the last few years. One could be forgiven for believing that this legislation is something of a new phenomena. For example, the Hazardous Waste Directive (HWD) and the European Waste Catalogue 1994 (EWC) laid some of the foundations for EU waste legislation. All EU member states were required to interpret these directives and to create a legislative and regulatory framework at a local level [17].

Waste legislation has therefore been implemented within member states in different ways and at different times based upon the legal interpretation of the directives while, at the same time, accounting for the different existing infrastructures available within each state. Storage and segregation of waste materials can often be overlooked, but it is a vital part of a compliance strategy. Although the hazardous waste and duty of care regulations cover responsibilities related to storage, there is no specific legislation. There are six key compliance areas relevant to the storage of waste chemicals.

Location and Contents of Storage Area

Storage areas should be located away from a watercourse and any foul or surface water drains. Where large volumes of waste are being stored, there should be a site storage plan in place outlining layout, locations, and segregation.

Infrastructure of Storage Area

The infrastructure of a storage area will depend on the type and volume of waste to be stored, but containment is the one common element to them all. The simplest method of creating a storage area is by creating a concrete wall within which waste is stored. However, there are now many portable or semiportable storage systems on the market that allow for total containment and an element of segregation [17].

Condition of Containers to be Stored

The container in which the waste is stored is the primary containment. Anything of a cardboard or fiberboard construction should not be stored in an area that is open to the elements as containment will be compromised.

Stock Rotation

Stock rotation is vital in any business and the same applies for the storage of waste.

Procedures for Storage and Segregation

Like many processes within an organization, it is important to document procedures for the storage and segregation of your waste. This could be in the form of a formal standard operating procedure (SOP), as part of the ISO 9001 quality structure, as a stand alone management system, or as part of an employee training matrix.

Procedures for Spillage Response

If a sizable spillage occurs, procedures should already be in place to provide adequate remedial action, including the correct absorbent materials, adequate personal protective equipment (PPE) for operatives, and technical expertise readily available.

Complex Criteria [17]

Manufacturers and fabricators in the FRP industry will almost certainly handle hazardous substances and produce hazardous waste. Many will, of course, already be familiar with their compliance requirements and will have resources in place to manage them. With hazardous waste, companies need to be aware of compliance criteria outside the normal regulatory frameworks provide by waste legislation. There can be no room for complacency. The risks are simply too great [17].

Next Generation Aircraft Wings

The next generation composite wing (NGCW) [18] initiative is being heralded as one of the most significant aircraft research and technology programs launched in the UK for several decades. Its aim is to develop technologies that will lead to the design and manufacture of aircraft that weigh less and are more environmentally friendly. Optimizing the weight and design of aircraft will improve their efficiency and performance, and burning less fuel will result in lower operating costs and lower gaseous emissions to the atmosphere. "Composite materials are aerospace's bright young things, offering huge potential for reducing CO_2 emissions and meeting EU emission targets," according to Neil Kirk, managing Director of Atkins' Aerospace business. "Atkins' engineers will be working to develop design principles that maximize strength to weight, and directly impact on fuel efficiency and overall emissions levels from aircraft." This NGCW program also takes into account the manufacturing facilities for the aircraft components. Morgan Professional Services (MPS), a provider of design, engineering, and project management services, has been tasked with developing an efficient, sustainable, and environmentally friendly factory of the future for Airbus, which meets the manufacturing process requirements for aircraft structures production [18].

Fuels

The problem with living in a dream world is that sooner or later reality will issue a wake-up call. For many years, if not decades, power companies in California have been trying to build new power plants. But the lawmakers and citizens refused to believe that they actually needed more power generation capacity, focusing instead on the CO_2 and other emissions from power plants. Power plants were renamed polluters, and the companies who wanted to build them demonized as greedy destroyers of the environment. In the real world, life is a matter of choices and trade-offs. Choosing to build a vibrant, growing economy leads to the need for choosing a source of power. That's when the trade-offs begin, because a growing economy is based on manufacturing, which needs power. In the far-off future, such power may be delivered by ocean waves, wind, and sun. But right now, the only realistic sources are fossil fuels and nuclear fission.

Natural gas is the cleanest fossil fuel, but federal laws and regulations have sharply curtailed both exploration and production. The same is true for oil, with the additional restriction of regulations that limit the building of processing facilities. The trade-off here is to lift the restrictions and allow drilling in pristine areas and offshore, at the price of a certain amount of risk to wildlife and possible spills. Is the trade-off worth it to grow the U.S. economy and prevent brownouts? Coal is the most plentiful and low-cost fossil fuel, but also is the one that emits the most CO_2. As CO_2 is considered by many to be the villain in the current global warming hysteria, coal-generated power is often rejected before its benefits can be enumerated. The trade-off here is to add a little more CO_2 to the atmosphere, to get cheaper, and more abundant energy. Nuclear fission is another viable choice, but today's restrictions and regulations mean that even if a nuclear power plant could generate power, new and more reliable designs have been developed and built around the world. It should also be considered whether it is worth reducing the restrictions and regulations to allow such plants to be built in the United States more quickly.

Reality has issued its wake-up call to California, in the form of recent brownouts and complete power outrages for parts of the state. It's a clear reminder that solutions to energy problems depend on actions taken based on realistic evaluations of benefits and disadvantages.

Clean Fuel Alternatives

As world demand increases for fossil fuels, alternative sources and new technology are mandatory and some of the concepts that are being researched and introduced are discussed below. A certain segment of our society erroneously believes we can drill our way to lower prices. Recent studies and research however, debunk such notions and indicate quite the opposite. For example, even if widespread offshore drilling is permitted, not only will it not bring down the price of gas by any meaningful amount, it will serve to encourage and increase our dependence on fossil fuels. Consider the fact that there are more than 6 billion people on our planet. Now take into consideration only approximately 12% of these people have access to powered vehicles. Clearly, the energy needs of the future cannot be met long-term with only fossil fuels. We need cleaner, renewable alternatives, and we need them now.

Alternatives do exist. No, I'm not referring to the methanol or ethanol approach or corn or other veggie (biodiesel) oils, as these solutions have proved too costly, cumbersome, and ineffective. Methanol and ethanol are temporary band-aid type solutions. Each is alcohol based. Alcohol is corrosive, therefore, mechanical or machine parts exposed to the chemical must be resistant to corrosion such a stainless steel or certain types of plastics (petroleum-based). Biodiesel fuels costs approximately $3.50 per gallon, not that much less than gasoline. And those residing in colder climates require higher concentration blends of biodiesel fuel additives to prevent the fuel from solidifying to that of a wax-like substance. Similarly, biodiesel storage tanks require heaters in most parts of the country.

Plans are underway that will lead us away from dependency on fossil fuels, eventually. The Department of Energy (DOE) demands the United States makes the transition to a hydrogen-based economy and President W. Bush (when in office) stated he wanted hydrogen-powered vehicles on the market by 2020. In addition, the President's Committee of Advisors on Science and Technology reported, "Successful application of fuel-cell technologies in automobiles will improve energy security and provide significant environmental benefits. A 10% market penetration could reduce U.S. oil imports by $130 million barrels per year" [19].

To date, pure hydrogen fuel cells appear to be the direction scientists are taking. Vehicles fueled by these cells emit practically zero pollutants and unlike oil or gas, we don't have to worry about spills and other environmental dangers. In contrast to the combustible engine, fuel cells provide power via electrochemical reactions. Heat and water are the only by-products generated by hydrogen when used as a fuel. The problem however, is that hydrigen-fueled vehicles will not be readily available until at least 2020, and that's being optimistic. There is no manufacturing or distribution system in place and hydrogen production is extremely expensive and energy consuming.

There's another approach that appears to be working fairly well in other parts of the world, such as Brazil. The country has more than 1.5 million vehicles in use powered by compressed natural gas (CNG). The number is expected to increase by 10% in 2009. Approximately 85% of America's natural gas is produced domestically. While some champion CNG as an alternative fuel the United States should explore, the truth of the matter regarding its use is a bit bleak. In order for a vehicle to travel reasonable distances, CNG must be compressed to 20.4–24.4 MPa (3000–3600 psi). At 24.4 MPa (3600 psi), CNG has approximately 1/3 the energy of gasoline and requires a larger, heavier, and much more expensive fuel tank. While considered a bargain compared to the price of gasoline, natural gas is also a nonrenewable source. Eventually, the correct combination of technology and alternative energies will emerge thereby introducing competition and placing an end to the monopolistic stranglehold big oil maintains on the United States and consumers.

Ethanol from Citrus Peels

With all the talk of going green heating up of late, the state of Florida is trying something different—going orange. The state is assisting private companies to produce alternative energy sources with a $25 million Farm to Fuel grant [20]. Some of that funding will be spent on technology that turns citrus peels into ethanol. Construction was scheduled to begin in 2008 on a brand new plant in Auburndale, Florida, which will produce ethanol from citrus peels. It is a process scientists with the USDA have been working on for more than a decade at a lab just outside of Winter Haven, Florida. The scientists have found a way to turn citrus peels into ethanol, which can be used as an alternative to fuel. USDA officials estimate using citrus peel technology could produce between 30 and 50 million gallons of ethanol each year.

The process starts with sterilizing the peels, then pressurizing them to extract limonene, which is used as a cleaning product. And that mushy substance eventually becomes the ethanol. Scientists add enzymes and yeast to the mush and then ferment it for 2 days. After being distilled, the end result is ethanol. "The Florida citrus industry produces between 3.5 and 5 million tons of wet waste per year," said B. Widmer, a PhD and USDA research chemist, "and there is no way you can landfill that" [20]. In January 2008, Florida Light and Power broke ground for another citrus peel ethanol plant in Hendry County, Florida.

Ethanol from Agricultural Waste

A biorefinery built to produce 1.4 million gallons of ethanol a year from cellulosic biomass opened in Jennings, Louisiana. The plant makes ethanol from agricultural waste left over from processing sugar cane. The facility will produce ethanol and be used to compare the production process commercially as well as validate cost and performance assumptions to prepare for the development of the first series of commercial plants. This phase would put the builder, Verenium, Cambridge, Massachusetts [21], on track to begin construction in 2009 on a 30 million gallon per year commercial plant, which will be the first of its kind, located in the southeastern United States.

Verenium's technology enables almost complete conversion of all the sugars found in cellulosic biomass. It is based on a combination of acid pretreatments, enzymes, and two types of bacteria to make ethanol from plant matter. This efficiency advantage, combined with the low input cost of cellulosic biomass, results in superior economics in the production of ethanol. The process [21] begins when the cane is ground up and cooked under

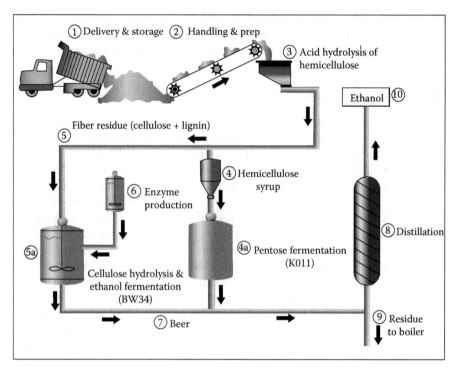

FIGURE 22.3
Cellulosic ethanol plant based on agricultural waste.

high pressure with a mild acid to hydrolyze the hemicellulose and separate it from the cellulose. The five-carbon sugars in hemicellulose are then fermented via genetically modified E. coli. The cellulose is broken down with enzymes and fermented with another type of bacteria. This bacteria also produces enzymes that break down cellulose, cutting in half the amount of enzymes needed from outside sources. The product is dilute ethanol, which is then distilled to make fuel, see Figure 22.3.

Algae and Biomass

In the ongoing quest for alternative, environmentally acceptable fuels, algae and other biomass materials are looking promising as feed stocks for processing and refining. However, finding the needed landmass to support this endeavor is problematic, according to a new report from the International Air Transport Association (IATA) [22]. The association called upon the industry to back the developing techniques aimed at growing vast fields of sustainable biomass materials, thereby solving the current shortage of feed stocks and avoiding competition with food production. Governments could play a greater role, IATA suggests, by subsidizing projects and enacting regulations.

High petroleum prices are bolstering the interest in alternative fuels initiatives. Efforts are being further driven by the airlines' commitment to environmentally safe energy use and the prospect that conventional sources of oil, many in unstable parts of the world, might be exhausted in several decades.

Algae plants are cited several times in the 68-page report as a feed stock receiving considerable attention. Algae are grown in photo bioreactors, open ponds, and in the open sea. Investment in the more effective bioreactors to produce enough algae to meet current aviation fuel demand would call for $74 billion to $2.4 trillion, depending on the process. Investment in open ponds for algae production, which would require higher initial outlays, has been pegged at $190 billion to $1.4 trillion. The Boeing Co. estimates that an area the size of Belgium (931,000 sq. km/359.4 sq. mi.) would be needed to grow algae in sufficient supply to meet current aviation fuel demand.

"Growing oil crops is well understood, but growing large amounts of sustainable oil crops is more challenging, because of the shortage of suitable land and fresh water," the report states. "The oil crops are not a substitute for the complete supply of oil in a sustainable way. Sustainability must be achieved and other biomass must be explored, for example, algae." Bio-diesel used by ground vehicles would compete with bio-jet. Subsidies for bio-diesel range from 0.26 to 1.05 per liter ($0.39–$1.60 per gal.). One challenge awaiting large-scale bio-jet production is the requirement that refineries be located within reasonable distances to the feedstock. A hub-and-spoke model for moving biomass would be efficient but investment in new refineries and transportation systems might be necessary. Nevertheless, the report states that the technical challenges to bio-jet production "have been overcome and many are optimistic that with continued investment in research and development it will be possible to overcome the commercial obstacles."

The report described as mature, the development of synthetic alternative fuels such as those derived from coal and natural gas through the Fischer-Tropsch process. The drop-in fuel, defined as indistinguishable from petroleum-based fuel, contains no sulfur and produces lower particulate emissions. However, it leaves a CO_2 footprint 1.5–2 times greater than conventional jet propellents. New processing facilities will require a high capital investment.

The U.S. Air Force has set early 2011 as the deadline for certification of its aircraft fleet for a 50-50 blend of Fischer–Tropsch fuel and JP-8 petroleum. It is advocating a mixed use of biomass and coal or natural gas to produce the synthetic fuel, which minimizes the output of CO_2. Two Fischer-Tropsch plants are scheduled to begin synthetic production in the next few years. Other renewable sources for refining biomass-derived jet fuel are forestry and agricultural crops and residues, sewage, industrial and animal residues, and municipal solid waste. Shell has begun a pilot facility on the island of Hawaii to grow marine algae for biofuel. In the Netherlands, Fred Kreuger, a professor at the Technical University at Delft, is working on growing an algae species that floats on water. IATA endorses the development of a road-map leading to the adoption of bio-jet. Officials say conversion to alternative fuels can only be accomplished after airlines have a thorough understanding of the economics and sustainability of bio-jet, taking into consideration production, refining, and transportation of these alternative fuels [22].

Biogas

India is the world's fifth largest and second-fastest-growing producer of greenhouse gas emissions. Generating electrical power by burning biogas (a mixture of methane and CO_2 produced by the bacterial degradation of organic matter) harnesses these greenhouse gases and reduces demand for fossil fuel-fired generating plants. To generate even more biogas energy, Capstone Turbine Corporation of Chatsworth, California, plans to install more microturbines in India in 2008–2009, building on lessons learned from its first installation, in Purilia, West Bengal, in 2006. The Purulia installation is at a dairy farm and consists of two 30 kilowatt microturbine systems, one of which is for backup. These can run either linked or independently of the power grid. In fact, when the grid is down, the diary is connected and that keeps it running. The microturbines use nickel alloys N06002, N07713, N07718, S301100, and S34700 in components such as the combustion chamber, spinning turbine, main rotor shaft, and recuperator housing, all of which run continuously with minimal maintenance. Austenitic stainless steels are the most cost-effective material for the systems that clean and compress the corrosive biogas prior to combustion in microturbines.

Biogas is mostly methane and CO_2, with hydrogen sulfide present as a contaminant. When hydrogen sulfide is mixed with water (with which biogas is always 100% saturated), it becomes a weak acid (hydrosulfuric acid). CO_2 forms a mild carbonic acid in the presence of water. Since biogas is very corrosive, a great deal of stainless steel is used in biogas conditioning and distributed generation. It has good corrosion resistance to those acids. Predominantly the structure uses S30400 and S30403, though sometimes the more corrosion-resistant S31600 and S31603 have been used. The S30403 might last 40–50 years whereas S31603 can last 100 years. A typical compression and cleaning system takes the gas drawn off the top of the sludge in a digestor, where manure has decayed for 20–30 days, and compresses and cleans it before it is burned.

In a simplified description of this process, the gas first passes through filters that remove the hydrogen sulfide. These tanks will typically be constructed of stainless steel plate 4.8 or 6.4 mm and are made to withstand negative pressure. The gas then passes through a compression system, where it is raised to pressures as high as 8.4 kg/cm². The piping used in this process is typically stainless steel. Next, it passes through heat exchangers to reduce the temperature to about 4°C (39.2°F), forcing more water out of the gas. The gas is then reheated to 27°C (80°F), reducing the relative humidity to 25%. Subsequently, the gas passes into stainless steel vessels where siloxane, a chemical used in lubricants and personal care products, is filtered out. Since siloxane turns into a glass-like substance at high temperatures, it must be removed before the gas is burned. After this step, the biogas is delivered to an end-use device, such as a turbine, internal combustion engine, or fuel cell.

Carbon Offsets and Biogas

Many airlines offer their customers a way to offset greenhouse gases produced by their flight. These so-called carbon offsets are small donations that are ultimately used to finance projects that will help reduce greenhouse gases in the world by an equivalent amount. Virgin Atlantic Airways, Ltd., for example, supports two projects with the money donated by customers. One is the reconstruction of a hydroelectric plant in Indonesia; the other, a project in India that generates electricity from biomass, such as sugarcane husks. In fact, many of the small biogas projects registered by the Clean Development Mechanism (CDM) in India burn renewable, alternative fuels in gas turbines to generate electricity. These turbines require nickel alloys for their efficient operation.

The CDM, which is an arrangement under the Kyoto Protocol allowing industrialized countries with a greenhouse gas reduction committment to invest in projects that reduce emissions in developing countries, has 835 registered projects, most of which are energy-related. Well over a third of all CDM's projects are in India (more than in any other country). India's participation in energy projects that reduce greenhouse gas emissions is a major development: According to the U.S. Agency for International Development, the country is the world's fifth largest (and second fastest growing) producer of greenhouse gases, and its power sector is the largest single contributor. USAID is helping India reduce its emissions with clean energy technologies and best practices. In 2003, the agency reported that biogas projects, along with other energy programs, helped prevent the production of 11.3 million tons of CO_2. As more projects are built to generate electricity from alternative fuel sources using gas turbines, fewer conventional power sources will be required, reducing global greenhouse gases.

Green Vision and Biogas

San Jose is moving closer to becoming the nation's first totally energy independent city. The CA city is pushing forward with its green vision of getting all its electrical power from clean, renewable sources, as well as diverting 100% of its waste from landfills and converting it into energy. In July 2009, the city council gave the green light to start negotiating plans that could lead to the nation's only organics-to-energy biogas facility. Renewable biogas, which contains methane, will help power the nation's 10th-largest city, which hopes to reduce its per capita energy use by 50% and get the remaining 50% from renewable sources. This project not only demonstrates San Jose's leadership in the production of renewable energy but will help in the city's economic development, zero waste, and energy goals for the city's green vision. After 3 years, the Biogas facility would start processing up to 150,000 tons of organic waste that would otherwise be destined for a landfill to create biogas in addition to high-quality compost that could be used to enrich local soils.

The biogas will be produced by the biological breakdown of food waste, as well as the organic share of the municipal solid waste system, in a process called dry anaerobic fermentation. The dry process, done in the absence of oxygen, is new to the United States. There are similar operations nationwide, but they involve wet waste, which is easier to recycle than dry waste. Dry waste is what usually ends up in landfills. The proposed new technology is already in use in 12 facilities in Germany and Italy. Thirteen more are planned for 2009–2010. The plant is planned to be built on a 40-acre site near the San Jose/Santa Clara Water Pollution Control Plant. The energy produced could be used to feed power to the water pollution plant, as well as sold as energy for the utility power grid.

Plans currently call for the creation of a fully integrated waste management system ecopark. These plans will be reviewed to determine if they adhere to the California Environmental Quality Act. The operation's impact on the habitat and neighborhood will be examined to make sure there is no imminent environmental threat because with any project that might seem too good to be true, there are often pitfalls. One challenge is dealing with leftovers with concentrated sludge containing pathogens, intact nutrients, and antibiotics. Poorly handled by-products could lead to leaks and runoffs that could be devastating to local plant and wildlife. Finally, another hope of this new technology is to create 25,000 clean tech jobs.

Switchgrass

The governor of TN is a big believer in alternative biofuels, although unlike many he doesn't think corn-based ethanol is the answer. He believes that with corn, you end up spending a barrel of oil to get a barrel and a half of ethanol. This doesn't address the issues of fossil fuel depletion or global warming. Instead, the governor is developing an ethanol program based on switchgrass, which can produce more biomass energy and grows in soil not suitable for agriculture, thus avoiding the current food-versus-fuel debate over corn. Better still, it is a high-yielding crop native to the state that is both drought- and flood-resistant. TN is a good state for growing switchgrass since it has good sunshine, good rain, and good soil. Additionally, the technical expertise exists in what it takes to produce cellulosic ethanol at the Oak Ridge National Laboratory (ORNL) and the UT School of Agriculture.

The state's biofuels experts are not likely to lack for resources. The DOE has invested $125 million at ORNL to build a biofuels research center that will focus on the science of converting switchgrass to ethanol. At the same time, the state has poured millions more into an initiative that rewards farmers for growing switchgrass and that helps pay for a biofuels refinery that will produce five million gallons of "grassline" per year. Ultimately, the goal is to produce one billion gallons a year and turn switchgrass into a major crop for the state.

Ethanol

The rise in the production of fuel ethanol has been both extraordinary and controversial. Seen as an important part of the solution for climate change and domestic energy security, fuel ethanol has also been blamed for the spike in food prices that took place in 2008. New policy developments have led to new technology developments. Ethanol is an increasingly common bio-fuel alternative to gasoline in many parts of the world. While fuel ethanol production in the United States is almost exclusively derived from corn, Brazil produces almost as much ethanol using sugar cane, and in the EU wheat is an important part of the feedstock mix. Ethanol is an attractive fuel because it comes from renewable sources. The use of ethanol as an additive to petroleum-based fuels can also result in cleaner burning with less emission of CO and particulates. For the United States, ethanol production has been encouraged as a way to decrease dependence on oil, and thereby improve national security.

The operating conditions for ethanol from the corn plant are not particularly corrosive to most stainless steels, pH ranges from 5.8 to 2 in most areas and, with a few exceptions, temperatures are relatively low. Pure ethanol is quite non-corrosive, but it is very hygroscopic; that is, absorbs water. Certain impurities from the bioethanol process, such as organic acids, sulfuric acid, and chloride-containing salt, will concentrate in the water phase, making it quite corrosive to carbon steel. A 400 million liter per year ethanol plant would typically require about 1600–2300 tons of stainless steel. Applications include equipment such as corn steepers, centrifuges, filters, mixers, dryers, and evaporators. Stainless steel tanks at various stages of production include liquefaction, fermentation, yeast slurry, beer well, washing, stillage, centrate surge, and syrup tanks. Most piping systems, including pumps and valves, are stainless steel.

Criticism of grain ethanol production grew in 2007–2008 as industry critics linked rising world food prices to the diversion of corn to ethanol production. Industry critics linked grain consumption and higher grain prices to the production of ethanol. Corn-based ethanol was also criticized for its life cycle environmental impacts. It was disputed whether corn-based ethanol as an automotive fuel increases rather than reduces greenhouse emissions, particularly when forests are converted into farm land. Although it is now suggested by some that high and rising grain prices were driven more by strong global economic growth and oil and gas prices than by the demand for ethanol, governments have been forced to examine the full life cycle impacts of the bio-fuels that are produced.

In 2007, the Energy Independence and Security Act (EISA) was passed in the U.S. EISA capped corn-based ethanol consumption in the United States at 69 billion liters (15 billion gallons) per year by 2015. This revised legislation continued to create the conditions for the U.S. bio-fuels industry to expand, while encouraging research and development into producing ethanol from alternative sources such as cellulose. Producing ethanol from cellulose is a new approach that may alleviate concerns about using forest land or food to produce fuel. Cellulosic ethanol can be produced from any plant material, including waste agricultural material such as corn stalks, potentially doubling yields and reducing its impact on climate change. Cellulosic biomass requires pretreatment before it can be converted into fermentable sugars. There is an array of pretreatment processes, many of them involving corrosive environments. Regardless of the technology used to break down the bio-mass in order to release the sugars, whether its acid hydrolysis or enzymatic hydrolysis, pretreatment tanks will be required as part of the production process.

Process technologies for the production of cellulosic ethanol are still in the development stage. Although the first commercial cellulosic plants are being built, large scale, cost-competitive, commercial production of cellulosic ethanol is not expected to get under way until the year 2011–2014. The ethanol industry, both in the United States and internationally, will require considerable quantities of stainless steel over the next 10–12 years. In the United States alone, the bio-fuels industry is expected to require about 500,000 tons of stainless. The bio-fuels industry in North America will remain a key end-use market for stainless steel; it is forecast to account for half of global production growth in bio-fuels over the next 20 years. Ethanol process technology companies like the way 304L performs and perceive it to be both reliable and cost-effective. With the advent of cellulosic ethanol, 316L stainless will certainly play an increasingly important role. The fuel mix used by societies in the twenty-first century will not resemble those of the late twentieth century. While technological progress and political priorities may favor certain alternatives at different times, what remains constant is the need for nickel-containing materials to provide the performance needed to produce them.

Furafuel

A chemical process called Furafuel, in which green waste is changed into a stable bio-crude oil, has reportedly been developed by CSIRO and Monash University, Australia. The process is based on low-value waste such as forest thinnings, crop residues, waste paper, and garden waste, significant amounts of which are currently dumped in landfills or burned. The plant wastes being targeted for conversion into bio-fuels contain a chemical known as lignocellulose, which is increasingly favored around the world as a raw material for the next generation of bio-ethanol. It is predominantly found in trees and is made up of cellulose; lignin, a natural plastic; and hemicellulose. By making changes to the chemical process, the developers have been able to create a concentrated bio-crude that is much more stable than any achieved elsewhere in the world.

Energy Technologies

According to the U.S. DOE, the United States currently uses about 20 million barrels of oil per day, at a cost of about $2 billion a week. Right now, the United States imports more than 55% of the oil that is consumed; that is expected to grow to 70% by 2025. Energy powers everything that operates in the United States, from businesses, to our homes, to our means of transportation. But the generation of power is the single largest source of air pollution in this country. Air pollution contributes to lung diseases that kill more than 330,000 Americans every year, according to the American Lung Association. And the hydrocarbons released from the use of fossil fuels are causing severe global warming, and could, ultimately, cause catastrophic climate changes. Therefore, with such statistics, the following is being done to reduce our dependence on fossil fuels, cut emissions from vehicles, and create a cleaner, more efficient energy-generation plan. Technologies such as hydrogen fuel cells, renewable energy sources, hybrid vehicles, and nanotechnology all have a major role to play in the future of energy and the environment.

Renewable Energy

The EPA estimates that CO_2 is responsible for one-half to two-thirds of our contribution to global warming. CO_2 is the major emission of fossil fuel use. Coal, oil, and natural gas, fossil fuels, are nonrenewable. They draw on finite resources that will eventually run out, become too expensive, or significantly damage the environment. Renewable energy, such as wind and solar energy, will not run out or seriously damage the environment.

Wind Power

According to the DOE, wind energy is the fastest-growing energy source in the world. It's also one of the lowest-priced renewable energy technologies, costing only 3–5 cents per kilowatt-hour. From North Dakota to Texas, there is enough wind power to power the entire U.S. power grid today. Wind power emits no CO_2, a greenhouse gas that is the number-one cause of global warming. Wind turbines use two or three long blades to collect energy in the wind and convert it to electricity. The diameter of the blades determines how much power the wind turbine can extract from the air. Wind power requires locations with high wind speeds such as plains, mountaintops, and offshore locations. To create enough electricity for a town or city, several turbines are placed together to create a wind farm. An Irish company called Airtricity has partnered with General Electric to build what will be the world's largest offshore wind farm on the Arklow sandbank, located off the east coast of Ireland. The wind farm will consist of giant, mega-turbines that will stand taller than a 30-story building. According to Airtricity, the turbines will supply electricity to about 3000 homes. Denmark currently receives 25% of its electricity from wind power, and Spain and Germany are close to that mark. By 2050, Europe expects to get 1/2 of its power from the wind.

Solar Power

Solar energy is continuously supplied to the Earth by the sun. Solar or PV power is generated when light is converted into a PV current. Photovoltaic systems consist of silicon wafers that, when hit by sunlight, undergo a chemical reaction that results in the release of electricity. According to the EPA, in the southwestern United States, sufficient solar energy falls on an area of 100 square miles to provide all of the country's electricity requirements.

Sandia National Laboratories (SNL), Albuquerque, New Mexico, and Stirling Energy Systems, Phoenix, Arizona, have built six solar dish-engine systems for electricity generation that provides enough grid-ready solar electricity to power more than 40 homes. The mini power plant produces up to 150 kW of grid-ready electrical power during the day. The goal was to construct 20,000 systems to be placed in one or more solar dish farms, providing electricity to southwest U.S. utility companies. Each unit operates automatically without operator intervention or on-site presence. Each morning at dawn, the unit starts up and operates throughout the day, tracking the sun, and responding to clouds and wind as needed. The solar dish generates electricity by focusing the sun's rays onto a receiver, which transmits the heat energy to an engine. The engine is a sealed system filled with hydrogen, and as the gas heats and cools, the pressure rises and falls. The change in pressure drives the pistons inside the engine, producing mechanical power, which in turn drives a generator and makes electricity. The units automatically shut down at sunset, and the system can be monitored over the Internet. Standalone units have already been demonstrated as an effective means of pumping water in rural areas.

Cars that Care

Currently, all major automakers are producing or working on hybrid electric vehicles (HEVs) and fuel cell vehicles. Hybrid cars, as the name implies, combine a gasoline engine with an electric motor. The first mass-produced HEV was the Toyota Prius in 1997. The current Prius is equipped with Toyota's Hybrid Synergy Drive powertrain, a full hybrid system that can operate in either gas or electric modes, as well as a mode in which both the gas engine and electric motor are in operation. Several years ago, Ford introduced the world's first full-hybrid sport-utility vehicle (SUV), the Ford Escape hybrid.

Honda also has been a leader in the HEV revolution with the Insight, introduced in 1999, and hybrid version of the Civic and Accord. The EPA has recognized the Honda Civic GX as the "cleanest car on Earth" with an internal combustion engine that runs only on natural gas. DaimlerChrysler in 2003 opened the first biogenous diesel fuel filling station in Germany. The fuel is largely CO_2 neutral, producing few pollutants through combustion. Called Biotrol, the fuel is produced from waste wood from forestry management sources. Hydrogen fuel cell cars can operate without a gasoline engine, using an electric motor that draws power from the fuel cell. Hydrogen is stored on the vehicle in a pressurized tank.

Nano Connection

Around 1985, Richard E. Smalley of Rice University invented the elongated carbon fullerene structures known today as nanotubes or bucky-tubes. The nanotube, and the exploding field of nanotechnology, can play a vital role in the world's energy challenge, according to Smalley. "Energy is the single most important challenge facing humanity today," Smalley stated. "We can find 'the New Oil,' the new technology that provides the massive clean, low-cost energy necessary before the century is over. Electricity will be the key," he said.

Smalley's vision includes every building, house, and business having its own local electrical energy storage device by 2050, and massive electrical power transmission over continental distances, which would enable electrical power to be transported from solar farms in New Mexico to New England. Here's where nanotechnology comes in, according to Smalley. "Nanotechnology in the form of single-walled carbon nanotubes, forming what we call the armchair quantum wire, may play a big role in this new electrical transmission system." A specific type of nanotube, called an armchair structure because it resembles a cross-section of an armchair, is a metallic conductor that can carry high current densities. The Earth's temperatures are rising at their fastest rate in 15,000 years, and there is more CO_2 in the atmosphere today than there has been in 45 million years. Whether it comes from the sun, the wind, the oceans, or nanotubes, a new method of affordable, clean energy must be developed.

Butanol

An Ohio State University team led by Shang-Tian Yang, professor of Chemical and Biomolecular Engineering [23], has found a way to double the production of the biofuel butanol, which might someday replace gasolene in automobiles. The process improves on the conventional method for brewing butanol in a bacterial fermentation tank. Normally, bacteria could only produce a certain amount of butanol, perhaps 15 grams of the chemical for every liter of water in the tank, before the tank would become too toxic for the bacteria to survive, explained Yang. The researchers developed a mutant strain of the bacterium Clostridium beijerinckii in a bioreactor containing bundles of polyester fibers. In that environment, the mutant bacteria produced up to 30 grams of butanol per liter.

Right now, butanol is mainly used as a solvent, or in industrial processes that make other chemicals. Once developed as a fuel, butanol could potentially be used in conventional automobiles in place of gasolene, while producing more energy than another alternative fuel, ethanol. "Today, the recovery and purification of butanol account for about 40% of the total production cost," explained Yang. "Because we are able to create butanol at higher concentrations, we believe we can lower those recovery and purification costs and make biofuel production more economical." Currently, a gallon of butanol costs approximately $3.00, a little more than the current price for a gallon of gasolene. The engineers are currently refining the technology and wish to further the technology within the industry.

Powering the Future with Advanced Nuclear Power Systems

Currently, 441 nuclear power reactors are in operation in 31 countries around the world. Generating electricity for nearly 1 billion people, they account for approximately 17% of worldwide electricity generation and provide half or more of the electricity in a number of industrialized countries. Nuclear power has an excellent operating record and generates electricity in a reliable, environmentally safe, and affordable manner without emitting noxious gases [24].

Concerns over energy resource availablity, climate change, air quality, and energy security suggest an important role for nuclear power in future energy supplies. While the current Generation II and III nuclear power plant designs provide a secure and low-cost energy supply in many markets, further advances in nuclear energy system design can broaden the opportunities for nuclear energy. The current fleet of Generation II and III nuclear reactors, generally designed, developed, and deployed in the 1970s, has done a marvelous job of safely and economically providing ~17% of the world's electricity (~20% in the United States). However, they operate with modest thermal efficiency, they rely upon active safety systems, and they generally operate without recycling fuel, thus utilizing relatively little of the vast reserves of fertile nuclear materials that are available. As a result, interest is growing in the development of advanced nuclear reactors.

U.S. Generation IV Priorities

Five advanced technology concepts are being pursued at varying levels of effort based on their technology status and potential to meet program and national goals.

- Two are thermal neutron spectrum systems: very high temperature reactor (VHTR) and supercritical water cooled reactor (SCWR), with coolants and temperatures that enable hydrogen or electricity production with high efficiency.
- Three are fast neutron spectrum systems: Gas-cooled (GFR), lead-cooled (LFR), and sodium-cooled (SFR) fast reactors. These will enable more effective management of actinides through recycling of most components in the discharged fuel. Most of the effort in the United States is devoted to the VHTR and the SFR [24].

The Very High Temperature Reactor (VHTR)

The VHTR system is based on a thermal neutron spectrum with graphite moderator and helium coolant. The VHTR system has coolant outlet temperatures approaching 1000°C (1832°F). The primary circuit is connected directly to a closed cycle gas turbine power conversion system for high efficiency (~48%) production of electricity. The VHTR is intended to be a high-efficiency system that can supply process heat to a broad spectrum of high-temperature and energy-intensive nonelectric processes as well as high-efficiency production of electricity. The system also has the flexibility to adopt different U/Pu and Th/U fuel cycles and minimize waste. The VHTR system is envisioned for hydrogen production and other process-heat applications because of its high-temperature capability.

Hydrogen may be produced by several processes. These include conventional electrolysis via electricity generated at high efficiency, high-temperature steam electrolysis by means of both electricity and high temperature heat, and thermochemical water-splitting processes requiring only heat. Production of hydrogen by CO_2-free nuclear processes has the potential to power our transportation economy as the supplies of petroleum-based fuels continue to shrink. Nuclear-produced hydrogen could also make synthetic-hydrocarbon fuels and power fuel-cell vehicles.

Sodium Cooled Fast Reactor

The sodium cooled fast reactor (SFR) system features a fast neutron spectrum and a closed fuel cycle for efficient conversion of fertile uranium and management of actinides. A full actinide recycled fuel cycle is envisioned with two options. One option is an intermediate size (150–500 MWe) sodium cooled reactor with a uranium-plutonium-minor actinide-zirconium metal alloy fuel [24]. The second option is a medium to large (500–1500 MWe) sodium cooled fast reactor with mixed uranium-plutonium oxide fuel. The SFR system is primarily envisioned for electricity production and actinide management.

Nanomaterials and Are They Environmentally Safe?

Nanotech sector leaders and analysts called for more funding for research into the environmental, health, and safety (EHS) impact of nanotechnology and one group released an inventory of existing EHS efforts [25].

Researchers and scholars have argued that the government should be more proactive in addressing the risks of nanotechnology, in particular nanomaterials, so that the public can be accurately informed about both the potential risks and rewards. "Even if studies showed every commercially relevant nanoparticle to be harmless in every real-world usage scenario, public skepticism about the safety of nanoparticles could still build and sharply limit the use of nanoparticles in products," said M. Nordan, Vice President of Lux Research.

Meanwhile, the Project on Emerging Nanotechnologies at the Woodrow Wilson International Center for Scholars unveiled an inventory of research into nanotechnology's potential environmental, human health, and safety effects. Wilson center scientists said particular areas needing support include investigating workplace safety issues like the risk of explosion in the production of nanopowders. They also said that virtually none of the research deals with future generations of nanomaterials; and that little funding is allocated to explore possible links between exposure to nanomaterials and diseases of the lung, heart, or skin.

Nanowaste: The Next Big Threat? [26]

For now, commercial nanotech is confined to nanomaterials, particles, and compounds that range from 1 to 100 nm (10^9 meters) in size. Nanoscale titanium-dioxide particles, for example, are used in cosmetics and sun blocks, and nanosized silica acts as a filler in several consumer products, including dental fillings. A recent discovery by researchers at Rice University, is making some people think twice about furthering their efforts in pursuing nanotechnology. In the past, scientists had thought that bucky balls were strongly hydrophobic and strictly insoluble in water. They believed that bucky balls dropped into the open environment could not be transported by water, but would simply stick to the soil and other organic materials. The Rice researchers were able to clump together the C60 molecules and they named them nano-C60. It easily travels in water if conditions are right and they also found that small concentrations of nano-C60 (20 parts/billion) killed half of the human liver and skin cells in lab samples. Other studies have shown that nano-C60 damages brain cells in fish and halts the growth of bacteria.

Scientists have claimed that they have no way to track nanomaterials like bucky balls or nanotubes in the environment. Even if they could be found, there is no way to remove them from the soil or water. In fact, there's no acceptable way to find or remove them from the human body. To make matters worse, no one has studied what the effects of nanotubes, bucky balls, and other newly created nanomaterials have on the environment or human physiology. The fear is that because nanomaterials are smaller than cells, they might enter and create havoc or bioaccumulate in smaller creatures, then work their way up the food chain in ever-increasing concentrations until they do cause problems for humans.

Some worry that nanomaterials could harm bacteria, the engine behind every ecosystem and food chain. With these unanswered questions, some groups want to halt nanotechnology until its safety can be proven. These same groups insist upon a moratorium on nanotech until an international body sets lab and manufacturing protocols for handling, using, and disposing of nanomaterials. Finally, the group wants countries to be informed about and prepared for any changes that new technologies such as nanotech or genetically modified foods might bring.

Although the prefix nano (Greek for dwarf) has only recently gained popularity, people have been creating and using nanomaterials for thousands of years. Medieval glaziers, for example, used nanometer-sized particles of gold and silver in their red and yellow stained glass. Diesel engines also emit nanomaterials in the soot they generate. There are also natural sources of nanomaterials, including forest fires that create bucky balls and volcanoes that put a variety of chemicals into the atmosphere in aerosol form. There's even a bacteria that digests iron and leaves nanosized particles of magnetite behind.

Scientists and engineers are not totally ignorant of the threat nanomaterials pose. There is a current theory on filters for small particles that relies on a concept called diffusional capture. It says that particles larger than 0.3 μm get stuck to fibers in the filter medium while smaller particles, including those in the nanometer range, get stuck to those particles clinging to the fibers.

Researchers also know that some nanomaterials break down naturally over time, much like plastics. Others are rapidly absorbed by the soil, which presumably reduces bioavailability and the threat of exposure to humans and other living creatures. Some nanomaterials are known to be safe based on their approval and use in medical applications. Numerous studies continue examining all aspects of how materials react, what happens when simple chemicals are added to nanomaterials, what happens to toxicity, what happens when nanomaterials are exposed to UV radiation from sunlight or other chemicals and life cycles and what other reactions we can expect. Nanomaterials are very expensive, costing thousands of dollars a pound for engineered versions of bucky balls and nanotubes. Companies are unlikely to view nanomaterials as waste products and thus, treat them with care, especially since they are also made in relatively small quantities.

Therefore, the debate goes on where some groups want to halt nanotech in its tracks until it is proven totally safe or acceptable standards are established, others want to move ahead and bring the benefits of the technology to society as soon as possible and turn a profit while doing so. In the middle, a

growing number of people want to investigate the possible dangers of nanomaterials without halting progress or banning anything without solid evidence of risk. There's also a growing consensus that nanomaterials should
not be released into the environment until more is known about their life
cycle outside the lab.

Most nanotech researchers urge a reexamination and restructuring of how
safety organizations like the EPA and OSHA define and treat nanomaterials.
One common recommendation is that the federal government spend more
money studying the health and environmental effects of nanomaterials.
Currently, about 6% of the federal nanotech budget goes into these efforts.
Industry leaders and social activists agree it should be more on the order of
10%. Some also believe companies looking to cash in on nanotech should
shoulder some of the R&D that will prove their products safe to manufacture, use, and dispose of.

Aircraft Emissions and Climate Change

The tough question remains. Is the aircraft industry doing enough to maintain public confidence in aviation's continued development? Or is it all just
veneer? It is not just window dressing, but does the public know that? If the
industry is to maintain the public trust, they will have to be highly ethi-
cal, fully transparent, and very honest with the consumer. They will have to
admit that today's aircrafts are still increasing CO_2 emissions to the environment. As Air Navigation Service Providers (ANSPs), the industry recognizes
that the fundamentals of aviation are forever changing. In addition to the
clear signs of climate change, world prices for fossil fuels, energy, raw materials, and food are shifting—changing the basic rules of (not just) society, but
especially for the aircraft industry.

If the industry wishes to see continued growth of aviation, it will have to
face some hard facts. Environmental performance is largely a by product
of efficient airspace design and optimized aircraft performance. To address
some of the most pressing airspace inefficiencies, the industry will have
to address basic airspace allocation to civil and military users. Similarly,
to optimize aircraft operations in all phases of flight, ANSPs must work
with regulators, aircraft manufacturers, airlines, airports, pilots, and engineers to optimize ground and flight operations to improve overall aircraft
performance.

The industry cannot minimize the environmental impact of aircraft operations by themselves. Solving their problems requires a balancing act among
many diverse interests. Nations, airports, airlines, ANSPs, and manufacturers
must focus on one thing: optimizing the performance of the entire aviation
system. To achieve this in a new economic reality, the industry will have to
change their ways, and start to think and act as one system. With a new economic reality taking hold, sometime in the not-too-distant future, aviation
growth could well stop. Ridiculous as this may sound, it could well result

from a combination of record oil prices, an extended credit crunch, increased taxation, market saturation, negative customer experiences due to the system breaking down, capacity shortfalls in the air and on the ground, and ever-inceasing security measures. But most importantly, growth in demand could slow as a result of a shift in public sentiment on climate change. Thus, the industry is threatened on two sides: the industry's own ability to maintain system reliability and performance, and public sentiment on climate change.

There was a major global warming legislation (America's Climate Security Act) introduced in the U.S. Senate in 2007, as well as other bills in 2009–2010. The legislation imposes an emissions cap on electrical utilities, manufacturing sources, transportation, and introduced a market-oriented cap-and-trade system, in which emission allowances could be bought and sold among firms. Hughes [27] describes key building blocks of global air traffic system modernization that has been deployed and could be utilized throughout the world to lower fuel burn, emissions, and noise while boosting runway capacity.

Soybeans and Composites

One of the first projects undertaken by the University of Delaware was to develop unsaturated polyester resin made from soybean oil [28]. In today's market, people look at using renewable resources for a variety of reasons, from marketing for "green" in building and construction, to alternative energy sources to replace oil. The composites industry is a part of that process. Soy-based polyurethanes are being evaluated in composites using the pultrusion process, and soy-based unsaturated polyester resins are being evaluated and used in a number of composite applications in the marine, construction, and agricultural industries. Epoxy resins will soon be made from glycerin, a by-product from biodiesel from soybeans, which will be a precursor to epichlorohydrin.

In the future, it is hoped that companies will look at biobased materials as a way to offset increases in material prices due to crude oil. Research is continually encouraged to find ways to incorporate soybeans into composites, primarily thermoset composites. However, involvement in incorporating these materials into thermoplastics and for fibers continues, while others are looking at fermenting the sugars (which are not easily digested by animals) in soybeans to make chemicals for further use in the composites area. One example is fumaric acid, which is used in the manufacture of unsaturated polyester resins. This can result in a usable chemical and a higher-quality food product.

Food/Bioplastics

In 1907, the Belgian-born chemist Leo Hendrik Baekeland ushered in a materials revolution with his invention of Bakelite, a plastic made from phenol and formaldehyde. Since then, plastics have permeated civilization, serving as the raw material for everything from packing to electronic devices, toys to prostheses, bicycle helmets to contact lenses. But the very indestructibility that makes plastic so attractive has dire consequences for our environment [29].

Professor Sergai Braun of The Hebrew University's Silberman Institute of Life Sciences recently developed a method for producing biodegradable plastic using cheap and plentiful by-products from the processing of protein-rich plants such as corn, canola, and soybeans. Professor Braun's technique utilizes a simple chemical reaction to produce a new building block for the construction of plastic. Materials made with this process has qualities similar to those of PP, yet retains the organic properties that render it highly degradable. The food oils and starches industry produces a surplus of vegetable proteins that can be used for manufacturing of environmentally friendly food packaging. In fact, just 6% of surplus vegetable protein production in the United States would be needed to produce plastic packaging for the entire country. Unlike conventional petroleum-based plastics, which persist in the environment for hundreds of years, those produced from plant sources should decompose in a controlled composting environment, such as a municipal facility, in under 90 days.

The growing use of bioplastics like those developed by Professor Braun promises to reduce the emission of greenhouse gases during the production process, reduce our dependence on oil, and reduce the amount of trash accumulating in the world's oceans. To quote Professor Braun, "Until now, the main barrier to the entry of biodegradable products into the plastics market was their high price. Biodegradable plastics, such as the popular PLA (polylactide acid plastic), cost five to ten times the price of ordinary packaging materials. Our invention is intended to change the market conditions and offer a cheap and available alternative that is also environmentally friendly." Perhaps the transformation of the industry spawned by Leo Baekeland has begun.

Energies

Renewable energy will be the world's fastest-growing source of electricity generation over the next two decades, although it will still make up a relatively minor portion of the global energy supply, according to the Energy Information Administration. The majority of that increase will come from the use of wind power and water power, or hydropower. The Presidential Administration of Barack Obama advocates a policy that would require 25% of U.S. electricity demand be met by renewable energy by 2025. But wood, coal, and oil continue to be the primary fuels that provide energy for the majority of human activities. Alternative technologies that utilize the natural energy sources of the Earth; such as wind, water, and the sun, continue to make progress.

Energy: Batteries

Discussions of cutting-edge energy always come down to a technology that is more than 200 years old: the battery. Modern versions are much more powerful than Alessandro Volta's pile of copper and zinc discs and brine-soaked cardboard. But they're still not up to twenty-first century tasks like powering the autos of 2020 or storing solar power for use at night. Many experts believe that batteries are the key to the nations' energy future. Energy storage is going to be vitally important. It will be important to match the production with the demand. That is easy to do when you have oil in the ground or coal that you can pile up, but you can't do that with electricity. You have to be able to store it somehow.

In the future, people are not only talking about batteries in cars, but batteries in power plants or in wind farms. There has to be different types of energy storage. You have to be able to store at the site, and you have to be able to store in a distributed way. And then when you talk about vehicles, you have to be able to store in a more compact and efficient way. So there will be all types of battery systems that are required. Will the batteries in our future be up to this challenge? Some people claim there is no doubt about it. Over the last couple of years there has been a lot of talk about how the battery is not ready for this. That's a fact: batteries and battery systems aren't ready for this. But that's because they haven't been required to be, there hasn't been a market. However, now people are moving beyond demonstration projects to commercialization. That is how this whole thing is going to move much more exponentially.

What caused the change? Was it $140-a-barrel oil? No one really knows. Since the automotive companies have been in crisis it now means it's time for a disruptive technology. The automotive industry has now moved beyond trying to protect the old technology, to a place where they want to be a part of the solution and now are a part of the race. The United States does not have a leadership position in the world in the consumer battery industry. Today it's in Korea. It's in Japan. And there's an awful lot of development in China. That's where the critical mass is. A partnership with the U.S. government to put manufacturing here and research and development in the United States—that is the proper thing to do. The payback on that is something that we're not going to see over the next few years exclusively.

Batteries are improving at the rate of roughly 8% a year. That doesn't seem to be the kind of disruptive technological advance that is needed, however, many believe that the technology has not been pushed hard enough. A true breakthrough advance is needed especially since this whole field has been underfunded and underfocused. If automobile companies can put vehicles on the road with real volume commitments, a step-up change is needed and can occur for batteries.

Current hybrid vehicles have both a gas engine and a battery-powered one. Many ask if it wouldn't make sense to have an all-electric vehicle? The hybridization of the fleet is moving very quickly in Europe, and you're going to see it in the United States very, very quickly. Then the plug-ins will come. You'll see the combustion engine get smaller, and then as confidence in the technology increases, and as the battery becomes more capable, you'll see the combustion engine go away, particularly for urban environments—people who have vehicles that they commute to work in, or that they use in their urban environments. Other vehicles will be available for other purposes and there is going to be a transformation that is unlike anything you've seen in the automotive industry for 50 years.

Advanced batteries are made with lithium. Most of the world's lithium reserves are found in Bolivia. Therefore, are we transferring dependence from unpredictable Middle Eastern oil to unpredictable Bolivian lithium? Experts claim that 97% of the batteries that are purchased today, lead-acid batteries, are recycled. So the lead that you get in your battery, you're just borrowing it for a little while. It's a highly recycled product. The same will occur with lithium batteries. In the future, Daimler's first vehicle lithium battery will be produced by Johnson Controls, the same supplier will produce the battery for Ford's first plug-in electric vehicle, and BMW's first lithium battery will also be built by Johnson Controls.

The higher energy and power density of Li-ion battery technology offers a significant reduction in weight and volume for HEV battery systems compared to lead acid and nickel metal hydride technologies. Recent demonstrations [30] by Saft's High Power Li-ion battery technology has shown a specific power of over 6000 W/kg under continuous discharge, and a pulse discharge of 8000–12,000 W/kg. This technology has been proven and implemented. The VL34P cell was recently developed to provide the dual performance of power and energy suited for military vehicle applications. This cell has been used in series hybrid applications to support battery-only operation, silent watch, and drive assist. The performance characteristics of the cell make it an ideal candidate for integration into battery systems for hybrid electric drive vehicles.

The cell incorporates improvements in many aspects of the cell design. The electrochemical design has been updated to improve the performance of the cell for high rates and low temperatures. It has been repackaged to improve the volumetric and gravimetric power and energy densities, while moving to lower-cost components and processes. Improved packaging and asembly methods have also allowed for a significant reduction in the internal resistance of the cell, resulting in improved power, improved energy, and lower heat generation. This allows for a uniform temperature distribution within the cell, and a means of effectively removing heat from the cell to improve cell life.

The VL34P cell has been shown to be well suited for use in military vehicle applications. The high power of the cell is able to support the charge and discharge power profiles of hybrid vehicles. The low cell resistance allows for simple cooling methodologies for the modules and batteries. Implementation of controls in the battery systems for vehicles provides an excellent means of effectively monitoring and integrating the battery within the vehicle controller. Iron phosphate is suitable for high-power applications that require an added level of redundant safety to extreme abuse. The requirements of the application must be considered, as there are tradeoffs in lower power and energy, poorer low-temperature performance, and less stability at high temperatures for storage.

Energy: Wind Turbines

Using smoke, laser light, model airplane propellers, and a campus wind tunnel, a team led by Johns Hopkins University researchers is trying to solve the airflow mysteries that surround wind turbines, an increasingly popular source of green energy [31]. The rise in oil prices and a growing demand for energy from nonpolluting sources has led to a global boom in construction of tall wind turbines that convert the power of moving air into electricity. The technology of these devices has improved dramatically in recent years, making wind energy more attractive. For example, Denmark is able to produce about 20% of its electric energy through wind turbines. But important questions remain: Could large wind farms, whipping up the air with massive whirling blades, alter local weather conditions? Could changing the arrangement of these turbines lead to even more efficient power production? The researchers from Johns Hopkins and Rensselaer Polytechnic Institute hope their work will help answer such questions.

"With diameters spanning up to 100 meters across, these wind turbines are the largest rotating machines ever built," said Charles Meneveau, a turbulence expert in Johns Hopkins' Whiting School of Engineering. "There's been a lot of research done on wind turbine blade aerodynamics, but few people have looked at the way these machines interact with the turbulent wind conditions around them. By studying the airflow around small, scale-model windmills in the lab, we can develop computer models that tell us more about what's happening to the atmosphere at full-size wind farms."

The researchers gather information on the interaction of the air currents and the model turbines by using a high-tech procedure called stereo particle-image velocimetry. First, they "seed" the air in the wind tunnel with a form of smoke—tiny particles that move with the prevailing airflow. Above the model turbines, a laser generates two sheet-like pulses of light in quick succession. A camera captures the position of the particles at the time of each flash. "When the images are processed, we see that there are two dots for every particle," said Meneveau. "Because we know the time difference between the two laser shots, we can calculate the velocity. So we get an instantaneous snapshot of the velocity vector at each point. Having these vector maps allows us to calculate how much kinetic energy is flowing from one place to another, in much greater detail than what was possible before."

As a result of this work the data could lead to a better understanding of real wind farm conditions. Meneveau pointed out that dense clusters of wind turbines also could affect nearby temperatures and humidity levels, and cumulatively, perhaps, alter local weather conditions. Highly accurate computer models will be needed to unravel the various effects involved. "Our research will provide the fluid dynamical data necessary to improve the accuracy of such computer models," Meneveau said. "We'd better know what the effects are in order to implement wind turbine technology in the most sustainable and efficient fashion possible" [31].

The United States now has the world's largest installed base of wind power, according to the World Wind Energy Association. More than 8300 megawatts of wind power were installed in 2008, expanding the nation's total wind-power-generating capacity by 50% in a single year.

Wind turbines convert the kinetic energy of the wind into mechanical energy, and then into electricity. A turbine is generally made up of rotor blades, a gearbox, and a generator. Because wind speeds fluctuate, operating the generator and the turbine in the most efficient way is difficult. Engineers at Purdue University and SNL have developed a technique that uses sensors and computational software to constantly monitor forces exerted on wind turbine blades, which are made primarily of fiberglass and balsa wood.

The engineers embedded sensors called uniaxial and triaxial accelerameters inside a wind turbine blade just as the blade was being built. The sensors measure two types of acceleration. One type, dynamic acceleration, results from gusting winds, while the other, called static acceleration, results from gravity and the steady background winds. Accurate measurement of both forms of acceleration is essential to estimate forces exerted on the blades. Sensor data in a smart system might be used to better control the turbine speed by automatically adjusting the blade pitch, while also commanding the generator to take corrective steps.

Electricity from the largest wind power farm on Indian land in the United States has begun flowing into California's power grid thanks to a partnership between GE Energy Financial Services and Babcock & Brown, the Australian investment and advisory firm that serves as the project sponsor. The wind farm comprises 25 turbines that each can generate two megawatts of electricity. After 8 months of construction and a month of testing, the turbines are feeding power into the San Diego Gas & Electric grid from the Campo Indian Reservation atop the Tecate divide.

A U.S. Government Agency recently ordered renewable energy tower systems, the Mojo, from Critical Solutions, Inc., Ashburn, Virginia. The Mojo is a road-ready, rapidly deployable trailer that includes a 26-ft telescoping tower, battery box, a complex set of electronics, as well as, four solar panels and a wind generator. A Mojo typically produces approximately 520 W of sustained solar power, and the wind turbines generate 350–600 W of sustained power. The U.S. Marine Corps is testing and evaluating the Mojo system for use in applications from sensor suites to communication.

The use of natural power sources eliminates the need for grid-based power, suiting a Mojo system for use in remote environments such as those in combat or border patrol situations. The systems can handle a variety of payloads to support missions such as intelligence, surveillance and reconnaissance, or communications. The balance of solar and wind power built into a Mojo enables it to operate through various environmental occurrences such as at night when solar power is not available or during weather when wind is absent. In the past several years, the use of carbon fibers has increased especially in its use in aircraft. Because of this growth in the use of carbon fibers, the wind energy systems engineers have looked at its use in wind turbine blades. As a result, wind energy systems and the aerospace industry encompassed the strongest market drivers for carbon fiber usage in 2007.

In terms of wind energy systems, turbine producer Vestas, headquartered in Randers, Denmark, predicts that the wind power share of global power consumption will increase up to 10% by 2020. The company utilizes carbon fiber in load-bearing spars on wind turbine blades, and has installed nearly 34,000 turbine systems to date. As of January 1, 2008, OEM Gamesa, of Bilboa, Spain, reports over 8000 MW of committed orders internationally for wind generators, and 3000 MW of completed installations. Wind power applications will drive growth volumes for large tow (24 k) carbon fiber producers and the global need for clean energy will again allow the producers for this end-use to continue to invest. In fact, the use of carbon fiber in wind energy blades is projected to be the second largest application after aerospace by 2010. Wind installations are growing at an accelerating rate, with a trend toward larger turbines with longer blades that use more carbon fiber content. In 2008, the estimate for installed new wind capacity is around 19,000 MWT, with a mix of turbine sizes. If the average size is 2.5 MWT, then 7600 turbines will be installed. The number of blades on that many systems is estimated at 22,800 and the amount of carbon fiber in each blade would range from zero to 1.5 tons.

Finally, wind power continued to be one of the most popular electricity generating technologies in the EU in 2007, making up 40% of the total new power installations. One reason for this increase in installations is that emission-free wind power can be brought on line quickly, and must play a major role in meeting climate protection targets. This is especially the case in the critical period between 2009 and 2020 when greenhouse gas emission must peak and begin to decline if the EU is to avoid the worst impacts of climate change. Wind energy has now become an important player in the world's energy markets. In terms of economic value, the global wind market is estimated to be worth about $36 billion per year in new generating equipment. Wind power is increasingly economically competitive with conventional sources of electricity. Increasing volatility in fossil fuel prices and increased concerns about energy security mean that wind power is often the most attractive option for new generation capacity, from any point of view.

The Purdue University blade is being tested on a research wind turbine at the U.S. Department of Agriculture's Agriculture Research Service Laboratory in Bushland, Texas. The research being pursued covers the application of a blade system for advanced, next-generation wind turbine blades that are more curved than conventional blades. Dong Energy is building and will run the Norwegian offshore wind farm Nygårdsfjell in the Narvik municipality in Northern Norway. The wind farm will have a capacity of between 25.3 MW and 33 MW. The Nygårdsfjell 2 wind farm represents yet another important step in the effort to triple production of renewable energy by 2020. It also contributes to achieving the vision of becoming a leading player in the renewable energy sector in Northern Norway and of reaching the goal of producing 300 GWh generated by wind power over the next 10 years.

Nygårdsfjell 2 will help to further develop DONG Energy's experience of setting up onshore farms in mountainous terrain, since DONG Energy has also erected the Storrun wind farm in similar terrain in Sweden, and runs the 6.9 MW Norwegian demonstration farm Nygårdsfjell 1, which has been operating since 2005. For the first time, offshore wind power is flowing from turbines in the North Sea to the German power grid. The consortium of EWE, E.ON, and Vattenfall has now successfully started up and adjusted for regular electricity generation, the first three of a total of 12 wind turbines at the Alpha Ventus wind farm. The turbines, with a nominal capacity of 5 MW, are located 45 km/28 miles north of the island of Borkum. It is anticipated that the wind farm's 12 turbines, five of which have already been completed, will all be in operation by the end of 2009. The already installed wind turbines are currently undergoing the so-called adjustment phase and as the name suggests, during this phase all the functions of the turbines are technically inspected and adjusted for subsequent long-term operation. This is comparable with making technical adjustments to the engine of a new car. The adjustment phase is followed by a period of trial operation. In this phase, the wind turbines are subjected to various test scenarios, such as operating under full load at different wind speeds.

With a booming wind energy industry, the question is now arising of how to deal with EOL turbines, and particularly the blades made of hard-to-cycle composites. There are several possible routes for the recycling of wind turbine blades. Today, EOL composites are generally shredded and sent to a landfill or incinerated. A number of organizations are trying to develop viable recycling alternatives, especially for carbon, which could have residual value and might pose an environmental hazard if disposed of inappropriately. There are already some answers. Reported in *Reinforced Plastics* magazine is a method of pyrolysis, in which the epoxy is thermally degraded, in essence burned away, in a reduced oxygen-combustion process. This technology is now being applied commercially, notably by UK recycler Milled Carbon. This company is a key player in a putative supply chain that could 1 day be disposing of thousands of tons of used carbon composite. There are indicators that microwave radiation can assist this process, enhancing the material recycling rate. Then there is the fluidized bed method, in which pre-fragmented waste material is fluidzed by having a high-temperature (around 550°C) fluid or gas passed through it from beneath in the presence of air, the ensuing pyrolysis and oxidation resulting in progressive separation of its components. These methods can, however, result in fiber shortening, some degradation of fiber properties, or both.

Now, though, researchers at the U.K. University of Nottingham [32] have raised hopes that a new system they have demonstrated on a small scale will facilitate the recovery, in near-virgin condition, of long high-modulus fibers that can be reused in significant structures. The answer is to chemically dissolve the epoxy matrix away. The Nottingham team's breakthrough has been to exploit the dissolving power of supercritical fluids, a class of solvent that has proved effective in other industries. The Nottingham researchers investigated supercritical water, carbon dioxide, and a number of organic solvents including ethanol, methanol, and acetone, before alighting on propanol, an affordable short-chain alcohol that is reasonably benign in its normal state. Propanol can successfully break down epoxy. When the team tested the effect of supercritical propanol on small samples of CFRP waste in both batch and semicontinuous reactors, the results were gratifying. The epoxy matrix was dissolved, leaving clean, unbroken fibers. Tests carried out on the fibers showed that these were almost as good as new, retaining a high proportion of their original strength and stiffness (up to 99% in some cases) and having a surface chemistry conducive to bonding with new matrix material. Fibers were left without any apparent damage even after total resin elimination. One indication of this was that the weight of recovered filaments matched that of the original fibers. The resin components decomposes into potentially useful lower molecular weight hydrocarbons, principally phenol.

In continuing work to optimize the process, the Nottingham experimenters concluded that temperature is the most critical variable, followed by pressure and flow conditions. They found that the temperature required, and hence process costs, can be reduced by using an alkali catalyst. One used was KOH (potassium hydroxide). The researchers claim that by using supercritical propanol, the following potential benefits can be gained; the reaction happens at manageable temperature and pressure, it recovers clean fibers in a condition that is close to virgin, and it also breaks down the matrix into useable chemicals. There are other aspects to be explored too. It is too early to say, for instance, whether the process, so far tried on polyacrylonitrile (PAN) carbon fibers, will be equally effective on pitch fibers.

It is known that fluidized bed, the other major recycling process that the Nottingham researchers have investigated, works well with mixed-materials, so it will be interesting to see how the supercritical fluid answer compares. It will also be interesting to find out how the process, or derivatives of it, measure up against resin chemistries including vinyl esters, phenolics, bismaleimides, and thermoplastics. More on recycling later in the chapter.

Energy Hydropower: (Utilizing The Ocean's Waves)

Of the several forms of hydropower, wave and tidal power are rising in popularity. Wave motion might yield more energy than tides, and its feasibility has been particularly investigated in Scotland and in the U.K. Wave and tidal power plant developer Aquamarine Power of Edinburg, Scotland plans to develop 1 gigawatt of wave and tidal power by 2020.

Aquamarine's Oyster wave power device is meant to be deployed at short depths of 10–12 meters, and is designed to capture the energy found in amplified surge forces in near-shore ocean waves. The system consists of a simple oscillating wave surge converter, or pump, fitted with double-acting water pistons. Each passing wave activates the pump, which delivers high-pressure water via a subsea pipeline to the shore. Its offshore component is a simple, highly reliable mechanical flap with minimal submerged moving parts. Onshore, the high-pressure water is converted to electrical power using conventional hydroelectric generators. Any excess energy is spilled over the top of the device's flap—its rotational capacity allowing it to duck under the waves. Oyster is unique in that it starts generating electricity in almost calm sea conditions and can continue generating during storms.

Another company developing wave energy technology is Wiltshire, UK-based Checkmate Seaenergy. Their Anaconda wave energy converter is a 200 meter long, water-filled distensible rubber tube that is anchored to the ocean's floor. The device is moored at the bow so that it faces waves head-on, and floats just beneath the ocean's surface in waters up to 50 meters deep. Anaconda is squeezed by passing waves, which form bulges in the tube and that travel down its length, gathering energy from the wave as it goes. The bulge wave then hits a hydraulic turbine at the stern, creating electricity. Since, Anaconda is mainly made of rubber, it offers a durable and cost-effective addition to wave energy technologies.

Energy: Solar

The U.S. solar energy industry grew about 9% in 2008, but the recession has cut demand for some solar installations. Solar energy projects will now be getting a financial boost of $117.6 million from the American Reinvestment and Recovery Act, which aims to accelerate commercialization of clean solar energy. Solar energy can be converted to electricity in a couple of ways. PV devices, or solar cells, change sunlight directly into electricity, whereas solar power plants indirectly generate electricity; the heat from solar thermal collectors is used to heat a fluid, which produces steam that is used to power a generator.

The U.S. DOE recently granted $3 million to Skyline Solar of Mountain View, California for solar PV research. The company manufactures a system called High Gain Solar (HGS), which uses metal reflectors to concentrate the sunlight onto monocrystalline silicon solar cells for electricity production. Skyline's HGS architecture delivers 10 times more energy per gram of silicon versus flat-panel systems in sunny locations. It is built primarily out of commodity materials with globally available manufacturing processes from the PV and automotive industries, thereby making it cost effective.

Parabolic troughs are used in solar thermal plants to concentrate light on tubes, heating up a fluid inside them that is then used to drive power-generating turbines. Skyline Solar replaced these tubes with narrow solar panels, adding a heat sink to keep them from getting too hot. Each HGS array consists of a reflective parabolic trough, or reflective rack, and four rows of HGS panels along its edges. The arrays are mechanically coupled together into long columns with adjacent units sharing mounting and tracking hardware. Single-axis tracking collects the maximum amount of light throughout the day, increasing overall energy production. Multiple columns are installed side-by-side to create large solar fields.

The reflective rack provides structural support for the overall array, and its top surface is covered with a thin, durable, metallic coating encased in oxide layers. This top surface is combined with a set of prefabricated struts and ribs underneath, forming a lightweight space frame. Skyline replaces most of the silicon found in traditional PV systems with highly reflective sheets of metal on the top of its reflective rack, and the arrays are built almost entirely of recyclable metal. The HGS systems are designed for large ground-mounted solar applications from 50 to 100 kilowatts up into the many megawatts.

Solar and wind-generating stations have been criticized from an aesthetic point of view, and a company called Solar Botanic of London plans to address this issue by building artificial trees that reap solar and wind energy. The company's plan involves three different energy-generation technologies: PVs, thermoelectrics (electricity from heat), and piezoelectrics (electricity from pressure). Solar Botanic's trees could be installed in areas where naturally growing groups of trees would previously have been used, such as along motorways, suburban streets, and parks.

Solar Botanic's essential technical element is the Nanoleaf, which is designed to capture the Sun's energy in PV and thermovoltaic cells. The Nanoleaves would rely on thin-film solar cells, perhaps made of copper indium gallium selenide, to convert sunlight into an electric current. The leaf stem and twigs comprise nanopiezovoltaic material—these tiny generators produce electricity kinetic energy caused by wind or falling raindrops. As the wind blows the layers of voltaic material in the stems, twigs and branches are moved, compressed, and stretched, creating electricity.

Energy: Geothermal

Geothermal energy provides clean, renewable base-load electricity available 24 hours a day, 7 days a week, and emits little or no CO_2. Engineered Geothermal Systems (EGS) projects produce electricity using heat extracted with engineered fluid-flow paths in hot rocks. The pathways are developed by stimulating them with cold water injected into a well at high pressure and using a variety of proprietary techniques developed by AltaRock Energy Inc., Sausalito, California. In the EGS power generation cycle, water is continuously injected down a well into the enhanced fractures, where it heats up as it flows through the rocks. The water is then brought to the surface in multilple production wells, and its heat is extracted to generate electricity in power plants. Finally, the water, depleted of its heat, is reinjected to be heated again in the fractures. The EGS concept provides the potential to remove the "dry hole" risk associated with conventional hydrothermal geothermal, which requires finding existing fractures that contain high flows of hot water. EGS would also allow the placement of geothermal development in sites without conventional geothermal resources.

Energy: Combinations

There are various technologies in progress as well as developments, which along with countless other advancements, are constantly evolving and undergoing further development. Skyline Solar has completed the construction of a grid-connected HGS demonstration project in San Jose, California. The system provides power for a transportation maintenance facility servicing over a million residents in Santa Clara County, California. Aquamarine's Oyster wave device produced electricity onshore on a full-scale test rig, proving that it can deliver electricity on a commercial scale. The output from a single pumping cyclinder delivered over 170 kW of electricity, demonstrating that a full-scale device, with two pumping cylinders, would deliver well in excess of the modeled output of 350 kW. A commercial farm of just devices could provide clean renewable energy to a town of 3000 homes. The Purdue University blade is being tested on a research wind turbine at the U.S. Department of Agriculture's Agriculture Research Service Laboratory in Bushland, Texas. The research being pursued covers the application of a blade system for advanced, next-generation wind turbine blades that are more curved than conventional blades.

Propellants and Rockets

The U.S. Air Force Office of Scientific Research (AFOSR) and NASA recently launched an environmentally friendly, safe propellant comprised of aluminum powder and water ice (ALICE). Alice has the potential to replace some liquid or solid propellants, and can be improved with the addition of oxidizers to become a potential solid rocket propellant on Earth. Away from this planet, on the Moon or Mars, ALICE can be manufactured in those locations instead of being transported at a large cost. When it is optimized, it could have a higher performance than conventional propellants. The success of the flight can be attributed to a sustained collaborative research effort on the fundamentals of the combustion of nanoscale aluminum and water over the last few years. ALICE has the consistency of toothpaste and can be fit into molds and cooled to −30°C 24 hours before flight. The propellant has a high burn rate and achieved a maximum thrust of 298 kg (650 lb) during the launch test.

Automobiles, Railcars, Racing Cars, Trucks

At a time when gas prices are high and resin prices are creeping up, marketing composites in the transportation industry can be challenging. It is especially tough if your customers are in the recreational vehicle, truck or automobile industry. However, there are other attractive niches in the transportation industry, and high gas prices are the reason these markets are looking to composites. Mass transportation is on the rise. More urban areas are installing a variety of transportation modes to move people in, out, and around the city, including light rails, streetcars, and aerial vehicles. With rising fuel prices, the demand to reduce weight is a high priority. Decision makers in all segments of the transportation industry—from cars to heavy rail trains and trailers—recognized the importance of taking out heavy materials and are considering substituting them with composites. This presents an opportunity for the composites industry, especially as it introduces new materials that compete with steel. The lighter weight and increased stiffness of cored composite materials are ideal for reducing weight in the transportation markets.

Hybrids

After two decades of high-octane growth in horsepower and heft, Detroit, Michigan is painfully rediscovering the eat-your-peas merits of fuel economy. General Motors, Ford, and Chrysler combined to lose $60 billion in their auto operations in years 2005, 2006, and 2007 and even more in 2008 as pain at the pump sent auto sales into the ditch. Detroit used to say that fuel economy was the last thing buyers asked for and the first thing they complained about. But lately concern about gas mileage has jumped into the top five reasons shoppers reject a car, according to new research from J.D. Power [33].

Now that mileage matters, the race is on to reverse a 20-year slide in fuel economy in America. And hybrids, for all their megawatt buzz, are not the only answer—they account for just 1.3% of the market, since most car buyers can't afford the $4000–$10,000 premium to own one. The real action in fuel economy is in conventional cars. And that's where the big fuel savings will come, since cars of all kinds account for 40% of America's consumption. (Homes consume 22% of our nation's energy, and they're attracting a lot of engineers' attention, too; in fact, one can get a 15–20% improvement in fuel economy from standard cars if customers demand it.) So the latest trend in automotive PR is to shine a spotlight on the guts of upcoming models, rather than just their curvy exteriors.

New cars are now exhibiting new V-6 engines, six-speed transmissions that improve fuel economy by 7%, and transmissions that continuously, but imperceptibly, modulate gears to boost mpg by 8%. Cars from Japan now generate fuel-saving technologies of up to 40 mpg. New breakthroughs include engines in which half the cylinders shut off at cruising speeds, boosting mileage by as much as 12%. Today's GM engineers went to school on their forefathers' flop to develop their new displacement on demand (DOD) V-8. And this time, the DOD is not DOA. The engineers replaced the old chunky mechanical controls with electronics that calibrate every fraction of a second whether to run on full or half-caf.

Hybrids, though, have taught the engineers that drivers like constant feed-back on their mileage. Most hybrids feature a mileage meter on the dash that lets you game the system. It turns out it helps to have the same information to calibrate your driving with DOD. Most models have a miser indicator on the dash. The dour economic news bodes better for plastics and composites, however, because the benefits that these materials offer in the face of an upward trend in metal prices, are exactly what the automotive industry needs. These benefits include reduced mass for improved fuel economy, and the ability to consolidate multiple parts into a single component for cost savings over metal-stamped assemblies. The fuel-cost scare was good news for the development of hybrid and electric vehicles, which have been in abundance at various 2009 International Auto Shows. GM has shown a prototype of the extended range electric Chevy Volt, slated for introduction in the United States in late 2010. The Volt uses 45.4 kg (100 lb) of thermoplastics, including composites in the hood and doors, plus unreinforced polymeric materials in the rear deck lid, roof, and fenders. Among the plastics used in the Volt is a hybrid thermoplastic having a continuous glass fiber reinforced sandwich composite for lightweight horizontal body panels [34].

Also there is the two-seat, all-electric Tesla Roadster Sport from California-based Tesla Motors. The company has begun taking orders for the sports car, which features a high power-to-weight ratio thanks to a skin of light-weight carbonfiber/epoxy compositethat, as the manufacturer tells it, took 2 years of design, prototyping, test, redesign, retest, meetings and arguments to develop. Tesla utilizes a carbon fiber mat in the RTM process to produce composite components for the car. Body panels are a sandwich made of two layers of carbon separated by core of glass and PP. The sandwich panels are encapsulated in epoxy resin. The car can reportedly go 0–60 mph in a blazing 3.7 seconds. British automaker Lotus, which helped Tesla Motors develop its vehicles, is expected to launch an extended-range electric car of its own in the near future. Lotus recently debuted its first all-new model since the Elise was introduced in 1995. The new Evora sports car employs a mostly aluminum chassis along with a composite roof that serves as stressed structural components and stressed composite composite body panels that add to the car's stiffness.

Future in Cars

Where are the futurists in the auto engine labs and where do they look for the next great leap forward—the wind-tunnel crew that celebrates more modest achievements. Lexus aero engineers crafted tiny plastic extensions onto the tail lights of the IS sedan to improve its aerodynamics by one hundreth of a point. Mercedes designers went swimming with the fishes to craft their Bionic car in the shape of a boxfish, which set a new automotive standard for slippery (when dry). Ford's aero engineers are considering installing shutters on front grilles that will close at high speeds to deflect air cleanly over the hood. Inside Ford's wind tunnel, a plastic air dam is seen under the Lincoln Zephyr's radiator to smooth the breezes beneath the car. Mileage gain: one-tenth mpg. The battle to get America's cars to zip, not guzzle, rides on such small victories.

You probably see more hybrid cars in advertisements than on the road. Although Toyota's Prius may be recognized by some and many other hybrids are in various stages of development and commercialization, the general perception is that hybrids not only can't compete seriously with conventional vehicles but also are more expensive. That perception is changing. As a result of rising gasoline prices, stringent environmental standards on emissions, continually improving technology and performance, and attractive tax incentives, hybrid cars are gaining market share—especially in Europe.

According to a 2007 study commissioned by the Nickel Institute on the importance of nickel to the EU economy, more than 50,000 hybrid vehicles were sold in the EU in 2006. By 2010, the hybrid market is expected to reach 4–5% of all new car sales. That translates into about 920,000 vehicles, two-thirds of which will contain nickel metal hydride (NiMH) batteries. The EU produces about a third of the world's new cars, and by 2020, the market share for hybrids in the EU is expected to triple to 15% of total sales.

Hybrid cars are classified as "micro," "mild," or "full" depending on the importance of battery power in the drive train of the vehicle relative to the internal combustion engine. In microhybrids, batteries play only an auxiliary role and are characteristically lead-acid batteries (LAB). Since the larger or dominant role played by batteries in mild and full hybrids requires greater electric storage capacity, power, efficiency, and recharging capacity, NiMH batteries are used in these vehicles. In comparison to other battery types, NiMH batteries provide 30–50% more charge capacity (power) per unit weight than a LAB and lasts 8–10 years compared with only 2–3 for a LAB. Furthermore, a NiMH battery can be recharged many thousands of times without first having to be fully discharged.

Nickel is used in NiMH batteries in both the positive and negative electrodes. The negative electrode is constructed of specific alloys such as lanthanum nickel ($LaNi_3$) or zirconium nickel ($ZrNi_2$) whereas the positive anode consists of a plate containing nickel hydroxide [$Ni(OH)_2$]. In other words, nickel is essential and substantial component of NiMH batteries. Since hybrids are more fuel-efficient than conventional internal combustion vehicles, they produce lower emissions, not only of CO_2 but CO, NO_x, and various hydrocarbons. They are also quieter because their internal combustion engine operates less than in a conventional vehicle.

In another application, ExxonMobil Corporation has partnered with QuestAir Technologies, Plug Power Inc., and Ben-Gurion University (BGU) of the Negev with plans to commercialize an on-vehicle hydrogen production system for use in a fuel cell-powered lift truck application [35]. Under the arrangements, Plug Power will seek to commercialize unique technologies developed by ExxonMobil, Questair Technologies, and Ben-Gurion University's Blechner Center for Industrial Catalysis and Process Development that take liquid fuels and convert them into hydrogen onboard the vehicle where it will be used in a fuel cell powertrain.

Currently, most prototype hydrogen vehicles on the road are powered by compressed or liquefied hydrogen that is delivered to distribution points and then stored onboard the vehicles at high pressures. The ExxonMobil system uses conventional fuels—gasoline, diesel, ethanol, or biodiesel—to produce hydrogen on demand, making that infrastructure unnecessary. Further, safety issues associated with transporting and storing hydrogen, both on and off the vehicle are avoided. "It is hoped to demonstrate significant infrastructure, logistics and cost advantages compared to other hydrogen vehicle systems, all the while reducing the impact on the environment," said Dr. Emil Jacobs, an ExxonMobil vice president for research and engineering. Ultimately, the system has the potential to provide up to 80% more fuel efficiency than today's internal combustion engines and to reduce CO_2 emissions by up to 45%. The immediate goal is to apply the system to lift trucks.

Also underway at the Blechner Center is work on a "second generation" biodiesel fuel. Existing biodiesel fuels are made by combining specific vegetable oils, like soybean or rapeseed, with methanol. Fuel produced by this process has different properties than crude-derived diesel, and its applications are limited. BGU is using a novel process to develop a biodiesel fuel from a variety of vegetable oils. Its composition, although similar to crude-derived diesel, contains major improvements, including low aromatics and no sulfur. It displays improved properties and excellent lubricity that translate into good performance in preliminary engine tests [35].

Formula 1 and Formula 3 Racing Cars

Barely a century after the advent of automobile racing, technology has developed so much that the sport is no longer simply about achieving raw speed. Although today's Formula 1 cars continue to break speed records, the challenge facing its engineers and designers is in how to achieve the most speed within a vast set of regulations outlining the acceptable materials, weight, and dimensions of the cars. Because of this never-ending game between the sport's governing body, known as F1A, and the engineers, and thanks to the sport's worldwide popularity, teams spend hundreds of millions of dollars each year producing technology that has no use outside Formula 1. As the series prepares for the opening race of the 2010 season, F1A is seeking a way not only to reduce costs and improve the show for spectators [36].

In a moment of foresight a year and a half ago, Max Mosley, the president of F1A, announced two new societal and business goals for the sport. It would aim to produce car technology to help improve the environment and to sell road cars by showcasing the manufacturers' environmentally friendly technology. The main effort and developments in Formula 1 are those that are directly helpful to the car industry and in particular things that are relevant to perhaps the biggest single issue that confronts the car industry worldwide, namely the reduction of the output of CO_2. This season marks the first step in such moves. In line with the EU objective for road-car fuel in 2010, racing fuel will have to contain 5.7% biofuel, which is made from organic matter. Another new rule eliminates a previously obligatory 10-minute period during qualifying when the cars did nothing but burn off fuel.

Next season, a device designed to save the car's kinetic energy during deceleration and then to use it in short power bursts—thus saving fuel—will be instituted. Other so-called clean technology systems, including one using heat from the engines to produce extra power, are in the works. A technical consultant, Tony Purnell, claims that racing could suffer from a public backlash if it did not develop environmentally cleaner cars. Additionally, what is ultimately at stake is trying to reduce CO_2 emissions by 50% while maintaining the power that highway drivers are accustomed to in their cars. In fact, if the challenge for the automotive industry, with Formula 1 spearheading the changes, is achieved then something worthwhile would be accomplished.

Following the recent turmoil in Formula 1 arising from the high costs of running competitive motor racing teams, the viability of motor racing is being critically questioned. As a result, the University of Warwick [37] unveiled the World First Formula 3 racing car that is powered by chocolate, steered by carrots, has bodywork made from potatoes, and can still do 125 mph around corners. There has been much speculation about the creation of this car but it has now been completed and is now ready to drive. It is the first Formula 3 racing car designed and made from sustainable and renewable materials, putting the world first by effectively managing the planet's resources. The car meets all the Formula 3 racing standards except for its biodiesel engine, which is configured to run on fuel derived from waste chocolate and vegetable oil. Formula 3 cars currently cannot use biodiesel. Components made from plants form the mainstay of the car's makeup, including a race specification steering wheel derived from carrots and other root vegetables, a flax fiber and soybean oil foam racing seat, a woven flax fiber bib, plant oil based lubricants, and a biodiesel engine configured to run on fuel derived from waste chocolate and vegetable oil. It also incorporates a radiator coated in a ground breaking emission destroying catalyst.

As OEMs focus on decreasing engine emissions, to meet future CO_2 gases, the WorldFirst project proves that if you are going to wholeheartedly embrace the "green is great" ethos you have to broaden your vision and have a strategy that stretches throughout the chain from the raw materials to the final disposal of the car. The project clearly demonstrates that automotive environmentalism can and should be about the whole package. This project proves that you can develop a working example of a truly "Green" motor racing car. The WorldFirst project expels the myth that performance needs to be compromised when developing the sustainable motor vehicles of the future.

Heavy Railcars

In the heavy railcar industry, continuous-fiber thermoplastic is a new material for producing panels for railcars, trucks, and refrigerated railcars. The benefit to the end user is a greener, higher impact, and scuff-resistant product. The consistent, sealed smooth surface is cleaner to work with, making this process more environmentally friendly.Compared to traditional thermosetting composite processes, using continuous-fiber thermoplastic material is easy, clean, and simple. The material is dry and easy to cut, with no special storage needed. There are no VOC emissions or hazardous chemicals.

Trucks

Have you ever wanted to make a vehicle run on garbage. A 42-year-old homebuilder, Dave Nichols, and auto shop owner from eastern CT modified his 1989 Ford-150 pickup truck to run on wood, leaves, cardboard and other "biomass" with a fuel system that he says expels virtually no pollution [38]. The technology is called gasification, and its been around since the 1800s, when it was used for street lamps and cooking. It even powered some vehicles during WWII, but faded away under oil's dominance.

Nichols and others say reviving gasification, which can also heat and power homes, has exciting possibilities, from reducing dependence on foreign oil to cutting pollution. It's a simple science from 130 years ago that can be used today to solve all of the current problems and it runs on potentially free fuel. This type of technology has to be developed and it has to be developed now. The new interest in gasification comes as the federal government and president presses to double the nation's use of renewable energy over the next 3 years, with $15 billion a year to be spent to develop solar power, wind power, advanced biofuels, fuel-efficient cars, and other technologies.

Gasification works by heating organic materials to high temperatures without flames. The resulting chemical reactions produce a hydrogen-hydrocarbon gas mixture in vapor form that is almost as potent as gasoline. Nichols pickup truck appears to run like any other and easily has reached 40 mph and above on local roads but it has no gas tanks. Nichols claims he can reach up to more than 80 mph. The only noticeable difference is a contraption—right behind the cab's rear window—that takes up some of the back and looks somewhat like a wood stove. Nichols says he has driven 10,000 miles without gas, including a trip when he loaded up the back of the pickup with about 181.4 kg (400 lb) of wood and drove 600 miles across Connecticut, then to New Hampshire and Boston before returning to home base. A pound of wood or other material will fuel the pickup truck for one to two miles, meaning that the truck costs about 8 cents a mile to fuel, compared to roughly 19 cents per mile if it used gasoline at today's prices. Therefore, this is real and not a game. Nichols has a goal to build a smaller version of the vehicle fueling system, so it could be more practical for cars.

Paints and Coatings

Efforts to institute environmentally responsible, cost-effective manufacturing are rapidly becoming more commonplace. One group of companies that is actively involved in this process—and in subsequently promoting the technologies and tactics that achieve positive results—is the U.S. paint and coatings industry. These paint and coating companies have demonstrated their ability to reduce the amount of waste generated during manufacturing and use fewer natural resources when making their products. At the same time, these companies achieved these environmental improvements without the purchase of expensive equipment or a massive retraining of employees. A look at how these companies instituted these changes can illustrate how manufacturing process improvements not only help the environment but also generate significant cost savings for manufacturers.

One such company was the Mesquite, Texas facility of Benjamin Moore & Company [39], known as the "Dallas Plant," which was designed from the ground up with pollution prevention, source reduction, and waste minimization in mind. The Dallas Plant targets a number of environmental improvements including: reducing point and nonpoint VOC emissions; controlling the generation of airborne particulate contaminants; reducing wash water; and reducing wastes. Specifically, point and nonpoint VOC emissions and generation of airborne particulates are 40% less than they would be at a similar plant that does not have the engineering controls designed into three key areas at the Dallas Plant:

- The bulk raw material delivery system
- The design and installation of mixing tank lids
- The design and installation of the remote vapor collection system

Reduction and control of the generation of airborne particulate matter and contaminants is accomplished through the use of a Torit Dust Collection System, engineering controls designed into mixing tanks, and use of slurry. The slurry system reduces particulates due to the fact that pigment is already predispersed and "wetted-in" upon addition to the batch.

Another facility is the Ace Paint Division [39] of Ace Hardware in Matteson, Illinois, which embarked on a program to reduce the cost of waste management, and set a goal of working toward zero discharge from its facility. The facility targeted three areas of concern, including: hazardous waste disposal from its oil base operation; nonhazardous waste; which was being compacted and landfilled; and latex nonhazardous waste, which was being treated and shipped out for further treatment and subsequently landfilled.

Companies that institute pollution prevention processes stand to benefit on a number of levels. In addition to contributing to significant environmental improvement, companies have enjoyed enhanced team spirit, improved dealer/vendor relations and savings that have run as high as $500,000. By recognizing deserving efforts such as those made by these companies and their programs it encourages other manufacturers to implement new pollution prevention techniques. The program demonstrates effective tactics and shows that pollution prevention can be successfully and cost-effectively incorporated into the manufacturing process. As new programs are developed and existing programs replicated, the paint and coatings industry is significantly reducing its waste generation and helping the environment, while enjoying the economic benefits of streamlined processes.

Architecture and Construction

What is SRI? Roofs after 2003 for large conference and convention centers had to meet minimum solar reflectance (the percentage of energy a surface reflects) and thermal emittance (the percentage of energy a material radiates after it is absorbed) values in order to earn LEED ppints for reducing heat islands. In 2006, LEED replaced these values with a single Solar Reflectance Index (SRI). In accordance with ASTM E 1980 standards, SRI is calculated from eight different parameters including wind speed, thermal emissivity, solar flux, and air temperature [40]. To earn LEED points, a low-sloped roof (equal or less than 2:12) has to have an SRI of at least 78. Steep-sloped roofs require an SRI of at least 29.

The real world is more complex than the SRI system suggests. There is a lot more going on than heat bouncing off a roof. A poor thermal conductor like stainless steel insulates and contributes to energy conservation. The SRI of stainless steel will also be more sustainable over the life of a building, compared with other roofing materials, because it is so durable and will be completely recyclable at the end of its life. LEED provides no explicit guidance for the use of stainless steel, though there are areas where stainless steel can help architects earn points; the reuse of building materials from existing structures, diversion of construction waste from disposal, and the recycled content of "new" stainless steel.

The high recycled content of austenitic stainless steel (ranging from 60 to 85% depending on geographic location) contributes significantly to the aggregate recycled content of materials used in new buildings. The Specialty Steel Industry of North America [40] reports that the average postconsumer recycled content of the 300 series stainless steel grades is about 75–85%.

Another area where the stainless steel industry could play a "LEED role" is in reduction of "heat islands," which refer to urban areas that are significantly warmer than their underdeveloped, rural surroundings. It is commonly thought that only specially painted roofs can meet those requirements, but recent tests funded by Contrarian Metal Resources in Pennsylvania, prove that this is not so: Contrarian's Invarimatte finish sheet stainless steel met the SRI requirement for steep-sloped roofs. The nondirectional matte surface finish used on the Pittsburgh Convention Center roof and the nonreflective finish was achieved by abrasive blasting called Architex. The roof sections are supported by cables or cantilevered at the wall ends, and form curves like those of the cables on the nearby bridge that spans the Allegheny River.

Net Composites and BRE are leading a new LINK collaborative research project funded through a renewable materials program [41]. The project has focused on addressing the environmental credentials of naturally derived construction materials. The work addresses the problem faced by the construction industry in that there is limited understanding of the environmental performance of these relatively new materials making like-for-like comparisons difficult. The project is focusing on environmental issues such as:

- Raw material supply—including crop production and land use
- Energy requirements for primary and secondary processing
- Durability of these naturally derived materials compared to conventional alternatives
- End-of-life (EOL) issues including: recyclability, composting, recovery, and reuse

The project team, comprising experts from the construction sector and natural materials industry, plans to produce information, guidance and tools that are crucial for the environmental understanding of these materials and their comparison to conventional alternatives. This will enable companies thruout the sector to make well-informed decisions about the likely benefits of crop-based materials without the cost and time requirements of conducting detailed life cycle impact studies. It will also allow the industry to deal with clients on a more informed basis, further stimulating the growth of materials derived from crops [41].

The 7th Annual Conference of the Network Group for Composites in Construction (NGCC) [42], held in Farnborough, UK, in May 2008, was aimed to increase awareness of the commercial benefits of using composites in construction and therefore assist in the future growth of this important market sector. The NGCC had identified that commercial pressures were both one of the main reasons for selecting composites in construction applications and also a potential barrier to their use, especially where the business case for composites was not adequately understood. It should be noted that NGCC is a membership-based organization promoting the use of FRP in construction.

One speaker, Professor Steve Denton from Parsons Brinkerhoff [42], highlighted several important issues that could easily be overlooked in assessing the business case for composites in construction, but which will be of increasing importance in the future. These include issues such as social and environmental impact, in addition to the more obvious ones of financial viability. Denton explained how the use of FRP composites for structures such as bridges can bring social benefits, for example by reducing delay and disruption to the road and rail networks because of quicker and easier installation. He also demonstrated how the use of FRP in major construction infrastructure is maturing and moving beyond pilot projects into higher volume commercial applications, selected for purely commercial reasons.

Bridge Construction

Christian Scholze from Fiberline Composites [42] presented an interesting graph that showed the energy consumed in producing various materials and the advantage demonstrated and the potential environmental benefits of FRP compared to other construction materials. The materials compared included FRP, PVC, aluminum, steel, and wood. Another recommendation suggested by Scholze was painting pultrusions used in external applications to provide additional environmental protection and a 30-year life to repainting is normally achieved. Scholze went on to site various footbridges [42] made entirely from FRP components, such as the Fiberline Bridge in Denmark that was erected in 1997, taking 18 hours to install on site. He also demonstrated how pultruded FRP decks are being used very successfully in combination with steel structures both for new bridges and to refurbish existing bridges by replacing corroded steel or rotten timber decks. The use of FRP enables rapid deck replacement, a reduction in dead load and reduced future maintenance. The use of FRP either for complete structure or in combination with steel, reduces the dead weight and enables large sections to be prefabricated.

Scholze concluded that while initial costs for a FRP bridge may be slightly higher than a steel equivalent, once the benefits of easier installation and reduced maintenance are taken into account, the total cost will be considerably lower for the FRP option.

FRPs in Construction

Dr. Sue Halliwell from NetComposites [42] presented the environmental case for FRPs in construction, which is becoming increasingly important in the materials selection process. She explained how both legislation and asset owners are demanding improved environmental performance and longer service life with reduced maintenance.

Addressing the U.K. building sector there is a target to reduce U.K. CO_2 emissions by 60% from 1990 levels by 2050. As 46% of carbon emissions are from buildings in use, the construction industry will clearly have a major role to play in meeting the target. Halliwell explained how FRPs could assist in meeting the environmental challenge as follows:

- Ideal for modular, factory-based production, leading to better QC
- Inherently thermal insulating, eliminating thermal bridging
- Resistant to passage of water vapor

- Joints can be vapor- and air-tight
- Thermal stability improves long-term performance in relation to air tightness
- Composite materials can be tailored to contain reflective coatings on strategic surfaces to reduce solar gain

Halliwell showed how environmental credentials may be assessed using techniques such as life cycle assessment and the Green Guide to Composites, which has an online tool available at www.netcomposites.com/composites-green-guide.asp. The Green Guide rates materials and processes from A (good) through to E (poor), and 12 different environmental impacts are individually scored and totalled to give an overall rating.

Finally, Halliwell addressed the critical issue of EOL options. She reported that, despite the development of recycling methods, the most common method of disposal of FRP is still landfill. It is expected that future waste management legislation and the increased cost of a landfill will increase the uptake of recycling and reuse and put more emphasis on minimizing waste. Halliwell explained that, at present, there is not a clear market for FRP composite recyclate and research is needed to develop markets and products for recycled FRP. Pilot projects have demonstrated that it is feasible to recycle FRP, for example by grinding to produce a filler for new materials, but a reliable supply chain must be established. It is also viable to incinerate FRP materials and recover energy due to their high calorific content [42].

Recycling

Ways to optimize the recycling of all materials in shredder residue, regardless of their source, are being studied by the U.S. Council for Automotive Research's (USCAR) Vehicle Recycling Partnership (VRP), composed of researchers from Chrysler, Ford Moror Co., and General Motors Corp. The VRP recently contracted with ECO2 Plastics Inc. to evaluate its proprietary polyethylene terephthalate plastic recycling technology.

The shredder residue plastics are cleaned in an environmentally friendly process that needs no water. Instead, a biodegradable solvent and liquid CO_2 remove the substances of concern so that the plastic can be more readily reused. Currently, more than 84% of each vehicle in the United States is recycled, with 95% of all vehicles going through the existing infrastructure. ECO2's proprietary recycling process addresses the plastics found in the unrecycled portion [43].

Today, many different types of industry are finding it increasingly difficult to dispose of production waste and waste materials. Manufacturers are facing an ever-increasing demand from both customers and government officials to provide alternatives to landfilling and incinerating these materials. In recent years, national governments and the EU have set out ambitious targets that require manufacturers to work toward a much higher level of recycling and to keep waste to an absolute minimum.

The Danish government had, for instance, set a target that by 2008 Danish industrial manufacturers should recycle at least 65% of their waste and surplus material, and the amount landfilled should not exceed 15%. Manufacturers, however, have not come close to meeting these targets. The Danish example is similar to developments in most other European countries, so it appears that there is a need for a different concept to speed up the process of recycling surplus and waste material. One way would be to impose financial sanctions on the industrial manufacturers that are not doing enough to meet the new requirements, but it is questionable whether or not these sanctions would have the desired effect. During the current economic downturn, it is probably best to avoid burdening the finances of manufacturers, if there are other options available.

The problem, however, lies not only with the industrial manufacturers. When these companies are unable to use their waste material in their normal production process, then other companies are needed that have a fresh perspective on the potential uses of the waste material. It appears to be impossible to make the industrial manufacturers thenselves find a complete solution for recycling their waste material, and perhaps, instead of punishing them, the resources would be better spent at developing and supporting companies that have new ideas of reusing this waste. For a number of years, the industry department at Danish company Barsmark AS has been developing, producing, and selling products made from recycled production waste, thus helping many industrial manufacturers with their waste problem [44].

In 2008 Barsmark and the company's production partner converted more than 8000 tons of recycled material into finished components with some unique properties. By 2018, the company's goal is to convert 25,000 tons of production waste into finished components and be sold to customers. Thus from 2009 to 2018 the company will have recycled more than 165,000 tons of material, which otherwise would have been landfilled or incinerated. To be able to obtain and use such a large amount of material, Barsmark is negotiating and creating various cooperative agreements with industrial manufacturers with usable surplus material, as well as new production partners in order to obtain the necessary production capacity. The two products in Barsmark Industry's [44] current product range are examples of the type of products that the company will focus on in the future.

PP-1700 is a sheet made from glass fiber mixed with epoxy made from waste material from the wind turbine industry. The material used for the production of the sheets is tested and evaluated to make sure that the PP-1700 is a homogeneous sheet. The sheet has a high strength-to-weight ratio as the material is used for manufacturing blades for wind turbines. The sheet is also corrosion-resistant, which makes it suitable for many different applications. Today, the PP-1700 sheet is primarily used as partition walls in pig sties, but it is also used as insert profiles in refrigerated trailers or for the strengthening of sandwich panels. PP-1700 is ideal for partition walls in pig sties because it fulfils almost every requirement that manufacturers of stable equipment and pig farmers have. The sheet is relatively rigid, has high yield strength, and can be cleaned with high pressure cleaners without damage. These qualities are extremely important when working with animals that weigh as much as 500 kg and that can really test the durability of the product.

The second product, PT-200, is made from waste polyurethane foam used, for example, as insulation material for refrigerators or polyester felt for car ceilings. As the foam originates from many different types of industry, it can contain trace amounts of other materials such as glass fiber or aluminum. This waste foam is converted into sheets by granulating the foam that is then pressed into a sheet. The sheet has the appearance of a standard chip board, but is 100% moisture resistant. Another benefit of this process is that the material wasted is kept at an absolute minimum, as any excess material is gathered and reused for producing new sheets. Similarly, old, used sheets and components can be recycled after a granulation process.

This material has been used in the transport industry. Some of Barsmark's customers have replaced the wooden components in their refrigerated trailers with components made from PT-200 sheet, thus avoiding their trailers gaining weight by absorbing water in the structure. Also, unlike wood, expansion of the component and formation of mold in this humid environment are no longer a concern. Barsmark customizes the components in its production facilities and they are delivered ready-to-use in the refrigerated trailers.

Other applications for PT-200 include: a core material in bathroom furniture, a core material in laboratory walls, and as insert profiles in glass and aluminum facades. These applications are characterized by a number of requirements: moisture resistance, resistance against the formation of mold and rot, stability of shape when exposed to moisture, low weight (approximately two-thirds that of MDF), and good insulating properties with a high level of screw extraction force. As it is possible to recycle old PT-200 sheets and components, there is also a high level of sustainability in the product.

Shredded Plastic Waste

Consider the mix of materials that go into household appliances and automobiles. According to researchers at Argonne National Laboratory [45], over 95% of the over 50 million vehicles scrapped globally each year enter a comprehensive recycling infrastructure. Nearly 75% of the weight of the vehicles is metals and the metals are profitably recycled through direct reuse, component remanufacturing, and scrap processing or shredding. Other components such as batteries, automotive fluids, some windshield glass, starters, alternators, and other dismantled parts are also recycled. It should also be noted that every time a ton of steel is recycled, it means that 1134 kg (2500 lbs) of iron ore, 453.6 kg (1000 lbs) of coal, and 18.1 kg (40 lbs) of limestone will not have to be mined.

The percentage of recycled materials from vehicles is about to go up. Argonne, working with the Vehicle Recycling Partnership (VRP) of the U.S. Council for Automotive Research (USCAR), a partnership described earlier in this chapter, and the American Chemistry Council—Plastics Division (ACC-PD), is developing technology to recover polymeric materials from shredder residue [45]. Scrap processors use giant 3000–8000 hp hammer mills to shred vehicles and other obsolete metal-containing products. Household appliances, industrial scrap, and demolition debris are all candidates for being turned into fist-sized chunks as a means of liberating the metals. Processing-unit operations vary, but the basic process involves air classification of the "lights" fraction followed by one or more stages of magnetic separation to recover the ferrous metals.

Trommels and screens are then used to remove particles smaller than about 0.4 cm (5/8 in.), followed by one or more stages of eddy-current separations to recover the nonferrous metals. Once the metals are gone, what's left is called shredder residue. It's typically a mix of polymers (plastics, rubber, and polyurethane foam), a "fines" fraction that includes metal oxides, glass, and dirt, as well as residual amounts of ferrous and nonferrous metals. For each ton of metal a shredding facility recovers there is roughly 226.8 kg (500 lb) of shredder residue. Unfortunately, this by-product has historically ended up in landfills. That's because recycling efforts were driven by the value of the metal, the single largest source of recycled ferrous scrap for the iron and steel industry is obsolete automobiles.

In the last 15 years, however, both automobiles and white goods have increasingly used polymers and composites. Existing separation technologies for polymer recovery rely on differences in density to separate solid particles. These methods will work on certain thermoplastics. But shredder residue content has overlapping densities and shapes that make it difficult to get enough material with the right purity for scrap processors.

Bulk Separation

Argonne's new process is designed to efficiently separate polymers of equivalent densities. Called froth flotation, it was originally developed to separate ABS from high-impact polystyrene (HIPS); a mixture of plastics that is typical of that recoverable from obsolete appliances. A pilot plant was built and confirmed the process economics and effectiveness of Argonne's froth-flotation process [45]. The pilot plant recovered ABS at purities in excess of 99% and at yields of more than 80%. ABS recovered from this operation was successfully used to injection mold automotive parts, thus confirming the feasibility of using obsolete post-consumer plastics (in some cases, plastics more than 15 years old) to meet the performance requirements of parts produced for this industry today. For a complete description including schematic diagrams of the froth-flotation and the Vinyloop processes see Hoffman [45].

Aerospace Recycling

The Boeing Co. (Seattle, Washington) and Alenia Aeronautica (Rome, Italy), have joined forces to help establish Italy's first composite recycling facility, which is located in Southern Italy [46]. Together with partners Milled Carbon (Birmingham, UK), Karborek (Puglia, Italy), and ENEA (Italian National Agency for New Technologies Energy and the Environment), the two companies will apply their expertise and work with academia to advance industry knowledge surrounding the recycling of composite airplane parts into reusable materials for manufacturing. The composites recycling facility, which is expected to be operational in late 2009, will be in Italy's Puglia region, near the Alenia Aeronautica manufacturing center. When fully operational, the center is expected to process an average of 1000 metric tons (1102 U.S. tons) of composite scrap annually.

Since high-value composite materials are playing an increasingly significant role in aviation's ability to develop lighter, more fuel efficient and environment friendly aircraft, the two companies are being proactive in developing technologies and capabilities today that will allow engineers to responsibly recycle the current precious resources, and help meet rising demand for high-quality composite material [46].

Recycling Glass

By dropping off glass jars in a city recycling container in Cheyenne, Wyoming, they won't end up being recycled any time soon. Their destination: a mound of glass at the city landfill, an ever-growing monument to the difficulty many communities across the country face in finding a market for a commodity that's too cheap for its own good. Thus, the city is stockpiling glass in a desperate search for a market. Cheyenne hasn't recycled the glass it collects, 9 tons a week, for years. Instead, the city has been putting it in the landfill, using it to surround the concrete-walled wells that pump toxic fluids out of the dump.

The economics of glass recycling have been marginal for some time. Nationwide, only about 25% of glass containers are recycled. That's compared to 31% of plastic containers, 45% of aluminum cans, and 63% of steel cans, according to the EPA [47]. Glass has piled up at the landfill serving Albuquerque, New Mexico, where officials in 2009 announced that a manufacturer of water-absorbing horticultural stones would eventually use up their stockpiles. New York City gave up glass recycling from 2002 to 2004 because officials decided it was too costly. The challenge is that the main ingredient in glass, sand, is plentiful and cheap—often cheaper than cullet, which is glass that has been prepared for recycling. Used glass must be sorted by color and cleaned before it can be crushed into cullet that is suitable for recycling into new containers. That contributes to much of the cost of recycling glass.

The other part of the cost is after cleaning, it must meet a specification of mixing it with sand, soda ash, and limestone. Last but not least are the transportation costs. The farther a community is from glass processors and container manufacturers, the more expensive it is to recycle it. One other possibility that some cities are considering is to purchase a glass pulverizer and to grind the glass into a fine consistency and use it in place of sand for road construction and at playgrounds.

Electronics Recycling

Electronics recycling comprises several EOL processes, from disassembly to data destruction to metals salvage and other steps. Some companies are one-stop-shops, smelting, grinding plastics, recovering reusable components, and reselling raw materials. Others partner with recyclers, perhaps performing data protection and initial disassembly, then handing over waste to dedicated processing centers. Beyond this are a myriad of brokers, specialists, and other processing/resale points that can make the reverse logistics of EOL more complicated than the traditional assembly/supply chain. It is through the multitiered network that broken computer parts become flower pots, for example.

For those about to enter the recycling field, either as a supervising partner or as an involved recycling company, the waters can be murky. Not only are there an estimated 1100 small recyclers in the United States, selling to a range of brokers, but the regulatory environment is fragmented and vague. Courtemanche [48] details which electronics recycling model may fit one's business.

Decomposing Plastics at Sea

Billions of pounds of plastic waste are floating in the world's oceans. Scientists are reporting that even though plastics are reputed to be virtually indestructible, they decompose with surprising speed and release potentially toxic substances into the water. "Plastics in daily use are generally assumed to be quite stable," said Katsuhiko Saido, a chemist with the College of Pharmacy, Nihon University, Chiba, Japan [49]. "We found that plastic in the ocean actually decomposes as it is exposed to the rain and sun and other environmental conditions, giving rise to yet another source of global contamination that will continue into the future." Saido's team found that when plastic decomposes it releases potentially toxic bisphemol A (BPA) and PS oligomer into the water, causing additional pollution. Plastics don't usually break down in an animal's body after being eaten, but the substances released from decomposing plastic are absorbed. BPA and PS oligomer can disrupt the functioning of hormones in animals and seriously affect reproductive systems.

A new method developed by the research team simulates the breakdown of plastic products at low temperatures, such as those found in oceans. The process involves modeling plastic decomposition at room temperature, removing heat from the plastic, and then using a liquid to extract BPA and PS oligomer. According to Saido, styrofoam is typically crushed into pieces in the ocean and finding these is no problem. But when the scientists were able to degrade the plastic, they discovered that three new compounds not found in nature formed: styrene monomer (SM), styrene dimer (SD), and styrene trimer (ST). Stryene is a suspected human carcinogen. BPA and PS oligomer are not found naturally, so they must have been created through the decomposition of the plastic. Trimer yields SM and SD when it decomposes from heat, so trimer also threatens living creatures.

As fishing stocks are decreasing at an alarming rate, instead of having fishing boats all over the world stand idle, why can't we have them adapted to retrieve plastics out of the sea and pay them a fee. The fees should come from each countries respective government. I would guess that residents of each country would back a campaign to do this and would welcome the challenge.

Recycling Carbon Fiber Composites

Disposing of carbon fiber composite will become a big issue as today's low-weight carbon fiber reinforced plastic (CFRP) structures reach the ends of their service lives. This is clear given that, globally, over 27,000 tons of carbon fiber tow are currently being produced each year. Grinding up the EOL material and sending it to landfill or incinerating it, as happens with low-value glass reinforced plastic (GRP), has little appeal due to the nondegradability of carbon thermosets and the health and safety risks they pose [32]. Add to this the fact that long-fiber, high-grade carbon is a valuable commodity, costing upward of $10,000 lbs per ton when new, and it would make sense to recover and recycle this if a means could be found that offered competitively priced material having minimally impaired mechanical properties and good surface chemistry. This poses the question of how to strip away the epoxy resin from fibers in such a way as to leave most of the original fiber properties undiminished.

There are already some answers. There is a method of pyrolysis, in which the epoxy is thermally degraded, in essence burned away, in a reduced-oxygen combustion process. Additionally, microwave radiation has been used as an assist to this process, enhancing the material recycling rate. Then there is the fluidized bed method, in which prefragmented waste material is fluidized by having a high-temperature (around 500°C [932°F]) fluid or gas passed through it from beneath in the presence of air, the ensuing pyrolysis and oxidation resulting in progressive separation of its components.

Several collaborative programs have been undertaken by academia and firms in the UK, Japan, and the US One program was High Value Composite Materials from Recycled CARbon Fibre (HIRECAR), which is aimed at finding ways to recycle carbon composite materials into car manufacture and other applications. A follow-on program is called AFRECAR (AFfordable

REcycled CARbon Fibres). AFRECAR aims not only to establish that a successful laboratory-scale process can be successfully scaled up to something able to deliver worthwhile amounts of recycled carbon, but also to find and develop uses for that carbon. This work was described earlier in this chapter under wind turbines and the University of Nottingham [32]. The other aspect of this program will explore the economics of recovering phenol and other hydrocarbons from the epoxy and whether these can be reused as chemical feedstock.

Finding and cultivating markets for recyclates is a key to building a viable recycling solution overall. The HIRECAR program focused largely on the use of recycled carbon fiber in bulk molding compound (BMC) for smaller, nonload-bearing components, and in sheet molding compound (SMC) where carbon fiber is rolled together with sheets of polymer. BMC and SMC are widely used by the automotive industry. Recycled carbon fiber is also being tested for use in tires, industrial injection molds, and sporting goods such as skateboards. Aero equipment suppliers are looking to exploit it for such items as aircraft baggage bins and galley carts. Another U.K. Government-funded program (REBrake) has been examining the use of recycled material in vehicle brakes. Flocked fiber can be used for conductive substrates, filter media, and decorative interiors. Short fibers can enhance the strength and durability of concrete used in construction.

The big apparent advantage of the supercritical fluid recovery method (developed by the University of Nottingham) is that it can retrieve the most valuable form of carbon fibers suitable for higher-grade applications including substantial engineered structures such as vehicle floor pans and secondary aerostructures. This is because of its ability to recover in impressive condition, long fibers that can subsequently be aligned, a key factor in making recycled fibers suitable for load-bearing applications. "If the material can be spun into continuous yarn with properties and costs competitive with those of new carbon fibers, it will expand the application of composites in vehicles and other structures where weight reduction is important," according to Dr. Ebby Shahidi, technical director of Advanced Composites Group. "Typical applications would include light body panels, chassis, etc., in the automotive sector, plus numerous components in marine, aerospace, and other industries."

The progress of the AFRECAR program [32] will be watched very closely by many companies, academia, engineers, and scientists as well as countries throughout the world. In fact. even legislators, driven by the environmental imperative, will insist on things happening. Being able to obtain a high-grade recycled product will make all the difference, greatly expanding the market for recyclate. It may also expand the market for carbon composites overall. Automotive manufacturers, for example, will be less concerned about adopting weight-saving composites in a legislative climate mandating that future vehicles will have to be up to 80% recyclable once they are confident that a means exists to recycle them cleanly and efficiently [50].

Recycling Traffic Lights

Recycled glass countertops made by Vetrazzo, LLC, Richmond, California, are said to be the most ecofriendly surface materials on the market. Vetrazzo believes that a green product should do more than use recycled materials or avoid release of dangerous airborne toxins. It should also solve an environmental problem. The company does that by creating a new market for waste glass, including glass that cannot be recycled into other products. Vetrazzo offers a green alternative to natural stone, like granite and marble. Each countertop panel is made from 250 kg (550 lb) of crushed recycled glass from traffic lights, windshields, and beverage bottles. Even the company's daily operations are designed with the planet in mind. Vetrazzo operates on a closed-loop water filtration system, saving approximately 670 m^3 (150,000 gal) per month. The factory has been 100% solar powered since 2008.

Cell Phones

When consumers upgrade to the latest model of cell phone, they rarely recycle their old handsets. If they did, we could reduce hazardous waste in landfills and save enough manufacturing energy to power more than 18,500 U.S. households for a year, according to the EPA, which estimates that just 10% of the 140 million cell phones replaced each year are recycled. Cell phones can contain dangerous substances like lead and chromium, as well as valuable metals like copper and easily reusable plastics. Industry officials have been trying for years without success to get consumers to recycle their old phones. Now, the government is getting involved. At least 19 states have enacted laws regarding the disposal of electronic devices (see the e-waste and recycling section), a trend the industry fears will lead to a patchwork of regulations, including some that would make manufacturers responsible for their products from cradle to grave. Some of the biggest cellphone makers—Nokia, Motorola, LG, Sony Ericsson, and others—are working with the EPA to encourage recycling.

Why are consumers reluctant to recycle cell phones? Apparently, size matters. It's not like the cellphone is taking up space in a garage or basement. Cell phones can be dropped off for recycling at virtually any service provider's location. But in spite of this situation, an environmental-advocacy group estimates that more than one billion unused cell phones have made their way into landfills or are collecting dust in America's closets and drawers.

Filters, Fumes, and Nanotechnology

Much has been written about the health risks of weld fumes, the various contaminate collection methods, and OSHA requirements/limits for worker exposure. Dust or fume collectors with pulse jet cartridge filters are the preferred method of cleaning the air because of their collection efficiency.

The particle size of weld fume is generally in the 0.3–0.7 micron size range. For hexavalent chromium [Cr(VI)], the OSHA limit is now 5 micrograms (0.005 mg/m³). For today's welding operations, nanofiber filtration offers an efficient and cost-effective way to go because weld fume and other matter smaller than one micron can be controlled.

If you look at any cartridge filter media through high magnification, you will see open spaces or holes. The smaller the holes, the better the media will be in capturing fine particulate. The best way to do this is to use the smallest fibers possible. A nanofiber is 1/1000 of a micron. Just how small are these fibers? Consider that there are 25,400 microns in an inch. The lower limit of visibility with the naked eye is 40 microns, and the average pore openings in your skin are 10 microns. As such, an extremely thin nanofiber surface layer on a cartridge filter is capable of capturing submicron particles [51]. Depending on the cartridge filter manufacturer, a very fine nanofiber layer will be between 0.3 and 1 micron thick and be placed over a cellulose substrate that is typically 0.0006 mm (0.016 in.) thick. This is equivalent to placing one sheet of copy paper (nanolayer) on top of 900–1000 sheets of copy paper (substrate).

Filters

The three main types of standard cartridge filters are cellulose, cellulose blended with a synthetic fiber (referred to as blended cellulose), and cellulose filters with a nanofiber layer. Some cellulose and blended cellulose filters consist of one homogeneous layer of media. These filters are true depth-loading filters. As air moves through the filter, particulate becomes embedded deep within the filter, restricting airflow and shortening filter life. Some cellulose or blended cellulose filters currently on the market have an outer layer of melt-blown fibers added to increase the filter's efficiency and life. The principle is the same as adding a nanofiber layer—to provide more surface loading so the filter is easier to pulse clean. The difference is in the diameter of melt-blown fibers versus nanofibers and the depth of the layer.

The most commonly used indicator of filter efficiency is the minimum efficiency reporting value (MERV). The higher the MERV rating (1–20), the better the filter is at removing particulate, especially very small particulate, from the air. The differences in filtering efficiencies between nanofiber, cellulose, and blended cellulose media were recently determined through independent lab testing. While MERV is an accurate measure of efficiency, filters should not be selected on just MERV alone. Other criteria, such as pressure drop, cleanability, compressed air usage, and filter life are important in determining a filter's total performance and life cycle cost. The best way to determine filter performance for your welding operation is to consult an expert, test the filters in your equipment, and ask for referrals from others in your industry [51].

Styrene Fumes in GRP Industry

Over the past 35 years the GRP industry, together with the unsaturated polyester resin producers, has put a huge amount of effort into reducing exposure to styrene. But up until now, a study has never been carried out into how successful these efforts have been [52].

Styrene

Styrene is the ideal monomer used for cross linking unsaturated polyester resins. Although alternative monomers have been extensively investigated but none can, on a broad scale, match the performance of styrene. A small part of the styrene monomer evaporates during processing into the atmosphere. Reducing worker exposure to styrene monomer has been a central focus of attention for the composites industry over recent decades. Realizing the styrene exposure problem, the resin industry quickly set about the task of developing low styrene emission (LSE) resins. The addition of LSE additives to the resin significantly reduces the emission of styrene, especially during the static phase of open mold processing.

As a result of the wax present in the additives, early LSE additives led to delamination problems, particularly in applications where the glass fiber reinforced laminates were subjected to high mechanical forces. Currently, LSE resins are widely used in open molding; the formulations have been greatly improved and the quality and value of LSE resins is undisputed. Other resin types were introduced, such as resins based on dicyclopentadiene (DCPD) as a raw material, which results in a slightly lower styrene content compared to standard unsaturated polyester formulations.

By modifying the molecular backbone of a resin, it is possible to reduce the styrene content further and developments continue into improved, lower styrene content resins. Whereas a styrene content of 40–45% was normal around 30 years ago, currently standard resins contain around 35% styrene monomer. In certain cases, resin formulations have been developed with a styrene monomer content as low as 20%, but the processing of these resins is more complicated, which reduces the possibilities for their more widespread use. In a review conducted by Industox Consult [52] in the Netherlands, it was found that the exposure data of styrene in the GRP industry retrieved for this review shows that the biological indicators of styrene confirm a decline in styrene exposure.

Scrubbing Out Sulfur

Pacific Northwest National Laboratory researchers have developed a reusable organic liquid that can pull harmful gases like CO_2 or SO_2 out of industrial emissions from power plants. Power plants could capture double the amount of harmful gases in a way that uses no water, less energy, and saves money.

"Power plants could easily retrofit to use our process as a direct replacement for existing technology," said David Heldebrant, PNNL's lead research scientist for the project [53].

Harmful gases such as CO_2 or SO_2 are called acid gases. The new scrubbing process uses acid gas-binding organic liquids that contain no water, and appear similar to oily compounds. These liquids capture the acid gases near room temperature. Scientists then heat the liquid to recover and dispose of the acid gases properly. These recyclable liquids require much less energy to heat, but can hold two times more harmful gases by weight than the current leading liquid absorbent used in power plants, which is a combination of water and monoethanolamine, a basic organic molecule that grabs the CO_2. The monoethanolamine component is too corrosive to be used without the excess water [53].

Solid-State Lighting

Scientists at Berkeley Lab have created nontoxic magnesium oxide nanocrystals whose size can be adjusted within just a few nanometers. The nanocrystals glow blue when exposed to ultraviolet light, and could be a bright candidate for lighting that consumes less energy and has a longer lifespan. The nanocrystals could also allow researchers to probe a key pathway in CO_2 sequestration. In its bulk form, magnesium oxide is a cheap, white mineral used in applications like insulating cables. Current routes for generating these alkaline earth metal oxide nanocrystals require processing at high temperatures, which causes uncontrolled growth or sticking of particles to one another—not a desirable outcome when the properties you seek are size-dependent. Vapor phase techniques, which provide size precision, are time and cost-intensive, and leave the nanocrystals attached to a substrate.

The scientists created the nanocrystals of magnesium oxide using an organometallic chemical synthesis route. Using a fundamentally new, unconventional mechanism for nicely controlling the size of these nanocrystals, an intriguing and surprising candidate for optical applications was realized. This efficient, bright blue luminescence could be an inexpensive, attractive alternative in applications such as bio-imaging or solid-state lighting.

Applications

Maglev System

A superconductive magnetic levitation transportation system performs short to long distance transport of cargo such as mail and foods [54]. The transportation vehicle is levitated on the magnetic levitation guide at the stable floated position, with the binding force by the pinning effect of the superconductor. The vehicle travels at high speed in the vacuum transportation passage by potential energy induced by the level difference between the start and the destination locations. Since external driving power is not needed and there is almost no pollution, the transportation system can help save energy and solve environmental problems.

Golf Balls

Arizona golf ball manufacturer, Dixon, has developed a line of environmentally responsible golf balls [55]. Dixon's Earth golf ball is green throughout and does not contain heavy metals like tungsten, cobalt, lead, or nonrenewable synthetic materials and compounds, often found in mainstream products. The ball has a graded-density core made from a proprietary polymer combination that preserves the playable properties of the ball. The ball is fabricated from new materials to ensure consistent playing properties but is 100% recyclable. Dixon works with a local company to recycle the balls into playgrounds, football field turf, and so on.

The golf balls are no less playable for being environmentally friendly. In testing, the Dixon Earth ball was found to play better than most two-piece golf balls and out-performed several three-piece, urethane cover balls. It plays the distance, spins, sticks on the greens, and has plenty of feel when putting. Each box of Dixon Earth golf balls includes a return mail pouch so golfers can mail used golf balls back to Dixon easily. Introduced in early 2008, the Dixon Earth golf ball conforms to USGA certification standards.

Ahead of Environmental Curve

Asian airlines are deftly handling the environmental debate, taking care of the warning signs that came out of Europe in 2007 to introduce timely and substantial measures to cut pollution and keep public opinion on their side [56]. While some European carriers were caught flat-footed by the sudden explosion of the environmental debate in that region of Europe in 2007, and found themselves scrambling to avoid punitive government policies, the Asian airline industry hasn't suffered an outbreak of antiaviation hysteria. For example, the region has been largely free of political pressure to specially punish airlines for their small but high-profile contribution to CO_2 emissions. There's also wide support among Asian airlines for substantial action, notably for a global emissions trading scheme. Under such a system, airlines would either cut their CO_2 pollution or pay other companies, probably in other industries, to do it for them.

Other efforts underway include instituting policies to reduce the weight of the aircraft, including monitoring long-term studies to evaluate the use of composite and lighter materials in the construction of cabins. A few major carriers are watching fuel consumption and also retiring older aircraft and introducing replacements that are smaller and, being new, more fuel efficient.

Wave Energy

A device that harnesses ocean waves [57] to generate electricity has been patented by M. Raftery. The advantage over existing technologies is a spring-rewind cable reel system. The cable reel system does not have limited stroke like the existing piston, linear magnet, and oscillating water column systems. A buoy or any other floating object can be used on the sea surface to drive the system. The generator housing can be powered to the sea floor, where it is safe from storms. The system is modular so thousands of units can be placed offshore, out of sight from the beach, and will produce enough power to replace nuclear or conventional fossil fuel power plants in the average seas.

Energy Clothes Dryer

Clothes dryers literally pump the heated and cooled air out of a home [58]. The dryer sits inside the home and intake air enters the dryer from within the home. The exhaust goes out of the home, pumping 12,000 cubic feet per hour of conditioned air out of the home. A recent invention design has an air intake inlet, which would take intake air from the exterior of the home. An internal duct inside the dryer takes the air from the outlet to the inside of the drying system. By taking intake air from outside of the structure, the dryer would no longer be pumping out the conditioned air from within the home. This is a green innovation that will greatly reduce the energy consumption needed to heat and cool the home. The system has been tested in very cold weather, as well as in hot, rainy weather with humidity at 100%.

Electrical Grids

The electric industry has been talking for decades about bringing the nation's antiquated, inefficient, glitch-prone energy grid into the computer age [59]. Now, with energy demand rising twice as fast as supply, it's finally happening, thanks to a rare alignment of interests, government, business, consumer, and environmental. Government and industry studies estimate that a modern digital energy grid could trim the country's power usage by 10%, reduce greenhouse gas emissions by 25%, and eliminate the need for $80 billion in new power plants. It's not a question of whether such a grid can be built, but when.

The basic idea is to replace a passive, analog electricity delivery system with one that is two-way and aware of what is happening to it at any moment. In other words, a smart grid. It's not going to be cheap. According to the Edison Electric Institute [59], utilities over the next 20 years will spend hundreds of billions of dollars on infrastructure improvements, including computers, sensors, and networking systems. This is a huge new market for Information Technology (IT) and networking companies, and there is no shortage of firms big and small scrambling for a piece of the business.

There are 1.4 billion electric meters in the world, about 10% of them in the United States, but many are the electromechnical type that hasn't changed much since it was invented in 1888. It costs roughly $300 apiece to replace them with a so-called smart meter. Southern California Edison will spend $1.7 billion over the next 4 years to equip 5.3 million homes with smart meters. The latest idea is a meter that could measure electrical usage and act as an Internet router, communicating wirelessly with both the utility and the end user. Utilities could monitor energy usage remotely and would know at the first sign of trouble what had gone wrong, without having to dispatch a truck to the site. Consumers, for their part, would know exactly what they were paying for electricity at any moment and could adjust their behavior during peak hours. This meter is being evaluated in 11 states and in Australia, currently (see Figure 22.4).

Some Ecofriendly Companies

Seattle Biodiesel [60] makes an alternative automobile fuel that many experts think could finally ease the nation's addiction to oil. Derived from vegetable oil, biodiesel can be blended with regular diesel or poured by itself into any conventional diesel car or truck. It produces relatively clean, almost sweet-smelling emissions. Biodiesel's obstacles have been its high price and the absence of a nationwide infrastructure to crush and refine oil-rich crops into usable fuel. Biofanatics usually have to drive to the back of a restaurant and beg for free waste oil to fill up their green machines. But is the Biodiesel plan

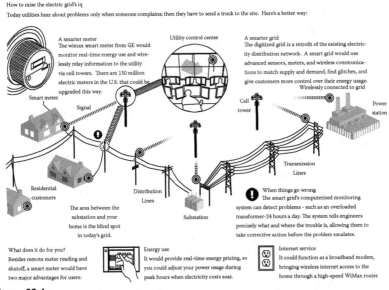

Figure 22.4
How to raise the electric grid's IQ.

to turn biodiesel into a viable national alternative? Their mission is to make a gas that is so cheap and plentiful that consumers don't even have to know it's not made from fossil fuels.

The company is now trying to create a local agricultural economy around biodiesel, using their new refinery to convince Washington farmers there's demand for feedstock such as canola and mustard seed. Seattle Diesel is now crushing and refining local crops, which means they can cut down on the expense of importing soybean oil from the Midwest. The company has scored some minor victories. A U.S. senator has used the plant as a backdrop to introduce legislation to boost the nation's biofuel production. Recently, for the first time in the region, pure biodiesel was cheaper than regular diesel. This price inversion hadn't been expected until 2009.

Hydrogenics produces fuel cells that fit together like LEGOs and extract electricity from the chemical reaction of hydrogen and water. It installed a hydrogen refueler inside a General Motors car-assembly plant in Canada to power two forklifts. And Hydrogenics technology started fueling a Purolator truck in Toronto and a transit bus in Winnipeg. Hydrogen fuel cells aren't just clean, they're also silent, and they produce water as a by-product. That's also creating defense applications that can capitalize on a stealth approach. Another upside: in desert settings like Iraq, the technology produces drinking water for soldiers. The U.S. Army recently started a year long trial using Hydrogenics to power a few of its Stryker light armored utility vehicles.

When this technology will move into mass markets is unclear. GM says it may happen in the auto industry over the next decade. (Hydrogenics helped develop a neighborhood car prototype that looks like a street-legal golf cart.)

Panda Development [60], has a plan to alleviate the world's environmental woes. It needs several hundred million dollars, a lot of corn, and 1 billion pounds of cow manure. Panda, a subsidiary of Dallas-based Panda Energy International, has a new building that will be utilized to produce ethanol (the cleaner but less efficient gas now used sparingly in U.S. automobiles, in part because it takes too much energy to produce.) Panda thinks it's found a solution. The company plans to collect truckloads of cow manure from feedlots. Gases produced by burning the bovine biomass will fuel Panda's plants, which in turn will convert corn to ethanol. Each plant will churn out 100 million gallons of ethanol a year. To be sure, it's unlikely that every car will eventually run on ethanol. But many are equipped to use fuel with at least some ethanol (mileage is slightly worse, but the fuel costs about 20–30 cents less per gallon than regular unleaded gas.)

STM Power [60], has developed a modern twist on the Stirling engine, including a "Fumes to Fuel" idea for Ford.

Candy lovers know that M&M's doesn't stick to your hands. The reason, palm oil, which is used in the candy's coating. One company that makes the palm oil used in M&M's, Aarhus of Port Newark, New Jersey, pays hundreds of dollars to send some of the 900 gallons of palm-oil waste that it generates daily to a landfill. But starting in 2006, the company hasn't been paying any disposal costs. Instead, it turns that palm oil into energy, using a new technology developed by an Ann Arbor, Michigan, company, STM Power.

STM's technology, which generates electricity on-site, is based on the Stirling cycle engine, which was invented in 1816 as a cooler alternative to the hot-burning steam engine. The process can harness the power of a variety of fuel sources, including methane gas and environmental pollutants to run an external combustion engine. The burning takes place outside the engine, and the heat is transferred to a small amount of hydrogen stored in tiny, semicircular tubes inside the Stirling engine. The heated hydrogen drives the pistons, to create new energy, and is then cooled and transferred back to be reheated, where the cycle repeats itself.

STM's units are expensive at $65,000 apiece, but each of its 1814 kg (4000-pound) power plants generates 55 kilowatts of electricity, enough to power 11 homes. Recently, STM shipped 31 of its units to a wide variety of customers, including EcoMEET Solutions in Tokyo, which is using gas from chicken manure as a fuel source for STM's technology. Ford Motor Co. started running one of STM's units for its truck plant in Wayne, Michigan. It's converting paint emissions containing VOCs into an electricity source, in a project it's calling "Fumes to Fuel." It's a great thing, since it turns an environmental issue on its head and now uses those environmental emissions to make electricity [60].

Imagine, for a moment, a future where electricity is generated not in big power plants but behind your home or in the basement of your office building. High-efficiency, low-polluting fuel cells the size of minivans provide all your power needs. They take hydrogen-rich natural gas or propane and then chemically strip the fuel of its electrons to produce electricity. One by-product is hydrogen, which can then be funneled into your new, clean-running hydrogen car. While the fuel cells run hotter than anything else in your home, perhaps as high as 1000°C (1832°F), then that's okay, too—the extra heat is captured and warms the water tank.

Ion America [60] is a quiet Silicon Valley start-up firm pursuing this coveted power paradigm, using solid oxide fuel cells. The company's first fuel cells have been undergoing testing and the prospects are exciting for some energy experts. The U.S. Navy gave Ion America a $2.7 million contract to test its fuel cells in submarines, where another by-product, oxygen, can be used as breathable air. A Chattanooga, TN, university installed a five-kilowatt prototype in 2007 and the city is asking for a more powerful, 100–200 kilowatt version to use in their downtown by 2009.

Some men root for their local sports teams, others for individual athletes. Personnel of Power Light owe allegiances to two solar-powered Mars rovers, Spirit and Opportunity [60], the rambling robots, designed to last just 90 days, continue to thrive in their 20th month of exploration on the Red Planet. Power Light people root for the rovers because they're symbols of the reliability and durability of harnessing energy from the sun.

A recent report by the renewable-energy group Worldwatch, found that grid-connected solar, the world's fastest-growing energy technology, expanded by 75% annually from 2000 to 2008. Power Light doesn't make the PV panels that capture the sun's energy, but it buys them from manufacturers in Japan and the United States. Then it integrates them into its electrical rooftop systems and takes the financial case for solar to the marketplace. Customers usually borrow to pay for the upfront costs, which are defrayed by state and federal subsidies aimed at reducing carbon emissions. Then they pay back the loan at a fixed rate.

Proponents of other sources of energy have trouble making the same consistent claims. Nuclear, gas, and coal-fired plants all have moving parts that need regular, and sometimes unscheduled, maintenance. PV technology is so reliable now that Power Light guarantees its panels for 25 years. In an uncertain energy market that degree of certainty is as welcome as sunshine.

Highly Efficient Plastic-Based Solar Cells

South Korean scientists said that they have created a highly efficient plastic-based power cell that can speed up commercial use of solar energy [61]. The team at the Gwangju Institute of Science and Technology (GIST), said the solar cells are designed to mimic the PV activities of plants, and reached an unprecedented energy efficiency rate of 6.2%. This is the highest number realized by any single-layer plastic, organic PV solar cell created in the world to date and should greatly help commercial use of power generation using sunlight. Energy efficiency indicates the percentage of sunshine that solar cells turn into electricity. The team used a new material that has open circuit voltage properties and titanium oxide has brought about this high efficiency. The energy efficiency reached was 17%, which is more than enough to start commercial power generation. Experts have said that an efficiency rate of 7% must be reached for plastic solar cells to become commercially viable. Conventional inorganic silicon-based solar cells used in homes have an efficiency rate of 7–8%, while very expensive panels placed on satellites have numbers reaching 15%.

Earth-Friendly Dog Waste Bags

Repell-em's [62] biodegradable dog waste bags will decompose 100% in both a compost or landfill environment. Therefore, people should stop using plastic dog waste bags that permanently add to our community landfills. The bags are durable and strong and 1.0 mil in thickness. Repell-em bags are thicker than other biodegradable dog waste bags and will not tear or fall apart easily. They have an all natural scent that repels insects, animals, and odors. These bags contain all natural scented oils that repel flies, moths, mosquitoes, dogs, cats, possums, skunks, and other pests. The red color also acts as a natural repellant because animals do not see the red spectrum of light well. Since the bags are of Earth-friendly materials and made from Earth-friendly bioplastic, these bags will decompose into all natural earth in a landfill or composter in 1–7 years. The bag material contains all natural biological microbes that will eat away at the material in both an aerobic and anaerobic environment. That means that the bags will decompose 100% whether in the presence of oxygen in a composter, or without oxygen under other trash in a landfill. The result is the return of all natural earth elements to the soil.

Architecture

In a recent feature article in *Materials Research Society's Bulletin*, Dr. Michelle Oyen explores potential uses of synthetic bone-like material. Oyen suggests that these materials will be too expensive to replace materials in typical construction and building applications, but can be developed for use in particularly demanding sections of advanced architecture as well as other specialist structural applications [63]. There is growing interest in materials and systems that imitate nature. Researchers are looking toward nature for inspiration because natural materials are composites, harnessing the best features of several different material types and combining them into a material that is more than the sum of its parts. Biomimetic materials synthesis aims to take the attractive features of a biological system and mimic either the material itself or the process that naturally occurs when the material is made.

Bone-like material could push the current limits of architecture, where the ideal material would exhibit exceptional mechanical properties, but also be very lightweight. Oyen says we should expect to see biosynthetic materials used in buildings of the future. She explains that "anywhere you have something heavy and brittle, like bricks or concrete, you might be able to use a bone-like material as a replacement where you would need less material (i.e., thinner and lighter sections) but still have excellent mechanical integrity." She suggests that the first architectural application of bone-like material could be domes or other larger vaulted structures.

Space Station

Orion Propulsion, Inc. [64], a Huntsville, Alabama-based company is hard at work producing flight hardware for another first-of-its-kind space venture. The propulsion system Orion is building is completely nontoxic, making it safer for the environment and the people working on the equipment. "The cool thing about these thrusters is we can test them here on earth and there won't be any hazard in testing them," said Tim Pickens, Orion's founder and CEO. The system that Orion Propulsion is building is part of the Ares I Manned Moon Landing Mission in 2015–2020 and supports fabrication and testing of the NASA crew launch vehicle's roll-control thrusters.

See-Through Solar Window Cells

New Energy Technologies, Inc., Burtonsville, Maryland, is developing tinted transparent glass solar windows capable of generating electricity by coating glass surfaces with the world's smallest known organic solar cells [65]. The technology uses an organic solar array, which achieves transparency through the creative use of conducting polymers that have the same desirable electrical properties as the world's most commercially popular semiconductor, silicon, yet boast a considerably better capacity to optically absorb photons from light and generate electricity. The ultra-small solar cells are fabricated using environmentally friendly hydrogen-carbon based materials, and successfully produce electricity from natural and artificial light. Unlike other solar technologies, New Energy's ultra-small solar cells generate electricity not only from the visible light spectrum found in sunlight but also by using the visible light found in artificial light, such as fluorescent lighting typically installed in offices and commercial buildings.

Clean-Up Technology

Researchers at North Carolina State University are demonstrating that trees can be used to degrade or capture fuels that leak into soil and groundwater. Through a process called phytoremediation, plants and trees remove pollutants from the environment or render them harmless. Dr. Elizabeth Nichols, a professor of Environmental Technology at NC State's Department of Forestry and Environmental Resources, and her team have partnered with state and federal agencies, the military, and industry to clean up a contaminated Coast Guard site in Elizabeth City, North Carolina. About 3000 trees were planted on the five-acre site, which stored aircraft fuel for the U.S. Coast Guard base from 1942 until 1991. Fuels have been released into the soil and ground water over time. Efforts to recover easily extractable fuel using a free product recovery system, or oil skimmers had stalled, so other remedial options were considered before choosing phytoremediation.

Phytoremediation uses plants to absorb heavy metals from the soil into their roots. The process is an alternative to standard clean-up methods currently used, which can be expensive and energy intensive. At appropriate sites, phytoremediation can be a cost effective and sustainable technology, according to Nichols. This technique is similar to one used 10 years ago where natural gas was injected into the ground at a chemical superfund site that caused the rapid reproduction of microbes in the soil that fed on the chemical contamination and acted as a very good remediation media.

Green Freezers from Ben and Jerry's

Think propane and butane are just for barbecuing? Think again: The common cooking fuels can also chill your drinks and ice cream with less energy and almost none of the global warming worries of current refrigerants. Some of the world's consumer product companies are promoting freezers and refrigerators in the United States that use propane, butane, and other coolants that don't trap heat in the atmosphere as much as freon and other conventional refrigerants. The new so-called hydrocarbon coolers, already popular in Europe, are being tested by Ben and Jerry's ice cream company at stores in the Washington and Boston areas. Meanwhile, G.E. is seeking approval to market a home refrigerator in the United States using a hydrocarbon refrigerant.

The new freezers take advantage of the way hydrocarbon gases absorb heat when they change from a liquid to a gas. It's the same process when a propane tank becomes cool to the touch when you're using it with a gas grill. The hydrocarbon refrigerant is compressed and expanded as it makes its way through the compressor and tubes surrounding the freezer. Unlike car exhaust or power plant pollution that's spewed directly into the air, the coolants used in most U.S. refrigerators today only enter the atmosphere when their compressors leak, or when appliances are thrown out and their refrigerant eventually escapes. If hydrocarbons are accidentally released into the atmosphere, their effect on trapping heat is about 1400 times less than conventional refrigerants, claims Pete Gosselin, director of engineering for Ben and Jerry's.

The fuels are flammable, of course, but current models only use the amount contained in two or three cigarette lighters. Electronic components are designed to prevent igniting a possible leak. The appliances cost about the same as similar conventional freezers and use about 10% less electricity. This is a big gain in terms of carbon footprint that 10% gain in electricity. Every kilowatt hour that comes in the wall, comes in with a certain amount of CO_2 footprint with it and if one can reduce 10% off that, then that's huge. The United States will be playing catch-up. Unilever, which has more than 2 million ice cream cabinets worldwide, including 100,000 in the United States now has more than 400,000 hydrocarbon-based units in Europe, Latin America, and Asia.

About 42,000 bottle vending machines using hydrocarbons or CO as refrigerant also have been installed in China, Europe, and Latin America by Coca-Cola, Carlsberg, and PepsiCo. McDonalds has opened two pilot restaurants in Denmark that don't use traditional refrigerants, according to the Refrigerants, Naturally Web site. The EPA, which allowed Ben and Jerry's to test the new coolers, has already completed a preliminary review on the freezers, as well as the new GE refrigerator. It expects to make a proposed rule on the machines available for public comment in late 2009 and a final decision could be issued in 2010.

Climate Change

As the United States lags on climate legislation, the U.N. Chief says China is poised to join the EU in claiming front-runner status among nations battling climate change. In fact, China is leaping ahead of the United States with domestic plans for more energy efficiency, renewable sources of power, cuts in vehicle pollution, and closures of dirty plants. China and India have announced very ambitious national climate change plans. In the case of China, so ambitious that it could well become the front-runner in the fight to address climate change. China is seeking to use 15% of its energy from renewable sources by 2020. China and the United States together account for about 40% of all the world's emissions of CO_2, methane, and other industrial warming gases. Japan has announced that its new goal is a 25% cut in greenhouse gas emissions from 1990 levels by 2020.

The United States has also announced a target of returning to 1990 levels of greenhouse emissions by 2020 [66].

Future

The world faces profound environmental challenges [67,68]; shortages of clean and accessible freshwater, degradation of terrestrial and aquatic ecosystems, increase in soil erosion, declines in fisheries, modifications in the chemistry of the atmosphere and, above all, rapid and substantial changes in climate. These changes are not isolated; they interact with each other and with natural variability in complex ways that cascade through the Earth's environment on local, regional, and global scales.

In the future, we must address the challenges that climate change will invariably present. We need the capability to monitor sources and sinks of greenhouse gases through this century and beyond. Concurrently, we need the capability to project, with a quantitative understanding of the uncertainties, the characteristics of climate change at least to the regional level and with far better temporal resolution than is currently available. Such projections are essential to help decision makers mitigate the many impacts of climate change on local and regional environments and populations. The times call for careful settings of priorities [68].

A July 2009 report "America's Future in Space Aligning the Civil Space Program with National Needs," set forth six strategic goals for guiding program choices and resource planning for U.S. civil space activities. The first of these is : "To re-establish leadership for the protection of Earth and its inhabitants through the use of space research and technology." The report also notes that the global perspective enabled by space observations is critical to monitoring climate change and testing climate models, managing Earth resources, and mitigating risks associated with natural phenomena. The report recommends that NASA and NOAA lead the formation of an international satellite-observing architecture capable of monitoring climate change and its consequences, and support the research needed to interpret and understand the data in time for meaningful policy decisions.

The challenge of climate change is growing and will not go away; it is not a "problem du jour." Sustained and aggressive actions are needed.

Commensurate funding is required to meet the climate challenge. So far, climate change, power supply security, and market liberalization have been the issues dominating many debates on green power and its promotion. But the EU's ambitious 2020 targets, expectations of a new global climate accord with even higher targets, and the increasingly pressing need to ensure sustainable energy supplies have shifted the goalposts, taking renewable energy right to the top of the energy policy agenda for many people. The question we must ask ourselves is no longer how we can best promote and market green power, but when green power will become the backbone of the electricity supply.

While the previous paragraphs discussed the environment of space there is also a concern for the earth. A new Tel Aviv University invention, a real-time Optical Soil Dipstick (OSD), provides a new diagnostic tool for assessing the health of the planet. Professor Eyal Ben-Dor, of TAU's Department of Geography, says his soil dipstick will help scientists, urban planners, and farmers understand the changing health of the soil. With climate change altering the planet, Ben-Dor explains that this dipstick could instantly tell geographers what parts of the United States are best, or worst, for farming. For authorities in California, it is already providing proof that organic farms are chemical-free, and it could be used to catch environmental industrial polluters.

Simple and inexpensive ways to test for soil health in the field are hard to come by. Soil maps of individual states are only compiled every 10 or 20 years, and each one costs millions. One testing process requires the use of a bulldozer, which dredges up large tracts of land to be sampled and analyzed in a laboratory. The OSD is a thin catheter-like device that is inserted into a small hole in the soil to give accurate and reliable information on the general health of the soil. Analyzing chemical and physical properties, the dipstick outputs its data to a handheld device or computer. The dipsticks can also be remotely and wirelessly networked to airplanes and satellites, providing the most comprehensive soil map of the United States [69].

The current global economic crisis has raised new questions. It's not that green power has become any less important. The increasingly apparent consequences of climate change have strengthened the will of political institutions to invest in renewable energy and energy efficiency. The U.S. administration is not the only government to recognize that investment in renewable energy is a good way of stimulating the economy. But do the markets, and the market participants themselves, also share this view? And will they act accordingly? How, in the tough environment of today, can we ensure that the necessary investment is made in sustainable energy? And what part do the state and the industrial, energy, and financial sectors have to play in all this? These questions raise many key issues and questions that I am posing for the future:

- What are the implications of the 2020 targets and any new directives from the different players' point of view?

- Are the existing promotional instruments and market mechanisms adequate to ensure long-term growth in green power or do we need more?

- How can renewable energy, and large-scale projects in particular,- be financed during the downturn?

- How will the various markets develop in technological and geo-graphic terms?

- What measures are required in terms of infrastructure, grids, and networks?

- What is the interplay of supply and demand in the green power market?

References

1. Rehl, C. The double edged sword for green initiatives, *Surface Mount Technology* October 22, 2008, p. 1 of 3.
2. Lopez, C.T. Army aims to reduce greenhouse gases, "carbon bootprint,", http://www.army.mil/-news/2009/04/06/19315, p. 1 of 2, May 6, 2009.
3. Turbini, L. J. Halogen: The latest green initiative, *Surface Mount Technology* June 26, 2008, p. 1 of 4.
4. Kluk, D., and de Krom, A. A rose is a ross is RoHS, *AM&P* (January 2006): 56–9.
5. Hoffman, J. M. Weather forecast for Europe: Hazy but green, *Machine Design* (March 9, 2006): 130–52.
6. *SMTA International Conference*, August 17–21, 2008, Orlando, FL.
7. Konrad, M. The environmental cost of green, *Surface Mount Technology* July 9, 2008, p. 1 of 3.
8. Norman, S. Solder dross recycling: A case study, *Surface Mount Technology* July 23, 2008, p. 1 of 4.
9. Abidh, G. Matching cleaning process to solder paste, *Surface Mount Technology* Jan. 8, 2009, p. 1 of 6.
10. Marsh, G. Composites that grow in fields, *Reinforced Plastics* (November 2008): 16–22.
11. Project will assess environmental impact of biomaterials, *Composites World*, http://www.compositesworld.com/news/project-will-asse..., 11/4/08, p. 1 of 2.
12. Developing biodegradable natural fibre composites, *Reinforced Plastics* (October 1, 2008): 1.
13. Biodegradable natural fibers announced, *Composites World*, Sept. 16, 2008.
14. Safety/environmental concerns, *2008 Sourcebook*, http://www.composites world.com, 32–3.
15. Hoffman, J. M. No more PVC? *Machine Design* (December 13, 2007): 24–5.
16. Wall, R. Waste management, *Aviation Week & Space Technology* (April 28, 2008): 44.

17. Oxford, P. Hazardous waste—A UK compliance overview, *Reinforced Plastics* (March 2007): 55–8.
18. Composites soar ahead in UK project, *Reinforced Plastics* (June 2008): 40–1.
19. Stone, W. Clean fuel alternatives, *C&N Publications, Inc.* (April 2009): 1, 21.
20. Largest wind farm on Indian land goes on line in California, http://www.geenergyservices.com; *AM&P* (February 2006): 24.
21. *AM&P* (August 2008): 21, http://www.verenium.com
22. Ott, J. Algae advances, *Aviation Week & Space Technology* (March 17, 2008): 66.
23. http://www.greendesignbriefs.com/component.content/article/5559, 8/23/2009, p. 1 of 3.
24. Schultz, K., Marder, J., and Rath, B. Advanced nuclear power systems, *AM&P* (November 2007): 45–8.
25. Forman, D. Nano industry requests cash for safety studies, *Small Times* (January/February 2006): 46–8.
26. Mraz, S. J. Nanowaste: The next big threat? *Machine Design* (November 17, 2005): 46–53.
27. Hughes, D. 2020 ATM today, *Aviation Week & Space Technology* (April 25, 2008): 52–5.
28. *Composite Manufacturing*, 09/08/2009, Q&A with United Soyboard D. Rust, 3 pages, info@acmanet.org
29. Plastics we can grow, *American Friends of Hebrew University, Newsletter* 5, no. 2 (Fall 2008): 1, 4.
30. Advanced lithium ion systems for military vehicles, http://www.defensetechbriefs.com/components/content/article/5556,9/9/2009, p. 2 of 3, and ARL–0071.
31. http://www.jhu.edu/news/home07/dec07/wind.html; 121/19/0731. Wind turbines produce "Green" energy, p. 1 of 4.
32. Marsh, G. Recycling carbon fibre composites, *Reinforced Plastics*, http://www.reinforced plastics.com/view/1426/recycling-carbon-fibre-composites/, September 9, 2009, p. 1 of 5.
33. Naughton, K. Go the extra mile, *Newsweek* (November 21, 2005): 50–4.
34. Stewart, R. Lightweighting the automotive market, *Reinforced Plastics* (March 2009): 14–21.
35. Exxonmobil to partner with BGU on new environment-friendly, fuel-efficient system, *IMPACT* (Winter/Spring, 2008): 4.
36. Spurgeon, B. An enterprise built for speed is shifting its focus to a green future, *New York Times* (February 2009): 9.
37. Formula 3 racing car powered by chocolate and steered by carrots, http://www2.warwick.ac.uk/newsandevents/pressreleases/racing_car, May 27, 2009, p. 1 of 3.
38. Conn. man uses old technology to run truck on wood, waste, *St. Petersburg Times* (June 28, 2009): 4A.
39. Shomon, M. J. Paint industry recognizes environmental improvements to manufacturing process, *SAMPE Journal* 32, no. 4 (July/August 1996): 42–3.
40. The greening of a convention center, *Nickel* 27, no. 3 (June 2008): 7–9.
41. Sustainability assessment to overcome barriers to renewable construction materials, *Biocompass*, http://www.biocompass.org.uk/, 11/11/2008, p. 1 of 1.

42. Kendall, D. The business case for composites in construction, *Reinforced Plastics* (July/August 2008): 20–7.
43. USCAR to optimize ways to recycle all automotive materials, *AM&P* (November 2007): 19.
44. Bennyson, R. Recycled materials are the future, http://www.reinforcedplastics.com/view/2095/recycled-materials-are-the-future, June 16, 2009, p. 1 of 4.
45. Hoffman, J. M. New life for shredded plastic waste, *Machine Design* (February 7, 2008): 55–8.
46. Boeing, Alenia launch composites recycling effort, *Composites World* July 22, 2008, p. 1 of 2.
47. Gruver, M. In Cheyenne, glass pile shows recycling challenges, http://www.baynews9.com/content/36/2009/9/27/526239.html, Sept 27, 2009, p. 1 of 3.
48. Courtemanche, M. eWaste: Which electronics recycling model fits your business? *Surface Mount Technology*, Sept 23, 2009, p. 1of 3; and Electro IQ.
49. Decomposing plastics at sea, *Green Design-Manufacturing*, http://www.greendesignbriefs.com/component.content/article/5661, Sept 15, 2009, p. 1 of 3.
50. Kasper, A. Recycling composites: FAQs, *Reinforced Plastics* (February 2008): 39.
51. Ravert, E. Nanofibers offer filtering efficiency and money savings, *Welding Journal* (November 2008): 34–6.
52. Kasper, A., and van Rooij, J. G. M. Trends in worker exposure to styrene in the Eyropean GRP industry, *Reinforced Plastics* (May 2007): 18–25.
53. Scrubbing out sulfur, *Green Design-Manufacturing*, http://www.greendesignbriefs.com/component/content/article/5600, August 23, 2009, p. 1 of 3.
54. http://link.abpi.net/l/php?20071127A4, November 27, 2007, p. 2 of 6.
55. Going green on the green, *AM&P* (December 2008): 4.
56. Perrett, B. Ahead of the curve, *Aviation Week & Space Technology*, (February 18, 2008): 89.
57. Raftery, M. Wave-energy-harnessing device, *NASA Tech Briefs* (April 2008): 22.
58. Blount, D. Energy-saving clothes dryer, *NASA Tech Briefs* (April 2008): 22.
59. *Fortune* (May 22, 2008): 30–2.
60. Romano, A. Ten eco-friendly companies, *Newsweek* (November 21, 2005): 5665.
61. Korean team develops highly efficient plastic-based solar cell, *Solid State Technology*, http://www.solid-state.com/display_news/177254/5/none/Korean_team_develops_highly_efficient_plasti. April 30, 2009, p. 1 of 3.
62. Be Earth-friendly when you walk your dog, http://www.biosmartbiodegradable.com/BiodegradableB..., October 1, 2008, p. 1 of 1.
63. Throw me a bone: Composite possibilities in architecture, *ACMA* (July/August 2008): 43, http://www.physorg.com
64. Berger, B. Startup tests propulsion system for commercial space station, *Space News* (June 30, 2009): 10.
65. See-thru solar window cells surpass thin-film and solar in artificial light, http://asm.inter-national.org/c.asp?777789&c8c39b99eb250e07&1, July 7, 2009, p. 1 of 4.
66. Heilprin, J. UN climate chief says China poised to lead, http://www.baynews9.com/content/36/2009/9/21/523543.html, September 21, 2009, p. 1 of 3.
67. http://www.senate.ca.gov, September 12, 2009, p. 1 of 2.

68. Moore, III, B. The challenge of understanding, monitoring, and managing the health of our planet, *Space News* (August 10, 2009): 19.
69. Optical dipstick assesses soil and overall planet health, http://www.green-designbriefs.com/component/content/article/5837, 10/13/2009, p. 1 of 3.

Bibliography

Acelo, R. Over-the-counter nano, *Small Times* (September/October 2007): 29–36.

A greener commute, *Parade* (November 23, 2008): 7.

Aircraft health monitors, *NASA Tech Briefs Insider*, August 7, 2007.

A new push for cellphone recycling, *Parade* (July 12, 2009): 9.

Beyond the surface, *Temple University Alumni Magazine* (Fall 2009): 25.

Biomaterials, http://www.biowerkstoff-kongress.de/, September 24, 2009, p. 1 of 3.

Borden, P. G. Laser applications in photovoltaics, *Photovoltaics World* (September/October 2009): 11–15.

Climate concerns turn city's smell into cash cow, http://www.baynews9.com/content/36/2009/10/17/534460.html, 10/17/2009, p. 1 of 3.

Could small springs beat batteries, *Green Design-Manufacturing*, http://www.green-designbriefs.com/component/content/article/5754, October 1, 2009, p. 1 of 3.

Dangers of e-waste, Parade.com/Intel.

Energy revamp would effect way we live and work in U.S., *St. Petersburg Times*, July 5, 2009.

Energy technologies for a clean future, *NASA Tech Briefs* (February 2005): 14–8, http://www.techbriefs.com

Environmentally compliant materials and processes, *Aeromat 2008 Conference and Exposition*, June 23–26, 2008, Austin Convention Center, Austin, Texas.

Ethanol on the front burner, *Machine Design* (November 20, 2008): 12–6.

EU and airlines and greenhouse gas emissions, http://link.abpi.net/l.php?20090818A14, August 18, 2009, p. 5 of 6.

Europe falls behind North America on clean energy investment, *Reinforced Plastics*, http://www.reinforcedplastics.com/view/3571/europe-falls-behind-north-america-on-clean-energy-invest..., August 28, 2009, p. 1 of 2.

eWaste: Is recycling a value-add EMS? *Surface Mount Technology-Electro IQ*, http://www.electroiq.com/index/surface-mount-technology/blogs/blog-display/s-blog/s-SMT/s-post987_...., September 23, 2009, p. 1 of 2.

eWaste: Turning the EOL burden into profitable revenue streams, *Surface Mount Technology- Electro IQ*, http://www.electroiq.com/index/surface-mount-technology/blogs/blog-display/s-blogs/s-SMT/s-post987_...., September 23, 2009, p. 1 of 3.

eWaste: Turning the EOL burden into revenue streams, *Surface Mount Technology*, August 18, 2009, p. 1 of 3.

eWaste: Which electronics recycling model fits your business? *Surface Mount Technology-Electro IQ*, September 16, 2009, p. 1 of 3.

Friedman, T. L. The people we have been waiting for, *New York Times* (January 2008): 11.

Geothermal heat extraction process takes advantage of new liquid, *Green Design-Manufacturing,* http://www.greendesignbriefs.com/component/content/article/5486, September 22, 2009, p. 1 of 3.

Gore, A. The climate for change, *New York Times* (November 9, 2008): 10.

Green Engineering, Select an application, http://www.ni.com/greenengineering/?metc = mtp7bv, April 23, 2008, p. 1 of 1.

Hardy, D. R., Rath, B. B., and Marder, J. Advanced coal combustion technologies, *AM&P* (April 2007): 30–33.

http://www.reinforcedplastics.v/com/view/3073/vedtas-launches-v112-3-mw-off-shore-wind-turbine/, September 17, 2009, p. 1 of 2.

Hwang, J. S. Part 8: Lead-free reliability for harsh environment electronics, *Surface Mount Technology,* http://smt.pennnet.com/Articles/Article_Display.cfm?ARTICLE_ID = 327374&p = 35&pc = E, May 8, 2008, p. 1 of 4.

Hydrogen car is here, a bit ahead of its time, *New York Times* (December 9, 2007): 11.

India's biogas boom, *Nickel* 23, no. 3 (June 2008): 4–5.

Kjaer, C. Wind energy exec says composite blades need to be lighter, *Composite Manufacturing, ACMA,* October 6, 2009, p. 1 of 2.

Linden, E. Catch the energy, *Parade* (April 20, 2008): 8.

http://link.abpi.net/l.php?20090421A14, April 21, 2009, p. 5 of 6.

Matheson, C. Better ways to go green, *Parade* (April 20, 2008): 8.

Mazur, S. Combining tin/lead and lead-free: A 5 step hybrid manufacturing process, *Surface Mount Technology,* September 2, 2009, p. 1 of 6.

New war on waste, *Fortune,* http://www.fortune.com/adsections, S1–5.

Nutcher, P. "LEED" the pack in green efforts, *Composite Manufacturing* (March 2008): 48.

Obama = green USA, *Reinforced Plastics,* http://www.reinforced plastics.com/articles/environment..., November 27, 2008, p. 1 of 2.

Obama urges Senate to pass energy bill, *St. Petersburg Times,* July 5, 2009.

Old traffic light lenses spiff up countertops, *AM&P* (December 2007): 3.

People, prosperity, and the planet (P3), http://www.greendesignbriefs.com/component/content/article/5836, 10/14/2009, p. 1 of 3.

Pulido, H. A., and De Sanctis, G. Conformal coating for microelectronics: A primer, *Surface Mount Technology,* http://smt.pennnet.com/Articles/Article_Display.cfm/ARTICLE_ID = 331528&p = 35&pc = , June 18, 2008, p. 1 of 5.

Regarding RoHS: Two years in, *Surface Mount Technology,* http://smt.pennnet.com/Articles/Article_Display.cfm?ARTICLE_ID = 332546&p = 35, June 26, 2008, p. 1 of 4.

Reinforced Plastics, http://www.reinforcedplastics.com/articles/environment..., December 19, 2007, p. 1 of 2.

Report focuses on green composites, *Composites World,* December 2, 2008, p. 1 of 1.

Researchers striving to cut waste when slicing silicon, *Green Design-Manufacturing,* http://www.greendesignbriefs.com/component/content/article/5580, August 23, 2009, p. 1 of 2.

Sequenced genome could enable more efficient biofuel production, http://www.greendesignbriefs.com/component/content/article/5835, 10/14/2009, p. 1 of 3.

Solar power outshining Colorado's gas industry, http://www.baynews9.com/content/36/2009/10/10/531675.html, 10/10/2009, p. 1 of 3.

Subrahmanian, K. P., and Dubouloz, F. Adhesives for bonding wind turbine blades, *Reinforced Plastics* (January/February 2009): 26–9.

Study: Composites in lithium-ion batteries show promise, *Composites World*, September 23, 2009, p. 1 of 1.

Thin-layer solar cells may bring cheaper green powder, *AM&P* (November 2007): 32.

UK invests more in coal than in marine renewables, *Reinforced Plastics*, http://www.reinforcedplastics.com/view/3549/uk-invests-more-in-coal-than-in-marine-renewables/, August 28, 2009, p. 1 of 2.

Understanding bio-materials, *Reinforced Plastics* (December 2008): 12.

van Gastel, S. The environmental impact of pick-and-place machines, *Surface Mount Technology*, April 2, 2009, p. 1 of 8.

Vestas launches V112, 3 MW offshore wind turbine, *Reinforced Plastics*, September 15, 2009.

Vianco, P. T. Environmental mandates and soldering technology: The path forward, *Welding Journal* (September 2007): 27–30.

Warming ocean triggers release of greenhouse gas, *Green Design-Manufacturing*, http://www.greendesignbriefs.com/components/content/article/5592, August 23, 2009, p. 1 of 2.

Index

A

Adhesive bonding
  abrasion, 19
  animal/vegetable origin, 12
  application, 13
  chemical treatment, 21
  corona discharge, 20–21
  flame treatment, 20
  joining methods
    riveting and bolting, 14
    stress distributions, 14
  joint design and adhesive selection,
    22–23
  laser, 21–22
  plasma treatment, 20
  pressure-sensitive tapes, 18
  pretreatments
    metals and ceramics, 17
    plastics, 18
  protein-based, 12
  selection, 16
  solvent wipe, 18–19
  surfaces preparation
    mechanical and physical
      properties, 15
  surface treatment, 13–14
  types
    ACAs/ACFs, 28–29
    ECAs, 26–27
    flexible, 24–25
    nanoglue, 23–24
    NCAs/NCFs, 29
    pressure-sensitive, 25
    waterproof bandage, 25–26
  wood, 12
Advanced combat helmet (ACH), 476
Aerospace aluminum applications,
    FSW, 128
  advantages, 123
  allowables approaches, 134
  aluminum alloys, 130
  assembled, 123
  benefits, 132

building blocks, 124
case study, 130
Code of Federal Regulations
  (FAR), 134
composition, 127
conductivity profiles, 128
corrosion, 132
  2000 and 7000 series
    aluminums, 132
  properties, 132
cost, 123
DCB stress corrosion, 131
distortion, 124
effects of, 126
exfoliation corrosion (EXCO) testing,
  128–129
  cross sections, 125–126, 130
extrusions, 126
747 Freighter barrier beam,
  132–133
hardness profile, 127–128
heat affected zone (HAZ), 125
  nugget area, 126
  PWA-T6 specimen, 128
integrated team approach, 124
key parameters, 133
kiss bonds and root side defects, 134
lazy "S" oxides strings, 133–134
material acreage, 132
parent material, 126
performance improvements, 123
potential defects, 133
process, 125
solid-state joining process, 123
stress corrosion cracking (SCC),
  125, 129
  results, 130
testing, 129
  specimen location, 128
  tensile and fracture toughness,
    127, 129
thermal and microstructural
    effects, 124